Chapter 3

p.m.f. probability mass

$P(\boldsymbol{x} = x)$ probability that the random variable $\boldsymbol{x}$ equals the value x

$E[\boldsymbol{x}]$ expected or average value of the random variable $\boldsymbol{x}$

$F(x)$ probability that $\boldsymbol{x} \leq x$, cumulative function

$g(x)$ any function of the random variable $\boldsymbol{x}$

$V[\boldsymbol{x}]$ variance of the random variable $\boldsymbol{x}$

z standardized random variable

p.d.f. probability density function

$f(x)$ height of the p.d.f. at $\boldsymbol{x} = x$

$P(x,y)$ probability that $\boldsymbol{x} = x$ and $\boldsymbol{y} = y$

$F(x,y)$ probability that $\boldsymbol{x} \leq x$ and $\boldsymbol{y} \leq y$

$P_x(x)$ marginal probability of $\boldsymbol{x}$

$P_{x|y}$ conditional probability of $\boldsymbol{x}$ given y

$C[x,y]$ covariance of $\boldsymbol{x}$ and $\boldsymbol{y}$

Chapter 4

p probability of success on a binomial trial

$P(a \leq \boldsymbol{x} \leq b)$ probability that $\boldsymbol{x}$ lies between a and b

λ Poisson mean and variance (lambda)

Chapter 5

z standardized normal random variable

$N(\mu,\sigma^2)$ a normal distribution with mean μ and variance σ^2

$\boldsymbol{T}$ random variable for exponential distribution

χ^2 random variable for chi-square distribution

Chapter 6

$\bar{x}$ sample mean (x-bar)

n number of sample values

s^2 sample variance

s sample standard deviation

$\mu_{\bar{x}}$ mean of all possible values of $\bar{x}$

$V[\bar{x}]$ variance of all possible values of $\bar{\boldsymbol{x}}$ ($\sigma^2_{\bar{x}}$)

$\sigma_{\bar{x}}$ standard error of the mean ($\sigma/\sqrt{n}$)

ν degree of freedom (nu = d.f.)

$\sqrt{(N - n)/(N - 1)}$ finite population correction factor

t random variable for student's $\boldsymbol{t}$-distribution

$\boldsymbol{F}$ random variable for $\boldsymbol{F}$-distribution

(Continued on back endpapers)

STATISTICAL METHODS

THIRD EDITION

DONALD L. HARNETT
Indiana University

ADDISON-WESLEY PUBLISHING COMPANY
Reading, Massachusetts
Menlo Park, California
London · Amsterdam
Don Mills, Ontario · Sydney

Library of Congress Cataloging in Publication Data

Harnett, Donald L.
Statistical methods.

Bibliography: p.
Includes index.
1. Mathematical statistics. I. Title.
QA276.H385 1982 519.5 81-3645
ISBN 0-201-03913-3 AACR2

Reprinted with corrections, June 1982

 Printed in the United States of America. Published simultaneously in Canada. Library of Congress Catalog Card No. 81-3645.

ISBN 0-201-03913-3
CDEFGHIJ-DO-89876543

To My Parents

PREFACE
TO THE THIRD EDITION

There has been a tremendous growth in recent years in the role played by quantitative analysis in many academic disciplines, particularly in the social and behavioral sciences. Because of this increasing importance of quantitative skills, college students are being encouraged (and often required) to take more mathematics during their academic career. Many schools now insist that students in business and economics, as well as liberal arts, take a course in college algebra and become familiar with at least the basic elements of both differential and integral calculus. Unfortunately, in the past, too many of these students then elected or were required to take a statistics course whose only prerequisite was "high school algebra." Recently, however, with more and more students taking the necessary mathematics, it has been possible to develop courses in statistics that build on the fundamentals of college algebra and differential and integral calculus. Taught in this fashion, statistics can provide a useful synergy: There is an improved understanding of algebra and calculus as the student learns about the theory and application of statistics. This book is written for the student who has had a background course in college algebra and some exposure to elementary calculus, in the hope of achieving this synergy.

Although the prerequisites for this text include knowledge of the elements of calculus, we have tried not to rely too heavily on the student's experience in this area. In general, the student is expected to be familiar with the notation and manipulation of simple mathematical functions, and to be able to follow the steps necessary for differentiating or integrating such functions. Calculus, to the extent to which it is used in this book, is reviewed in Appendix A, which also includes simple problems students can work out to refresh their memory on the processes of integration and differentiation. Appendix A also includes a review of summation notation, as well as a review of variables and functions. In the text itself, the use of complicated proofs and mathematical notation has been minimized. Proofs, where necessary, are given in simple terms, and in some cases relegated to a footnote. The more advanced sections (and exercises) in each chapter are preceded by an asterisk. These starred sections can be omitted without loss of continuity.

While this book is designed as an introductory text, it will also benefit more advanced students who have taken only a course in descriptive statistics. Because the text provides the instructor with great flexibility in the assignment of topics (including the optional starred sections), it can be used for a one- or two-semester course at either the undergraduate or the graduate level. The first eight chapters cover the basic elements of classical statistical analysis, including descriptive statistics, probability theory, probability distributions, sampling, estimation, and hypothesis testing. Chapter 9 presents the fundamentals of statistical decision theory, while Chapters 10 through 15 discuss topics of more special interest in statistics, namely regression, correlation, time series and index numbers, analysis of variance, and nonparametric statistics. The materials presented in Chapters 9, 10, 14, and 15 do not require familiarity with the material in the other chapters in this group, so the instructor may omit, or rearrange the sequence of, chapters as desired. Chapter 13 can also be used independently, once the prerequisite Chapter 10 has been covered.

This third edition has almost twice as many problems as the second edition. Problems have been placed at the end of sections of material, as well as at the end of each chapter. As before, the "Exercises" at the end of each chapter are designed to present the student with problems that are a bit more challenging in their level of difficulty, or at least require some independent work involving concepts not explicitly covered in the book. The multiple-choice questions at the ends of some chapters represent questions typical of those that appear on the CPA or CMA certification exams. A solutions manual is available to instructors by writing to Addison-Wesley, Reading, MA., 01867.

The present text was designed to be slightly more advanced than *Introductory Statistical Analysis* (Harnett/Murphy, 2nd ed., Addison-Wesley, 1980). Although both books incorporate much of the same materials, this book assumes prior knowledge of some elementary calculus, while the Harnett/Murphy book does not assume calculus. In addition, this text covers slightly more material on ANOVA, includes more material on nonparametric statistics, and generally has more difficult problems. For readers who are interested in additional problems *with solutions*, we suggest the *Student Workbook for Introductory Statistical Analysis* (Addison-Wesley, 1980). With only minor rearranging, that workbook is most appropriate for use with the present text.

This book would not have been possible, at least in its present form, without the patient assistance and advice of Professor Robert L. Winkler. His most tangible contribution was the authorship of two chapters (Chapters 9 and 14).

Equally important, however, were his comments and suggestions through all stages of the development of the remaining thirteen chapters. Professor Winkler's review of this material contributed greatly to whatever merit is found in these chapters. The responsibility for errors, of course, remains mine.

In addition to the contributions of Professor Winkler, I received help and encouragement during the preparation of this text from many of my friends, colleagues, and students, among them Professors William Perkins, Ira Horowitz, Wain Martin, Victor Cabot, and James Murphy. Professor William Zani provided a sounding board for many of my ideas during the development of the book; Mary O'Leary and Jeanie Tate did an admirable job when faced with the unenviable task of typing manuscript from my handwriting. Steve Schlickman worked with me for nearly a year on the book, developing during that time many of the exercises at the ends of the chapters. Credit also belongs to the staff of Addison-Wesley, who helped in many different ways. I am grateful to the Library Executor of the late Sir Ronald A. Fisher, F.R.S., to Dr. Frank Yates, F. R. S., and to Longman Group Ltd. London, for permission to reprint the Table of Critical Values of the *t*-Distribution from their book *Statistical Tables for Biological, Agricultural and Medical Research* (6th edition, 1974). Last, but certainly not least, I owe a debt of gratitude to my wife, Janet, to my son, Kendall, and to my daughter Kristina, who, having lived through three editions, seem little thc worse for wear.

Bloomington, Indiana D.L.H.
December 1981

CONTENTS

* Starred sections represent more advanced topics, and may be omitted without loss of continuity.

1

INTRODUCTION AND DESCRIPTIVE STATISTICS

"Statistical thinking will one day be as necessary for efficient citizenship as the ability to read and write."

H. G. WELLS

1.1 STATISTICS AND STATISTICAL ANALYSIS

Statistical techniques are put to use in one form or another in almost all branches of modern science, and in many other fields of human activity as well. As Solomon Fabricant said over 25 years ago, "The whole world now seems to hold that statistics can be useful in understanding, assessing, and controlling the operations of society." Progress in our society can be measured by a variety of numerical indexes. Statistics are used to describe, manipulate, and interpret these numbers.

Beginning Applications of Statistics

Although the origins of statistics can be traced to studies of games of chance in the 1700's, it is only in the past sixty years that applications of statistical methods have been developed for use in almost all fields of science—social, behavioral, and physical. Most early applications of statistics consisted primarily of data presented in the form of tables and charts. This field, known as *descriptive statistics*, soon grew to include a large variety of methods for arranging, summarizing, or somehow conveying the characteristics of a set of numbers. Today, these techniques account for what is certainly the most visible application of statistics—the mass of quantitative information that is collected and published in our society every day. Crime rates, births and deaths, divorce rates, price indexes, the Dow-Jones average, and batting averages are but a few of the many "statistics" familiar to all of us.

In addition to conveying the characteristics of quantifiable information, descriptive measures provide an important basis for analysis in almost all academic disciplines—especially in the social and behavioral sciences, where human behavior generally cannot be described with the precision possible in the physical sciences. Statistical measures of satisfaction, intelligence, job aptitude, and leadership, for example, serve to expand our knowledge of human motivation and performance. In the same fashion, indexes of prices, productivity, gross national product, employment, free reserves, and net exports serve as the tools of management and government in considering policies directed toward promoting long-term growth and economic stability.

The Use of Statistics in Decision-Making

Despite the enlarging scope and increasing importance of descriptive methods over the past several hundred years, these methods now represent only a minor, relatively unimportant portion of the body of statistical literature. The phenomenal growth in statistics since the turn of the century has taken place mainly in the field called *statistical inference* or *inductive statistics*. This field is concerned with the formulation of generalizations, as well as the prediction and estimation of relationships between two or more variables. The terms inferential and inductive analysis are used here because this aspect of statistics involves drawing conclusions (or "inferences") about the unknown characteristics of certain phenomena on the basis of only limited or imperfect information. Generally, this involves drawing conclusions about a set of data (called a population) based on values observed in a sample drawn from that population. From this sample information, statistical inference can often derive the quantitative information necessary for deciding among alternative courses of action when it is impossible to predict exactly what the consequence of each of these will be.

The process of drawing conclusions from limited information is one familiar to all of us, for almost every decision we face must be made without knowing with certainty the consequences. In deciding to watch television tomorrow rather than study, you may, at least subconsciously, be inferring that your grades will not suffer as a result of this decision. You probably have given considerable time and thought to your choice of a major in college, but here again your decision must be made on the basis of the limited amount of information that can be provided by aptitude tests, guidance counselors, and the advice of your parents and friends. If you make a poor choice you may suffer the loss of considerable time and money. Similar problems are faced in business. Should a new product be introduced? What about plant expansion? How much should be spent on advertising? The economic advisor to government policy-makers must choose among various alternative recommendations for preventing unemployment, improving the trade balance, dampening inflationary spirals, and increasing production and income. In general the best choices can only be "inferred" from less-than-perfect information about future events. As a result, such decisions often are made under conditions that expose the decision-maker to considerable risk. The process of making decisions under these circumstances is usually referred to as decision-making under uncertainty, or as decision-making under risk.*

Statistics as a decision-making tool plays an important role in the areas of research and development, and guidance and control in a wide variety of fields. Both government and industry, for instance, participate in the development, testing, and certification of new drugs and medicines, a process that often requires a large number of statistical tests (and decisions) concerning the safety and effectiveness of these drugs for public use. Similarly, the psychologist, the lawyer, and any person who makes decisions involving uncertain factors such as human behavior often will base decisions on data of a statistical nature. Since complex decision situations almost always call for some type of statistical analysis—formal or informal, explicit or implicit—it is difficult to overemphasize the importance of inferential statistics to the decision-maker. In fact, *statistics is often defined as the set of methods for making decisions under uncertainty.*

Definition of the Statistical Population

In statistical inference problems, the set of all values under consideration—that is, all pertinent data—is customarily referred to as a *population* or *universe.*

* Although the two terms "decision-making under uncertainty" and "decision-making under risk" often are given slightly different interpretations in decision-theory literature, we shall consider them synonymous in this book.

In general, any set of quantifiable data can be referred to as a population if that set of data consists of ***all values of interest****.*

For example, in deciding on the choice of a major you may want to make inferences about the set of grades you can expect to receive in the courses in a certain field; or perhaps you may want to estimate the salaries being earned by people with degrees in this field. Similarly, a business manager may want to know how many customers might buy a new product considered for possible production, or what increase in sales to expect from the implementation of a particular advertising campaign. A government policy-maker may want to estimate the changes in demand for food that would result from increased welfare payments to the handicapped or a relaxation of restrictions on imported products. A legislator may need to infer the level of state revenue available from a special exercise tax on tobacco products or a sales tax so that a reliable budget can be prepared. In each of these cases the set of *all relevant values* constitutes a population. The IQs of all students in the sophomore class at a certain school could represent a population—as could the IQs of all students in a university, the past levels of advertising expenditures for a given firm, the profits of all companies in the U.S.A., or the monthly level of energy consumed by families with incomes under $10,000.

In making decisions, we naturally would prefer to have access to as much information as possible about the relevant population or populations. One can avoid the possibility of making an incorrect inference only when all the information about a population is available. Unfortunately, it is usually impossible or much too costly to collect all the information concerning the population associated with a practical problem. Consequently, inferences (and the resulting decisions) must be made on the basis of limited or imperfect information about the population. The function of statistics as an aid to the decision-maker is to help him or her to decide (a) what information is needed for a particular type of decision, and (b) how this information can best be collected and analyzed for use in making the decision.

In trying to decide what information about a population is necessary for making a decision we shall be referring to certain numerical characteristics that distinguish that population. These numerical characteristics, called *parameters*, describe specific properties of the population.

Numerical characteristics of populations are referred to as ***population parameters****, or simply* ***parameters****.*

For instance, one parameter of the population "executive salaries in the steel industry" is the "average salary in that industry," since this measure describes the *central tendency* of all salaries in that population. Another parameter would be a measure of the *spread* or *variability* of all salaries. A precise definition of these parameters will be given in Sections 1.3 and 1.4. For now, it is important to note that the task of determining the exact value of a population parameter may be quite difficult. This may be due to the inconvenience or impracticality of collecting the necessary data. If we are interested in the population "executive salaries," for example, it may not be possible even to identify all the executives in a given industry, much less find their salaries. Suppose that we are interested in executive salaries in the future, say one year from now. The parameters for this population, "executive salaries one year from now," may be impossible to evaluate, not only because of the reluctance of executives to disclose their salaries but also because many of these people may not know what their salaries will be one year from now.

Use of Samples from a Population

Since it is often impossible or impractical to determine the exact value of the parameters of a population, the characteristics of a given population are commonly judged by observing a sample drawn from all possible values.

A sample is a subset of a population.

The individual values contained in a sample are often referred to as *observations*, and the population from which they come is sometimes called the *parent population.* We may, for example, take a sample of 100 executives, determine their current salaries, and on the basis of these observations, make statements about different characteristics (parameters) of the population of all executive salaries—such as the average salary, or the variability of salaries in a certain industry, or perhaps the average salary in that industry a year from now.

As another example, suppose a quality-control engineer is responsible for ensuring the reliability of electrical components produced in some production process. Testing each and every item may be prohibitively expensive. Or it may be impossible if the inspection process destroys the components. Consider the problem of producing a stereo cartridge designed to last 1000 hours of playing time. An inspector might test each cartridge for 1000 hours or until it becomes defective; but then what would be left for the manufacturer to sell? The solution to this problem lies in determining the reliability of all items produced (a population parameter) by inspecting only a subset (i.e., a sample) of the items. In order to make decisions about a population on the basis of a sample, we will calculate certain numerical characteristics of a sample.

Numerical characteristics of samples are referred to as ***sample statistics*** *or simply* ***statistics***.

If 100 executives are sampled, their average salary is a sample statistic. A measure of the spread of these 100 salaries represents another sample statistic. Figure 1.1 illustrates these relationships. As we will study extensively later in this text, sample statistics are most often used to either:

a) make estimates about certain population parameters, or
b) test hypotheses (or assumptions) about certain population parameters, or
c) determine the optimal decision in a context of uncertainty.

In Fig. 1.1 the upper arrow represents the process of estimating population parameters. The lower arrow represents hypothesis testing.

The methods for relating populations and samples will be described in greater detail beginning in Chapter 6. Before reaching Chapter 6, however, we will devote considerable attention to the concepts and techniques that form a foundation for these methods. It is important to establish here the major purpose for which these concepts and techniques will be used—*to aid in drawing inferences and testing hypotheses about population parameters on the basis of sample statistics, and to make decisions based on knowledge of the reliability of statistical estimates.*

In order to be able to discuss the objectives described above, we will need to learn more about how a population or a sample can be described.

Fig. 1.1 Statistical terms.

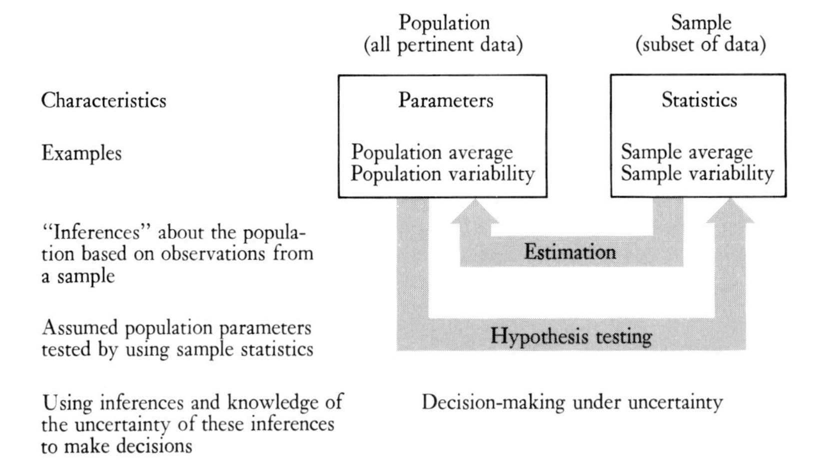

1.2 DEFINING A POPULATION

Graphical Forms

Graphs and charts, the most popular and often the most convenient means for presenting data, are usually employed when a visual representation of all or a major portion of the information is desired. Although there are many alternative methods for presenting data in this form, only a few will be discussed here. The "pie chart," for example, is a familiar device for describing how a given quantity is subdivided. In Fig. 1.2 the relevant quantity is the U.S. Federal budget dollar, and the subsets represent where the dollar comes from and where it goes.

Another popular descriptive device is the chart measuring changes over time in some index, such as the U.S. energy demand shown in Fig. 1.3 and the Alcoa advertisement shown in Fig. 1.4. Our final illustration of a graphical form (Fig. 1.5) is derived from an example in Darrell Huff's delightful book *How To Lie With Statistics* (Norton, 1954). This example depicts the increase in the number of cows between 1860 (eight million) and 1936 (24 million). In drawing pictures of cows to represent this growth, one naturally would be inclined to draw the 1936 cow three times as long as the 1860 cow. Of course, a cow three times as long looks rather peculiar unless it is also three times as high. But if the 1936 cow is three times as high and three times as wide, it is

Fig. 1.2 Shift in budget priorities.

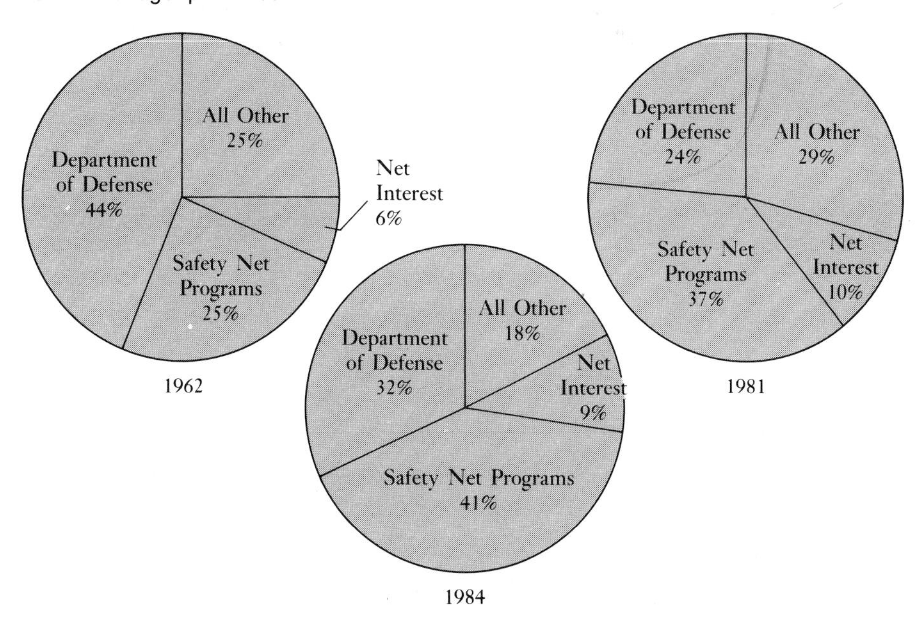

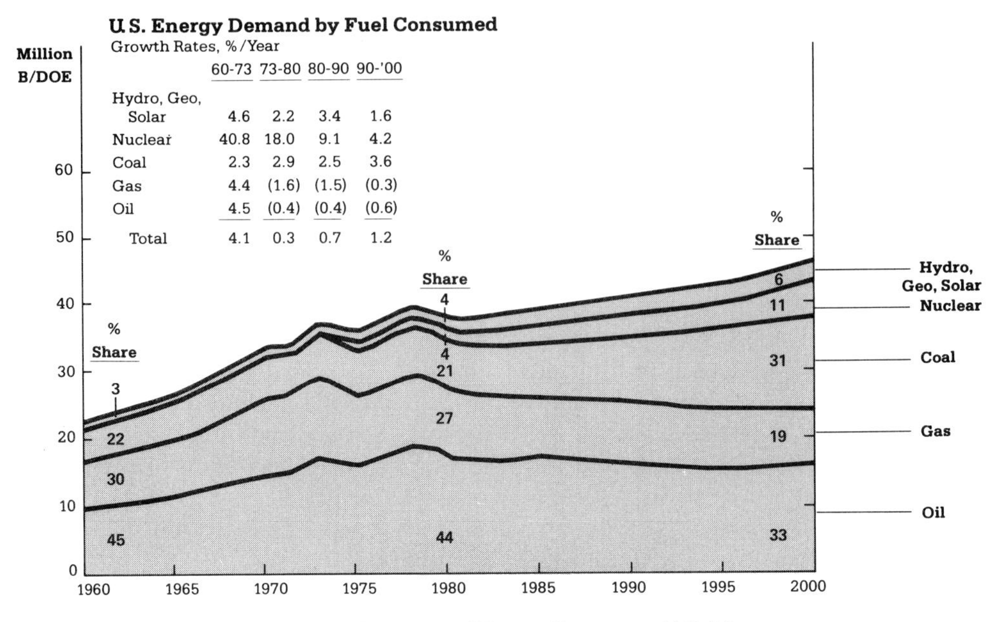

Fig. 1.3 U.S. energy demand. (Courtesy of Exxon Company, U.S.A.)

Fig. 1.4 Alcoa ad in *Time*, July 21, 1980.

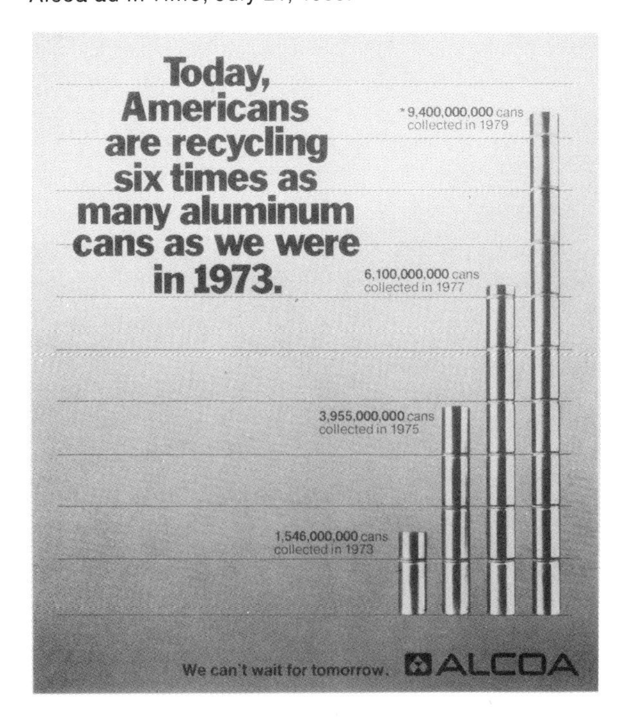

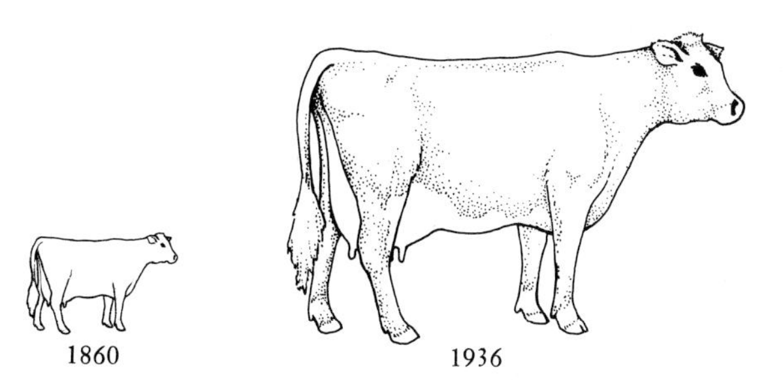

Figure 1.5

nine times as large as the 1860 cow in terms of area. Such a figure would seriously misrepresent the true growth.

Business journals and news magazines often adopt a somewhat whimsical device to avoid this difficulty. Interested primarily in presenting information in a form understandable to the layman (not the statistician), such journals would depict growth in the number of cows by showing one cow to represent 1860 and three cows to represent 1936—thus, each would represent eight million animals. Similarly, growth in auto production from, let's say, ten million to 27 million cars could be indicated by lopping off the motor and front wheels of the third car. This truncation is less gruesome in the case of the car than in the case of the cow.

Frequency Distributions

Although graphs and charts such as those shown in Figs. 1.2. and 1.3 often serve a very useful function, they are not appropriate for most purposes of statistical analysis and decision-making because they provide only a representation of the information and not the actual data themselves. Statistical purposes usually require that the data be presented in a form that gives a more precise indication of the information at hand. Some method is needed that will summarize or describe large masses of data without loss or distortion of the essential characteristics of the information, and also make the data easier to present. One such method is the arrangement of data into what is called a *frequency distribution* or *frequency table.* In constructing a frequency distribution it is first necessary to divide the data into a limited number of different categories or classes, and then to record the number of times (the frequency) an observation falls (or is distributed) into each class. As an illustration of the construction of a frequency distribution, suppose the 40 numbers in Table 1.1 represent the number of new cars sold in a two-month period by 40 different Chevrolet dealers.

A wide choice of classes or categories could be used to summarize these new-car sales. Statisticians have developed certain guidelines for the construction of such categories, which can be illustrated in terms of the data from Table 1.1.

Table 1.1

63	68	71	74	76	78	81	84	85	89
66	70	73	75	76	79	82	84	85	90
67	71	73	75	76	79	82	85	86	92
68	71	74	75	77	79	84	85	86	94

1. Classes are generally chosen so that the magnitude of each class, called the *class interval*, is the same for all categories. Otherwise, interpretation of the frequency distribution may be difficult. For example, grouping the sales data in Table 1.1 into unequal categories such as 60–64, 65–74, 75–89, 90–95 is ill-advised. Comparisons between the number of observations in different categories would be misleading since the size of the categories is not consistent.
2. The number of classes used should probably be *fewer than* 20 (for ease of handling and to ensure sufficient compacting of the information) and *at least* 6 (to avoid loss of information due to grouping together widely diverse data).
3. Open-ended intervals should be avoided. Too much information is lost if categories such as "<65" or ">90" are used. If a few extreme values do not conveniently fit into frequency categories they should be listed separately. The best frequency distribution for the other values should be determined after excluding these extremes.
4. Categories should be defined so that no single observation could fall into more than one category. For example, categories such as 65–70 and 70–75 are ambiguous, since where to place sales of exactly 70 cars is not clear. In this particular case one could use instead the classes 64.5–69.5 and 69.5–74.5. Note that the overlap at 69.5 causes no problem because car sales can only be integers (i.e., one cannot sell a fraction of a car).
5. The midpoints of each category should be representative of the values assigned to that category. This is important because these midpoints, called *class marks*, will be used in the calculation of summary measures.

Following these basic rules, suppose that we break the data in Table 1.1 into seven classes of equal width: 59.5–64.5, 64.5–69.5, . . . , 84.5–89.5, and 89.5–94.5. The boundaries of each of these classes are referred to as their *upper* and *lower class limits*. In this example, the lower class limits are 59.5, 64.5 . . . , 89.5, while the upper class limits are 64.5, 69.5, . . . , 94.5. Table 1.2 shows the frequency distribution of sales in these classes and, in addition, gives the relative frequency of each class. *Relative frequency* is determined by dividing the frequency of each class by the total number of observations and expressing the result as a decimal. Thus the first relative frequency in Table 1.2 is $\frac{1}{40} = 0.025$;

Table 1.2

Class	Frequency Distribution	Relative Frequency
59.5–64.5	1	0.025
64.5–69.5	4	0.100
69.5–74.5	8	0.200
74.5–79.5	11	0.275
79.5–84.5	6	0.150
84.5–89.5	7	0.175
89.5–94.5	3	0.075
Sum	40	1.000

the second is $\frac{4}{40} = 0.100$; and so forth. Note that the sum of all relative frequencies must equal 1.000. Calculation and tabulation of these statistics makes it clear that sales of 74.5–79.5 new cars occurred with the greatest frequency.

Visual Representation of Data

While it is often useful to arrange the values in a data set from smallest to largest (as in Table 1.1), or to classify data into categories and determine frequencies and relative frequencies (as in Table 1.2), many analysts prefer a pictorial representation. Perhaps the most common type is the graph in which the classes are plotted on the horizontal axis and the frequency of each class is plotted on the vertical axis. This type of graph is called a *histogram* or *bar graph*. Figure 1.6 represents the histogram for the frequency distribution in Table 1.2.

A helpful addition to the histogram is the *frequency polygon*, constructed by drawing a straight line between the midpoints (class marks) of adjacent class intervals. The frequency polygon for the data in Fig. 1.6, indicated by the dashed line through the class marks, serves to smooth a set of values. The advantage of

Fig. 1.6 Histogram and frequency polygon of data in Table 1.2.

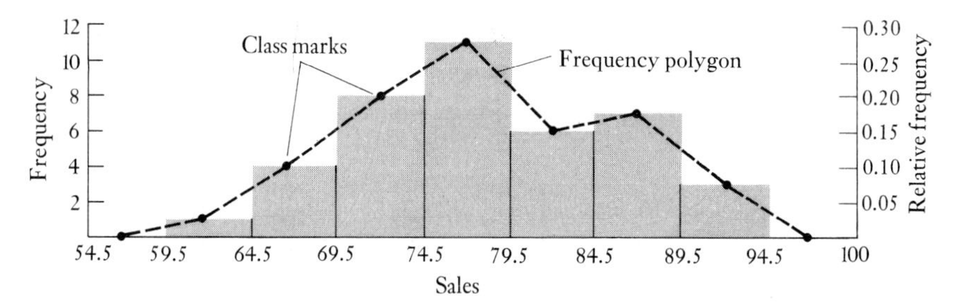

using smoothed approximations of discrete data is that working with these approximations is generally much easier. The reader should verify at this point that in Fig. 1.6 the graph of the *relative* frequencies is the same as that of the absolute frequencies except that the values for the vertical scale are different (relative frequencies are shown on the right side of Fig. 1.6). Note in Fig. 1.6 that the frequency polygon is closed off at either end by extending it to the horizontal axis (midpoint of next class).

Cumulative Frequency Distributions

Another important method of presenting a data set is the table of cumulative frequencies, or table of *cumulative relative frequencies.* Table 1.3 applies this method to the data from our car sales example. Cumulative frequencies represent the *sum* of the frequencies, from the lowest to the highest class. For example, by the end of the second class (at 69.5) the cumulative frequency of car sales is $1 + 4 = 5$. Similarly, at the end of the 74.5–79.5 class the cumulative frequency is $1 + 4 + 8 + 11 = 24$. At this point the cumulative relative frequency is $\frac{24}{40} = 0.600$. This value indicates that 60 percent of all car sales were less than or equal to 79.5. The cumulative relative frequency for the highest class, 89.5–94.5 in this case must equal 1.00, since all values certainly fall in this class or some lower class.

Just as a graph of the frequencies of a set of values provides a visual description of the original data, so a graph of the cumulative frequency or the cumulative relative frequency provides visual information about cumulative values. Note in Fig. 1.7 that cumulative relative frequencies can be smoothed in a fashion similar to relative frequencies; in this case the polygon is called an *ogive*, and the lines are drawn between points at the *beginning* of each class.

The distributions we have just seen in Tables 1.2 and 1.3 could lead to many questions for the regional manager of Chevrolet. For example, why are

Table 1.3

Class	Frequency	Cumulative Frequency	Cumulative Relative Frequency
59.5–64.5	1	1	0.025
64.5–69.5	4	5	0.125
69.5–74.5	8	13	0.325
74.5–79.5	11	24	0.600
79.5–84.5	6	30	0.750
84.5–89.5	7	37	0.925
89.5–94.5	3	40	1.000

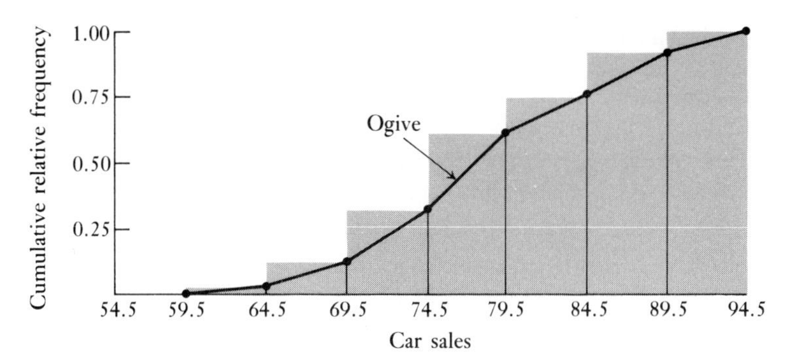

Fig. 1.7 Cumulative relative frequency.

sales not the same for each dealership? Are sales higher in specific weeks during the two-month period? Perhaps sales are related to the number of local radio or television advertisements in those particular weeks. If it is possible to determine such relationships, could local dealers increase the overall average sales by changing advertising campaigns? Perhaps sales can be promoted by hiring an extra employee or by keeping a larger inventory of new cars.

Although these questions concerning this small operation do not call for earth-shattering decisions, it is evident that statistical analysis of available data can be useful in finding answers. However, if the data represent compounded portfolio yields by the 40 largest insurance companies, similar kinds of questions could be very important to the financial management of a single company, or to stockholders, bankers, and brokers in general. The same type of statistical analysis developed in this text using simple examples also applies to a wide range of very important managerial and governmental decision-making problems.

Summary Measures

Thus far we have concentrated on describing an entire set of observations, either graphically or by means of a frequency distribution. In many cases, however, having one or more descriptive measures that summarize the data in some quantitative form is preferable to working with all observations. In particular, we are usually most interested in the two measures mentioned earlier, the *central location* of the data and the *variability* or spread of the observations. Such characteristics of a data set are called *summary measures.* As we indicated in Fig. 1.1, when these summary measures apply to an entire population, they are called parameters; when the data represent a sample drawn from a population, they are called statistics. Their usefulness is obvious if one considers the difficulty of making a logical presentation of the meaning and interpretation of a given data set. Simple intuitive or "naked-eyeball"

Table 1.4 Common Summary Measures

Central Location	Variability
Arithmetic mean	Standard deviation
Median	Variance
Mode	Range
Geometric mean	Interquartile range

analysis of the values can be misleading and may easily miss some important implications. Furthermore, presenting such analysis of large data sets is tedious and boring for the listener or reader. If the important and most useful information in a data set can be condensed into a few summary measures, then comprehension and comparison of various features of different populations or samples become much easier. In decision-making problems (for example, in statistical analysis of pollution abatement systems or the welfare program), summary presentation of this nature is called data reduction. All the information in a data set that is useful for a particular purpose is "reduced" into a single measure, such as a reliability measure or an average payment.

As we have said, the two types of summary measures most often used in statistical inference and decision-making are the central location and the variability of the data. There are a number of different ways of measuring these two characteristics, as shown in Table 1.4. Some of these terms are perhaps already familiar to you while others are new, technical terms. Although each is useful for certain purposes, this text will emphasize the two most common and useful measures in statistical inference, namely, the *arithmetic mean* and the *standard deviation.*

1.3 CENTRAL LOCATION

The single most important measure describing numerical information is the location of the *center* of the data. The term *central location* may refer to any one of a number of different measures including the mean, the median, and the mode. Each of these measures is appropriate for certain descriptive purposes, but completely inappropriate for others.

The Mode

The mode is defined as the value that occurs most often or, equivalently, the point (or class mark) corresponding to the value with the highest frequency.

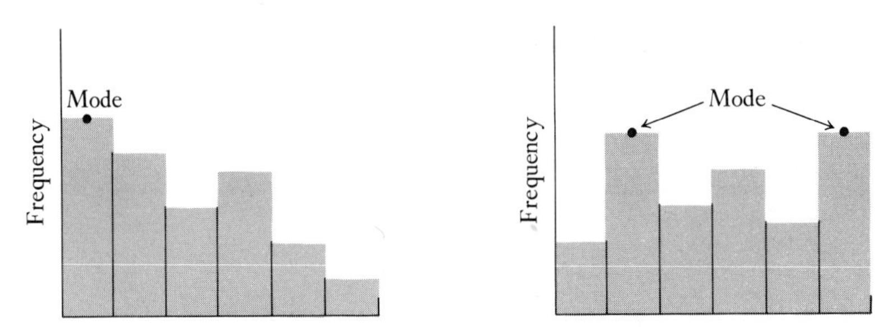

Fig. 1.8 The mode as a measure of central location.

For the sales shown in Table 1.1, the mode is 85 cars, since 85 occurs more often than any other number. For the grouped data of Fig. 1.6, the mode is at the class-mark score of 77, which represents the class with the highest frequency. Note that the mode may be a poor measure of central location, since the most frequently occurring value may not appear near the center of the data. Furthermore, the mode need not even be unique. Consider the frequency distributions shown in Fig. 1.8. The mode of the first distribution is located at the lowest class and certainly cannot be considered representative of central location. The second distribution has two modes, neither of which appears to be representative of the central location of the data. For these reasons, the mode has limited use as a measure of central location for decision-making. For descriptive analysis, however, the mode is useful because it represents the most frequently occurring value.

The Median

The median is defined as the middle value in a set of numbers arranged in order of magnitude.

When it is desirable to divide the data into two groups, each group containing exactly the same number of values, the median is the appropriate point of division. Finding the median of a set of numbers is not difficult when these numbers are arranged in ascending or descending order. If the number of values in the data set (N) is odd, the middle value can be determined by counting off, from either the highest or lowest value, $(N + 1)/2$ numbers; the resulting number divides the data into the two desired groups and thus represents the median. For example, in a list of five values the median is found by counting down (or up) $(5 + 1)/2 = 3$ values; in a list of seven values, the median is found by counting down (or up) $(7 + 1)/2 = 4$ values.

Table 1.5

	Data Set A	Data Set B	Data Set C	Data Set D
Observations	2, 2, 3, 4, 8, 10, 13	−5, 8, 8, 9, 10, 12	2, 3, 4, 4, 4, 7	11, 9, 26, 11, 10, 11
Median	4	8.5	4	11
Mode	2	8	4	11
Mean	6	7	4	13

When N is even, there are two middle values, and the median is usually defined as the number halfway between these two values. The median of six values is thus halfway between the third and fourth numbers. In Table 1.1, for instance, there are 40 numbers, so the median must be halfway between the twentieth and the twenty-first. Since the twentieth value (from the lowest end) is 77 and the twenty-first value is 78, the median in Table 1.1 is 77.50. Note in the examples given in Table 1.5 that no matter how many items there are, the median always has a value such that the *number of values* on each side of the median *is equal* when all the values are arranged in numerical order. The only time this rule causes some confusion (but, technically, still holds) is when there are several values equal to the median, as in Examples C and D in Table 1.5. In data set D we have not arranged the data in numerical order, to emphasize that the data don't always come in order.

The Mean

By far the most common measure of central location is the *arithmetic mean*, commonly referred to as the *average*. Calculating the mean (the word "arithmetic" is usually omitted) of a set of numbers is a relatively straightforward process, one that most of us learned long before knowing anything else about statistics.

The mean of a set of numbers is the sum of all values being considered divided by the total number of values in the set.

The symbol μ (the Greek letter mu) is used to represent the mean of a population.* Perhaps the mean IQ of all members of the sophomore class at your college is $\mu = 123.4$, or the mean starting salary of all graduating seniors may

* In statistics it is traditional to let Greek letters represent the characteristics of a population. A subscript is added to the Greek letter in those cases where two or more populations are being studied. For example, μ_x represents the population mean of all x-values, while μ_y represents the population mean of all y-values.

be $\mu = \$19{,}573$. As another example, the mean number of television sets per household in the United States is supposed to be $\mu = 1.91$. In most statistical problems the value of the population parameter μ is unknown and must be estimated on the basis of a sample statistic, such as the sample mean.

In order to develop a formula for the mean of a population, we continue to denote the number of values in the population by the letter N and let x_1 equal the first value in the population, x_2 equal the second value, and so forth, with x_N equaling the last value. The mean of these N values is their arithmetic sum divided by N, as shown in Formula (1.1).

Population mean:

$$\mu = \frac{x_1 + x_2 + \cdots + x_N}{N} = \frac{1}{N}\sum_{i=1}^{N} x_i. \tag{1.1}$$

To illustrate the use of Formula (1.1), we will find the mean of the values in Table 1.1. In this case $N = 40$, $x_1 = 63$, $x_2 = 66$, and $x_N = x_{40} = 94$:

$$\mu = \frac{63 + 66 + 67 + \cdots + 94}{40} = \frac{3128}{40} = 78.20.$$

Thus, the average number of new cars sold by the 40 dealers is 78.20.

The mean is shown in row four for each data set in Table 1.5. Note from these additional examples that the mean may be less than, greater than, or equal to the median. For the data from Table 1.1, the mean (78.20) is larger than the median (77.50) but smaller than the mode (85). The mean of a data set may be thought of as the *point of balance* of the data, analogous to the center of gravity for a distribution of mass in physics.

The Mean of a Frequency Distribution

The formula presented for the mean [Formula (1.1)] is based on the assumption that each value of the data set is given separately. Often, however, it is much easier to manipulate large amounts of data by grouping them into a *frequency distribution.* Columns 1 and 2 of Table 1.6 give an example of such a distribution: the starting monthly salaries (reported to the nearest 100 dollars) of 250 recent college graduates.

One way to find the mean of this population would be to sum all 250 values separately (8 values of \$1700 + 23 values of \$1800 + $\cdots$ + 11 values of \$2200), and then divide by 250. This is the procedure presented earlier in Formula (1.1). But most of us learned long ago that multiplication is easier than repeated addition; hence, we should take advantage of the fact that there are only six *different* values in Table 1.6, not 250. In other words, instead of adding \$1700

Table 1.6 Calculating the Mean for a Frequency Distribution of Monthly Salaries of Recent College Graduates

(1) Salary x_i	(2) Frequency f_i	(3) $x_i f_i$	(4) Relative Frequency f_i/N	(5) $x_i\left(\frac{f_i}{N}\right)$
1700	8	13,600	0.032	54.4
1800	23	41,400	0.092	165.6
1900	75	142,500	0.300	570.0
2000	90	180,000	0.360	720.0
2100	43	90,300	0.172	361.2
2200	11	24,200	0.044	96.8
Sum	250	492,000	1.000	1968.0

eight times, we can use the product 8($1700) = 13,600. Similar products are used for every value of x_i, as shown in column (3) of Table 1.6. The sum of these products for all six values divided by 250 yields μ.

$$\mu = \frac{492{,}000}{250} = \$1968.$$

The mean or average starting salary is thus $1968.

To obtain a formula for the above process, let x_i represent the ith value of x, and let f_i represent the frequency associated with that value. If there are c different values of x_i (i.e., c different rows), then μ is:

Population mean for frequency distribution:

$$\mu = \frac{1}{N}\sum_{i=1}^{c} x_i f_i. \tag{1.2}$$

We must emphasize that Formula (1.2) is merely a more general way of writing Formula (1.1). If each value of x_i occurs with a frequency of $f_i = 1$ (as in Table 1.1), then Formula (1.2) is identical to Formula (1.1), and the value of c will equal the value of N.

It is convenient in view of our forthcoming development of means in Chapter 3 to rewrite Formula (1.2) in a slightly different (but equivalent) form, placing N inside the sum sign.

Population mean for frequency distribution:

$$\mu = \sum_{i=1}^{c} x_i \left(\frac{f_i}{N} \right) \tag{1.3}$$

The use of Formula (1.3) is demonstrated by column (5) of Table 1.6. Note that the sum of this column yields $\mu = 1968$, the same value we calculated above by using Formula (1.2).

Comparison of the Mode, Median, and Mean

The arithmetic mean is the most widely used measure of central location. Its disadvantage for descriptive purposes is that it is affected more by extreme values than the median or the mode because it takes into account the difference among all values, not merely their rank order (as does the median) or their frequency (as does the mode). A recent cartoon illustrated this problem quite well by depicting a small town worker commenting to a reporter that "the average yearly income in this town is $100,000—there's one person making a million, and ten of us workers making $10,000."

Use of the median requires knowledge not only of the frequency of the values in a data set, but also of their ranking, so that these values can be ordered and the middle value obtained. Suppose each of 50 pro-football rookies can be rated I, II, III, IV, or V according to his performance in a tryout camp and his past record. A ranking of I indicates that the player is a guaranteed all-pro, while a ranking of V indicates that the player should investigate other career options quickly. Consider, for example, the data in Table 1.7.

The mode for these data is rank II, while the median occurs at rank III (since the twenty-fifth and twenty-sixth values both lie in this rank). It would be inappropriate to calculate a mean, since the differences between ranks are not precisely known, nor can these differences be assumed to be equal.

In contrast to the example above, economic and business problems generally involve data in which the differences among values are known—income measures, output quantities, retained profits, prices, and interest rates. The same factor that makes the mean inappropriate for frequency data and ranked data is its special advantage in these cases—it is a more reliable measure of

Table 1.7 Ranking of Pro-Football Rookies

Rank	I	II	III	IV	V
Frequency	3	17	14	6	10

central location because it requires more knowledge about the population, namely the difference between each value in the data set.

In concluding this section, we must point out that the mean, median, and mode are not the *only* measures of central location. Another type of mean, the *geometric mean*, is especially useful in certain types of problems in business and economics.* We will not present such measures here, since we wish to emphasize the use of the arithmetic mean.

Define: Population, parameter, sample statistic, histogram, frequency polygon, class marks, relative frequency, mean, median, mode, central location, ogive.

Problems

1.1 Explain what you understand to be the primary purpose of descriptive statistics. Give an example from a recent newspaper or magazine of some use of descriptive statistics.

1.2 Assume that you are given a list of the starting salaries of 50 students from last year's graduating class in your university.

a) Describe a situation in which these 50 values represent a population.

b) Describe another situation in which these 50 values represent a sample.

c) If you are given just the 50 values, is it possible for you to determine whether or not these values represent a population? If you are told that their mean is $\mu = \$18,873$, how do you now know that these values represent a population?

1.3 A $25 sweater is on sale for $15 and a $75 coat is on sale for $60. Find the average percent decrease in price for the two items.

1.4 The following 16 values represent the asking price for homes listed for sale in an upper-middle class neighborhood (in thousands of dollars).

68, 83, 47, 51, 91, 89, 99, 73,
62, 58, 91, 66, 75, 84, 77, 69

a) Construct a histogram representing this population, using classes of 39.5–49.5, 49.5–59.5, etc. Draw the frequency polygon.

b) Construct the cumulative relative frequency distribution. What percent of the population of prices is less than $70,000?

c) Find the mean and the median.

d) Does this population have a unique mode? If so, is this mode a good measure of central location?

* The *geometric mean* is a particularly appropriate measure of the central location of data expressed in relative terms, such as rates of change or ratios (e.g., change in the Consumer and Producer price indexes). The geometric mean gives equal weight to changes of equal *relative* importance. For example, if an index is doubled in value, this change is weighted equally to a change that halves the value of this index.

1.5 In a recent study, the Federal Aviation Administration (FAA) tabulated the ages of 301 adult passengers who were flying with children under 12 years of age.

Age:	18–22	23–27	28–32	33–37
Number:	60	80	50	40
Age:	38–42	43–47	48–52	>90
Number:	30	20	20	1

a) Draw the histogram for these data, excluding the one 97-year-old person. Draw the frequency polygon.

b) Show the relative frequencies on your histogram in part (a).

c) Construct the cumulative relative frequency distribution for the data in part (a). What percent of this population is younger than 33 years?

d) In which class does the median fall? (Include the 97-year-old in this case.) Would the midpoint of this class be a good guess for the median? If not, what age would you use as an approximation to the median?

e) What is the mode of this distribution?

f) Calculate a mean for these data, including the 97-year-old. (*Hint:* Use the midpoints 20, 25, . . . , 50 to represent all values in a class.)

1.6 Records were kept on the number of absences of 100 assembly-line workers during a one-month period.

Number of Sick Days	0	1	2	3	4
Frequency (f)	47	33	14	5	1

a) What are the median and mode of this population?

b) Find the mean of this population.

c) Sketch the relative frequency distribution and the cumulative relative frequency distribution.

1.7 There are a number of ways to "lie" with statistics when using index numbers. To illustrate, suppose you are preparing a report on how much more it costs now for textbooks and materials for a statistics course, relative to 1970. Let's say that textbook prices have doubled since 1970 and the calculator that you have to buy has halved in price.

a) Assign both the textbooks and calculator an index of 100 for 1970. The present-day indexes will thus be 200 and 50, respectively. Comparing the average index for 1970 vs. the present, has the average index increased or decreased?

b) Follow the same procedure as indicated in part (a) except now let the two present-day indexes equal 100. Has the average index increased or decreased from 1970 to the present? Comment on why your answer to (a) differs from your answer to (b).

1.8 Publishers Clearing House recently announced a contest in which the following amounts (among others) would be awarded.

Award	3900	2600	1300	1200	1000	500	250	100	50	25
Frequency of Award	1	3	5	5	6	72	50	250	500	1750

a) What are the median and mode of this population?

b) Find the mean of this population.

c) If one million people enter this contest, what is the average amount each person will win?

1.9 One line of General Motors cars recently sold 850,000 units. The following table gives the percent of these sales in ten different categories, each category representing the fuel economy (mpg) of a particular model.

mpg	20.97	25.39	21.02	20.43	19.94	18.46	17.79	18.92	17.81
% of Sales	2.05	25.28	8.42	32.20	3.52	12.99	1.20	8.24	6.10

a) Find the mean fuel economy (mpg) for this line of car.

b) Find the median and mode of this population

1.10 Sales by the Colgate Palmolive Company for 1978–80 (in millions of dollars) are shown in Figs. 1.9 and 1.10. Comment on why Figs. 1.9 and 1.10 differ. Which one gives the "true" picture of sales? What possibilities do you see here for "lying" with statistics?

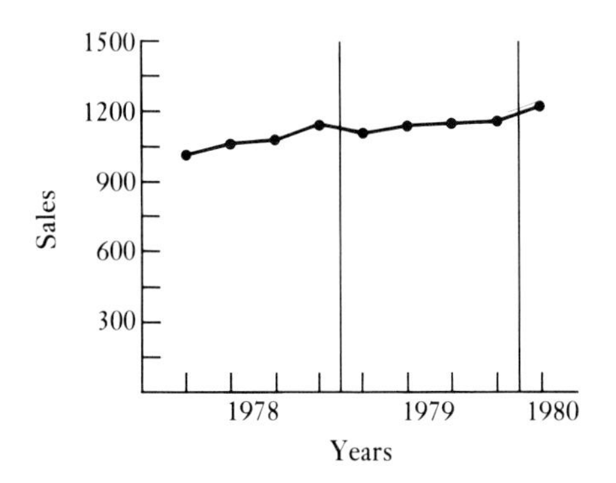

Figure 1.9

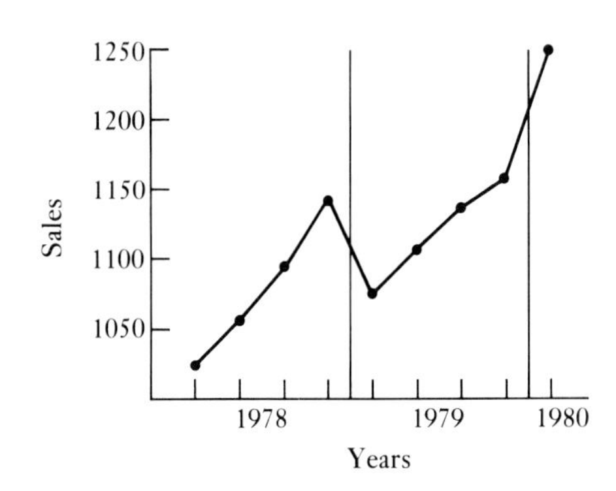

Figure 1.10

1.11 Draw a "pie chart" (comparable to Fig. 1.2) using the following data on the projected energy demand (by consuming sector) for the year 2000.

Sector	New Energy	Industrial	Transportation	Residential/ Commercial
Percent of Demand	8	37	20	35

1.12 A survey recently indicated the percent, by functional field, of MBAs expected to be hired by energy-related companies over the next five years. Present the data given below in the form of a "pie chart."

Accounting:	9.6%	Planning:	8.9%
Finance:	42.4%	Personnel:	5.2%
Marketing:	13.8%	Information systems:	4.9%
Operations management:	8.4%	Other:	6.8%

1.13 Suppose that a certain gasoline producer sponsors a mileage economy test involving 30 cars. The cumulative frequencies of the number of miles per gallon, x, recorded to the nearest gallon, are given below.

x	$\sum f$
25	8
26	17
27	24
28	30

Find the average miles per gallon for this population by using Formula (1.2) and Formula (1.3).

1.14 Twenty communities provide information on the vacancy rate in local apartments. Find the mean of the following population of vacancy rates, using both Formula (1.2) and Formula (1.3). [*Hint:* Use the middle value for each vacancy rate.]

Frequency	Vacancy Rate
10	3–7%
6	8–12%
4	13–17%

1.4 MEASURES OF DISPERSION

Measures of central location usually do not give enough information to provide an adequate description of the data because variability or spread is ignored. An individual who makes a judgment on the mean alone may be compared to the person whose head is in a freezer and whose feet are in an oven declaring "On the average, I feel fine." Another measure that indicates how spread out or *dispersed* the data are is needed.

Since there are a number of ways to measure spread, let us consider some of the properties that a good measure should have. A good index of spread should be independent of the central location of the observations; that is, independent of the mean of the data. This property implies, in effect, that if a constant were added to (or subtracted from) each value in a set of observations, this transformation would not influence the measure of spread. In addition, to be most useful, a measure of spread should take into account *all* observations in its calculation, rather than just a few selected values such as the highest and lowest. Finally, a good measure should reflect the typical spread of the data and it should be convenient to manipulate mathematically.

The Range

One simple example of a measure of spread is the *range*, which is defined as the *absolute difference* between the highest and lowest values. The range for the data in Table 1.1 is the absolute difference between the highest value, 94, and the lowest value, 63, or

$$|94 - 63| = 31.$$

Similarly, the ranges of the four examples in Table 1.5 (on page 16) are

$$|13 - 2| = 11, \quad |12 - (-5)| = 17, \quad |7 - 2| = 5, \quad \text{and} \quad |26 - 9| = 17,$$

respectively. The range has the advantages of being independent of the measure of central location and being easy to calculate. It has the *disadvantages* of ignoring all but two values of the data set and not necessarily giving a *typical* measure of the dispersion, since a single extreme value changes the range radically.

Deviations About the Mean

We now begin development of a measure of variability that has all the desirable properties mentioned at the beginning of this section. First, recall that we want our measure to be independent of the mean of the data. In other words, the value of μ should have no influence on the value of our measure of variability. This objective is accomplished by working always with data sets that have the *same* mean. That is, if all the populations that we wish to describe have the same mean, then the value of μ cannot influence our measure of variability. It is obvious, of course, that not all data sets have the same mean to begin with. We have to *transform* the values in each set in such a way that the transformed means are all equal.

The transformation used in statistics is designed to yield a data set always having $\mu = 0$. To make $\mu = 0$ is quite simple: Merely subtract the mean of the original data set from each value in that set. Each resulting number, called a *deviation* and denoted $(x - \mu)$, indicates how far and in which direction the original number lies from the mean. For example, the deviation $(x - \mu) = 5$ reflects a value of x five units above the mean, and the deviation $(x - \mu) = -7$ indicates a value of x seven units below the mean.

The sum of the deviations $(x - \mu)$ will always equal zero.* Similarly, the average of the deviations $(x - \mu)$ will always equal zero. Consider, for example, the five values in Table 1.8 representing the price per share of five different common stocks you are considering purchasing. The mean of these prices is $\mu = 10$. If we subtract μ from each value of x, the mean of the new set of values, labeled $x - \mu$, must equal zero. Any set of numbers can be transformed in this fashion into a set of deviations with a mean of zero.

To formalize the above process, consider a population with N values x_i, $i = 1, 2, \ldots, N$, and mean μ. When these N values are transformed into

* The fact that the $\sum_{i=1}^{N} (x_i - \mu)$ must equal zero can be proved as follows:

$$\sum_{i=1}^{N} (x_i - \mu) = \sum_{i=1}^{N} x_i - \sum_{i=1}^{N} \mu = \frac{N}{N} \sum_{i=1}^{N} x_i - N\mu = N\mu - N\mu = 0.$$

Table 1.8

	x	Deviations $x - \mu$
	4	-6
	8	-2
	10	0
	13	$+3$
	15	$+5$
Sum	50	0
Mean	10	0

deviations from the mean, the new values are $x_i - \mu$ for $i = 1, 2, \ldots, N$, and the sum of these deviations must be zero. That is,

$$\sum_{i=1}^{N} (x_i - \mu) = 0. \tag{1.4}$$

Since the transformation $(x_i - \mu)$ gives all sets of data a common central location (i.e., they all have means of zero), measures of dispersion defined in terms of deviations about the mean have the desirable property of being independent of central location.

While the sum of the deviations about μ is advantageous because it takes into account *all* the observations and is independent of the mean, this sum clearly cannot be our measure of variability since its value always equals zero. We will see in the following section that if we *square* each deviation before we sum, the resulting measure no longer equals zero and is relatively easy to manipulate mathematically. If we take the *average* of these squared deviations (i.e., divide their sum by N), our measure also will reflect the typical spread of the data.*

Standard Deviation and Variance

Consider the *average squared deviations* about the mean. Squaring the deviations avoids the problem inherent in ordinary deviations about the mean (namely that their sum always equals zero). Indeed, as we indicated above, this index

* Another possibility for measuring variability is to average the sum of the absolute deviations about the mean, which is

$$\frac{1}{N}\sum |x - \mu| = \text{mean absolute deviation.}$$

This measure has all of the desirable properties described above, except for the fact that it is not convenient to manipulate mathematically.

meets all the properties of a good measure of spread; *thus the average squared deviation is the traditional method for measuring the variability of a data set.* Since it uses the deviations about the mean, it is independent of central location. It uses every value in the data set and is reasonably easy to compute mathematically. Furthermore, it is very sensitive to any change in the values—even a single change of one value in a set of 100 would result in a different measure of variability.

This "average squared deviation" measure is called the *variance* and is denoted for a population by the symbol σ^2, which is the square of the lower case Greek letter *sigma*. It is defined as follows:

Population variance:
$$\sigma^2 = \frac{1}{N}\sum_{i=1}^{N}(x_i - \mu)^2. \tag{1.5}$$

In calculating a variance using Formula (1.5), one first calculates each deviation; these deviations are then squared, the squared deviations are then summed, and finally the sum is divided by N.

Very often, the square root of the variance, denoted σ and called the *population standard deviation,* is used in place of (or in conjunction with) the population variance to describe variability. The standard deviation is usually more convenient than the variance for *interpreting* the variability of a data set, since σ^2 is in squared units while σ is in the same units as the original data. The population standard deviation is defined as follows:

Population standard deviation:
$$\sigma = \sqrt{\frac{1}{N}\sum_{i=1}^{N}(x_i - \mu)^2}. \tag{1.6}$$

To illustrate the calculation of a population variance and standard deviation, we will assume that the values of x in Table 1.9 represent the number of tape recorders assembled by ten different workers on an assembly line over the past month. That is, worker 1 assembled 115 recorders, worker 2 assembled 122 recorders, etc. The mean number of recorders assembled, at the bottom of the first column, is

$$\mu = \frac{1200}{10} = 120 \text{ (recorders).}$$

The deviations about the mean are shown in the second column. Note that the sum of these deviations equals zero, as it must. The third column of values

Table 1.9

	x	$x - \mu$	$(x - \mu)^2$
	115	−5	25
	122	+2	4
	129	+9	81
	113	−7	49
	119	−1	1
	124	+4	16
	132	+12	144
	120	0	0
	110	−10	100
	116	−4	16
Sum	1200	0	436
Mean	$120 = \mu$	0	$43.6 = \sigma^2$

gives the *squared* deviations about the mean, the sum of which is 436. Hence, the *average squared deviation* (the variance) of this population is:

Recorders variance:

$$\sigma^2 = \frac{1}{N}\sum_{i=1}^{N}(x_i - \mu)^2 = \frac{436}{10}$$

$$= 43.6 \text{ (recorders)}^2.$$

While the variance meets the criteria desired for a good measure of dispersion, the standard deviation is a better measure of the typical size of a deviation from the mean because this measure is in the same units as the original data (while, as said earlier, the variance is in squared units). For the data in Table 1.9 the standard deviation is

$$\sigma = \sqrt{\text{Variance}} = \sqrt{43.6} = 6.60 \text{ (recorders)}.$$

In general, a precise interpretation of values of σ and σ^2 is difficult because variability depends so highly on the unit of measurement. For instance, variability of income in the U.S. is certainly larger when measured in dollars than when measured in thousands of dollars. In all cases, as the spread of a population increases, the values of σ^2 (and σ) will also increase. On the other hand, if $\sigma^2 = \sigma = 0$, this means there is no variability at all to the data (all x-values are the same and equal to their mean; that is, x is a constant).

One rule of thumb that often provides a good *approximation* to the spread of a set of observations states that

---

***about 68 percent** of all values will fall within one standard deviation to either side of the mean, and **about 95 percent** of all values will fall within two standard deviations to either side of the mean.**

In other words, the interval from $(\mu - \sigma)$ to $(\mu + \sigma)$, which we will write as $(\mu \pm \sigma)$, often will contain about 68 percent (or about $\frac{2}{3}$) of all population values. Similarly, the interval $(\mu - 2\sigma)$ to $(\mu + 2\sigma)$, that is $(\mu \pm 2\sigma)$, often will contain about 95 percent of all the population values. We must emphasize that these rules are only approximations and do not necessarily hold for any one discrete example. For other examples, the approximations may be quite good. Consider, for instance, the distribution of IQs in the U.S. measured on the Stanford-Binet test, which has a mean of $\mu = 100$ and a standard deviation of $\sigma = 16$. Studies of this population indicate about 68 percent of all U.S. IQs fall between 84 and 116 (which is $\mu \pm 1\sigma$) and approximately 95 percent of all IQs fall between 68 and 132 (which is $\mu \pm 2\sigma$). It is not difficult, for example, to show for the data on tape recorders (Table 1.9) that 60 percent of the population (six of the ten values) falls in the interval $(\mu \pm \sigma)$ and 100 percent of the data falls in the interval $(\mu \pm 2\sigma)$. These intervals are shown in Table 1.10. Recall that $\mu = 120.0$ and $\sigma = 6.60$.

As a final example of the process of calculating and interpreting the variance and standard deviation of a population, consider once again the car-sales

Table 1.10

Interval	Values Within Interval	Actual Percent of Population Within Interval	Rule of Thumb
$\mu \pm \sigma = 120 \pm 6.60$ $= \begin{cases} 113.4 \text{ to} \\ 126.6 \end{cases}$	115, 116, 119, 120, 122, 124	60%	68%
$\mu \pm 2\sigma = 120 \pm 2(6.60)$ $= \begin{cases} 106.80 \text{ to} \\ 133.20 \end{cases}$	110, 113, 115, 116, 119 120, 122, 124, 129, 132	100%	95%

* In Chapter 5 we will show that this rule of thumb is based on the assumption that the population has a symmetrical bell-shaped distribution called the *normal distribution*. A more general rule for such approximations is given in Problem 1.48.

data in Table 1.1. We previously calculated the mean of this population to be $\mu = 78.20$. We now calculate the forty deviations from this mean so that σ^2 and σ can then be determined:

$$\begin{aligned}
63 - 78.20 &= -15.20 \\
66 - 78.20 &= -12.20 \\
67 - 78.20 &= -11.20 \\
68 - 78.20 &= -10.20 \\
&\vdots \\
92 - 78.20 &= 13.80 \\
94 - 78.20 &= 15.80
\end{aligned}$$

The mean of these deviations must be zero (since their sum is zero). The value of σ^2 (or variance) is

$$\begin{aligned}
\sigma^2 &= \frac{1}{N}\sum_{i=1}^{N}(x_i - \mu)^2 \\
&= \frac{1}{40}[(-15.20)^2 + (-12.20)^2 + (-11.20)^2 + \cdots + (13.80)^2 + (15.80)^2] \\
&= 55.36 \text{ (cars)}^2,
\end{aligned}$$

and the standard deviation is

$$\sigma = \sqrt{55.36} = 7.44 \text{ (cars)}$$

If the rule of thumb described earlier holds, then $\mu \pm \sigma$ should contain about 68 percent of the car-sales data, and $\mu \pm 2\sigma$ should contain about 95 percent of all these values. Checking these intervals against the values in Table 1.1 gives the following results:

Interval	Percent of Values
$\mu \pm \sigma = 78.20 \pm 7.44 = 70.76$ to 85.64	70%
$\mu \pm 2\sigma = 78.20 \pm 2(7.44) = 63.32$ to 93.08	95%

Hence, had we known only that $\sigma = 7.44$ and $\mu = 78.20$ for this particular set of observations, we could have given a good description of the variability of the data as well as its central location.

Considering all the measures of central location and dispersion that we have presented, the two measures most often useful in statistical inference and decision-making are the *mean* and the *standard deviation.* These are common household terms to any statistician, as they are used every day in helping to make decisions based on statistical analysis of data sets. The mean is precisely the balance point of all the values. The standard deviation is the typical (or standard) size of the difference (deviation) between the individual values of the population and the mean of the population. As such it provides a good insight into the extent of variability in the data set, especially when the rules of thumb apply. The reader should keep in mind that the variance and the standard deviation do *not* represent two different ways of measuring the variability of a population. Since σ is merely the square root of σ^2, these two measures reflect the *same information* about variability, but are expressed in different units. The standard deviation is easier to interpret because it is not in squared units, but it is more difficult to manipulate mathematically than the variance because of the square-root sign.

Population Variance for a Frequency Distribution

In Formula (1.5) we calculated the variance of a population, assuming that all frequency values were equal to one (i.e., $f_i = 1$). Formula (1.5) can be generalized to take into account frequencies other than one in exactly the same manner in which the formula for μ was generalized. Again, we assume that there are c different values of x (that is, c rows). A squared deviation $(x_i - \mu)^2$ is calculated for every row. Each squared deviation is then multiplied by its frequency. Dividing the sum of the products by N we get:

Population variance for a frequency distribution:

$$\sigma^2 = \frac{1}{N}\sum_{i=1}^{c}(x_i - \mu)^2 f_i. \tag{1.7}$$

Formula (1.7) is illustrated by the first four columns of Table 1.11 for our salary example,* which originated in Table 1.6 (on page 18). Using Formula (1.7) and the sum in column 4, we get

$$\sigma^2 = \frac{1}{250}(3{,}004{,}000) = 12{,}016 \text{ (dollars)}^2.$$

As in the case of the formula for computing the mean, it is convenient to modify Formula (1.7) by moving the value $(1/N)$ inside the sum sign. This yields

* The inquisitive reader may wonder why the sum of column (3) $[(x_i - \mu)]$ in Table 1.11 does not equal zero, as Formula (1.4) seems to indicate that it should. The reason is that each deviation occurs more than once, according to the frequencies in column (2). If each deviation is multiplied by its frequency and the products are summed, their total will be zero.

Table 1.11 Calculating the Mean for a Frequency Distribution of Monthly Salaries of Recent College Graduates

(1) Salary (x_i)	(2) Frequency, f	(3) $(x_i - \mu)$	(4) $(x_i - \mu)^2 f_i$	(5) Relative Frequency (f_i/N)	(6) $(x_i - \mu)^2 \left(\frac{f_i}{N}\right)$
$1700	8	−268	574,592	0.032	2,298.368
1800	23	−168	649,152	0.092	2,596.608
1900	75	−68	346,800	0.300	1,387.200
2000	90	32	92,160	0.360	368.640
2100	43	132	749,232	0.172	2,996.928
2200	11	232	592,064	0.044	2,368.256
1968 = μ	250 = Sum		3,004,000	1.000	12,016.000

the following (equivalent) formula:

Population variance for a frequency distribution:

$$\sigma^2 = \sum_{i=1}^{c} (x_i - \mu)^2 \left(\frac{f_i}{N}\right) \tag{1.8}$$

The fact that Formula (1.8) is equivalent to Formula (1.7) can be seen by examining the sum of the values in column (6) in Table 1.11:

$$\sigma^2 = \sum_{i=1}^{c} (x_i - \mu)^2 \left(\frac{f_i}{N}\right) = 12{,}016 \text{ (dollars)}^2.$$

This is the same value obtained previously by using Formula (1.7). The standard deviation is the square root of this value, $\sigma = 109.6$, and it is measured in the original units (dollars).

Short-Cut Formula for Variance

Statisticians, like most of us, really don't enjoy performing a lot of burdensome calculations. Perhaps that is one reason why a short-cut formula was developed for calculating variances. Formula (1.9) is merely the result of manipulating Formula (1.7) algebraically, and therefore is *exactly* equivalent to Formulas (1.7) and (1.8). It will always give the same value for σ^2 as these formulas (except for rounding errors).* The advantage of the short-cut formula is that in many problems it makes the process of calculating σ^2 somewhat easier (especially when μ is not an integer), as it does not require finding each deviation and squaring it. This short-cut formula is:

Short-cut formula for calculating σ^2:

$$\sigma^2 = \left(\sum_{i=1}^{c} x_i^2 \frac{f_i}{N}\right) - \mu^2. \tag{1.9}$$

To illustrate the use of Formula (1.9), let us return to the grouped data on car sales in Table 1.3. Although this table was derived from the original (ungrouped) data in Table 1.1, we will assume that the 40 individual values of x_i in Table 1.1 are not known. When only grouped data are available, then the midpoint of each class is used to represent all population values in that class. For example, we will assume that all four values of x_i in the second class in Table 1.3 (64.5–69.5) fall at the midpoint of this class, which is 67. Thus, for grouped data the symbol x_i represents the midpoint of the ith class; for ungrouped data, the symbol x_i represents the ith value in the population. All the data necessary for calculating σ^2 by the short-cut formula are given in Table 1.12.

The sum of the values in the last column of Table 1.12 represents the first term in Formula (1.9). The sum of the values in the fifth column (78.25) represents the mean of the grouped data (note that the mean of the grouped data is slightly

Table 1.12 Calculation of σ^2 by Using Short-Cut Formula (1.9) and Data from Table 1.3

(1)	(2)	(3)	(4)	(5)	(6)	(7)
Class	Midpoint (x)	f_i	f_i/N	$x_i\left(\frac{f_i}{N}\right)$	x_i^2	$x_i^2(f_i/N)$
59.5–64.5	62	1	0.025	1.550	3844	96.100
64.5–69.5	67	4	0.100	6.700	4489	448.900
69.5–74.5	72	8	0.200	14.400	5184	1036.800
74.5–79.5	77	11	0.275	21.175	5929	1630.475
79.5–84.5	82	6	0.150	12.300	6724	1008.600
84.5–89.5	87	7	0.175	15.225	7569	1324.575
89.5–94.5	92	3	0.075	6.900	8464	634.800
		$N = 40$	1.000	$78.25 = \mu$		6180.250

* Proof of this fact is left as an exercise for the reader in Problem 1.45.

higher than the mean of the ungrouped data, 78.20). The variance is thus

$$\sigma^2 = \sum_{i=1}^{c} x_i^2(f_i/N) - \mu^2 = 6180.250 - (78.25)^2 = 57.1875\ (\text{cars})^2.$$

The reader may wish to verify that this same value of σ^2 will result if Formula (1.8) is used instead of the short-cut formula. Note also that the current value of σ^2 (=57.1875) is slightly larger than the value calculated on page 29, $\sigma^2 = 55.36$. This difference occurs because we used grouped data to calculate the current value.

Before presenting other measures, we will discuss why the characteristics of both central location and variability are used in most statistical analysis. Some examples of simple decision problems will illustrate why one characteristic without the other is, in general, insufficient.

Inadequacy of Central Location Measures Alone

Consider the case in which a person must choose between two sales jobs, each having potential earnings of $25,000 a year. One company representative says that their sales persons earning $25,000 work an average of 30 hours per week. The other prospective employer says the average for similar employees is 50 hours. One might decide on this basis to work for the first company with the "average" work week of 30 hours. Be careful! The average hours worked per week to earn $25,000 in the first company may be as low as 30 because the company president has hired some relatives who hardly work at all each week while the typical salesperson is working 65 hours per week. It is necessary to know more about the *distribution* of the data set "hours worked per week" before this information can be used intelligently in making a decision.

Inadequacy of Variability Measures Alone

Consider the job choice given above, but suppose that an important factor in the decision is the current earnings of employees who five years ago had starting positions similar to the one now being offered. Assume that the data for the first company indicate that the range from the lowest to the highest salary is $15,000; the data for the second company indicate a much larger range of $100,000. The job with the second company might appear preferable because it seems to have a higher potential for rewards. Suppose, however, that the data for the first company also show that these five-year employees earn an average of $35,000, while those with the second company earn an average of only $30,000. Perhaps the only people in the second company earning over $30,000 are a few privileged salespersons getting unearned commissions on government contract purchases. From this example we see that a better decision can be made when one knows more than just the variability of the data.

Define: dispersion, range, deviation, variance, standard deviation.

Problems

1.15 Find the variance for the population described in Problem 1.4.

1.16 Find the variance for the population described in Problem 1.5.

1.17 a) Find the variance of the number of absences per worker in Problem 1.6, using Formula (1.7).

b) Verify your answer to (a) by using the short-cut formula.

c) What percent of the population falls within the interval $\mu \pm \sigma$? Within the interval $\mu \pm 2\sigma$?

1.18 a) Find the standard deviation for the population in Problem 1.9.

b) Find the percent of the population that falls within the $\mu \pm \sigma$ and $\mu \pm 2\sigma$ intervals.

1.19 The following frequency distribution represents the price-earnings ratio for a population of 20 common stocks.

Class	Frequency	Class Mark	fx	fx^2
1–5	4	3	12	36
6–10	8	8	—	—
11–15	3	—	—	—
16–20	5	—	90	—

a) Find the mean of this population by completing the table above.

b) Find the variance, using Formula (1.9).

1.20 Find the variance for the population in Problem 1.14 using all three of the equivalent variance formulas given in the text [Formulas (1.7), (1.8), and (1.9)].

1.21 The Graduate Management Admissions Test (GMAT) is often required for admission to graduate schools of business. Suppose you learn that the population of test scores is bell-shaped and that approximately 95% of all test scores fall between 280 and 680.

a) What value of μ and σ would you estimate for this population based on the information given?

b) What percent of the population would you estimate to fall between 380 and 580? What percent would you estimate to be greater than 580?

1.22 The risk in purchasing a common stock is often measured by variance of the prices the stock might assume over a specified interval of time. Consider the following stocks, each with a predicted mean and variance for one year from now.

Stock	A	B	C	D
Mean	10	200	100	100
Variance	4	16	50	100

a) Which stock has the largest risk?
Which stock has the lowest risk relative to its price?

b) Based on these estimates, what is the lowest price you might expect for stock B?

1.23 Statistical analysis has become increasingly important in equal employment opportunity litigation. For example, in examining the hiring practices of a company it is often necessary to define characteristics of the population from which workers could be drawn. In one study, the following commuting distances were determined for workers in one metropolitan area.

Distance Traveled to Work	Fraction of Workers
less than 1 mile	0.12
1.0–1.9 miles	0.10
2.0–3.9 miles	0.19
4.0–5.9 miles	0.17
6.0–9.9 miles	0.18
10.0–14.9 miles	0.12
15 or more	0.12
	1.00

a) Calculate the mean and variance of this population using the midpoint of each class (use 20 miles as the midpoint of the last class).

b) Comment on why this particular classification of distances is not convenient from a statistical viewpoint.

1.5 OTHER DESCRIPTIVE MEASURES

While the mean and the standard deviation are the most common descriptive measures, there are a number of other measures that give additional information about the characteristics of a data set. This section is devoted to describing, rather briefly, a few of these measures.

Percentiles, Deciles, and Quartiles

The summary measures described thus far all use just a single number to describe certain characteristics of a population. In some circumstances it may be helpful to use *more* than one number to describe a data set. For example, a company recruiter visiting a college campus may be interested in learning more than just the mean or median grade point average for all graduating seniors. This person may want to know the average of those members of the graduating class who form the upper 10 percent, the upper 20 percent, and so forth. Percentiles, deciles, and quartiles are useful in this circumstance in that they divide a data set into a specified number of groups, each containing the same number of values. *Percentiles* divide the data into 100 equal parts, each representing one percent of all values. The 90th percentile, for example, is that value which has 90 percent of all values below it and 10 percent above it. Thus, a student scoring higher than 95 percent and lower than five percent of all students on the college board exams is said to have scored in the 95th percentile. Percentiles can be *determined exactly* from a table of cumulative relative frequencies of ungrouped data and *approximated* from a table of grouped data. From Table 1.3, for example, we can estimate only that the value of the 50th percentile (the median) is somewhere between 74.5, representing a cumulative relative frequency of 0.325, and 79.5, representing a cumulative relative frequency of 0.600.

The ogive given in Fig. 1.7 is useful for approximating a value for the median. Since the median is (by definition) the 50th percentile, we need to find the point on the horizontal axis corresponding to $\sum f/N = 0.50$. Figure 1.11 (a modified version of Fig. 1.7) shows a horizontal line from 0.50 to the ogive; the dotted vertical line from the ogive to the horizontal axis represents our approximation to the median. We can, by interpolation, give an exact value for this approximation by first noting that the amount of relative frequency in the class containing the median (74.5 − 79.5) is 0.600 − 0.325 = 0.275. The distance from 0.325 to 0.500 (the median) is 0.500 − 0.325 = 0.175. Hence, the horizontal line at 0.500 is

$$\frac{0.500 - 0.325}{0.600 - 0.325} = \frac{0.175}{0.275} = 0.636,$$

or 63.6 percent of the distance from 0.325 to 0.600. By the rules of similar triangles, the median must also be 63.6% of the distance between 74.5 and

Fig. 1.11 Determining a median.

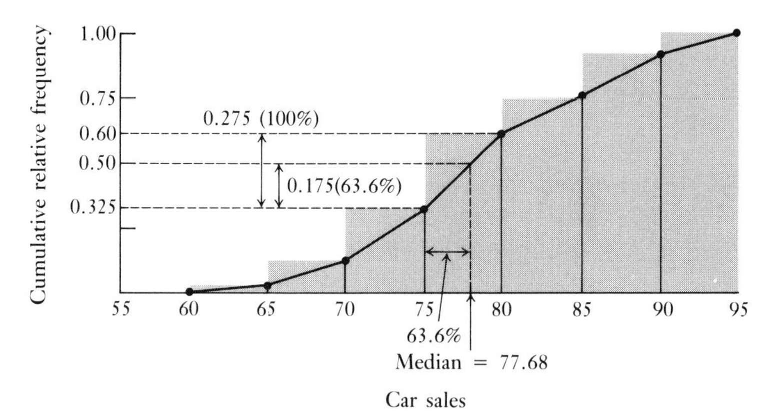

79.5. That is,

$$\text{median} = 74.5 + 0.636(79.5 - 74.5) = 77.68.$$

Note how well this value corresponds to the actual (ungrouped data) median of 77.50 calculated earlier.

Quartiles and deciles are defined in much the same fashion as percentiles: *quartiles* divide the data into four equal parts, while *deciles* divide the data into ten equal parts. The *first* quartile value is that point which exceeds one-fourth and is exceeded by three-fourths of the observations. Only three quartile values are necessary to divide the data into four parts. Likewise, nine decile values divide a set of observations into ten equal parts. The *fifth decile* and the *second quartile* values are equivalent to the median. Determination of quartile values for *ungrouped* data is discussed below.

Calculation of Quartiles—Ungrouped Data

For many populations it may be difficult to determine a unique number corresponding to each quartile value. When this happens there are a number of rather arbitrary methods for specifying a single point. We shall describe one such method by applying it to find quartiles for *ungrouped data.* Assume that there are N numbers ranked in an array from lowest to highest and we wish to determine the three quartile values q_1, q_2, and q_3. The *ranks* corresponding to the first, second, and third quartile values are as follows:

$$\begin{aligned} \text{Rank of the first quartile value} &= (1/4)N + \tfrac{1}{4}, \\ \text{Rank of the second quartile value} &= (1/2)N + \tfrac{1}{2} = \text{Median,*} \\ \text{Rank of the third quartile value} &= (3/4)N + \tfrac{1}{4}. \end{aligned}$$

For example, the data of Table 1.1 consist of 40 observations. The rank of the first quartile value is thus

$$(\tfrac{1}{4})40 + \tfrac{1}{4} = 10.25.$$

This means that the first quartile value is located beyond the 10th-ranked value by one-fourth the distance to the 11th-ranked value; i.e., one-quarter of the way between the 10th and 11th values. Since ranks 10 and 11 are both 73, the first quartile value is $q_1 = 73$. The second quartile value corresponds to the rank $(\frac{1}{2})N + \frac{1}{2} = 20.50$, or $q_2 = 77.5$, which is the median. Finally, since $(\frac{3}{4})N + \frac{1}{4} = 30.25$, and the 30th rank is 84 while the 31st rank is 85, the third quartile value is $q_3 = 84.25$.

Once the first and third quartile values have been found, it is easy to calculate the 50-percent midrange, for this value is merely the difference $q_3 - q_1$.

* This formula for the median is equivalent to $(N + 1)/2$, presented in Section 1.3.

In fact, the ranks at q_1 and q_3 were defined so that the difference between these ranks would exactly equal 50 percent of the values in the population:

$$(\tfrac{3}{4}N + \tfrac{1}{4}) - (\tfrac{1}{4}N + \tfrac{1}{4}) = \tfrac{1}{2}N.$$

The 50-percent midrange is often called the *interquartile range* since it *gives the range between the first and third quartiles.* For the car-sales example,

$$q_3 = 84.25 \qquad \text{and} \qquad q_1 = 73.00;$$

hence, the interquartile range is

$$q_3 - q_1 = 11.25.$$

The interquartile range is a useful measure because it is relatively easy to interpret and not too difficult to calculate. In many circumstances it can also be used to give a rough approximation to the standard deviation of a set of values. The interquartile range includes 50 percent of all values, whereas plus or minus one standard deviation includes approximately 68 percent of all values. Thus, if we multiply the interquartile range by the ratio 68/50 the result should be a good approximation to the interval of length 2σ between $\mu - \sigma$ and $\mu + \sigma$. In the car-sales example, for instance, multiplying $q_3 - q_1 = 11.25$ by 68/50 yields 15.3; this is not a bad estimate of the true interval between $\mu \pm \sigma$, which is $2(7.44) = 14.88$.

Shapes of Distributions

Having a method for describing the *shape* of a frequency distribution is often more helpful than just being able to describe the central location or spread of a set of values. Most of the distributions representing real-world problems are called *unimodal* distributions, implying that they have only one peak, or *mode.* A distribution with two peaks is called a *bimodal* distribution. Often, distributions with more than one mode actually reflect the combination of two or more *separate* kinds of data into a single set of values.

Consider, for example, the frequency distribution shown in Fig. 1.12 representing the frequency of sales of television sets for a large department store, in intervals of \$100. What Fig. 1.12 actually represents is *two* unimodal distributions: one reflecting the sales of black-and-white television sets, and the

Fig. 1.12 Television set sales.

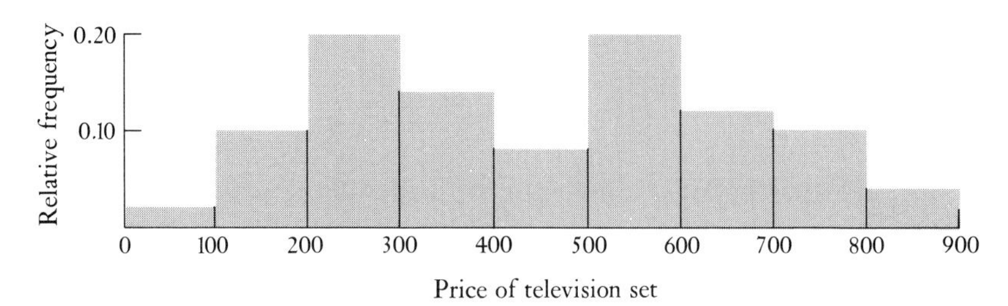

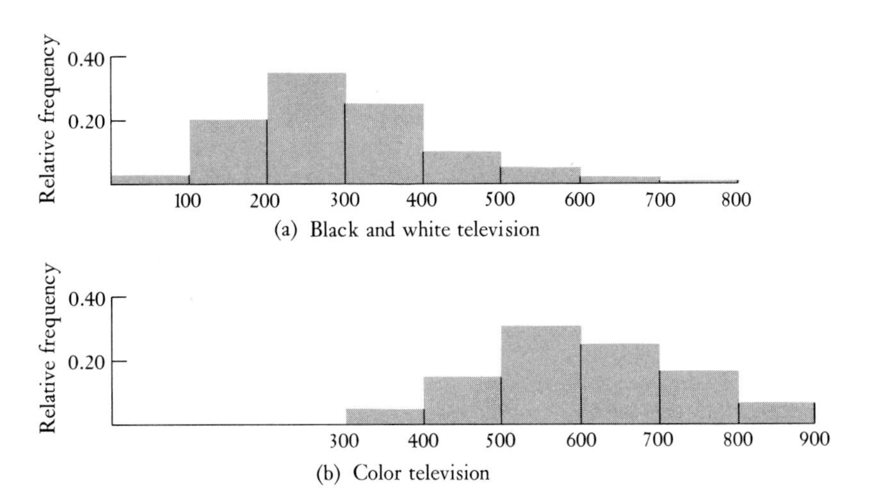

Fig. 1.13 Television set sales.

other representing the sales of color television sets. If we make this distinction and plot the resulting frequency distributions, the two distributions in Fig. 1.13 are obtained. Note that the distribution in Fig. 1.13(a) has a fairly long "tail" to the right, a characteristic common to many distributions representing data in the behavioral and social sciences, especially income distributions.

A distribution is *symmetric* if it has the same shape on both sides of its median. Imagine folding the picture of a distribution in half at its median. To be symmetric, the two halves must match perfectly—they must be "mirror images" of one another. Figure 1.13(b) represents a distribution with a relatively *symmetrical* shape. For all symmetric distributions the median equals the mean. The mode will also equal the median if the distribution is unimodal. Figure 1.14 shows three symmetric distributions.

A distribution that is not symmetric, but rather has most of its values either to the right or to the left of the mode, is said to be *skewed*. If most of the values

Fig. 1.14 Symmetric distributions.

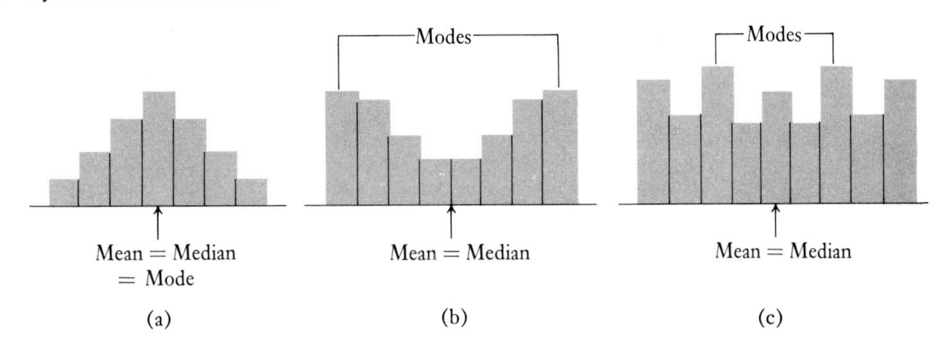

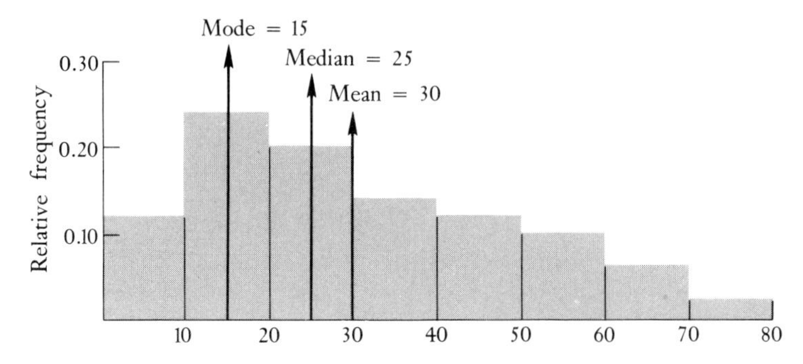

Fig. 1.15 The relationship between the mean, the median, and the mode for a distribution with positive skewness.

of a distribution fall to the *right* of the mode, as in Fig. 1.13(a), this distribution is said to be skewed to the right or *skewed positively*. A distribution with the opposite shape, with most values to the *left* of the mode, is said to be skewed to the left, or *skewed negatively*. Note in Fig. 1.14 how the lack of symmetry in a distribution affects the relationship between the mean, the median, and the mode. For a completely symmetrical unimodal distribution, such as Fig. 1.14(a), these three values must all be equal. As the distribution becomes skewed positively, the mode remains at the value representing the highest frequency, but the median and the mean move to the right, as in Fig. 1.15.

Various memory aids make it easy to remember the direction and the effect of skewness on the measures of central location in a unimodal distribution. Two useful ones are:

1. The order of magnitude of the central location measures is *alphabetical* in a *negatively* skewed distribution (Mean < Median < Mode) and *reversed* in a *positively* skewed distribution (such as Fig. 1.15).
2. If a distribution is stretched sideways, the direction of stretch is the direction of skewness. A distribution with its right side stretched out in the positive (increasing numerical value) direction has positive skewness. A distribution with its left tail stretched out in the negative (decreasing numerical value) direction is negatively skewed.

Kurtosis

Another descriptive measure of the shape of a distribution relates to its flatness or peakedness. The term applied to this characteristic of shape is *kurtosis*. The flat distribution with short broad tails illustrated in Fig. 1.16 is called *platykurtic*. A very peaked distribution with long thin tails is called *leptokurtic*. Measures of kurtosis (and skewness) are important to mathematical statisticians in their study of the theoretical properties of distributions. Although summary measures for kurtosis exist, we will not present them in this text, as measures of kurtosis are of relatively little use in elementary applications of statistics.

Fig. 1.16 Platykurtic curves have short tails like a platypus, while leptokurtic curves have long tails like kangaroos, noted for "lepping" (after W. S. Gosset).

1.6 SUMMARY

In this chapter we have focused on describing the central location and variability of populations as aids to decision-making. The two measures of central location that are of lesser importance are the:

Median: the middle value of the population.
Mode: the most frequently occurring value.

The population mean (μ) was presented as the most important measure of central location, and the variance (σ^2) was presented as an important measure of variability. These two parameters were defined in several ways, the most important of which are:

$$\mu = \sum_{i=1}^{c} x_i \left(\frac{f_i}{N}\right),$$

$$\sigma^2 = \sum_{i=1}^{c} (x_i - \mu)^2 \left(\frac{f_i}{N}\right) = \sum_{i=1}^{c} x_i^2 \frac{f_i}{N} - \mu^2.$$

In most practical problems the standard deviation (σ) is a more easily interpreted measure of variability, since it is expressed in the same units as the original data. The rules of thumb for interpreting σ are:

$\mu \pm \sigma$ will often contain approximately 68% of the population,
$\mu \pm 2\sigma$ will often contain approximately 95% of the population.

Define: percentile, decile, quartile, bimodal distribution, symmetric and skewed distributions, kurtosis.

Problems

1.24 A leading manufacturer guarantees its quartz watch will be correct within ± 10 seconds each month. Suppose we know that the population of errors by all such watches is bell-shaped, the mean error is zero ($\mu = 0$), and the variance of errors is $\sigma^2 = 100$ (seconds)^2 per month.

a) What percent of the population of watches would you expect to not meet the guarantee?

b) Assume only watches with an error of more than ± 20 seconds will be returned for repair. If 100,000 watches are sold with this guarantee, how many would you expect to be returned for repair?

1.25 Given the following grade distribution on a statistics quiz, find the mean and variance of x. Use the midvalue of each class (the class mark) as the value of x.

Grade	Frequency	Class Mark
96–100	4	98
91–95	5	93
86–90	6	⋮
81–85	5	
76–80	1	
71–75	4	
66–70	2	
61–65	5	

1.26 In a study for discrimination between men and women in the hiring practices of a certain firm, the following starting salaries were paid to nine men and nine women hired for jobs specifying "no previous experience necessary."

	Males	Females	Hourly Rate
File Clerk	2	7	4.28
Messenger	1	0	4.32
Shipping Clerk	2	1	4.41
Assembler—Type I	3	1	4.60
Assembler—Type II	1	0	4.67

a) Find the average wage paid to the men and the average to women. Do you think the difference is large enough to indicate discrimination?

b) Find the average for all 18 new workers.

1.27 Assume that the 250 population values in Table 1.11 came from data with class intervals \$1650–\$1750, \$1750–\$1850, etc. (Don't worry about the overlap of classes at \$1750, \$1850, \$1950, \$2050, and \$2150.) Also, assume that the frequencies in each class are evenly distributed throughout the class (i.e., the 8 values in the first row fall evenly between \$1650 and \$1750, the 23 values in the second row fall evenly between \$1750 and \$1850).

a) Find the values of q_1, q_2, and q_3 based on the above assumptions.

b) Find the value of the sixth decile (i.e., that value which has 60 percent of the population below it).

c) Find the value of the 90th percentile.

1.28 The population "kitchen employees in seven local restaurants" is 2, 9, 1, 3, 10, 3, 2.

a) Find the three quartile values for this population. What is the interquartile range?

b) Calculate σ^2 for these values. Does multiplying the interquartile range by 68/50 yield a good approximation to 2σ?

c) Is this population symmetric or is it skewed? If skewed, is it skewed to the right or left?

1.29 a) What does it mean if a math score on the SAT is reported as being in the "85th percentile?"

b) If your college grades are in the "third quartile," what does this mean?

1.30 The following frequency distribution shows the *cumulative relative* frequency of weekly sales of the Snoopy Hat Corporation for last year (52 weeks).

Sales	**Cumulative Relative Frequency**
0–\$5,999	4/52
\$6,000–11,999	10/52
12,000–17,999	20/52
18,000–23,999	36/52
24,000–29,999	48/52
30,000–35,999	52/52

a) Draw the histogram for these data.

b) Compute the mean and the mode.

c) Compute the variance.

d) Sketch the cumulative relative frequency distribution.

1.31 The following annual starting salaries were offered to 16 students about to receive their college degrees:

\$17,500	\$14,900	\$16,200	\$16,500
15,400	16,400	15,800	15,500
14,600	17,600	16,800	16,000
15,600	16,400	14,400	14,100

a) Find the mean, the median, the variance, and the interquartile range for these 16 observations.

b) Use a class interval of $500 to form a frequency distribution for the data. Construct a histogram for this distribution, and draw the frequency polygon connecting the class marks (start with the class $14,001–14,500). Is this distribution skewed right, symmetric, or skewed left?

c) Form a cumulative relative frequency distribution, and plot this distribution. What percent of the graduates will earn more than $15,500? What percent will earn less than $16,001?

d) Use the frequency distribution you calculated for part (b) to find the mean of the grouped data.

1.32 For the data given in Exercise 1.30, compute

a) the sum of the absolute deviations about the mean;

b) the standard deviation;

c) the percent of observations falling within one standard deviation from the mean, and the percent falling within two standard deviations from the mean.

1.33 Ten residents of Metropolis report the following incomes:

$30,500	$24,150	$22,505	$26,245	$25,570
$26,600	$34,800	$31,325	$29,170	$39,000

a) Find the mean, standard deviation, and median of these data.

b) Compare the variability and skewness of this distribution with a regional income distribution that reports a mean of $30,000, standard deviation of $3,000, and median of $28,000.

1.34 How are the mean, the median, and the mode related in a completely symmetrical and unimodal frequency distribution? How are they related in positively skewed and in negatively skewed unimodal distributions? Sketch several distributions to illustrate your answer.

1.35 a) The number of days a certain machine was "down" last year for five different repairs was: 3, 5, 1, 8, and 3. Show that the mean and standard deviation are 4.0 and 2.37, respectively.

b) Assume that another population has the same values given in part (a), plus the two additional values 2 and 6. Find the mean and show that this population has a smaller standard deviation.

1.36 The number of fatal automobile accidents in ten metropolitan cities for a one-year period was: 35, 14, 6, 18, 14, 27, 19, 7, 12, and 14. Find the mean, mode, median, range, the quartile values, and the interquartile range.

Exercises

1.37 Suppose that Village Pantry, a neighborhood convenience grocery store, reports that 223 items were returned last year. On the other hand, Kroger, the nearby national chain supermarket, reports 1571 returns. Does this indicate that Village Pantry is doing a better job than Kroger servicing its customers? Comment.

1.38 What is statistical inference? Why is statistical inference important in the social and behavioral sciences? Give several examples of the use of statistical inference in your major field of study in college.

1.39 Find an example (from a newspaper, magazine, etc.) of using only *means* to draw some conclusion. Examine the argument closely and explain the inadequacy of it, or explain how knowledge of a variability measure could strengthen or change the argument.

1.40 Acquire 50 observations on a variable of interest to you and your fellow students (e.g., wage rates, apartment rentals, football statistics, anatomical measurements). Construct a frequency distribution and a cumulative frequency distribution for your data. Sketch both distributions.

1.41 Use interpolation to find the median for Problem 1.30.

1.42 A movie producer holds a preview of a new movie and asks viewers for their reaction. By age groups, the following results are obtained:

	Age Group			
	Under 20	**20–39**	**40–59**	**60 and over**
Liked the movie	140	75	50	10
Disliked the movie	60	50	50	20

Using some diagrams or summary measures, argue toward which age groups the firm should aim its advertising campaign for the movie.

1.43 Throckmorton Jones, manager of Shark Loan, Inc., has kept a record of the frequency of the *time between arrivals* of customers at his loan office. These data, shown below, indicate that the time-interval between consecutive arrivals was between zero and 20 minutes on 50 occasions, between 20 and 40 minutes on 33 different occasions, etc. Since there were 150 customers during this period, there are 150 interarrival times (the time to the first customer is counted as one interarrival time).

Minutes Between Customers (t)	**Frequency (f)**
$0 \le t < 20$	50
$20 \le t < 40$	33
$40 \le t < 60$	22
$60 \le t < 80$	15
$80 \le t < 100$	11
$100 \le t < 120$	8
$120 \le t < 140$	5
$140 \le t < 160$	3
$160 \le t < 180$	2
$180 \le t < 200$	1
	150

a) Construct a histogram of relative frequencies and a cumulative relative frequency distribution for the interarrival times. Draw the polygon for these distributions.

b) Find the mode, the median, the mean, and the standard deviation (use class marks of 10, 30, 50, . . .). Use interpolation to find the median.

c) What is the value of the interquartile range? What percent of the observations fall within $\mu \pm 1\sigma$? What percent fall within $\mu \pm 2\sigma$?

d) Based on these data, how many customers would you estimate for Shark Loan next week if the office is open five days a week, ten hours a day?

1.44 Assume that you are responsible for auditing last month's accounts receivable for the 10,000 credit accounts on the Easy Charge Company. The company has furnished you with the following summary data for this population (this is a slight abstraction of 12,223 actual credit balances).

Balance due	Frequency	Balance due	Frequency
$0– 99.99	3123	$600–699.99	180
100–199.99	2085	700–799.99	90
200–299.99	1927	800–899.99	53
300–399.99	1355	900–999.99	25
400–499.99	743	$1000 and over	19
500–599.99	400		

a) Draw the histogram and frequency polygon for these data.

b) As an auditor, are you satisfied with the manner in which the company summarized the data?

c) Use the midpoint of each class and the associated frequencies to calculate the mean of this population. Assume that all 19 values in the eleventh class fall at $1050.

d) Calculate σ and σ^2 for this population. What percent of the population lies within $\mu \pm \sigma$ and $\mu \pm 2\sigma$?

e) Assuming that the population values are evenly spread throughout each class, calculate the median of this population.

f) Making the same assumptions as in part (e), calculate q_1, q_2, q_3, and the interquartile range. Also calculate the ninth decile (ninetieth percentile).

g) Is this distribution skewed to the right or left?

1.45 Prove that Formula (1.9) is mathematically equivalent to Formula (1.8).

1.46 An important statistic to consider when using a statistical sampling audit plan is the population variability. The population variability is measured by the:

a) sample mean

b) standard deviation

c) kurtosis

d) estimated population total minus the actual population total.

1.47 The frequency distribution of employee years of service for Henry Enterprises is right-skewed (i.e., it is not symmetrical). If a CPA wishes to describe the years of service of the typical Henry employee in a special report, the measure of central tendency that the CPA should use is the:

a) standard deviation b) arithmetic mean

c) mode d) median.

1.48 *Tchebysheff's Theorem* states that at least $(1 - 1/K^2)$ of any population will lie within K standard deviations of their mean, where K is any number ≥ 1.0. For example, using $K = 2$, this theorem says that at least $(1 - \frac{1}{4}) = \frac{3}{4}$ of any population will lie within $\mu \pm 2\sigma$. Similarly, if $K = 3$, then at least $(1 - \frac{1}{9}) = \frac{8}{9}$ of any population will lie within $\mu \pm 3\sigma$. K need not be an integer.

a) What percent of any population will lie between $\mu \pm 4\sigma$?

b) Show, for Problem 1.36, that at least $\frac{3}{4}$ of the population lies between $\mu \pm 2\sigma$.

c) Show, for Table 1.1, that at least $\frac{8}{9}$ of the population lies between $\mu \pm 3\sigma$.

d) Is Tchebysheff's Theorem consistent with the rule of thumb presented in Section 1.4?

1.49 A record was kept of the dollar value of sales over 50 weeks by a person selling surgical supplies.

Sales	f
\$0–999	11
1000–1999	15
2000–2999	18
3000–3999	6

Divide each value of x by 1000, and then calculate μ and σ^2. Find the mean and variance of the *original* data using these values of μ and σ^2.

1.50 a) Find the median for Problem 1.49 using interpolation.

b) Find the interquartile range for Problem 1.49 using interpolation.

1.51 The *coefficient of variation* is defined as a standard deviation divided by a mean. This measure indicates the percentage that σ is of μ, and thus permits comparison of the relative variability of one population to the relative variability in another population. For example, if two populations have the same standard deviation, $\sigma = 100$, but $\mu_1 = 1000$ and $\mu_2 = 2000$, then the coefficient of variations are, respectively, $100/1000 = 0.10$, and $100/2000 = 0.05$. The first population has a percentage of variability (10%) twice as large as the second population (5%) even though their standard deviations are the same.

a) If test scores on the Graduate Record Exam (GRE) have $\mu = 500$ and $\sigma = 80$, whereas scores on the Graduate Management Admissions test have $\mu = 540$ and $\sigma = 100$, do these tests have similar relative variability? Explain.

b) Calculate the coefficient of variation for males and females in Problem 1.26. Is the percentage of variability about equal?

c) Suppose you are trying to decide between three investments that all require the same capital outlay. The average return and variance of returns for the three are:

1	2	3
$\mu = 2000$	$\mu = 1000$	$\mu = 1500$
$\sigma = 1200$	$\sigma = 500$	$\sigma = 700$

Using the coefficient of determination as a measure of the "risk" of each investment, which is the most risky? Which is least risky? Which one yields the highest average return?

1.52 On May 4, 1981, a *Time* magazine article reported the average (for nine schools) starting salary for MBAs to be $27,500. If you know that salaries are unimodal with a mode of $26,000, and $\sigma = \$2000$, sketch an approximation of the distribution of salaries.

GLOSSARY

descriptive statistics: methods concerned with arranging, summarizing, or somehow conveying the characteristics of a set of numbers.

statistical inference: making generalizations, predictions, or conclusions about characteristics of a population based on the characteristics of a sample drawn from that population.

decision making under uncertainty (or risk): decision-making process in which there is uncertainty about what outcome will result from a particular action.

population: all relevant values of interest in a particular context.

parameters: certain characteristics of a population, such as its central location or variability.

sample: a subset of a population.

sample statistic: a numerical characteristic of a sample.

frequency distribution: a table or graph describing the number of observations for each class of a data set.

frequency polygon: a series of straight lines connecting the midpoints of each class in the graph of a frequency distribution.

ogive: a series of straight lines connecting the class limits in a graph of a cumulative frequency distribution.

mode: that value which occurs most frequently.

median: the middle value in a set of values (population or a sample).

population mean: the average of a set of N values:

$$\mu = \frac{1}{N}\sum_{i-1}^{N} x_i \quad \text{or} \quad \mu = \sum_{i=1}^{c} x_i\left(\frac{f_i}{N}\right).$$

range: difference between highest and lowest values in the set of values.

population variance (σ^2): the average of the squared deviations of values of x from their mean.

$$\sigma^2 = \left(\frac{1}{N}\right)\sum_{i=1}^{c}(x_i - \mu)^2 f_i.$$

$$\sigma^2 = \sum_{i=1}^{c} x_i^2(f_i/N) - \mu^2.$$

standard deviation (σ): the square root of the variance

$$\sigma = \sqrt{\text{variance}} = \sqrt{\sigma^2}.$$

rule of thumb: $\mu \pm 1\sigma$ includes about 68% of the values;
$\mu \pm 2\sigma$ includes about 95% of the values.

quartiles: those values which separate a data set into four equal parts.

q_1 = value which exceeds one-fourth and is exceeded by three-fourths of the observations. Formula is $N/4 + \frac{1}{4}$.

q_2 = value which divides data into two equal parts (the median). Formula is $N/2 + \frac{1}{2}$.

q_3 = value which exceeds three-fourths and is exceeded by one-fourth of the observations. Formula is $3N/4 + \frac{1}{4}$.

interquartile range: difference between third quartile value (q_3) and the first quartile value (q_1).

bimodal distribution: a distribution with two distinct modes.

symmetric distribution: a distribution with the same shape on both sides of the median.

skewed distribution: any distribution that is not symmetric. Skewed to the right (positively) means that most of the values are to the right of the mode. Skewed negatively is the opposite.

deciles: nine values that separate a data set into ten equal parts.

kurtosis: a measure of the shape of a distribution.

2

PROBABILITY THEORY: DISCRETE SAMPLE SPACES

"The theory of probabilities is nothing more than good sense confirmed by calculations."

PIERRE LAPLACE

2.1 INTRODUCTION

As we mentioned in Chapter 1, most problems of statistics involve elements of uncertainty, since it is usually not possible to determine in advance the characteristics of an unknown population or to foresee the exact consequences which will result from each course of action in a decision-making context. A necessary part of an analytical approach to these problems must thus involve evaluations of just how likely it is that certain events have occurred or will occur. "How likely is it that a given sample accurately reflects the characteristics of a certain population?" or "What is the chance that a given consequence will occur following a certain decision?" are examples of inquiries into the probability of an event or set of events of this nature.

A probability is a number between 0 and 1 which indicates how likely it is that a specific event or set of events will occur.

If an event has a probability of 0, then this event will *never* occur. On the other hand, an event with a probability of 1 will occur with *absolute certainty*. Finding the value between 0 and 1 which depicts how likely is a given outcome, a specific event, or a sequence of events, forms a fundamental part of almost all types of statistical analysis; and probability theory provides the foundation for the methods of this analysis.

The origins of probability theory date back to the 1600's, when the mathematicians Blaise Pascal and Pierre Fermat became interested in games of chance. Although Pascal and Fermat corresponded regularly about problems involving elements of chance, not until over 100 years later did this new branch of mathematics find many applications beyond the French gambling houses of the seventeenth century. The work of Karl Gauss and Pierre Laplace was especially important during the late 1700's in extending probability theory to problems of the social sciences and actuarial mathematics. Laplace commented that "It is remarkable that a science which began with the consideration of games of chance could have become the most important object of human knowledge." Despite the contributions made in the seventeenth through nineteenth centuries, most modern-day statistics was developed in the past fifty years. R. A. Fisher, J. Neyman, E. S. Pearson, and A. Wald are some of the more prominent researchers who have contributed to the phenomenal growth of statistics in recent years.

2.2 THE PROBABILITY MODEL

Often you may wish to specify a probability associated with some situation. Perhaps you want to know the probability that no automobile accidents will occur in a given location over a certain time period, or the probability that the price of dairy products will decrease, or the probability that candidate X will be elected senator, or the probability that your favorite person finally will phone—right after you have gone out to get something to eat. To specify the probability associated with a given situation, it is extremely important to define what *experiment* underlies this situation, and what outcomes can result from this experiment. As we will demonstrate in this chapter and Chapter 3, making probability assessments is simplified by using a *probability model* as the framework for your mental construction of the problem. Once the nature of the problem is clearly formulated in this way, the solution is routinely obtained by using specified *probability laws and formulas*.

The Experiment

The first component in the probability model is the definition of the experiment. In a probability context *an experiment is any situation capable of replication under essentially stable conditions*. The repetitions may be feasible and be performed (such as flipping a coin), or they may be abstract and theoretically conceivable. For example, investing $1000 in the stock market could be an experiment, even though you intend to do it only once. You could imagine repeating such an investment over and over again many times and consider theoretically the chances that you would make capital gains or losses, or earn a yield of more or less than ten percent, or achieve some other result. Driving a car from New York to Phoenix might be an experiment. You may plan to do it only once, but you may be interested in the theoretical chances of making the trip without tire or engine trouble. You are imagining abstractly many replications of the trip under similar circumstances. Even taking an accounting exam might be considered an experiment. If you imagined many students taking the exam with similar preparation, you might be interested in the chances that more than half will pass the exam or that you will get an A or a B on the exam. Clearly, an experiment is defined very broadly here, to correspond to *any situation involving uncertainty*, whether it actually recurs many times or whether the repetitions are hypothetical.

When an experiment is defined, it is extremely important to specify *all* the procedures associated with the experiment. For example, it would be ambiguous to define an experiment as "draw two cards from a deck of 52," since this definition leaves doubt as to whether or not the first card drawn is replaced before the second card is drawn. Since probability values often depend on which of these conditions holds, many experiments involving several steps must be defined as either "with replacement" or "without replacement."

Sample Spaces—The Outcomes of an Experiment

The different outcomes of an experiment are often referred to as *sample points*, and the set of all possible outcomes is called the *sample space*. As we will describe in the following sections, sample spaces are often distinguished according to whether they are *discrete* or *continuous*. Continuous sample spaces generally involve outcomes that are *measured* (such as length, time, weight, mass) and the number of sample points is always infinite. Discrete sample spaces may contain a finite or an infinite number of points, all of which can be separated and counted.

Suppose an experiment is defined to be "observe one day's weather." Although we could define numerous types of sample spaces, let us assume only two sample points can occur each day: (rain, no rain). Note that if the experiment is defined to be "observe two days' weather," then the sample space consists of the four points

(rain, rain), (rain, no rain), (no rain, rain), (no rain, no rain).

In this weather example, the sample space is *discrete* (because the outcomes can be separated from one another and counted) and *finite* (because the number of possible outcomes is limited). Suppose, however, that we define the experiment as "observe daily weather until the first rain appears." In this case the number of sample points is infinite, since there is no limit to the number of outcomes. Such a sample space is still discrete because the number of outcomes (i.e., the number of days before it rains) can be separated and counted. Thus, a discrete sample space may contain either a finite or an infinite number of outcomes.

Examples of discrete sample spaces that contain a *finite* number of outcomes include the number of minority candidates hired from a list of candidates for a certain job, the number of electronics companies that submit bids on a government contract, or the number of defectives in a shipping lot of a given size. Problems that involve a discrete sample space having an *infinite* number of outcomes include the quoted dollar-value of a common stock (it can vary from 0 to infinity, in units of one-eighth of a dollar), or the number of flaws per unit of material from a textile plant (which can be *any* positive integer).

In contrast to a discrete sample space is the concept of a *continuous sample space*. A sample space is continuous if the number of possible outcomes is infinite and uncountable. An experiment involving the time it takes a lightbulb to burn out represents a continuous sample space because the outcome of the experiment could be *any* real number from zero to the upper bound (such as 10,000 hours). There is obviously an infinite number of outcomes possible here, and no way to separate and count them. Generally, we are dealing with continuous sample spaces when the data involved are obtained by *measurement* rather than by counting. Thus, a set is continuous if *any* value within an interval can occur, such as any value between 0 and 1, or any value between 200 and 1000, or any value from 0 to infinity. The net weight of a box of packaged cereal, the length of a fish caught in a trout stream, the average speed of the winning car at the Indianapolis 500, or the distance a car can travel on a gallon of gas are all examples of outcomes in a continuous sample space.

Most applications of probability theory involve experiments with a finite number of outcomes, although the number is often large enough so that it makes little difference (for practical purposes) if it is assumed to be infinite. Furthermore, many experiments involving only a discrete set of outcomes can be *approximated* by a continuous set. The advantage of such approximations is that they often *simplify* the derivation of certain statistical results.

We must hasten to add that it is not always clear from the statement of an experiment exactly what outcomes are relevant. For example, in the experiment "take an accounting exam," one student may define the sample space to be all possible *numerical* grades (perhaps a number between 0 and 100), another student may be interested only in the *letter* grade (A, B, C, D, or F), while a third student may define the sample space as simply {pass, fail}. Similarly, in observing weather over two days, we may have no interest in the *order* of the outcomes. In this case the relevant sample space would be the three points:

(two days without rain), (one day rain, one day no rain),

(two rainy days).

When working with continuous sample spaces, it may be advantageous to *group* the sample space into a small number of discrete outcomes. For example, in working with the time it takes a lightbulb to burn out, we might group the outcomes into the following three sets: (1) time is less than 100 hours, (2) time is between 100 and 1000 hours, and (3) time exceeds 1000 hours. To specify the probabilities associated with a given situation it is important that the relevant outcomes be carefully defined.

In defining the sample space of an experiment, one must be sure that it is not possible for two or more outcomes to occur in the same replication of the experiment. Outcomes defined in this manner are said to be *mutually exclusive.* For example, the two outcomes {rain, no rain} are mutually exclusive for the experiment "observe one day's weather," but they are *not* mutually exclusive for the experiment "observe two days' weather" (since in two days we could have one day of each). Similarly, the student who defines the outcomes of an accounting exam as "A," "Pass," and "Fail" has not specified a mutually exclusive set because both "A" and "Pass" could happen at the same time.

A second requirement in defining the sample space of an experiment is that *the list of outcomes must be exhaustive;* that is, no possible outcomes can be omitted. If a flipped coin can stand on its edge, then this outcome must be added to the sample space. If our lightbulb could last longer than the specified upper bound (such as 10,000 hours), then this possibility must be added to the sample space. In general, when one is constructing a probability model, it is extremely important that the outcomes associated with the sample space (that is, the sample points) be *both mutually exclusive and exhaustive.*

To summarize briefly, we have defined the following terms and their part in every probability model.

A) An experiment—any situation capable of replication under stable conditions.

B) Outcomes of an experiment (sample space)
   1. *Discrete (separable and countable outcomes)*
      a) finite (an upper limit on number);
      b) infinite (no upper limit on number).
   2. *Continuous (nonseparable)*
      The number of outcomes is infinite.

C) Event—some subset of the outcomes of an experiment.

D) Mutually exclusive and exhaustive events—events that do not overlap (are mutually exclusive) and account for (exhaust) all possible outcomes of the experiment.

There are two additional components of every probability model: (1) a random variable, and (2) a probability function. These two components of the probability model are extremely important in describing the probability of the entire set of events of interest in a given experiment; they will be discussed in detail in Chapter 3.*

2.3 SUBJECTIVE AND OBJECTIVE PROBABILITY

Since there is some disagreement, even among authorities, about the interpretation of probability, it is advisable to begin a discussion of the fundamental concepts of the theory of probability by describing the two major viewpoints or interpretations. Both of these interpretations deal with determining the probability of the occurrence of an event. They differ in the methods they prescribe for determining this probability. The first and more traditional viewpoint defines *probability as the relative frequency with which an event occurs over the long run.* Probabilities determined by the long-run relative frequency of an event are usually referred to as *frequency probabilities*, or *objective probabilities* (since they are determined by "objective evidence" and would have the same value regardless of *who* did the determination). The second interpretation of probability assigns probabilities based on the decision-maker's subjective estimates, using prior knowledge, information, and experience as a guide. This approach, in which probabilities are referred to as *subjective* probabilities, recently has gained considerable importance in statistical theory, largely because of the influence of such statisticians as L. J. Savage, R. Schlaifer, and H. Raiffa.

Suppose you believe the chances are one in four that you will earn an "A" in statistics, or that the odds are eight to one against the Pittsburgh Pirates' winning the National League Pennant next year. These are subjective evaluations, in which *your* personal opinion about the probability of these events need not agree with those of your statistics professors or those of the manager of the Pittsburgh Pirates, or with anyone else's, for that matter. You might, for example, believe that the odds of a tail appearing on the flip of a certain coin are less than 50–50, while everyone else thinks the chances of a head and a tail are equal. You are, of course, entitled to your own opinion; presumably, however, if we flipped this coin enough times, your opinion and everyone

* Some instructors may wish to cover the portion of Chapter 3 relating to random variables at this point.

else's should begin to coincide. In other words, given a sufficient amount of data about the past occurrences of an event, we would expect one's subjective opinion to agree fairly closely with the long-run relative frequency of that event.

The problem (and controversy) in the "frequency" approach to estimating probabilities is that in many real-world problems, there may be little or no historical data available on which to base an estimate of the probability of an event. In such cases, only a subjective probability can be determined; and this probability may differ even among experts who have similar technical knowledge and identical information. For example, various space scientists in 1960 gave different estimates of the probability that men would walk on the moon within that decade. Different weathermen observing the same climatic conditions may give different probabilities that rain will occur on a given day. Fortunately, the rules and operations governing probability theory are the same whether the number itself is generated by an objective or subjective approach.

A simple example of the concept of probability as a long-run relative frequency is an experiment in which a coin is tossed over and over again, where the outcome of each toss is observed and recorded. For the first few tosses the proportion of heads may fluctuate rather wildly; after a sufficiently large number of tosses, however, it should begin to stabilize about one particular value. The more tosses observed, the closer should the proportion of heads approach this long-run or limiting value. Just such an experiment was performed by the statistician J. E. Kerrick during his internment in Denmark during World War II; Kerrick tossed a coin 10,000 times and obtained 5067 heads; the frequency distribution is shown in Fig. 2.1.

Note that in Fig. 2.1 the relative frequency of heads begins to stabilize, after quite a few tosses, about one particular value (which in this case appears

Fig. 2.1 Relative frequency of heads in Kerrick experiment.

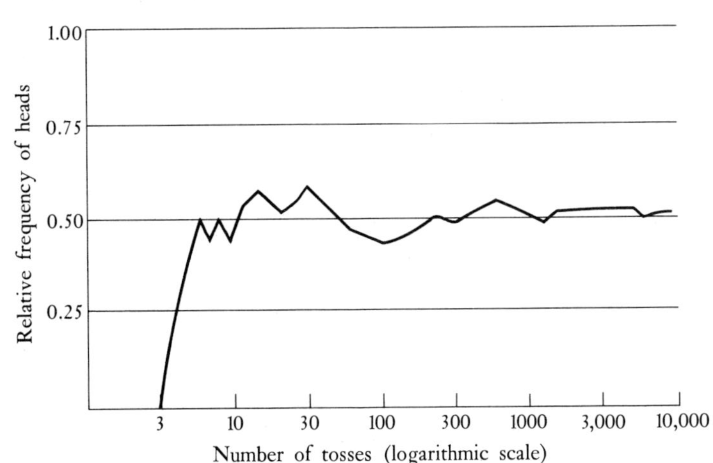

to be about 0.50). One would expect, as the number of tosses becomes even larger, that the relative frequency of heads will come closer and closer to this particular value until, at the limit (an infinite number of tosses), they are equal. The value at which they are equal, sometimes called the *limit of relative frequency*, represents what we have called long-run relative frequency; and it is this concept of the long-run relative frequency of an event which is generally implied when referring to an "objective probability" or to "the frequency approach to probability." Thus, a probability of heads equal to 0.50 in the Kerrick experiment means that *in the long run* the expected proportion of heads is 50 percent. In most practical problems, of course, the limit of relative frequency can only be estimated, since it is seldom possible to observe a sufficient number of observations to determine this limit precisely.

Expected Frequency

At times it is convenient to determine another interpretation of probability, that represented by the *expected* or *theoretical long-run relative frequency* of an event. In such cases, the probability of an event depends on assumptions about the conditions underlying the occurrence of this event. For example, if the coin in the experiment by Kerrick had been assumed to be "fair" (i.e., not biased to heads or tails), then the *theoretical long-run relative frequency* of heads for this coin would have been, by definition, 0.50. The underlying conditions or assumptions about an experiment (such as tossing a coin) represent part of what is called the *experimental model.* Making assumptions of this nature is useful in testing to determine if theoretical probability values differ from actual (or observed) probabilities, whether these values are subjective or objective. Notice that Kerrick's observed value differs very little from the theoretical value of 0.50.

Probability Axioms

The set of *all possible outcomes* of interest in an experiment corresponds to what we called a population (all values of interest) in Chapter 1. This set is often referred to as the *sample space* (denoted by the letter S) since it represents the region (or "space") corresponding to all possible sample results.

We can now present the two basic properties necessary for defining the probability of an event. If $P(E_i)$ is the probability of event E_i in a sample space S, then both of the following properties must hold.

Property 1.	$1.0 \geq P(E_i) \geq 0$	(2.1)
Property 2.	$P(S) = 1.0$	

It is easy to recognize that these properties are consistent with our previous examples. The first one says that the probability of an event can never be less than zero (which represents an impossibility) nor greater than one (which represents a certainty). The second one says that the events E_i comprising the sample space must be exhaustive—it must be a certainty that one of the mutually exclusive events in the sample space will take place in each replication of the experiment.

Much of the historical development of probability can be traced to analysis of problems in which only a finite number of outcomes may take place, and in which each of these outcomes is assumed to have the same chance of occurring. In such situations, if any one of N "equally likely" outcomes can take place, then the probability of any one outcome occurring is $1/N$. For example, suppose Publishers Clearing House mails out announcements of a contest to 200,000 people, stating that the grand prize winner will be selected "at random" from the list of 200,000 people. Random selection in this context means that each person is equally likely to be selected; the probability of being the grand prize winner is thus

$$P(\text{grand prize winner}) = \frac{1}{200{,}000} = 0.000005^*$$

When the number of equally likely outcomes in a given problem is relatively small, it is often possible to determine the probability of each outcome by counting the total number of outcomes (N), and then calculating $1/N$. In Section 2.11 we will discuss several rules that aid in this counting process. Before doing so, we now present a special rule, which can be used when all sample points are equally likely.

The probability of an event is the ratio of the number of outcomes comprising this event to the total number of equally likely outcomes in the sample space.

$$P(\textit{event}) = \frac{\textit{No. of outcomes comprising the event}}{\textit{Total no. of equally likely outcomes in sample space}}.$$

We will give numerous examples of this rule throughout this chapter. For now, let's just consider a simple situation where you would like to determine the probability that a woman will be selected to fill the next Supreme Court vacancy. Suppose there are four female candidates and six male candidates, and all ten people are assumed to have the same chance of being selected. In

* Probability values can be stated as either fractions or decimals. For the most part, in this chapter the fractional form will be more convenient.

this situation,

$$P(\text{woman selected}) = \frac{\begin{array}{c}\text{No. of outcomes where}\\ \text{"woman selected"}\end{array}}{\begin{array}{c}\text{No. of equally likely}\\ \text{outcomes in the experiment}\\ \text{"replace a Supreme Court Judge"}\end{array}} = \frac{4}{10} = 0.40.$$

To further illustrate this rule, consider the procedure followed by a well-known investment service, which rates common stocks in terms of both (1) short-term growth potential and (2) long-term growth potential. Short-term growth is rated as either 1, 2, or 3 while long-term growth is rated as either 1, 2, 3, or 4 (in both cases, higher ratings imply better potential). The sample space for this experiment "classify a common stock" can be illustrated as in Fig. 2.2.

Figure 2.2 illustrates the 12 sample points for classifying a stock. If all 12 sample points are equally likely (they may not be), then the probability of any *one* occurring is $1/N = 1/12$.

Now suppose we define the following two events.

E_5 = sum of two ratings equals 5

E_3 = either long-term ≥ 3 or short-term $= 3$, or both.

To find $P(E_5)$ and $P(E_3)$ we add the equally likely outcomes associated with each one. For example, the three outcomes that are equal to five are

(4, 1), (3, 2), and (2, 3),

where the first number in each set corresponds to long-term growth and the second corresponds to short-term growth. Since each one of these sample points is assumed to have a probability of $\frac{1}{12}$,

$$P(E_5) = \frac{3}{12}.$$

Fig. 2.2 The sample space for stock classifications.

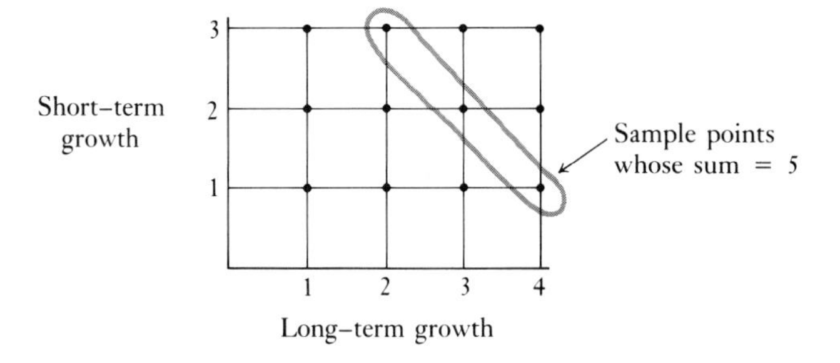

Similarly, the following eight outcomes comprise the event E_3:

$$(4, 1), \quad (4, 2), \quad (4, 3), \quad (3, 1), \quad (3, 2), \quad (3, 3), \quad (2, 3), \quad \text{and} \quad (1, 3).$$

Hence, $P(E_3) = \frac{8}{12} = \frac{2}{3}$.

We will elaborate on the rules for computing the probability of an event in the next several sections.

2.4 COUNTING RULES

Many probability experiments involve two or more steps, each of which can result in one of a number of different outcomes. To calculate probabilities in such experiments, we often need to first determine the total number of possible outcomes. For example, suppose three different people are asked whether they own any common stock. Now there are three steps (interviewing three people) and each step has two outcomes [own stock (S) or not own stock (NS)]. The fact that there are eight different outcomes in this experiment is easily seen if we construct a "tree diagram" (Fig. 2.3).

Basic Counting Rule

In the stock survey example above there are three steps (people) and each step has two possible outcomes (S or NS). The total number of outcomes in this

Fig. 2.3 Tree diagram for the stock/no stock experiment.

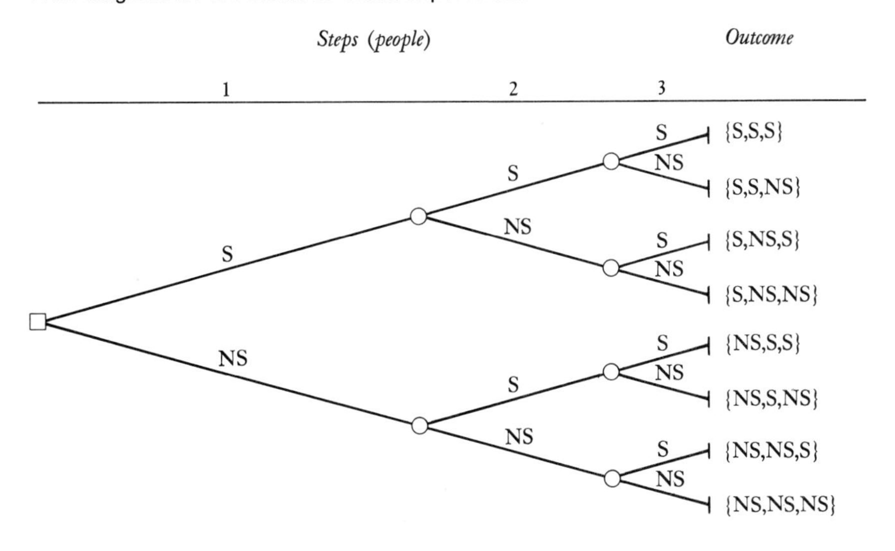

experiment (which we denote by the letter N) is

$$N = 2 \cdot 2 \cdot 2 = 8.$$

To generalize this type of calculation, suppose we denote the number of outcomes in the first step of an experiment as n_1, the number of outcomes in the second step as n_2, and so forth, with n_k denoting the number of outcomes in the last (or kth) step. The *basic counting rule* states that the total number of outcomes (N) equals the product of the number of outcomes in each step.

Basic counting rule:

Total number of sample points $N = n_1 \cdot n_2 \cdots n_k.$ (2.2)

In our stock survey example, $n_1 = 2$, $n_2 = 2$, and $n_3 = 2$; hence, $N = n_1 \times n_2 \times n_3 = 2 \times 2 \times 2 = 8$. As another illustration of Formula (2.2), consider a company that is going to select one of four different pricing policies, and then select one of three different advertising packages. In this example, $n_1 = 4$ (pricing) and $n_2 = 3$ (advertising). The total number of outcomes of this experiment is (by Formula 2.2):

$$N = n_1 \times n_2 = 4 \times 3 = 12.$$

Define: Subjective and objective probability, sample spaces and sample outcomes, $P(E_i)$, basic counting rule.

Problems

2.1 Consider the experiment "I take a statistics course for credit."

a) Describe the outcomes of the experiment. [*Note:* Your answer may differ from another person's.] Is this sample space finite or infinite, discrete or continuous?

b) Is the probability P(my grade = A) a subjective or objective probability?

c) Determine values of P(A grade), P(B grade), P(C grade), P(D grade), and P(F grade). Show that these values meet the two basic probability properties if no other grades are allowed, such as plus or minus grades or an "incomplete."

2.2 Describe an experiment for each of the following situations (do not use examples mentioned in the text):

a) the sample space is discrete and finite;

b) the sample space is discrete and infinite;

c) the sample space is continuous.

2.3 Indicate whether the sample space is finite or infinite, discrete or continuous, if the experiment involves:

a) the number of grains of sand in the world;

b) the number of years before a woman is first elected President of the U.S.A.;

c) the percent by which the price of a stock might decrease, if percents are restricted to integer values;

d) the percent by which the price of a stock might decrease, if percents can be any decimal value;

e) the amount of rainfall in New York City in April.

2.4 Indicate whether the following events are (1) mutually exclusive and/or (2) exhaustive:

a) the events "stock market goes up" and "stock market doesn't go up";

b) in one draw from a deck of 52, the events "draw an ace" and "draw a spade";

c) the events "ace of spades on first draw" and "ace of spades on second draw" in two draws from a deck of cards (with replacement);

d) repeat (c), but without replacement;

e) the events "receive an A in statistics," "pass statistics," and "fail statistics";

f) the events "make dean's list" and "earn an A in statistics."

2.5 Suppose the Publishers Clearing House contest mentioned in Section 2.3 is giving out one grand prize, 500 second prizes, and 1000 third prizes. There are 50,000 people who might win, and no one can win more than one prize. What is the probability you will win if you are one of the 50,000 people?

2.6 Four different students take a certain course "pass or fail."

a) Describe the sample space for this experiment. Draw the tree diagram.

b) How many different outcomes are there if it is important which students pass and which ones fail? How many outcomes are there if one is interested only in the number of passes and failures?

c) If "pass" and "fail" are considered equally likely, what are $P(0 \text{ pass})$, $P(1 \text{ passes})$, $P(2 \text{ pass})$, $P(3 \text{ pass})$, and $P(4 \text{ pass})$? Do these values sum to 1.0?

2.7 In one draw from a deck of cards, what is the probability of the event "face card" (an ace is not counted as a face card)?

2.8 Suppose you are planning on taking a survey of two out of five previously selected local retail establishments.

a) Describe the sample space for the experiment, "select two establishments without replacement." [*Hint:* Label the establishments 1, 2, 3, 4, 5, and list all pairs; note that order is *not* important here.]

b) What is the probability that establishment 1 will be included in the survey?

c) What is the probability that either 1 or 2 will be included in the survey?

2.9 Identify the following as either subjective or objective probabilities:

a) the probability that you will win the grand prize in a Publishers Clearing House sweepstakes;

b) the probability that two or more people in your statistics class have the same birthday;

c) the probability that you will earn an "A" grade in statistics;

d) the probability that unemployment in the U.S. will average less than six percent next year;

e) the probability that a Republican will be elected in the next presidential election.

2.10 Tennis rackets are often classified according to their (1) power and (2) control. Describe the sample space for a classification involving three power levels (1, 2, 3) and two control levels (1, 2). If all sample points are equally likely, what is the probability of each point? What is the probability, assuming equally likely outcomes, that power will be 2 or higher and control will equal 2?

2.11 An important task of certain business and governmental agencies involves assigning workers to specific tasks (or machines). For example, the Department of Labor assigns OSHA (Occupational Safety & Health Act) inspectors to inspect employer locations for safety. Suppose a city has three inspectors and two "first-priority" locations (where a catastrophe or fatality has occurred), and must now assign one inspector to one of the two locations.

a) Draw the decision tree for the experiment "assign one inspector to one first-priority location." [*Hint:* Label inspectors as A, B, C and the locations as 1, 2.] What is the probability of each sample point, assuming random assignment?

b) Use your decision tree to calculate the probability that A is assigned to 1 (A-1), or B is assigned to 2 (B-2).

2.5 PROBABILITY RULES

One can often determine the probability of an event from knowledge about the probability of one or more other events in the sample space. In this section we will discuss rules for finding the probability of complementary and conditional events, the probability of the union of two events, and the joint probability of two events.

Basic Definitions. Before describing these rules we present a few definitions. If we designate A and B as two events of interest in a particular experiment, then the following definitions hold:

1. $P(\bar{A})$ = Probability that A does *not* occur in one replication of the experiment.

   $P(\bar{A})$ is called the probability of the *complement* of A.

2. $P(A|B)$ = Probability that A occurs *given that* B has taken place (or will take place).

   $P(A|B)$ is called the *conditional probability* of A given B.

3. $P(A \cap B)$ = Probability that *both* A *and* B occur in one replication of the experiment.

   $P(A \cap B)$ is called the probability of the *intersection* of A and B or the *joint* probability of A and B.

4. $P(A \cup B)$= Probability that *either* A *or* B *or both* occur in one replication of the experiment.

   $P(A \cup B)$ is called the probability of the *union* of A and B.

Each of these definitions is elaborated in the sections that follow.

Perhaps the simplest way to form a new event from a given event is to take the *complement* of that event. For example, the complement of the event A, which is denoted by $\bar{A}$ (read A-bar), contains all the points in the sample space that are *not part* of A. If S denotes the total sample space, the $P(\bar{A})$ is defined as follows:

Probability rule for complements: $P(\bar{A}) = 1 - P(A)$

To illustrate this rule, suppose the sample space S is the set of points in Fig. 2.2 (on page 60) and E_5 = points whose sum = 5. Then

$$P(\bar{E}_5) = \text{points whose sum does not equal 5}$$

$$= 1 - P(E_5) = 1 - \frac{3}{12} = \frac{9}{12}.$$

As a further illustration of the probability rule for complements, if the probability that the Pittsburg Pirates will win the National League pennant is 0.25, then the probability that they will *not* win it is

$$1 - 0.25 = 0.75.$$

Some of our friends (who are Cubs fans) say that a probability of winning of0.25 is too high for the Pirates, reminding us that this probability is a *subjective* probability.

Two complementary events must be *exhaustive* because they take into account (or exhaust) all possible events. In addition, complementary events must always be *mutually exclusive* because none of the sample points in A can be a sample point in $\bar{A}$.

Conditional Probability

Suppose that we are interested in determining whether some event A occurs "given that" or "on the condition that" some other event B has already taken place (or will take place in the future). Such a conditional event is read as "A given B," and is usually written as $(A \mid B)$, where the vertical line is read as "given."

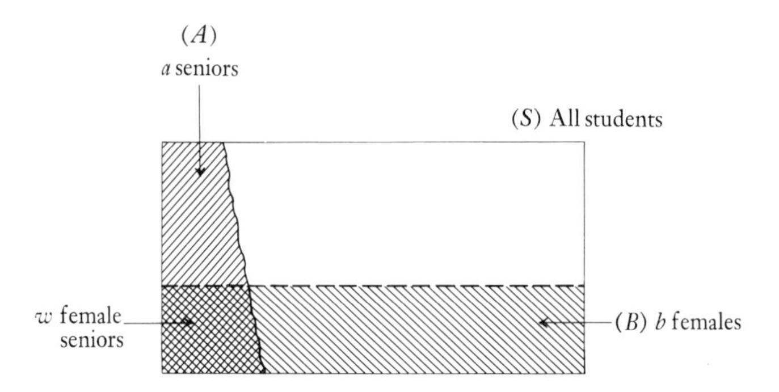

Fig. 2.4 Venn diagram of overlapping events.

The Venn diagram* (Fig. 2.4) can be useful in illustrating a conditional probability. Assume that the sample space under consideration is all 10,000 students in a university, and that event A represents the $a = 2000$ students who are seniors, while B represents the $b = 3500$ students who are females. Suppose also that $w = 800$ of these 3500 females are seniors.

The conditional event $A|B$ represents those students who are seniors selected from those who satisfy the condition of being *female*. The probability of conditional events of this nature is often of interest in specific sampling problems. For example, if a student is selected at random, and given that the selected student is female, what is the probability that the student is also a senior? We denote this probability by $P(A|B)$ and determine its value by calculating the frequency of students who are female and seniors (w) relative to the total number of females (b). This relative frequency is

$$P(A|B) = \frac{w}{b} = \frac{800}{3500} = 0.228.$$

Similarly, $P(B|A)$ can be expressed as the question, "given that a selected student is a senior, what is the probability that the student is female?" The answer is

$$P(B|A) = \frac{w}{a} = \frac{800}{2000} = 0.40.$$

Let us now write $P(A|B)$ in a slightly more convenient form. By dividing w and b by N (the total number of sample points) we obtain

$$P(A|B) = \frac{w}{b} = \frac{w/N}{b/N}.$$

* Named after logician J. Venn (1834–1923).

Another way of writing b/N is $P(B)$. Similarly, w/N can be written as $P(W)$. Recall from our earlier definition that an event W, which represents the occurrence of "both a senior (A) and a female (B)," is the *intersection of A and B*; thus, $P(W) = P(A \cap B)$. Putting these facts together, the probability that a randomly selected student is a senior, *given that* this person is a female, is:

$$P(A|B) = P(\text{Senior}|\text{Female}) = \frac{P(A \cap B)}{P(B)} = \frac{w/N}{b/N} = \frac{w}{b} = \frac{800}{3500} = 0.228.$$

Similarly, the probability that a randomly selected student is a female *given that* this person is a senior is

$$P(B|A) = \frac{P(A \cap B)}{P(A)} = \frac{w/N}{a/N} = \frac{w}{a} = \frac{800}{2000} = 0.40.$$

We can now formalize the definition of a conditional probability.

Conditional probability of A, given B:

$$P(A|B) = \frac{P(A \cap B)}{P(B)}.$$

Conditional probability of B, given A: (2.3)

$$P(B|A) = \frac{P(A \cap B)}{P(A)}.$$

To illustrate Formula (2.3) using our stock-rating example, let's calculate $P(\text{short-term} = 2|\text{long-term} = 3)$. From Fig. 2.2 we know there are 12 sample points, and hence $P(\text{short-term} = 2 \cap \text{long-term} = 3) = \frac{1}{12}$. Also, since there are four long-term categories, $P(\text{long-term} = 3) = \frac{1}{4}$. Thus,

$$P(\text{short-term} = 2|\text{long-term} = 3) = \frac{P(\text{short-term} = 2 \cap \text{long-term} = 3)}{P(\text{long-term} = 3)}$$

$$= \frac{\frac{1}{12}}{\frac{1}{4}} = \frac{4}{12} = \frac{1}{3}.$$

This result is easily verified using Fig. 2.2, as there are three (equally likely) outcomes associated with the column corresponding to the event "long-term = 3."

In our next example of conditional probabilities, we emphasize that it is not necessary to know the total number of sample points (N) in order to use Formula (2.3). Suppose that in a production process two parts of a particular product are produced simultaneously. Let D_1 denote the fact that the first part is defective, and D_2 denote the fact that the second part is defective. From past production records it is known that the probability that part 1 is defective is $P(\mathrm{D}_1) = 0.15$. Also, it is known from these records that the probability that *both* parts are defective is

$$P(\mathrm{D}_1 \cap \mathrm{D}_2) = 0.05.$$

The conditional probability that part 2 is defective, *given that* the first part is defective, can be determined by using Formula (2.3):

$$P(\mathrm{D}_2 | \mathrm{D}_1) = \frac{P(\mathrm{D}_1 \cap \mathrm{D}_2)}{P(\mathrm{D}_1)} = \frac{0.05}{0.15} = \frac{1}{3}.$$

$P(\mathrm{D}_1 | \mathrm{D}_2)$ cannot be calculated in this example unless we know the value of $P(\mathrm{D}_2)$.

As a final illustration of conditional probabilities, consider a physician advising an overweight patient who smokes on the probability of a heart attack. In this example, there are two previous conditions, smoking and being overweight. The appropriate conditional probability in this case can be written as follows

$$P(\text{Heart attack} | \text{smokes and is overweight}).$$

Now, if we let A = (heart attack) and B = (smokes and is overweight), then formula (2.3) yields the following relationships.

$$P(\text{Heart attack} | \text{smokes} \cap \text{overweight}) = P(A|B) = \frac{P(A \cap B)}{P(B)}.$$

Presumably, the physician can estimate the above (subjective) probabilities based on a variety of factors (patient's age and weight, how many cigarettes are smoked, family history, etc.). One study suggests that for certain age groups the incidence of heart attacks is four times larger for people who both smoke and are overweight. For these people, the probability of a heart attack in a given year for smokers who are overweight is

$$P(\text{Heart attack} | \text{smokes and is overweight}) = 0.048.$$

An objective problem involving similar conditional values will be presented in the next section.

Probability of an Intersection

In studying conditional probabilities in the last section, we indicated that $P(A \cap B)$ represents the probability of the intersection of A and B—that is, the probability that both A and B take place in one replication of an experiment. For example, the crosshatched area in Fig. 2.4 represents the probability that a student is both a senior and a female. Since there are 800 female seniors, $P(\text{senior} \cap \text{female}) = 800/10{,}000 = 0.08$. Several additional intersections were presented in the last section, namely

$$P(\mathrm{D}_1 \cap \mathrm{D}_2) = 0.05 \quad \text{and} \quad P(\text{short-term} = 2 \cap \text{long-term} = 3) = \frac{1}{12}.$$

We now want to develop a formula for the probability of an intersection. To do so, one merely has to solve Formula (2.3) for $P(A \cap B)$. The resulting formula is called the *general rule of multiplication* and provides a method for finding the probability of an intersection.

General rule of multiplication:

$$P(A \cap B) = P(A)P(B|A) = P(B)P(A|B). \tag{2.4}$$

The first part of Formula (2.4) can be interpreted as follows: The probability that *both* A and B take place is given by two occurrences—first, event A takes place, with probability $P(A)$, and then event B takes place on the condition that A has already occurred, with probability $P(B|A)$. The probability that both occurrences take place is the *product* of these two probabilities, or $P(A)P(B|A)$. For example, let us again consider the production problem previously presented, where we assumed that the probability of receiving one defective component is $P(\mathrm{D}_1) = 0.15$. Suppose that we also know from past records that the conditional probability $P(\mathrm{D}_2|\mathrm{D}_1) = \frac{1}{3}$. These two probabilities can be used to determine $P(\mathrm{D}_2 \cap \mathrm{D}_1)$, as follows:

$$P(\mathrm{D}_2 \cap \mathrm{D}_1) = P(\mathrm{D}_1)P(\mathrm{D}_2|\mathrm{D}_1) = (0.15)(\tfrac{1}{3}) = 0.05.^*$$

The probability that the intersection of two (or more) events occurs in a given experiment is often referred to as the *joint probability* of these events; the term "joint probability" implies that the events under consideration take place in the same replication of an experiment. Depending on the nature of the experiment, events that occur jointly do not necessarily take place at identical points in calendar or clock time. For example, in our marketing policy problem, the result of the pricing and advertising decision may be considered a joint occurrence even though the two decisions are not made simultaneously. Similarly, suppose NBC is interviewing two randomly selected U.S. Senators. The process of interviewing one senator and then another from the total of 100 senators can be thought of as a joint occurrence even though the two interviews are not simultaneous. If R_1 = Republican on the first interview and R_2 = Republican on the second interview, then $P(R_1 \cap \mathrm{R}_2)$ is the joint prob-

* If the values of $P(\mathrm{D}_2)$ and $P(\mathrm{D}_1|\mathrm{D}_2)$ are known, then we could calculate $P(\mathrm{D}_2 \cap \mathrm{D}_1)$ as follows:

$$P(\mathrm{D}_2 \cap \mathrm{D}_1) = P(\mathrm{D}_2)P(\mathrm{D}_1|\mathrm{D}_2).$$

This approach must also yield $P(\mathrm{D}_2 \cap \mathrm{D}_1) = 0.05$.

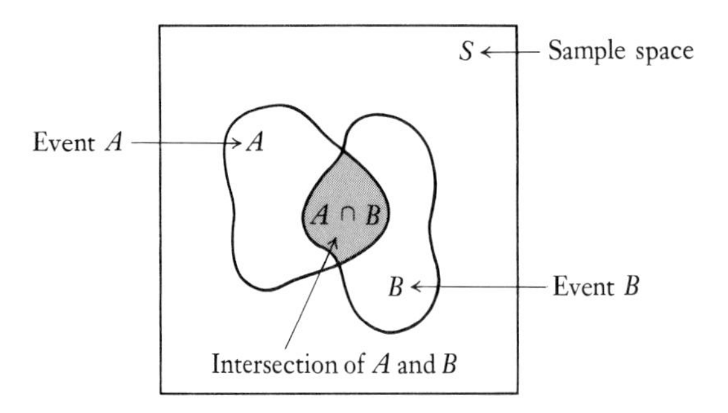

Fig. 2.5 The intersection of A and B (shaded portion).

ability that both are Republicans. If the Senate has 48 Republicans and 52 Democrats, then $P(\mathrm{R}_1) = 48/100 = 0.48$. After interviewing one Republican senator (who cannot be interviewed again), there remain 99 senators, 47 of whom are Republicans; hence,

$$P(\mathrm{R}_2 | \mathrm{R}_1) = \frac{47}{99},$$

and, by Formula (2.4),

$$P(\mathrm{R}_1 \cap \mathrm{R}_2) = P(\mathrm{R}_1)P(\mathrm{R}_2 | \mathrm{R}_1) = \frac{48}{100} \times \frac{47}{99} = 0.228.$$

Because the expressions $P(E_1 \cap E_2)$ and $P(E_2 \cap E_1)$ do not imply any ordering, over time, of the two events E_1 and E_2, but only that these events both occur in a single trial of the experiment, $P(E_1 \cap E_2)$ must equal $P(E_2 \cap E_1)$. Also, writing $P(E_2 | E_1)$ does not imply that E_1 precedes E_2 in chronological order. For instance, it is just as legitimate to determine the probability that two cars on a two-lane highway will "pass safely" (event A) given that one driver is "legally drunk" (event B) by $P(A | B)$, as it is to determine the probability that one of the drivers is legally drunk given that the two cars pass safely, $P(B | A)$.* In general, $P(A | B)$ does not equal $P(B | A)$.

To illustrate graphically the intersection of two events, in the Venn diagram shown in Fig. 2.5 we let A and B represent events of some sample space S. The intersection of A and B equals the shaded portion of these two events, labeled $A \cap B$.

Although the notation becomes rather cumbersome and the analysis more complex, it is possible to extend the rules for conditional and joint probabilities to apply to problems involving three or more events. To appreciate the concepts involved, consider a single relatively simple example—that of

* In some problems we probably won't want to consider both $P(E_1 | E_2)$ and $P(E_2 | E_1)$. For example, If R_1 = Republican on first interview and R_2 = Republican on second interview, the probability $P(\mathrm{R}_1 | \mathrm{R}_2)$ does not make any intuitive sense.

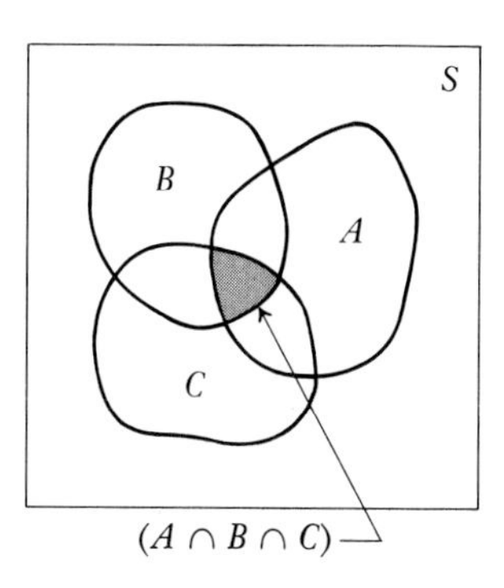

Fig. 2.6 The intersection of *A*, *B*, and *C* (shaded portion).

determining the probability that a random sample of three automobiles, selected without replacement from a list containing 10 J-cars and 10 K-cars, will consist of exactly three K-cars. This probability can be broken down into three occurrences: (1) select a K-car on the first draw (K_1), (2) select a K-car on the second draw *given* that a K-car was selected on the first draw ($K_2|K_1$), and (3) select a K-car on the third draw *given* that a K-car was selected on *both* the first and second draws ($K_3|K_2 \cap K_1$). The product of the probabilities of these three occurrences gives the joint probability of $K_1 \cap K_2 \cap K_3$.

$$P(K_1 \cap K_2 \cap K_3) = P(K_1)P(K_2|K_1)P(K_3|K_2 \cap K_1)$$

$$= \left(\frac{10}{20}\right)\left(\frac{9}{19}\right)\left(\frac{8}{18}\right) = \frac{2}{19}.$$

Figure 2.6 illustrates the intersection of three events (A, B, C) as the shaded area in which the sample points satisfy all three events.

Probability of a Union

The probability of the union of two events A and B, written $P(A \cup B)$, is the probability that either A occurs, or B occurs, or both A and B take place. The union of events A and B is illustrated by Fig. 2.7. The probability $P(A \cup B)$ is found by identifying the proportion of sample points that are included either in A or in B, or in the intersection of $(A \cap B)$.

The crosshatched area in Fig. 2.4 on page 66 also represents a union—in this case the union of the events *senior* and *female*. Recall from that example that there were $a = 2000$ seniors, $b = 3500$ females, $w = 800$ female seniors, and $N = 10{,}000$ students. To determine the probability that a randomly selected student is either a senior or a female (or both), we cannot merely add

$$\frac{a}{N} + \frac{b}{N} = \frac{2000}{10{,}000} + \frac{3500}{10{,}000},$$

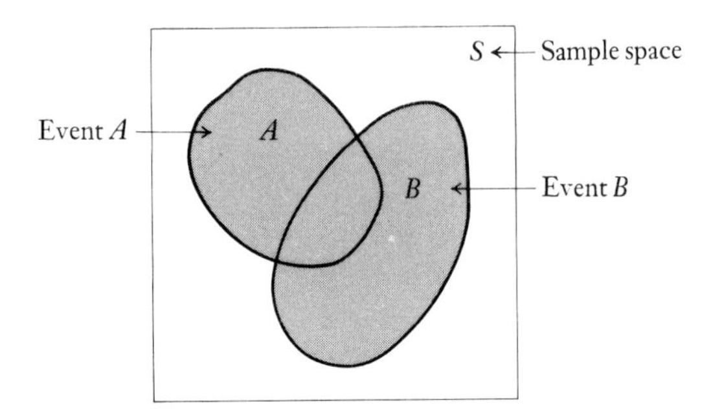

Fig. 2.7 The union of A and B (shaded portion).

because this sum *double-counts* the females who are seniors. In other words, that sum is too large by the amount

$$\frac{w}{N} = \frac{800}{10{,}000}.$$

The correct probability is

$$\begin{aligned} P(\text{senior} \cup \text{female}) &= P(\text{senior}) + P(\text{female}) - P(\text{senior} \cap \text{female}). \\ &= \frac{a}{N} + \frac{b}{N} - \frac{w}{N} \\ &= \frac{2000}{10{,}000} + \frac{3500}{10{,}000} - \frac{800}{10{,}000} \\ &= 0.20 + 0.35 - 0.08 = 0.47. \end{aligned}$$

This example demonstrates what is called the *general rule of addition.* If we substitute the letter A for senior and B for female in the formulation above, then the general rule of addition can be written as follows:

> *General rule of addition:*
>
> $$P(A \cup B) = P(A) + P(B) - P(A \cap B). \tag{2.5}$$

Formula (2.5) can be used to find the probability of the union of A and B in our stock-rating problem, where we define

A = long-term growth is rated 3 or higher, and

B = short-term growth is rated 3.

From Fig. 2.2 on page 60 we can show that $P(A) = \frac{2}{4}$ and $P(B) = \frac{1}{3}$. Also, we calculated earlier that $P(A \cap B) = \frac{2}{12}$. Thus,

$$P(A \cup B) = P(A) + P(B) - P(A \cap B)$$
$$= \frac{2}{4} + \frac{1}{3} - \frac{2}{12} = \frac{8}{12}.$$

This probability agrees with the result we calculated (less formally) in Section 2.1. You should also be able to count in Fig. 2.2 the eight sample points where either long-term growth ≥ 3, or short-term growth $= 3$.

In our production process example, suppose that we now assume that $P(D_2) = 0.10$, in addition to the values we previously assumed, namely $P(D_1 \cap D_2) = 0.05$ and $P(D_1) = 0.15$. These values can be used to calculate $P(D_1 \cup D_2)$.

$$P(D_1 \cup D_2) = P(D_1) + P(D_2) - P(D_1 \cap D_2)$$
$$= 0.15 + 0.10 - 0.05 = 0.20.$$

Define: Probability of a complement, joint probabilities, general rule of addition, conditional probabilities, general rule of multiplication, intersection, and union.

Problems

2.12 A study on the probability that a randomly selected person smokes cigarettes divided the U.S. population into three age groups: under 30 (<30), between 30 and 50 (30–50), and over 50 (>50). Half those under 30 were found to smoke.

a) If $P(<30) = \frac{1}{2}$, find the probability that a randomly selected person is under 30 and smokes.

b) If $P(\text{Smokes}|<30) = \frac{1}{2}$, $P(\text{Smokes}|>50) = \frac{1}{2}$, and $P(\text{Smokes}|30\text{–}50) = \frac{1}{4}$, does this indicate independence or dependence between age and smoking?

c) If $P(30\text{–}50) = \frac{1}{4}$ and $P(>50) = \frac{1}{4}$, find $P(\text{S}) = P(\text{Smokes})$.

d) Replace the probability symbols in the following table with their appropriate values.

	<30	30–50	>50	
Smokes	$P(\text{S} \cap <30)$	$P(\text{S} \cap 30\text{–}50)$	$P(\text{S} \cap >50)$	$P(\text{S})$
Doesn't smoke	$P(\bar{\text{S}} \cap <30)$	$P(\bar{\text{S}} \cap 30\text{–}50)$	$P(\bar{\text{S}} \cap >50)$	$P(\bar{\text{S}})$
	$P(<30)$	$P(30\text{–}50)$	$P(>50)$	

e) Find $P(\text{S} \cup >50)$.

2.13 The following data describe certain characteristics of the students enrolled at a university.

	Men	Women	Over 21
Freshmen	1325	1100	125
Sophomores	1200	900	175
Juniors	900	850	325
Seniors	725	775	950
Graduates	1350	875	2225

a) How many students are enrolled in this university?

b) What is the probability that a student selected "at random" will be a woman?

c) What is the probability that a student selected at random will be a senior?

d) Calculate $P(\text{sophomore} \cap \text{male})$ and $P(\text{sophomore} | \text{male})$

2.14 As an exercise to test your understanding of the concepts discussed in this chapter, answer the following questions about a friend who is in your statistics class (or yourself).

a) Estimate the probability that this person will receive a grade of A on the next exam [denote this as $P(\text{AE})$].

b) What is $P(\overline{\text{AE}})$?

c) Estimate the probability that this person will earn an A in the course (denoted as $P(\text{AC})$.

d) Estimate $P(\text{AC}|\text{AE})$.

e) Estimate $P(\text{AE}|\text{AC})$. Must this answer be the same as for $P(\text{AC}|\text{AE})$? Explain.

f) Determine $P(\text{AC} \cap \text{AE})$ and $P(\text{AC} \cup \text{AE})$ based on your answers to parts (a) through (e).

g) Draw a Venn diagram representing the events AE and AC.

2.15 The Easy Charge Company of Exercise 1.44 has presented the following table representing a breakdown of customers according to the amount they owe and whether a cash advance has been made.

Amounts Owed by Customers	Cash Advance	No Cash Advance
\$0– 99.99	229	2894
\$100–199.99	378	1707
\$200–299.99	501	1426
\$300–399.99	416	939
\$400–499.99	260	483
\$500 or more	289	478
Total customers	2073	7927

a) Find $P(\text{Cash advance})$ and $P(\overline{\text{Cash advance}})$.

b) Find $P(\text{Cash advance}|\text{Amount owed} < \$100)$.

c) Find $P(\text{Amount owed} < \$100|\text{Cash advance})$.

d) Find $P(\text{Amount owed} = \$100 - \$199.99 \cap \overline{\text{Cash advance}})$.

e) Find $P(\text{Amount owed} = \$100 - \$199.99 \cup \overline{\text{Cash advance}})$.

2.6 SPECIAL CASES OF PROBABILITY RULES

Formulas (2.4) and (2.5) are *general* formulas that are appropriate for all types of events. There are, however, special cases of these formulas that can make calculating certain probabilities easier. We will examine these special rules for the case of mutually exclusive events and for independent events.

Mutually Exclusive Events

When the events A and B are mutually exclusive, the danger of double counting is eliminated. That is, if A and B are mutually exclusive events, then there is *no overlap* of sample points in A and B that can be counted twice. The intersection $A \cap B$ in this special case has probability zero, $P(A \cap B) = 0$. We can thus rewrite Formula (2.5) as shown next.

> *Special case of addition rule (given mutually exclusive events):*
>
> $$P(A \cup B) = P(A) + P(B). \tag{2.6}$$

The Venn diagram in Fig. 2.8 illustrates the relationship between the sets A and B when these sets are mutually exclusive.

Determining the probability that *either* the Braves *or* the Cubs will win the National League pennant this year is an example of the type of problem that involves Formula (2.6). These events are mutually exclusive because it is not possible for both teams to win. (Some people would say that the probability that *either* will win is close to zero.) Another example is that of determining the probability that NBC interviews a senator from Indiana or a senator from New York when randomly selecting one person from the U.S. Senate. Since

Fig. 2.8 Illustration of the special rule of addition for mutually exclusive events:

$$P(A \cap B) = 0,$$

so

$$P(A \cup B) = P(A) + P(B).$$

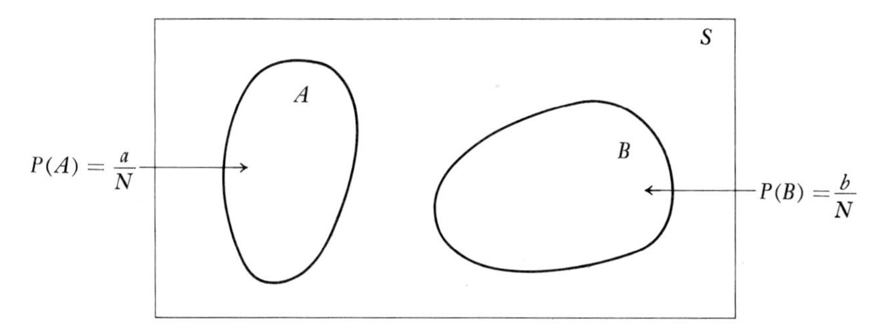

$P(\text{Ind.}) = P(\text{N.Y.}) = \frac{1}{50}$, the probability of a union is

$$P(\text{Ind.} \cup \text{N.Y.}) = \frac{1}{50} + \frac{1}{50} = \frac{1}{25} = 0.04.$$

Formulas (2.5) and (2.6) can be extended to three or more events in a union, although the notation and formulas become tedious when (2.5) is extended. In the case of (2.6), if the events $E_1, E_2, \ldots, E_k$ are *mutually exclusive* [i.e., for all i, j, $P(E_i \cap E_j) = 0$], then the probability of their union is simply the sum of the probabilities of each event because there is no intersection, and thus none of the sample points is counted twice. That is, the extension of Formula (2.6) is

Special rule for union of k mutually exclusive events:

$$P(E_1 \cup E_2 \cup \cdots \cup E_k) = \sum_{i=1}^{k} P(E_i).$$

Special Case of Independence

Two events are said to be *independent* if the occurrence (or nonoccurrence) of either event cannot influence (or be influenced by) the occurrence (or nonoccurrence) of the other event. That is, the probability of any given event must be unaffected by the other event in order for independence to hold. Dependence implies just the opposite—that the occurrence of one event *is* influenced by the occurrence of one or more other events being considered. For example, the probability that you make the dean's list is highly dependent on the occurrence or nonoccurrence of the event "receive an A in statistics." On the other hand, the probability that you receive an A in statistics is independent of whether or not the Pirates win the National League pennant this year.* Medical scientists, for example, are continually researching diseases to determine their independence of or dependence on a variety of factors. The probability of getting cancer, for instance, has been reported to be related to (dependent on) whether one smokes, the quality of the air one breathes, and even the food one eats.

In many problems it may be difficult to determine whether or not two or more events are independent or dependent. As a current example, note how much controversy has been generated as to whether or not the presence of the artificial sweetener saccharine in soft drinks is independent of the occurrence of bladder cancer in humans. Similarly, a local retail store may wish to study

* Note that independence here implies *both* that an A grade does not influence the standing of the Pirates, *and* the Pirates' fate in the pennant race does not influence your probability of an A grade.

whether sales are dependent on numerous factors, such as the state of the economy, the price of a competitor's product, the weather, and the amount of advertising. To illustrate dependence versus independence, recall our example about students in a university (see Fig. 2.4 on page 66), where there were 3500 females and 800 seniors in a university of 10,000 students. Suppose we define the following for a sample of one person:

$$f = \text{person is a female},$$
$$s = \text{person is a senior}.$$

Are the events f and s independent or dependent? To answer this question we need to determine whether the occurrence of f is influenced by s, or vice versa. In Section 2.5, we calculated

$$P(\text{senior}|\text{female}) = \frac{800}{3500} = 0.228.$$

There were 2000 seniors in our problem, hence

$$P(\text{senior}) = \frac{2000}{10{,}000} = 0.200.$$

Thus, knowing that the person selected is a female (f) *changes* the probability this person is a senior from 0.200 to 0.228; hence, the event f *does* influence the occurrence of the event s, and we then know that these two events must be dependent. For s and f to have been independent, the following relationship would have to hold true:

$$P(\text{senior}|\text{female}) = P(\text{senior}).$$

In many statistical problems, it may not be clear whether the events of interest are independent or dependent. In such cases either (1) it is not important or it is impossible to determine whether the events are independent or dependent, (2) one is trying to *prove* whether the events are independent or dependent, or (3) the experimenter can *assume* that the events are independent because of the way the experiment is defined. These three situations are elaborated below.

1. When it is not important or it is impossible to determine whether the events are independent or dependent, then the formulas derived thus far [(2.3), (2.4), and (2.5)] should be used. *These formulas hold for both independent and dependent events.*
2. In some problems it is important to attempt to *prove* whether events are independent or dependent. In real-world problems, this is important when one is trying to establish the relationship between events (such as cigarette smoking and cancer). From a statistical point of view, it is more convenient to work with events that are independent, for then we can use simplified versions of Formulas (2.3), (2.4) and (2.5). If A and B are independent, then

$P(A)$ must be unaffected by whether or not B occurs, and $P(B)$ must be unaffected by whether or not A occurs. The rule for proving the independence of two events A and B is as follows:

A and B are independent if either of the following two relationships holds true:

$$P(A|B) = P(A) \qquad \text{or} \qquad P(B|A) = P(B). \tag{2.7}$$

If $P(A|B) = P(A)$ then it must be true that $P(B|A) = P(B)$, and conversely.*

Since independence is the complement of dependence, to verify that two events are dependent, we need only show that one of the two relationships in (2.7) does *not* hold true; that is,

A and B are dependent if either one of the following two relationships holds true:

$$P(A|B) \neq P(A) \qquad \text{or} \qquad P(B|A) \neq P(B).$$

Suppose we use the rules described above to determine whether or not independence exists in our production-process example. To determine whether D_1 and D_2 are independent or dependent, we must look at one or both of the following relationships:

$$P(D_1|D_2) \text{ versus } P(D_1) \qquad \text{and/or} \qquad P(D_2|D_1) \text{ versus } P(D_2).$$

If *either* relationship can be shown to be an equality, then D_1 and D_2 are independent. Similarly, if either relationship can be shown *not* to be an equality, then D_1 and D_2 must be dependent. Remember from the statement on page 73 that the value of $P(D_2)$ was stated to be $P(D_2) = 0.10$. Also remember that on page 68 we gave the following conditional probability, $P(D_2|D_1) = \frac{1}{3}$. Since

$$P(D_2|D_1) = \tfrac{1}{3} \text{ does not equal } P(D_2) = 0.10,$$

the events D_1 and D_2 cannot be independent; in other words, they must be dependent.

As an example of independent events, let us consider the experiment "draw one card from a deck of 52." Events A and B are defined as follows: A = a spade is drawn and B = an ace is drawn. To prove that A and B are independent, we need to show that either

$$P(\text{spade}|\text{ace}) = P(\text{spade}) \qquad \text{or} \qquad P(\text{ace}|\text{spade}) = P(\text{ace}).$$

* Proof of this is left as an exercise for the reader.

Since there are 13 spades and 4 aces in a deck of 52 cards,

$$P(\text{spade}) = \tfrac{13}{52} = \tfrac{1}{4} \qquad \text{and} \qquad P(\text{ace}) = \tfrac{4}{52} = \tfrac{1}{13}.$$

As everyone knows, there is one ace of spades in a deck of cards; hence $P(\text{ace} \cap \text{spade}) = \frac{1}{52}$.

Using these results and Formula (2.3),

$$P(\text{spade}|\text{ace}) = \frac{P(\text{spade} \cap \text{ace})}{P(\text{ace})} = \frac{\frac{1}{52}}{\frac{4}{52}} = \frac{1}{4}.$$

Now since

$$P(\text{spade}|\text{ace}) = \tfrac{1}{4} = P(\text{spade}),$$

we have proved independence of the events ace and spade in one draw from a deck of cards. The reader should try proving $P(\text{ace}|\text{spade}) = P(\text{ace})$ to see that this relationship also holds (as it must if "ace" and "spade" are independent).

3. In some experiments the events are *assumed* explicitly (or at least implicitly) to be independent. For instance, the outcomes of the slot machines in Las Vegas are generally assumed to be independent; this implies that the result of pulling a lever has no influence on the results of any other "pulls." In Chapter 6, we will assume that sample observations are taken "randomly" (independently), meaning that the outcome of one sample observation cannot influence the outcome of any other sample observation.

Probability Rules for Independent Events

The most important effect of assuming independence is that it greatly simplifies the calculation of joint probabilities. This can be shown by substituting Formula (2.7) into the general rule of multiplication [Formula (2.4)]. The result is the special case of the multiplication rule.

If A and B are independent, then the joint probability of A and B can be determined by using the following formula:

Special case of multiplication rule:

$$P(A \cap B) = P(A)P(B) = P(B)P(A). \tag{2.8}$$

Let us assume that the probability that you will earn an A in statistics is $P(\text{A grade}) = 0.50$ and the probability that the Pirates win the National League pennant is $P(\text{Pirates win}) = 0.25$. If these events are assumed to be independent (e.g., you are not taking statistics during the summer and have season tickets to the Pirates' games), then

$$P(\text{A grade} \cap \text{Pirates win}) = P(\text{A grade})P(\text{Pirates win}) = (0.50)(0.25) = 0.125.$$

As another example, suppose a soft-drink manufacturer uses two machines in its capping process. Defective caps result from one machine with probability $P(\text{D}_1) = 0.001$, and defective caps result from the second machine with probability $P(\text{D}_2) = 0.003$. Assuming independence, the probability of a defective cap being found on both of two randomly selected bottles, each from a different machine, is

$$P(\text{D}_1 \cap \text{D}_2) = P(\text{D}_1)P(\text{D}_2) = (0.001)(0.003) = 0.000003.$$

There are thus three chances in a million that both selected bottles have defective caps.

From Formula (2.8) we can easily derive a rule for calculating joint probabilities when there are more than two events. If a sequence of k events $E_1, E_2, \ldots, E_k$ are assumed to be independent, then the joint probability $P(E_1 \cap E_2 \cap \cdots \cap E_k)$ can be determined as follows:

Special multiplication rule for k independent events:

$$P(E_1 \cap E_2 \cap \cdots \cap E_k) = P(E_1)P(E_2) \cdots P(E_k). \tag{2.9}$$

To illustrate the use of Formula (2.9), suppose a survey is being taken in a city where 40% of the registered voters are supposed to be opposed to a certain bond issue $[P(O) = 0.40]$. If five registered voters are selected randomly, what is the probability that all five will be opposed? Using Formula (2.9),

$$P(O_1 \cap O_2 \cap O_3 \cap O_4 \cap O_5) = (0.40)(0.40)(0.40)(0.40)(0.40) = (0.40)^5 = 0.01024.$$

This result says that only slightly over 1% of the time will all five people surveyed oppose the issue, assuming the probability of opposition is 0.40, and that events are independent.

The Distinction Between Mutually Exclusive Events and Independence

Beginning students of statistics often confuse the concepts of independence and mutually exclusive events. It is vital that the reader understand the difference between these concepts and the implications of each in probability theory.

Thus, we present the following two general statements:

1. Events that are mutually exclusive must be dependent, but dependent events need not be mutually exclusive.
2. Events that are not mutually exclusive may be either independent or dependent; however, events that are independent cannot be mutually exclusive.

To illustrate the first of these two statements, consider the following three events:

E_1 = high government spending,
E_2 = low government spending,
E_3 = low unemployment.

The events E_1 and E_2 are mutually exclusive; they are also dependent, since if one happens, this influences the probability of the other happening (it cannot). Thus we have demonstrated the first part of statement 1. Now E_1 and E_3 are generally regarded as dependent events (high government spending lowers unemployment), but they certainly are not mutually exclusive. This demonstrates the second part of statement 1.

To illustrate the first part of statement 2, suppose a survey is taken of 100 top economists to determine how many favor tighter controls over foreign imports. Let's assume 40 of these economists favor tighter controls, and define the following two events.

T_1 = first economist sampled favors tighter controls,
T_2 = second economist sampled favors tighter controls.

These two events are not mutually exclusive. They are dependent events if sampling is without replacement [then $P(T_2|T_1) = \frac{39}{99} \neq P(T_2) = \frac{40}{100}$]. If sampling had been with replacement (which is not usually the case), then

$$P(T_2|T_1) = P(T_2) = \frac{40}{100},$$

and T_2 and T_1 are independent.

Suppose that two events, A and B, are assumed to be independent; then we know that

$$P(A|B) = P(A) = \frac{a}{N} \qquad \text{and} \qquad P(B|A) = P(B) = \frac{b}{N}.$$

On the other hand, if A and B are mutually exclusive events (as shown in Fig. 2.8 on page 75), then

$$P(A|B) = 0 \qquad \text{and} \qquad P(B|A) = 0.$$

Thus, if A and B are independent they cannot be mutually exclusive because

$$P(A) = \frac{a}{N} \neq 0 \qquad \text{and} \qquad P(B) = \frac{b}{N} \neq 0.^*$$

To summarize,

Mutually exclusive implies dependence, and independence implies not mutually exclusive, but no other simple relationships hold true.

Define: Special rule of addition, dependence and independence, special case of multiplication rule.

Problems

2.16 In Problem 2.15 are the events "cash advance" and amount owed \$0–\$99.99 independent?

2.17 If age and sex are assumed to be independent in Problem 2.13, what is the probability that a randomly selected student is a male over 21? What is $P(\text{male} \cup \text{female})$?

2.18 During the 1979 Three Mile Island nuclear problems, at least three separate problems contributed to the near disaster: (1) a relief valve stuck; (2) a pressurizer gave a false reading; and (3) an auxiliary feedwater pump was left closed when it should have been open. Suppose the probability that each one of these three events occurs is 0.01, and they are assumed to be independent. What is the joint probability that all three will occur jointly? Would you conclude, given these values, that the TMI incident was just "bad luck"?

2.19 A machine is known to produce defective components with a probability of $p = 0.05$.

a) Assume defective components are produced independently of one another. If we examine three components, what is the probability all three will be defective?

b) If you find three defectives, as in part (a), what might you conclude about the machine?

2.20 Consider the pie chart in Fig. 2.9, featured in *Time* magazine on August 11, 1980.

a) According to this chart, what is the probability that a randomly selected person who is unemployed will be either adult or white, $P(W \cup A)$. What is $P(W \cap A)$?

* We assume that $a > 0$ and $b > 0$. If either a or b equals zero then it makes little sense to talk about independence or dependence since the event never occurs.

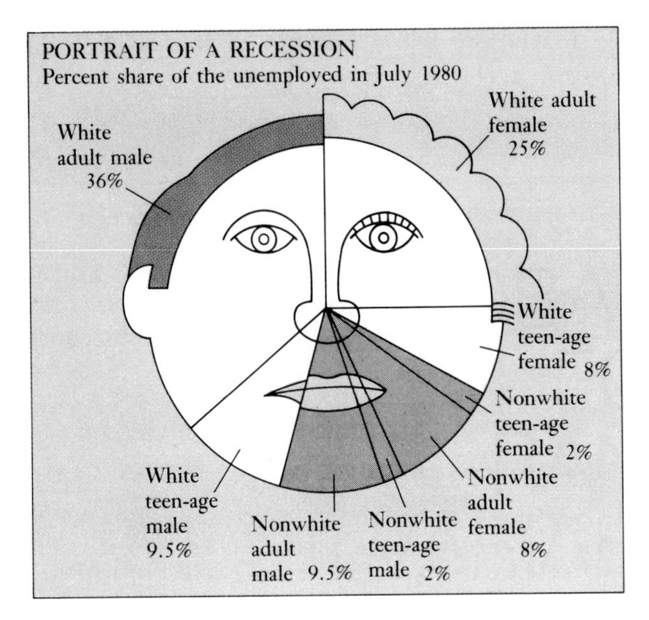

Fig. 2.9 Portrait of a recession. (Adapted by permission from *Time* chart by Nigel Holmes. Copyright 1980 Time Inc. All rights reserved.)

b) Given that a person is an adult, what is the probability that this person is white? What is $P(W)$?

c) Is the event W independent of the event A? Explain, using your answer to part (b).

d) Find the value of $P(A)P(W)$. Use this value and your results from part (a) to determine whether A and W are independent.

2.21 Suppose on the Management Accounting Examination there are three multiple-choice questions on probability. Each question has four possible answers.

a) Can the MAE administrators be 99-percent confident that a student who randomly picks answers will not answer all three correctly?

b) What is the probability that a person picking randomly will not get any answer correct?

2.22 An August 1980 Gallup survey in the *Wall Street Journal* reported on hours worked by 780 chief executives as shown in Fig. 2.10 on page 84.

a) Among large firms, what is the probability that a randomly selected executive will be working more than 60 hours per week?

b) Suppose you interview four randomly selected chief executives in small firms. What is the probability that all four will be working less than 50 hours per week, using the data above?

2.23 One part of the trial of the Great Train Robbery ($7,000,000 stolen) took place in January, 1964. In it, Dr. I. Holden assessed the probability of finding two different paints on the soles of one pair of shoes identical to two paints found at the robbers' hideout. His answer was that if there are 1000 different colors of paint (one firm

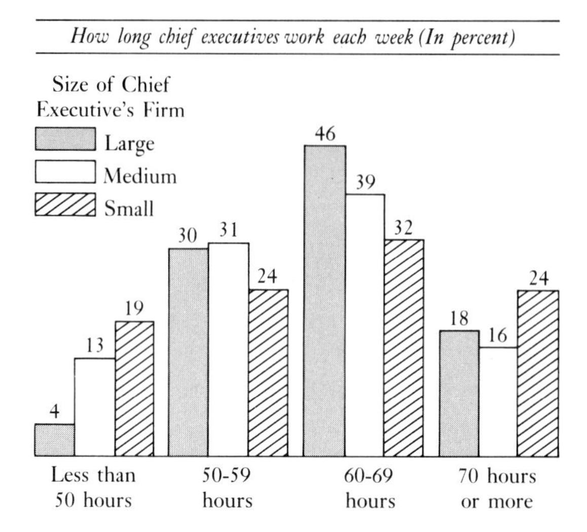

Fig. 2.10 Executive work week. (Reprinted by permission of *Wall Street Journal.* © Dow Jones & Company, Inc. All rights reserved.)

advertised that number), and if we assume the probability of getting one particular color on one's shoe is 1/1000, then the probability that the paint on a person's shoes would exactly match the two colors is 0.000001. Comment on any assumptions or difficulties you see with this analysis.

2.7 MARGINAL PROBABILITY

In a number of circumstances it is convenient to think of a single event as always occurring jointly with other events, where these other events are assumed to influence the probability of the first occurring. For instance, it may be helpful not only to identify defective items coming off a production process, but also to specify exactly which machines, or which workers, produced these defectives. Insurance companies are interested in not only the amount of damage associated with each automobile accident, but among other things, the city where the accident took place, and the age and sex of the driver. In such situations, the probability of the event in question (e.g., the probability of a defective item across all machines and all workers, or the probability of an accident involving at least x dollars across all cities and drivers) may not be known directly, but perhaps this value can be determined by summing its chance of occurrence in combination with the other relevant factors identified in the problem.

Suppose a bookstore is concerned not only with the number of books it sells in a given week, but also with the number of hardback books relative to the number of paperbacks. Assume that Table 2.1 represents the number of copies of three bestsellers (A, B, and C) sold in their paperback and hardback versions

Table 2.1

	Book			Marginal totals
	A	B	C	
Hardbacks	25	20	5	50
Paperbacks	125	80	45	250
Marginal totals	150	100	50	300

during one week. Out of the 300 books sold that week, 50 were sold in a hardback version. Hence, for the week in question the probability that a customer chose the hardback version is $P(\mathrm{H}) = \frac{50}{300} = \frac{1}{6}$. $P(\mathrm{H})$ in this context is called a *marginal probability* since its value is determined from information given in the "margin" of a table of frequencies. The (marginal) probability that book A was bought is $P(\mathrm{A}) = \frac{150}{300} = \frac{1}{2}$, while $P(\mathrm{B}) = \frac{100}{300} = \frac{1}{3}$, and $P(\mathrm{C}) = \frac{50}{300} = \frac{1}{6}$. Note that a marginal probability is actually nothing more than an unconditional probability, and that the sum of the three unconditional probabilities, $P(\mathrm{A}) + P(\mathrm{B}) + P(\mathrm{C})$, must equal *one.*

The important point of this section is that a marginal probability can be derived without knowing the marginal totals if certain other joint or conditional probabilities in the situation are known. In the present problem, for example, we can calculate the marginal probability $P(\mathrm{H})$ if we know the joint probability that H occurs with events A, B, and C. That is, since H must occur with one of these events, the marginal probability $P(\mathrm{H})$ can be written as the sum of the probability that H intersects with A, B, or C. Thus,

$$P(\mathrm{H}) = P(\mathrm{H} \cap \mathrm{A}) + P(\mathrm{H} \cap \mathrm{B}) + P(\mathrm{H} \cap \mathrm{C}).$$

From Table 2.1 we see that H occurs jointly with A, B, and C with probabilities

$$P(\mathrm{H} \cap \mathrm{A}) = \tfrac{25}{300}, \qquad P(\mathrm{H} \cap \mathrm{B}) = \tfrac{20}{300}, \qquad \text{and} \qquad P(\mathrm{H} \cap \mathrm{C}) = \tfrac{5}{300}.$$

Hence,

$$\begin{aligned} P(\mathrm{H}) &= P(\mathrm{H} \cap \mathrm{A}) + P(\mathrm{H} \cap \mathrm{B}) + P(\mathrm{H} \cap \mathrm{C}) \\ &= \tfrac{25}{300} + \tfrac{20}{300} + \tfrac{5}{300} = \tfrac{50}{300} = \tfrac{1}{6}. \end{aligned}$$

Thus we see that the marginal probability $P(\mathrm{H})$ is the sum of the joint probability of this event together with all events with which it could possibly be associated (i.e., books A, B, and C).

To provide a more general statement of the above result, suppose we let A represent the event of interest in a problem of the above type (e.g., a hardback), and let $E_1, E_2, \ldots, E_j, \ldots, E_N$ represent a set of mutually exclusive events, one of which must always occur jointly with A (e.g., the books). The value of $P(A)$ is given by summing the joint probability $P(A \cap E_j)$ over all possible values of j $(0 \le j \le N)$.

Marginal probability of Event A:

$$P(A) = \sum_{j=1}^{N} P(A \cap E_j). \tag{2.10}$$

Formula (2.10) is not usually the most convenient form of expressing a marginal probability. Rather, the joint probability $P(A \cap E_j)$ is usually written in its equivalent form given by the general rule of multiplication (Formula (2.4)). Using this formula we can write $P(A \cap E_j) = P(E_j)P(A|E_j)$; hence

Marginal probability of Event A:

$$P(A) = \sum_{j=1}^{N} P(A \cap E_j) = \sum_{j=1}^{N} P(E_j)P(A|E_j). \tag{2.11}$$

Let us now rewrite the marginal probability $P(\text{H})$ in our book example in terms of Formula (2.11):

$$P(\text{H}) = P(A)P(\text{H}|A) + P(B)P(\text{H}|B) + P(C)P(\text{H}|C).$$

To find $P(\text{H})$ in this fashion, we need to know the three conditional probabilities $P(\text{H}|A)$, $P(\text{H}|B)$, and $P(\text{H}|C)$. These values are seen in Table 2.1 to be

$$P(\text{H}|A) = \frac{25}{150} = \frac{1}{6},$$

$$P(\text{H}|B) = \frac{20}{100} = \frac{1}{5},$$

$$P(\text{H}|C) = \frac{5}{50} = \frac{1}{10}.$$

Since we also know that $P(A) = \frac{150}{300}$, $P(B) = \frac{100}{300}$, and $P(C) = \frac{50}{300}$, $P(\text{H})$ can be calculated as follows:

$$P(\text{H}) = P(A)P(\text{H}|A) + P(B)P(\text{H}|B) + P(C)P(\text{H}|C)$$

$$= \frac{150}{300} \cdot \frac{25}{150} + \frac{100}{300} \cdot \frac{20}{100} + \frac{50}{300} \cdot \frac{5}{50} = \frac{1}{6}.$$

As another example of the use of Formula (2.11), consider the problem of estimating the probability that Carl Yastrzemski (Boston Red Sox baseball player) will get a hit on a randomly selected time at bat, if it is known that in the past the relative frequency of his getting a hit off a righthanded pitcher was 0.315, and the relative frequency of getting a hit off a lefthanded pitcher was

0.262.* If 0.315 is taken as an estimate of the probability he gets a hit given a righthanded pitcher [i.e., $P(\text{Hit} \mid R) = 0.315$], and 0.262 as the probability of a hit given a lefthanded pitcher [i.e., $P(\text{Hit} \mid L) = 0.262$], and we assume the probability that Yastrzemski will face righthanded and lefthanded pitchers to be $P(R) = 0.75$ and $P(L) = 0.25$, then

$$\begin{aligned} P(\text{Hit}) &= P(R)P(\text{Hit} \mid R) + P(L)P(\text{Hit} \mid L) \\ &= 0.75(0.315) + 0.25(0.262) = 0.302. \end{aligned}$$

2.8 APPLICATION OF PROBABILITY THEORY: AN EXAMPLE

Suppose a manufacturer producing delicate electronic components cannot determine whether a component is defective without tearing the component apart and, in the process, destroying the usefulness of that component. However, there is a machine that its makers claim will help detect defective components. This machine, which indicates only that the component appears to be good (+) or that it appears to be defective (−), is not infallible; that is, it will sometimes indicate positive when the component is defective, and sometimes indicate negative when the component is good. To determine the ability of the machine to distinguish between good and defective items, 800 randomly selected components were first tested on the machine and then torn apart to see whether they were good or defective. The results of this research are shown in Table 2.2. The data in Table 2.2 can be used to make estimates of a number of different probability values.

Intersections First note that four joint probabilities representing the *intersection* of two events can be determined directly from the data in Table 2.2: the joint probability of a good item *and* a positive test equals the relative

Table 2.2

		State of Component		Marginal totals
		Good	Defective	
Result of test	+	448	60	508
	−	112	180	292
Marginal totals		560	240	800

* Data based on Yastrzemski's major-league career record.

frequency of *both* events occurring together, which is

$$P(\text{Good} \cap \text{Test } +) = \frac{448}{800} = 0.560$$

$$P(\text{Good} \cap \text{Test } -) = \frac{112}{800} = 0.140$$

$$P(\text{Def.} \cap \text{Test } +) = \frac{60}{800} = 0.075$$

$$P(\text{Def.} \cap \text{Test } -) = \frac{180}{800} = \frac{0.225}{1.000}$$

Note that the probabilities of the above four *mutually exclusive* and *exhaustive* events sum to one.

Marginals The four probabilities calculated above can now be used to calculate other probabilities in the combined forms of the sets discussed earlier. For instance, although the marginal (unconditional) probability of a good item can be determined from Table 2.2 to be $P(\text{Good}) = \frac{560}{800} = 0.700$, its value can also be determined via Formula (2.1-1):

$$\begin{aligned} P(\text{Good}) &= P(\text{Good} \cap \text{Test } +) + P(\text{Good} \cap \text{Test } -) \\ &= 0.560 + 0.140 = 0.700 \end{aligned}$$

The probability $P(\text{Test } +)$ can be determined in a similar fashion:

$$\begin{aligned} P(\text{Test } +) &= P(\text{Good} \cap \text{Test } +) + P(\text{Def.} \cap \text{Test } +) \\ &= 0.560 + 0.075 = 0.635. \end{aligned}$$

Unions The probability of the union of two events, such as the probability of *either* a good component *or* a positive test, can be determined by using the general rule of addition (Formula (2.5)):

$$\begin{aligned} P(\text{Good} \cup \text{Test } +) &= P(\text{Good}) + P(\text{Test } +) - P(\text{Good} \cap \text{Test } +) \\ &= 0.700 + 0.635 - 0.560 = 0.775. \end{aligned}$$

The manufacturer in our problem is primarily interested in how well this testing device works—in other words, if it reads + on a given component, how often will the component be good? Similarly, if it reads −, how often will the component be defective? These values can be determined from Table 2.2 by noting that out of the 508 components which tested +, 448 were good; hence

$$P(\text{Good} \mid \text{Test } +) = \frac{448}{508} = 0.882.$$

Similarly, of the 292 components which tested −, 180 were defective, so that

$$P(\text{Def.}|\text{Test} -) = \frac{180}{292} = 0.616.$$

These values can formally be determined by using the formula for conditional probability:

$$P(\text{Good}|\text{Test} +) = \frac{P(\text{Good} \cap \text{Test} +)}{P(\text{Test} +)} = \frac{0.560}{0.635} = 0.882;$$

$$P(\text{Def.}|\text{Test} -) = \frac{P(\text{Def.} \cap \text{Test} -)}{P(\text{Test} -)} = \frac{0.225}{0.365} = 0.616.$$

The value of $P(\text{Good}|\text{Test} -)$ must equal

$$1 - P(\text{Def.}|\text{Test} -) = 1 - 0.616 = 0.384,$$

since this event and $P(\text{Def.}|\text{Test} -)$ are complementary.

The results of the machine test in this problem are obviously not independent of the state of the component. If they were independent, $P(\text{Good}|\text{Test} +)$ would have to equal $P(\text{Good})$, and $P(\text{Def.}|\text{Test} -)$ would have to equal $P(\text{Def.})$, which is not the case. But *how much better off* is the manufacturer by knowing the results of the test? The manufacturer is in good shape if the test is positive, as then a guess that the component is good will be correct 88.2 percent of the time. After a negative indication, however, the manufacturer who guesses the component to be defective will be correct only 61.6 percent of the time. Fortunately, a positive test will occur most often (with probability 0.635), and a negative test will occur only relatively infrequently (with probability 0.365). Multiplying 0.882 by 0.635 and 0.616 by 0.365, we obtain, as the sum of these products, 0.785. This is the probability of a correct assessment, assuming the manufacturer always accepts the result of the machine test. This weighting is just a common-sense use of the formal rule for adding up all the joint occurrences of these events in which the test indicates the correct state of the component:

$$\begin{aligned} P(\text{Good} \cap \text{Test} +) + P(\text{Def.} \cap \text{Test} -) &= 0.882(0.635) + (0.616)(0.365) \\ &= 0.785. \end{aligned}$$

If we now knew how much this machine costs (to buy and operate), and how much the firm's revenue will increase by being better able to distinguish between good and defective components, then we could determine whether the machine is worth purchasing. Without this machine, the manufacturer who presumes all components are good will be correct 70 percent of the time [since $P(\text{Good}) = \frac{560}{800} = 0.70$].

2.9 BAYES' RULE

One of the most interesting (and controversial) applications of the rules of probability theory involves estimating unknown probabilities and making decisions on the basis of new (sample) information. Statistical decision theory is a new field of study which has its foundations in just such problems. Chapter 9 investigates the area of statistical decision theory in some detail; this section describes one of the basic formulas of the area, *Bayes' rule*.

An English philosopher, the Reverend Thomas Bayes (1702–1761), was one of the first to work with rules for revising probabilities in the light of sample information. Bayes' research, published in 1763, went largely unnoticed for over a century, and only recently has attracted a great deal of attention. His contribution consists primarily of a unique method for calculating conditional probabilities. The so-called "Bayesian" approach to this problem addresses itself to the question of determining the probability of some event, E_i, *given that* another event, A, has been (or will be) observed; i.e., determining the value of $P(E_i|A)$. The event A is usually thought of as sample information, so that Bayes' rule is concerned with determining the probability of an event given certain new information, such as that acquired from a sample, a survey, or a pilot study. For example, a sample output of 3 defectives in 20 trials (event A) might be used to estimate the probability that a machine is not working correctly (event E_i); or you might use the results of your first exam in statistics (event A) as sample evidence in estimating the probability of receiving an "A" in this course (event E_i).

Probabilities before revision by Bayes' rule are called *a priori* or simply *prior probabilities*, because they are determined *before* the sample information is taken into account. Prior probabilities may be either objective or subjective values.

A probability which has undergone revision in the light of sample information (via Bayes' rule) is called a *posterior probability*, since it represents a probability calculated *after* this information is taken into account. Posterior probabilities are always conditional probabilities, the conditional event being the sample information. Thus, by using Bayes' rule, a prior probability, which is an unconditional probability, becomes a posterior probability, which is a conditional probability. In order to calculate such posterior probabilities, we will first derive Bayes' rule for the general problem of determining $P(E_i|A)$.

Recall that earlier in this chapter we indicated that a joint probability can be written in two different conditional forms (see Formula (2.4)). This implies that we can also write the joint probability $P(E_i \cap A)$ in a similar fashion,

$$P(A)P(E_i|A) = P(E_i \cap A) = P(E_i)P(A|E_i).$$

If the above formula holds, then it must also be true that

$$P(A)P(E_i|A) = P(E_i)P(A|E_i).$$

We can now solve for $P(E_i|A)$ directly by dividing both sides by $P(A)$:

$$P(E_i|A) = \frac{P(E_i)P(A|E_i)}{P(A)}. \tag{2.12}$$

The numerator of Formula (2.12) represents the probability that A and E_i will both occur, while the denominator is the probability that A alone will occur. But $P(A)$ can be written in more meaningful terms for this type of problem as a marginal probability. The probability $P(A)$, written as a marginal probability, is given by Formula (2.11) as $P(A) = \sum_{j=1}^{N} P(E_j)P(A|E_j)$. Substituting this value into Formula (2.12) yields Bayes' rule:

Bayes' rule:

$$P(E_i|A) = \frac{P(E_i)P(A|E_i)}{\sum_{j=1}^{N} P(E_j)P(A|E_j)} \tag{2.13}$$

$$= \frac{P(E_i)P(A|E_i)}{P(E_1)P(A|E_1) + P(E_2)P(A|E_2) + \cdots + P(E_N)P(A|E_N)}$$

Remember, Bayes' rule is just another way of writing a conditional probability, for its numerator is a form of $P(A \cap E_j)$, and its denominator is $P(A)$. The probabilities $P(E_j)$ are the prior probabilities. It is important to note that the event E_i is *one* specific event contained in the set of all possible events denoted as E_j, where $j = 1, 2, \ldots, N$. In other words, the value of i must be one of the numbers $1, 2, \ldots, N$. The variable j is just an index of summation. Probabilities of the form $P(A|E_j)$ are called *likelihoods* since they indicate how "likely" it is that the specified sample result (A) will occur *given* that the event has taken (or will take) place.

Battery Manufacturer Example As an example of Bayes' rule ,suppose a manufacturer of automobile batteries produces two types of batteries (regular and heavy-duty) at each of three plants (A, B, and C). Plant A produces 300 batteries a day, two hundred of which are regular, and 100 heavy-duty. Plant B produces 200 batteries each day, 50 regular and 150 heavy-duty. Plant C produces 50 batteries of *each* type in a day. What is the probability that a randomly selected heavy-duty battery came from plant B? That is, what is $P(\text{Plant B}|\text{Heavy-duty})$? A diagram of the plants and their production of batteries might help. (See Table 2.3).

Since we want to calculate $P(\text{Plant B}|\text{Heavy-duty})$, "heavy-duty" represents the sample information (event A) and "plant B" represents the event of

Table 2.3

	Regular Batteries				Heavy-Duty			Total
Plant A	50	50	50	50		50	50	300
Plant B				50	50	50	50	200
Plant C				50			50	100
Total daily plant production								600

interest (E_i). The prior probabilities of all the events E_j (i.e., the plants) are

$$P(\text{Plant A}) = \frac{300}{600} = \frac{1}{2},$$

$$P(\text{Plant B}) = \frac{200}{600} = \frac{1}{3},$$

$$P(\text{Plant C}) = \frac{100}{600} = \frac{1}{6}.$$

The likelihoods are also easily obtained by looking at the diagram above:

$$P(\text{Heavy-duty}\,|\,\text{Plant A}) = \frac{100}{300} = \frac{1}{3},$$

$$P(\text{Heavy-duty}\,|\,\text{Plant B}) = \frac{150}{200} = \frac{3}{4},$$

$$P(\text{Heavy-duty}\,|\,\text{Plant C}) = \frac{50}{100} = \frac{1}{2}.$$

We can now use this information, and Bayes' rule, to calculate $P(\text{Plant B}\,|\,\text{Heavy-duty})$, which we abbreviate as $P(\text{B}\,|\,\text{HD})$.

$$P(\text{B}\,|\,\text{HD}) = \frac{P(\text{B})P(\text{HD}\,|\,\text{B})}{P(\text{A})P(\text{HD}\,|\,\text{A}) + P(\text{B})P(\text{HD}\,|\,\text{B}) + P(\text{C})P(\text{HD}\,|\,\text{C})}$$

$$= \frac{(\frac{1}{3})(\frac{3}{4})}{(\frac{1}{2})(\frac{1}{3}) + (\frac{1}{3})(\frac{3}{4}) + (\frac{1}{6})(\frac{1}{2})} = \frac{1}{2}$$

The posterior probability of plant B is thus $\frac{1}{2}$. This should seem reasonable, since the prior probability was $P(\text{Plant B}) = \frac{1}{3}$, and the sample information (heavy-duty) favors plant B; hence, we would expect its probability to increase. Note that this result is also logical in view of Table 2.3. There are six boxes representing production of heavy-duty batteries, and plant B is associated with three of them; hence, we can see that the posterior probability must be $\frac{1}{2}$. Bayes' rule has merely given us the ratio of the number of events of interest (heavy-duty batteries from plant B) to the total number of sample outcomes (heavy-duty batteries).

We leave it as an exercise for the reader to calculate the other posteriors in this example, such as $P(\text{A}|\text{HD})$, $P(\text{C}|\text{HD})$, $P(\text{A}|\text{Regular})$, etc.

Components Example As a final example of Bayes' rule, let us again consider the problem facing the manufacturer of delicate electronic components. Recall that $P(\text{Good}) = 0.70$ and $P(\text{Def.}) = 0.30$. These are the manufacturer's prior probabilities of a good or a defective component. Once more, we assume that the testing device described in Section 2.8 is being considered. This time, however, we do not want to assume that all the data in Table 2.2 are available. It is perhaps more reasonable to assume that the device has been tested on some components known to be good and some known to be defective. For instance, suppose all we know from Table 2.2 is that 80% of the good components yield a + reading, and 20% yield a − reading. Similarly, for the defective component, all we know is that 75% of the time the test reads −, and 25% of the time it reads +. That is,

$$P(+|\text{Good}) = 0.80, \qquad P(-|\text{Good}) = 0.20,$$

and

$$P(+|\text{Def.}) = 0.25, \qquad P(-|\text{Def.}) = 0.75.$$

All these values are consistent with those in Table 2.2. They are the likelihoods for this problem.

As we indicated in Section 2.8, the probabilities of interest to the manufacturer are those which indicate *how often* the test is correct—that is, $P(\text{Good}|+)$ and $P(\text{Def.}|-)$. These values can be calculated by Bayes' rule:

$$P(\text{Good}|+) = \frac{P(\text{Good})P(+|\text{Good})}{P(\text{Good})P(+|\text{Good}) + P(\text{Def.})P(+|\text{Def.})}$$

$$= \frac{(0.70)(0.80)}{(0.70)(0.80) + (0.30)(0.25)} = 0.882.$$

Similarly,

$$P(\text{Def.}|-) = \frac{P(\text{Def.})P(-|\text{Def.})}{P(\text{Def.})P(-|\text{Def.}) + P(\text{Good})P(-|\text{Good})}$$

$$= \frac{(0.30)(0.75)}{(0.30)(0.75) + (0.70)(0.20)} = 0.616.$$

The two remaining probabilities in this problem can also be calculated by Bayes' rule. However, it is much easier to find these values by the complementary law of probability.

$$P(\text{Def.}|+) = 1 - P(\text{Good}|+) = 1 - 0.882 = 0.118,$$
$$P(\text{Good}|-) = 1 - P(\text{Def.}|-) = 1 - 0.616 = 0.384.$$

A good exercise for the reader would be to verify these values by using Bayes' rule.

Problems

2.24 Two hundred marketing strategies were classified as "very effective," "moderately effective," or "not effective" in conjunction with three pricing strategies (I, II, III) as shown below:

Marketing Strategies	Pricing Strategies			
	I	II	III	Totals
Very effective	20	50	30	100
Moderately effective	20	20	20	60
Not effective	20	10	10	40
Totals	60	80	60	200

a) Convert these data into a table showing a joint probability in each cell.

b) Use Formula (2.11) to calculate the marginal probability $P(\text{Very effective})$.

c) Use Bayes' rule to calculate the posterior probability $P(\text{Pricing strategy II}|\text{Very effective})$.

2.25 Suppose that I acquire a ski resort. On a given weekend operation, the probability that I make a profit if the weather is "favorable" is $\frac{3}{4}$. If the weather is "unfavorable," the probability that I make a profit is $\frac{1}{8}$. Assume the forecast is for a $\frac{2}{5}$ chance of "favorable" weather.

a) What is the probability that I will make a profit from the weekend operation?

b) Suppose that on Monday I tell you that I made a profit; find the probability that the weather on the preceding weekend was "favorable."

2.26 A man goes fishing for the first time. He has three types of bait, only one of which is correct for the type of fishing he intends to try. The probability that he will catch a fish if he uses the correct bait is $\frac{1}{3}$. If he uses the wrong bait, his chances of catching a fish are $\frac{1}{5}$.

a) What is the probability that he will catch a fish?

b) Given that the man caught a fish, what is the probability that he used the correct type of bait?

2.27 In a certain city it is known that one-fourth of the people leave their keys in their cars. The police chief estimates that five percent of the cars with keys left in the ignition will be stolen, but that only one percent of the cars without keys left in the ignition will be stolen. What is the probability that a car stolen in this city had the keys in the ignition?

2.28 A student recognizes five potential questions he may be asked on a quiz. However, he only has time to study one of them thoroughly, and he selects this one randomly. Suppose the probability that he passes the test if this selected question appears is 0.90, but the probability that he passes the test if one of the other four questions appears is only 0.30. The test contains only one question and it is one of these five.

a) What is the probability that he will pass the test?

b) Suppose that you see the student next week and he has passed the test. What is the probability that the question he selected to study was in fact the one on the quiz?

2.29 Use Bayes' rule to find the probability in Problem 2.13, $P(\text{male}|\text{senior})$.

2.30 Use Bayes' rule on the data in Problem 2.12 to calculate the probability that a randomly selected person smokes, given that the person <30.

2.31 The Easy Charge Company described in Exercise 1.44 has released the following data on the number of customers who were given cash advances last month.

Amounts Owed by Customers	**Cash Advance**	**No Cash Advance**
\$0– 99.99	229	2894
\$100–199.99	378	1707
\$200–299.99	501	1426
\$300–399.99	416	939
\$400–499.99	260	483
Over \$500	289	478
Total customers	2073	7927

a) Use Formula (2.11) to calculate the marginal probability $P(\text{Cash advance})$.

b) Use Bayes' rule to calculate the posterior probability $P(\text{Balance due} < \$100|\text{Cash advance})$.

2.32 Suppose that you have three urns, each filled with red and blue marbles. The first urn contains one red and three blue marbles, the second contains two red and two blue, and the third contains three red and one blue. You select an urn at random, and then randomly select a marble from this urn.

a) If the marble drawn is red, what is the probability that you have drawn from the first urn? What is the probability that you drew from the second urn? From the third urn?

b) Repeat part (a), assuming that the marble drawn was blue.

2.33 Given the following table of survey information:

	Salary ($000)	Years of College	
		At Least 2	None
(A_1)	20–22.9	30	50
(A_2)	23–25.9	50	40
(A_3)	26–28.9	20	10
		100	100

a) Illustrate the use of Bayes' rule by finding the probability of selecting a non-college respondent, presuming that the one selected is in the highest salary bracket indicated.

b) Are salary and years of college independent in this problem?

2.34 Consider two types of economic stabilization policies—fiscal policy (controlled by Congress) and monetary policy (controlled by the Federal Reserve Board). Assume the policy decisions made by these two bodies are independent of one another, and that the action of either group is correct 80 percent of the time and incorrect 20 percent of the time. Finally, assume that the probabilities that the economy follows a generally stable growth pattern due to (or in spite of) these policy actions are:

$P(\text{Stable growth}\,|\,\text{Neither acting correctly}) = 0.40;$

$P(\text{Stable growth}\,|\,\text{Both acting correctly}) = 0.99;$

$P(\text{Stable growth}\,|\,\text{Only 1 acting correctly}) = 0.70.$

a) Use the independence assumption to calculate:

$P(\text{Neither acting correctly}),$

$P(\text{Both acting correctly}),$

$P(\text{Only 1 acting correctly}).$

b) You are given the sample information that growth is stable for a particular period. Use Bayes' rule to calculate:

$P(\text{Only 1 acting correctly}\,|\,\text{Stable growth}),$

$P(\text{Both acting correctly}\,|\,\text{Stable growth}),$

$P(\text{Neither acting correctly}\,|\,\text{Stable growth}).$

Check to see if these three probabilities sum to 1.0.

2.35 Suppose that a questionnaire is sent to rural households with probability $P(\text{R}) = 0.50$ and to urban households with probability $P(\text{U}) = 0.50$, where R stands for "rural" and U for "urban." Households are divided into low-income (L) and high-income (H). Furthermore, the following conditional probabilities are known.

$P(\text{H}\,|\,\text{R}) = 0.20$, $P(\text{L}\,|\,\text{R}) = 0.80$ $\qquad$ $P(\text{H}\,|\,\text{U}) = 0.40$, $P(\text{L}\,|\,\text{U}) = 0.60$

Now suppose that the location code has been omitted on one of the questionnaires received so that it is not known whether it is from a rural or urban household. Our prior probabilities suggest that the probability is 0.5 that it comes from a rural household and 0.5 that it comes from an urban household. Suppose that analysis of the responses in the questionnaire shows that it was obviously completed by a high-income household. How does this new information affect our probabilistic knowledge of whether it came from a rural or an urban household?

2.10 PROBABILITIES OF REPEATED TRIALS

Some of the examples given in this chapter have involved experimental situations in which more than two events can occur and where the events in question are combinations of the occurrence of simple events. Determining probabilities in such problems generally depends on a knowledge of the *number* of events in such combinations, a knowledge that may not be easily obtained from simply listing all the points in the sample space. Determining the number of events for such experiments involves a discussion of permutations and combinations.

Permutations A *permutation* is an arrangement or ordering of outcomes. The number of such arrangements or orderings is often important in determining the probability of some set of particular events.

The number of permutations of n outcomes is the maximum number of different ways these n outcomes can be arranged or ordered.

Permutations are involved in probability questions related to the ranking of a set of items, such as investment opportunities, or college football teams. For example, suppose we are interested in the number of ways three candidates (A, B, and C) for election to a particular office might be ranked by the voters. Six different rankings or permutations are possible, as shown in Table 2.4. If these permutations are all equally likely, the probability that any *one* will occur is $\frac{1}{6}$. If there were four candidates instead of three, we could show (but won't) that there are 24 different permutations. As the number (n) of candidates gets larger, we soon get tired of listing all possibilities, and need an easier way to determine the number of permutations.

Consider the number of different ways of filling each rank in Table 2.4. There are three ways to fill the first rank (with A, B, or C). Now suppose the

Table 2.4 Permutations of A, B, C

		1	2	3	4	5	6
	1	A	A	B	B	C	C
Rank	2	B	C	A	C	A	B
	3	C	B	C	A	B	A

first rank is filled, say with A; then the second rank can be filled in two ways (with either B or C). Once the second rank has been filled, there is only *one* way to fill the third rank. The number of permutations is thus the total number of ways of filling position *one* (which in this case is 3) times the ways of filling position *two* (2), times the ways of filling position three (1), or

$$(3)(2)(1) = 6.$$

Similarly, with four objects to rank, the number of permutations is

$$(4)(3)(2)(1) = 24.$$

This type of reasoning can be extended to any number of objects; if there are n objects, there are n ways to fill the first position, $(n - 1)$ ways to fill the second, and so forth, the product of these terms being the total number of permutations:

$$n(n - 1)(n - 2) \cdots (1).$$

Usually the symbol $n!$ is reserved for this type of product, where ! is to be read as "factorial," and where n can be any integer greater than or equal to zero. For example, "5 factorial" is

$$5! = (5)(4)(3)(2)(1) = 120,$$

and "10 factorial" is

$$10! = (10)(9) \cdots (1) = 3{,}628{,}800.$$

By definition, $0! = 1$. Now, if we have n objects and want to determine the numbers of permutations of *all n* of these objects (as we did above), this number, which is denoted by the symbol ${}_nP_n$, is called the *number of permutations of n objects taken n at a time.*

> *Permutations of n objects taken n at a time:*
>
> $${}_nP_n = n! \tag{2.14}$$

Permutation problems often do not involve all n objects at the same time, but rather some *subset* of these objects. Suppose we designate the number of objects in the subset by the letter x. These problems are concerned with the number of permutations of n objects when only x of these objects are considered at a time (i.e., x positions to fill). In general, there are x terms to multiply when there are x positions to fill. To illustrate, consider the permutations of four objects (A, B, C, and D) considered two ($x = 2$) at a time, as shown in Table 2.5.

Table 2.5 Permutations of A, B, C, D (taken x = 2 at a time)

		1	2	3	4	5	6	7	8	9	10	11	12
Rank	1	A	A	A	B	B	B	C	C	C	D	D	D
	2	B	C	D	A	C	D	A	B	D	A	B	C

As before, there are $n\,(=4)$ ways of filling the first position, and $n - 1\,(=3)$ ways of filling the second position; but now we have only two positions to fill; hence the total number of permutations in this case is $n(n-1)$, or $(4)(3) = 12$. In general there will always be x terms to multiply when there are x positions to fill. The *last* term in the progression $n(n-1)(n-2)\cdots$ must therefore always be $(n - x + 1)$ in order to have only x numbers to multiply (since there are x numbers between $n - x + 1$ and n). Letting ${}_nP_x$ represent the number of permutations of n objects considered (or taken) x at a time, then,

$$ {}_nP_x = n(n-1)(n-2)\cdots(n-x+1). \tag{2.15}$$

Formula (2.15) is not the most common formula representing ${}_nP_x$; there is a more convenient (although equivalent) form in which to express this relationship.*

Permutations of n objects taken x at a time:

$$ {}_nP_x = \frac{n!}{(n-x)!} \tag{2.16}$$

To illustrate the use of Formula (2.16), suppose a list of ten investments for a business firm is presented to the board of control, and each member is asked to rank the five projects considered to represent the best opportunities. How many different rankings exist? Here $n = 10$, $x = 5$, and

$$ {}_nP_x = {}_{10}P_5 = \frac{10!}{(10-5)!} = 30{,}240.$$

* Formulas (2.15) and (2.16) can easily be shown to be equivalent by expanding the latter, and then cancelling common terms:

$$ {}_nP_x = \frac{n(n-1)(n-2)\cdots(n-x+1)\cancel{(n-x)}\cancel{(n-x-1)}\cdots\cancel{(1)}}{\cancel{(n-x)}\cancel{(n-x-1)}\cdots\cancel{(1)}}$$

$$ = n(n-1)(n-2)\cdots(n-x+1).$$

One hopes there is enough agreement within the board of control so that a much smaller number of different rankings will actually be submitted.

Combinations The number of permutations of a set of objects depends on how many ways these objects can be ordered. But perhaps one cannot or does not want to be concerned with order; for instance, the order in which voters are surveyed on public issues is generally assumed to be unimportant, as is the order in which cards are received in a bridge hand. In these problems, interest usually centers on the number of ways a specific *combination* of objects can occur, where two sets of objects are identical if they contain exactly the same elements, no matter how these objects are arranged. Consider once again four objects (A, B, C, D) taken two at a time; there are only *six different combinations* of these objects, since sets with the same elements, such as AB and BA, cannot be counted twice.

Combinations					
1	2	3	4	5	6
A	A	A	B	B	C
B	C	D	C	D	D

The number of combinations of n objects considered (or taken) x at a time is written as ${}_nC_x$, *and equals the maximum number of different sets which can be collected using x out of the n objects.*

There are always fewer combinations than permutations for a given n and x, since different orderings *do not count* for combinations, but *do* for permutations. But how many fewer combinations are there? We need to reduce the number of permutations by a factor which depends on how many arrangements are identical, such as AB and BA for $x = 2$, and ABC, ACB, BAC, BCA, CAB and CBA for $x = 3$. Note from these examples that the number of permutations is twice as large as the number of combinations for $x = 2$, and six times as large for $x = 3$. The number of permutations is twice as large in the former case because, for each set of two different objects, there are exactly two permutations, only *one* of which counts as a combination. The number of permutations can be seen to be six times as large in the latter case by the same type of reasoning—there are exactly six permutations for each set of three objects, only *one* of which counts as a combination. Listed below are the four different combinations of letters A, B, C, D, and the six permutations associated with each one of these combinations.

Combinations	Permutations
ABC	ABC, ACB, BAC, BCA, CAB, CBA
ABD	ABD, ADB, BAD, BDA, DAB, DBA
ACD	ACD, ADC, CAD, CDA, DAC, DCA
BCD	BCD, BDC, CBD, CDB, DBC, DCB

These examples can be generalized by noting that the number of permutations will always be larger than the number of combinations by a factor equal to the number of different ways of arranging the x objects taken x at a time. This number is ${}_xP_x = x!$, which means that ${}_nP_x$ will always be larger than ${}_nC_x$ by a factor of $x!$; thus,

$$ {}_nP_x = x!({}_nC_x) \qquad \text{or} \qquad {}_nC_x = \frac{{}_nP_x}{x!}. $$

Now, by means of Formula (2.16), we know that ${}_nP_x = n!(n - x)!$, so ${}_nC_x$ can be written as follows

Combinations of n objects taken x at a time:

$$ {}_nC_x = \frac{n!}{x!(n - x)!}. \tag{2.17} $$

Formula (2.17) is often abbreviated as

$$ {}_nC_x = \binom{n}{x}, $$

where the term in the parentheses is not a fraction, but merely a different way of denoting the number of combinations of n objects taken x at a time. Suppose we want to form a committee of five people, where there are ten people who could be asked to serve on this committee. For this problem we need to know the number of combinations of $n = 10$ objects, taken $x = 5$ at a time, or ${}_{10}C_5 = \binom{10}{5}$:

$$ {}_{10}C_5 = \binom{10}{5} = \frac{10!}{5!(10 - 5)!} = \frac{(10)(9)(8)\cdots(1)}{[(5)(4)\cdots(1)] \times [(5)(4)\cdots(1)]} = 252. $$

Table 2.6 presents a summary of the probability rules presented thus far.

Table 2.6 Summary of Probability Rules

Rule Name	Formula	General Rule	Rule for Mutually Exclusive Events	Rule for Independence
Complements		$P(\bar{A}) = 1 - P(A)$	—	—
Conditional probability	(2.3)	$P(A \mid B) = \dfrac{P(A \cap B)}{P(B)}$	$P(A \mid B) = 0$	$P(A \mid B) = P(A)$
Joint probability	(2.4)	$P(A \cap B) = P(B)P(A \mid B) = P(A)P(B \mid A)$	$P(A \cap B) = 0$	$P(A \cap B) = P(A)P(B)$
Probability of a union	(2.5)	$P(A \cup B) = P(A) + P(B) - P(A \cap B)$	$P(A \cup B) = P(A) + P(B)$	$P(A \cup B) = P(A) + P(B) - P(A)P(B)$
Marginal probability	(2.11)	$P(A) = \sum P(A \cap E_i) = \sum P(E_j)P(A \mid E_j)$	—	—
Bayes' rule	(2.13)	$P(E_i \mid A) = \dfrac{P(E_i)P(A \mid E_i)}{\sum_j P(E_j)P(A \mid E_j)}$.	—	—

*2.11 EXTENSIONS OF THE RULES FOR PERMUTATIONS AND COMBINATIONS (OPTIONAL)

One of the most common complicating factors in many problems involving permutations and combinations is that often the total set of n objects will contain some objects which are exactly alike—i.e., they cannot be distinguished one from another. In this case, some modifications of the formulas presented in the last section are necessary. We first consider permutations, then combinations.

Permutations Let's assume that instead of all n objects being distinguishable, only k different types of objects are distinguishable ($k < n$). Furthermore, we assume that of these k types of objects, there are n_1 of the first kind alike, n_2 of the second kind alike, and so forth, with n_k of the kth kind alike. Also, $n = n_1 + n_2 + \cdots + n_k$. The number of permutations of these n objects is:

Permutations of n objects, n_1 alike, n_2 alike, . . . , n_k alike:

$$\frac{n!}{n_1!n_2!\cdots n_k!} \qquad (2.18)$$

The numerator of (2.18) is the total number of permutations of n objects taken n at a time, since ${}_nP_n = n!$; but n_1 of these n objects are exactly alike, so we can't count their $n_1!$ permutations as different orderings. To avoid counting them as different orderings, $n!$ needs to be divided by $n_1!$ Similarly, to avoid counting as different permutations the orderings of $n_2, n_3, \ldots, n_k$, we need to divide $n!$ by $n_2!, n_3!, \ldots, n_k!$ The result is Formula (2.18).

To illustrate the use of Formula (2.18) consider the supervisor of a job shop who is faced with scheduling seven major jobs over the next month. Out of the seven jobs, four are of the same type, two are of another kind, and the remaining one is of a third type. The supervisor wants to calculate the number of different ways the three types of jobs can be scheduled. In this case $n = 7$, $n_1 = 4$, $n_2 = 2$, and $n_3 = 1$. Thus, the number of permutations is

$$\frac{n!}{n_1!n_2!n_3!} = \frac{7!}{4!2!1!} = 105.$$

One special case of Formula (2.18) is particularly important, when $k = 2$ and $n_1 = x$, $n_2 = n - x$. Substituting these values into Formula (2.18) gives the following:

> *Number of permutations when $n_1 = x$ are alike, $n_2 = n - x$ alike:*
>
> $$\frac{n!}{x!(n-x)!}. \tag{2.19}$$

Note that this formula, which gives the number of *permutations*, is identical to Formula (2.17), which gives the number of *combinations*. The fact that these two formulas are identical causes some confusion among beginning students, for it is common to denote the number of permutations represented by (2.19) by the combinational symbol $\binom{n}{x}$.

Combinations The rule for combinations can be extended in much the same manner as we did above for permutations, by assuming that we have a total of n objects, with n_1 of these objects being alike, n_2 alike, . . . , and n_k alike, where $n = n_1 + n_2 + \cdots + n_k$. In this case we are interested in determining the number of combinations possible if, out of the n_1 objects of the first kind, x_1 are taken at a time, and out of the n_2 of the second kind, x_2 are taken at a time, and so forth, with x_k being taken at a time out of the n_k of the kth kind. Now we know from the rule for combinations that $\binom{n_1}{x_1}$ is the number of combinations of n_1 taken x_2 at a time, . . . , and $\binom{n_k}{x_k}$ is the number of combinations of n_k taken x_k at a time. By the basic counting rule, the total number of combinations is the product of each one of these terms:

> *Number of combinations of n_1 taken x_1 at a time, n_2 taken x_2 at a time, etc.:*
>
> $$\binom{n_1}{x_1}\binom{n_2}{x_2}\cdots\binom{n_k}{x_k}$$

To illustrate this formula, you might recall that Bloomington and Indianapolis were the only two Indiana cities among 17 finalists for an "All American City" award in 1981. Suppose that five awards are to be made. How many combinations of winners will include *exactly one* Indiana city? In this situation, there are $n_1 = 2$ cities from Indiana, of which $x_1 = 1$ must be picked, and $n_2 = 15$ other cities, of which $x_2 = 4$ must be picked. The total number of combinations is

$$\binom{n_1}{x_1}\binom{n_2}{x_2} = \binom{2}{1}\binom{15}{4} = 2730.$$

Hypergeometric Distribution Now that we know the number of ways of drawing x_1 out of n_1, x_2 out of n_2, etc., it is not difficult to determine the *probability* that a random draw of

$$(x_1 + x_2 + \cdots + x_k) \text{ objects}$$

out of a total of $(n_1 + n_2 + \cdots + n_k)$ objects will give this result. This probability is given by what is called the hypergeometric distribution.

Hypergeometric probability distribution:

$$P(x_1 \text{ out of } n_1, \ldots, x_k \text{ out of } n_k) = \frac{\binom{n_1}{x_1}\binom{n_2}{x_2}\cdots\binom{n_k}{x_k}}{\binom{n_1 + n_2 + \cdots + n_k}{x_1 + x_2 + \cdots + x_k}} \quad (2.20)$$

The numerator of this formula is merely the number of combinations we calculated above. The denominator is the total number of combinations of $n_1 + n_2 + \cdots + n_k$ objects taken $x_1 + x_2 + \cdots + x_k$ at a time.

For our "All American City" example, $n_1 = 2$, $n_2 = 15$, $x_1 = 1$, $x_2 = 4$. Assuming a random selection of five cities, the probability of exactly one Indiana city is

$$P(x_1 = 1 \text{ out of } n_1 = 2, \quad \text{and} \quad x_2 = 4 \text{ out of } n_2 = 15) = \frac{\binom{2}{1}\binom{15}{4}}{\binom{17}{5}}$$

$$= \frac{2730}{6188} = 0.441.$$

Probability in More Complicated Situations Many problems in statistics involve probabilities that are determined by using, simultaneously, a number of the rules presented in this chapter. Some of these probabilities can be determined in several different ways, as we will illustrate shortly. One useful rule in determining the probability of an event which is composed of many different occurrences all of which have the same probability, is to find the probability of each occurrence, and then multiply by the number of such occurrences:

$$P(\text{Event}) = P(\text{Each occurrence}) \times (\text{Number of relevant occurrences}) \quad (2.21)$$

To illustrate how probability is calculated in more complicated situations, suppose that ten companies have submitted designs to the U.S. Government for production rights to a new aircraft for the Air Force. The government is about to announce the names of the three companies who will be given grants to build a model from their design. The three winners will be announced on three separate days (to maximize press coverage). All ten companies are assumed to have an equal chance at winning one of the three awards.

Suppose you own stock in four of the ten companies. You want to determine (1) the probability that your companies will win all three awards, (2) the probability that *two* of your companies will win.

Question 1. We can determine this probability in a very straightforward fashion by using the general rule of multiplication. Let's denote by y_1 the event one of your firms wins the award announced on the first day, and y_2 and y_3 the event one of your firms wins on each of days two and three. The probability in question is $P(y_1 \cap y_2 \cap y_3)$.

$$P(y_1 \cap y_2 \cap y_3) = P(y_1)P(y_2 | y_1)P(y_3 | y_1 \cap y_2).$$

Your four firms have a probability of winning on the first day of $P(y_1) = \frac{4}{10}$. If y_1 occurs, you have three firms left out of nine; hence, $P(y_2 | y_1) = \frac{3}{9}$. Similarly, $P(y_3 | y_1 \cap y_2) = \frac{2}{8}$. Hence,

$$P(y_1 \cap y_2 \cap y_3) = \left(\frac{4}{10}\right)\left(\frac{3}{9}\right)\left(\frac{2}{8}\right) = \frac{1}{30}.$$

Now, the same answer can be obtained by first calculating how many different combinations of three winners can be selected from 10 companies. This is given by the rule for combinations, where $n = 10$ and $x = 3$:

$${}_{10}C_3 = \frac{10!}{3!(10 - 3)!} = 120.$$

There are thus 120 different ways of selecting three winners. Since we assumed that all companies have equal chances of winning, the probability of occurrence of each one of these ways is the same, namely 1/120. But how many of these ways have *three* of your firms included? The answer to this is given by the number of combinations of three winners taken out of four companies:

$${}_4C_3 = \frac{4!}{3!(4 - 3)!} = 4.$$

Using Formula (2.21) we have

$$P(\text{Event: 3 winners}) = P\begin{pmatrix}\text{Each occurrence} \\ \text{of 3 winners}\end{pmatrix}\begin{pmatrix}\text{Number of relevant} \\ \text{occurrences}\end{pmatrix}$$

$$= \left(\frac{1}{120}\right)(4) = \frac{1}{30}.$$

Question 2. Two approaches are again possible in answering the next question, "What is the probability that *exactly* two of your firms are winners?" Let's continue with the approach above, where we know that there are ${}_{10}C_3 = 120$ different ways of selecting three winners. We now need to determine the total number of these ways which include *two* winners from your four firms, and one *other* winner. The number of combinations of two of your firms being selected out of your four is ${}_4C_2 = 4!/(2!2!) = 6$. The number of ways of selecting one firm out of the remaining six is ${}_6C_1 = 6$. By the basic rule of counting discussed at the beginning of this section, the total number of combinations of two of your firms and one other is thus $({}_4C_2)({}_6C_1) = (6)(6) = 36$.

$$P(\text{Event: 2 winners}) = P\begin{pmatrix}\text{Each occurrence of}\\ \text{2 out of 3 winners}\end{pmatrix}\begin{pmatrix}\text{Number of relevant}\\ \text{occurrences}\end{pmatrix}$$

$$= \left(\frac{1}{120}\right)(36) = \frac{3}{10}.$$

Another way of determining this probability is to first determine the probability of your two firms winning in *some specified order*, and then multiplying this probability by the number of occurrences of such orderings. For example, let's calculate the probability that your firms win on the first and second days, but another firm wins on the third day (which we designate as O_3). This probability can be calculated by the general rule of multiplication.

$$P(y_1 \cap y_2 \cap O_3) = P(y_1)P(y_2 | y_1)P(O_3 | y_2 \cap y_1)$$

$$= \left(\frac{4}{10}\right)\left(\frac{3}{9}\right)\left(\frac{6}{8}\right) = \frac{1}{10}.$$

Now, the number of ways of having exactly two winning firms on three announcement days is

$${}_3C_2 = \frac{3!}{2!(3-2)!} = 3.$$

Thus,

$$P(\text{Event: 2 winners}) = P\begin{pmatrix}\text{Each occurrence of}\\ \text{2 out of 3 winners}\end{pmatrix}\begin{pmatrix}\text{Number of relevant}\\ \text{occurrences}\end{pmatrix}$$

$$= \left(\frac{1}{10}\right)(3) = \frac{3}{10}.$$

It is important to point out now that probability problems of the above type are generally not easy for most students to solve. It takes lots and lots of practice to feel at ease with such problems. We therefore urge you to work as many of the Review Problems at the end of the chapter as possible, and don't become discouraged! The reader who has a good understanding of the use of the concepts and rules presented in this chapter will have an important foundation for the discussion of probability distributions in the next two chapters, and for statistical inference and decision-making under uncertainty.

Multinomial Probability Distribution In this section we want to derive a general expression for Formula (2.21), which is repeated below.

$$P(\text{Event}) = P(\text{Each occurrence}) \times \begin{pmatrix}\text{Number of relevant}\\ \text{occurrences}\end{pmatrix}.$$

Again assume n objects, n_1 alike, n_2 alike, . . . , and n_k alike, with $n = n_1 + n_2 + \cdots + n_k$. We now assume that the probability of receiving these objects is not the same—those objects of the first type have a probability p_1, those of the second type have a probability p_2, etc., with p_k being the probability of the kth kind. The sum of these probabilities is unity—that is, $p_1 + p_2 + \cdots + p_k = 1$. Now, let's determine the probability of one particular order of the n objects, where the n_1 objects occur first, the n_2 second, and so forth.

$$P(\text{One occurrence}) = \underbrace{p_1 p_1 \cdots p_1}_{n_1 \text{ times}} \underbrace{p_2 p_2 \cdots p_2}_{n_2 \text{ times}} \cdots \underbrace{p_k p_k \cdots p_k}_{n_k \text{ times}}$$
$$= p_1^{n_1} \cdot p_2^{n_2} \cdots p_k^{n_k}.$$

From Formula (2.18), we already know the number of occurrences of this type of probability. The product of this probability and (2.18) is called the *multinomial distribution.*

Multinomial distribution:

$$P(n_1 \text{ of 1st type}, \ldots, n_k \text{ of } k\text{th type}) = \frac{n!}{n_1!n_2! \cdots n_k!} p_1^{n_1} p_2^{n_2} \cdots p_k^{n_k}.$$

Suppose we know that 10 blood donors are going to be giving blood in the next hour. The probability of Type O is 0.50, of Type A is 0.20, of Type AB is 0.15, and of Type B is 0.15. What is the probability that the $n = 10$ donors will be made up of $n_1 = 5$ of blood type O, $n_2 = 3$ of blood type A, $n_3 = 2$ of blood type of AB, and $n_4 = 0$ of type B?

$$P(5 \text{ of type 1, 3 of type 2, 2 of type 3 and 0 of type 4})$$
$$= \frac{10!}{5!3!2!0!} (0.50)^5 (0.20)^3 (0.15)^2 (0.15)^0 = 0.14.$$

Define: permutations, n factorial, combinations, hypergeometric distribution, multinomial distribution.

Problems

2.36 A club has 14 male members and 10 female members. If a committee of eight is to be chosen randomly from the membership of this club, what is the probability that one-half of the female members will be on the committee?

2.37 Assume that a store selling televisions has three black-and-white sets that are identical, and two identical color sets. If the store owner wants to arrange them on a shelf, how many different permutations are there of the five sets?

2.38 How many different permutations are there of the ten letters in the word "statistics"?

2.39 Assume that Doc Counsilman, former Olympic swimming coach, has seven possible swimmers for a four-person relay team.

a) If the order of the swimmers for the free-style relay is unimportant, how many different relay teams are possible?

b) Assume for the medley relay team that order is important; how many different teams are possible?

c) Swimmer Mark S. is one of the seven swimmers. If the assignment of the swimmers is random, how many of the teams in part (a) will include Mark S.? How many of the teams in part (b) will include Mark S. ?

2.40 Sears advertises that the customer can select one of the possible frequency codes on an automatic garage-door opener by setting nine switches in one of two positions (e.g., + or −).

a) How many different codes might a burglar have to try before being assured of finding the correct one (the answer "all of them" is not sufficient)? What is the probability that the burglar will be right on one randomly selected try?

b) Suppose the burglar knows that eight of the nine switches are set on + and one on −. What is now the probability of his being correct on one randomly selected try?

2.41 Wendy's hamburger chain advertises that you can order your hamburger with any one of 256 different combinations of toppings. How many different toppings will you get if you order your burger with "everything," assuming each topping (such as tomato) is either on or off?

2.42 a) The President of the U.S.A. is scheduled to visit three different countries. In how many different orders can these three countries be visited? Draw the tree diagram.

b) In how many different orders can the President visit three countries if there are ten possible countries that could be visited?

c) How many different combinations of three countries could be visited if there is a list of ten possible countries to visit?

2.43 A novice auto mechanic, who has decided that one of three problems is the cause of an engine malfunction, is going to try three corresponding solutions. Label the three solutions as A, B, C, and then list the different orders (permutations) in which they could be tried.

2.44 A radio repairman wants to replace a defective tube in an old radio. He has seven tubes in the repair kit, but only two of them will work. He selects the tubes at random one after another without replacement. What is the probability that the repairman will have to try three or more tubes before finding a good one?

2.45 a) A sportswriter is considering how to rank the ten teams in a conference. How many different ways can the teams be ranked if ties are not considered?

b) How many different combinations are possible if only the first three places (first, second, and third) are considered?

c) How many different rankings are possible if only the first three places are considered?

2.46 Suppose the Certified Management Accountancy administrators announce a list of five essay questions, two of which will be selected randomly to be on the next exam.

You randomly pick two questions and study for only those two. You know you will pass the exam only if at least one of the two questions you study is selected by the instructor. List the sample space for the experiment "administrators pick two questions without replacement." What is the probability that you will pass?

2.47 Many business firms are reluctant to transmit internal data over from one computer terminal to another without proper security procedures. One method for securing data has been to code the data at its entry point, with the decode method known only to receiving personnel. Suppose a coding scheme is suggested in which the customer can select any one of 2^{56} different codes. As president of this company would you consider this enough codes, knowing that a "spy" could use a computer program capable of trying 1000 codes a second in an effort to break your code?

2.48 a) In dealing from a deck of 52 cards, how many bridge hands of 13 cards are possible?

b) How many different bridge hands are there containing all four aces?

c) What is the probability of a bridge hand containing all four aces?

2.49 Suppose eight salespeople of a company are cited for outstanding sales during the past year. Five of these people are married. Suppose four of the eight are selected at random to receive a week's vacation for two in Hawaii. What is the probability that at least three of those selected will be married?

2.50 Suppose that your instructor in statistics announces that the final exam will consist of five questions, which will be randomly selected from a list of ten questions handed out one week before the exam. In order to pass the exam, a student must be able to answer at least four out of the five exam questions selected. What is the probability that a student who can answer only eight of the ten questions will pass the exam?

2.51 Use the multinomial distribution to determine the probability of receiving, in seven draws with replacement from a deck of cards, two black cards, four hearts, and the jack of diamonds.

2.52 An insurance company classifies drivers as "high," "moderate," or "low" risk. The percent of drivers in each category is 20, 30, and 50%, respectively. If 10 derivers are picked randomly from company files, what is the probability that there will be no more than two of the "high risk" and no more than two of the "moderate risk" drivers?

2.53 Decisions regarding governmental expenditures for public welfare often require a choice among various projects whose relative merits involve strong disagreement. Suppose public input is being gathered on 10 separate projects, as follows: social service (2), public transportation (1), arts and leisure time (3), education (1), and business and job development (3).

a) Determine the number of sample points for the experiment "rank all 10 projects." If all outcomes are equally likely, what is the probability of each outcome?

b) Suppose three projects are picked, at random, to be winners. What is the probability there will be at least one social service project and at least one arts and leisure project among the three winners?

2.54 The world's largest air cargo carrier, Flying Tiger, has a fleet of 18 DC–8's (in addition to seven Boeing 747's). On the average, each aircraft is staffed by eight crews, and each crew has three pilots.

a) If there are 144 different DC–8 crews (18×8), how many ways can one group of eight crews be selected?

b) Write an expression for the number of ways of selecting a group of eight crews for all 18 of the DC–8's. (*Hint:* pick one group of eight, then another group of eight, etc.)

c) If eight crews are assigned to one specific aircraft, how many ways can you pick one crew, and one head pilot, to fly this aircraft to given location?

Exercises

2.55 A group of four golfers stumbled across a nest of yellow jackets and two of them were stung. Three of the golfers broke 100 for the round, and all players either were stung, or broke 100, or both.

a) What is the probability that a player broke 100 and was stung?

b) Given that a player has been stung, what is the probability that he broke 100?

c) Given that a player broke 100, what is the probability this person was stung?

2.56 a) Show that when $x = n$, the formula for permutations is equivalent to the basic counting rule.

b) Show that when $x = 0$ or $x = n$, ${}_nC_x$ always equals 1.

2.57 A statistics class has 23 students registered for the course. What is the probability that at least two students will have the same birthday? (*Hint:* Approximate the answer using logarithms.)

2.58 Every once in a while there is an article in the newspaper about a bridge hand in which one of the four players received 13 cards of the same suit. Set up the combinational problem to determine the probability of such an occurrence.

2.59 Show that

$$\binom{n+1}{r} = \binom{n}{r-1} + \binom{n}{r}$$

2.60 In drawing two cards without replacement from a deck of cards, what is the probability of drawing a king on the second draw, given that either a king or an ace was drawn on the first draw?

2.61 Given the following set of weekly wages in dollars for six employees: 272, 300, 288, 295, and 278. If two of these employees are to be selected at random to serve as labor representatives, what is the probability that at least one will have a wage lower than the average?

2.62 a) In a poker hand, what is the probability of receiving four of a kind (i.e., four with the same face value out of five cards)? Ignore the order in which the cards are received.

b) What is the probability of receiving a full house in a poker hand (i.e., three cards with one face value and two cards with another face value)?

2.63 Assume that you are in a gambling casino. Find and compare your chances of getting at least one six if you roll one die four times with your chances of getting at least one double six if you roll two dice 24 times.

2.64 How many different basketball lineups can be made from a team of ten men if all ten men can play any position? How many lineups are possible if the team contains two centers, four guards, and four forwards, and the lineup must include one center, two guards, and two forwards?

2.65 Shark Loans (from Chapter 1) has recorded both the activity level of the local economy (either Hi, Med, or Low) and the mean interarrival times (0–20 min, 20–60 min, 60–200 min) for its customers over the past 150 weeks.

Time Interval Between Customers		State of Economy Hi	Med	Low
A	0–20	30	12	8
B	20–60	30	21	4
C	60–200	30	12	3

Based on these data:

a) Find $P(\text{Hi})$, $P(\text{Med})$, $P(\text{Low})$, $P(A)$, $P(B)$, and $P(C)$.

b) Find $P(A|\text{Hi})$, $P(A|\text{Med})$, $P(A|\text{Low})$.

c) Use Formula (2.11) to find $P(A)$.

d) Are the events A, B, C independent of the state of the economy?

e) Suppose Shark would like to revise the probabilities $P(\text{Hi})$, $P(\text{Med})$, $P(\text{Low})$ in light of sample evidence (S_1). Find $P(\text{Hi}|S_1)$, $P(\text{Med}|S_1)$, and $P(\text{Low}|S_1)$ given that $P(S_1|\text{Hi}) = 0.05$, $P(S_1|\text{Med}) = 0.10$, and $P(S_1|\text{Low}) = 0.40$.

2.66 Prove that if $P(A|B) = P(A)$, then $P(B|A) = P(B)$.

GLOSSARY

experiment: any operation capable of replication under essentially stable circumstances; a situation involving uncertainty.

sample space: outcomes of an experiment; may be discrete and finite, discrete and infinite, or continuous (infinite).

event: a subset of the outcomes of an experiment.

mutually exclusive outcomes: two or more possible results of an experiment, any one of which, if it occurs, rules out the occurrence of any other.

exhaustive outcomes: outcomes that account for ("exhaust") the entire sample space.

probability value: a number between 0 and 1 that indicates how likely an event is to occur.

probability of an event (using equally likely outcomes): ratio of the number of outcomes comprising the event to the total number of outcomes in the sample space.

objective probability: a probability value determined by "objective" evidence, often relative frequencies.

subjective probability: a probability value determined by an individual based on this person's knowledge, information, and experience.

basic counting rule: total number of sample points

$$N = n_1 \cdot n_2 \cdot n_3 \cdots n_k.$$

***n* factorial:** $n! = n(n-1)(n-2)\cdots(1)$.

permutations of *n* objects taken *x* at a time:

$$_nP_x = \frac{n!}{(n-x)!}.$$

combinations of *n* objects taken *x* at a time:

$$_nC_x = \frac{n!}{x!\,(n-x)!} = \binom{n}{x}.$$

probability of an event:

$$P(\text{event}) = P(\text{one sample point}) \times (\text{number of relevant sample points}).$$

two basic properties: 1. $1.0 \geq P(E_i) \geq 0$

2. $P(S) = 1.0 = \sum_{\text{All } i} P(E_i)$

joint probability (intersection—both A and B occur in one replication of an experiment): $P(A \cap B) = P(A|B)P(B) = P(B|A)P(A)$.

additive probability (union—either A or B or both occur in one replication of an experiment): $P(A \cup B) = P(A) + P(B) - P(A \cap B)$.

conditional probability of A given that B has taken (or will take) place: $P(A|B) = P(A \cap B)/P(B)$.

complementary probability (A will not take place in one replication of an experiment): $P(\bar{A}) = 1 - P(A)$.

independence: one event does not influence the probability of another. If independent, $P(A|B) = P(A)$ and $P(A \cap B) = P(A)P(B)$.

marginal probability: $P(A) = \sum_j P(E_j)P(A|E_j)$.

Bayes' rule:

$$P(E_i|A) = \frac{P(E_i)P(A|E_i)}{\sum_j P(E_j)P(A|E_j)} = \frac{P(\text{event})P(\text{sample info.}|\text{event})}{\sum_j P(\text{event}_j)P(\text{sample info.}|\text{event}_j)}.$$

3

PROBABILITY THEORY: RANDOM VARIABLES

"Chance favors only the prepared mind."

LOUIS PASTEUR

3.1 INTRODUCTION

Now that we have studied the rules for associating a probability value with a single event, or with a combination of events in an experiment, we can proceed to the next logical problem, that of describing the probability of *all* events in a given experiment. This problem becomes especially important in decision-making contexts; we can seldom properly evaluate a course of action on the basis of one or two outcomes, but rather should consider all possible consequences of the action. Most of us face problems of this nature regularly; for instance, in deciding whether to take a particularly hard elective course (or whether to study tonight or watch television), you would certainly want to consider how this decision will affect your chances of receiving not only the grade of "A," but also grades of "B," "C," "D," or "F." Similarly, the speculator in the stock market should (but does not always) consider the probability of losing significant proportions of the investment, as well as of making a profit.

The outcomes of an experiment are said to take place *randomly*, since they cannot, by definition, occur in any particular order or pattern. Such variables, whose values thus cannot be known in advance by the person conducting the experiment, are called *random variables*. In tossing a pair of dice, for example, the sum of the two faces showing on each toss represents a random variable unless it is *known* exactly which outcome will occur. Similarly, the grade you receive in statistics can be considered a random variable if it is not known in advance (e.g., before the course begins) exactly which grade you will receive. In other words, something is left to chance, such as how you feel the day of the final exam, the whims of your professor, etc. On the other hand, the amount of money you earned last year is not (to you at least) a random variable, since presumably you know the exact amount—there is no uncertainty connected with it. However, your income is probably a random variable to your statistics instructor since its value is not known to this person.

Random Variable

Give an experiment and a set of *mutually exclusive* and *exhaustive* outcomes, it is common to consider questions about the probability of the occurrence of any one or more of these outcomes by using the following definition of a *random variable*.

A ***random variable*** *is a well-defined rule for assigning a numerical value to all possible outcomes of an experiment.*

This means that the symbols used in Chapter 2 to designate the outcomes of an experiment—"heads," "defective," etc.—are now going to be replaced with numbers. A random variable is a rule designating a number to be associated with each outcome of the experiment.

The outcomes of some experiments readily meet this definition of a random variable because they are already well-defined numbers. For example, the number of hours that a given lightbulb might last is a well-defined number; the number of defectives that could occur in a lot of transistors is a well-defined number; and the potential yield on an investment of $1000 is a well-defined number. In other cases, the outcomes of an experiment may be qualitative. For example, the outcome of coin-flipping is heads or tails, and the outcome of taking a course could be a grade of A, B, C, D, or F. In these instances the probability model must specify exactly what numerical value corresponds to each qualitative outcome. Registrars at many colleges do this for grades by letting A = 4, B = 3, C = 2, D = 1, and F = 0. In tossing a coin, one common way to define a random variable is to let heads = 1 and tails = 0. There may be less agreement in attempting to define a random variable for the experiment "drive from New York to Phoenix." In this case the sample space would need to be converted to some consistent measure, such as the *number of dollars* required for automobile repairs.

In working with *continuous* sample space, it is sometimes convenient to reduce the sample space to just a few discrete points. For example, the yield on a $1000 investment might be classified as falling into one of just a small number of intervals (such as 0 to 2.0, 2.1 to 4.0, etc.); and the dollar amount of repairs on a trip to Phoenix could be classified as either less than $50, between $50 and $100, or over $100. In all these examples we have a random variable only when numerical values are assigned to the outcomes of the experiment by a well-defined rule.

In making the assignment of numerical values to the outcomes of an experiment, we will denote random variables by letters in *boldface* type, such as $\boldsymbol{x}$, $\boldsymbol{y}$, $\boldsymbol{z}$, or sometimes by subscripted boldface letters such as $\boldsymbol{x}_1, \boldsymbol{x}_2, \boldsymbol{x}_3$. *Specific* values of such random variables will be denoted by letters in lightface type, such as x, y, z, or perhaps x_1, x_2, x_3. Thus, the designation $\{\boldsymbol{x} = x\}$ is read as "the random variable $\boldsymbol{x}$ takes on the value x." The following examples will illustrate this notation.

1. *Experiment*: Flipping a coin once
   Outcomes: Two discrete outcomes, heads and tails
   Sample space: Discrete and finite
   Random variable: Define $\{\boldsymbol{x} = 1\}$ if heads occurs and $\{\boldsymbol{x} = 0\}$ if tails occurs.

   Although any values may be used to give numerical labels to the outcomes in an experiment such as this one, zero and one are especially convenient mathematically in many situations involving just two outcomes. Since the variable $\boldsymbol{x}$ in this case gives a well-defined rule for assigning numerical values to the experiment, $\boldsymbol{x}$ is a *random variable*.

2. *Experiment*: Taking an exam
   Outcomes: Grades A, B, C, D, F
   Sample space: Discrete and finite
   Random variable: Define $\{\boldsymbol{y} = 4\}$ if the grade is A,
   $\{\boldsymbol{y} = 3\}$ if the grade is B,
   $\{\boldsymbol{y} = 2\}$ if the grade is C,
   $\{\boldsymbol{y} = 1\}$ if the grade is D, and
   $\{\boldsymbol{y} = 0\}$ if the grade is F.

   The familiar four-point grade system is simply an assignment of numbers to a grade measure. Since the variable $\boldsymbol{y}$ gives a well-defined rule for assigning numbers to the outcomes of this experiment, $\boldsymbol{y}$ is a random variable.

3. *Experiment*: Driving a car from New York to Phoenix
   Outcomes: Various car troubles that might be encountered on trip
   Sample space: Discrete (infinite but countable)
   Random variable: Define z = nearest number of dollars paid for repairs,

$$\{z = 0, 1, 2, 3, \ldots\}$$

   The random variable z in this case is discrete, and it is also infinite since there is no limit on the amount of repairs. Realistically, however, there is some upper bound to the value of z, perhaps equal to the cost of the car if it is a total loss due to an accident. Also, this probability model assumes no negative values for z, since we doubt that anyone can find a "Tom Sawyer" mechanic willing to pay for the chance to do the needed repairs.

4. *Experiment*: Investing $1000 in a common stock
   Outcomes: Values of yield or rate of return
   Sample space: Continuous (always infinite)
   Random variable: Define x = value of yield, $(-\infty < x < +\infty\}$.

   A *continuous random variable* is obtained from a continuous sample space whenever a single value of x is assigned to each outcome in the sample space. Thus, since a yield can be *any* positive number (or a negative number), x must be continuous.

5. *Experiment*: Investing $1000 in a common stock

   In this example we simplify Experiment 4 somewhat by grouping the various yields into different classes. For example, we might let one class represent all yields between 0 and 2%, another represent 2.1 to 4.0%, etc. This simplification results in the following probability model:

   Outcomes: Class intervals of yields
   Sample space: Discrete and infinite (there is no limit on the number of classes)
   Random variable: Define x = the midpoint or some representative value (the class mark) of the yields in each class interval.

Assigning values of the random variable so they equal the class marks makes it much easier to find the mean and variance of the probability distribution under study. We will describe the process of finding means and variances of probability distributions later in this chapter. As a final note in this section, we should point out that the numerical value assigned to an outcome in an experiment need not be unique to that outcome. That is, several different outcomes may be assigned the same numerical value. This fact is easily seen in the experiment about driving to Phoenix, for in this case there are certainly many different outcomes (car troubles) that would lead to the same value of the random variable (that is, lead to the same dollar value of cost).

Probability Distributions

Once an experiment and its outcomes have been clearly stated, and the random variable of interest has been defined, then the probability of the occurrence of any value of the random variable can be specified. Let us present some new examples. Suppose 140 students are registered for a certain course, and they are to be divided randomly into four sections. The number of students assigned to each section is determined by the size of the available classrooms as follows:

Section	Class Size
1	25
2	45
3	40
4	30
Total	140

If you are one of the students involved in this assignment, you could view this process as an experiment with *four outcomes*. The sample space is discrete and finite; a random variable $\boldsymbol{x}$ may be defined to have values equal to the section number, $\boldsymbol{x} = 1, 2, 3$, or 4. The probability that you will be assigned to any one section can be determined and denoted by the symbol $P(x)$. For example, the probability of your being assigned to Section 1 is denoted $P(\boldsymbol{x} = 1)$ or $P(1)$. The probability that you will be assigned to Section 1 is simply the proportion of assignments that are made to Section 1 relative to the total number of students, which is

$$P(1) = \tfrac{25}{40} = 0.179.$$

This is the *least probable* outcome of the experiment. What value of $\boldsymbol{x}$ has the highest probability? That is, what section assignment is the most probable outcome? Clearly, you have the highest chance of being assigned to Section 2, since this section will have the most students. We find

$$P(2) = \tfrac{45}{140} = 0.321.$$

Continuing in this manner we can find the probability of each possible value for the random variable $\boldsymbol{x}$. When this is done, we have obtained the *probability distribution* for $\boldsymbol{x}$. Table 3.1 and Fig. 3.1 depict the probability distribution for the random variable in this assignment problem.

In this example only four discrete values of $\boldsymbol{x}$ have a positive probability. All other values of $\boldsymbol{x}$ have a probability of occurring equal to zero, indicating that they are impossible. Also, the sum of the probabilities of all values of $\boldsymbol{x}$ is

Table 3.1 Probability Distribution for x in the Section Assignment Example

Outcome	Value of x	$P(x)$
Section 1	1	$\frac{25}{140} = 0.179$
Section 2	2	$\frac{45}{140} = 0.321$
Section 3	3	$\frac{40}{140} = 0.286$
Section 4	4	$\frac{30}{140} = 0.214$
Sum		1.000

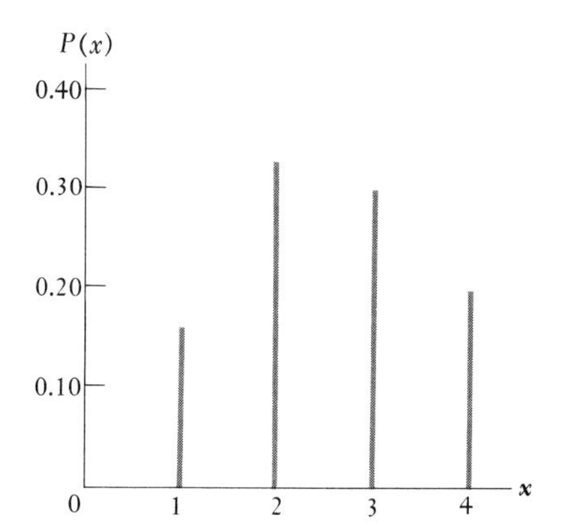

Fig. 3.1 Graph of the discrete probability distribution for **x** given in Table 3.1.

equal to 1.0, indicating that these four are the only possible outcomes; we know with certainty that one of them will occur in each student assignment.

The construction of a probability distribution is not always as simple as in the previous example (where the random variable had a unique value for each section). To illustrate a more complicated example, consider the experiment of throwing a pair of "fair" (i.e., perfectly balanced) six-sided dice a single time, and observing the total number of dots facing up. The sample space is discrete and finite, as illustrated in Fig. 3.2.

Let us define a random variable $\boldsymbol{x}$ so that it has values equal to *the sum of the dots* facing up on the pair of dice. That is, $\boldsymbol{x}$ can assume any integer value from 2 to 12, depending on the outcome of the roll of the dice. The probability

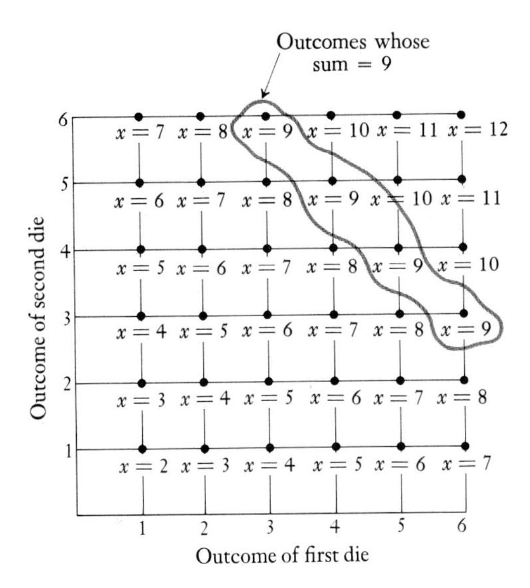

Fig. 3.2 Sample space for throwing a pair of dice.

of any value of $\boldsymbol{x}$ is given by the number of sample points for which the sum of the dots equal x, divided by the *total number* of sample points. For example, let us find $P(\boldsymbol{x} = 9)$, which is usually shortened to $P(9)$. We can observe in Fig. 3.2 the number of sample points corresponding to this value of $\boldsymbol{x}$. Out of all 36 outcomes, those satisfying the condition $\{\boldsymbol{x} = 9\}$ are the four ordered pairs

$$(3, 6), \qquad (4, 5), \qquad (5, 4), \qquad \text{and} \qquad (6, 3)$$

where the first entry is the outcome of the first die, and the second entry is the outcome of the second die. Thus,

$$P(9) = \tfrac{4}{36}.$$

Similarly, we could find the probability that the value of $\boldsymbol{x}$ would be 3. The ordered pairs (1, 2) and (2, 1) are the only sample points satisfying $\{\boldsymbol{x} = 3\}$, and so

$$P(3) = \tfrac{2}{36}.$$

Figure 3.3 is a graph of the probability distribution of the random variable $\boldsymbol{x}$ = sum of dots.

The determination of a probability distribution completes the process of describing what is called the *probability model.* Figure 3.4 summarizes this model. First, the experiment must be stated clearly, so that one can specify the relevant sample space. A random variable is then associated with this sample space, which makes it possible to define the probability distribution.

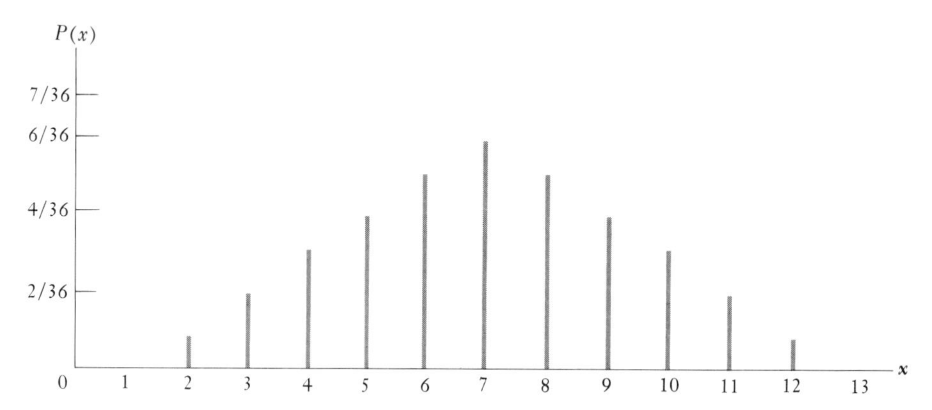

Fig. 3.3 Probability distribution of x = sum of dots face up after tossing a pair of dice.

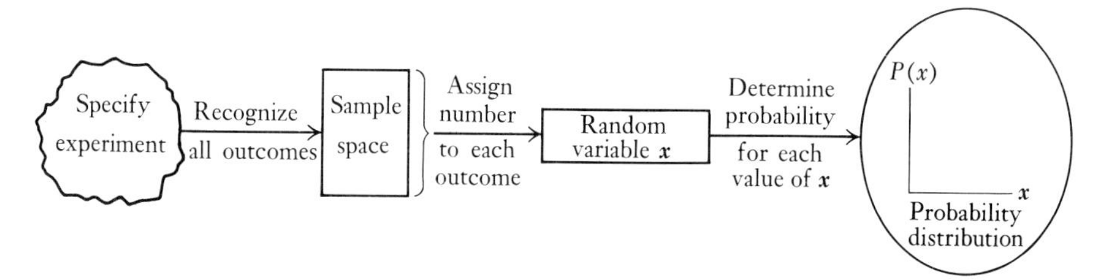

Fig. 3.4 The probability model.

3.2 DISCRETE RANDOM VARIABLES

A probability distribution involving only discrete values of x *is usually called a* ***probability mass function.***

A probability mass function (abbreviated p.m.f.) is usually described in one of three ways: (1) by a graph, such as Fig. 3.3, (2) by a table of values, such as Table 3.1, or (3) by a formula.* The name "mass function" derives from the fact that all outcomes associated with the value of a discrete random variable can be represented on a graph by a vertical line whose height (or *mass*) indicates the probability of that value.

To illustrate the concept of describing a mass function by graph, table, and formula, consider the problem of determining how long a certain grocery item might sit on the shelf before being sold. For one relatively low-selling

* The reader is referred to the review of functions in Appendix A.

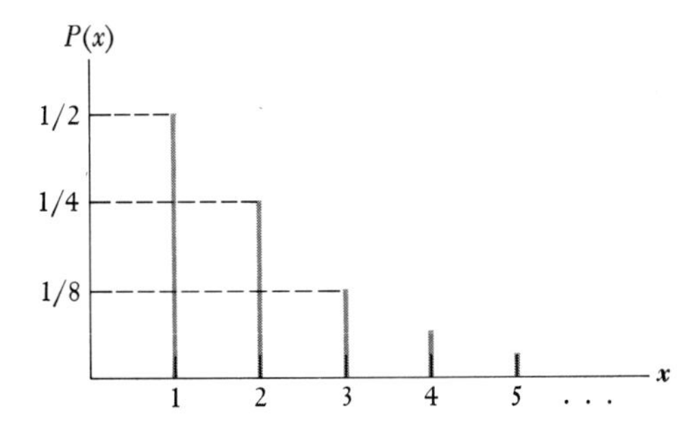

Fig. 3.5 Probability mass function for day of sale of grocery item.

item, the estimate is that there is a 50–50 chance this item will be sold on any given day. Suppose we let $\boldsymbol{x}$ = the day on which the item is sold. Since there is a 50–50 chance that the item is sold on the first day, $P(\boldsymbol{x} = 1) = \frac{1}{2}$. If $\{\boldsymbol{x} = 2\}$, this means the item was not sold on day one, and *is* sold on day two. Thus, $P(\boldsymbol{x} = 2) = \frac{1}{2} \cdot \frac{1}{2} = \frac{1}{4}$. Similarly $P(\boldsymbol{x} = 3) = \frac{1}{2} \cdot \frac{1}{2} \cdot \frac{1}{2} = \frac{1}{8}$. A graph of the p.m.f. for this experiment, and a table of its values are shown in Fig. 3.5 and Table 3.2, respectively.

Using a graph and/or a table in this problem is inconvenient because of the fact that although $\boldsymbol{x}$ is discrete, an infinite number of x values is possible. That is, the item might not be sold for a very large number of days. Thus, Fig. 3.5 and Table 3.2 are somewhat unsatisfactory because a series of dots is used to indicate the values of $\boldsymbol{x}$ and their probability after $\{\boldsymbol{x} = 5\}$. A formula, on the other hand, can be used to *explicitly* specify how $P(x)$ and x are related for all values of $\boldsymbol{x}$. In this case the relationship is easily verified to be the following:

$$P(x) = \begin{cases} (\frac{1}{2})^x & \text{for } x = 1, 2, 3, \ldots, \infty. \\ 0 & \text{otherwise.} \end{cases}$$

By substituting $x = 1$, $x = 2, \ldots, x = 5$ into this formula, the reader can verify the values in Table 3.2 and Fig. 3.5.

Table 3.2

x	$P(x)$
1	$\frac{1}{2}$
2	$\frac{1}{4}$
3	$\frac{1}{8}$
4	$\frac{1}{16}$
5	$\frac{1}{32}$
⋮	⋮

By now the reader should recognize the similarity between the concept of relative frequency, as described in Chapter 1, and the concept of a probability distribution. The difference, technically, is that relative frequencies often are only frequencies based on the outcomes of one or more replications of an experiment, whereas a probability distribution can be viewed as the theoretical long-run relative frequency for all conceivable replications. The same properties defined in Chapter 2 for probability values can now be specified in terms of random variables.

Properties of all probability mass functions

Property 1: $0 \leq P(\mathbf{x} = x) \leq 1;$

Property 2: $\sum_{\text{All } x} P(\mathbf{x} = x) = 1.$

Cumulative Mass Function (c.m.f.)

The concept of *cumulative* relative frequency and the graphical representations thereof, which were introduced in Chapter 1, also have their counterparts in the study of probability. A *cumulative mass function* (c.m.f.) describes how probability accumulates in exactly the same fashion as the cumulative column in Table 1.3 (on page 12) describes how relative frequency accumulates—by *summing* all the relative frequency values. The value of the cumulative mass function at any given point x is usually denoted by the symbol $F(x)$, where $F(x)$ is the *sum of all values* of the probability mass function for all values of the random variable $\mathbf{x}$ that are *less than or equal to* x. That is,

Cumulative mass function at the point x:

$$F(x) = P(\mathbf{x} \leq x) = \sum_{\mathbf{x} \leq x} P(x). \tag{3.1}$$

For our grocery-item example, the value of $F(x)$ when $x = 1$ is $F(1) = 0.50$ because $P(\mathbf{x} \leq 1) = 0.50$. Similarly, $F(2) = 0.75$ because $P(\mathbf{x} \leq 2) = 0.50 + 0.25 = 0.75$. As with a probability mass function, a cumulative mass function must be defined for *all* values of the random variable. This is usually accomplished by means of either a graph, a table, or a formula. The table and graph for the grocery example are shown in Table 3.3 and Fig. 3.6. A

Table 3.3

x	$F(x)$
1	0.500
2	0.750
3	0.875
4	0.938
5	0.969
⋮	⋮

c.m.f. graph will always look like a series of steps (a "step-function") going up from zero to one as the value of $\boldsymbol{x}$ increases.

Figure 3.6 is perhaps a better way to illustrate the c.m.f. than Table 3.3, for the former emphasizes the fact that $F(x)$ is defined for *all* values of $\boldsymbol{x}$ from negative infinity to positive infinity. That is, the cumulative mass function $F(x)$ is defined for *any* value of $\boldsymbol{x}$ between positive and negative infinity, not just the integer values listed in Table 3.3. As should be clear from Fig. 3.6, $F(x) = 0$ for all values of $\boldsymbol{x}$ from minus infinity up to $x = 1$. At $x = 1$ the value of $F(x)$ becomes 0.50. Similarly, at $x = 2$ the value of the function becomes 0.75. Note that for *any* number between $x = 2$ and $x = 3$, $F(x) = 0.75$. Suppose that we arbitrarily pick a number, say $x = 2.45$. From Fig. 3.6, $F(2.45)$ can easily be seen to be

$$F(2.45) = 0.750.$$

The value $F(3) = 0.875$ is interpreted as meaning that there is a probability of 0.875 that this grocery item will be sold in three days or less.

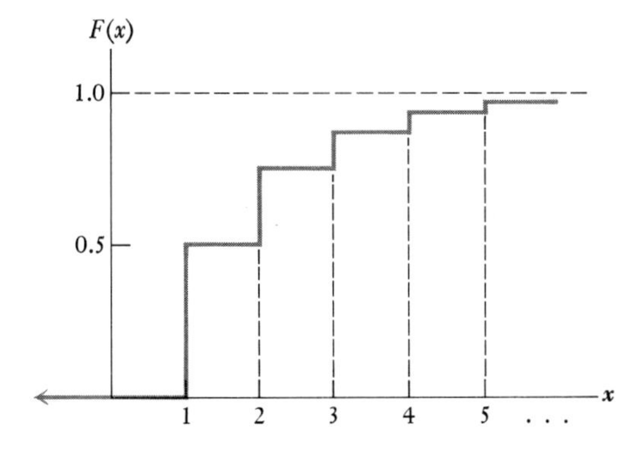

Fig. 3.6 The cumulative mass function for day of sale of a grocery item.

Examples. Many other examples of discrete probability functions can be cited. Consider a situation involving the possibility of exploring for oil at one of five sites. From past experience, it is known that an average of two of every five wells will be dry, two will not be commercially feasible, and one will be a success. Thus, the experiment of choosing one site has three different outcomes. Now, suppose we let a random variable $\boldsymbol{x}$ assume values of -1, 0, and $+10$ for the three outcomes, respectively. These x-values may reflect some management view of the net payoff (or loss) associated with the different outcomes. The values of the probability mass function and the cumulative mass function for this random variable are easily determined; they are given in Table 3.4.

Table 3.4 The p.m.f. and c.m.f. for the Oil Exploration Example

	Value of x	-1	0	$+10$	All Other Values
p.m.f.	$P(x)$	$\frac{2}{5}$	$\frac{2}{5}$	$\frac{1}{5}$	0
c.m.f.	$F(x)$	$\frac{2}{5}$	$\frac{4}{5}$	1.0	

As always, $0 \leq P(x) \leq 1.0$, and $F(x)$ is a function of x that never decreases. Only *three* values of $\boldsymbol{x}$ have positive probabilities, so this random variable is discrete. For any value less than -1, the cumulative value is zero. Now let us consider values larger than $x = -1$. For example,

$$F(4) = \sum_{x \leq 4} P(x).$$

Since the interval $x \leq 4$ includes $x = -1$ and $x = 0$, the nonzero probabilities to be included are

$$F(4) = P(-1) + P(0) = \tfrac{2}{5} + \tfrac{2}{5} = \tfrac{4}{5}.$$

At the value $x = 4$, we have accumulated four-fifths of the probability values. Indeed, the value of $F(x)$ is $\frac{4}{5}$ for all values, $0 \leq x < 10$. At the point $x = 10$, the value of $F(x)$ steps up to 1.0.

As another example, consider the volume of take-out sales at a donut shop drive-in window. Customers may buy a single donut or packages of 2, 4, 6, or a dozen donuts. In this experiment involving sales to a given customer, there are five discrete outcomes. Suppose that we let the random variable $\boldsymbol{x}$ represent the number of donuts sold to a customer, where the values of $P(x)$ and $F(x)$ determined from the relative frequencies of actual sales are given as follows:

x	**1**	**2**	**4**	**6**	**12**
$P(\mathbf{x} = x)$	0.08	0.27	0.10	0.33	0.22
$F(x) = P(\mathbf{x} \leq x)$	0.08	0.35	0.45	0.78	1.00

We see that for this discrete set of values, $0 \leq P(x) \leq 1.0$ for each x, and

$$\sum_{\text{All } x} P(x) = 1.0,$$

so that these values satisfy the two properties of a probability mass function presented earlier. The definition of a cumulative mass function can be used to find a number of probabilities that might be of interest to the manager of the donut shop, such as:

1. The probability of selling four donuts or fewer to a given customer

   $P(\mathbf{x} \leq 4) = F(4) = 0.45.$

2. The probability of selling packages of more than two donuts, $P(\mathbf{x} > 2)$. This probability can be easily determined by using the complementary law of probability,

   $P(\mathbf{x} > 2) = 1 - P(\mathbf{x} \leq 2).$

   Since $P(\mathbf{x} \leq 2) = F(2)$, the solution is

   $P(\mathbf{x} > 2) = 1 - F(2) = 1.0 - 0.35 = 0.65.$

3. The probability of selling packages in which the number of donuts is less than or equal to 6 but more than 1, $P(1 < \mathbf{x} \leq 6)$. As shown in the p.m.f. table,

   $P(\mathbf{x} \leq 6) = F(6) = 0.78.$

   But, since $F(6)$ *includes* the probability $P(\mathbf{x} \leq 1) = F(1)$, we must *subtract* this value from $F(6)$. Hence,

   $P(1 < \mathbf{x} \leq 6) = F(6) - F(1) = 0.78 - 0.08 = 0.70.$

4. The probability of selling packages of 12 donuts or *fewer* is

   $P(\mathbf{x} \leq 12) = F(12) = 1.0.$

As you can see from the above examples, the cumulative mass function is useful in determining probability values for various types of events. Since this function will be used frequently in problems of statistical inference, the student should thoroughly understand this concept and all of the above examples and hints for interpretation before proceeding. A good exercise for the reader would be to sketch the function $F(x)$ for this example of the sales of donuts, and determine the answers to items 1 through 4 above by using your graph.

Some Further Questions

Perhaps the reader has become curious about the parameters of the distribution of donut sales. Recall from Chapter 1 that the important features of *any* distribution (which we now know includes probability distributions) are the measures of *central location* and *dispersion*. Knowing the mean number of donuts per sale, for example, would be of value in deciding how many donuts to produce.

A measure of the dispersion of sales is also important when one is interested in deviations from the expected or theoretical average payoff. For instance, what is the chance that the shop will sell all its donuts because of a succession of sales of one-dozen-donut packages? For planning purposes, one might want to know how many packages of each type should be made so that the chance of selling out after 500 sales is less than 0.05. These and various other questions necessary for decision-making could be posed. In the next section, the measures of central tendency and dispersion for a probability distribution are discussed. To repeat an earlier remark, while the problems and questions used here as examples to guide and motivate your study of statistics may not seem worth the effort, remember that the same concepts, measures, formulas, and methods can be and are applied to important decision-making situations involving not pennies but millions of dollars; not tossing dice but choosing portfolio holdings; not flipping coins but controlling production processes; not counting donuts, but making pricing decisions in an oligopoly market; and other significant endeavors. Some of the problems or exercises will illustrate these widespread and important uses.

3.3 EXPECTED VALUE: THE DISCRETE CASE

By means of the discussion presented thus far, we can now determine the probability of any single event of an experiment, or describe the probability of the entire set of outcomes associated with a given experiment. Yet this information may not be concise enough for most decision-making contexts. Recall that we had the same problem in Chapter 1, when it was not sufficient merely to present all the data, but in addition several characteristics of these data were given—the most important of which were the mean and the variance. The same type of measures are also useful in describing probability distributions, but in this case we must speak not of an *observed* mean or an *observed* variance, but of the mean or variance which would be *expected* to result, on the average, from the random variable under consideration. These values are thus given the name *expectations*, or *expected values*.

The expected value of a discrete random variable is found by multiplying each value of the random variable $\boldsymbol{x}$ *by its probability and then summing over all values of* $\boldsymbol{x}$.

The letter E usually denotes an expected value and must be followed by brackets giving the random variable of interest; $E[\boldsymbol{x}]$, for instance, represents the expected value of the random variable $\boldsymbol{x}$. Using this notation, the expected value of a discrete random variable $\boldsymbol{x}$ is:

Expected value of $\boldsymbol{x}$ *(discrete case):* $$E[\boldsymbol{x}] = \sum_{\text{All } x} xP(x). \qquad (3.2)$$

The expected value of a random variable is nothing more than the arithmetic mean of that variable—i.e., it is the center of gravity, or the balancing point of the values of $\boldsymbol{x}$ "weighted" by their probability. Thus, $E[\boldsymbol{x}]$ is the mean of the population of $\boldsymbol{x}$ values, where the relative frequency of each value of $\boldsymbol{x}$ is $P(x)$ and we can write:

Population mean: $$\mu = E[\boldsymbol{x}] = \sum_{\text{All } x} xP(x).$$

Note that this definition of μ corresponds very closely to our definition in Chapter 1 of the mean of a population for grouped data (Formula (1.3)). The major difference is that in Chapter 1 each value of $\boldsymbol{x}$ was "weighted" by its relative frequency f_i/N, while in this chapter the weight is the relative frequency $P(x)$. As we pointed out previously, f_i/N is the relative frequency of just one replication (or a few replications) of an experiment. while $P(x)$ can be thought of as the expected relative frequency of an infinite number of replications of the experiment.

As an example of the calculation of an expected value, suppose a stockbroker is trying to estimate the price at which a certain stock will sell a year from now. The broker lets $\boldsymbol{x}$ = price in one year, and has established (based on various objective and subjective factors) the following probability mass function.

x = Price in One Year	$P(x)$
$80.	0.05
85.	0.10
90.	0.20
95.	0.20
100.	0.25
105.	0.10
110.	0.10
	1.00

The broker wishes to know what value to expect for $\boldsymbol{x}$, on the average, in this experiment. In other words, if we repeated this experiment many times, what would be the average of all the x values? What we want to determine is the *expected value* of $\boldsymbol{x}$, or $E[\boldsymbol{x}]$.

One method to *approximate* the mean value in any experiment is to replicate the experiment many times, add up the observed numbers, and divide by the number of observations; but such a procedure is often impractical, if not impossible, and gives only an approximation of the desired value. Fortunately, there is no need to replicate an experiment if the probability mass function is known, for we then already know the *expected relative frequency* of each event. For instance, if the experiment above were repeated 100 times, we would expect the price of $x = 80$ to occur 5 times, since

$$P(\boldsymbol{x} = 80) = 0.05.$$

The "weight" we assign to $x = 80$ is thus 0.05, or equivalently, $\frac{5}{100}$. Similarly, the weight for $x = 85$ is 0.10 (or $\frac{10}{100}$) since $P(\boldsymbol{x} = 85) = 0.10$. By substituting the values from our stock example into Formula (3.2), we can calculate the expected price of the stock next year as follows:

$$\begin{aligned} E[\boldsymbol{x}] &= \sum_{\text{All } x} xP(x) \\ &= 80(0.05) + 85(0.10) + 90(0.20) \\ &\quad + 95(0.20) + 100(0.25) + 105(0.10) + 110(0.10) \\ &= 96.00. \end{aligned}$$

The expected, or average, price of the stock in one year is thus $96.00. Another way of saying the same thing is to state that the mean (or balance point) of this probability mass function is $\mu = \$96.00$.

As another example, consider the donut-sales situation with $\boldsymbol{x}$ and $P(x)$ defined as in the previous section. What is the expected value of $\boldsymbol{x}$ in this example? That is, what is the balance point of the probability mass function for the average number of donuts per sale? Using Formula (3.2), we obtain:

$$\begin{aligned} E[\boldsymbol{x}] &= \sum_{\text{All } x} xP(x) \\ &= 1(0.08) + 2(0.27) + 4(0.10) + 6(0.33) + 12(0.22) \\ &= 5.64 \text{ donuts.} \end{aligned}$$

This means that, over an infinite number of sales, the *average* number of donuts sold at one time would be 5.64. The expected number of donuts sold in a given number of sales, N, equals $NE[\boldsymbol{x}]$. For example, in 500 sales, the total number of donuts expected to be sold would be

$$500 \times 5.64 = 2820 \text{ donuts.}$$

Thus, if 250 dozen donuts (3000) were produced to cover 500 sales, then the shop could expect a surplus of

$$3000 - 2820 = 180 \text{ donuts.}$$

The number 180 represents only a theoretical value, since one can be sure of this surplus only as an average amount over an infinite number of days' sales under similar conditions. Over any finite number of sales, the actual average sale may deviate from the 5.64 donuts average.

Expected Value of a Function of a Random Variable

Not only can we take an expectation of a simple random variable, but we can also take an expectation of *any function* of a random variable. For instance, instead of finding the mean of the random variable $\boldsymbol{x}$, we might be interested in determining the expected value of $\boldsymbol{x}^2$, or of $\log \boldsymbol{x}$, or of $e^{\boldsymbol{x}}$. If $\boldsymbol{x}$ is a random variable, then these functions of $\boldsymbol{x}$ are also random variables, and their expected values can be determined. Suppose we let $g(\boldsymbol{x})$ represent the random variable whose value is $g(x)$ when the value of $\boldsymbol{x}$ is x. The expected value of $g(\boldsymbol{x})$ is defined as follows:

Expected value of $g(\boldsymbol{x})$: $$E[g(\boldsymbol{x})] = \sum_{\text{All } x} g(x)P(x). \tag{3.3}$$

The only difference between Formulas (3.3) and (3.2) is that in (3.3), $P(x)$ are the weights of values of $g(\boldsymbol{x})$ rather than values of $\boldsymbol{x}$.

One of the most important expectations to calculate in statistics is the one in which $g(\boldsymbol{x}) = \boldsymbol{x}^2$; that is, $E[\boldsymbol{x}^2]$. This expectation often makes calculating the variance of $\boldsymbol{x}$ a much easier task.

To illustrate the use of Formula (3.3) with $g(\boldsymbol{x}) = \boldsymbol{x}^2$, we will suppose that in our stockbroker example we are interested in determining $E[\boldsymbol{x}^2]$, which is the average or expected value of $\boldsymbol{x}^2$. That is, what would be the average value if every value of $\boldsymbol{x}$ in the probability mass function were squared? The function would look like this:

x^2	80^2	85^2	90^2	95^2	100^2	105^2	110^2
$P(x)$	0.05	0.10	0.20	0.20	0.25	0.10	0.10

We calculate the average value of $\boldsymbol{x}^2$ by substituting x^2 for $g(x)$ in Formula (3.3), as follows:

$$\begin{aligned} E[\boldsymbol{x}^2] &= \sum_{\text{All } x} x^2 P(x) \\ &= (80^2)(0.05) + (85^2)(0.10) + (90^2)(0.20) + (95^2)(0.20) \\ &\quad + (100^2)(0.25) + (105^2)(0.10) + (110^2)(0.10) \\ &= 9280 \text{ (dollars squared).} \end{aligned}$$

The average squared value of $\boldsymbol{x}$ is thus $E[\boldsymbol{x}^2] = 9280$.*

The value of $E[\boldsymbol{x}^2]$ can also be calculated for our donut-sales example as follows:

x^2	1^2	2^2	4^2	6^2	12^2
$P(x)$	0.08	0.27	0.10	0.33	0.22

$$\begin{aligned} E[\boldsymbol{x}^2] &= \sum_{\text{All } x} x^2 P(x) \\ &= (1^2)(0.08) + (2^2)(0.27) + (4^2)(0.10) + (6^2)(0.33) + (12^2)(0.22) \\ &= 46.32 \text{ (donuts squared).} \end{aligned}$$

The Variance of a Random Variable

In the same way that the variance of a population is defined as the average squared deviation of the population values from their mean (μ), so the *variance* of a random variable can be defined in terms of the expected squared deviation

* Note that $E[\boldsymbol{x}^2] = 9280$ does not equal $(E[\boldsymbol{x}])^2 = 96^2 = 9216$. In general, $E[\boldsymbol{x}^2] \neq (E[\boldsymbol{x}])^2$.

of the values of $\boldsymbol{x}$ around their expected value $E[\boldsymbol{x}]$. We denote this variance of the random variable $\boldsymbol{x}$ by the symbol $V[\boldsymbol{x}]$, which is defined as follows:

$$V[\boldsymbol{x}] = \sigma^2 = E[(\boldsymbol{x} - \mu)^2].$$

Since the squared-deviation term within the brackets $[(\boldsymbol{x} - \mu)^2]$ is a function of the random variable $\boldsymbol{x}$, it can be written as $g(\boldsymbol{x}) = (\boldsymbol{x} - \mu)^2$. This means that because we know how to find $E[g(\boldsymbol{x})]$ from Formula (3.3), we can now find $E[g(\boldsymbol{x})] = E[(\boldsymbol{x} - \mu)^2]$. Making the substitution for $g(\boldsymbol{x}) = [(\boldsymbol{x} - \mu)^2]$ in Formula (3.3), we obtain:

Variance of $\boldsymbol{x}$:

$$V[\boldsymbol{x}] = \sigma^2 = E[(\boldsymbol{x} - \mu)^2] = \sum_{\text{All } x} (x - \mu)^2 P(x). \qquad (3.4)$$

Formula (3.4) is the traditional way of defining the variance of a discrete random variable. It can also be used to compute a standard deviation, since the standard deviation, denoted by σ, is always the square root of the variance:

Standard deviation of $\boldsymbol{x}$: $\quad \sigma = \sqrt{V[\boldsymbol{x}]}. \qquad (3.5)$

In order to illustrate the use of Formula (3.4), suppose we again use the probability distribution for the donut-sales situation described earlier. The probability mass function for this example is illustrated in Fig. 3.7.

Since we know, for this example, that the expected value is

$$E[\boldsymbol{x}] = \mu = 5.64,$$

Fig. 3.7 Probability mass function for sales of donuts.

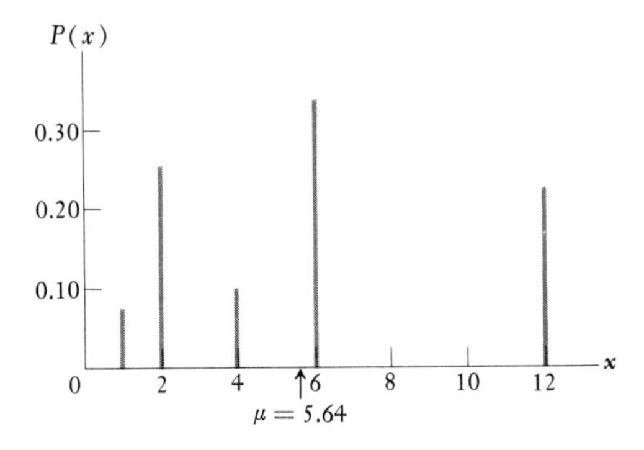

we substitute this value in Formula (3.4) to calculate $V[\mathbf{x}]$:

$$V[\mathbf{x}] = \sigma^2 = \sum_{\text{All } x} (x - 5.64)^2 P(x)$$

$$= (1 - 5.64)^2 0.08 + (2 - 5.64)^2 0.27 + (4 - 5.64)^2 0.10 + (6 - 5.64)^2 0.33 + (12 - 5.64)^2 0.22,$$

and

$$V[\mathbf{x}] = 14.51 \text{ (donuts squared)}.$$

The standard deviation is $\sigma = \sqrt{V[\mathbf{x}]} = 3.81$ donuts.

The value $\mu = 5.64$ is the *center* of this probability distribution, in the sense that it is the *expected number of donuts in each sale*. The value $\sigma = 3.81$ indicates the expected size of the spread of the distribution of sales of donuts.

It is important to emphasize at this point that the reader should always check the reasonableness of calculations such as $E[\mathbf{x}]$ and $V[\mathbf{x}]$. For instance, if $E[\mathbf{x}]$ is not near where you would expect the center of gravity to be, then you should double-check your calculations. A good way to check the reasonableness of $V[\mathbf{x}]$ is to use the rule of thumb for variances presented in Chapter 1. This rule says that if the probability distribution is fairly symmetrical and unimodal, then 68% and 95% represent good approximations to the percent of the distribution falling in the intervals $\mu \pm \sigma$ and $\mu \pm 2\sigma$, respectively. For probability distributions, these rules mean that approximately 68% and 95% of the *probability* should be within $\mu \pm \sigma$ and $\mu \pm 2\sigma$, respectively. Using the above results for μ and σ, and Fig. 3.7, these intervals can be shown to contain:

Interval	**Probability in Interval**
$\mu \pm \sigma = 5.64 \pm 3.81 = 1.83$ to 9.45	$[0.27 + 0.10 + 0.33 = 0.70]$
$\mu \pm 2\sigma = 5.64 \pm 2(3.81) = 0$ to 13.26	$[0.08 + 0.27 + 0.10 + 0.33 + 0.22 = 1.00]$

We see that the actual percents agree fairly closely with the rule of thumb. The reader should remember, however, that had the distribution been sufficiently asymmetric, the percentages falling into these intervals might have been further from the 68% and 95% values.

While Formula (3.4) is theoretically correct, it has a major disadvantage for computational purposes. Specifically, if the mean μ is not an integer, but has three or more decimals, then subtraction of μ from each value of $\mathbf{x}$ can be tedious. Also, squaring the deviations then gives numbers with at least *six* decimals. Finally, if the number of values taken on by $\mathbf{x}$ is quite large, all these subtractions, squares, and the ultimate summation require many steps and this raises great potential for computational errors.

To circumvent these objections to Formula (3.4), there is a computational formula for the variance of a probability distribution very similar to the short-cut formula for σ^2 in Chapter 1. The present formula requires that one first calculate $E[\mathbf{x}]$ and $E[\mathbf{x}^2]$. This formula is equivalent to Formula (3.4).*

Equivalent formula for the variance of $\mathbf{x}$:

$$V[\mathbf{x}] = \sigma^2 = E[\mathbf{x}^2] - (E[\mathbf{x}])^2 \tag{3.6}$$

Formula (3.6) might be remembered as the *mean of squares minus the squared mean.* This is a general formula that applies to any random variable, discrete or continuous. It will be used again many times throughout this book.

Suppose that we apply Formula (3.6) to the donut-sale problem to verify that the value of $V[\mathbf{x}] = 14.51$, as was calculated earlier. Recall that we already know that

$$E[\mathbf{x}] = 5.64 \qquad \text{and} \qquad E[\mathbf{x}^2] = 46.32.$$

Substituting these values into Formula (3.6) yields $V[\mathbf{x}]$.

$$V[\mathbf{x}] = E[\mathbf{x}^2] - (E[\mathbf{x}])^2 = 46.32 - (5.64)^2 = 14.51.$$

The reader should verify, using both Formulas (3.4) and (3.6), that the variance for the stock example (on page 130) is

$$\sigma^2 = V[\mathbf{x}] = 64.$$

The result $\sigma = \sqrt{64} = 8$ is reasonable since

$$\mu \pm \sigma = 96 \pm 8 = 88 \text{ to } 104,$$
$$\mu \pm 2\sigma = 96 \pm 2(8) = 80 \text{ to } 112,$$

and these intervals contain, respectively, 65% and 100% of the values of $\mathbf{x}$ in the probability mass function (sufficiently close to the 68% and 95% rule of thumb).

Define: random variable, probability mass function (p.m.f.), cumulative mass function (c.m.f.), expected value, $E[g(\mathbf{x})]$, $V[\mathbf{x}]$.

---

* The equality between Formulas (3.4) and (3.6) is easily demonstrated as follows:

$$\begin{aligned}
E[(\mathbf{x} - \mu)^2] &= \textstyle\sum(x - \mu)^2 P(x) \\
&= \textstyle\sum(x^2 - 2x\mu + \mu^2)P(x) && \text{(by expansion)} \\
&= \textstyle\sum x^2 P(x) - 2\mu \sum x P(x) + \mu^2 \sum P(x) && \text{(by summing each term)} \\
&= E[\mathbf{x}^2] - 2\mu E[\mathbf{x}] + \mu^2 && (\textstyle\sum P(x) = 1 \text{ by definition}) \\
&= E[\mathbf{x}^2] + (-2 + 1)(E[\mathbf{x}])^2 && (\text{since } \mu = E[\mathbf{x}]) \\
&= E[\mathbf{x}^2] - (E[\mathbf{x}])^2.
\end{aligned}$$

Problems

3.1 Sketch the cumulative mass function for the random variable $\boldsymbol{x}$ of the donut-sales example discussed earlier, where $\boldsymbol{x}$ is the number of donuts sold to a customer. What is the median number of donuts sold?

3.2 A dealer selling farm machinery estimates that monthly sales of a new grain harvester will be $x = 1, 2, 3$, or 4 according to the following p.m.f.:

$$P(x) = \begin{cases} \frac{1}{10}x & \text{for } x = 1, 2, 3, 4 \\ 0 & \text{otherwise} \end{cases}$$

a) Sketch this function and show that it satisfies the two properties necessary if $P(x)$ is to be a p.m.f.

b) Sketch the c.m.f.

c) Find $E[\boldsymbol{x}]$ and $V[\boldsymbol{x}]$ for this function.

3.3 A statistics professor announces that final grades will consist of 20% A's, 30% B's, 30% C's, 10% D's and 10% F's.

a) Sketch the p.m.f. and c.m.f., letting $A = 4$, $B = 3$, $C = 2$, $D = 1$, and $F = 0$.

b) Let $\boldsymbol{x}$ = numerical grade received, and calculate $E[\boldsymbol{x}]$ and $V[\boldsymbol{x}]$.

c) Calculate $\mu \pm \sigma$ and $\mu \pm 2\sigma$ for this problem. Is the percent of values of $\boldsymbol{x}$ within these intervals close to the rule of thumb?

3.4 A manufacturer can ship either 4000 or 12,000 boxes of spark plugs to an automotive outlet in Germany. Suppose we let $\boldsymbol{x}$ = sales of spark plugs, in units of 1000 boxes. The manufacturer estimates that the following p.m.f. accurately describes sales:

$$P(x) = \begin{cases} \frac{3}{x} & \text{for } x = 4 \text{ or } 12 \\ 0 & \text{otherwise} \end{cases}$$

a) Sketch this p.m.f., verifying that it meets the two necessary conditions for a mass function.

b) Sketch the c.m.f.

c) Find the mean and the variance of expected sales.

3.5 Reconsider the oil exploration example on page 126.

a) Sketch the p.m.f. and the c.m.f.

b) Find $E[\boldsymbol{x}]$.

c) Find $V[\boldsymbol{x}]$ first by using Formula (3.4) and then by using Formula (3.6).

3.6 In 1980, A & P sponsored an "Old Fashioned Bingo" game in which they published the following odds for one visit to an A & P market.

Prize	Odds	Prize	Odds
$1000	1 in 416,666	$5	1 in 1,562
100	1 in 41,666	1	1 in 114
10	1 in 3,125		

a) Find the expected value of the payoff for one visit.

b) Find the variance of the payoff for one visit.

c) What is the probability that you will not win a single prize in 20 consecutive visits?

3.7 A large eastern construction company classifies its laborers as either 3 (unskilled), 2 (semi-skilled) or 1 (skilled). Based on past records, they have determined the p.m.f. for rating new workers to be;

$$P(x) = \begin{cases} x^2/14 & \text{for } x = 1, 2, 3 \\ 0 & \text{otherwise} \end{cases}$$

a) Sketch this p.m.f. and the corresponding c.m.f. What proportion of workers are classified as unskilled?

b) Find $E[x]$ and $V[x]$.

3.8 A hardware store is participating in a promotion in which each customer receives a card, and each card contains a hidden coin. For every 100 cards, there are 10 nickels, 10 dimes, 15 quarters, 15 half dollars, and 50 souvenir coins (no monetary value).

a) For a customer with one card, what is the probability the coin will be a dime?

b) What is the probability a customer with two coins will have 50 cents or more? Assume that the content of all cards is independent.

c) Sketch the p.m.f. and c.m.f. for this distribution.

d) What is the average amount this store will give away to each customer?

e) What is the variance of the amount this store will give away to each customer?

3.9 Show, for Problem 3.7, that $(E[x])^2 \neq E[x^2]$.

3.10 In a campus chest bazaar, a booth offers the chance to throw a dart at a balloon. If you break the balloon, you receive a prize equal to the amount hidden behind the balloon. Suppose that each balloon is equally likely to be hit, and that the average chance of a hit is $\frac{1}{2}$ for all expected participants. The awards are distributed as follows:

40% have payoff of 5¢	20% have payoff of 25¢
30% have payoff of 10¢	10% have payoff of \$1.00

If 15¢ a dart is charged, what is the booth's expected return for 500 darts thrown?

3.4 CONTINUOUS RANDOM VARIABLES

Probability Density Functions

So far we have examined experiments involving only a discrete set of outcomes, thus limiting ourselves to discrete probability values. As we indicated earlier, however, an outcome set can be continuous as well as discrete; this implies that the random variable in an experiment must be able to assume a continuous form. Continuous random variables are especially convenient to work with, so much so that even when the set of outcomes is discrete it is often advantageous to use a *continuous approximation* to these values. Fortunately, most probability theory is basically the same for discrete and continuous random variables, and the formulas presented in Chapter 2 hold for both cases.

Probability functions defined in terms of a continuous random variable are usually referred to as *probability density functions* (abbreviated p.d.f.), or simply as *density functions*. It is helpful to think of a density function as the frequency polygon for the histogram of a discrete probability function involving a large number of events. Figure 3.8 illustrates what we mean by this, using yearly sales data for a basic statistics text.

Notice how the frequency polygon in Fig. 3.8 serves to smooth the discrete probability values. Now, suppose we decrease the width of the classes, from a class interval of 1000 to a class interval of say 500, or 250, or even to a class interval of 1. In Fig. 3.9, using class intervals of 500 and then of 250, we illustrate the change in the frequency polygon as the width of the interval decreases and the number of intervals increases correspondingly—namely, that it begins to look more and more like a smooth, continuous function. The limiting form of the frequency polygon, as the width of the class interval goes to zero, will be a function with no bends or angles (i.e., a function for which the first derivative exists for all values of $\boldsymbol{x}$).

As the width of the class interval goes to zero, the *number* of classes (or events) under consideration must increase until, at the limit (i.e., number of classes $= \infty$), there are an *infinite* number of such events between any two values of $\boldsymbol{x}$. Because of this fact, one of the important rules of continuous random variables is the following:

For continuous random variables, the probability that any one specific value of $\boldsymbol{x}$ *takes place equals zero.*

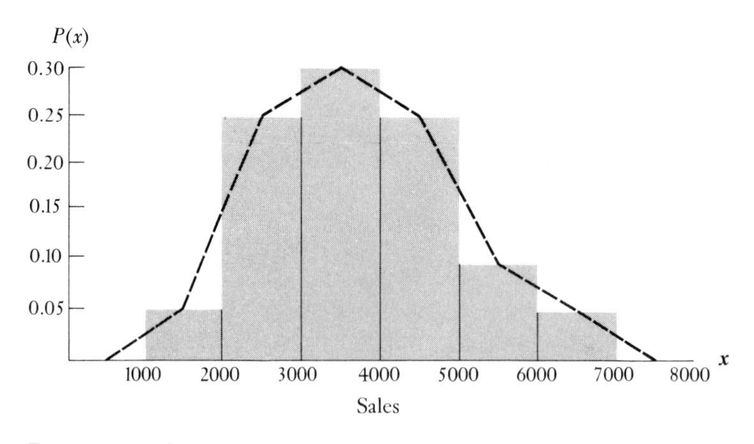

Fig. 3.8 Frequency polygon.

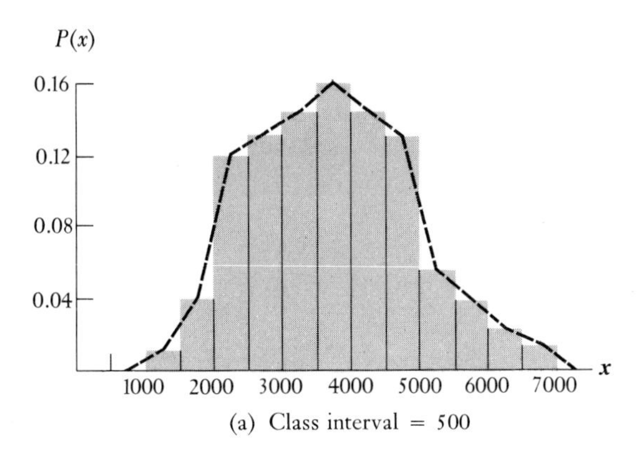

(a) Class interval = 500

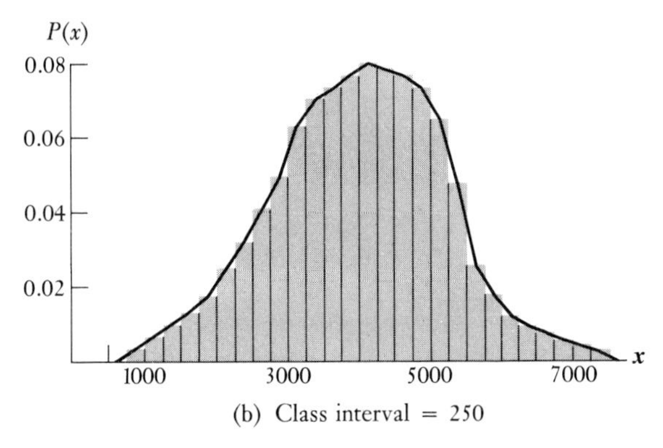

(b) Class interval = 250

Fig. 3.9 Frequency polygon for two different class sizes.

This means that we cannot talk about the probability of a single sample point (it equals zero!); instead, probability must always be defined *over an interval* (between two points). Now, suppose we want to evaluate $P(a \leq \mathbf{x} \leq b)$, the probability that the random variable $\mathbf{x}$ falls between the two points a and b.* If we had only a finite number of outcomes, this value would be given by summing the probability of each of the events. As the number of events becomes larger and larger, however, the frequency polygon becomes a better and better approximation to these values (see Fig. 3.10) until, at the limit, it exactly describes the relative frequency of the random variable $\mathbf{x}$. Note in Fig. 3.10 that, in using a continuous function to approximate a probability mass function, the sum of the probability "spikes" in part (a) of that figure is approximated by the *area* under the curve in part (b). This curve, which we have labeled $f(x)$, is the probability density function.

* Note that, since the probability that any one value of $\mathbf{x}$ will occur equals zero, the following is implied:

$$P(a \leq \mathbf{x} \leq b) = P(a < \mathbf{x} \leq b) = P(a \leq \mathbf{x} < b) = P(a < \mathbf{x} < b).$$

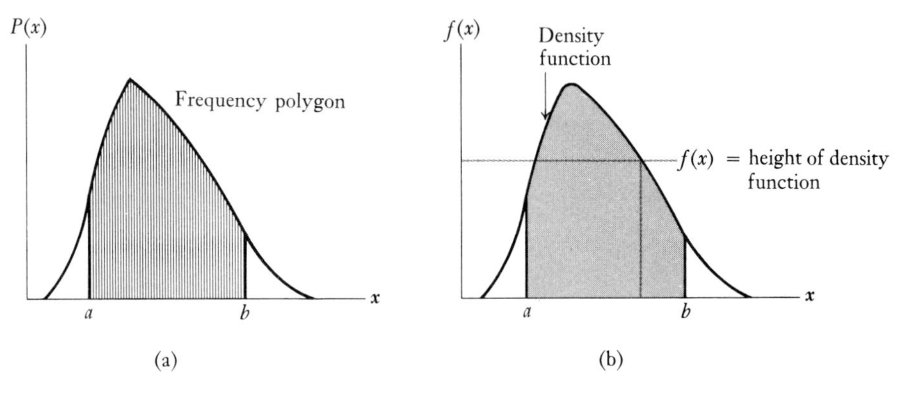

Fig. 3.10 The continuous approximation to $P(a \leq \mathbf{x} \leq b)$.

The symbol $f(x)$ will be used throughout this book to denote a probability density function, as distinguished from $P(x)$, which represents the probabilities involved in either a density function or a mass function. While the values of $P(x)$ can be specified either by a list of all such values or by means of a formula, the values of $f(x)$ must always be given by a formula since $f(x)$ is a continuous function. In this light, the reader is cautioned against making the serious error of interpreting values of $f(x)$ as probabilities. The value of $f(x)$ represents the *height* of the density function at the point x.

In order to be able to evaluate the probability $P(a \leq \mathbf{x} \leq b)$ for part (b) in Fig. 3.10, we need to determine the total area under $f(x)$, the density function, from a to b. Recall from your study of calculus that this area can be written as the integral from a to b, or $\int_b^a f(x)\,dx$. Thus,

$$P(a \leq \mathbf{x} \leq b) = \begin{pmatrix}\text{Area under } f(x) \\ \text{from } a \text{ to } b\end{pmatrix} = \int_a^b f(x)\,dx.$$

A number of formulas for $f(x)$ will be investigated later in this chapter and in Chapter 4. For now, the diagrams in Fig. 3.11. should suffice to give some insight into the types of function we will be discussing in detail later in this book. In the first diagram the density function is seen to be a constant, equal to 2.0 for values between $x = 1.0$ and $x = 1.5$, and equal to zero for all other values of $\mathbf{x}$. In this case the random variable $\mathbf{x}$ might represent the number of bushels of wheat the USSR buys from the U.S.A. in a given year (in millions). Perhaps the U.S.A. won't sell less than 1.0 million or more than 1.5 million; hence, $\{1.0 \leq \mathbf{x} \leq 1.5\}$.

In the second diagram, the function $f(x)$ is the straight line $\frac{1}{4} + \frac{1}{8}x$ for values between $x = -2$ and $x = 2$ (and zero otherwise). We might magine $\mathbf{x}$ in this

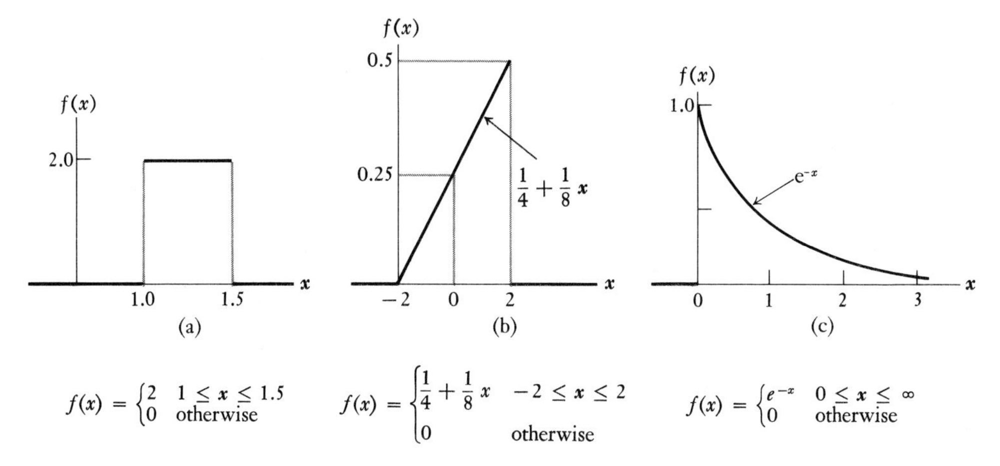

Fig. 3.11 Continuous probability functions.

situation to be the *error* which could occur in certain estimates. For example, in Fig. 3.11(b) $\boldsymbol{x}$ could be the error in estimating the number of cubic yards of concrete needed to complete an office building (i.e., there could be an over-estimation or underestimation by as much as two cubic yards).

In the third diagram, $f(x)$ is a decreasing function for x between zero and infinity.* This type of function is often used in situations where x represents the *time between* certain events.

Probability values determined by integrating a density function must naturally conform to the same probability rules that apply to values determined from a probability mass function. That is, all probabilities must be greater than or equal to zero, and total probability must equal one. This means that the function $f(x)$ can never assume a value less than zero, for if it did, the probability determined by integrating over the negative portion of such a function could also assume a value less than zero. On the other hand, it is *not* necessary for $f(x)$ to always assume values less than or equal to one; this is because the fact that the height of a function is greater than one does not necessarily imply that its total area is also greater than one [see Fig. 3.11(a)]. The two properties of a probability density function are thus:

Properties of all probability density functions:

1. $f(x) \geq 0,$ (density function always positive)
2. $\int_{\text{All } x} f(x)\,dx = 1$ (area under $f(x)$ equals 1)

(3.7)

* In Fig. 3.11(c), the symbol e denotes the nonrepeating, nonterminating decimal $e \cong 2.71828\ldots$

Cumulative Distribution Functions

Before being more precise about the properties of $f(x)$, we need to define the cumulative function associated with a probability density function. This function is called a *cumulative distribution function* (c.d.f.) or, in its more common form, just a *distribution function.* As in the discrete case, this function represents the *probability that the random variable* $\mathbf{x}$ *assumes a value less than or equal to some specified value.* Instead of summing discrete probabilities, however, we now must *integrate* over the relevant range of the probability density function. The same symbol employed in the discrete case, $F(x)$, is used to denote a cumulative distribution function. That is, $F(x) = P(\mathbf{x} \leq x)$. Now, assume $f(x)$ is defined for all values of $\mathbf{x}$ greater than or equal to minus infinity (for some of these values, of course, it may be that $f(x) = 0$). The cumulative distribution function $F(x)$ is defined as follows:

Cumulative distribution function:

$$F(x) = P(\mathbf{x} \leq x) = \int_{-\infty}^{x} f(x)\,dx. \tag{3.8}$$

$$= \text{area under } f(x) \text{ up to the point } x.$$

Fortunately, it is usually not necessary to integrate a given density function $f(x)$ every time one wishes to determine a new value of $F(x)$. Rather, we can often find a *general* formula for $F(x)$ by integrating Formula (3.8) without specifying any particular value of $\mathbf{x}$. For example, the formula for $F(x)$ in Fig. 3.11(a) for values between $x = 1.0$ and $x = 1.5$ is found by integrating $f(x)$ from 1.0 up to x, as follows:

$$F(x) = \int_{1.0}^{x} (2)\,dx = \Big[2x\Big]_{1.0}^{x} = 2x - 2.$$

The general formula for this example is thus:

$$F(x) = \begin{cases} 1.0, & \text{if } x \geq 1.5, \\ 2x - 2, & \text{if } 1.0 \leq x \leq 1.5, \\ 0, & \text{if } x \leq 1.0. \end{cases}$$

Let's check the $2x - 2$ portion of this formula by trying these values: $F(1.0)$, $F(1.25)$, and $F(1.5)$:

$$F(1.0) = 2(1.0) - 2 = 0,$$

$$F(1.25) = 2(1.25) - 2 = 0.50,$$

$$F(1.5) = 2(1.5) - 2 = 1.0.$$

These three values agree with the probability values we know they should assume.

Formulas for the other two functions in Fig. 3.11 can be calculated in the same manner. The reader should verify each of the integrals.

$$\text{If } f(x)=\begin{cases}\frac{1}{4}+\frac{1}{8}x & -2\le x\le 2\\ 0, & \text{otherwise,}\end{cases} \quad \text{then } F(x)=\begin{cases}1, & \text{for } x\ge 2,\\ \frac{1}{16}x^2+\frac{1}{4}x+\frac{1}{4} & \text{for } -2\le x\le 2\\ 0, & \text{for } x\le -2\end{cases}$$

$$\text{If } f(x)=\begin{cases}e^{-x}, & 0\le \mathbf{x}\le\infty,\\ 0, & \text{otherwise,}\end{cases} \quad \text{then } F(x)=\begin{cases}1-e^{-x}, & \text{for } x\ge 0,\\ 0, & \text{for } x\le 0.\end{cases}$$

The cumulative distribution for all three functions in Fig. 3.11 is graphed in Fig. 3.12.

Note that we can *prove* for these examples that they are proper probability density functions by showing that Properties 1 and 2 both hold. Property 1, which is $f(x)\ge 0$, is easily seen to hold by looking at the graph of the functions in Fig. 3.11. The integrations necessary for Property 2 are completed below:

$$\text{a) for } f(x)=\begin{cases}2, & 1.0\le x\le 1.5,\\ 0, & \text{otherwise,}\end{cases} \qquad \int_1^{3/2}(2)\,dx=\Big[2x\Big]_1^{3/2}=1;$$

$$\text{b) for } f(x)=\begin{cases}\frac{1}{4}+\frac{1}{8}x & -2\le x\le 2\\ 0, & \text{otherwise,}\end{cases} \qquad \int_{-2}^{2}\left(\tfrac{1}{4}+\tfrac{1}{8}x\right)dx=\Big[\tfrac{1}{4}x+\tfrac{1}{16}x^2\Big]_{-2}^{2}=1;$$

$$\text{c) for } f(x)=\begin{cases}e^{-x}, & 0\le \mathbf{x},\\ 0, & \text{otherwise.}\end{cases} \qquad \int_0^{\infty}(e^{-x})\,dx=\Big[-e^{-x}\Big]_0^{\infty}=1.$$

The function $F(x)$ is especially useful in evaluating the probability that the random variable $\mathbf{x}$ falls in the interval between $\mathbf{x}=a$ and $\mathbf{x}=b$ (where $a<b$).

Fig. 3.12 Cumulative distribution functions.

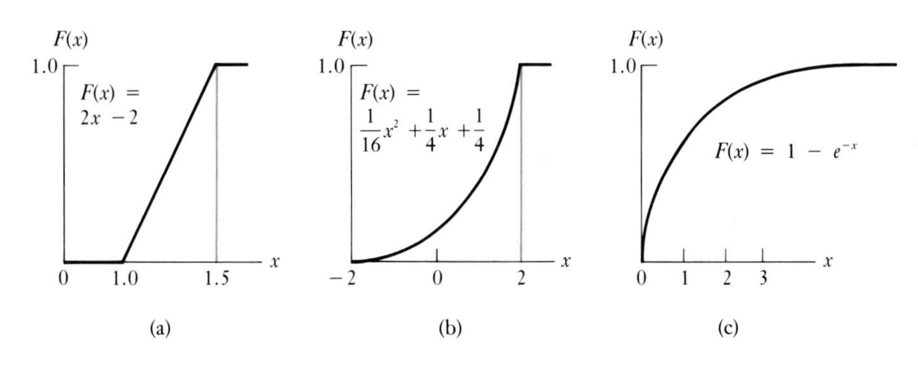

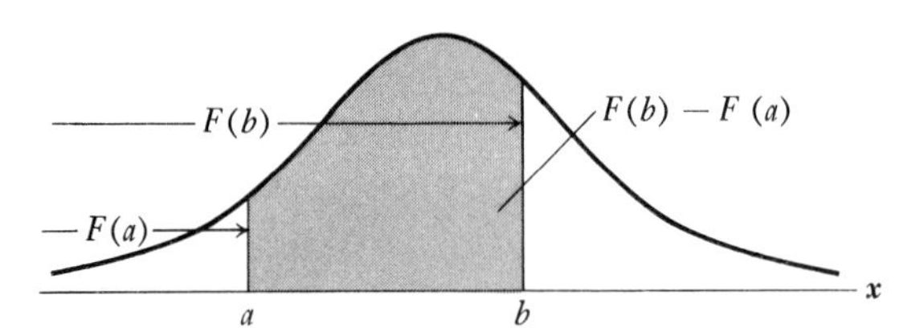

Figure 3.13

To do this, we use the following rule:

$$P(a \leq \mathbf{x} \leq b) = F(b) - F(a). \tag{3.9}$$

This formula should be intuitively appealing, since $F(b)$ is the probability from minus infinity up to the point $\mathbf{x} = b$, and $F(a)$ is the probability from minus infinity up to the point $\mathbf{x} = a$. Subtracting the two gives the area between a and b, as shown in Fig. 3.13.

Suppose for the function in Fig. 3.11(c) we want to find $P(1 \leq \mathbf{x} \leq 2)$. This probability can be evaluated as follows (recall for that function, $F(x) = 1 - e^{-x}$):

$$\begin{aligned} P(1 \leq \mathbf{x} \leq 2) &= F(2) - F(1) \\ &= [1 - e^{-2}] - [1 - e^{-1}]. \end{aligned}$$

Since $e^{-2} = 0.135$ and $e^{-1} = 0.368$,

$$P(1 \leq \mathbf{x} \leq 2) = [1 - 0.135] - [1 - 0.368] = 0.233.$$

It is important to remember that $f(x)$ is not a probability value, but the height of the density function. On the other hand, $F(x)$ is a probability value, namely $P(\mathbf{x} \leq x)$.

3.5 APPROXIMATING A DISCRETE RANDOM VARIABLE BY A CONTINUOUS RANDOM VARIABLE

To illustrate the process of approximating discrete probability values with a continuous function, suppose a mail-order book club is interested in the pattern with which its subscribers pay for the books they order. At the same time that a member's order is filled, the customer is sent a bill on which payment is due within five weeks of the shipping date. In analyzing recent records of this book

Table 3.5

(1) Week Payment Was Received x	(2) Number of Payments Received f	(3) Relative Frequency $f/N = P(x)$	(4) Cumulative Relative Frequency $F(x)$	(5) $0.08x - 0.04$
1	3,940	0.039	0.039	0.040
2	12,012	0.120	0.159	0.120
3	20,133	0.201	0.360	0.200
4	27,852	0.279	0.639	0.280
5	36,063	0.361	1.000	0.360
Sum	100,000	1.000		

club, it was found that only about 16 percent of all customers return their payment within the first few weeks; most people wait until weeks four or five to send in their money. Table 3.5 shows the number of payments received in each of the five weeks for the past 100,000 orders. We will explain column (5) of this table in a moment.

Discrete Case

Suppose we now plot the mass and cumulative functions describing, for each of the five weeks, the probability that a randomly selected customer will make the payment. Figure 3.14 shows the graph of these functions. The tops of the probability lines in this figure form a fairly straight line which has a slope of

Fig. 3.14 Probability functions representing mail-order payments.

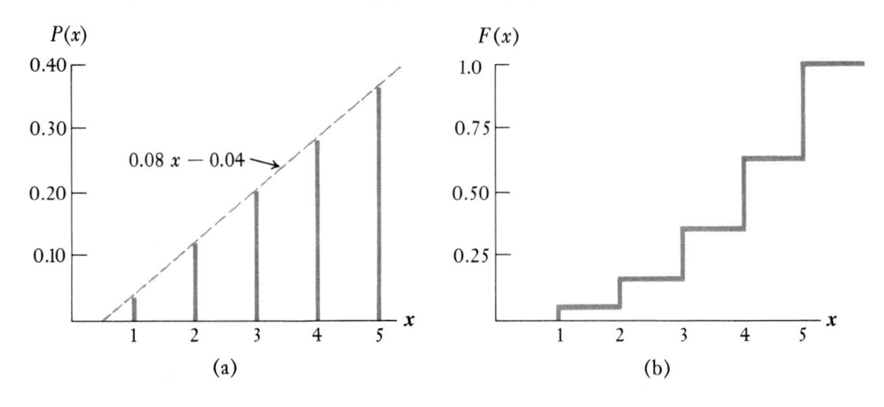

0.08 and a vertical intercept of -0.04. This relationship is shown by the dotted line in Fig. 3.14. An equation which represents the line connecting the tops of the probability values can thus be written as

$$P(x) = 0.08x - 0.04.$$

This equation quite accurately describes $P(x)$ for the discrete values $x = 1, 2, 3, 4$, and 5, as can be seen by comparing columns (3) and (5) in Table 3.5.

Continuous Case

The probabilities shown in Table 3.5 assume that the book club can distinguish only which week ($x = 1, 2, 3, 4$, or 5) during which a customer's payment was received. But suppose we want a continuous approximation which assumes that a payment can be received at any value of $\boldsymbol{x}$ between 0 and 5. That is, we want to assume that $\boldsymbol{x}$ is a continuous random variable which can assume positive probabilities in the interval $0 \leq \boldsymbol{x} \leq 5$. To make this approximation we need a density function which yields probabilities that are comparable to those in the discrete case—namely, 0.04 that the payment was received *between* the shipping date and the end of the first week $[P(0 \leq \boldsymbol{x} \leq 1) = 0.04]$, a probability of 0.12 that the payment was received in the second week. $[P(1 \leq \boldsymbol{x} \leq 2) = 0.12]$, and so forth, with $P(4 \leq \boldsymbol{x} \leq 5) = 0.36$. Although we omit the details, it is not hard to determine that a function giving this approximation is the following:

$$f(x) = \begin{cases} 0.08x, & 0 \leq \boldsymbol{x} \leq 5, \\ 0, & \text{otherwise.} \end{cases}$$

The density function and cumulative functions for this example are shown in Fig. 3.15.

To show that this density function does, in fact, satisfy our needs, we have calculated the appropriate probabilities as follows:

$$P(0 \leq \boldsymbol{x} \leq 1) = \int_0^1 0.08x\,dx = \Big[0.04x^2\Big]_0^1 = 0.04,$$

$$P(1 \leq \boldsymbol{x} \leq 2) = \int_1^2 0.08x\,dx = \Big[0.04x^2\Big]_1^2 = 0.12,$$

$$P(2 \leq \boldsymbol{x} \leq 3) = \int_2^3 0.08x\,dx = \Big[0.04x^2\Big]_2^3 = 0.20,$$

$$P(3 \leq \boldsymbol{x} \leq 4) = \int_3^4 0.08x\,dx = \Big[0.04x^2\Big]_3^4 = 0.28,$$

$$P(4 \leq \boldsymbol{x} \leq 5) = \int_4^5 0.08x\,dx = \Big[0.04x^2\Big]_4^5 = 0.36.$$

It is also not difficult to show that $f(x) = 0.08x$ satisfies both Properties 1 and 2 for continuous frequency functions, as its value is always positive when $0 \leq \boldsymbol{x} \leq 5$, and $\int_0^5 0.08x\,dx = [0.04x^2]_0^5 = 1$.

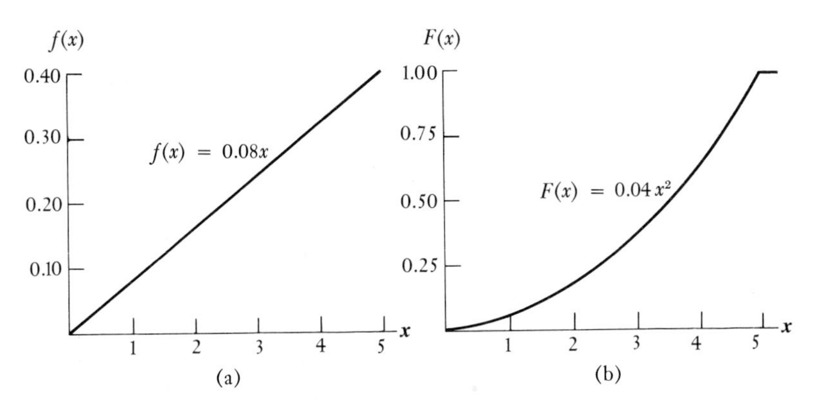

Fig. 3.15 Density and cumulative functions for mail-order payments.

3.6 EXPECTED VALUE: THE CONTINUOUS CASE

The formulas on expected value discussed thus far apply only to discrete data. Formulas for the continuous case follow directly from Formula (3.3) by substituting an integral sign for the summation notation and using $f(x)\,dx$ instead of $P(x)$. The continuous analogy to Formula (3.3) is thus:

Expected value of $g(\mathbf{x})$ *(Continuous case):*

$$E[g(\mathbf{x})] = \int_{\text{All } x} g(x)f(x)\,dx. \tag{3.10}$$

There are two special cases of Formula (3.10) which are of interest, that giving the mean of the random variable $\mathbf{x}$, which is $\mu = E[\mathbf{x}]$, and that giving the variance, which is $\sigma^2 = V[\mathbf{x}] = E[(\mathbf{x} - \mu)^2]$. These formulas, given below, are comparable to Formulas (3.2) and (3.4) for the discrete case.

Expected value of $\mathbf{x}$ *(Continuous case):*

$$E[\mathbf{x}] = \mu = \int_{\text{All } x} xf(x)\,dx.$$

Variance of $\mathbf{x}$ *(Continuous case):*

$$E[(\mathbf{x} - \mu)^2] = \sigma^2 = \int_{\text{All } x} (x - \mu)^2 f(x)\,dx.$$

(3.11)

As before, the variance of $\boldsymbol{x}$ is denoted as either σ^2 or $V[\boldsymbol{x}]$, and the standard deviation is denoted as σ, where $\sigma = \sqrt{V[x]}$. To calculate a variance it is again convenient to use the "mean square – square mean" relationship presented in Formula (3.6). This formula is repeated below.

$$V[\boldsymbol{x}] = E[\boldsymbol{x}^2] - (E[\boldsymbol{x}])^2. \tag{3.6 repeated}$$

We can illustrate the process of using Formula (3.11) by recalling the density function used to approximate the discrete data in the book-club problem:

$$f(x) = \begin{cases} 0.08x, & 0 \leq x \leq 5, \\ 0, & \text{otherwise;} \end{cases}$$

$$\mu = E[\boldsymbol{x}] = \int_0^5 x(0.08x)\,dx = \int_0^5 0.08x^2\,dx = \left[\frac{0.08x^3}{3}\right]_0^5 = 3.33;$$

$$E[\boldsymbol{x}^2] = \int_0^5 x^2(0.08x)\,dx = \int_0^5 0.08x^3\,dx = \left[\frac{0.08x^4}{4}\right]_0^5 = 12.50,$$

$$\sigma^2 = E[(\boldsymbol{x} - \mu)^2] = E[\boldsymbol{x}^2] - (E[\boldsymbol{x}])^2 = 12.50 - (3.33)^2 = 1.41.$$

Note that the mean and variance for this density function are not identical to the values $\mu = 3.80$ and $\sigma^2 = 1.36$ obtained in the discrete analysis. The mean is lower and the variance larger in the continuous case because of the way the data are aggregated—all customers paying within a given week are counted as paying at the *end* of that week in the discrete case, while customers are assumed to be paying *throughout* the five-week period in the continuous case. The discrete probability values begin at $x = 1$, the continuous values at $x = 0$; hence, the latter set of outcomes has a lower mean and is more spread out.

The standard deviation for the density function in this problem is $\sigma = \sqrt{1.41} = 1.19$. We can check the reasonableness of the value $\sigma = 1.19$ by seeing whether approximately 95% of the area under the curve is contained in

$$\begin{aligned} \mu \pm 2\sigma &= 3.33 \pm 2(1.19) \\ &= 0.95 \text{ to } 5.71. \end{aligned}$$

We need only find the area under the curve from 0.95 to 5.00, however, because $f(x) = 0$ when $x \geq 5$ (by definition):

$$\begin{aligned} P(\mu - 2\sigma \leq \boldsymbol{x} \leq \mu + 2\sigma) &= P(0.95 \leq \boldsymbol{x} \leq 5.00) \\ &= \int_{0.95}^{5.00} f(x)\,dx = \int_{0.95}^{5.00} 0.08x\,dx \\ &= \left[0.04x^2\right]_{0.95}^{5.00} = 0.964. \end{aligned}$$

The area under the curve is 0.964, which is close to our approximating value, 0.950.

3.7 EXPECTATION RULES*

The concept of mathematical expectation, or expected value, will prove so useful in the coming chapters that the consideration of a few of the important properties of expectations will be beneficial at this time. In most cases these rules will be presented without proof, and it is not necessary that they be memorized. They will, however, be referred to in subsequent sections and be useful in solving problems.

Rule 1: $E[k] = k$;
The expected value of a constant is the constant itself.

Rule 2: $V[k] = 0$;
The variance of a constant is zero.

Rule 3: $E[k\mathbf{x}] = kE[\mathbf{x}]$;
The expected value of the product of a constant times a variable is the product of the constant times the expected value of the variable.

Rule 4: $V[k\mathbf{x}] = k^2V[\mathbf{x}]$;
The variance of the product of a constant times a variable is the product of the ***square*** *of the constant times the variance of the variable.*

Rule 5†: $E[a \pm b\mathbf{x}] = a \pm bE[\mathbf{x}]$;
The expected value of the quantity $(a + b\mathbf{x})$ [*or* $(a - b\mathbf{x})$] *is a plus (or minus) b times the expectation of* $\mathbf{x}$.

Rule 6‡: $V[a \pm b\mathbf{x}] = b^2V[\mathbf{x}]$
The variance of the quantity $(a + b\mathbf{x})$ [*or* $(a - b\mathbf{x})$] *equals* b^2 *times the variance of* $\mathbf{x}$.

Note in Rules 4 and 6 that the constant is squared on the righthand side. This is because variances involve squared deviations and any constant times $\mathbf{x}$ will itself be squared in calculating the variance.

All six of these rules are illustrated by the use of the probability distributions in Table 3.6. Some additional rules for expected values are illustrated in Section 3.9.

Column (1) of Table 3.6 demonstrates Rule 1; it also demonstrates Rule 2, since there is no variability to a constant. Rule 3 is demonstrated by the comparison of columns (4) and (7). $E[\mathbf{x}] = 3$ in column (4). If each value of

* This section and all the remaining sections in this chapter may be omitted without loss of continuity. If Sections 3.8 and 3.9 are omitted, they may be studied when regression and correlation are covered, beginning in Chapter 10.

† We will prove a more general version of Rule 5 in Section 3.9.

‡ We will prove a more general version of Rule 6 in Section 3.9.

Table 3.6 Example of Rules for Expectations

	(1)	(2)	(3)	(4)	(5)	(6)	(7)	(8)
	k	x	$P(x)$	$xP(x)$	$2x$	$P(2x)$	$(2x)P(2x)$	$5 + 2x$
	2	3	$\frac{1}{3}$	$\frac{3}{3}$	6	$\frac{1}{3}$	$\frac{6}{3}$	11
	2	2	$\frac{1}{3}$	$\frac{2}{3}$	4	$\frac{1}{3}$	$\frac{4}{3}$	9
	2	4	$\frac{1}{3}$	$\frac{4}{3}$	8	$\frac{1}{3}$	$\frac{8}{3}$	13
Mean	2			3			6	11

$\boldsymbol{x}$ is multiplied by 2, the result is column (5), which we have labeled $2\boldsymbol{x}$. The expected value of the variable $2\boldsymbol{x}$ is calculated in column (7), and is seen to be $E[2\boldsymbol{x}] = 6$. Thus, we have demonstrated Rule 3,

$$E[k\boldsymbol{x}] = kE[\boldsymbol{x}].$$

To verify Rule 4, let us calculate the variance of the values in column (5) by using Formula (3.4), with $E[2\boldsymbol{x}] = 6$. Note that multiplying x by 2 doesn't change its probability—hence, $P(x) = P(2x)$.

$$V[2\boldsymbol{x}] = \sum(2x - 6)^2P(2x) = (6 - 6)^2(\tfrac{1}{3}) + (4 - 6)^2(\tfrac{1}{3}) + (8 - 6)^2(\tfrac{1}{3})$$
$$= 0 + \tfrac{4}{3} + \tfrac{4}{3} = \tfrac{8}{3}.$$

If Rule 4 holds, then $V[2\boldsymbol{x}] = (2)^2V[\boldsymbol{x}]$. The variance of $\boldsymbol{x}$ is

$$V[\boldsymbol{x}] = \sum(x - 3)^2P(x) = (3 - 3)^2(\tfrac{1}{3}) + (2 - 3)^2(\tfrac{1}{3}) + (4 - 3)^2(\tfrac{1}{3})$$
$$= 0 + \tfrac{1}{3} + \tfrac{1}{3} = \tfrac{2}{3}.$$

Thus, we have

$$(2)^2V[\boldsymbol{x}] = 4(\tfrac{2}{3}) = \tfrac{8}{3} = V[2\boldsymbol{x}],$$

and Rule 4 is verified.

Column (8) in Table 3.6 can be used to demonstrate the rule for expectations of the form $a + b\boldsymbol{x}$, which are called linear transformations of the variable $\boldsymbol{x}$. First, let's find thc mean of column (8), which, by Rule 5, should be

$$E[5 + 2\boldsymbol{x}] = 5 + 2E[\boldsymbol{x}].$$

From our analysis thus far we know that $E[\boldsymbol{x}] = 3$; hence,

$$E[5 + 2\boldsymbol{x}] = 5 + 2E[\boldsymbol{x}] = 5 + 2(3) = 11.$$

Also, by visual examination the mean of column (8) can be seen to be 11; thus we have verified Rule 5.

According to Rule 6, the variance of the values in column (8) should be

$$V[5 + 2\boldsymbol{x}] = (2)^2V[\boldsymbol{x}] = 4V[\boldsymbol{x}]$$

We previously calculated $V[\mathbf{x}]$ to equal $\frac{2}{3}$. This means that the variance of column (8), by Rule 6, is

$$V[5 + 2\mathbf{x}] = 4[\tfrac{2}{3}] = \tfrac{8}{3}.$$

To verify this variance, let us now calculate $V[5 + 2\mathbf{x}]$ directly from column (8), remembering that $P[5 + 2x] = P(x) = \frac{1}{3}$ and $E[5 + 2\mathbf{x}] = 11$.

$$\begin{aligned} V[5 + 2\mathbf{x}] &= \sum (x - 11)^2 P(x) \\ &= (11 - 11)^2(\tfrac{1}{3}) + (9 - 11)^2(\tfrac{1}{3}) + (13 - 11)^2(\tfrac{1}{3}) \\ &= \tfrac{8}{3}. \end{aligned}$$

This verifies Rule 6.

Expectation of $(a + bx)$

Rules 5 and 6 are often useful in simplifying the calculations of a mean and a variance. For example, we can solve Rule 5 for $E[\mathbf{x}]$ as follows:

$$E[\mathbf{x}] = \frac{E[a + b\mathbf{x}] - a}{b}. \tag{3.12}$$

Suppose we use Formula (3.12) to calculate the mean of the data in the first column of Table 3.7 (assume these values occur with equal probability). Rather than working with the original data, it is considerably easier if we reduce the data to a simpler form. For example, let's divide each value of x by 1000, and then subtract 2500 from each number, as shown in the second two columns in Table 3.7. The values of a and b in Formula (3.12) are thus $b = \frac{1}{1000}$ and $a = -2500$. Since the expectation of the new set of numbers (in column 3) is 2, we can calculate $E[\mathbf{x}]$ as follows:

$$\begin{aligned} E[\mathbf{x}] &= \frac{E[-2500 + x/1000] - (-2500)}{1/1000} \\ &= \frac{2 + 2500}{1/1000} = 2{,}502{,}000. \end{aligned}$$

Table 3.7

x	$x/1000$	$-2500 + x/1000$
2,483,000	2,483	−17
2,519,000	2,519	19
2,511,000	2,511	11
2,495,000	2,495	−5
$E[\mathbf{x}] = ?$	$E[-2500 + \mathbf{x}/1000] = 2$	

Variance of $(a + bx)$

Rule 6 implies that adding (or subtracting) the constant a to a random variable does *not* change its variance. However, changing the *scale* of a random variable by multiplying it by the constant b changes the variance by the square of that constant. For the data in Table 3.7, if you calculate the variance of the transformed values, the result is $V[-2500 + \mathbf{x}/1000] = 195$. Hence, $V[\mathbf{x}]$ can be computed as follows:

$$V[a + b\mathbf{x}] = b^2 V[\mathbf{x}];$$

Therefore,

$$V[\mathbf{x}] = \frac{1}{b^2} V[a + b\mathbf{x}] = \frac{1}{b^2} V[-2500 + \mathbf{x}/1000]$$

$$= \frac{1}{(1/1000)^2}(195) = 195{,}000{,}000.$$

The standard deviation of these data is $\sigma = \sqrt{195{,}000{,}000} = \$13{,}964$.

Expectations Involving $\mathbf{z} = (\mathbf{x} - \mu)/\sigma$

Two additional expectations, both involving the function $(\mathbf{x} - \mu)/\sigma$, are especially useful in statistics. By using this function, the random variable $\mathbf{x}$ is said to be "transformed" by subtracting μ from each value of $\mathbf{x}$, and then by dividing each resulting deviation $(\mathbf{x} - \mu)$ by σ. The letter $\mathbf{z}$ is usually used to denote this transformation. We will write the transformation as $\mathbf{z} = (\mathbf{x} - \mu)/\sigma$ and call it a "standardization" of the variable $\mathbf{x}$.

The standardization $\mathbf{z} = (\mathbf{x} - \mu)/\sigma$ transforms any variable $\mathbf{x}$ into a new variable $\mathbf{z}$, which has a mean of zero and a variance of one.

The new variable $\mathbf{z}$ is important for the comparison of distributions having different means and variances. We will describe in detail in Chapter 5 the usefulness of this standardization. For now, you might imagine someone who has taken one admissions test in which the average score is 500 and the variance is 100, and taken another admissions test in which the average is 600 and the variance is 150. Standardizing these tests would allow us to compare this person's scores on the two tests.

Our interest in this chapter is merely in demonstrating that the expected value of $\mathbf{z} = (\mathbf{x} - \mu)/\sigma$ is zero, and its variance is 1.0. It is easy to show that $E[\mathbf{z}] = E[(\mathbf{x} - \mu)/\sigma] = 0$. Intuitively, we know that the mean is the balance point of any distribution. Some values of $\mathbf{x}$ are above μ and some are below μ,

Table 3.8

	(1) x	(2) $x-\mu$	(3) $(x-\mu)^2$	(4) $(x-\mu)/\sigma = z$	(5) Squared Deviations of z
	13	-7	49	$-\frac{7}{4} = -1.75$	$(-1.75)^2 = 3.0625$
	21	1	1	$\frac{1}{4} = 0.25$	$(0.25)^2 = 0.0625$
	25	5	25	$\frac{5}{4} = 1.25$	$(1.25)^2 = 1.5625$
	19	-1	1	$-\frac{1}{4} = -0.25$	$(-0.25)^2 = 0.0625$
	22	2	4	$\frac{2}{4} = 0.50$	$(0.50)^2 = 0.2500$
Sum	100		80	0	5.000
Mean $\mu_x =$	20		$\sigma_x^2 = 16$ $(\sigma = 4)$	$\mu_z = 0$	$\sigma_z^2 = 1.000$

so that the weighted average of all the values of x is μ. Thus, the weighted average of all the differences between the values of x and μ must be zero. Since zero divided by any constant, including the standard deviation σ, is still zero, it is reasonable that $E[z] = 0$. For the interested reader, we formally prove $E[z] = 0$ and $V[z] = 1$ in a footnote.* An example illustrating these properties is given below.

Consider the probability distribution in Table 3.8, in which each value of x is assumed to have the same probability ($\frac{1}{5}$). The mean and variance of the x

---

* First, we find $E[z]$:

$$E[z] = E\left[\frac{x-\mu}{\sigma}\right] = \frac{1}{\sigma}E[x-\mu] \qquad \begin{cases}\text{Since } \sigma \text{ is a known constant} \\ \text{(Rule 3)}\end{cases}$$

$$= \frac{1}{\sigma}(E[x] - \mu) \qquad \text{Rule 5, since } \mu \text{ is a constant}$$

$$= \frac{1}{\sigma}(\mu - \mu), \qquad E[x] = \mu.$$

Therefore,

$$E[z] = 0.$$

Thus, we have shown that the mean of the standard measure z is zero.

We can find the variance of z as follows:

$$V[z] = V\left[\frac{x-\mu}{\sigma}\right] = \frac{1}{\sigma^2}V[x-\mu] \qquad \begin{cases}\text{Since } \sigma \text{ is a known} \\ \text{constant (Rule 4)}\end{cases}$$

$$= \frac{1}{\sigma^2}(\sigma^2), \qquad \begin{cases}\text{Since } \mu = \text{constant} \\ \text{(Rule 6), and } V[x] = \sigma^2.\end{cases}$$

Therefore,

$$V[z] = 1.$$

values in column (1) are shown to be

$$E[\boldsymbol{x}] = \mu = 20 \qquad \text{and} \qquad V[\boldsymbol{x}] = \sigma^2 = 16.$$

The $z = (x - \mu)/a$ values are given in column (4). These values are shown to have a mean of zero at the bottom of that column. The variance of the variable $\boldsymbol{z} = (\boldsymbol{x} - \mu)/\sigma$ is calculated in column (5), and is shown to equal 1.00.

Define: random variable, probability mass function, expected value, cumulative mass function, standardization, probability model, variance, probability density function (p.d.f.), cumulative distribution function (c.d.f.), and linear transformations.

Problems

3.11 The random variable $\boldsymbol{x}$ = time of arrival of the first customer at a certain store (where $\boldsymbol{x}$ is in hours) is defined as follows:

$$f(x) = \begin{cases} 2x & \text{for } 0 \le x \le 1, \\ 0 & \text{otherwise} \end{cases}$$

a) Sketch the p.d.f. and show that it meets the two properties of all density functions.

b) What is $P(\boldsymbol{x} \le \frac{1}{2})$? Find and interpret $F(1)$.

3.12 a) Find $E[\boldsymbol{x}]$ and $V[\boldsymbol{x}]$ for the p.d.f. in Problem 3.11.

b) Find the median of the function. (*Hint:* Half of the area lies to the left of the median.)

c) Find the area between $\mu \pm 2\sigma$ and compare this area with the 95% rule of thumb.

3.13 a) Using Problem 3.11, write down the value of $F(x)$ for the following: $x = 0$, $x = \frac{1}{2}$, $x = 0.707$, $x = 1.0$, $x = 3.7$.

b) Find a formula for $F(x)$ and sketch this function.

3.14 From Phase I to Phase II, a glass furnace takes between an hour and $1\frac{1}{2}$ hours ($1 \le \boldsymbol{x} \le \frac{3}{2}$) to reach operating temperature. The actual time required follows the p.d.f.:

$$f(x) = \begin{cases} 2x - \frac{1}{2} & 1 \le x \le \frac{3}{2}, \\ 0 & \text{otherwise} \end{cases}$$

a) Graph $f(x)$ and show that it meets the two properties of all density functions.

b) Find $F(x)$ and graph this function.

c) Find $P(\frac{5}{4} \le x \le \frac{3}{2})$.

d) Find the area between $\mu \pm 1\sigma$ and $\mu \pm 2\sigma$ and compare these values to the rule of thumb.

3.15 With the normalization of diplomatic relationships between the U.S. and China, numerous business opportunities have opened up. Suppose a U.S. manufacturer has decided to export a certain type of personal computer to China. The manufacturer estimates the annual demand ($\boldsymbol{d}$) can be represented by the following p.d.f., where $\boldsymbol{d}$

is in thousands of units:

$$f(x) = \begin{cases} (d - 30)/450 & \text{for } 30 \le d \le 60 \\ 0 & \text{otherwise} \end{cases}$$

a) Sketch this function and verify that the area under $f(x)$ equals 1.0.

b) Find $P(\boldsymbol{d} > 45)$ and $F(50)$.

c) Find the expected number of sales.

3.16 Consider the following p.d.f., where $\boldsymbol{x}$ = amount of snow (in feet) in Indianapolis in December.

$$f(x) = \begin{cases} \frac{4}{3} - x^2 & 0 \le x \le 1, \\ 0 & \text{otherwise} \end{cases}$$

a) Sketch this p.d.f. and show that it meets the two conditions of all density functions.

b) What is the probability of more than 6 inches of snow?

3.17 Find a formula for $F(x)$ for Problem 3.16, and use this formula to find $F(\frac{1}{3})$.

3.18 A study of the length of quarter-mile outdoor tracks found that they varied from 5 inches too short to 5 inches too long. An expert has used the following p.d.f. to describe the error in the case of a randomly selected track.

$$f(x) = \begin{cases} \frac{1}{10} & \text{for } -5 \le x \le 5 \\ 0 & \text{otherwise} \end{cases}$$

a) Sketch this p.d.f. and show that its area is 1.0.

b) What is the probability that a track is more than 3 inches too long? What percent of tracks in this population will be more than 2 inches over or under the correct length?

3.19 A Penn State professor has determined that someone always falls asleep during the 1:00–2:00 P.M. class (too many big lunches). The professor insists that the following p.d.f. accurately describes the moment the first student falls asleep ($\boldsymbol{x}$ = time, in hours).

$$f(x) = \begin{cases} 3x^2 & 0 \le x \le 1 \\ 0 & \text{otherwise} \end{cases}$$

a) Sketch this function, showing that it meets the two p.d.f. conditions.

b) Find the formula for $F(x)$. Use this formula to find $F(\frac{1}{4}) = P(\boldsymbol{x} \le \frac{1}{4})$.

3.20 The dollar values of daily sales by a certain small store for the first ten days of the month are 175, 188, 196, 202, 194, 215, 188, 194, 196, and 202.

a) Find the mean of this population (each value has probability $\frac{1}{10}$).

b) Find σ^2 and σ.

c) What percent of the observations fall within $\mu \pm \sigma$ and $\mu \pm 2\sigma$?

d) Repeat parts (a) and (b), subtracting 175 from each observation. How does this subtraction affect μ? How does it affect σ^2?

e) Standardize these ten values by calculating $(x - \mu)/\sigma = z$ for each value.

f) Repeat parts (a) and (b) using the values from part (e). How does division by σ affect the variance?

3.21 An entrepreneur is faced with two investment opportunities that each require an initial outlay of \$10,000. The estimated return on investment $\mathbf{x}$ will be either \$40,000, \$20,000, or \$0, with probabilities of 0.25, 0.50, and 0.25, respectively. For investment $\mathbf{y}$ the returns will be \$30,000, \$20,000, or \$10,000 with probabilities of one-third in each case.

a) Compute $E[\mathbf{x}]$ and $E[\mathbf{y}]$.

b) Check your calculations in (a) by first dividing each value of $\mathbf{x}$ and $\mathbf{y}$ by 10,000, then calculating the mean of the two new sets of numbers, and finally determining $E[\mathbf{x}]$ and $E[\mathbf{y}]$ by multiplying the results by 10,000.

c) Compute $V[\mathbf{x}]$ and $V[\mathbf{y}]$. Try to check your answer by again dividing by 10,000. What is the relationship between $V[\mathbf{x}]$ and $V[\mathbf{x}/10{,}000]$ and $V[\mathbf{x}]/10{,}000$?

---

*3.8 DISCRETE BIVARIATE PROBABILITY (OPTIONAL)

In some situations an experiment involves outcomes which are related to two (or more) random variables. This section covers the theory of such probability functions, which are called *multivariate probability functions*. In this book only the case of *bivariate probability functions* (two variables) will be covered.

The function describing the probability of two random variables is often called a *joint probability function*. For the two random variables $\mathbf{x}$ and $\mathbf{y}$, the joint probability that $\mathbf{x} = x$ and $\mathbf{y} = y$ is written as follows:

> *Joint probability that* $\mathbf{x} = x$ *and* $\mathbf{y} = y$:
>
> $P(\mathbf{x} = x \text{ and } \mathbf{y} = y) = P(x, y).$

As we will show shortly, most of the univariate rules discussed thus far have comparable rules in the bivariate (or multivariate) case. For example, the two properties of all probability functions presented in Section 3.2 have direct counterparts for joint probability functions:

Property 1: $0 \leq P(x, y) \leq 1;$

Property 2: $\sum_{\text{All } y} \sum_{\text{All } x} P(x, y) = 1.$

These and other properties of joint probability functions will be discussed as we progress through this section.

To illustrate a joint probability distribution, let's examine the results of a study designed to investigate the relationship between the number of jobs a college graduate holds in the first five years after graduation ($\mathbf{x}$), and the number of increases in responsibility (that is, $\mathbf{y}$ = number of promotions). Two hundred

Table 3.9 Frequencies for Jobs—Promotions Study

		Number of Promotions (y)				Marginal Total
		1	2	3	4	
Number of jobs (x)	1	20	30	24	12	86
	2	10	14	20	10	54
	3	8	4	28	20	60
Marginal total		38	48	72	42	200

recent college graduates of comparable age and undergraduate background were surveyed and then classified according to the number of jobs and promotions they received in their first five years out of college. The results of this study are given in Table 3.9. Now, we want to translate these data on frequencies to relative frequencies (or probabilities, as we will interpret them). For example, $P(\mathbf{x} = 2, \mathbf{y} = 3) = P(2, 3)$ is the joint probability that one person drawn randomly from this population had two jobs and was promoted three times in the five years. We see that 20 people out of the total of 200 had these characteristics; hence $P(2, 3) = 20/200 = 0.10$. The remaining joint probabilities, which are calculated in a similar fashion, are shown in Table 3.10.

Analogous to a cumulative probability function for a single random variable is the cumulative joint probability function. This function is denoted as $F(x, y)$ and defined as:

Cumulative joint probability: $F(x, y) = P(\mathbf{x} \leq x \quad \text{and} \quad \mathbf{y} \leq y).$

Table 3.10 Probabilities for Jobs—Promotions Study

		Number of Promotions (y)				Marginal Total
		1	2	3	4	
Number of jobs (x)	1	0.10	0.15	0.12	0.06	0.43
	2	0.05	0.07	0.10	0.05	0.27
	3	0.04	0.02	0.14	0.10	0.30
Marginal total		0.19	0.24	0.36	0.21	1.00 $= \sum_y \sum_x P(x, y)$

The value of $F(2, 3) = P(\boldsymbol{x} \leq 2 \text{ and } \boldsymbol{y} \leq 3)$ in Table 3.10 can be seen to equal $P(1, 1) + P(1, 2) + P(1, 3) + P(2, 1) + P(2, 2) + P(2, 3) = 0.59$. Note that $F(x, y)$, like $F(x)$, can assume values only between 0 and 1.0.

Marginal Probability

The concept of a marginal probability that we will use here is the same as that discussed in Chapter 2 except that now in using abbreviations we must be careful not to confuse the marginal probability $P(\boldsymbol{x} = x)$ with the marginal probability $P(\boldsymbol{y} = y)$. For instance, it is not clear if $P(2)$ refers to the former or the latter case. To make this distinction we will abbreviate $P(\boldsymbol{x} = 2)$ as $P_x(2)$ and abbreviate $P(\boldsymbol{y} = 2)$ as $P_y(2)$. With this notation, and using Formula (2.11) as a reference, we can write the marginal probability of $\boldsymbol{x}$ and $\boldsymbol{y}$ as follows:

Marginal probability of $\boldsymbol{x}$: $$P_x(x) = \sum_{\text{All } y} P(x, y);$$

Marginal probability of $\boldsymbol{y}$: $$P_y(y) = \sum_{\text{All } x} P(x, y).$$

Table 3.10 can be used to illustrate these probabilities.

$$P_x(1) = \sum_{\text{All } y} P(1, y) = P(1, 1) + P(1, 2) + P(1, 3) + P(1, 4)$$

$$= 0.10 + 0.15 + 0.12 + 0.06 = 0.43;$$

$$P_y(3) = \sum_{\text{All } x} P(x, 3) = P(1, 3) + P(2, 3) + P(3, 3)$$

$$= 0.12 + 0.10 + 0.14 = 0.36.$$

Thus we can conclude that the probability is 0.43 that a person randomly selected from this population will have had exactly one job, while 0.36 is the probability a person will have had exactly three promotions.

Conditional Probability

A conditional probability for two random variables $\boldsymbol{x}$ and $\boldsymbol{y}$ is defined in the same manner. This concept was defined in Chapter 2, but again we have to be careful about notation. Suppose we let $P_{x|y}(x|y)$ denote the conditional probability $P(\boldsymbol{x} = x | \boldsymbol{y} = y)$, and $P_{y|x}(y|x)$ denote $P(\boldsymbol{y} = y | \boldsymbol{x} = x)$.

Conditional probability of $\boldsymbol{x}$, *given* y:

$$P_{x|y}(x|y) = P(\boldsymbol{x} = x | \boldsymbol{y} = y) = \frac{P(x, y)}{P_y(y)}.$$

Conditional probability of $\boldsymbol{y}$*, given* x*:*

$$P_{y|x}(y|x) = P(\boldsymbol{y} = y|\boldsymbol{x} = x) = \frac{P(x, y)}{P_x(x)}.$$

The first formula can be used to calculate $P_{x|y}(2|3)$ from Table 3.7:

$$P_{x|y}(2|3) = P(\boldsymbol{x} = 2|\boldsymbol{y} = 3) = \frac{P(2, 3)}{P_y(3)}.$$

Since $P(2, 3) = 0.10$ and $P_y(3) = 0.36$,

$$P_{x|y}(2|3) = \frac{0.10}{0.36} = 0.278.$$

In other words, if it is known that a person had three promotions, the probability that this person had exactly two jobs is 0.278.

Independence

Just as we were able in Chapter 2 to determine whether or not two events are independent, so in the present context we can determine whether or not two random variables are independent. The test for independence in the two cases is very similar. In order for two random variables to be independent, all the joint probability values must equal the product of their marginal values. That is, if $\boldsymbol{x}$ and $\boldsymbol{y}$ are independent, then the following relationship must hold for all values of $\boldsymbol{x}$ and $\boldsymbol{y}$.

Joint probability of $\boldsymbol{x}$ *and* $\boldsymbol{y}$ *if independent:*

$$P(x, y) = P_x(x)P_y(y). \tag{3.13}$$

If the above relationship does not hold for *all* possible combinations of $\boldsymbol{x}$ and $\boldsymbol{y}$, then these values are not independent. Only one violation is necessary. For the data in Table 3.9, it is easily shown that *none* of the pairs $\boldsymbol{x}$ and $\boldsymbol{y}$ satisfies Formula (3.13). For example, $P(1, 2) = 0.15$, but this value is not equal to the product

$$P_x(1)P_y(2) = (0.43)(0.24) = 0.1032.$$

Thus, we can conclude that the number of jobs and promotions a person has had are *not* independent.

*3.9 BIVARIATE EXPECTATIONS (OPTIONAL)

We have discussed the important measures of the mean, $\mu = E[\boldsymbol{x}]$, and the variance, $V[\boldsymbol{x}] = \sigma^2$ for a single random variable. These same measures can be used to describe similar concepts in bivariate problems. Suppose we write a function of two random variables $\boldsymbol{x}$ and $\boldsymbol{y}$ as $g(\boldsymbol{x}, \boldsymbol{y})$. In general, the expectation of such a function is given by a direct extension of Formula (3.3).

Expected value of $g(\boldsymbol{x}, \boldsymbol{y})$:

$$E[g(\boldsymbol{x}, \boldsymbol{y})] = \sum_{\text{All } y} \sum_{\text{All } x} g(\boldsymbol{x}, \boldsymbol{y})P(x, y). \tag{3.14}$$

For example, we may want to find the expected value of the product of $\boldsymbol{x}$ times $\boldsymbol{y}$, in which case

$$g(\boldsymbol{x}, \boldsymbol{y}) = x \cdot y \qquad \text{and} \qquad E[\boldsymbol{x} \cdot \boldsymbol{y}] = \sum_{y} \sum_{x} (x \cdot y)P(x, y).$$

Similarly, if $g(\boldsymbol{x}, \boldsymbol{y}) = x + y$, then $E[x + y] = \sum_y \sum_x (x + y)P(x, y)$. We now investigate these two special cases of Formula (3.14).

Expectations of x · y

To illustrate Formula (3.14) when $g(\boldsymbol{x}, \boldsymbol{y}) = x \cdot y$, consider a lumber yard which sells plywood paneling in two lengths, 4 ft and 8 ft, and in three different widths, 2 ft, 4 ft, and 6 ft. The lumber yard is interested in determining the average amount of paneling sold in terms of area (square ft). That is, they want to determine $E[\boldsymbol{x} \cdot \boldsymbol{y}]$, where $\boldsymbol{x}$ = length and $\boldsymbol{y}$ = width. By the basic counting rule of Chapter 2, there are $n_1 = 2$ times $n_2 = 3$, or $n_1 \cdot n_2 = 6$ different arrangements of widths and lengths sold. The distributions $P(x)$, $P(y)$, and $P(x, y)$ for the sale of these six combinations, based on company records, are given in Table 3.11. The final column shows the average square feet of paneling sold is $\mu = 30.80$.

Table 3.11 Plywood Paneling

	x	$P(x)$	$xP(x)$	y	$P(y)$	$yP(y)$	x, y	$P(x, y)$	Area $x \cdot y$	$E[x \cdot y] = (x \cdot y)P(x, y)$
	4	0.20	0.80	2	0.15	0.30	(4, 2)	0.05	8	0.40
	8	0.80	6.40	4	0.55	2.20	(4, 4)	0.05	16	0.80
				6	0.30	1.80	(4, 6)	0.10	24	2.40
							(8, 2)	0.10	16	1.60
							(8, 4)	0.50	32	16.00
							(8, 6)	0.20	48	9.60
Sum		1.00	7.20		1.00	4.30		1.00		30.80

Expectation of $x \cdot y$ When x and y Are Independent

One special case of the expectation of $\boldsymbol{x} \cdot \boldsymbol{y}$ is worth noting—the case in which the variables $\boldsymbol{x}$ and $\boldsymbol{y}$ are independent.

> *Expectation of* $\boldsymbol{x} \cdot \boldsymbol{y}$*, assuming independence:*
>
> $$E[\boldsymbol{x} \cdot \boldsymbol{y}] = E[\boldsymbol{x}]E[\boldsymbol{y}]. \qquad (3.15)$$

We will illustrate Formula (3.15) by considering the choices made by new students regarding university housing and laundry services. Each student may choose to live in a single, double, or twin-double (suite) dormitory room, denoted by $\boldsymbol{x} = 1, 2,$ or 4-person rooms. A second choice is available between two laundry plans, allowing for \$4 or \$9 worth of laundry per week. The \$4-per-week plan is included in regular dormitory fees. The \$9-per-week plan has an additional fee. The laundry plan selected is denoted by $\boldsymbol{y} = 4$ or 9. The probability distributions of $P(x)$, $P(y)$, and $P(x, y)$, based on the students' choices, are given in the matrix below. These probabilities illustrate that dormitory-room selections and laundry-plan choices are independent (i.e., statistically unrelated decisions) for these new students, since $P(x, y) = P_x(x)P_y(y)$ for all x and y.

x \ y	Laundry 4	9	$P(x)$
1	0.24	0.16	0.40
Room 2	0.12	0.08	0.20
4	0.24	0.16	0.40
$P(x)$	0.60	0.40	$1.00 = \sum\sum P(x, y)$

The fact that $\boldsymbol{x}$ and $\boldsymbol{y}$ are independent in this example can also be shown by using some of the rules of expectations presented earlier in this chapter. Table 3.12 shows a number of expectations derived from this example, which will be applied extensively throughout the remainder of this chapter.

From columns (3) and (6) of Table 3.12 we see that

$$E[\boldsymbol{x}]E[\boldsymbol{y}] = (2.40)(6.00) = 14.40,$$

which is the same result as that given in column (9),

$$E[\boldsymbol{x} \cdot \boldsymbol{y}] = 14.40.$$

This result verifies that $\boldsymbol{x}$ and $\boldsymbol{y}$ are indeed independent.

Table 3.12 Room Choice/Laundry Example

(1) x	(2) $P(x)$	(3) $xP(x)$	(4) y	(5) $P(y)$	(6) $yP(y)$	(7) (x, y)	(8) $P(x, y)$	(9) $E[x \cdot y] =$ $(x \cdot y)P(x, y)$
1	0.40	0.40	4	0.60	2.40	(1, 4)	0.24	(4)(.24) = 0.96
2	0.20	0.40	9	0.40	3.60	(1, 9)	0.16	(9)(.16) = 1.44
4	0.40	1.60				(2, 4)	0.12	(8)(.12) = 0.96
						(2, 9)	0.08	(18)(.08) = 1.44
						(4, 4)	0.24	(16)(.24) = 3.84
						(4, 9)	0.16	(36)(.16) = 5.76
Sum	1.00	2.40		1.00	6.00		1.00	14.40

Covariance of x and y

At this point we introduce another measure of variation that is very important in most statistical analysis—the *covariance of* $\boldsymbol{x}$ *and* $\boldsymbol{y}$ (which we denote as $C[\boldsymbol{x}, \boldsymbol{y}]$). The covariance of two random variables is a measure of how they vary together (how they "covary"). If we let

$$\mu_x = E[\boldsymbol{x}] \qquad \text{and} \qquad \mu_y = E[\boldsymbol{y}],$$

then

$$\textit{Covariance of } \boldsymbol{x} \textit{ and } \boldsymbol{y}: \qquad C[\boldsymbol{x}, \boldsymbol{y}] = E[(\boldsymbol{x} - \mu_x)(\boldsymbol{y} - \mu_y)]. \tag{3.16}$$

Like a variance, a covariance is somewhat difficult to interpret. If high values of $\boldsymbol{x}$ (high relative to μ_x) tend to be associated with high values of $\boldsymbol{y}$ (relative to μ_y), and low values associated with low values, then $C[\boldsymbol{x}, \boldsymbol{y}]$ will be a large positive number.* If the covariance is a large negative number, this means that low values of one variable tend to be associated with high values of the other, and vice versa. If two variables are independent, then $C[\boldsymbol{x}, \boldsymbol{y}] = 0$ (i.e., they are not related).

To calculate a covariance, we could use the definition of $E[g(\boldsymbol{x}, \boldsymbol{y})]$ in Formula (3.14). Rather than do this, we present an equivalent formula which is much more convenient computationally:

$$C[\boldsymbol{x}, \boldsymbol{y}] = E[\boldsymbol{x} \cdot \boldsymbol{y}] - E[\boldsymbol{x}]E[\boldsymbol{y}]. \tag{3.17}$$

* This is because when $(\boldsymbol{x} - \mu_x)$ is +, $(\boldsymbol{y} - \mu_y)$ tends to be +, and when $(\boldsymbol{x} - \mu_x)$ is −, $(\boldsymbol{y} - \mu_y)$ tends to be −; hence the sign of $(\boldsymbol{x} - \mu_x)(\boldsymbol{y} - \mu_y)$ tends to be positive.

The reader may recognize this formula as just a variation of the mean square − square mean relationship. In fact, in all of our formulas dealing with the covariances, if one merely substitutes the variable $\mathbf{x}$ wherever the variable $\mathbf{y}$ occurs, the result is a variance formula. For example, $C[\mathbf{x}, \mathbf{y}]$ becomes $C[\mathbf{x}, \mathbf{x}]$, which, by Formulas (3.16) and (3.4), is seen to be exactly the same as $V[\mathbf{x}]$.

Now, let's use Formula (3.17) to calculate $C[\mathbf{x}, \mathbf{y}]$ for the data in Table 3.11. From that table, $E[\mathbf{x}] = 7.20$, $E[\mathbf{y}] = 4.30$, and $E[\mathbf{x} \cdot \mathbf{y}] = 30.80$:

$$\begin{aligned} C[\mathbf{x}, \mathbf{y}] &= E[\mathbf{x} \cdot \mathbf{y}] - E[\mathbf{x}]E[\mathbf{y}] \\ &= 30.80 - (7.20)(4.30) = -0.16. \end{aligned}$$

The covariance in this case is close to zero, which means that the relationship between the length ($\mathbf{x}$) and width ($\mathbf{y}$) of the lumber sold is not very strong.

If $\mathbf{x}$ and $\mathbf{y}$ are independent, we know from Formula (3.15) that $E[\mathbf{x}, \mathbf{y}] = E[\mathbf{x}]E[\mathbf{y}]$. Substituting this relationship into Formula (3.17) yields a result that has already been stated; *when* $\mathbf{x}$ *and* $\mathbf{y}$ *are independent, then their covariance is zero,**

$$C[\mathbf{x}, \mathbf{y}] = E[\mathbf{x}]E[\mathbf{y}] - E[\mathbf{x}]E[\mathbf{y}] = 0.$$

This relationship can be demonstrated by applying Formula (3.17) to the data in Table 3.12, where $\mathbf{x}$ and $\mathbf{y}$ are independent combined random variables.

Expectations of $a\mathbf{x} + b\mathbf{y}$

Another special case of $E[g(\mathbf{x}, \mathbf{y})]$ that has practical importance is when

$$g(\mathbf{x}, \mathbf{y}) = a\mathbf{x} + b\mathbf{y},$$

where a and b are constants.

Expected value of $(a\mathbf{x} + b\mathbf{y})$:†

$$\begin{aligned} E[a\mathbf{x} + b\mathbf{y}] &= \sum_{\text{All } y} \sum_{\text{All } x} (ax + by)P(x, y) \\ &= aE[\mathbf{x}] + bE[\mathbf{y}]. \end{aligned} \qquad (3.18)$$

We will illustrate Formula (3.18) by considering the special case in which the constants a and b both equal $+1.0$. This makes our algebra easier while still

* The converse of this relationship does not necessarily hold true.

† The reader may recognize that Rule 5 on page 149 is merely a special case of Formula (3.18), in which one of the variables is a constant.

maintaining the essence of the process. Thus, we will demonstrate that

$$E[x + y] = E[x] + E[y]. \tag{3.19}$$

Consider a company that specializes in sending "care" packages to college students. These packages consist of a fruit box plus a "free gift," which is always either a glass mug or a metal coin bank, alternately placed in successive packages. The company is currently trying to analyze the total weight of their packages in order to better control mailing costs. The fruit boxes come in three weights, 2, 5, and 10 pounds. The mug weighs one pound and the coin bank weighs $\frac{1}{2}$ pound. The company is interested in determining the average weight of each package, or $E[x + y]$, where x = weight of the fruit and y = weight of the gift. The probability distribution of $P(x)$, $P(y)$, and $P(x, y)$, based on sales records, is given in Table 3.13.

From the final column of Table 3.13 we see that the average weight of the "care" packages is 5.40 pounds. This result is also obtained by summing the totals of columns (3) and (6):

$$E[x] + E[y] = 4.65 + 0.75 = 5.40.$$

Generally, it is easier to find $E[x + y]$ by summing $E[x] + E[y]$. As an exercise, the reader may use the data in Table 3.12 (page 162) to show that

$$E[x + y] = 8.40.$$

Variance of $(ax + by)$

Our final discussion in this chapter involves calculating $V[ax + by]$, where a and b are constants. Although the formula given below is presented without proof, its derivation is not difficult using the concepts provided in this chapter.

Table 3.13 Package Weight Example

(1) Weight of Fruit (x)	(2) $P(x)$	(3) $xP(x)$	(4) Weight of Gift (y)	(5) $P(y)$	(6) $yP(y)$	(7) (x, y)	(8) $P(x, y)$	(9) $(x + y)$	(10) $(x + y)P(x, y)$
2	0.45	0.90	$\frac{1}{2}$	0.50	0.25	$(2, \frac{1}{2})$	0.20	2.5	0.500
5	0.35	1.75	1	0.50	0.50	$(2, 1)$	0.25	3.0	0.750
10	0.20	2.00				$(5, \frac{1}{2})$	0.15	5.5	0.825
						$(5, 1)$	0.20	6.0	1.200
						$(10, \frac{1}{2})$	0.15	10.5	1.575
						$(10, 1)$	0.05	11.0	0.550
Sum	1.00	4.65		1.00	0.75		1.00		5.400

> *Variance of* $(ax + by)$:*
>
> $$V[ax + by] = a^2V[x] + b^2V[y] + 2abC[x, y]. \qquad (3.20)$$

We will not try to illustrate Formula (3.20) for various values of a and b, but again will consider the special case where a and b both equal 1.0. That is, we will illustrate the simpler rule,

$$V[x + y] = V[x] + V[y] + 2C[x, y],$$

by applying it to the data in Table 3.11. The values of $E[x^2]$ and $E[y^2]$ for the data from that table (on page 160) can be shown to be

$$E[x^2] = 54.4 \quad \text{and} \quad E[y^2] = 20.2.$$

We have already shown that $E[x] = 7.2$ and $E[y] = 4.3$. Thus,

$$V[x] = E[x^2] - (E[x])^2 = 54.4 - (7.2)^2 = 2.56,$$

and

$$V[y] = E[y^2] - (E[y])^2 = 20.2 - (4.3)^2 = 1.71.$$

Since we previously calculated $C[x, y] = -0.16$, we can write

$$\begin{aligned} V[x + y] &= V[x] + V[y] + 2C[x, y] \\ &= 2.56 + 1.71 + 2(-0.16) = 3.95. \end{aligned}$$

Independence

As stated earlier, when x and y are independent, then $C[x, y] = 0$; hence

> *Special case for independent random variables* x *and* y:†
>
> $$V[ax + by] = a^2V[x] + b^2V[y]. \qquad (3.21)$$

Note that Formula (3.21) implies that if $a = 1$ and $b = \pm 1$, then

$$V[x + y] = V[x - y] = V[x] + V[y].$$

---

* The reader should note that Rule 6 presented on page 149 is merely a special case of Formula (3.20), in which one of the variables is a constant. Also, the sign of the last term in (3.20) will be negative if either a or b is negative (but not both), or if the covariance is negative.

† Formula (3.21) holds whenever $C[x, y] = 0$, not merely when x and y are independent.

Table 3.14 Bivariate Expectation Formulas

Function	Formulas	Special Case of Independence
1. Mean of $(x \cdot y)$	$E[x \cdot y] = \sum\sum (x \cdot y)P(x, y)$	$E[x] \cdot E[y]$
2. Covariance of (x and y)	$C[x, y] = E[x \cdot y] - E[x]E[y]$	0
3. Mean of $(ax + by)$	$E[ax + by] = aE[x] + bE[y]$	No change
4. Variance of $(ax + by)$	$V[ax + by] = a^2V[x] + b^2V[y] + 2abC[x, y]$	$a^2V[x] + b^2V[y]$

Expressed in words, the variance of the sum or difference of two variables that are independent always equals the *sum* of the two variances calculated separately. The reader can verify this fact by using the data in Table 3.12 (on page 162) to show that

$$V[x + y] = V[x - y] = 7.84.$$

A summary of the bivariate expectation formulas from this section is provided in Table 3.14.

Define: Expectation rules, linear transformations, joint probability function, bivariate expectations, covariance, independence, conditional probability function, marginal probability function, cumulative joint mass function

*3.10 CONTINUOUS BIVARIATE PROBABILITY (OPTIONAL)

A major portion of this chapter has been devoted to probability rules involving two or more *discrete* random variables. Similar rules can be developed for probability density functions, for we are often interested in the probability that two or more *continuous* random variables will be combined in specified ways. For instance, a stockbroker may want to estimate the probability that prices on *both* the New York Stock Exchange *and* the American Stock Exchange will decrease between 10 and 20 percent in the next month, or a building contractor may wish to estimate the probability that in the next year *either* material costs or labor costs or both will increase more than five percent. As we pointed out earlier in this chapter, however, it is usually the entire probability density function rather than the probability of isolated events alone which helps to evaluate a particular course of action in a decision-making context. This section covers the theory of probability density functions which involve more than one variable; such density functions are called *multivariate* density functions.

Although in this book only the case of *bivariate* density functions (two variables) will be covered, the reader should be able to generalize most of the concepts discussed, or may wish to consult a text on mathematical statistics.

Consider a bivariate density function involving the random variables $\boldsymbol{x}_1$ and $\boldsymbol{x}_2$. For instance, we might be interested in the weight ($\boldsymbol{x}_1$) and the height ($\boldsymbol{x}_2$) of American males, or $\boldsymbol{x}_1$ might represent the temperature and $\boldsymbol{x}_2$ the humidity at a specified time and date in selected U.S. cities. Similarly, $\boldsymbol{x}_1$ and $\boldsymbol{x}_2$ might be the length and width of a certain industrial product. For problems of this nature, the first step is to describe the relationship between $\boldsymbol{x}_1$ and $\boldsymbol{x}_2$ by defining a density function which gives the probability that $\boldsymbol{x}_1$ assumes any one of a given set of values at the *same time* that $\boldsymbol{x}_2$ assumes any one of another set of values. This function, which indicates the relative frequency of the joint occurrence that the random variable $\boldsymbol{x}_1$ assumes the specific value x_1 *at the same time* that the random variable $\boldsymbol{x}_2$ assumes the specific value x_2, is called the *joint density function of* $\boldsymbol{x}_1$ *and* $\boldsymbol{x}_2$ and is written as $f(x_1, x_2)$. Figure 3.16 shows the joint density function of $f(x_1, x_2) = 4$ for $0 \leq \boldsymbol{x}_1 \leq \frac{1}{2}$ and $0 \leq \boldsymbol{x}_2 \leq \frac{1}{2}$, and $f(x_1, x_2) = x_1 + x_2$ for $0 \leq \boldsymbol{x}_1 \leq 1$ and $0 \leq \boldsymbol{x}_2 \leq 1$.

Probability in the bivariate case is not determined by the *area* under the joint density function $f(x_1, x_2)$, but must be measured by the *volume* under this function. Volume, as you may recall from your study of calculus, is measured by taking the *double* integral of the function under investigation, integrating with respect to first one variable and then the other. Thus, the volume under the joint density function $f(x_1, x_2)$ equals $\iint f(x_1, x_2)\,dx_1\,dx_2$. Suppose that we want to determine the probability that $\boldsymbol{x}_1$ lies between two points a and b, and at the same time, that $\boldsymbol{x}_2$ lies between the points c and d. This probability is given by the volume under $f(x_1, x_2)$ between these limits, or $\int_c^d \int_a^b f(x_1, x_2)\,dx_1\,dx_2$. Consider the second example shown in Fig. 3.16, and assume that we would like to determine the probability that $\boldsymbol{x}_1$ lies between 0

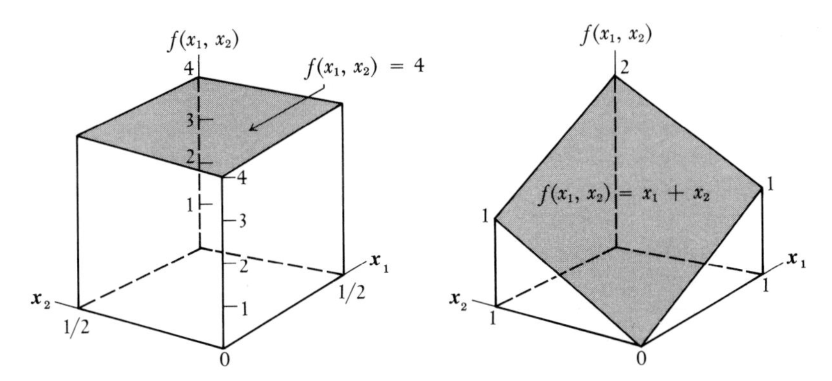

Fig. 3.16 Multivariate density functions.

and 1 and $\boldsymbol{x}_2$ lies between 0 and $\frac{1}{2}$:

$$P(0 \le \boldsymbol{x}_1 \le 1 \quad \text{and} \quad 0 \le \boldsymbol{x}_2 \le 1/2)$$

$$= \int_0^{1/2} \int_0^1 f(x_1, x_2)\,dx_1\,dx_2$$

$$= \int_0^{1/2} \int_0^1 (x_1 + x_2)\,dx_1\,dx_2 = \int_0^{1/2} \left[\frac{x_1^2}{2} + x_1 x_2\right]_0^1 dx_2$$

$$= \int_0^{1/2} \left(\frac{1}{2} + x_2\right) dx_2 = \left[\frac{x_2}{2} + \frac{x_2^2}{2}\right]_0^{1/2} = \frac{3}{8}.$$

The *joint distribution function* of a bivariate density function is written as $F(x_1, x_2)$ and denotes the probability that the first random variable, $\boldsymbol{x}_1$, assumes a value less than x_1, and the second random variable, $\boldsymbol{x}_2$, assumes a value less than x_2. In other words, it is the volume contained under $f(x_1, x_2)$ up to the points $\boldsymbol{x}_1 = x_1$ and $\boldsymbol{x}_2 = x_2$. This probability can be denoted as follows:

$$F(x_1, x_2) = \int_{-\infty}^{x_2} \int_{-\infty}^{x_1} f(x_1, x_2)\,dx_1\,dx_2.$$

This joint distribution function definition can be used to test whether the functions described by Fig. 3.16 satisfy Property 2, as they must if they do, in fact, represent density functions. Note that these functions satisfy Property 1, since $f(x_1, x_2) \ge 0$ over their defined ranges. To test whether the probability of the universal set equals one we must integrate over all possible values of x_1 and x_2. The value of

$$\int_{\text{All } x_2} \int_{\text{All } x_1} f(x_1, x_2)\,dx_1\,dx_2$$

equals one for both functions, as we show below.

1. $f(x_1, x_2) = 4; 0 \le \boldsymbol{x}_1, \boldsymbol{x}_2 \le \frac{1}{2}$;

$$\int_0^{1/2} \int_0^{1/2} (4)\,dx_1\,dx_2 = \int_0^{1/2} \left[4x_1\right]_0^{1/2} dx_2 = \int_0^{1/2} 2\,dx_2$$

$$= \left[2x_2\right]_0^{1/2} = 1.$$

2. $f(x_1, x_2) = x_1 + x_2; 0 \le \boldsymbol{x}_1 \le 1, 0 \le \boldsymbol{x}_2 \le 1$;

$$\int_0^1 \int_0^1 (x_1 + x_2)\,dx_1\,dx_2 = \int_0^1 \left[\frac{x_1^2}{2} + x_1 x_2\right]_0^1 dx_2$$

$$= \int_0^1 \left(\frac{1}{2} + x_2\right) dx_2$$

$$= \left[\frac{x_2}{2} + \frac{x_2^2}{2}\right]_0^1 = 1.$$

The probability relationships studied in Chapter 2 can now be shown to have their counterparts for bivariate density functions. Rather than present the relationships necessary for all types of bivariate analysis, we shall merely illustrate several of the most important ones.

First, consider the continuous analogy to Formula (2.11), which describes marginal probability. Instead of summing the joint probability $P(A \cap E_j)$ over all values of E_j to evaluate $P(A)$, we now integrate the joint density function $f(x_1, x_2)$ over all values of $\mathbf{x}_2$ to determine $f(x_1)$. Hence, the *marginal density function of* $\mathbf{x}_1$ is defined as

$$f(x_1) = \int_{\text{All } x_2} f(x_1, x_2)\, dx_2.$$

Thus, to derive the marginal density function of one variable, the effect of the other variable must be removed by integrating over all values of that variable. To illustrate, suppose we let $f(x_1, x_2) = x_1 x_2$, where $0 \leq \mathbf{x}_1 \leq 2$ and $0 \leq \mathbf{x}_2 \leq 1$, and determine $f(x_1)$ and $f(x_2)$:

$$f(x_1) = \int_0^1 x_1 x_2\, dx_2 = \left[\frac{x_2^2}{2} x_1\right]_0^1 = \tfrac{1}{2} x_1,$$

$$f(x_2) = \int_0^2 x_1 x_2\, dx_1 = \left[\frac{x_1^2}{2} x_2\right]_0^2 = 2x_2.$$

Once the marginal density functions are known it is a simple matter to determine conditional functions; the conditional density function of x_2 given x_1 must equal the joint density function, $f(x_1, x_2)$ divided by the marginal density function, $f(x_1)$ (see Formula (2.3)), or

$$f(x_2 | x_1) = \frac{f(x_1, x_2)}{f(x_1)}.$$

In the example above, where $f(x_1, x_2) = x_1 x_2$ and $f(x_1) = \frac{1}{2}x_1$, the conditional density function of x_2 given x_1 must equal

$$f(x_2 | x_1) = \frac{f(x_1, x_2)}{f(x_1)} = \frac{x_1 x_2}{\frac{1}{2}x_1} = 2x_2,$$

and

$$f(x_1 | x_2) = \frac{f(x_1, x_2)}{f(x_2)} = \frac{x_1 x_2}{2x_2} = \frac{1}{2} x_1.$$

Note the significance of these results. The conditional density functions in each case exactly equal the marginal density functions we calculated earlier; that is, $f(x_2 | x_1) = f(x_2) = 2x_2$ and $f(x_1 | x_2) = f(x_1) = \frac{1}{2}x_1$. This will always happen if, as we discussed in Chapter 2, the variables (or events, in Chapter 2) are *independent*. A more general statement concerning independence also follows from the discussion in Chapter 2 (see Formula 2.8), since it can be shown that if the random variables x_1 and x_2 are independent, then their joint frequency function can always be factored into a function involving only x_1 times a function involving only x_2; that is,

$$f(x_1, x_2) = f(x_1)f(x_2).$$

The variables x_1 and x_2 are independent in the above example, and hence $f(x_1, x_2)$ can be factored: $f(x_1, x_2) = f(x_1)f(x_2) = (x_1)(x_2)$.

Define: Joint probability, marginal probability, bivariate expectations, covariance, bivariate probability, multivariate density function.

Problems

3.22 Reconsider the stock market example on page 130. Divide each value of x by 5, and then subtract 15 from each of the resulting values.

a) Calculate $E[y]$ and $V[y]$ for this new variable $(x/5) - 15 = y$.

b) Based on the fact that $E[x] = 96$ and $V[x] = 64$ show how you could have predicted the values in part (a) using the expectation rules.

3.23 a) Given a random variable x with mean 10 and variance 9, find the expected value and variance of the random variable y where $y = 12 + 2x$.

b) Given: a random variable x with $E[x] = 5$ and $V[x] = 9$, and another random variable y with $E[y] = 10$ and $V[y] = 25$. The variables x and y are independent.

1. Find $E[x \cdot y]$, $E[x + 2y]$, and $E[13 - 2x]$.
2. Find $V[x - y]$, $V[x + 2y]$, and $V[13 - 2x]$.
3. What is the value of $C[x, y]$?

3.24 The data in the following table represent the sales of cars by one of the major automobile companies for each of the past four years.

$P(x)$	x
0.250	1,751,000
0.250	1,389,000
0.250	1,512,000
0.250	1,456,000
1.000	$E[x] = ?$

a) Calculate the four values of $y = a + bx$, where $a = -1500$ and $b = \frac{1}{1000}$.

b) Calculate $E[x]$ by calculating $E[y] = E[a + bx]$. The value of $E[x]$ is then derived by solving Rule 5 of Section 3.7 for $E[x]$:

$$E[x] = \frac{E[a + bx] - a}{b}.$$

c) Calculate $V[x]$ by finding $V[a + bx]$ first, and then by solving Rule 6 for $V[x]$:

$$V[x] = \frac{V[a + bx]}{b^2}.$$

3.25 Some people claim that (x) the severity of a winter can be predicted in the fall by (y) the amount of fuzz on a caterpillar. A study of this phenomenon was conducted using three grades for the severity of winter (1 = mild, 2 = average, 3 = harsh) and three categories of the density of caterpillar fuzz $\{y = 1, 2, 3\}$. The following joint probability table reflects the results of this study.

		x(winter)			
		1	**2**	**3**	
	1	0.10	0.05	0.05	0.20
y(fuzz)	**2**	0.10	0.20	0.10	0.40
	3	0.05	0.15	0.20	0.40
		0.25	0.40	0.35	

a) Find $P_x(2)$, $P_y(2)$, and $P_{x|y}(3|2)$.

b) Would you conclude that x and y are independent or dependent from these data? Explain!

c) Find $E[x]$ and $E[y]$.

d) Find $E[x \cdot y]$ and $E[x + y]$.

e) Find $C[x, y]$.

f) Find $E[5x - 3y]$.

g) Find $V[5x - 3y]$.

3.26 A recent study involving 46,000 women of child-bearing age in the United Kingdom suggested that women who both smoke and use birth control pills run a higher risk of death from diseases of the circulatory system. The following table indicates the annual mortality rate per 100,000 women from such diseases based on this study.

Number of Deaths

	Pill Users	**Nonusers**
Nonsmokers	13.5	3.0
Smokers	39.5	8.9

a) How many variables are there in this study? Name them.

b) What kinds of probabilities would you want to consider to determine whether smoking increases the risk of death for pill users?

c) What can one conclude from the data in this table about the relationship between using the pill, smoking, and the mortality rate? [Be careful!]

d) What additional data or analysis would you suggest to supplement these data?

3.27 a) For the data of Table 3.12 show that $E[x + y] = 8.40 = E[x] + E[y]$.

b) For the data of Table 3.12 show that $V[x - y] = 7.84 = V[x] + V[y]$.

3.28 a) Assume that each pair of values of x and y shown below occurs with equal probability (that is, $\frac{1}{4}$ each).

x	5	2	3	6
y	6	6	6	6

1) Find $E[x]$, $E[y]$, $V[x]$, and $V[y]$.

2) Find $E[x + y]$ and $E[x - y]$.

3) Are x and y independent or dependent? Explain.

4) Find $V[x + y]$ and $C[x, y]$.

b) Repeat the above calculations for the following data:

x	5	2	3	6
y	8	4	5	7

Explain (intuitively) why $C[x, y]$ is larger for part (b) than for part (a).

3.29 A company selling water softeners has three models ($y = 1, 2, 3$) and will sell each model with a guarantee of either 1, 5, or 10 years ($x = 1, 5$, or 10). The joint probability function $P(x, y)$, is given by the following table:

y \ x	1	5	10
1	$\frac{1}{20}$	$\frac{2}{20}$	$\frac{3}{20}$
2	$\frac{4}{20}$	0	$\frac{3}{20}$
3	$\frac{2}{20}$	$\frac{4}{20}$	$\frac{1}{20}$

a) Write down the cumulative joint mass function $F(x, y)$.

b) Find the marginal probability functions for x and y. What is the value of $P_y(2)$?

c) Find $E[x]$ and $E[y]$.

d) Calculate $P_{x|y}(10|1)$.

e) Determine whether x and y are independent.

3.30 Use the information in Problem 3.29 to calculate

a) $E[x \cdot y]$.

b) $E[x + y]$ and $E[x - y]$.

c) $C[x, y]$.

d) $V[x + y]$ and $V[x - y]$.

3.31 Given the following joint density function:

$$f(x) = \begin{cases} \frac{1}{6}(3x_1 + 9x_2), & \text{for } 0 \le x_1, x_2 \le 1, \\ 0, & \text{otherwise.} \end{cases}$$

a) Find the cumulative function $F(x_1, x_2)$. What is the value of this function at $x_1 = \frac{1}{2}$, $x_2 = \frac{1}{2}$?

b) Show that $f(x)$ satisfies the properties of a probability function.

c) Determine the joint probability of $0 \le x_1 \le \frac{1}{2}$ and $\frac{1}{2} \le x_2 \le 1$.

d) Find the marginal density function for $\boldsymbol{x}_1$ and for $\boldsymbol{x}_2$.

e) Find the conditional distribution of $\boldsymbol{x}_2$ given $\boldsymbol{x}_1$.

f) Find the probability that $\boldsymbol{x}_2$ will take on a value between 0 and $\frac{3}{4}$, given that $x_1 = 0$.

3.32 Let $\boldsymbol{y}$ be a random variable that assumes the value 1 with probability p, and the value 0 with probability $(1 - p)$. Find $E[\boldsymbol{y}]$ and $V[\boldsymbol{y}]$.

3.33 a) Let $\boldsymbol{x}$ be a random variable representing the number of heads that appear when a fair coin is tossed four consecutive times. Let $\boldsymbol{y} = 2\boldsymbol{x} - 4$. Is $\boldsymbol{y}$ a random variable? If so, describe it.

b) Find $E[\boldsymbol{x}]$ and $V[\boldsymbol{x}]$.

c) Using the probability distribution for $\boldsymbol{y}$ described in part (a), find $E[\boldsymbol{y}]$ and $V[\boldsymbol{y}]$.

d) Using the results in part (b) and the expectation rules, find $E[\boldsymbol{y}]$ and $V[\boldsymbol{y}]$. Compare your answers here to your results in part (c).

Exercises

3.34 Prove that a random variable $\boldsymbol{y}$ and a constant k are always statistically independent.

3.35 Given: $\boldsymbol{x}$ is a discrete random variable with values chosen at random from the set $\{1, 2, 3, 4, 5, \text{and } 6\}$; $\boldsymbol{y}$ is a random variable with values at least as large as $\boldsymbol{x}$ and not greater than 6.

a) Determine the joint probability function of $\boldsymbol{x}$ and $\boldsymbol{y}$.

b) Find the conditional probabilities $P(\boldsymbol{y} | x = 3)$.

c) Find the expected value of $\boldsymbol{y}$.

3.36 Suppose that you are offered the opportunity to participate in an experiment in which a fair coin is tossed until the first tail appears. If a tail appears on the first toss you will be paid \$2 for participating. If the first tail appears on the second toss, you will be paid \$4 for participating. You will be paid \$8 if the first tail appears on the third toss, and \$16 if the first tail appears on the fourth toss; in other words, your payment is increased by a factor of 2 for each head that appears, the game ending with the appearance of a tail.

a) How much would you be willing to pay to participate in this game? (If you said less than \$2 you don't understand the game.)

b) Determine the expected value for this game. Are you willing to pay more or less than the expected value of the game? Explain why.

3.37 In Chapter 4 we will study a mass function called the *Poisson probability distribution.*

$$P(x) = \begin{cases} \dfrac{e^{-\lambda}\lambda^x}{x!}, & \text{for } x = 0, 1, 2, \ldots, \infty, \text{ and } \lambda = \text{constant } \geq 0, \\ 0, & \text{otherwise.} \end{cases}$$

a) Show that this function meets the two properties of all p.m.f.

$$\left(\textit{Hint: } \sum_{x=0}^{\infty} \lambda^x/x! = e^{\lambda}.\right)$$

b) Find $E[\boldsymbol{x}]$ and $E[\boldsymbol{x}^2]$, and use these results to find $V(\boldsymbol{x})$.

3.38 In Chapter 5 we will study a density function called the *exponential function*:

$$f(x) = \begin{cases} \lambda e^{-\lambda x}, & \text{for } x \geq 0 \text{ and } \lambda = \text{constant } \geq 0, \\ 0, & \text{otherwise.} \end{cases}$$

a) Prove that this function meets the two conditions of all density functions.

b) Find the mean and variance of $f(x)$. (*Hint:* see Appendix A.)

c) Graph the p.d.f. and c.d.f.

3.39 A bag contains 7 white and 3 black balls. Four players (A, B, C, and D) agree to take turns drawing balls from this bag (without replacement) until a white ball is drawn. If the first player (A) draws the white ball he receives $25. If he draws a black ball, he must put $50 into the "pot." The game continues in this manner until a white ball is drawn. Find the expected value to each of the four players, assuming the first $25 has been "donated" by someone else.

3.40 Prove that

$$V[ax + by] = a^2V[x] + b^2V[y] + 2abC[x, y].$$

3.41 a) Prove that Rule 5 is a special case of Formula (3.18).

b) Prove that Rule 6 is a special case of Formula (3.20).

3.42 Use Formula 3.16 to verify that, for the data in Table 3.11, $C[x, y] = -0.16$.

3.43 a) Prove that

$$C[x, y] = E[x \cdot y] - E[x]E[y].$$

b) For the following joint probability distribution, show that $C[x, y] = 0$, but that the variables x and y are not independent.

		y: −1	y: 0	y: 1
x	0	0	.25	0
	1	.25	0	.25
	2	0	.25	0

3.44 The kth moment about a is defined to be

$$E[(x - a)^k].$$

a) Find the 1st moment about $a = 0$, assuming

$$f(x) = \begin{cases} \frac{1}{2}(x + 1), & -1 \leq x \leq 1, \\ 0, & \text{otherwise.} \end{cases}$$

b) Find the 2nd moment about $a = \mu$ for the p.d.f. defined in part (a). What is another name for this 2nd moment?

3.45 Consider two stocks, x and y, which have the following joint distribution of their returns (perhaps computed over the last year).

		Return of x in %: 4	6	8
Return of y in %	4	0	0	.4
	6	0	.2	0
	8	.2	.2	0

a) Compute $E[\mathbf{x}]$, $E[\mathbf{y}]$, $V[\mathbf{x}]$ and $V[\mathbf{y}]$. Which stock appears to be more attractive?

b) Consider a portfolio that consists of one-half of stock x and one-half of stock y. Using the rules for linear combinations of variables, find $E[0.5\mathbf{x} + 0.5\mathbf{y}]$ and $V[0.5\mathbf{x} + 0.5\mathbf{y}]$. Compare the attractiveness of this combined portfolio with that of investing entirely in x or entirely in y.

c) Confirm your results in part (b) by actually tabulating the distribution of the new portfolio.

GLOSSARY

random variable: a well-defined rule for assigning a numerical value to all possible outcomes of an experiment.

probability distribution: a specification (usually by a graph, a table, or a function) of the probability associated with each value of a random variable.

p.m.f. or probability mass function: a discrete probability distribution.

p.d.f. or probability density function: a continuous probability distribution.

cumulative mass function (c.m.f.): a summation of the values of a p.m.f., starting at the lower limit and going up to and including a specified value of the random variable. $F(x) = P(\mathbf{x} \leq x)$.

cumulative distribution function (c.d.f.): the integral of a p.d.f. starting at the lower limit and ending at some specified value of the random variable. $F(x) = P(\mathbf{x} \leq x)$.

expected value of x or $E[x]$: the weighted mean or average of the random variable $\mathbf{x}$. $E[\mathbf{x}] = \sum xP(x)$ or $E[\mathbf{x}] = \int xf(x)dx$.

variance of x or $V[x]$: the expection of the squared deviations of a random variable about its mean μ. $V[\mathbf{x}] = E[(\mathbf{x} - \mu)^2] = E[\mathbf{x}^2] - \mu^2 = E[\mathbf{x}^2] - (E[\mathbf{x}])^2$.

standardization of x: the transformation $(\mathbf{x} - \mu)/\sigma = \mathbf{z}$.

***joint probability that $\mathbf{x} = x$ and $\mathbf{y} = y$:** $P(x, y)$.

***conditional probability that $\mathbf{x} = x$ given that $\mathbf{y} = y$:** $P_{x|y}(x|y)$.

***covariance of x and y:** $C[\mathbf{x}, \mathbf{y}] = E[\mathbf{x} \cdot \mathbf{y}] - E[\mathbf{x}]E[\mathbf{y}]$.

***expectation of a weighted sum of combined random variables:** $E[a\mathbf{x} + b\mathbf{y}] = aE[\mathbf{x}] + bE[\mathbf{y}]$.

***variance of a weighted sum of combined random variables:** $V[a\mathbf{x} + b\mathbf{y}] = a^2V[\mathbf{x}] + b^2V[\mathbf{y}] + 2abC[\mathbf{x}, \mathbf{y}]$.

***independence:** $E[\mathbf{x} \cdot \mathbf{y}] = E[\mathbf{x}] \cdot E[\mathbf{y}]$; this implies a special case of item 12, $C[\mathbf{x}, \mathbf{y}] = 0$, and item 14, $V[a\mathbf{x} + b\mathbf{y}] = a^2V[\mathbf{x}] + b^2V[\mathbf{y}]$.

* These items are described in sections that may be omitted without loss of continuity.

4

DISCRETE PROBABILITY DISTRIBUTIONS

"Lest men suspect your tale untrue, keep probability in view."

JOHN GAY

4.1 INTRODUCTION

While it is often useful to determine probabilities for a specific random variable, or for more than one random variable, many apparently different situations in statistical inference and decision-making involve the same type of probability functions. In such instances it is useful to apply the theory of probability functions described in Chapter 3, to obtain some *general* results about a probability distribution, such as its mean and variance, rather than re-deriving these characteristics in each special case which has different numbers. It would be quite discouraging to go through the process of formulating a new mass or density function and deriving its mean and variance every time we are concerned with a slightly different experiment. Fortunately, there are enough similarities between certain types, or families, of apparently unique experiments to make it possible to develop formulas representing the general characteristics of these experiments. We shall discuss three discrete distributions in this chapter, the *binomial*, the *hypergeometric*, and the *Poisson*. Three continuous distributions, the normal, the exponential, and the chi-square distributions, are discussed in Chapter 5. These distributions are among the most widely known and used distributions in statistics. Two other distributions that have applications across a wide variety of fields are discussed in Chapter 6 in the context of sampling theory.

4.2 THE BINOMIAL DISTRIBUTION

Many experiments share the common element that their outcomes can be classified into one of two events. For instance, the experiment "toss a coin" must result in either a *head* or a *tail*; the experiment "take an exam" can be considered to result in the outcomes *pass* or *fail*; a production process may turn out items which are either good or defective; and the stock market in general goes either up or down. In fact, it is often possible to describe the outcome of many of life's ventures in this fashion merely by distinguishing only two events, "success" and "failure." Experiments considered as involving just two possible outcomes play an important role in one of the most widely used discrete probability distributions, the *binomial distribution*.

Several generations of the Bernoulli family, Swiss mathematicians of the 1700's, usually receive credit as the originators of much of the early research on probability theory, especially that involving problems characterized by the binomial distribution. Therefore the Bernoulli name has now come to be associated with this class of experiment, and each repetition of an experiment involving only two outcomes (e.g., each toss of a coin) is called a *Bernoulli trial*. For the purposes of probability theory, interest centers not on a single Bernoulli trial but rather on a series of *independent*, *repeated* Bernoulli trials. That is, we are interested in more than one trial. The fact that these trials must be "independent" means that the results of any one trial cannot influence the results of any other trial. In addition, when a Bernoulli trial is "repeated," it means that the conditions under which each trial is held must be an exact replication of the conditions underlying all other trials, implying that the probability of the two possible outcomes cannot change from trial to trial. The status of items from a production process (such as "good" or "defective") could represent a Bernoulli process, as could the responses to a marketing survey (e.g., male/female, like/dislike, or purchase/not purchase).

Binomial Parameters

In a binomial distribution the probabilities of interest are those of receiving a certain number of successes, x, in n ***independent*** *trials, each trial having the* ***same probability****, p, of success.*

Note that the two assumptions underlying the binomial distribution, namely independent trials and a constant probability of success, are met by what we have called independent repeated Bernoulli trials; and so by definition the binomial distribution is appropriate in an experiment involving these trials. The binomial distribution is completely described by the values of n and p, which are referred to as the "parameters" of this distribution. The word parameter in this context means the same as it did in Chapter 1—it refers to a characteristic of a *population*. In the binomial distribution n is the parameter "number of trials," and p is the parameter "probability of success on a single trial." Given specific values of n and p, one can calculate the probability of any specified number of successes, as well as determine other characteristics of the binomial distribution, such as its mean and variance.

To illustrate a situation in which the binomial distribution applies, suppose a production process is producing solid-state components that are classified as either "good" or "defective." When the process is not working correctly, there is a *constant* probability, $p = 0.10$, that a component will be defective. In this situation the number of defectives (x) can range anywhere from zero up to the total number of objects examined (n). The binomial distribution can be used to determine the probability for any specified value of x and n. For example, we may want to ask, "What is the probability that a random sample of four will result in one defective?" or "What is the probability that there will be two or fewer defectives in a sample of 4?" The use of the word "random" in this context implies independence among the items sampled. We shall calculate these probabilities later in this section.

The Binomial Formula

We can calculate the probabilities in a binomial situation by using the probability rules developed in Chapter 2. Recall that at that time we presented the relationship

$$P(\text{Event}) = \begin{pmatrix}\text{Number of relevant} \\ \text{occurrences}\end{pmatrix} P(\text{One occurrence})$$

In a binomial problem we are interested in calculating the probability of exactly x successes in n repeated Bernoulli trials, each having the same probability of success, p. That is, we want x successes and $(n - x)$ failures. To calculate such probabilities, it is necessary to find the probability of *one* occurrence of this type, and then multiply this probability by the number of such occurrences. Since it doesn't make any difference which occurrence we investigate first, let's (arbitrarily) take the one in which the x successes occur first, followed by the $(n - x)$ failures. If we let S = success and F = failure, then this particular ordering can be represented as follows:

$$\underbrace{\text{S S} \cdots \text{S}}_{x \text{ successes}}\ \underbrace{\text{F F} \cdots \text{F}}_{\substack{n - x \\ \text{failures}}}$$

To determine the joint probability of this particular sequence of successes and failures, recall that all trials are assumed to be independent. This means that the joint probability of these n events, which we write as

$$P(\mathrm{S} \cap \mathrm{S} \cap \cdots \cap \mathrm{S} \cap \mathrm{F} \cap \mathrm{F} \cap \mathrm{F} \cap \cdots \cap \mathrm{F}),$$

equals the *product* of the probabilities of each event,

$$P(\mathrm{S} \cap \mathrm{S} \cap \cdots \cap \mathrm{S} \cap \mathrm{F} \cap \mathrm{F} \cap \cdots \cap \mathrm{F}) = \underbrace{P(\mathrm{S})P(\mathrm{S}) \cdots P(\mathrm{S})}_{x \text{ successes}} \underbrace{P(\mathrm{F})P(\mathrm{F}) \cdots P(\mathrm{F})}_{(n-x) \text{ failures}}.$$

It is customary to simplify the above notation by letting $P(\mathrm{S}) = p$ and $P(\mathrm{F}) = q$ where q must equal $(1 - p)$ because the two events, success and failure, are *mutually exclusive* and *exhaustive*. The desired joint probability is thus

$$P(\mathrm{S} \cap \mathrm{S} \cap \cdots \cap \mathrm{S} \cap \mathrm{F} \cap \mathrm{F} \cap \cdots \cap \mathrm{F}) = \underbrace{p\,p \cdots p}_{x \text{ successes}} \underbrace{q\,q \cdots q}_{(n-x) \text{ failures}}.$$

Since there are x p-values and $(n - x)$ q-values, the above joint probability can be written as

$$P(\mathrm{S} \cap \mathrm{S} \cap \cdots \cap \mathrm{S} \cap \mathrm{F} \cap \mathrm{F} \cap \cdots \cap \mathrm{F}) = p^x q^{n-x}.$$

Let us now relax the assumption that the x successes occur in a specified order. Recall that the joint probability of a series of independent events does not depend on the *order* in which they are arranged; for example

$$P(E_1 \cap E_2) = P(E_2 \cap E_1).$$

Hence, $p^x q^{n-x}$ represents the probability not only of our one arrangement, but of *any* possible arrangement of x successes and $(n - x)$ failures. The question at this point is how many different occurrences there are of x successes and $(n - x)$ failures. The answer is the number of combinations of n objects, taken x at a time, or ${}_nC_x = \binom{n}{x}$.* Therefore, multiplying the probability of each occurrence, $p^x q^{n-x}$, by the total number of such occurrences, ${}_nC_x$, gives the probability of x successes in n trials. This product is the binomial probability mass function of an experiment involving n repeated independent Bernoulli trials, where the probability of a success on each trial is a constant, p.

Binomial distribution:

$$P(x \text{ successes in } n \text{ trials}) = \begin{cases} \binom{n}{x} p^x q^{n-x} & \text{for } \begin{cases} x = 0, 1, 2, \ldots, n \\ n = 1, 2, \ldots, \end{cases} \\ 0 & \text{otherwise.} \end{cases} \qquad (4.1)$$

* If you read Section 2.8 in Chapter 2, you may recognize that the answer to this question is given by the number of permutations of n objects of which there are x alike (success) and $(n - x)$ of the other kind alike. The total number of such permutations is given by Formula 2.18, as $n!/(x!(n - x)!)$. But another way of writing $n!/(x!(n - x)!)$ is ${}_nC_x = \binom{n}{x}$; hence the total number of occurrences equals the number of *combinations of* n objects, taken x at a time.

Table 4.1 Binomial Probabilities for $n = 4$, $p = 0.10$

x	$\binom{n}{x}$	$p^x q^{n-x}$	$P(x) = \binom{n}{x} p^x q^{n-x}$
0	$\binom{4}{0} = 1$	$(0.10)^0(0.90)^4 = 0.6561$	$P(0) = 0.6561$
1	$\binom{4}{1} = 4$	$(0.10)^1(0.90)^3 = 0.0729$	$P(1) = 0.2916$
2	$\binom{4}{2} = 6$	$(0.10)^2(0.90)^2 = 0.0081$	$P(2) = 0.0486$
3	$\binom{4}{3} = 4$	$(0.10)^3(0.90)^1 = 0.0009$	$P(3) = 0.0036$
4	$\binom{4}{4} = 1$	$(0.10)^4(0.90)^0 = 0.0001$	$P(4) = 0.0001$
			Sum $= 1.0000$

To illustrate the use of Formula (4.1), let's answer several of the questions posed earlier in this section about the number of defectives in a sample of $n = 4$ from the output of a production process in which the probability of a defective is $p = 0.10$. Define a success in this case to be the identification of a defective item, so that $\mathbf{x}$ = the number of defectives.* The probability that exactly one component will be defective ($\mathbf{x} = 1$) out of a sample of four ($n = 4$), when $p = 0.10$ (and hence $q = 0.90$), is

$$\begin{aligned} P(\mathbf{x} = 1) &= \binom{n}{x} p^x q^{n-x} \\ &= \binom{4}{1}(0.10)^1(0.90)^3 \\ &= \frac{4!}{1!3!}(0.10)(0.90)(0.90)(0.90) = 0.2916. \end{aligned}$$

The probability that there are exactly *two* defectives ($\mathbf{x} = 2$) in a sample of $n = 4$ can be calculated to be

$$P(\mathbf{x} = 2) = \binom{4}{2}(0.10)^2(0.90)^2 = 0.0486.$$

Indeed, the probability of any number of defectives from 0 to 4 may be determined in the same way. Table 4.1 presents these values.

* We shall show in a moment that as long as one is consistent throughout the problem, it makes no difference whether $\mathbf{x}$ = the number of defectives, or $\mathbf{x}$ = the number of good items.

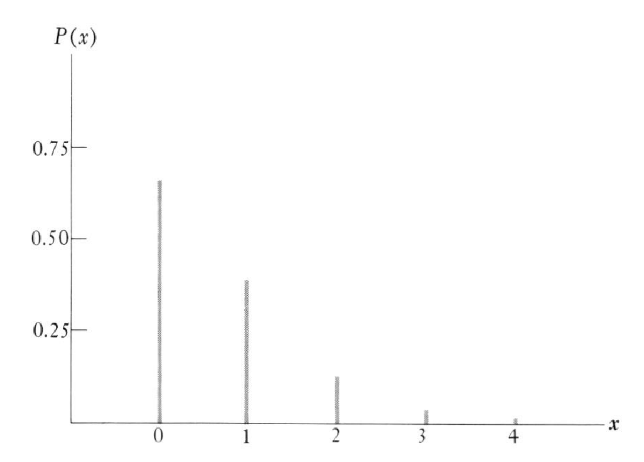

Fig. 4.1 Binomial distribution for $n = 4$, $p = 0.10$.

From Table 4.1 it is easy to answer our second question posed earlier, concerning the probability of two or fewer defectives when $n = 4$. This value is:

$$\begin{aligned} P(\mathbf{x} \leq 2) &= P(0) + P(1) + P(2) \\ &= 0.6561 + 0.2916 + 0.0486 \\ &= 0.9963. \end{aligned}$$

A graph of all the values for $n = 4$ appears in Fig. 4.1.

The calculations in Table 4.1 are not very complex, but the difficulty could readily become overwhelming if n becomes much larger. Consider a more realistic situation where $n = 20$ items are sampled, and the question posed earlier is now raised again—what is the probability of finding exactly two defectives, when $p = 0.10$? Using Formula (4.1), we have:

$$\begin{aligned} P(\mathbf{x} = 2) &= \binom{n}{x} p^x q^{n-x} = \binom{20}{2} (0.10)^2 (0.90)^{18} \\ &= \frac{20 \times 19 \times (18!)}{2 \times 1 \times (18!)} (0.10)(0.10)\underbrace{(0.90) \cdots (0.90)}_{18 \text{ terms}}. \end{aligned}$$

Fortunately it is not necessary to carry out such calculations, for tables of binomial probabilities for various values of n and p are readily available. Table I in Appendix B gives the probability for a number of the more commonly referred to values of n, for values of p from 0.01 to 0.99. The probability of $P(\mathbf{x} = 2)$ for the above problem can be seen to equal 0.2852, by referring to the set of probabilities headed $n = 20$ and finding the values corresponding to $\mathbf{x} = 2$ on the lefthand margin and $p = 0.10$ across the top. (The decimal points have all been omitted from Table I.) Note that the entire probability distribution for $n = 20$ and $p = 0.10$ is given in Table I, although probability values smaller than 0.00005 are rounded to 0.0000 in this table.

Another important fact to notice about Table I is its symmetry. Values of p less than 0.51 are found across the top of each set of numbers, while values of p greater than 0.50 are found across the bottom of each set (in which case the values of $\boldsymbol{x}$ are read from the righthand margin). This symmetry in Table I results from the fact that the probability of x successes when p is the probability of success exactly equals the probability of $(n - x)$ failures when the probability of a failure is $(1 - p)$. Thus, in our defective-components example, if $n =$ the number of *good* items in the sample (rather than the number of defectives), and hence the probability of a success equals 0.90 (rather than 0.10), then the probability of 18 good items out of 20 is read from exactly the same point in the table as was the probability of two defectives, $P(\boldsymbol{x} = 18) = 0.2852$.

Note from Table 4.1 that the binomial distribution for $n = 4$ and $p = 0.10$ satisfies the two properties described in Chapter 3 for all probability mass functions—namely, that

$$1 \geq P(x) \geq 0 \qquad \text{for all } x,$$

and

$$\sum_{\text{All } x} P(x) = 1.^*$$

Mean and Variance of the Binomial

Since the binomial distribution is characterized by the value of the two parameters, n and p, one might anticipate that the summary measures of the mean and standard deviation also can be determined in terms of n and p. For example, it should appear reasonable that the mean number of successes in any given experiment must equal the number of trials (n) times the probability of successes on each trial (p). If, for example, the probability that a process

* To show that the function

$$P(x) = \binom{n}{x} p^x q^{n-x}$$

meets the two properties of probability described in Chapter 3 for all values of n and p, first note that $\binom{n}{x}$, p^x, and q^{n-x} must all be positive; hence their product will always be positive. To prove that $\sum P(x) = 1$, recall the binomial theorem (refer to any good book on algebra), which allows us to state the following:

$$\textit{Binomial Theorem:} \qquad \sum_{x=0}^{n} \binom{n}{x} p^x q^{n-x} = (p + q)^n.$$

By substituting $(1 - p)$ for q in the righthand side of the above relationship, we can prove that the second axiom necessary for defining a probability mass function holds true:

$$\sum_{x=0}^{n} \binom{n}{x} p^x q^{n-x} = (p + q)^n = [p + (1 - p)]^n = 1.$$

produces a defective item is $p = 0.10$, then the mean number of defectives in 20 trials is $20(0.10) = 2$; the mean number in 50 trials is $50(0.10) = 5$; and the mean number in 100 trials is $100(0.10) = 10$. Thus, the *mean* number of successes in n trials is np:

*Binomial mean:** $\quad \mu = np.$ (4.2)

The variance (and standard deviation) of the binomial distribution are also functions of the parameters n and p, as follows.

Binomial variance:† $\quad \sigma^2 = npq;$

Binomial standard deviation: $\quad \sigma = \sqrt{npq}.$ (4.3)

The variance in 20 Bernoulli trials of a process producing defectives with probability $p = 0.10$ is thus

$$npq = 20(0.10)(0.90) = 1.80;$$

the standard deviation is

$$\sqrt{npq} = \sqrt{1.80} = 1.34.$$

---

* To more formally derive this mean, assume that the variable $\boldsymbol{x}$ assumes a value of 1 for each success and a value of 0 for each failure, in an experiment consisting of n independent repeated Bernoulli trials. The number of successes in n trials can now be determined merely by summing these n values of $\boldsymbol{x}$. The mean number of successes is given by taking the expectation of this sum or $E[\boldsymbol{x} + \boldsymbol{x} + \cdots + \boldsymbol{x}]$. By the rules on expectations presented in Section 3.7, $E[\boldsymbol{x} + \boldsymbol{x} + \cdots + \boldsymbol{x}]$ can be simplified as follows:

$$E[\underbrace{\boldsymbol{x} + \boldsymbol{x} + \cdots + \boldsymbol{x}}_{n \text{ values}}] = \underbrace{E[\boldsymbol{x}] + E[\boldsymbol{x}] + \cdots + E[\boldsymbol{x}]}_{n \text{ values}} = nE[\boldsymbol{x}].$$

The expected value $E[\boldsymbol{x}]$ equals the sum of the product of each possible value that $\boldsymbol{x}$ can assume ($\boldsymbol{x} = 1$, $\boldsymbol{x} = 0$) times the probability of each of these values, $P(1) = p$ and $P(0) = q$. Hence, using Formula (3.5) in Chapter 3, we find:

$$E[\boldsymbol{x}] = \sum_{x=0}^{1} xP(x) = (0)q + (1)p = p.$$

Thus, p is the expected value on one trial, and since there are n trials, $E[\boldsymbol{x}] = np$.

† To derive the variance of the binomial distribution, first note that:

$$E[\boldsymbol{x}^2] = \sum_{x=0}^{1} x^2P(x) = (0^2)q + (1^2)p = p.$$

We can now use the mean-square − square-mean relationship as follows:

$$V[\underbrace{\boldsymbol{x} + \boldsymbol{x} + \cdots + \boldsymbol{x}}_{x \text{ terms}}] = \underbrace{V[\boldsymbol{x}] + V[\boldsymbol{x}] + \cdots + V[\boldsymbol{x}]}_{n \text{ terms}} = n\,V[\boldsymbol{x}];$$

$$n\,V[\boldsymbol{x}] = n\{E[\boldsymbol{x}^2] - (E[\boldsymbol{x}])^2\} = n\{p - p^2\} = np(1 - p) = npq.$$

Note from the table below that the intervals $\mu \pm \sigma$ and $\mu \pm 2\sigma$ for this problem contain slightly more of the total probability (obtained from Table I) than the rule of thumb given in Chapter 1 (68 and 95 percent) would indicate.

Interval	Percent of Probability
$\mu \pm \sigma = 2.00 \pm 1.34 = 0.66$ to 3.34	75%
$\mu \pm 2\sigma = 2.00 \pm 2(1.34) = 0$ to 4.68	97%

The Shape of the Binomial Distribution

The shape of the binomial distribution depends on its parameters, n and p. It will be useful to consider three different combinations of n and p:

1. When n is small and p is large (that is, $p > \frac{1}{2}$);
2. When n is small and p is also small (that is, $p < \frac{1}{2}$);
3. When $p = \frac{1}{2}$ and/or n is large.

1. *When n is small and $p > \frac{1}{2}$.* To illustrate this case, the distribution for $n = 5$, $p = 0.80$, is shown in Fig. 4.2. As Fig. 4.2 illustrates, the binomial distribution when n is small and $p > \frac{1}{2}$ will be skewed to the left, or *negatively* skewed.
2. *When n is small and $p < \frac{1}{2}$.* A typical illustration of this case is the binomial distribution for $n = 4$ and $p = 0.10$, shown in Table 4.1 and pictured in Fig. 4.1. This distribution is skewed to the right, or *positively* skewed, as all binomial distributions will be when n is small and $p < \frac{1}{2}$.

Fig. 4.2 Binomial distribution for $n = 5$, $p = 0.80$.

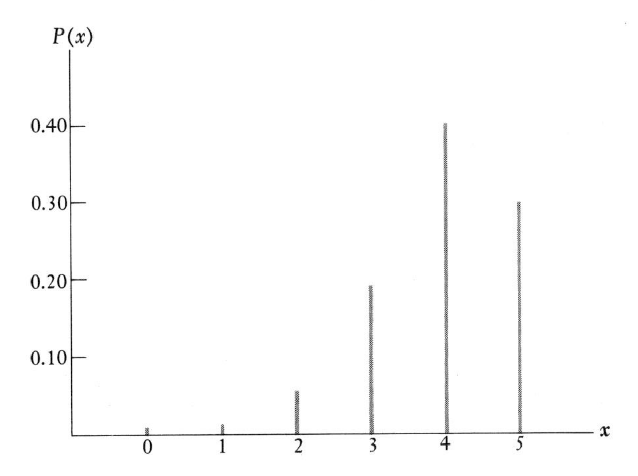

3. *When $p = \frac{1}{2}$ and/or n is large.* When $p = 0.50$, the binomial distribution will always be a symmetrical distribution with its mean equal to its median. This can be seen in a simple example by considering $n = 2$, and $p = 0.50$. The three values of $\mathbf{x}$ in this case are

$$P(\mathbf{x} = 0) = 0.25, \qquad P(\mathbf{x} = 1) = 0.50, \qquad \text{and} \qquad P(\mathbf{x} = 2) = 0.25.$$

Note that in this example, median = mean = mode = 1 (the mode equals the mean and median when n is even and $p = \frac{1}{2}$). An important fact about the binomial is that even when $p \neq \frac{1}{2}$, the shape of the distribution takes on a more and more symmetrical appearance the larger the value of n. Figure 4.3 illustrates this fact. Note in Fig. 4.3(a) and (b) that even though n is as small as 20, the distributions of $p = 0.20$ and $p = 0.40$ are fairly symmetrical in appearance. For $n = 100$ and $p = 0.30$, shown in Fig. 4.3(c), the distribution has a very symmetrical "bell" shape.

Fig. 4.3 The binomial distribution: $P(x) = \binom{n}{x} p^x q^{n-x}$.

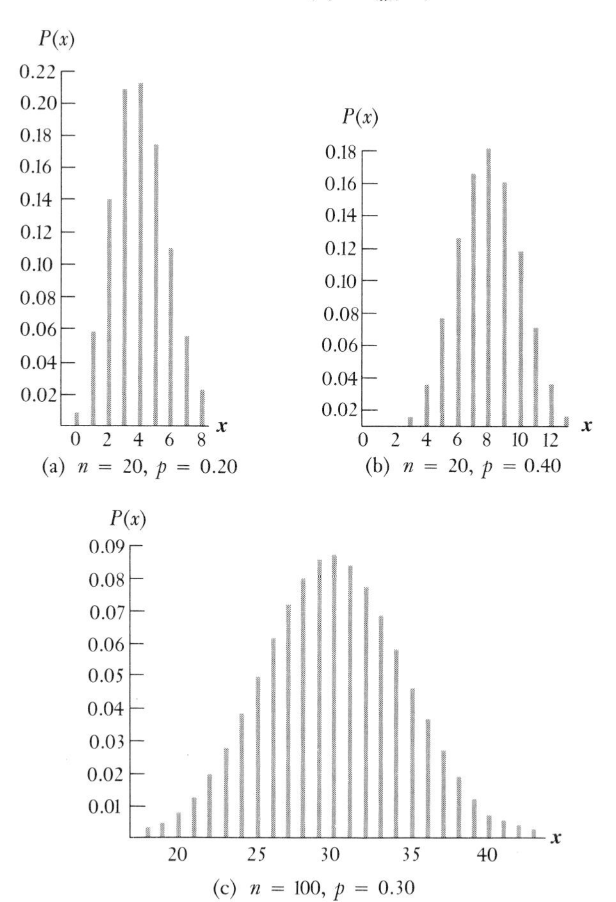

Applications of the Binomial Distribution

Example 1. The binomial distribution is useful as the basis for investigating a number of special problems in statistical inference and decision-making. To take a simple example, suppose a U.S. senator is concerned about how to vote on several bills involving unemployment and inflation. One group of concerned citizens claims that 90% of the population considers inflation a more serious problem; another group claims that the population is equally divided, 50% considering inflation more serious and 50% considering unemployment more serious. Since it is impossible or at least impractical for the senator to learn the preferences of the entire population, suppose a small sample is collected by telephoning (at random) 15 households and asking their priorities. Assume that in ten of these households someone is home, and willing to express a preference, with the following results:

Inflation the More Serious	Inflation/Unemployment Equally Serious	Total
7	3	10

We can determine which group's claim seems more reasonable on the basis of this sample by using the binomial distribution. The view of the first group is that $p = 0.90$, as opposed to $p = 0.50$ for the second group. Note that if the first group is correct, the mean number of households indicating inflation as more serious (in a sample of 10) would be $\mu = np = 10(0.90) = 9$. If the second group is correct, the mean would be $\mu = np = 10(0.50) = 5$. The observed result, seven indicating inflation, falls right in the middle of these two means. However, let's calculate the probability of exactly seven successes ($x = 7$ indicating inflation priority) in a sample of $n = 10$ households, using both $p = 0.90$ and $p = 0.50$.

$$\text{If } p = 0.90\text{: } P(\mathbf{x} = 7) = \binom{10}{7}(0.90)^7(0.10)^3 = 0.0574;$$

$$\text{If } p = 0.50\text{: } P(\mathbf{x} = 7) = \binom{10}{7}(0.50)^7(0.50)^3 = 0.1172.$$

We thus see that the probability of finding seven households answering inflation is twice as likely if $p = 0.50$ as when $p = 0.90$, indicating that this sample tends to support the equally serious group.

We must quickly point out that our conclusion in the above analysis is very sensitive. Suppose the sample had yielded just one more inflation preference, so that 8 out of 10 indicated inflation is a higher priority. In this case $P(\mathbf{x} = 8) = 0.1937$ when $p = 0.90$, and $P(\mathbf{x} = 8) = 0.0439$ when $p = 0.50$. This means that the inflation claim is more reasonable. To avoid such sensitive results in statistical analysis, larger samples are usually taken. Furthermore, probabilities are usually not calculated for a *single* outcome, such as $x = 7$, but for a *set* of outcomes, such as $P(\mathbf{x} \leq 7)$ or $P(\mathbf{x} \geq 7)$. For example, the reasonableness of the hypothesis that $p = 0.90$ when $x = 7$ is usually determined by asking how likely it is to have 7 or *fewer* households when $p = 0.90$. This probability is $P(\mathbf{x} \leq 7) = 0.0702$. Similarly, the reasonableness of the unemployment claim is determined by asking how likely it is to find 7 *or more* households when $p = 0.50$. This probability is $P(\mathbf{x} \geq 7) = 0.1719$. Thus, it is over twice as likely to get seven or more households when $p = 0.50$ as it is to get seven or less viewers when $p = 0.90$, indicating again that we should support the claim that unemployment and inflation are equally serious problems.

The following procedure is generally followed in testing hypotheses using a discrete distribution such as the binomial.

> *If the expected value exceeds the observed value, calculate the probability that* $\mathbf{x}$ *is less than or equal to the observed value* [*e.g., if* 9 *is expected and* 7 *is observed, calculate* $P(\mathbf{x} \leq 7)$]. *If the observed value exceeds the expected value, calculate*
>
> $$P(\mathbf{x} \geq \textit{observed value}).$$

Example 2. As another example of the use of the binomial in decision-making, let's assume that the production process described earlier in this chapter is malfunctioning, and will require either a minor or a major adjustment. If the defective rate is ten percent ($p = 0.10$), then only a minor adjustment is necessary; if the number of defectives has jumped to 25.0 percent ($p = 0.25$), then a major adjustment is necessary. The problem at this point is how to decide, on the basis of a random sample of size $n = 20$, whether the process requires a minor or a major adjustment. This decision is not without risks, however, for we assume it is costly to make the wrong decision—i.e., to make a minor adjustment to a process needing a major adjustment, or to make a major adjustment to process needing only minor adjustments.

Suppose, for the moment, that *four* defectives is established as the decision point concerning repairs—if there are four or more defectives, then a major adjustment is made; with three or less defectives, minor repairs are made. This decision rule will lead to an *incorrect* decision if $p = 0.10$ and the number of defectives is $\mathbf{x} \geq 4$ (since then a major adjustment will be made). Another type of incorrect decision will be made when $p = 0.25$ and the number of defectives is $\mathbf{x} \leq 3$ (since then a minor adjustment will be made). To determine how likely it is to make these errors, we need to know the probability of $\mathbf{x}$ defectives in a sample of $n = 20$ when $p = 0.10$ and when $p = 0.25$. Table 4.2 provides the appropriate values.

Table 4.2 Binomial Probabilities for Production Process Example

Number of Defectives (x)		Decision Rule	If $p = 0.10$: $P(x) = \binom{20}{x}(0.10)^x(0.90)^{20-x}$	If $p = 0.25$: $P(x) = \binom{20}{x}(0.25)^x(0.75)^{20-x}$
0		Minor repairs	0.1216	0.0032 (0.2251 for x = 0–3)
1		Minor repairs	0.2702	0.0211
2		Minor repairs	0.2852	0.0669
3	Decision	Minor repairs	0.1901	0.1339
4	point	Major repairs	0.0898 (0.1331 for x = 4–20)	0.1897
5		Major repairs	0.0319	0.2023
6		Major repairs	0.0089	0.1686
7		Major repairs	0.0020	0.1124
8		Major repairs	0.0004	0.0609
9		Major repairs	0.0001	0.0271
10		Major repairs	0.0000	0.0099
11		Major repairs	0.0000	0.0030
12		Major repairs	0.0000	0.0008
13		Major repairs	0.0000	0.0002
14–20		Major repairs	0.0000	0.0000
Sum			1.0000	1.0000

The probability of making the two types of errors is shown in Table 4.2. First, if $p = 0.10$, the probability of receiving four or more defectives is $P(\mathbf{x} \geq 4) = 0.1331$; this means that, under this decision rule, the probability of *incorrectly* making a major adjustment (when only a minor one is needed) is 0.1331. The probability of incorrectly making a minor adjustment when $p = 0.25$ is $P(\mathbf{x} \leq 3) = 0.2251$.

These probabilities of making an incorrect decision depend on the fact that $\mathbf{x} = 4$ was the decision point selected. For example, if the decision point between a major and a minor adjustment is set at $\mathbf{x} = 3$ rather than $\mathbf{x} = 4$, then the appropriate probabilities can again be determined from Table 4.2.

Decision point of $x = 3$:

If $p = 0.10$, *the probability of incorrectly making a major adjustment is* $P(\mathbf{x} \geq 3) = 0.3232$.

If $p = 0.25$, *the probability of incorrectly making a minor adjustment is* $P(\mathbf{x} \leq 2) = 0.0912$.

For a given value of n, one of these types of errors (e.g., incorrectly making a major or a minor adjustment) can be made smaller only if the other is allowed to become larger. Just what decision rule to use, such as $\boldsymbol{x} = 3$ or $\boldsymbol{x} = 4$, depends largely on the costs associated with making these errors. We shall examine this subject in more detail in Chapters 8 and 9.

Problems

4.1 Six products are selected at random from a very large group of products of a certain kind. If 40% of the products are defective, what is the probability that no more than four of the six selected will be defective?

4.2 It is established that a fighter-bomber will hit its target $\frac{3}{4}$ of the time. Suppose military tacticians want a certain crucial target area destroyed. If five planes are assigned to strike this same target area one time each, what is the probability that it will be hit two or more times? Assume independence among strikes by different planes.

4.3 Suppose that a certain business submits bids for construction projects, and that there are always four other companies bidding on the same projects. Assume that the long-run chances of this firm being the low bidder are one out of five. What is the probability that this business will be awarded the contract as the low bidder in exactly two of the next four projects? Explain what it means to assume independent trials in this case.

4.4 If $\frac{2}{3}$ of the students on a certain campus are lower-division students (freshman and sophomores), what is the probability that five students randomly selected will include exactly three lower-division students?

4.5 a) Sketch the binomial p.m.f. and the binomial c.m.f. for $n = 8$ and $p = 0.40$ (use Table I).

b) What is the probability $P(\boldsymbol{x} = 7)$ for the above parameters? What is the probability $P(\boldsymbol{x} \geq 7)$?

c) Find the mean and variance of this distribution.

4.6 A bakery in Los Angeles was fined for distributing consistently underweight loaves of bread. Suppose that this bakery claims that the authorities just took a "bad sample," and that in reality 50% of their loaves are either exactly the correct weight or slightly overweight.

a) If, in a random sample of 100 loaves, 69% are underweight, would you support the bakery's claim? Explain.

b) Suppose y is the number of loaves that are underweight in a sample of 100. How large does y have to be before you start disbelieving the bakery's claim?

4.7 A national survey indicated that the average realtor sells one residential dwelling for every twenty homes shown to prospective buyers.

a) If a realtor shows 100 homes, what is the expected number of sales? What is the variance of sales if $n = 100$? Assume sales are independent of one another.

b) What is the probability that the realtor in (a) will sell one or fewer homes in 100 showings?

c) If a local realtor sells only one home after 100 showings, would you conclude this person is performing "below average"? Explain.

4.8 Eight helicopters were part of the April 25, 1980, ill–fated attempt by the U.S. to rescue the Iranian hostages. At least six helicopters had to be operational for the mission to continue.

a) Suppose that the probability of failure for each helicopter during the mission is 0.10, and failures are independent of one another. What is the probability that three or more of the helicopters will fail?

b) In the April 25th mission three helicopters failed, causing the rescue to be terminated. Would you attribute this termination to "bad luck"? Explain.

c) Sketch the p.m.f. for this binomial experiment. Show μ and σ on your graph.

4.9 The Big Rivers Electric Cooperative at Henderson, Kentucky, had a new peak demand of 983 megawatts last July 16th. Suppose Big Rivers sells power to the Tennessee Valley Authority (TVA) whenever demand is less than 900 megawatts. Big Rivers estimates that demand over the next several years will exceed 900 megawatts on only 5% of the 50 days between July 1st and August 19th.

a) What is the average number of days Big Rivers can expect to sell power to TVA, using the binomial distribution? What is the most and the fewest days they can typically expect to sell?

b) Do the assumptions of the binomial appear reasonable in this situation?

4.10 In a recent ad in the *Wall Street Journal,* Volvo stated their car is ". . . made well enough to have a life expectancy of 17.9 years in Sweden." If we accept this statement as true, could we further state that there is a 50–50 chance that a given Volvo's life expectancy will exceed 17.9 years in Sweden? Explain.

4.11 A study by one automobile manufacturer indicated that one out of every four new automobiles required engine work under their new-car guarantee, with an average cost of $50 per repair. How much would you expect repairs to cost for sales of 100 new cars under this guarantee? For 100 cars, what is the maximum you would expect?

4.12 Eisner Supermarket recently ran a promotional campaign in which customers can win a prize by scratching off a card the one covering, out of six, that has a prize pictured behind it. One store manager is suspicious of a flaw in the game because he observed 100 randomly selected customers, and there were 24 winners. Should the manager be worried?

4.3 BINOMIAL PROPORTIONS

All the binomial problems studied thus far involved the variable x, where x represents the *number* of successes in n Bernoulli trials. In this section, we will show that any one of these problems could have been solved by using the variable x/n, where x/n represents the *proportion* of successes in n Bernoulli trials. For all practical purposes, *it makes no difference in solving a problem whether x is used, or x/n is used.*

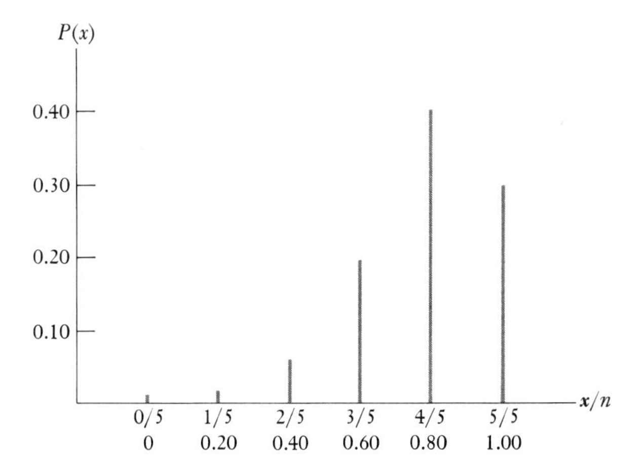

Fig. 4.4 Binomial distribution for $n = 5$, $p = 0.80$, using x/n as the variable.

To illustrate the statement above, suppose we redo Fig. 4.2, which shows the binomial distribution for $n = 5$, $p = 0.80$. Instead of using x as our horizontal axis, this time (in Fig. 4.4) we use x/n as the horizontal axis.

The important aspect of Fig. 4.4 is that it looks identical to Fig. 4.2 except for the values on the horizontal axis. In other words, dividing each value of $\mathbf{x}$ by n doesn't change any of the probability values. The mean and variance of $\mathbf{x}/n$ will be different from $\mu = np$ and $\sigma^2 = npq$, but we can easily derive the new values using the algebra of expectations.* These new values are:

$$\textit{Mean of } \mathbf{x}/n = E[\mathbf{x}/n] = p$$

$$\textit{Variance of } \mathbf{x}/n = V[\mathbf{x}/n] = \frac{pq}{n}$$

* The proofs are:

$$E[\mathbf{x}/n] = \frac{1}{n} E[\mathbf{x}] \qquad \text{by Rule 3 of Section 3.7}$$

$$= \frac{1}{n}(np) \qquad \text{since } E[\mathbf{x}] = np$$

$$= p.$$

$$V[\mathbf{x}/n] = \frac{1}{n^2} V[\mathbf{x}] \qquad \text{by Rule 4 of Section 3.7}$$

$$= \frac{1}{n^2}(npq) \qquad \text{since } V[\mathbf{x}] = npq$$

$$= \frac{pq}{n}.$$

Since Table I is presented in terms of the *number* of successes (x), rather than the *proportion* of successes (x/n), a problem involving x/n is usually solved by transforming it to the comparable problem involving x. Consider the following problem, which illustrates this point.

A newspaper has suggested that a certain politician will (perhaps should) win only 40% of the votes in an upcoming election. To test this assertion, the politician takes a random sample of 50 voters. Should the politican doubt the newspaper's "usually reliable sources" if 54% of these 50 voters say they will vote for the politician?*

Remember from Section 4.2 that in problems of this nature we generally don't calculate the probability of one specific value (like 54%), but rather the probability that the proportion is 54% *or larger*, given that $n = 50$. That is, we want to calculate

$$P\left(\frac{x}{n} \geq 0.54\right).$$

Since $n = 50$, $x/n \geq 0.54$ is equivalent to $x \geq 50\,(0.54)$, or $x \geq 27$. Thus

$$P\left(\frac{x}{n} \geq 0.54\right) = P(x \geq 27).$$

This problem is easily solved by using Table I for $n = 50$, $p = 0.40$, and $x \geq 27$. Be sure to recognize that p in this problem is 0.40 (and not 0.54), because 40% is the assertion we are considering. From Table I,

$$P(x \geq 27) = 0.0154 + \cdots + 0.0001 = 0.0314.$$

Thus, in random samples of this nature, only a little over 3% of the time would one expect 27 or more "successes" in 50 trials. The politician would appear to have good reason to doubt the newspaper's assertion and to believe that the true population proportion is larger than 0.40.

Would the newspaper's assertion still be in doubt if, say, 48% of 50 people said they intend to vote for the politican? In this case

$$P\left(\frac{x}{n} \geq 0.48\right) = P(x \geq 24) = 0.0978,$$

which means there is almost a 10% chance that this result (48%) could occur if the newspaper is correct. Perhaps 0.10 is too high a probability for the politician to risk criticizing the newspaper for its 40% assertion. In Chapter 8 we will study the process of trying to decide how low a probability has to be before one can reject the assertion on which this probability was calculated. *One quite arbitrary rule often used is that the probability must be less than 0.05 in order to reject the assertion.*

In our politician example, the result could have been in the *opposite* direction—perhaps only 22% of the sample say they intend to vote for this

* Ignore the fact that some people may not vote as they say they will.

candidate. The appropriate calculation now is for the probability that $\mathbf{x}/n$ will be 0.22 *or less*. From Table I, $n = 50$ and $p = 0.40$,

$$P\left(\frac{\mathbf{x}}{n} \leq 0.22\right) = P(\mathbf{x} \leq 11) = 0.0056.$$

There is thus less than a 1% chance of having 11 or fewer "yes" votes when the true population proportion of "yes" voters is 40%. Since this probability is quite low (much less than the 5% rule mentioned above), a sample of 22% would again give the politician cause to doubt the newspaper's assertion and to worry that the true percentage of "yes" voters in the population may even be less than 40%.

Define. Bernoulli trial, independent trials, binomial parameters, binomial proportion.

4.4 HYPERGEOMETRIC DISTRIBUTION (OPTIONAL)

The binomial distribution assumes that the value of p is constant from trial to trial. This assumption generally implies that sampling is either with replacement, or from an infinitely large population. *When sampling is without replacement from a finite population the hypergeometric distribution is appropriate.*

Let's assume that we are sampling from a finite population which contains k different types of objects. Furthermore, we assume the population has n_1 objects of the first kind, n_2 of the second type, and so forth, with n_k of the last (kth) kind. The total number of objects in the population is thus

$$n = n_1 + n_2 + \cdots + n_k.$$

Now suppose that we want to determine the probability of drawing a sample which includes exactly x_1 objects of the first kind, x_2 of the second kind, . . . , and x_k of the kth kind. We use the hypergeometric formula.*

Hypergeometric distribution:

$P(x_1 \text{ out of } n_1, x_2 \text{ out of } n_2, \ldots, x_k \text{ out of } n_k)$

$$= \frac{\binom{n_1}{x_1}\binom{n_2}{x_2}\cdots\binom{n_k}{x_k}}{\binom{n_1 + n_2 + \cdots + n_k}{x_1 + x_2 + \cdots + x_k}}. \tag{4.4}$$

* Readers who studied Section 2.11 will recognize this formula as identical to (2.20).

To illustrate the use of the hypergeometric distribution, imagine a production process in which electrical components are produced in lots of fifty. If this process is working correctly, there will be no defective items among the 50 produced. Suppose, however, that the process has, in fact, been malfunctioning and exactly five components in a particular lot are defective. What is the probability that exactly one of these defectives will appear in a sample of four randomly selected components? Note that we are sampling without replacement from a finite population ($n = 50$). Hence the hypergeometric distribution is appropriate. Now, what is the probability that a sample of 4 contains exactly 1 defective and 3 good components? The probability that the sample contains $x_1 = 1$ of the $n_1 = 5$ defective components, and $x_2 = 3$ out of the $n_2 = 45$ good components, is given by Formula (4.4).

$$P(1 \text{ defective and } 4 \text{ good}) = \frac{\binom{5}{1}\binom{45}{3}}{\binom{50}{4}} = \frac{\dfrac{5!}{4!1!}\dfrac{45!}{42!3!}}{\dfrac{50!}{46!4!}} = 0.308$$

Similarly, one might want to calculate the probability of receiving *two or less* defectives in a sample of 4:

$$P(2 \text{ or less defectives}) = P(0 \text{ def.}) + P(1 \text{ def.}) + P(2 \text{ def.})$$

$$= \frac{\binom{5}{0}\binom{45}{4}}{\binom{50}{4}} + \frac{\binom{5}{1}\binom{45}{3}}{\binom{50}{4}} + \frac{\binom{5}{2}\binom{45}{2}}{\binom{50}{4}}$$

$$= 0.647 + 0.308 + 0.043 = 0.998.$$

Problems

4.13 A Chicago-based company proudly advertises that, on the average, three out of every five MBA students it hires are still working for the company five years later.

a) Assume that this company hired 10 MBAs last year. Sketch the p.m.f. for the number of these MBAs who will still be working for the company in five years, assuming the binomial distribution holds. Indicate the mean and standard deviation on your sketch.

b) If only three of these MBA students are still working for the company in five years, would you conclude that this company is no longer retaining 60 percent of the MBAs?

4.14 A research study reported in the *Los Angeles Times* concluded that "... children of alcoholics are more likely to have alcohol problems than children of nonalcoholics ..." Consider a study of 50 U.S. men aged 23–45, all of whom had been adopted by a nonalcoholic couple before they were six weeks old. One or both of the natural parents of these men had been alcoholics; 18 percent of the 50 men became alcoholics. If the rate of alcoholism among U.S. men born of nonalcoholic parents is 5 percent, is the conclusion above justified? What precautions are necessary in reaching such a conclusion?

4.15 A study of the population of workers available in the Detroit area for certain jobs in the automotive industry indicated that 20 percent are black, while 80 percent are nonblack. If one company hires 100 new workers for these jobs and 12 percent are black, would you conclude that the company is discriminating against blacks?

4.16 How does the hypergeometric distribution differ from the binomial distribution? Under what circumstance is each one appropriate?

4.17 Suppose that a university committee of six is to be selected from a list consisting of five men and five women.

a) If the committee is to be chosen by random selection, what is the probability there will be three women and three men?

b) What is the probability that a majority of the committee will be women?

4.18 Suppose that in a production run of ten units, three are defective. A sample of three units is to be randomly drawn from the ten. What is the probability of receiving *at least* one defective unit if the samples are drawn (a) with replacement? (b) without replacement?

4.19 Go back and work (or rework) Problems 2.46, 2.49, and 2.50 from Chapter 2.

4.20 Brand B Aspirin Company decides to survey 10 out of 100 doctors who were randomly selected from a large population. If 50 percent of these 100 doctors actually prefer Brand B, what is the probability that the results of the survey will show that "9 out of 10 doctors surveyed prefer Brand B"?

4.21 The Department of Transportation (DOT) has announced that it will award a grant for the study of traffic accidents to each of four different universities. Twenty universities have applied for a grant. If Stanford, Indiana University, and the University of North Carolina are among the 20 applicants, what is the probability that these three universities will all be winners, assuming the selection is made randomly?

4.22 Redo Problem 4.20 using a binomial approximation. In this case, let the probability that a doctor prefers Brand B be $p = 0.50$, and calculate $P(\mathbf{x} = 9)$ assuming $n = 10$.

4.23 Redo Problem 4.21 using the binomial distribution.

4.24 Assume that 2073 of the 10,000 customers of the Easy Charge Company from Problem 2.31 received a cash advance during September. You take a random sample of 100 of their 10,000 customers.

a) Write down the hypergeometric expression representing the probability that 21% of the sample will be customers who received a cash advance.

b) Find the binomial probability that best approximates the hypergeometric probability represented in part (a).

4.25 A special congressional investigation committee of eight people was selected randomly from a Senate composed of 55 Democrats and 45 Republicans.

a) Write down the hypergeometric expression for the probability the committee consists of 7 Democrats and 1 Republican.

b) Approximate the probability in part (a) using the binomial distribution.

c) Sketch the p.m.f. for $x = 0, 1, 2, \ldots, 8$ where x is the number of Republicans. Should the Democrats complain that they have been unfairly represented on the committee if five of the committee members are Republicans? Explain.

4.26 A continual problem with the commercial airlines is estimating the number of seats to reserve for "nonsmokers." If each person requesting nonsmoking cannot be accommodated, then FAA regulations state that the entire plane may be limited to nonsmoking.

a) Suppose that a fully booked plane has 10 seats remaining, five in nonsmoking and five in smoking sections. What is the probability that there will not be enough nonsmoking seats if $p = 0.40$ is the probability a person insists on a nonsmoking seat, and requests are independent?

b) How many seats would you reserve for nonsmokers if you want to honor every nonsmoking request on 99 percent of all flights? Assume $p = 0.40$, $n = 300$, and independence.

4.27 The dean at the Harvard Business School has been asked to select 3 students for a certain committee. Twenty first-year students volunteer to be on this committee, as do thirty second-year students. If the dean selects the three students randomly from the total group of 50 volunteers, what is the probability that one first year and two second year students will be selected?

4.5 THE POISSON DISTRIBUTION

Another important discrete distribution, the Poisson distribution, has recently found fairly wide application, especially in the area of operations research. This distribution was named for its originator, the French mathematician S. D. Poisson (1781–1840), who described its use in a paper in 1837. Its rather morbid first applications indicated that the Poisson distribution quite accurately described the probability of deaths in the Prussian army resulting from the kick of a horse, as well as the number of suicides among women and children. More recent (and useful) approximations include many applications involving arrivals at a service facility, or requests for service at that facility, as well as the rate at which this service is provided. A few of the many successful approximations to which the Poisson distribution has been applied include problems involving the number of arrivals or requests for services per unit time at tollbooths on an expressway, at checkout counters in a supermarket, at teller windows in a bank, at runways in an airport, by maintenance men in a repair shop, or by machines needing repair.

In examples of the above nature, the Poisson distribution can be used to determine the probability of x occurrences (arrivals or service completions) per unit time if four basic assumptions are met. First, it must be possible to divide the time interval being used into a large number of small subintervals in such a manner that the probability of an occurrence in each of these subintervals is very small. Second, the probability of an occurrence in *each* of the subintervals must remain constant throughout the time period being considered. Third, the probability of two or more occurrences in each subinterval must be small enough to be ignored. And finally, an occurrence (or nonoccurrence) in one interval must not affect the occurrence (or nonoccurrence) in any other subinterval—i.e., the occurrences must be independent.

Consider arrivals at a bank per hour, and suppose we can divide a given hour into intervals of one second, where the probability that a customer arrives during each second is very small and remains constant throughout the one-hour period. Furthermore, assume that only one customer can arrive in a given second (e.g., the door is large enough to admit only one person), and that the number of arrivals in a given time period is independent of the number of arrivals in any other time period (e.g., customers don't turn away because of long lines). Under these circumstances, the number of arrivals in the one-hour period meets the four basic assumptions, and thus it is not unreasonable to assume the Poisson distribution holds. These assumptions and how they fit the bank example are summarized below.

Assumption	Bank Example
1. Possible to divide time interval of interest into many small subintervals.	1. Can divide the hour into subintervals of one second each.
2. Probability of an occurrence remains constant throught time interval.	2. The hour is one for which we have no reason to suspect an uneven flow of customers.
3. Probability of two or more occurrences in a subinterval is small enough to be ignored.	3. Impossible for two or more people to enter the bank simultaneously (i.e., in the same second).
4. Independence of occurrences.	4. Arrivals at the bank are not influenced by the length of the lines.

Of these four assumptions, numbers (1) and (3) are general enough to apply to almost any setting involving arrivals over time.* The assumptions that occurrences are constant over time and independent, however, are much less likely to be met in potential applications of the Poisson distribution. Nevertheless, the Poisson does seem to apply in a surprisingly large variety of different situations.

* The Poisson distribution can also be applied to problems involving the number of occurrences of a random variable for a given unit of *area*, such as the number of typographical errors on a page, the number of white blood cells in a blood suspension, or the number of imperfections in a surface of wood, metal, or paint.

Examples of the Poisson distribution such as those given above are concerned with the probability of x occurrences (arrivals or service completions) *per unit of time*. The only parameter necessary to characterize a population described by the Poisson distribution is the *mean rate* at which events take place in each unit of time. We shall use the Greek letter lambda (λ) for this parameter. Lambda can thus be defined as the mean rate of occurrence for any convenient unit of time, such as one minute, ten minutes, an hour, a day, or even a year. A value of $\lambda = 2.3$, for example, could indicate that there are, on the average, 2.3 requests for service in a particular bank every 10 minutes. For practical applications, the mean rate at which events occur must be determined empirically. That is, λ must be known in advance, such as on the basis of a previous study of the situation. Once λ is known, the frequency function for the Poisson distribution can be used to determine the probability that exactly x occurrences, or events, take place in the specified interval of time. The Poisson distribution is defined as follows:*

Poisson distribution:

$$P\begin{pmatrix}x \text{ occurrences in a} \\ \text{given time unit}\end{pmatrix} = \begin{cases} \dfrac{e^{-\lambda}\lambda^x}{x!} & \text{for } \begin{cases} x = 0, 1, 2, \ldots, \\ \lambda > 0 \end{cases} \\ 0 & \text{otherwise} \end{cases} \tag{4.5}$$

To illustrate the use of Formula (4.5), suppose the bank in our discussion above knows from past experience that, between 10 A.M. and 11 A.M. each day, the mean arrival rate of customer is $\lambda = 60$ customers (per hour). Since we've assumed that arrivals are constant during a given time interval, this rate is equivalent to an arrival rate of $\lambda = 1$ customer per minute. Now, suppose the bank wants to determine the probability that exactly two customers ($\boldsymbol{x} = 2$) will arrive in a given one-minute time interval between 10 and 11 A.M. Substi-

* The reader interested in proving that the Poisson distribution meets the two properties defining a probability distribution should note that $e^{-\lambda}$, λ^x, and $x!$ must always be greater than or equal to zero; hence $P(x) \geq 0$. Proving that $P(x)$ meets the second property requires the assumption that the following relationship holds true:

$$e^{\lambda} = \sum_{x=0}^{\infty} \frac{\lambda^x}{x!}.$$

This expansion of e^{λ} is known as a Maclaurin series, and its proof can be found in any good intermediate book on calculus. We use it below to prove that $P(x) = 1$:

$$\sum_{x=0}^{\infty} P(x) = \sum_{x=0}^{\infty} \frac{e^{-\lambda}\lambda^x}{x!} = e^{-\lambda} \sum_{x=0}^{\infty} \frac{\lambda^x}{x!} = e^{-\lambda}(e^{\lambda}) = 1.$$

tuting $\lambda = 1$ and $\boldsymbol{x} = 2$ into Formula (4.5) yields:

$$P(2 \text{ arrivals}) = \frac{e^{-1}1^2}{2!} = \frac{1}{2}e^{-1}.$$

Since $e^{-1} = 0.3679$,

$$P(2 \text{ arrivals}) = 0.3679/2 = 0.1839.$$

Similarly, they might want to calculate P(2 or less arrivals):

$$\begin{aligned} P(2 \text{ or less arrivals}) &= P(0) + P(1) + P(2) \\ &= \frac{e^{-1}1^0}{0!} + \frac{e^{-1}1^1}{1!} + \frac{e^{-1}1^2}{2!} \\ &= 0.3679 + 0.3679 + 0.1839 \\ &= 0.9197. \end{aligned}$$

As was the case for the binomial, Poisson probabilities have been extensively tabled, so that one can avoid the task by using Formula (4.5) to calculate such values. Table II in Appendix B gives these probabilities for selected values of λ from $\lambda = 0.01$ to $\lambda = 20.0$. The probability values for the above example are shown in Table II, under the heading $\lambda = 1.0$. These values are graphed in part (a) of Fig. 4.5; parts (b) and (c) of that figure show the probability mass function for $\lambda = 3.8$ and $\lambda = 10.0$.

Characteristics of the Poisson Distribution

As can be seen in the graphs in Fig. 4.5, the Poisson is a discrete mass function which is always skewed to the right (since $\boldsymbol{x}$ cannot be lower than zero, but may be any positive integer). When λ is not too close to zero, however, the shape of the Poisson distribution will often have a very symmetrical appearance, as shown in Fig. 4.5(c).

As we indicated above, λ is the only parameter of the Poisson distribution. Because of this fact, both the mean and the variance of the Poisson must be a function of λ. We know already that the mean number of occurrences in a Poisson distribution is λ. Hence,

Mean of Poisson: $\mu = \lambda$.

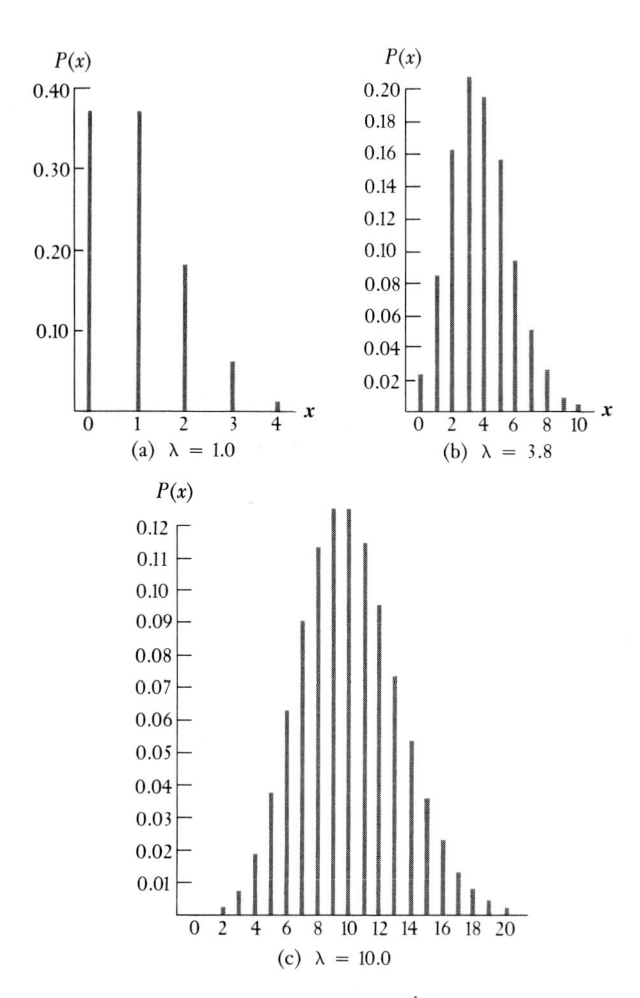

Fig. 4.5 The Poisson distribution: $P(x) = \dfrac{e^{-\lambda}\lambda^x}{x!}$.

Not only is the mean of a Poisson distribution denoted by λ, but the variance of the Poisson distribution also equals λ. That is,

> *Variance of Poisson:* $\sigma^2 = \lambda.$

Thus, if a value of $\lambda = 1.0$ indicates that customers are arriving at a bank at an average rate of 1 customer per minute, then the *variance* of the number of arrivals in each minute is also 1.0.

As another illustration of the Poisson distribution, suppose it has been suggested to the manager of a large supermarket that arrivals at the checkout counters might follow a Poisson distribution during certain periods of time. The manager decides to investigate this possibility by first checking the reasonableness of the four basic assumptions listed earlier, especially assumptions number 2 (constant probability) and 4 (independence). Let's assume the manager has decided to use 2 to 4 P.M. Monday–Friday, as the time interval, and all four assumptions appear reasonable for this time interval.

The next step this manager might take is to collect some observations concerning the number of arrivals within the two-hour time from 2 to 4 P.M. Rather than count the number of customers for many different *two-hour* time periods, the manager has recorded arrivals in each of 100 randomly selected *one-minute* periods from 2 to 4 P.M. This procedure is reasonable since arrivals are assumed to be constant over the two-hour period; hence it makes no difference, theoretically at least, what time unit is used. The number of customers arriving per minute (x) in this study ranged from 0 to 10, as shown in Table 4.3. For example, the first row indicates that only once did zero customers arrive in the one-minute periods; on eight occasions exactly one customer arrived; on 19 occasions exactly two customers arrived, and so forth.

Recall that, for the Poisson distribution, both μ and σ^2 are equal. Hence, in checking to see if the data in Table 4.3 do follow a Poisson distribution, the manager might first wish to see if $\mu = \sigma^2$. Column 4 in this table gives the

Table 4.3 Supermarket Example

(1) Arrivals (x_i)	(2) Observed Frequency f_i	(3) $\frac{f_i}{N}$	(4) $x_i \frac{f_i}{N}$	(5) x^2	(6) $x^2 \frac{f_i}{N}$
0	1	0.01	0.00	0	0
1	8	0.08	0.08	1	.08
2	19	0.19	0.38	4	.76
3	23	0.23	0.69	9	2.07
4	17	0.17	0.68	16	2.72
5	15	0.15	0.75	25	3.75
6	8	0.08	0.48	36	2.88
7	3	0.03	0.21	49	1.47
8	3	0.03	0.24	64	1.92
9	2	0.02	0.18	81	1.62
10	1	0.01	0.10	100	1.00
	100	1.00	3.79		18.27

appropriate values for calculating μ (via Formula (1.8) in Chapter 1). We see that $\mu = 3.79$.

The variance of the data in Table 4.3 can be calculated using the mean-square–square-mean relationship, where the appropriate value of $\sum x^2(f_i/N)$ is given in Column 6 of Table 4.3:

$$\sigma^2 = \sum x^2 \frac{f_i}{N} - \mu^2 = 18.27 - (3.79)^2 = 3.91.$$

We see, for this population, that μ and σ^2 are very nearly equal in value (3.79 vs. 3.91). Theoretically they are supposed to be *exactly* equal, but in just 100 time periods we would not expect the observed value of either μ or σ^2 to precisely equal the theoretical value.

The closeness of μ and σ^2 in this case is encouraging enough for the store manager to make the most important test, namely, to see if the observed relative frequencies correspond to the theoretical relative frequencies from the Poisson. To make this comparison, suppose that the manager assumes that $\lambda = 3.8$ (note that it is not clear from the data whether $\lambda = 3.79$ or 3.91, or some other value). The choice of $\lambda = 3.8$ is somewhat arbitrary. Now, if the number of arrivals in these 100 one-minute periods does, in fact, follow a Poisson distribution with $\lambda = 3.8$, we would expect the observed frequencies in Column 3 of Table 4.3 to closely correspond to the probabilities of the Poisson distribution $P(\mathbf{x}) = e^{-3.8}(3.8)^x/x!$. These values, shown in Table II, Appendix B, under the heading $\lambda = 3.8$, are reproduced in Table 4.4.

We see in Table 4.4 that the observed relative frequencies correspond quite well to the Poisson values for $\lambda = 3.8$. A good exercise for the reader at this point would be to verify, for the probabilities shown in the last column of Table 4.4, that $\mu = \sigma^2 = 3.8$.

Since the probabilities in the last column in Table 4.4 completely describe the Poisson distribution for $\lambda = 3.8$, we can determine from these values other probabilities which might be of interest. For instance, the probability of *two or fewer* customers arriving in a given one-minute interval is

$$P(\mathbf{x} \leq 2) = \sum_{x=0}^{2} \frac{e^{-3.8}(3.8)^x}{x!} = 0.2689.$$

Similarly, the probability of *eight or more* customers in a one-minute period equals one minus the cumulative value of seven or less customers:

$$P(\mathbf{x} \geq 8) = 1 - P(\mathbf{x} \leq 7) = 1 - \sum_{x=0}^{7} \frac{e^{-3.8}(3.8)^x}{x!} = 0.04.$$

Probability values such as these might be of use to the supermarket manager in making a number of decisions, especially those concerned with the number of checkout counters to have open during the 2 to 4 P.M. period.

Table 4.4 Observed vs. Theoretical Poisson Probabilities

Arrivals (x)	Observed Relative Frequency	Poisson Values $P(x) = \frac{e^{-3.8}(3.8)^x}{x!}$
0	0.010	0.0224
1	0.080	0.0850
2	0.190	0.1615
3	0.230	0.2046
4	0.170	0.1944
5	0.150	0.1477
6	0.080	0.0936
7	0.030	0.0508
8	0.030	0.0241
9	0.020	0.0102
10	0.010	0.0039
11	0.000	0.0013
12	0.000	0.0004
13	0.000	0.0001
Sum	1.000	1.0000

4.6 POISSON APPROXIMATION TO THE BINOMIAL*

Recall that we mentioned that binomial probabilities may be tedious and time-consuming to compute. Fortunately, under certain circumstances, binomial probabilities can be approximated quite accurately by the Poisson distribution. Whenever n is large, and p is not close to $\frac{1}{2}$, the Poisson provides a good approximation.

In order for one probability distribution to approximate another, the two distributions must have similar characteristics. In particular, we would want the means of the two distributions to be equal, and the two variances to be equal as well, or at least almost equal. Thus, if the Poisson distribution, which has a mean of $\mu = \lambda$, is to approximate a binomial distribution, whose mean is $\mu = np$, then these two values should be set equal. That is,

$$\text{(Mean of Poisson)} \quad \lambda = np \quad \text{(Mean of binomial).}$$

* This section and the next may be omitted without loss of continuity.

What about the variances? Recall that the variance of the binomial is npq, and the variance of the Poisson is λ. Thus, we would like $\lambda = npq$. But this is impossible since we have already specified that $\lambda = np$. Hence, if $\lambda = np$, the variance of the Poisson will never equal the variance of the binomial since npq never exactly equals np except in the trivial cases when $n = 0$ or $p = 0$. If, however, p is close to zero, then $1 - p = q$ will be close to 1, and when q is close to 1, np will be correspondingly close to npq. That is, if q approximately equals $1 (q \cong 1)$, then

$$V[\text{Binomial}] = npq \cong np(1) = np = \lambda = V[\text{Poisson}].$$

Just equating means and variances does not ensure that one distribution will be a good approximation to another one; for this case, however, it can also be shown that, for a constant value of λ (that is, $\lambda = np =$ a constant), the Poisson is, in fact, the limit of the binomial as n goes to infinity. That is,

$$\lim_{n \to \infty} \binom{n}{x} p^x q^{n-x} = \frac{e^{-\lambda} \lambda^x}{x!} \qquad \text{(for } \lambda = np \text{ a constant).}$$

In other words, the Poisson distribution can be thought of as a binomial distribution in which the number of trials, n, is infinitely large. This relationship guarantees that the Poisson will yield a better and better approximation to the binomial as n becomes larger and larger ($n \to \infty$). Note that as n becomes larger, the value of p in the relationship $\lambda = np$ must become smaller and smaller, since we assumed λ is a constant. This verifies what we said above—that when p is small (that is, $q \cong 1$), then the Poisson will give a good approximation to the binomial. Just how small p needs to be depends on how "good" an approximation one desires. One rule of thumb requires $p \leq 0.10$ and $n \geq 30$ for a good approximation; but the example below shows that reasonably good approximations can be achieved under less stringent conditions.

Suppose we are interested in approximating the probabilities in Table 4.2, which gives the probability of receiving between 0 and 13 defectives in a sample of 20 items, with $p = 0.10$. Letting $\lambda = np$ implies that $\lambda = 20(0.10) = 2.0$. The binomial and Poisson probabilities for these parameters are given in Table 4.5. Note how closely they agree, despite the fact that n is less than 30.

We have said that in order for the Poisson to provide a good approximation to the binomial, the value of p must be "small," and q close to 1.0. But you should recall that, because of the symmetry in the binomial distribution, it makes no difference which event is labeled "success" and which is labeled "failure." Thus, if p happens to be close to 1.0 in a particular problem, rather than close to 0.0, all one needs to do is interchange terms, letting p represent the probability of the event formerly associated with q; or exactly the same effect can be accomplished by setting $\lambda = nq$, instead of setting $\lambda = np$. Thus, our condition for the Poisson to be a good approximation to the binomial is: *the value of p is not close to* $\frac{1}{2}$. As n becomes larger, the approximation gets better.

Table 4.5 Approximating the Probabilities in Table 4.2

x	$P(x) = \binom{20}{x}(0.10)^x(0.90)^{20-x}$	$P(x) = \dfrac{e^{-2.0}(2.0)^x}{x!}$
0	0.1216	0.1353
1	0.2702	0.2707
2	0.2852	0.2707
3	0.1901	0.1804
4	0.0898	0.0902
5	0.0319	0.0361
6	0.0089	0.0120
7	0.0020	0.0034
8	0.0004	0.0009
9	0.0001	0.0002

*4.7 DERIVATION OF THE POISSON MEAN AND VARIANCE (OPTIONAL)

In this section we shall prove that the mean and the variance of the Poisson distribution both equal λ. First, by the definition of an expected value,

$$E[\mathbf{x}] = \sum_{x=0}^{\infty} x \frac{\lambda^x e^{-\lambda}}{x!}.$$

Note that the term corresponding to $x = 0$ has a value of zero, so we can rewrite the above summation as $\sum_{x=1}^{\infty} x(\lambda^x e^{-\lambda}/x!)$. Now, rewrite $x!$ as $x(x - 1)!$, and cancel the x in the numerator with the x in the denominator, leaving

$$E[\mathbf{x}] = \sum_{x=1}^{\infty} \frac{\lambda^x e^{-\lambda}}{(x - 1)!}.$$

Since λ and $e^{-\lambda}$ are constants, we can take them outside the summation sign:

$$E[\mathbf{x}] = \lambda e^{-\lambda} \sum_{x=1}^{\infty} \frac{\lambda^{x-1}}{(x - 1)!}.$$

If we let $y = x - 1$, then the term

$$\sum_{x=1}^{\infty} \frac{\lambda^{x-1}}{(x - 1)!} = \sum_{y=0}^{\infty} \frac{\lambda^y}{y!}$$

can be seen to correspond to a Maclaurin series, where

$$e^{\lambda} = \sum_{y=0}^{\infty} \frac{\lambda^y}{y!}.$$

Hence,

$$E[\boldsymbol{x}] = \lambda e^{-\lambda}(e^{\lambda}) = \lambda.$$

Deriving $V[\boldsymbol{x}]$ requires much the same manipulation as described above. In following the proof given below, the reader should take special note of the limits of summation, which change when the term being summed equals zero. Also, it will be observed in step 2 that $(x - 1 + 1)$ is substituted for x. Such a substitution is for mathematical convenience only: $(x - 1 + 1)$ obviously equals x. We first compute $E[\boldsymbol{x}^2]$, and then derive $V[\boldsymbol{x}]$:

$$E[\boldsymbol{x}^2] = \sum_{x=0}^{\infty} (x^2)\frac{\lambda^x e^{-\lambda}}{x!} = \sum_{x=1}^{\infty} x\frac{\lambda^x e^{-\lambda}}{(x-1)!}$$

$$= \sum_{x=1}^{\infty} \frac{(x-1+1)\lambda^x e^{-\lambda}}{(x-1)!} = \sum_{x=1}^{\infty} \frac{(x-1)\lambda^x e^{-\lambda}}{(x-1)!} + \sum_{x=1}^{\infty} \frac{\lambda^x e^{-\lambda}}{(x-1)!}.$$

Factoring $\lambda^2 e^{-\lambda}$ from the first term, and $\lambda e^{-\lambda}$ from the second, we obtain:

$$E[\boldsymbol{x}^2] = \lambda^2 e^{-\lambda} \sum_{x=2}^{\infty} \frac{\lambda^{x-2}}{(x-2)!} + \lambda e^{-\lambda} \sum_{x=1}^{\infty} \frac{\lambda^{x-1}}{(x-1)!}.$$

By the Maclaurin expansion, we now have

$$E[\boldsymbol{x}^2] = \lambda^2 e^{-\lambda}(e^{\lambda}) + \lambda e^{-\lambda}(e^{\lambda}) = \lambda^2 + \lambda.$$

It follows from the above results that

$$V[\boldsymbol{x}] = E[\boldsymbol{x}^2] - (E[\boldsymbol{x}])^2 = (\lambda^2 + \lambda) - \lambda^2,$$

or

$$\sigma^2 = \lambda.$$

The Poisson distribution thus has the unusual property of having its mean equal to its variance.

Problems

4.28 Use Formula (4.5) to determine the probabilities associated with the Poisson distribution for $\lambda = 2.0$. Sketch both the mass and the cumulative distributions. Check your answers by using Table II of Appendix B. What are the mean and the variance of this distribution?

4.29 A barbershop has on the average ten customers between 8:00 and 9:00 each morning that it is open. Customers arrive according to the Poisson distribution.

a) What is the probability that the barbershop will have exactly ten customers between these hours on a given morning?

b) What is the probability that the barbershop will have more than twelve customers?

c) What is the probability that the barbershop will have fewer than six customers?

4.30 In an airport, an average of 8.5 pieces of baggage per minute are handled, following a Poisson distribution. Find the probability of 10 pieces of baggage being handled in a selected minute of time.

4.31 Suppose in a textile manufacturing process, an average of 2 flaws per 10 running yards of material have appeared. What is the probability that a given ten-yard segment will have 0 or 1 defects, if the number of flaws follows a Poisson distribution?

4.32 Suppose on the average 2.3 telephone calls per minute are made through a central switchboard, according to a Poisson distribution. What is the probability that during a given minute 2 or more calls will be made?

4.33 A State Farm Insurance agent processes an average of 10 claims each working day, following a Poisson distribution.

a) Sketch the Poisson p.m.f., indicating on your sketch both the mean and standard deviation.

b) Sketch the c.m.f. for $\lambda = 10$.

4.34 Use Table I to determine the binomial probabilities for $n = 10$ and $p = 0.20$, and then use Table II to find the Poisson approximation to these probabilities. Comment on how good the Poisson approximation is in this case. Graph both distributions on the same sheet of paper.

4.35 If 10 cards are drawn with replacement from a deck of cards, what is the probability of receiving five aces? (Let $P(\text{Ace}) = p = \frac{4}{52} \cong 0.08$.) What is the Poisson approximation to this probability? Compare the variance of the binomial distribution for this problem with the variance of the Poisson approximation.

4.36 In a production process the probability of a good item is binomially distributed with $p = 0.90$.

a) Find the probability of exactly 9 good items out of 10.

b) Approximate the answer above by using the Poisson distribution.

c) Repeat parts (a) and (b), assuming that you want to determine the probability of 8 or more successes.

4.37 a) Find what percent of the Poisson distribution lies within $\mu \pm 1\sigma$ and $\mu \pm 2\sigma$ when $\lambda = 1$. How do these values compare with the rule of thumb given in Chapter 1?

b) Repeat part (a) for $\lambda = 4$ and $\lambda = 9$, and comment on why you think the percents are getting closer to the rule of thumb.

4.38 In litigation involving drugs or exposure to radiation (such as the thalidomide or Love Canal cases), the question arises as to reasonable expectations about how often abnormalities can be expected to occur. For example, in a study of 300 pregnant women taking a drug thought to be harmless, 24 gave birth to a premature baby (less than 5 lbs. 4 oz., or born more than three weeks early). If the national average for premature babies is 6 percent, would you suspect this drug of being harmful? Explain, using a Poisson approximation to the binomial.

Exercises

4.39 If a manufacturing process is working correctly, only 10 percent of the items produced will be defective. You take a sample of 7 items.

a) What is the probability that 3 or more of these will be defective?

b) If 3 items were defective would you, as a quality-control inspector, take any action? Why or why not?

4.40 Suppose that a baseball player has averaged four official times at bat in each of 150 games and hit 50 home runs. Based on his past performance, what is the probability that in game number 151 this player will hit at least one home run in his first four official times at bat? Are the assumptions of the binomial distribution reasonable for this application? Explain.

4.41 The expected value of x in a series of repeated Bernoulli trials is defined as follows:

$$E[\mathbf{x}] = \sum_{x=0}^{n} x({}_nC_x)p^x q^{n-x}.$$

a) Use the above relationship to prove that the mean of the binomial distribution equals np.

b) In the same manner, prove that the variance of the binomial distribution equals npq.

4.42 A classical example of the Poisson distribution resulted from a study of the number of deaths from horse kicks in the Prussian Army from 1875 to 1894. The data for this example are:

Deaths per Corps (per year)	Observed Frequency
0	144
1	91
2	32
3	11
4	2
5 and over	0
Total	280

a) Fit a Poisson distribution to these data. (*Hint*: Note that there were 196 deaths from the 280 observations; hence the mean death rate was $\frac{196}{280} = 0.700$.) How good does the Poisson approximation appear to be?

b) Do the assumptions of the Poisson distribution seem reasonable in this problem? Explain.

4.43 Using the Poisson distribution in the third column of Table 4.4, show by calculation that its mean and variance are both equal to 3.8.

4.44 a) Assume that the customers of *Shark Loans* (Exercise 1.43) arrive at the loan office at a rate of $\lambda = 12$ customers per day. Graph the probability mass function for all values of $\mathbf{x}$ between 5 and 18.

b) Suppose the people at Shark have noticed that approximately 50 people walk by their office each day. If we assume [from part (a)] that there is a constant probability that each of these people will enter, the binomial distribution can be used to describe the arrival rate.

1) What is the appropriate value of p if we assume 12 of the 50 people enter? What will μ and σ^2 be for this binomial distribution?

2) Superimpose on your graph from part (a) the probability mass function of the binomial for values of x between 4 and 18.

4.45 Out of 10 salemen, seven (call them group A) make sales on 20% of their calls. The other three (call them group B) make sales on 50% of their calls.

a) What is the average percentage of sales to calls for all ten salesmen?

b) Suppose the sales manager selects three of these salesmen at random to assign to a new territory. What is the probability that exactly two of them will be from group A?

c) Suppose we follow one of the salesmen in group A on his next five calls, each of which is considered independent of any of the others. What is the probability that he makes more than one but less than four sales in those five calls?

4.46 The following bridge hand almost broke up the 1955 world championship tournament because it occurred twice in the space of a few hours.

Spades:	A, K, 9, 5	Hearts:	Q, 8, 4
Diamonds:	J, 7, 3	Clubs:	10, 6, 2

a) Write down an expression representing the probability of the occurrence of this hand, assuming the cards are dealt randomly.

b) Determine a nonprobabilistic explanation as to why such a bridge hand might have occurred twice in the 1955 world championship.

4.47 A Certified Public Accountant has been asked by a client, a department store, to assist in determining the effects on customer service of eliminating a clerk in one department. The probability of a customer's arriving for service is the same at all moments in time regardless of what has happened in previous moments. If the Certified Public Account analyzes this queueing (waiting-line) problem mathematically, the frequency distribution generally used would be the

a) normal   b) binomial

c) hypergeometric   d) Poisson

4.48 A discrete distribution is generated by picking, randomly, an integer between 0 and 9. If each number has a probability of being selected of $\frac{1}{10}$, find $E[x]$ and $V[x]$.

4.49 The Pascal (or negative binomial) distribution is appropriate when one is interested in determining the probability that n Bernoulli trials will be required to produce r successes. This probability is

$$P(n) = \frac{(n-1)!}{(r-1)!(n-r)!} p^r(1-p)^{n-r} \qquad \text{for } n \geq r.$$

For example, suppose an advertising agency is trying to evaluate the effects of a television commercial advertising swimming pool chlorine. A caller is assigned to randomly contact residential dwellings in Phoenix, where 40 percent of the homes have pools.

a) What is the probability that it will take 10 or less calls to find exactly 5 homes with pools (assume all calls are answered politely).

b) What is the average number of calls required to have five successes if $E[n] = r/p$.

c) What is the variance of the number of calls in part (a) if $V[\mathbf{n}] = r(1 - p)/p^2$.

GLOSSARY

binomial distribution: discrete probability mass function (p.m.f.) involving independent, repeated, Bernoulli trials. (A Bernoulli trial is a replication of an experiment in which one of two mutually exclusive and exhaustive outcomes must take place.)

binomial formula: $P(x) = {}_nC_x p^x q^{n-x}$.

binomial parameters: $\mu = E[\mathbf{x}] = np$;

$\sigma^2 = V[\mathbf{x}] = npq$.

expectations for a binomial proportion: $E[\mathbf{x}/n] = p$, $V[\mathbf{x}/n] = pq/n$.

hypergeometric distribution: discrete probability mass function (p.m.f.) involving sampling without replacement from a finite population. Involved with calculating probability such as x_1 out of n_1 *and* x_2 out of n_2.

hypergeometric p.m.f.:

$$P(x_1 \text{ out of } n_1, \ldots, x_k \text{ out of } n_k = \frac{\binom{n_1}{x_1} \cdots \binom{n_k}{x_k}}{\binom{n_1 + \cdots + n_k}{x_1 + \cdots + x_k}}.$$

Poisson distribution: discrete p.m.f. involving events which take place relatively infrequently when only a small subinterval of time is being considered.

Poisson p.m.f.:

$$P(x) = \begin{cases} \dfrac{e^{-\lambda}\lambda^x}{x!} & x = 0, 1, 2, \ldots \\ 0 & \text{otherwise.} \end{cases}$$

Poisson parameters: $E[\mathbf{x}] = \lambda$;

$V[\mathbf{x}] = \lambda$.

5

CONTINUOUS PROBABILITY DISTRIBUTIONS

"In graphing the data they fell,
In line with a swoop and a swell.
Experimentally normal,
Their pattern was formal,
Their shape emulating a bell."

ROBERT LAMBORN

5.1 INTRODUCTION

In this chapter we continue presenting a number of the more commonly used probability distributions, but now our discussion will include only continuous distributions. Remember, if a random variable is continuous between two points, then *any* number between these points is at least theoretically possible. For practical purposes, we usually can't measure such variables with very great accuracy. For example, a swimmer's time in the 100-yard freestyle can, theoretically, be any number between 0 and infinity. But even with electronic timing devices we know that times can be recorded up to only about $\frac{1}{1000}$ of a second. Hence, while such a variable is theoretically continuous, for practical purposes it is discrete. We will see in this chapter that it is often more convenient to manipulate continuous variables than discrete variables.

Three continuous variables are discussed in this chapter, the *normal*, the *chi-square*, and the *exponential*. Two additional continuous distributions, the ***t*-*distribution*** and the ***F*-*distribution*** are discussed in Chapter 6 in the context of sampling theory.

5.2 THE NORMAL DISTRIBUTION

Scientists in the eighteenth century noted a predictable regularity to the frequency with which certain "errors" occur, especially errors of measurement. Suppose, for example, a machine is supposed to roll a sheet of metal to a width of exactly $\frac{5}{16}$ in.; while this machine produces sheets that are $\frac{5}{16}$ in. wide on the average, some sheets are in "error" by being slightly too wide, others by being slightly too narrow. Experiments producing errors of this nature were found to form a symmetrical distribution which was originally called the "normal curve of errors." The continuous probability distribution which such an experiment approximates is usually referred to as the *normal distribution*, or sometimes the *Gaussian distribution*, after an early researcher, Karl Gauss (1777–1855).

The normal distribution undoubtedly represents the most widely known and used of all distributions. Because the normal distribution approximates many natural phenomena so well, this distribution has developed into a standard of reference for many probability problems. In addition, the normal distribution has the convenient property that, under certain conditions, it can be used to approximate the binomial distribution and the Poisson distribution. The normal distribution is so important in the theory of statistics that a considerable portion of the sampling, estimation, and hypothesis-testing theory we will study in the rest of this book is based on the characteristics of this distribution.

Characteristics of the Normal Distribution

The normal distribution is a continuous distribution in which x can assume any value between minus infinity and plus infinity ($-\infty \leq x \leq \infty$). Two parameters describe the normal distribution, μ, representing the mean, and σ, representing the standard deviation.* The normal density function contains two constants, π (where $\pi = 3.1415\ldots$) and e (where $e = 2.7182\ldots$). The normal density function is given by Formula (5.1). Note that in labeling this formula we use the symbol $N(\mu, \sigma^2)$, which is a traditional designation for a normal curve with mean μ and variance σ^2.

* Beginning students of statistics are sometimes confused by the fact that for the normal distribution, the general symbols μ and σ are used to represent the specific parameters of this distribution rather than some different ones such as n and p in the binomial, or λ in the Poisson. This is merely a tradition, since the normal is the most commonly used distribution.

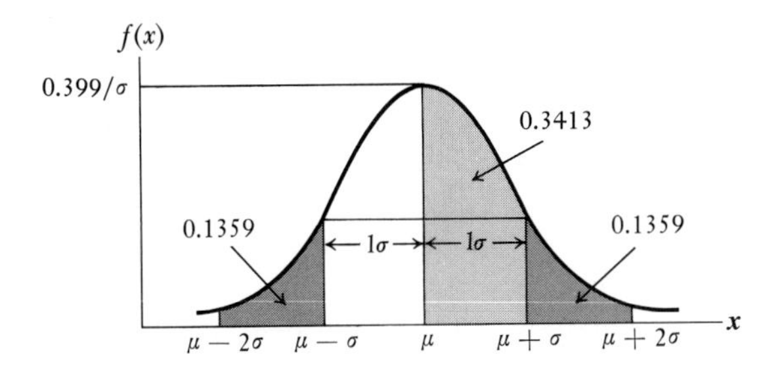

Fig. 5.1 Normal distribution with mean μ and standard deviation σ.

Normal density function, $N(\mu, \sigma^2)$*:*

$$f(x) = \frac{1}{\sigma\sqrt{2\pi}} e^{-(1/2)[(x-\mu)/\sigma]^2} \qquad \text{for } -\infty \leq x \leq \infty. \tag{5.1}$$

The curve described by Formula (5.1) is a completely symmetrical, bell-shaped p.d.f., whose graph is shown in Fig. 5.1.

Note that, in Fig. 5.1, the area under the curve from μ to $\mu + 1\sigma$ is 0.3413. Thus, $P(\mu \leq \mathbf{x} \leq \mu + 1\sigma) = 0.3413$. By symmetry,

$$P(\mu - 1\sigma \leq \mathbf{x} \leq \mu + 1\sigma) = 2(0.3413) = 0.6826.$$

We can also see that

$$P(\mu + 1\sigma \leq \mathbf{x} \leq \mu + 2\sigma) = 0.1359,$$

and hence,

$$P(\mu - 2\sigma \leq \mathbf{x} \leq \mu + 2\sigma) = 0.6826 + 0.1359 + 0.1359 = 0.9544.$$

Had we included a few additional values in Fig. 5.1, we could also show that*

$$P(\mu - 3\sigma \leq \mathbf{x} \leq \mu + 3\sigma) = 0.9974.$$

* Since probabilities in the continuous case are given by integrating the p.d.f. over the appropriate intervals, we can thus write:

$$P(\mu - \sigma < \mathbf{x} < \mu + \sigma) = \int_{\mu-\sigma}^{\mu+\sigma} \frac{1}{\sigma\sqrt{2\pi}} e^{(-1/2)(x-\mu)^2}\,dx = 0.6826;$$

$$P(\mu - 2\sigma < \mathbf{x} < \mu + 2\sigma) = \int_{\mu-2\sigma}^{\mu+2\sigma} \frac{1}{\sigma\sqrt{2\pi}} e^{(-1/2)(x-\mu)^2}\,dx = 0.9544;$$

and

$$P(\mu - 3\sigma < \mathbf{x} < \mu + 3\sigma) = \int_{\mu-3\sigma}^{\mu+3\sigma} \frac{1}{\sigma\sqrt{2\pi}} e^{(-1/2)(x-\mu)^2}\,dx = 0.9974.$$

Note that the probability values for $\mu \pm \sigma$ and $\mu \pm 2\sigma$ exactly equal the rule-of-thumb values specified in Chapter 1 for the area included in these intervals (68 and 95 percent). This is not merely a coincidence, for *the rule of thumb we have been using throughout this book to interpret the size of a standard deviation was based on the normal distribution.* The extent to which the intervals we have considered previously have *differed* from our rule of thumb reflects the fact that these distributions have *not* been normal distributions.

It is important to remember that all normal distributions have the same bell-shaped curve pictured in Fig. 5.1, regardless of the values of μ and σ. The value of μ merely indicates where the center of the "bell" lies, while σ indicates how spread out (or wide) the distribution is. Note that in Fig. 5.1 the height of the density function at the point $x = \mu$ is $0.399/\sigma$. This fact can be derived from Formula (5.1) by substituting μ for x, and then noting that

$$f(x) = f(\mu) = \frac{1}{\sigma\sqrt{2\pi}} e^{(-1/2)(\mu-\mu)^2} = \frac{1}{\sigma\sqrt{2\pi}} e^0.$$

Since $e^0 = 1$ and $1/\sqrt{2\pi} = 0.399$,

$$f(\mu) = \frac{0.399}{\sigma}.$$

Thus if $\sigma = 1.0$, then $f(\mu) = 0.399/1.0 = 0.399$. When $\sigma = 0.5$, $f(\mu) = 0.399/0.5 = 0.798$, and if $\sigma = 1.5$, $f(\mu) = 0.399/1.5 = 0.266$. The normal distributions corresponding to these three values of σ are shown in Fig. 5.2.

It is easy to see, from Figs. 5.1 and 5.2, that Property 1 of all p.d.f.'s is satisfied for the normal distribution, since $f(x) \geq 0$ for all $-\infty \leq x \leq \infty$. Proving Property 2, that $\int_{-\infty}^{\infty} f(x)\,dx = 1$, is more difficult since it involves switching to polar coordinates. This proof, and proving that $E[x] = \mu$ and $E[(x - \mu)^2] = \sigma^2$ for the normal distribution, are beyond the scope of this book.

As we have described, probabilities involving the normal distribution can be determined by integrating the normal density function over the appropriate interval. Unfortunately, this integration is not an easy process, so we will need

Fig. 5.2 Three normal distributions with different standard deviations.

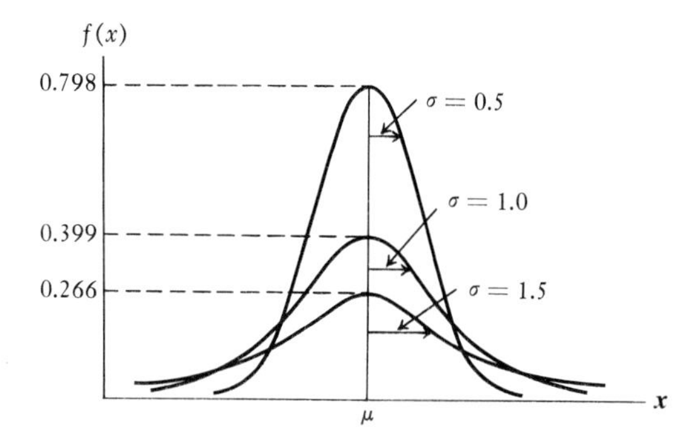

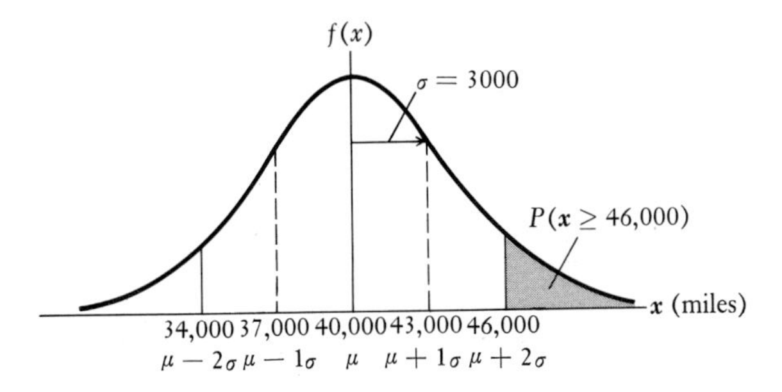

Fig. 5.3 Tire-life example.

to develop a convenient means for determining such probabilities. However, even with the limited information present thus far about the normal, it is possible to solve certain problems, such as the following one.

Tire-life Example Suppose the random variable $\boldsymbol{x}$, representing the tread life (in miles) of a certain new radial tire, is normally distributed with mean, $\mu_x = 40{,}000$ miles, and standard deviation, $\sigma_x = 3{,}000$ miles $[N(40{,}000, 3000^2)]$. This information is sufficient to completely determine the probability of any event concerning the values of $\boldsymbol{x}$. For example, the probability that a randomly selected tire lasts more than 40,000 miles is $P(\boldsymbol{x} > 40{,}000) = \frac{1}{2}$, since half of the probability in a normal distribution lies on each side of the mean. On the other hand, suppose we calculate the probability that a tire of this model selected at random has a tread life of between 34,000 and 46,000 miles—$P(34{,}000 \leq \boldsymbol{x} \leq 46{,}000)$. We see in Fig. 5.3 that 46,000 is exactly *two standard deviations* above the mean,

$$46{,}000 - 40{,}000 = 6{,}000 = 2\sigma_x.$$

Similarly, 34,000 is exactly two standard deviations *below* the mean. The probability $P(34{,}000 \leq \boldsymbol{x} \leq 46{,}000)$ is thus equivalent to $P(\mu - 2\sigma \leq \boldsymbol{x} \leq \mu + 2\sigma)$, which we calculated earlier (see Fig. 5.1) to be

$$P(\mu - 2\sigma \leq \boldsymbol{x} \leq \mu + 2\sigma) = 0.9544.$$

Now let's move to a slightly more complicated problem—determining the probability that a tire will last at least 46,000 miles, or $P(\boldsymbol{x} \geq 46{,}000)$.* From Fig. 5.3 we see that $P(\boldsymbol{x} \geq 46{,}000)$ is equivalent to $P(\boldsymbol{x} \geq \mu + 2\sigma)$. But $P(\boldsymbol{x} \geq \mu + 2\sigma)$ is exactly *half* of the area *not* included in $P(\mu - 2\sigma \leq \boldsymbol{x} \leq \mu + 2\sigma)$, since the normal is symmetric. In other words $P(\boldsymbol{x} \geq \mu + 2\sigma)$ is one-half of the *complement* of $P(\mu - 2\sigma \leq \boldsymbol{x} \leq \mu + 2\sigma) = 0.9544$. That is,

$$P(\boldsymbol{x} \geq \mu + 2\sigma) = \tfrac{1}{2}(1.0000 - 0.9544) = \tfrac{1}{2}(0.0456) = 0.0228.$$

By the symmetry of the normal distribution, we also know that

$$P(\boldsymbol{x} \leq \mu - 2\sigma) = 0.0228 = P(\boldsymbol{x} \leq 34{,}000).$$

* Remember that $P(\boldsymbol{x} > 46{,}000) = P(\boldsymbol{x} \geq 46{,}000)$ because $P(\boldsymbol{x} = 46{,}000) = 0$.

To carry this analysis one step further, we now have enough information to calculate, directly,

$$P(\boldsymbol{x} \leq 37{,}000) = P(\boldsymbol{x} \leq \mu - 1\sigma).$$

From Fig. 5.1 we know the probability that $\boldsymbol{x}$ falls between $\mu - 2\sigma$ and $\mu - 1\sigma$ is 0.1359. If we add to this value the probability we calculated above,

$$P(\boldsymbol{x} \leq \mu - 2\sigma) = 0.0228,$$

the result is

$$P(\boldsymbol{x} \leq \mu - 1\sigma) = 0.0228 + 0.1359 = 0.1587.$$

Finally, for this example, suppose we are interested in the probability that tread life will fall between 44,500 and 47,500 miles or $P(44{,}500 \leq \boldsymbol{x} \leq 47{,}500)$. The answer can be determined in two ways. Using calculus, one could integrate the normal p.d.f. with $\mu_x = 40{,}000$ and $\sigma_x = 3{,}000$ from 44,500 to 47,500; or one could use a table of areas already calculated for the precise parameters, $\mu_x = 40{,}000$ and $\sigma_x = 3{,}000$. Both of these methods are unsatisfactory, however, the first one because it is too tedious to evaluate a new integral every time one investigates a different set of parameters or new $\boldsymbol{x}$-values, and the second one because no such tables exist (there obviously can't be tables listing the infinite number of values of μ and σ).

5.3 STANDARDIZED NORMAL

Values of $\boldsymbol{x}$ for the normal distribution are usually described in terms of *how many standard deviations* they are away from the mean. The value $x = 200$, for example, has little meaning unless we know in what units $\boldsymbol{x}$ was measured (e.g., feet, miles, pounds). On the other hand, the statement that $\boldsymbol{x}$ is one standard deviation larger (or smaller) than the mean can be given a very precise interpretation, since it is always meaningful to talk of $\boldsymbol{x}$ being a certain number of standard deviations above (or below) the mean, no matter what value σ assumes or on what scale the variable $\boldsymbol{x}$ is measured. Now, if $\boldsymbol{x}$ is measured in terms of standard deviations about the mean, it is natural to describe probability values in the same terms—that is, by specifying the probability that $\boldsymbol{x}$ will fall within so many standard deviations of the mean. There are three commonly encountered intervals, the first two of which we have referred to often in the last two chapters: $\mu \pm \sigma$, $\mu \pm 2\sigma$, and $\mu \pm 3\sigma$.

Treating the values of $\boldsymbol{x}$ in a normal distribution in terms of standard deviations about the mean has the advantage of permitting all normal distributions to be compared to one common or *standard* form. In this standard form, different values of μ and σ no longer generate completely different curves, since $\boldsymbol{x}$ is measured only about μ, and all distances away from μ are in terms of multiples of σ. In other words, it is easier to compare normal distributions having different values of μ and σ if these curves are transformed to one common form, which is called the *standardized normal.* The standardized normal, by definition, has a mean of zero ($\mu = 0$) and a standard deviation of one ($\sigma = 1$). Note that if the standard deviation is one, the variance must also be one since $\sigma^2 = 1.0$ when $\sigma = 1.0$.

This process of standardization gives a hint toward the best method of attack in answering questions concerning a normal probability distribution. Instead of trying to directly solve a probability problem involving a normally distributed random variable x with mean μ and standard deviation σ, an indirect approach is used. We first convert the problem to an equivalent one dealing with a normal variable measured in standard deviation units, called a standardized normal variable. A table of standardized normal values (Table III) can then be used to obtain an answer in terms of the converted problem. Finally, by converting back to the original units of measurement for x, we can obtain the answer to the original problem. Figure 5.4 gives a schematic outline of this method of solving probability problems.

Recall that we discussed in Chapter 3 the process of transforming a random variable x (whether normally distributed or not), with mean and standard deviation σ, into a standardized measure with mean zero and standard deviation one. The appropriate transformation was shown to be $z = (x - \mu)/\sigma$. The mean of the variable z was shown in Section 3.7 to be $E[z] = 0$ and the variance was shown to be $V[z] = 1.0$.

Although we have used the letter z to represent a random variable with mean zero and variance one, this variable traditionally also represents the standardized form of the normal distribution. The letter z is thus generally associated with the normal distribution $N(0, 1)$. That is,

Standardized normal random variable:

$$z = \frac{x - \mu}{\sigma} \quad \text{is} \quad N(0, 1). \tag{5.2}$$

The density function for z can be derived by substituting z for $(x - \mu)/\sigma$, and $\sigma = 1$, into Formula (5.1). The resulting p.d.f. is denoted as $f(z)$.

Fig. 5.4 Problem-solving tactic using the standardized normal.

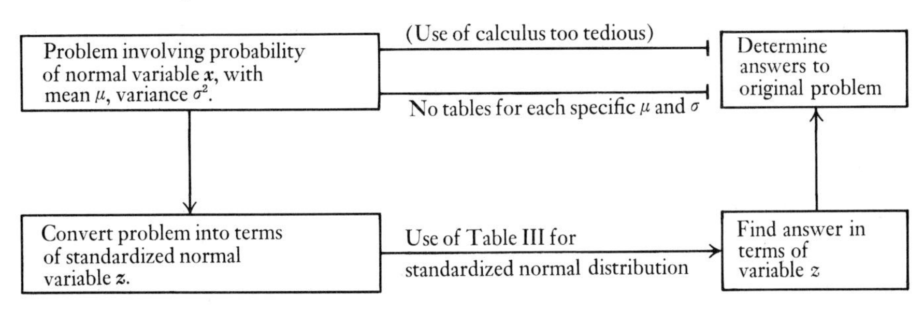

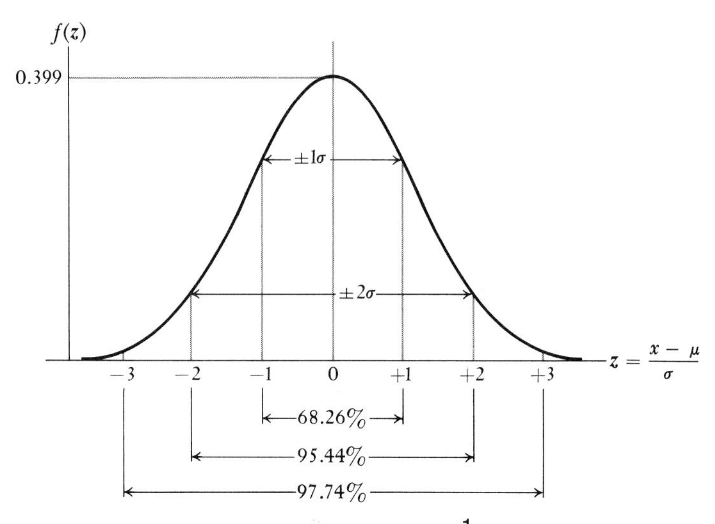

Fig. 5.5 Standardized normal distribution: $f(z) = \frac{1}{\sqrt{2\pi}} e^{-(1/2)z^2}$.

Standardized normal density function:

$$f(z) = \frac{1}{\sqrt{2\pi}} e^{(-1/2)z^2}. \tag{5.3}$$

This function is shown graphically in Fig. 5.5. The reader should compare Fig. 5.5 with 5.1 to verify that the former is merely a special case of the latter, where $\mu = 0$ and $\sigma = 1$.

The interpretation of $\boldsymbol{z}$-values is relatively simple. Since $\sigma_{\boldsymbol{z}} = 1$, a value of $\boldsymbol{x}$ is, for example, two standard deviations away from the mean whenever $z = \pm 2$; likewise, if $z = \pm 1.56$, the corresponding x-value is exactly 1.56 standard deviations away from the mean (or $|x - \mu| = 1.56\sigma$).

To illustrate the use of the standardized normal, let's consider the tire-life problem posed at the end of the last section—namely, $P(44{,}500 \leq \boldsymbol{x} \leq 47{,}500)$. In transforming this probability into an equivalent one in standardized-normal form, we need to apply the transformation $\boldsymbol{z} = (\boldsymbol{x} - \mu)/\sigma$ to *each* part of the expression in parenthesis. For example, the value 44,500 is transformed into its equivalent form by subtracting $\mu_x = 40{,}000$ from it, and then dividing by $\sigma_x = 3000$; the value 47,500 is transformed in exactly the same manner. Finally, we can think of the variable $\boldsymbol{x}$ being transformed in the same way, since the new variable, $\boldsymbol{z}$, equals $(\boldsymbol{x} - \mu)/\sigma$. Thus,

$$P(44{,}500 \leq \boldsymbol{x} \leq 47{,}500) = P\left(\frac{44{,}500 - 40{,}000}{3000} \leq \frac{\boldsymbol{x} - \mu}{\sigma} \leq \frac{47{,}500 - 40{,}000}{3000}\right)$$

$$= P(1.5 \leq \boldsymbol{z} \leq 2.5).$$

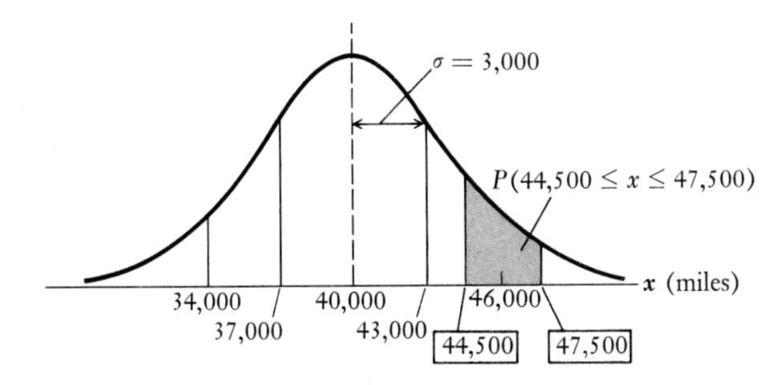

Tire life: $\mu = 40{,}000$, $\sigma = 3{,}000$

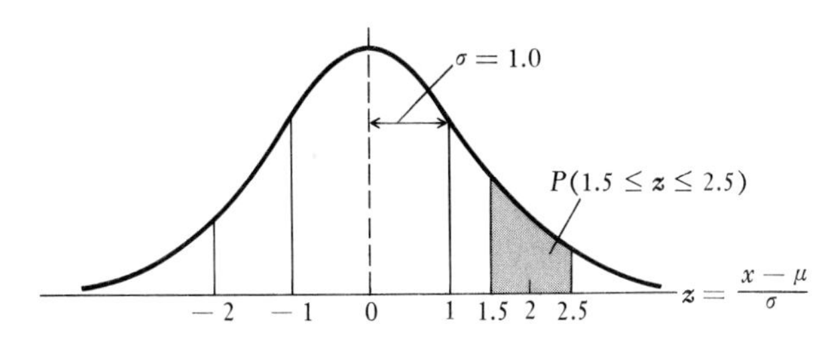

Standardized normal: $\mu = 0$, $\sigma = 1.0$

Fig. 5.6 Tire-life example.

Figure 5.6 illustrates the equivalence between:

$$P(44{,}500 \leq \mathbf{x} \leq 47{,}500) \qquad \text{and} \qquad P(1.5 \leq \mathbf{z} \leq 2.5)$$

As we indicated previously, probabilities involving z-values are usually determined using tables of standardized normal values, such as those shown in Table III, Appendix B. In our discussion of the standardized normal distribution and Table III, we will make use of the cumulative distribution function $F(z) = P(\mathbf{z} \leq z)$. This function is shown in Fig. 5.7. As we pointed out in Chapter 1, a cumulative function can be used to calculate percentiles. For example, $x = 1$

Fig. 5.7 Cumulative **z**-values.

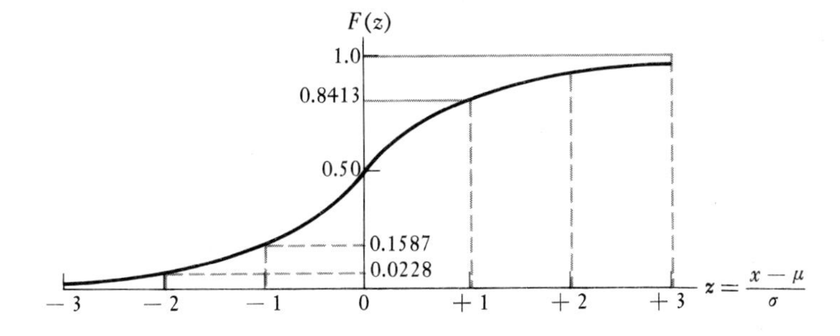

represents the 84.13th percentile, since $F(1) = 0.8413$. Note that we have plotted z-values only from -3 to $+3$, since very little area lies beyond these limits. At $z = 0$ (the mean of $\mathbf{z}$), the value of $F(z)$ must be 0.50, since $z = 0$ represents the median (as well as the mode and the mean) of the $\mathbf{z}$-values. Most of the other values in Fig. 5.7 should be familiar to you by now. For example, $F(-2) = 0.0228$. This value agrees with the one calculated in the example on tire life, as we saw then that $P(\mathbf{x} \leq \mu - 2\sigma) = 0.0228$. The value $F(-1) = 0.1587$ should also appear familiar, as this is the same value we calculated in the tire-life problem for $P(\mathbf{x} \leq \mu - 1\sigma)$.

Since the normal distribution is completely symmetrical, tables of $\mathbf{z}$-values usually include only *positive* values of z. Thus, the lowest value in Table III is $z = 0$, and the cumulative probability at this point is $F(0) = P(\mathbf{z} \leq 0) = 0.50$. Table III gives values of $\mathbf{z}$ to two decimal points, up to the point $z = 3.49$. The values of $\mathbf{z}$ to one decimal are read from the left margin in Table III, while the second decimal is read across the top. The body of the table gives the values of $F(z)$.

To illustrate the use of Table III, we will consider four basic rules. The reader should try to understand (visualize) these rules, not necessarily memorize them.

Rule 1: *$P(\mathbf{z} \leq a)$ is given by $F(a)$ when a is positive.*

We can illustrate this rule by determining the probability $P(\mathbf{x} \leq 45{,}000)$ in the tire-life problem:

$$P(\mathbf{x} \leq 45{,}000) = P\left(\frac{\mathbf{x} - \mu}{\sigma} \leq \frac{45{,}000 - 40{,}000}{3000}\right)$$

$$= P(\mathbf{z} \leq 1.66) = F(1.66).$$

The value of $F(1.66)$ is found in Table III by looking for 1.6 along the left margin and 0.06 across the top. The intersection of these values shows $F(1.66) = 0.9515$. Thus, $P(\mathbf{x} \leq 45{,}000) = 0.9515$.

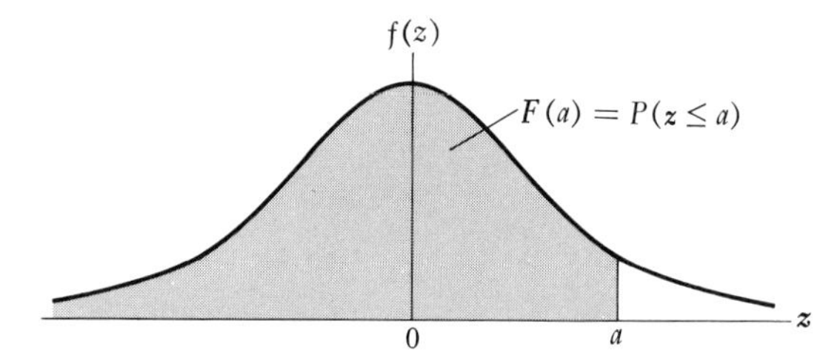

Rule 2: *$P(\mathbf{x} \geq a)$ is given by the complement rule as $1 - F(a)$.*

This rule is a direct application of the complement rule of Chapter 2. To illustrate it, consider the tire-life problem in which we calculated $P(\mathbf{x} \geq \mu + 2\sigma) =$

0.0228. This probability is equivalent to $P(z \geq 2.0)$, which, by complement rule, equals $1 - F(2.00)$. From Table III, $F(2.00) = 0.9772$; hence,

$$P(z \geq 2.0) = 1 - 0.9722 = 0.0228.$$

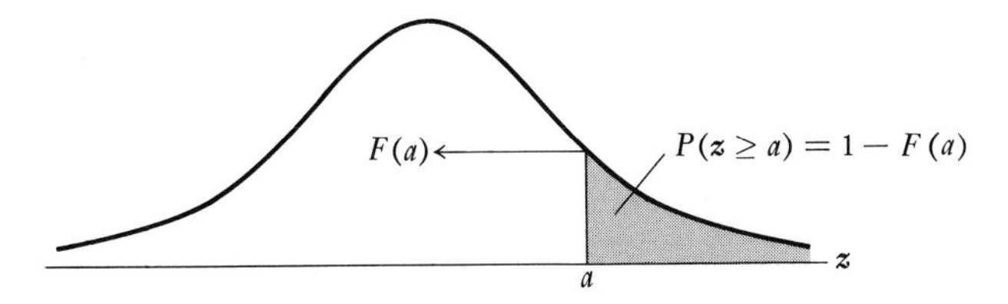

Rule 3: *$P(z \leq -a)$ is given by* $1 - F(a)$.

Because the normal distribution is completely symmetrical, $P(z \leq -a) = P(z \geq a)$. Since, by Rule 2, we know that $P(z \geq a) = 1 - F(a)$, we also know that $P(z \leq -a) = 1 - F(a)$. For example, let's return to the tire-life problem of calculating $P(x \leq 37{,}000) = P(z \leq -1.00)$. From Table III, $F(1.00) = 0.8413$; hence,

$$P(x \leq 37{,}000) = 1 - 0.8413 = 0.1587$$

(which agrees with our earlier result).

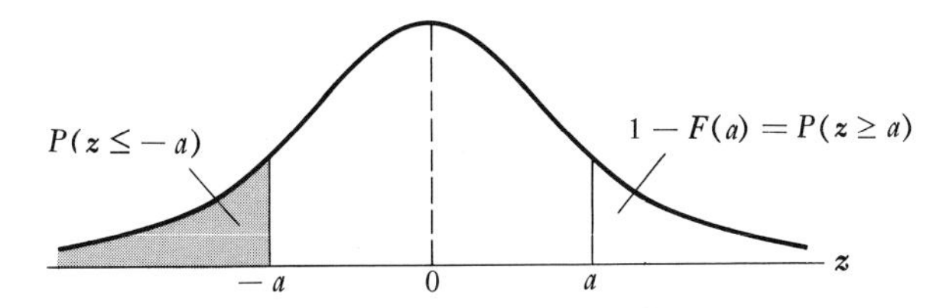

Rule 4: *$P(a \leq z \leq b)$ is given by $F(b) - F(a)$.*

This rule is the same as Formula (3.9) in Chapter 3. To illustrate it, suppose we solve the problem at the start of this section, $P(1.5 \leq z \leq 2.5)$. From Table III,

$$F(2.5) = 0.9938 \qquad \text{and} \qquad F(1.5) = 0.9332.$$

Hence,

$$P(1.5 \leq z \leq 2.5) = 0.9938 - 0.9332 = 0.0606.$$

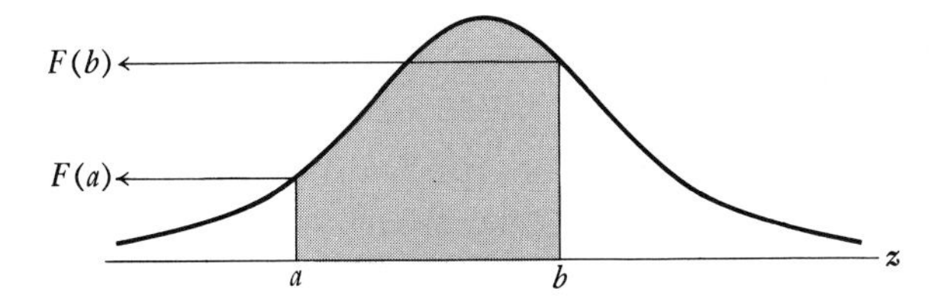

There are many other types of problems that can be solved using the standardized normal distribution. To illustrate several of these problems, we conclude this section with the following example.

Calculating a Probability Interval

Suppose we want to solve a probability problem involving an interval $a \leq \boldsymbol{x} \leq b$ where both a and b are unknown. For example, in the tire-life problem, one might wish to determine two values a and b such that the probability is 0.95 that a randomly selected tire will fall between these values;

$$P(a \leq \boldsymbol{x} \leq b) = 0.95.$$

It should be readily apparent that there is an infinitely large number of such intervals, depending on how the values of a and b are selected. If our interval is to include 95 percent of the area under the curve and exclude 5 percent, we could exclude all 5 percent above b, or exclude all 5 percent below a, or exclude part of it below a and part above b. In many cases the best way to split the percent to be excluded among the two tails will be specified in the problem. If it is not specified, then *it is generally agreed* that:

---

*The best way to split the percent to be excluded is in a manner which makes the interval from a to b as small as possible. The smallest interval is obtained by excluding equal amounts in both the upper and lower tails of the distribution.**

This result, known as the Neyman–Pearson theorem, is proven in more advanced books on statistics. It tells us that the best way to split our 5 percent is to have 2.5 percent in each tail. Let's now use this result to solve the tire-life problem, $P(a \leq \boldsymbol{x} \leq b) = 0.95$.

To solve this problem let's first standardize it as much as possible, by subtracting $\mu_x = 40{,}000$ from each term, and then dividing by $\sigma_x = 3000$.

$$P(a \leq \boldsymbol{x} \leq b) = P\left(\frac{a - 40{,}000}{3000} \leq \frac{\boldsymbol{x} - \mu_x}{\sigma_x} \leq \frac{b - 40{,}000}{3000}\right) = 0.95$$

or

$$P\left(\frac{a - 40{,}000}{3000} \leq \boldsymbol{z} \leq \frac{b - 40{,}000}{3000}\right) = 0.95. \qquad (5.4)$$

We can't solve this expression yet. However, suppose we try to find, using Table III, the appropriate z-values; that is, let us try to find an interval $c \leq$

* As we will discuss later, this nontechnical statement is true only for unimodal, symmetrical distributions such as the normal. For other types of distributions, further conditions must be specified.

$z \leq d$ such that

$$P(c \leq z \leq d) = 0.95.$$

From Table III, $F(1.96) = 0.9750$, which means that

$$P(z \geq 1.96) = 1.000 - 0.975 = 0.025.$$

By symmetry $P(z \leq -1.96) = 0.025$. Since each of the values ± 1.96 excludes an area of 0.025 on the normal distribution, the interval *between* these values must include an area of $1 - 0.025 - 0.025 = 0.95$. That is,

$$P(-1.96 \leq z \leq 1.96) = 0.95. \tag{5.5}$$

By comparing equations (5.4) and (5.5) it is apparent that

$$\frac{a - 40{,}000}{3000} = -1.96 \qquad \text{and} \qquad \frac{b - 40{,}000}{3000} = +1.96.$$

Solving for a and b yields $a = 34{,}120$ and $b = 45{,}880$, so we can now write the appropriate intervals as:

$$P(34{,}120 \leq \mathbf{x} \leq 45{,}880) = 0.95.$$

The reader should verify that 34,120 to 45,880 is the *smallest* interval possible, by trying several other possible splits of the 5 percent to be excluded. Figure 5.8 shows the normal distribution for this example.

Many natural phenomena tend to result in normal distributions, such as length, height, and breadth of animals or plants; medical counts of sugar, white-blood cells, incidence of inner ear disease; and behavioral, emotional, or psychological measures of human actions, aptitudes, or abilities. Also, the distribution of measured errors (or degree of perfection) in production processes of many kinds tends to be normal; these include errors from a specified standard in diameters of pistons, cylinders or gun barrels, weight of packaged products, and even lengths of yardsticks.

However, the family of probability problems where the standardized normal z is useful extends even beyond these many instances. Many other distributions

Figure 5.8

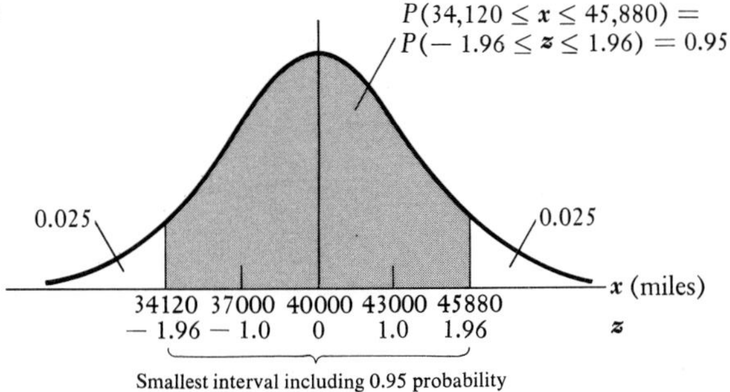

applying to other types of problems tend to be normal distributions under certain conditions. One example that is discussed in the next section is the binomial distribution. Finally, as we will see in Chapters 6 through 8, the standardized normal distribution is of primary importance in problems of statistical inference dealing with means of samples from a population whose distribution is unknown.

Define: Gaussian distribution, $F(z)$, $N(\mu, \sigma^2)$.

Problems

5.1 Explain what is meant by the phrase "standardization of a variable." Why do we standardize variables?

5.2 Suppose that the expected life of a certain brand of water softener is $N(10, 2)$, measured in years. Sketch the p.d.f.

5.3 Sketch the *cumulative* distribution function for the p.d.f. given in Problem 5.2.

5.4 Assume that a manufacturer of tennis balls produces a ball whose bounce from six feet is normally distributed with a mean of 42″ and a variance of 4″.

a) Sketch this p.d.f.

b) How often will a ball bounce more than 46″? How often will it bounce less than 40″?

c) How often will two consecutive bounces be less than 40″ if bounces are independent?

5.5 For Problem 5.4, find values of a and b such that:

a) $P(\mathbf{x} \leq a) = 0.95$

b) $P(\mathbf{x} \geq b) = 0.95$

c) $P(a \leq \mathbf{x} \leq b) = 0.95$

5.6 The number of hours per week of lost work due to employees' illness in a certain automobile assembly plant is approximately normally distributed, with a mean of 60 hours and a standard deviation of 15 hours. For a given week, selected at random, what is the probability that;

a) The number of lost work hours will exceed 85 hours?

b) The number of lost hours will be between 45 and 55 hours?

c) The number of lost work hours will be exactly 60?

5.7 A certain disease is contagious for an average of 10 days, with a standard deviation of 2 days. How long would you isolate a person with this disease if you want to be 99-percent certain the person is not contagious?

5.8 A credit card company has found its average balance per customer is $N(250, 2500)$.

a) Find the probability that a balance is over \$300 [i.e., $P(\mathbf{x} \geq \$300)$]. Find $P(\mathbf{x} \leq \$150)$.

b) Find the value of a such that the probability of a customer's balance being less than this amount is 0.05.

c) Determine values of a and b such that

$P(a \leq \mathbf{x} \leq b) = 0.99.$

d) What is the probability that eight out of ten randomly selected balances will be less than \$300?

5.9 Suppose the lengths of fish caught in a certain lake are $N(12'', 16'')$.

a) If $\mathbf{x}$ = length, find $P(\mathbf{x} > 16)$, $P(\mathbf{x} > 24)$, and $P(8 \leq \mathbf{x} \leq 16)$.

b) Find values of a and b such that $P(a \leq \mathbf{x} \leq b) = 0.9544$.

5.10 Use the standarized normal distribution to find, for Problem 5.9:

a) $P(\mathbf{x} \geq 13.6)$, $P(\mathbf{x} \geq 7.6)$, and $P(8.4 \leq \mathbf{x} \leq 14.4)$.

b) If all fish less than 6 inches must be thrown back, what is the probability that a catch will have to be returned?

c) What is the probability a fish will be either shorter than 6″ or longer than 16″?

5.11 a) If the income in a community is normally distributed, with a mean of \$19,000 and a standard deviation of \$2000, what minumium income does a member of this community have to earn in order to be in the top 10 percent? What is the maximum income one can have and still be in the middle 50 percent?

b) If $\mathbf{x}$ = income, find values of a and b such that

$P(a \leq \mathbf{x} \leq b) = 0.9500.$

5.12 The average IQ of the U.S. population, as measured by the Stanford-Binet test, is supposed to be $N(100, 16^2)$.

a) What is the probability that a randomly selected person will have an IQ greater than 140?

b) What percent of the U.S. population should have IQs less than 80?

c) What is the probability that a randomly selected member of your statistics class has an IQ of less than 100?

5.13 A traveling circus has found that its average attendance per performance is 6000 people, with a standard deviation of 1500. Assume that the attendance is normally distributed.

a) Suppose that a town asks the circus to come, but the only available site in town holds only 8000 people. What is the probability that this site will not be large based on past experience?

b) If the circus loses money on 20% of its performances, what attendance must it have to break even for a given performance? Assume that the profit it makes for a given performance depends only on the number of people attending the performance.

5.14 Suppose that the Nielson ratings indicate a certain television show has an average weekly audience of 18 million viewers. The standard deviation of audience size is estimated at 3 million viewers. For any given week, what is the lowest number of viewers you would expect to be watching? What is the largest number you would expect to be watching? Assume a normal distribution.

5.15 Assume that the scores on the Graduate Management Admission Test (GMAT) are $N(540, 100^2)$, and scores on the Graduate Record Exam (GRE) are $N(500, 80^2)$. Who has performed better—a student with a 600 on the GRE or a student with a 660 on the GMAT?

5.4 NORMAL APPROXIMATION TO THE BINOMIAL

In studying the binomial distribution, we saw that when the number of trials (n) is large, this distribution can be tedious to calculate. Fortunately, it is often possible in this situation to use the normal distribution to *approximate* the binomial. Remember that in Fig. 4.3 we saw that when n is as small as 20, the binomial has a fairly symmetrical (bell-shaped) appearance, even when p is not very close to $\frac{1}{2}$. In general, the larger the value of n, and the closer p is to $\frac{1}{2}$, the better the normal will approximate the binomial. Just how large n needs to be depends on how close p is to $\frac{1}{2}$, and on the precision desired, although fairly good results are usually obtained when $npq \geq 3$.

To use a normal distribution to approximate the binomial, it is necessary to let the mean of the normal have the same value as the mean of the binomial, $\mu = np$, and let the standard deviation of the normal equal the standard deviation of the binomial, $\sigma = \sqrt{npq}$. Then probabilities for any values of $\mathbf{x}$ can be approximated by converting to the standardized normal variable $\mathbf{z}$ as follows,

Normal approximation to binomial when $npq \geq 3$:

$$\mathbf{z} = \frac{\mathbf{x} - \mu}{\sigma} = \frac{\mathbf{x} - np}{\sqrt{npq}}. \tag{5.6}$$

Substituting np for μ and $\sqrt{npq}$ for σ does not ensure that the values of $\mathbf{z}$ resulting from Formula 5.6 will correspond to a standardized normal distribution with $\mu = 0$ and $\sigma^2 = 1$. It can be shown, however, that as n gets larger and larger, the ratio $(\mathbf{x} - np)/\sqrt{npq}$ does, in fact, become a better and better approximation to the standardized normal.*

One additional factor must be considered in using the normal to approximate the binomial, namely, that a discrete distribution involving only integer values (the binomial) is being approximated by a continuous distribution (the

* In other words, it can be shown that

$$\lim_{n \to \infty} \frac{\mathbf{x} - np}{\sqrt{npq}} = N(0, 1).$$

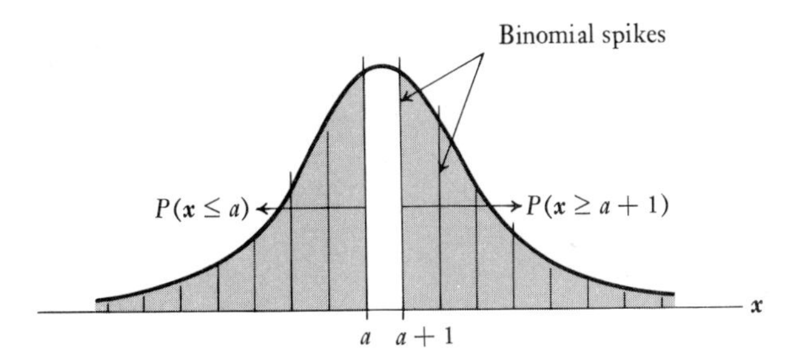

Figure 5.9

normal) in which $\mathbf{x}$ can take on any value between negative and positive infinity. The problem that can arise in this situation is illustrated in Fig. 5.9.

For the binomial "spikes," we see that $P(\mathbf{x} \leq a)$ and $P(\mathbf{x} \geq a + 1)$ will sum to 1.0 whenever a is an integer. But if we sum the area under the normal curve corresponding to $P(\mathbf{x} \leq a)$ and $P(\mathbf{x} \geq a + 1)$, this area does *not* sum to 1.0 because the area from a to $(a + 1)$ is missing.

The usual way to handle this problem is to *associate one-half of the interval* from a to $(a + 1)$ *with each adjacent integer*. The continuous approximation to the probability $P(\mathbf{x} \leq a)$ would thus be $P(\mathbf{x} \leq a + \frac{1}{2})$, while the continuous approximation to $P(\mathbf{x} \geq a + 1)$ would be $P(\mathbf{x} \geq a + \frac{1}{2})$. This adjustment is called a *correction for continuity*. Notice that the effect of the correction for continuity is to enlarge each endpoint in the binomial problem by one-half of a unit.

A Sample Calculation

Suppose we want to calculate the probability that in a random sample of 100 people, 60 or more will indicate that they favor a certain legislative proposal when the true population proportion of those favoring this proposal is $p = 0.64$. The parameters necessary for determining the binomial probability $P(60 \leq \mathbf{x} \leq 100)$ are $n = 100$ and $p = 0.64$. For a normal approximation, we correct for continuity by enlarging the endpoints 60 and 100 by one-half unit each. The appropriate probability is $P(59.5 \leq \mathbf{x} \leq 100.5)$, with parameters

$$\mu = np = 64 \qquad \text{and} \qquad \sigma = \sqrt{npq} = \sqrt{100(0.64)(0.36)} = 4.8.$$

The values of these two probabilities are:

Binomial probability:

$$P(60 \leq \mathbf{x} \leq 100) = \sum_{x=60}^{100} \binom{100}{x} (0.64)^x (0.36)^{100-x}$$

$$= 0.8263 \qquad \text{(from Table I);}$$

Normal approximation:

$$P(59.5 \le \mathbf{x} \le 100.5) = P\left(\frac{59.5 - np}{\sqrt{npq}} \le z \le \frac{100.5 - np}{\sqrt{npq}}\right),$$

$$P\left(\frac{59.5 - 64}{4.8} \le z \le \frac{100.5 - 64}{4.8}\right) = P(-0.94 \le z \le 7.6)$$

$$= F(7.6) - F(-0.94)$$
$$= 1.000 - [1.00 - F(+0.94)]$$
$$= F(0.94) = 0.8264 \quad \text{(from Table III).}$$

The approximation in this case is very good, as in this case $npq = 23.04$, which is much larger than the suggested condition $npq \ge 3$.

Consider one additional example, that of calculating the probability that the Dow-Jones average will go down on 4 out of 12 randomly selected days assuming $p = 0.30$ is the (constant) probability it will go down on any given day. With $n = 12$ and $p = 0.30$, the appropriate binomial probability is:

Binomial probability:

$$P(\mathbf{x} = 4) = \binom{12}{4}(0.30)^4(0.70)^8 = 0.2310.$$

Using the normal approximation, we can approximate $P(\mathbf{x} = 4)$ as $P(3.5 \le \mathbf{x} \le 4.5)$. In this case, $\mu = np = 12(.3) = 3.6$, and $\sigma = \sqrt{npq} = \sqrt{2.52} = 1.59$.

Normal approximation:

$$P(3.5 \le \mathbf{x} \le 4.5) = P\left(\frac{3.5 - np}{\sqrt{npq}} \le z \le \frac{4.5 - np}{\sqrt{npq}}\right)$$

$$= P\left(\frac{3.5 - 3.6}{1.59} \le z \le \frac{4.5 - 3.6}{1.59}\right)$$

$$= P(-0.06 \le z \le 0.57)$$

$$= F(0.57) - F(-0.06)$$

$$= 0.7157 - [1.0 - F(+0.06)]$$

$$= 0.7157 - [1.0 - 0.5239] = 0.2396.$$

The normal approximation to the binomial is quite good in this case, even though $npq = 12(0.3)(0.7) = 2.52$ is smaller than 3. Figure 5.10 gives the histogram for the binomial distribution with $n = 12$, $p = 0.30$; superimposed on this histogram is a normal distribution with $\mu = 3.6$ and $\sigma = 1.59$. Notice how good the fit is despite the small sample size.

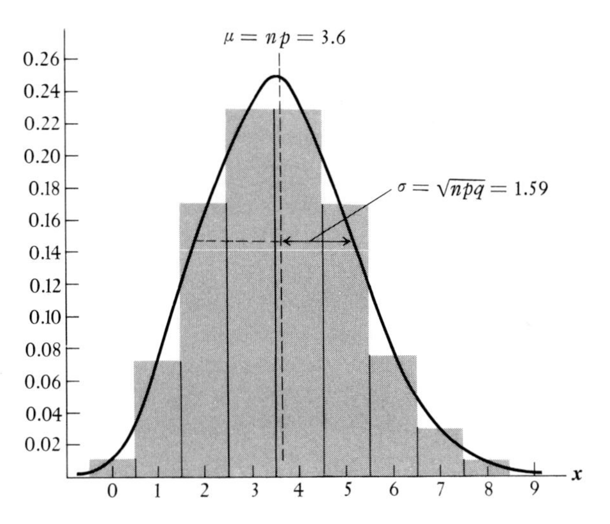

Fig. 5.10 Normal approximation to the binomial ($n = 12$, $p = 0.30$; $\mu = 3.6$, $\sigma = 1.59$).

In these examples comparing the exact probability of the binomial with the approximate probability of the normal with a continuity correction, the use of the binomial may seem simpler since the selected values of n and p could be found in Table I. Remember the point of these comparisons, however. When the value of n is large and is not in Table I, or the value of p has three or more decimal places and is not in Table I, then the normal approximation can and should be used. It will be quite accurate if $npq \geq 3$.

Normal Approximation Using Proportions

Recall from Chapter 4 that binomial problems can be phrased in terms of the *proportion* of successes as well as the *number* of successes. For solving problems by the normal approximation method, the following formula based on proportions is equivalent to Formula (5.6) on page 228.

Normal approximation to binomial using proportions:

$$z = \frac{\frac{x}{n} - p}{\sqrt{\frac{pq}{n}}}. \tag{5.7}$$

Formula (5.7) should be recognized as the standardization of the variable $\boldsymbol{x}/n$, as in Chapter 4 we showed that the variable $\boldsymbol{x}/n$ has a mean of $E[\boldsymbol{x}/n] = p$ and a variance of $V[\boldsymbol{x}/n] = pq/n$.

To demonstrate the equivalence of Formulas (5.6) and (5.7), let us reconsider the legislative proposal problem from the previous subsection. Suppose that we wish to approximate the probability that the *proportion* of voters in a sample of 100 falls between 60% and 70% when $p = 0.64$, or

$$P\left(0.60 \leq \frac{\boldsymbol{x}}{n} \leq 0.70\right).$$

Again, we have to correct for continuity by enlarging the interval by one-half unit (.005) at both ends. The lower limit thus becomes 0.595 and the upper limit 0.705, and the new probability is

$$P\left(0.595 \leq \frac{\boldsymbol{x}}{n} \leq 0.705\right).$$

This probability can now be standardized by using Formula (5.7), where $p = 0.640$ and $\sqrt{pq/n} = \sqrt{(0.64)(0.36)/100} = 0.048$.

$$\begin{aligned} P\left(\frac{0.595 - 0.640}{0.048} \leq \frac{(\boldsymbol{x}/n) - p}{\sqrt{pq/n}} \leq \frac{0.705 - 0.640}{0.048}\right) &= P(-0.94 \leq z \leq 1.35) \\ &= F(1.35) - F(-0.94) \\ &= 0.9115 - [1 - F(0.94)] = 0.7379. \end{aligned}$$

The reader may wish to verify this result using Formula (5.6). This verification will demonstrate that a binomial problem can be approximated using either the number of successes [Formula (5.6)] or the proportion of successes [Formula (5.7)].

5.5 EXPONENTIAL DISTRIBUTION (OPTIONAL)

Another important continuous distribution, the *exponential distribution*, is closely related to a discrete distribution discussed previously, the Poisson. Both the Poisson and the exponential distribution have found many applications in operations research, especially in studies of queueing (waiting-line) theory. These two distributions are related in such applications by the fact that if events (e.g., requests for service or arrivals) are assumed to occur according to a Poisson probability law, then the exponential distribution can be used to determine the probability distribution of the time which elapses *between* such events. For example, if customers arrive at a bank in accordance with a Poisson distribution, the exponential may be used to determine the probability distribution of the time between these arrivals. The time it takes to be serviced (called the service time) in these models is another application of the exponential distribution.

The exponential distribution is a continuous function which has the same parameter, λ, as the Poisson. Lambda, as before, represents the mean rate at which events (arrivals or service completions) occur. Thus a value of $\lambda = 3.0$ might imply that service completions occur, on the average, at the rate of 3.0 per minute (or any other time unit). If a telephone line handles an average of 20 customers per hour, then λ, defined as the mean number of customers being served by the telephone facilities, is $\lambda = 20$ (per hour) or $\lambda = 1/3$ (per minute). Similarly, $\lambda = 3.8$ might imply, as it did in Section 4.5, that on the average 3.8 customers arrive at a checkout counter in a supermarket every minute.

One major assumption necessary for applying the exponential distribution to applications involving service facilities is that the time between arrivals (if $\lambda =$ the arrival rate) or the time for completing the service (if $\lambda =$ the service rate) is usually relatively short (e.g., my wife notwithstanding, phone conversations are usually rather brief). The longer the time interval becomes, the *less* likely it is that the service completion (or the next arrival) will take that long or longer. Suppose we let the random variable $\boldsymbol{T}$ represent the amount of time between service completions, or between arrivals. As $\boldsymbol{T}$ becomes larger and larger, the value of $f(T)$ for the exponential becomes smaller and smaller. In fact, as can be seen in the graph of the exponential distribution in Fig. 5.11, $f(T)$ approaches zero as $\boldsymbol{T}$ goes to infinity.

Note that the exponential distribution, similar to the Poisson, assumes a value other than zero only when $\boldsymbol{T}$ is greater than or equal to zero and when λ is greater than zero. The vertical intercept of the function shown in Fig. 5.11 is seen to equal λ, which means that $f(0) = \lambda$. These relationships characterize the exponential distribution, whose density function is:

Exponential distribution:

$$f(T) = \begin{cases} \lambda e^{-\lambda T} & \text{for } 0 \leq T \leq \infty, \lambda > 0, \\ 0 & \text{otherwise.} \end{cases} \tag{5.8}$$

Fig. 5.11 The exponential distribution: $f(T) = \lambda e^{-\lambda T}$.

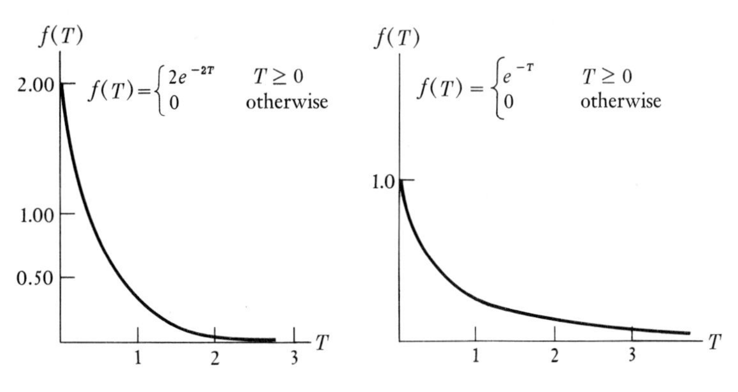

Mean and Variance of the Exponential

Remember that if we interpret λ as the mean arrival (or service) rate, then the exponential gives the probability distribution of the time *between* arrivals (or between service completions).* It should thus not be very surprising to learn that the mean of the exponential distribution is $1/\lambda$. For example, suppose the mean service rate of a cashier in a bank equals one-half customer every minute, or $\lambda = \frac{1}{2}$. Since it takes, on the average, one minute to serve half a customer, it takes two minutes to serve one customer; hence, the mean time between service completions is $1/\lambda = 2$. Similarly, if the mean number of arrivals at a certain airport is $\lambda = 20$ per hour, then the mean time between arrivals will be

$$\frac{1}{\lambda} = \frac{1}{20} = 0.05 \text{ hours}$$

(or one airplane approximately every 3 minutes). As was true for the Poisson, the mean and the variance of the exponential are both functions of λ.

The mean and variance of the exponential are shown in Formula (5.9).†

$$\begin{array}{ll} \textit{Exponential} & \mu = \dfrac{1}{\lambda}; \\[2ex] \textit{Exponential Variance:} & \sigma^2 = \dfrac{1}{\lambda^2}. \end{array} \tag{5.9}$$

* To prove that the exponential distribution meets the two axioms defining a probability density function, first note that $f(T)$ is easily seen to be always positive or zero. Showing that the second axiom holds merely requires integrating $f(T)$ from 0 to infinity:

$$\int_{\text{All } T} f(T)\,dT = \int_0^\infty \lambda e^{-\lambda T}\,dT = \Big[-e^{-\lambda T}\Big]_0^\infty = e^0 = 1.$$

† Deriving the mean and the variance for the exponential distribution requires integrating $f(T)$ by parts (see Appendix A):

$$E[\boldsymbol{T}] = \int_0^\infty (T)\lambda e^{-\lambda T}\,dT = \Big[-Te^{-\lambda T}\Big]_0^\infty - \int_0^\infty -e^{-\lambda T}\,dT$$

$$= \Big[-Te^{-\lambda T}\Big]_0^\infty - \left[\frac{1}{\lambda}e^{-\lambda T}\right]_0^\infty = \frac{1}{\lambda}.$$

$$E[\boldsymbol{T}^2] = \int_0^\infty (T^2)\lambda e^{-\lambda T}\,dT = \Big[-T^2e^{-\lambda T}\Big]_0^\infty - \int_0^\infty -2Te^{-\lambda T}\,dT$$

$$= \Big[-T^2e^{-\lambda T}\Big]_0^\infty + 2\left[-\frac{1}{\lambda}Te^{-\lambda T}\right]_0^\infty + 2\int_0^\infty \frac{1}{\lambda}e^{-\lambda T}\,dT$$

$$= \Big[-T^2e^{-\lambda T}\Big]_0^\infty + 2\left[-\frac{1}{\lambda}Te^{-\lambda T}\right]_0^\infty - 2\left[\frac{1}{\lambda^2}e^{-\lambda T}\right]_0^\infty$$

$$= \frac{2}{\lambda^2}.$$

$$V(\boldsymbol{T}) = E[\boldsymbol{T}^2] - (E[\boldsymbol{T}])^2 = \frac{2}{\lambda^2} - \left(\frac{1}{\lambda}\right)^2 = \frac{1}{\lambda^2}.$$

Since the exponential and the Poisson distributions can both be applied to problems of arrivals at a service facility, suppose we reconsider the example in Section 4.4, using the exponential function to describe the time between arrivals at a supermarket where the mean arrival rate is $\lambda = 3.8$. By substituting $\lambda = 3.8$ in Formula (5.7) we get

$$f(T) = 3.8e^{-3.8T}.$$

Since $\lambda = 3.8$ for this example, the mean time between arrivals is

$$\frac{1}{\lambda} = \frac{1}{3.8} = 0.263 \quad \text{(minute);}$$

the variance of the time between arrivals is

$$\frac{1}{\lambda^2} = \frac{1}{(3.8)^2} = 0.069 \text{ (minutes}^2\text{)}.$$

Now, suppose we wanted to calculate the probability that the time between arrivals is less than one minute, $P(\mathbf{T} \leq 1)$. One way to find such a probability is to integrate the exponential function to find the area under the curve over the interval in question. Thus, one way to find the probability $P(\mathbf{T} \leq 1)$ is to evaluate the integral shown below (we will use an easier approach in a moment).

$$P(\mathbf{T} \leq 1) = \int_0^1 3.8e^{-3.8T}\,dT = \left[-e^{-3.8T}\right]_0^1$$
$$= 1 - e^{-3.8} = 0.978.$$

Similarly, one might want to determine the probability that the time between two consecutive arrivals will be between one-half and one minute:

$$P(\tfrac{1}{2} \leq \mathbf{T} \leq 1) = \int_{1/2}^1 3.8e^{-3.8T}dT = \left[-e^{-3.8T}\right]_{1/2}^1$$
$$= 0.150 - 0.022 = 0.128.$$

Fortunately, such integrations are not usually necessary, since tables have been prepared which permit evaluation of probabilities involving the exponential distribution. As was the case for the normal distribution, the cumulative function is presented in Appendix B (Table IV). Since there are an infinite number of different combinations of λ and T which might appear in such a table, Table IV presents values of $F(T)$ only for selected values of the product of λ times T, or λT. For example, if one is interested in determining the probability $P(\mathbf{T} \leq 0.5)$ when $\lambda = 3.8$, then it is necessary to first multiply the critical value of $T(=0.5)$ times the value of $\lambda(=3.8)$, which is $(0.5)(3.8) = 1.9$. The value of $F(T)$ corresponding to $\lambda T = 1.9$ can then be found in Table IV to be $F(T) = 0.850$. Thus,

$$P(\mathbf{T} \leq 0.5) = F(0.5) = 0.850.$$

By the complementary rule, we know that $P(\mathbf{T} \geq 0.5) = 1 - 0.850 = 0.150$.

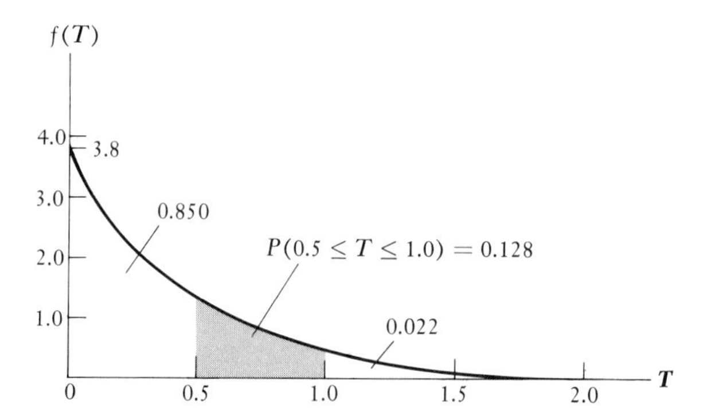

Fig. 5.12 Example of exponential probability problem with $\lambda = 3.8$.

Similarly, if one wants to determine the probability that an arrival will occur between one-half and one minute when $\lambda = 3.8$, Table IV can be used to find

$$P(0.5 < \boldsymbol{T} < 1) = F(1.0) - F(0.5).$$

We already know that $F(0.5) = 0.850$. The value for $F(1.0)$ is found in Table IV under $\lambda(T) = 3.8(1.0) = 3.8$, and equals $F(3.8) = 0.978$. Thus, $P(0.5 \leq \boldsymbol{T} \leq 1.0) = F(1.0) - F(0.5) = 0.978 - 0.850 = 0.128$, which agrees with our earlier result. This example is illustrated in Fig. 5.12.

Calculations of exponential probabilities can be especially useful in the queueing problems mentioned earlier. If, in addition to studying the pattern of arrivals at our supermarket checkout counter, we had also investigated the service time of the cashiers, then we could develop a relationship which indicates the probability that our cashiers will not be busy for a period of $\boldsymbol{T}$ minutes and the probability that the customers will have to wait more (or less) than $\boldsymbol{T}$ minutes. Ideally, such an investigation could lead to an analysis of the benefits of keeping a customer waiting versus the cost of hiring (or firing) another cashier, the result being a staffing policy designed to balance these costs and benefits. (Too often, it seems, the arrival rate *exceeds* the service rate for extended periods of time in many supermarkets.)

5.6 THE CHI-SQUARE DISTRIBUTION (OPTIONAL)*

A number of important probability distributions are closely related to the normal distribution. One of the most widely used of these related distributions is a continuous p.d.f. called the *chi-square* distribution. The chi-square distri-

* This section contains a formal introduction to the chi-square distribution. The material here forms a good introduction to Sections 6.9 and 6.10, but it is not a prerequisite to these sections.

bution gets its name because it involves the *square* of normally distributed random variables, as we will explain below.

Up to this point we have used examples involving just a single normal variable x. And we saw how this variable can be transformed into an equivalent z-variable in standardized normal form by letting $z = (x - \mu)/\sigma$. Now, let's assume we want to investigate the combined properties of more than just one standardized normal variable, where these variables are assumed to be independent of one another. You might imagine a number of machines (or workers) each producing (independently) a product whose length is normally distributed with μ and variance σ^2 [that is, $N(\mu, \sigma^2)$]. Furthermore, let's assume that the lengths from each machine have been standardized by using the transformation $z = (x - \mu)/\sigma$. The lengths produced by the different machines thus form a set of independent standardized normal variables,

$$z_1 = \frac{x_1 - \mu_1}{\sigma_1}, \qquad z_2 = \frac{x_2 - \mu_2}{\sigma_2}, \qquad \text{and so forth.}$$

It will be convenient to denote the *number* of variables in this situation (i.e., the number of machines or workers) by the letter ν (the Greek letter nu), and to take the *square* of each of these nu variables. We thus have a set of ν independent variables, as follows:

$$z_1^2 = \frac{(x_1 - \mu_1)^2}{\sigma_1^2}, \qquad z_2^2 = \frac{(x_2 - \mu_2)^2}{\sigma_2^2}, \qquad \ldots, \qquad z_\nu^2 = \frac{(x_\nu - \mu_\nu)^2}{\sigma_\nu^2}.$$

You may recall from Chapter 3 that we can form a new random variable by combining two or more random variables. For the present case, we want to take the *sum* of the variables $z_1^2, z_2^2, \ldots, z_\nu^2$. This sum is denoted by the letter χ^2 (the square of the Greek letter chi, or "chi-square"). Usually a subscript is added to the χ^2 symbol to denote the fact that there are ν values being summed. That is,

$$\chi_\nu^2 = z_1^2 + z_2^2 + \cdots + z_\nu^2 = \sum_{i=1}^{\nu} z_i^2.$$

The distribution of the random variable χ^2, which is referred to as the chi-square distribution, is important in a number of different contexts. For example, if z_i represents a deviation about the mean of the length of a certain product (in standardized units), then z_i^2 is the square of this deviation. And $\chi_\nu^2 = \sum_{i=1}^{\nu} z_i^2$ is the sum of the squared deviations of the ν different machines producing this product. Since a variance is also defined in terms of the sum of a set of squared deviations about a mean, it shouldn't surprise you that one of the most important applications of the χ^2 distribution involves variances. We will describe, more completely, this application of the χ^2 distribution in Chapter 6.

Properties of the χ^2 Distribution

The chi-square distribution is a family of density functions having a single parameter, ν, which is called the *number of degrees of freedom*. The degrees of freedom associated with a particular chi-square distribution completely determines the characteristics of its density function, $f(\chi^2)$. For example, the mean and variance of the χ^2 distribution are both related to ν, as follows:

$$\textit{Chi-square mean} \quad = E[\chi^2_\nu] = \nu$$

$$\textit{Chi-square variance} = V[\chi^2_\nu] = 2\nu.$$

The shape of the chi-square distribution is highly skewed to the right when ν is small. Consider, for example, the situation where $\nu = 1$, in which case

$$\chi^2_1 = z^2_1 = \frac{(x_1 - \mu_1)^2}{\sigma^2_1}.$$

The mean of this distribution is $E[\chi^2_1] = 1$, and its variance is $V(\chi^2_1) = 2$. Since the values of χ^2_1 will all be positive, and most of them will be relatively close to 1.0 (because we divided by σ^2_1), χ^2_1 has the shape shown in Fig. 5.13 corresponding to $\nu = 1$. The χ^2 distributions for $\nu = 2$, 4, 6, and 11 are also shown in this figure.

As ν becomes larger and larger, the shape of the χ^2 distribution becomes more and more symmetrical in appearance. Consider, for example,

$$\chi^2_{11} = z^2_1 + z^2_2 + \cdots + z^2_{11}.$$

In this case the mean is $E[\chi^2_{11}] = 11$ and $V[\chi^2_{11}] = 22$. The chi-square distribution for $\nu = 11$ is shown in detail in Fig. 5.14.

Fig. 5.13 The chi-square distribution for various values of ν.

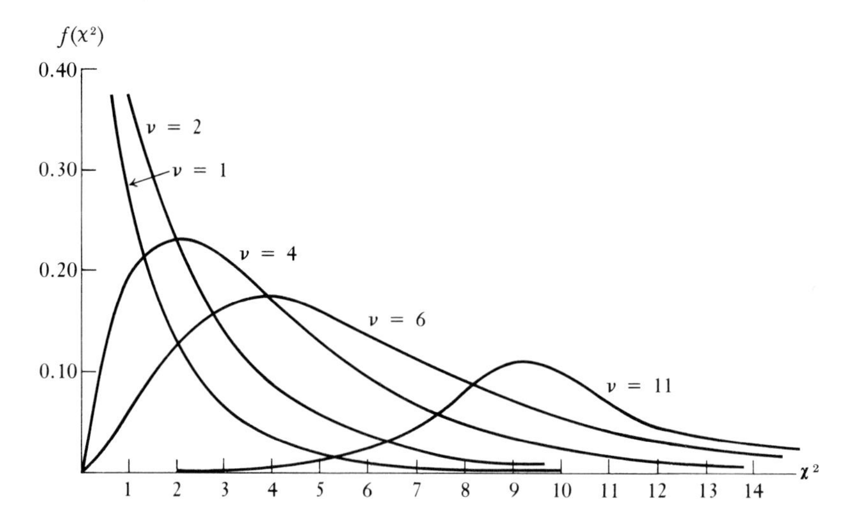

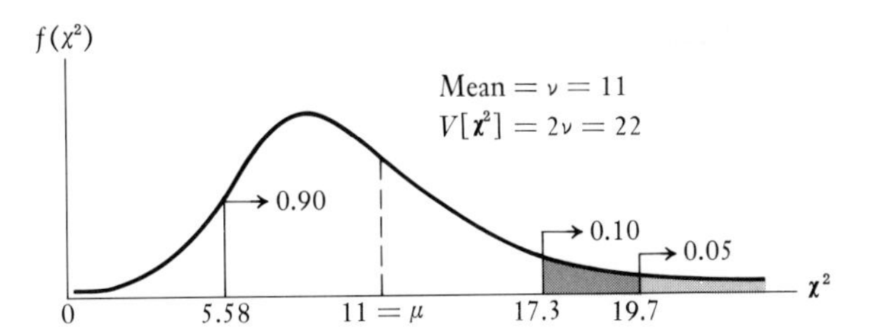

Fig. 5.14 Chi-square distribution for $\nu = 11$.

We have shown several probability values in Fig. 5.14. For example, we see that when $\nu = 11$,

$$P(\chi^2_{11} \geq 17.3) = 0.10, \qquad \text{and} \qquad P(\chi^2_{11} \geq 19.7) = 0.05.$$

Notice in this figure how symmetrical the distribution of χ^2 is for $\nu = 11$, almost like a normal distribution. It can be shown, in fact, that the limit of the χ^2 distribution as $\nu \to \infty$ is the normal.

Use of χ^2 Table—Examples

As was the case with the normal distribution, tables of the χ^2 distribution are readily available. In this case, however, the task of listing values of the distribution is more complicated because the chi-square distribution is different for each value of ν. For this reason it is common to list, in χ^2 tables, only those values of χ^2 which correspond to probabilities most often used. In addition, chi-square tables often give the value of $P(\chi^2 \geq \chi^2)$ rather than values of the cumulative function, as was the case for the normal distribution. The χ^2 table in this book (Table V, Appendix B) lists the values of χ^2 for ten different probability values (across the top of the table) and for various values of ν up to $\nu = 30$. When $\nu \geq 30$, the formula at the bottom of the table gives the normal approximation to the χ^2 distribution.

First Example. Suppose we want to determine the probability that the value of a chi-square random variable is greater than 18.0 when there are 11 degrees of freedom—that is, find $P(\chi^2_{11} \geq 18.0)$. From Fig. 5.14 we can easily see that $P(\chi^2_{11} \geq 18.0)$ must be larger than 0.05 (since 18.0 is less than 19.7), and it must be smaller than 0.10 (since 18.0 is larger than 17.3). Thus, we can write:

$$0.10 > P(\chi^2_{11} \geq 18.0) > 0.05.$$

This relationship also could have been determined by using Table V. First, we need to select the row corresponding to $\nu = 11$, and then find the columns

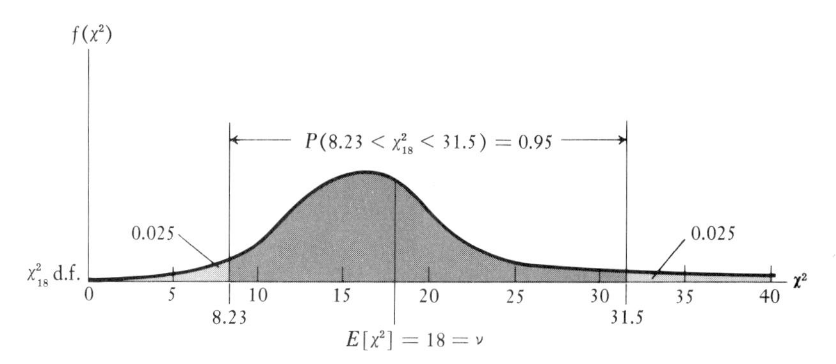

Figure 5.15

giving values of χ^2 which bracket the number 18.0. In this case, we see that

$$P(\chi_{11}^2 \geq 17.3) = 0.10 \qquad \text{and} \qquad P(\chi_{11}^2 \geq 19.7) = 0.05;$$

hence $P(\chi_{11}^2 > 18.0)$ must fall between 0.10 and 0.05. In other words, when $v = 11$ a value of χ_{11}^2 of 18.0, or larger, will occur between 5 and 10 percent of the time.

Second Example. As a second example, suppose we want to find the narrowest interval where the probability of χ^2 falling within the interval is 0.95, assuming that the number of degrees of freedom is 18. In other words, we want to find values a and b such that $P(a < \chi_{18}^2 < b) = 0.95$. To get the narrowest interval, we exclude an equal probability on the left and the right ends of the distribution, namely, 0.025 on each tail.* In terms of Table V, this means that we must find values a and b such that:

$$P(\chi^2 \geq b) = 0.025 \quad \text{and } P(\chi^2 \leq a) = 0.025 \left[\text{or } P(\chi^2 \geq a) = 0.975\right].$$

Using the row in Table V for $v = 18$, we find $P(\chi^2 \geq b) = 0.025$ when $b = 31.5$, and $P(\chi^2 \geq a) = 0.975$ if $a = 8.23$. Therefore,

$$P(8.23 \leq \chi_{18}^2 \leq 31.5) = 0.95,$$

and the range of the interval is as narrow as possible while still including 95 percent of the area under $f(\chi_{18}^2)$. Note that the distance from the two endpoints to the mean is not the same, since the chi-square distribution is not symmetric. It is the amount of *area* under $f(\chi^2)$ outside each endpoint (0.025) which is made equal in order to solve the problem. Figure 5.15 shows the χ^2 distribution for this example.

Define: binomial approximation, correction for continuity, exponential distribution, chi-square distribution.

* This statement is correct only if v not small ($v \geq 10$); otherwise, the chi-square distribution is quite skewed.

Table 5.1 Summary of Probability Distributions

Probability Distribution	Parameters	Characteristics	Mass or Density Function Given in Formula	Mean	Variance	Reference Sections	Probability Table
Discrete							
Binomial $\boldsymbol{x}$	$0 \leq p \leq 1$ $n = 0, 1, 2, \ldots$	Skewed unless $p = 0.5$, family of distributions	(4.1)	np	npq	4.2, 4.3	$P(\boldsymbol{x}_{\text{binomial}})$ in Table I
Poisson $\boldsymbol{x}$	$\lambda > 0$	Skewed positively, family of distributions	(4.5)	λ	λ	4.5	$P(\boldsymbol{x}_{\text{Poisson}})$ in Table II
Continuous							
Normal $\boldsymbol{x}$	$-\infty < \mu < +\infty$ $\sigma > 0$	Symmetrical, family of distributions	(5.1)	μ	σ^2	5.2	—
Standardized normal $\boldsymbol{z}$	—	Symmetrical, single distribution	(5.3)	0	1	5.3	$F(z)$ in Table III
Exponential $\boldsymbol{T}$	$\lambda > 0$	Skewed positively, family of distributions	(5.3)	$1/\lambda$	$1/\lambda^2$	5.5	$F(T)$ in Table IV
Chi-square $\boldsymbol{\chi}^2$	$v \geq 1$	Skewed positively, family of distributions	(See Table V)	v	$2v$	5.6	$P(\boldsymbol{\chi}^2 \leq \chi^2)$ in Table V

Problems

5.16 Senator Fogbound claims that 75% of his constituents favored his voting policies over the past year. In a random sample of 50 of these people, only 50% favored his voting policies. Is this enough evidence to make the senator's claim strongly suspect? Use a normal approximation.

5.17 Approximate your answer to Problem 4.14 on page 195 by using the normal approximation.

5.18 A recent article in the *Wall Street Journal* indicated that last year approximately 30% of all new cars sold were not made in the U.S.A. Suppose you take a random survey of 400 people who recently have purchased a new car, and find that 100 bought a foreign car. What might you conclude from this study, using a normal approximation to the binomial?

5.19 Approximate your answer to Problem 4.15 on page 196 by using the normal distribution.

5.20 If a pair of fair dice is tossed 49 times, what is the probability of receiving either a 7 or an 11 on 17 or more of these tosses? (Use a normal approximation and remember that $P(7 \text{ or } 11)$ for a pair of fair dice is $\frac{2}{9}$.) Would you recommend that a visitor to a gambling casino complain of loaded dice if this person observed exactly a 7 or 11 exactly 17 times?

5.21 Describe how the Poisson and the exponential distributions are related. What assumptions underlie these distributions? Given the exponential distribution whose parameter, λ, equals 3.0:

a) Graph the probability density function and the cumulative probability distribution.

b) What is the mean and the variance of this distribution?

c) What percent of the area of this distribution lies within $\pm$ one standard deviation of the mean? Within $\pm$ two standard deviations of the mean?

5.22 Suppose that you observe the following service times by a teller in a bank: $\frac{1}{2}$, 1, $\frac{1}{2}$, 6, 1, 3, where all times are in minutes. Assume that service times are exponentially distributed.

a) What is the mean service time, given these observations? What is the variance?

b) Graph the exponential distribution for this application.

c) What is the probability that service will take longer than two minutes?

5.23 Assume that the interarrival time at a certain tollbooth is exponentially distributed with a mean interarrival time of $\frac{1}{2}$ minute.

a) What is the value of λ for this problem?

b) What is the variance of the interarrival times?

c) What is the probability that the time between two consecutive arrivals will be between 0 and 1 minute?

d) What is $P(\boldsymbol{T} > 2.0)$?

5.24 A bank can service its customers, on the average, at the rate of four customers per six-minute period. Assume the number of customers serviced to be Poisson-distributed.

a) What is the probability that this bank will be able to service six or more customers in a six-minute period?

b) What is the probability that a customer will take longer than three minutes?

c) What is the probability that a customer will take between three and six minutes?

5.25 A barbershop has an average of ten customers between 8:00 and 9:00 each morning that it is open. Customers arrive according to a Poisson distribution.

a) What is the probability that the barbershop will have exactly 10 customers between these hours on a given morning?

b) What is the probability that the time between consecutive arrivals (customers) will exceed six minutes?

c) What is the probability that the time between consecutive arrivals will fall between 3 and 6 minutes?

5.26 Airplanes land at the Purdue University airport at the rate of one every 30 minutes, following a Poisson distribution.

a) What is the probability that the time between arrivals will be less than 15 minutes?

b) What is the probability that the time between arrivals will be greater than three quarters of an hour?

5.27 Recall the data in Chapter 1 for the Shark Loan Company (Problem 1.43).

a) Graph the exponential p.d.f. with $\lambda = 0.02$. Compare this graph with the polygon for Problem 1.43. Do they seem to correspond well?

b) Elaborate on your answer to part (a) by finding $P(0 \leq T \leq 20)$, $P(20 \leq T \leq 60)$, and $P(T \geq 60)$, and then comparing these answers to the table of values given in Exercise 2.65.

c) What are the mean, median, and standard deviation of an exponential distribution with $\lambda = 0.02$? Are these answers consistent with values for Problem 1.43(b)?

5.28 a) What parameter characterizes the chi-square distribution?

b) What is the mean and variance of a chi-square distribution?

c) Sketch the chi-square distribution for $v = 6$. Indicate on your sketch the region cutting off 5% of the righthand tail of this distribution.

5.29 Find values of a and b such that $P(a \leq \chi^2 \leq b) = 0.95$, assuming that $v = 8$. Sketch this area on a graph.

5.30 Suppose that a random variable is known to be chi-square distributed with parameter $v = 24$.

a) What is the mean and the variance of the chi-square distribution for this parameter?

b) What is the probability that the value of χ^2 will exceed 43.0? What is $P(\chi^2 \geq 33.2)$? What is $P(\chi^2 \geq 9.89)$?

c) Use your answers to parts (a) and (b) to draw a rough sketch of the chi-square distribution for $v = 24$.

d) Superimpose on your sketch for part (c) a graph of the normal distribution for $\mu = 24$ and $\sigma^2 = 48$. How closely do the two distributions agree?

Exercises

5.31 The *uniform* or *rectangular* distribution can be defined in terms of its cumulative distribution function, as follows:

$$F(x) = \begin{cases} \dfrac{x - a}{b - a}, & \text{for } a \leq x \leq b, \\ 0, & \text{for } x < a, \\ 1, & \text{for } x > b. \end{cases}$$

a) Derive the uniform density function for $a \leq x \leq b$ by differentiating $F(x)$ with respect to $\boldsymbol{x}$.

b) Sketch both the density function and the cumulative function for the uniform distribution.

c) Show that the uniform distribution satisfies the two properties of all probability functions.

d) Find the mean and the variance of this distribution.

5.32 How many questions would a professor have to put on a multiple-choice exam, where there are four choices for each question, in order to be 99.9 percent sure that a student who makes a random guess on each question misses at least one-half of the questions? (Use the normal approximation to the binomial without correcting for continuity.)

5.33 a) Prove that the standardized normal density function reaches its maximum height at $z = 0$. (*Hint:* Set the first derivative equal to zero.)

b) Prove that the points $z = \pm 1$ for the standardized normal density function are inflection points. (*Hint:* Set the second derivative equal to zero.)

5.34 A continuous p.d.f. not presented in this chapter is the *Beta distribution*. This distribution has two parameters, r and n.

$$f(x) = \begin{cases} \dbinom{n-1}{r-1} x^{r-1}(1 - x)^{n-r-1} & 0 \leq x \leq 1, \\ 0 & \text{otherwise.} \end{cases}$$

a) Discuss the similarities and differences between this distribution and the binomial distribution.

b) Graph this distribution for $r = 1, n = 2$.

c) Determine $E[\boldsymbol{x}]$.

d) If $r \leq n/2$, will the distribution be positively or negatively skewed?

5.35 In Chapter 6 we will show that the random variable $(v\boldsymbol{s}^2/\sigma^2)$ follows a $\boldsymbol{\chi}^2$ distribution (that is, $\boldsymbol{\chi}^2 = v\boldsymbol{s}^2/\sigma^2$), where $\boldsymbol{s}^2$ is a sample variance.

a) If $E[\boldsymbol{\chi}^2] = v$, what is $E[\boldsymbol{s}^2]$?

b) If $V[\boldsymbol{\chi}^2] = 2v$, what is $V[\boldsymbol{s}^2]$?

c) Find the probability that $\boldsymbol{s}^2 \geq 20$, when $\sigma^2 = 10$ and $v = 8$, by calculating

$$P\left(\chi^2 \geq \frac{vs^2}{\sigma^2}\right).$$

5.36 One of the most important characteristics of the exponential distribution is called the "memoryless property." This property states that $P(\boldsymbol{T} > a + b | \boldsymbol{T} > a) = P(\boldsymbol{T} > b)$, where a and b are times. For example, suppose that the average time between arrivals of an empty taxicab at a certain New York City corner is 4 minutes, following an exponential distribution.

a) If you have been waiting for a cab for 3 minutes, what is the probability that you will wait at least two more minutes?

b) Prove the memoryless property.

6

SAMPLING AND SAMPLING DISTRIBUTIONS

"You don't have to eat the whole ox to know that the meat is tough."

SAMUEL JOHNSON

6.1 INTRODUCTION

This chapter begins the task of relating the probability concepts studied in the past four chapters to the objective of statistics stated in Chapter 1: to draw inferences about population parameters on the basis of sample information.

Recall from Chapter 1 that a major portion of statistics is concerned with the problem of estimating population parameters, or testing hypotheses about such parameters. If we could take a "census" (i.e., examine all items in the population), then the value of the population parameters would be known. Unfortunately, a census of a population is usually not feasible because of monetary and/or time limitations. Hence, we must rely on observing a subset (or *sample*) of the items in the population, and use this information to make estimates or test hypotheses about the unknown parameters. The process of making estimates will be covered in Chapter 7, while the subject of hypothesis testing will be described in Chapter 8. In these subsequent chapters we will usually assume that a sample has been taken. It is therefore important that we describe in this chapter how a sample is taken, and what type of information can be drawn from a sample.

There are four basic questions that must be asked about samples and the process of inference:

1. What are the least expensive methods for collecting samples which best ensure that the samples are representative of the parent population?
2. What is the best way to describe sample information usefully and clearly?
3. How does one go about drawing conclusions from samples and making inferences about the population?
4. How reliable are the inferences and conclusions drawn from sample information?

By describing various *sample designs*, we will answer first the question of how sample information can be most efficiently collected.

A sample design is a procedure or plan, specified before any data are collected, for obtaining a sample from a given population.

6.2 SAMPLE DESIGNS

Nonsampling Errors

The primary requisite for a "good" sample is that it be representative of the population that one is trying to describe. There are, of course, many ways of collecting a "poor" sample. One obvious source of errors of misrepresentation arises when the *wrong population is sampled inadvertently*. The 1936 presidential election poll conducted by the now defunct *Literary Digest* remains a classic example of this problem. The *Literary Digest* predicted, on the basis of a sample of over two million names selected from telephone directories and automobile registrations, that Landon would win an overwhelming victory in the election that year. Instead, Roosevelt won by a substantial margin. The sample collected by the *Digest* apparently represented the population of predominantly middle- and upper-class people who owned cars and telephones; it misrepresented the general electorate, however, and Roosevelt's support came from the lower-income classes, whose opinions were not reflected in the poll. Telephone surveys even in the 1980's can misrepresent certain populations. Current estimates indicate that 15 percent of the U.S. adult population either do not have phones or have unlisted numbers.

Modern polls also can be imprecise despite the aid of computers and elaborate interviewing techniques. During the 1980 Presidential election, for example, most polls indicated a close race between Carter and Reagan. Instead, Reagan won by a margin of 51% to 41%. Most of the polls missed Reagan's actual voting percentage by at least four points, which, in the case of the George Gallup poll, was the largest error factor it has ever had in a presidential election. In retrospect, the public opinion industry has suggested that since a large number of voters apparently moved to Reagan during the last two days of campaigning, more last-minute polls should have been taken. One reason for a lack of such last-moment polls is the cost—a major national survey involving at least 1500 people will cost approximately $25,000.

Perhaps the most widely publicized poll is the A. C. Nielsen Company's rating of television viewing in the U.S. The Nielsen organization, which is the largest marketing research firm in the world, bases its ratings on a sample of 1170 TV homes. This sample is based on a random sample of locations selected from a Census Bureau list so as to give a geographical spread across the country. The ratings, which may have an error of as much as 2.6 points,* can mean life or death for a TV show, and affect billions of advertising dollars a year. Advertising agencies may pay over $300,000 a year for the Nielsen service, a price which is small when compared to the $200,000 (or more) a minute that is often charged for commercials on top-rated shows.

A potential source of error in sampling, especially in surveys of public opinion, comes from *response bias*. Poorly worded questionnaires or improper interview techniques may elicit responses that do not reflect true opinions. Kinsey's research on sex practices, for example, received widespread criticism for reporting responses to questions to which most people are fairly sensitive. Such responses are, therefore, likely to be distorted from the truth. Similarly, it is amazing how the economic well-being of certain college alumni can vary over the interval between the annual homecoming reunion and the annual fund-raising drive.

These types of error are called *nonsampling errors*. Nonsampling errors include all kinds of "human errors"—mistakes in collecting, analyzing, or reporting data; sampling from the wrong population; and response bias. If the researcher incorrectly adds a column of numbers, this represents a nonsampling error just as much as does the failure of a respondent to provide truthful information on a questionnaire.

* Their rating plus or minus 2.6 points would bracket the correct rating about 95% of the time.

Sampling Errors

In addition, even in well-designed and well-executed samples, there are bound to be cases in which the sample does not provide a true representation of the population under study, simply because samples represent only a portion of a population. In such cases the information contained in the sample may lead to incorrect inferences about the parent population; that is, an "error" might be made in estimating the population characteristics based on the sample information. Errors of this nature, representing the differences that can exist between a sample statistic and the population parameter being estimated, are called *sampling errors.* Sampling errors obviously can occur in all data-collection procedures except a complete enumeration of the population (a census).

One primary objective in sample design is to minimize both sampling and nonsampling errors. Errors are costly, not only in terms of the time and money spent in collecting a sample, but also in terms of the potential loss implicit in making a wrong decision on the basis of an incorrect inference from the data. An inaccurate public-opinion survey, for instance, could cost a politician votes if a campaign design is based on inferences from these data. Similarly, investment in real estate or stocks might cause an investor to lose a considerable amount of money if the (sample) information that led to a particular investment proved incorrect.

Note that it is the *decisions* resulting from incorrect inferences that may be costly, not the incorrect inferences themselves; hence, it is customary to refer to one objective of sampling as that of *minimizing the cost of making an incorrect decision* (*error*). But reducing the costs of making an incorrect decision usually implies increasing the cost of designing and/or collecting the sample. For example, additional effort (or money) devoted to designing a questionnaire, identifying the correct population, or collecting a larger sample usually results in a more representative sample. We can therefore state that the primary objective in sample design is to *balance the costs of making an error and the costs of sampling.*

Designing an optimal sampling procedure may not be easy. For one reason, the elements of a given population may be extremely difficult to locate, gain access to, or even identify. For example, it may be impractical, if not impossible, to identify the population elements of "color television owners" in a particular city in the United States. Another obvious difficulty already mentioned is cost; budget constraints, for example, may force one to collect fewer data or to be less careful about collecting these data than ideal designs would dictate. Also, the costs of making an incorrect decision may be very hard to specify. A discussion of all the problems inherent in sample design, especially those concerned with nonsampling errors and the costs of making an incorrect decision, falls outside the scope of this book. Therefore, we shall concentrate our attention on the problem of determining which sample designs most effectively minimize sampling errors.

Probabilistic Sampling

Often, the first criterion for a good sample is that each item in the population under investigation have an *equal and independent chance* to be part of the sample; also, it is often advantageous that each set of n items have an equal probability of being included. Samples in which every possible sample of size n (i.e., every combination of n items from the N in the population) is equally likely are referred to as *simple random samples*. These are the types of samples that we have used implicitly throughout our discussion of probability theory to illustrate the use of formulas and probability functions.

Simple random sampling requires that one have access to all items in the population. For a small population of elements that are easy to identify and sample, this procedure normally gives the best results. However, simple random sampling of a large population may be difficult, perhaps even impossible, to implement; at best, it will be quite costly. For this case, a more practical procedure must be designed, even though it also will be more restrictive.

In another popular sampling plan, called *systematic sampling*, a random starting point in the population is selected, and then every kth element encountered thereafter becomes an item in the sample. For example, every 200th name in a telephone directory might be called in order to survey public opinion. This method is *not* equivalent to simple random sampling because every set of n names does not have an equal probability of being selected. Bias will result under systematic sampling if there is a *periodicity* to the elements of the population. For instance, sampling sales in a supermarket every seventh day certainly will result in a sample that represents only the sales of a single day, say Monday, rather than the weekly pattern.

Using a Random Number Table

Designing a sample in which each set of items in the population has an equal probability of being selected usually requires carefully controlled sampling procedures. The stereotype of drawing slips of paper from a goldfish bowl may satisfy the requirements of a simple random sample, but there is no practical way to be certain of this. The 1969 Selective Service draft lottery was highly criticized for using essentially a "goldfish bowl" technique without adequate mixing. A more systematic approach to ensuring randomness is to select a sample with the aid of a table of random numbers. In such a table, each digit between 0 and 9 is called a *random digit*; here, the word random implies that all of these digits have the same long-run relative frequency (i.e., the same probability of occurring), and the occurrence or nonoccurrence of any number is independent of the occurrence or nonoccurrence of all other numbers, or of all sets of n other numbers. In a table of random numbers, random digits are usually combined to form numbers of more than one digit. For example, random digits taken in pairs will result in a set of 100 different numbers (00 to 99), each with a probability of occurring of $\frac{1}{100}$, each independent of other numbers formed in the same manner. Likewise, in a table of random numbers consisting of groups of three random digits, each of the 1000 numbers between 000 and 999 will have a probability of occurring of $\frac{1}{1000}$ and will be independent of the remaining numbers. Table VI is a page of random numbers from a book published by the Rand Corporation, containing one million random digits.

A table of random numbers can be used in the following manner to select a simple random sample of n items from a finite population of size N. First, a unique number between 0 and N must be assigned to each of the N items in the population. The table of random numbers is then consulted. The first n numbers encountered (starting at *any* point in the table and moving systematically across rows or down columns) which are less than N constitute a set of n random numbers. The n elements corresponding to these n numbers form the random sample.

To illustrate this procedure, we will use Table VI to select a sample of four items from a population of 75 elements. A random selection of a starting point in the table is customary; let us arbitrarily start with the number which is tenth from the bottom in the first column, 09237. Since our population has less than 100 elements, we need to look at only two digits of each number. Suppose we use the first two digits. Then the first item in our sample becomes item number 09. Reading down, the next three items are 11, 60, and 71 (we skip 79 because we have only 75 elements in the population). Note that if our sample were to contain five items, the number 09 would have occurred again. In most practical situations it may not be possible or desirable to sample the same item twice. Its usefulness may be destroyed by the first sample, as would be the case in testing the tread life of a tire or measuring the yield from a new seed variety at an agricultural testing station. The usual procedure in this situation is to let the next element on the list (e.g., item 63) take its place. When duplicated items are discarded, it must be recognized that sampling is now taking place without replacement, rather than with replacement.

Certain special terms are used by the accounting profession to designate specific types of sampling objectives. For example:

Attribute sampling: used when the auditor wants to estimate population proportion (p). For example, a CPA may want to estimate the proportion of accounts receivable that exceed some specified dollar value.

Variable sampling: used when the auditor wants to estimate a population mean (μ). For instance, a CPA may wish to estimate the mean dollar amount owed by all customers.

Acceptance sampling: used to determine whether or not an error rate is large enough to reject the entire population. One might want to determine whether the number of defectives in a sample of items is too large to accept the entire lot.

Discovery sampling: a special case of attribute sampling used to locate relatively rare occurrences. For example, an auditor might be looking for critical errors and fraud. This approach is not used for estimating population characteristics.

Sampling with Prior Knowledge

Two important random sampling plans depend on prior knowledge about the population: stratified sampling and cluster sampling.

Stratified Sampling The use of *stratified sampling* requires that a population be divided into homogeneous classes or groups, called *strata*. Each stratum is then sampled according to certain specified criteria. The advantage of this procedure is that if homogeneous subsets of the population can be identified, then only a relatively small number of observations is needed to determine the characteristics of each subset. It can be shown that:

> *The optimal method of selecting strata is to find groups with a large variability between strata, but with only a small variability within strata.*

We illustrate stratified sampling by considering the task of determining the majority political preference in a given city. Assume that it is known, from previous surveys and elections, that political preferences in this town tend to correspond to various income levels. For instance, upper-income families tend to have similar opinions, as do middle-income and lower-income families. Assume further that, in this particular city, it is well known that the upper-income families and the lower-income families will have less variability of opinion within their respective groups than will the middle-income group. It may be that upper-income families in general will favor a fiscally conservative candidate, and lower-income families will favor a candidate who promises increased city services, while middle-income families will be less predictable.

A *proportional* stratified sampling plan selects items from each stratum in proportion to the size of that stratum. This procedure ensures that each stratum in the sample is "weighted" by the number of elements it contains. If the category upper-income families includes ten percent of the voting population, then a proportional stratified sampling plan will randomly select ten percent of the sample from this group. Many times, however, a more efficient procedure is to select a *disproportionate* stratified sample. A plan of this nature collects more than a proportionate amount of observations in those strata with the most variability; e.g., the middle-income group in the above example. In other words, by allocating a disproportionate amount of effort (time, money, etc.) to those groups whose opinions are most in doubt, one often obtains a maximum amount of information for a given cost. Similarly, if it is more costly to sample from a particular stratum, one may elect to take fewer items from that stratum.

Cluster Sampling *Cluster sampling* represents a second important sampling plan in which the population is subdivided into groups in an attempt to design an efficient sample. The subdivisions or classes of the population in this case are called clusters, where each cluster, ideally, has the same characteristics as the parent population. If each cluster is assumed to be representative of the population, then the characteristics of this population can be estimated by (randomly) picking a cluster and then randomly sampling elements from within this cluster. Sampling within a cluster may take any of the forms already discussed and may even involve sampling from clusters within a cluster (called two-stage cluster sampling). The criterion for the selection of optimal clusters is exactly opposite to that for strata:

> *There should be little variability between clusters, but a high variability (e.g., representation of the population) within each cluster.*

Cluster sampling can be illustrated by extending our previous example. Assume that we now want to sample political preferences in all cities of the United States rather than just one. In this case a simple random sample would probably be very difficult and expensive, if not impossible, to collect; instead, it may be that a number of cities adequately represent the population of all cities, and it would then be sufficient to sample from just one of these cities. Within the chosen city one could use simple random sampling, stratified sampling, or systematic sampling, or one could break the city into smaller clusters. In cluster sampling there is always the danger that a cluster is not truly representative of the population; a geographical bias, for example, may exist when one city is used to represent political preferences in all cities of the United States.

Double, Multiple, and Sequential Sampling

One of the most important decisions in any sampling design involves selecting the *size* of the sample. Usually size is determined in advance of any data collection, but in some circumstances this may not be the most efficient procedure. Consider the problem of determining whether a shipment of 5000 items meets certain specified standards. It would be too expensive to check all 5000 items for their quality, so a sample is drawn and each item in the sample is tested for quality. Rather than take one large sample, of perhaps 100 items, a preliminary random sample of 25 items could be drawn and inspected. It may often be unnecessary to examine the remaining 75 items, for perhaps the entire lot can be judged on the basis of these 25 items. If a high percentage of the 25 components were defective, the conclusion drawn would probably be that the quality of the entire lot may not be acceptable. A low percentage of defectives may lead to accepting the lot. Values other than these extremes may also lead to acceptance or rejection of the entire lot. Nevertheless, there will usually be a range in which there is doubt about the quality of the entire lot. For example, it may be normal to have one or two defectives in a sample of 25; more than three, however, may lead one to suspect the entire lot, but not necessarily to reject it. An additional sample, perhaps the remaining 75 items, could then be taken, and the lot judged on the basis of all 100 items.

Samples in which the items are drawn in two different stages, such as in the sequential fashion described above or from a cluster within a cluster, represent a process referred to as *two-stage sampling.* Virtually all important samples, and certainly all large-scale surveys, represent one form or another of multiple-stage sampling, and the sample design is usually not simple to plan. The major advantage of double, multiple, and sequential sampling procedures obviously depends on the savings that result when fewer items than usual must be observed. These procedures are especially appropriate when sampling is expensive, as when inspection destroys the usefulness of a valuable item or when travel expenses of the survey team would be high.

Nonprobabilistic Sampling

In some sense all nonprobabilistic sampling procedures represent *judgment samples*, in that they involve the selection of the items in a sample on the basis of the judgment or opinion of one or more persons. Judgment sampling is usually employed when a random sample cannot be taken or is not practical. It may be that there is not enough time or money to collect a random sample; or perhaps the sample represents an exploratory study where randomness is not too important. On the other hand, when the number of population elements is small the judgment of an expert may be better than random methods in picking a truly representative sample. For example, you are using judgment sampling when you ask a friend's opinion about a movie, or about a particular college course. Similarly, "representative" individuals or animals are often chosen to participate in experiments, and accountants frequently select "typical" weeks for auditing accounts.

In *quota sampling*, each person gathering observations is given a specified number of elements to sample. This technique is used often in public-opinion surveys, in which the interviewer is allocated a certain number of people to interview. The decision as to exactly whom to interview is usually left to the individual doing the interview, although certain guidelines are almost always established. With well-trained and trustworthy interviewers, this procedure can be quite effective and can be carried out at a relatively low cost. Great danger exists, however, that procedures left to the interviewers' judgment and convenience may contain many unknown biases not conducive to a representative sample. Quota sampling is used often to obtain market research data or to survey for political preferences. In some of these situations a quota sample can be thought of as a special form of stratified sampling, in which interviewers are sent out and told to obtain a specified number of interviews from each stratum.

The least representative sampling procedure selects observations on the basis of convenience to the researcher; i.e., a *convenience sample.* "Street-corner surveys," in which the interviewer questions people as they go by, seems to be a favorite method of local TV news reporters for collecting public opinions. This method obviously cannot be considered very likely to yield a representative sample; more often, the results are biased and quite unsatisfactory. Convenience sampling is not widely used in circumstances other than preliminary or exploratory studies, or where representativeness is not a crucial factor.

The following table itemizes the sampling procedures described in Section 6.2.

Probabilistic	**Nonprobabilistic**
Simple random sampling	Judgment sampling
Stratified sampling	Quota sampling
Cluster sampling	Convenience sampling
Multiple/sequential sampling	

6.3 SAMPLE STATISTICS

As we have indicated previously, the usual purpose of sampling is to learn something about the population being sampled. In selecting a sampling design, the primary considerations are the importance of the information to be gathered and the desired degree of accuracy of what is learned about the population. In view of these purposes, it is important that we structure the problem of taking a sample and analyzing the sample results in terms of the concepts of probability presented in Chapters 2 through 5.

Assume that we are planning to take a sample of n observations in order to determine the characteristics of some random variable $\boldsymbol{x}$. The process of taking a sample from this population can be viewed as an experiment, and the observations that may occur in such an experiment make up the sample space. Suppose we let the random variables $\boldsymbol{x}_1, \boldsymbol{x}_2, \ldots, \boldsymbol{x}_n$ represent the observations in this sample. That is, the random variable $\boldsymbol{x}_1$ represents the observation which occurs first in a sample of n observations, x_2 represents the second observation, and so forth. In simple random sampling, every item in the population has an equal chance of being the observation that occurs first, so in this case the sample space for x_1 would be the entire population of x-values. It is important to remember that $x_1, x_2, \ldots, x_n$ are all *random variables*, and each of these variables has a theoretical probability distribution. Under simple random sampling, the probability distribution of each of the random variables $x_1, x_2, \ldots, x_n$ will be identical to the distribution of the population random variable x (since the marginal distribution of x_i is the same as that of the population).

Once we have collected a random sample of n observations, we have one value of x_1, one value of x_2, and so forth, with one value of x_n. We now need to discover how to learn more about the characteristics of the random variable x (the population) by making use of the sample values of $x_1, x_2, \ldots, x_n$. In general, the population parameters of interest are usually those described in Chapter 1—i.e., the summary measures such as central location, dispersion, skewness, or kurtosis. It is intuitively appealing (and mathematically provable) that the best estimate of a population parameter is given by a comparable sample measure (a sample *statistic*). For example, we will see that the best estimate of the central location of a population is a measure of the central location of a sample; and the best estimate of the population dispersion is a measure of the dispersion of the sample. *Thus, a sample statistic is used as an estimate of a population parameter.** Many different sample statistics can be used to estimate the population parameters of interest. Generally, the population parameters of most interest are the mean and variance, since these two measures are so useful in describing a distribution and so necessary in decision-making. Consequently, the most useful sample statistics are those that provide the best information about these two parameters, namely the sample mean and the sample variance. As we indicated earlier, there are a number of different sampling procedures that one may use to estimate population parameters. The particular procedure selected will have an effect on one's ability to make inferences about the underlying population. We will assume, at least initially, simple random sampling. In the development to follow, the reader should bear in mind that a similar development could be presented for sample statistics other than the mean and variance.

* A sample statistic can be defined as a function of some (or all) of the n random variables $x_1, x_2, \ldots, x_n$. That is, a sample statistic is a random variable that is based on the sample values of $x_1, x_2, \ldots, x_n$. This means that there is a theoretical probability distribution associated with every sample statistic. For example, suppose we let $R = x_{\max} - x_{\min}$ be the *range* of the values in a sample. In this case, the sample statistic R is a random variable which is a function of only two values in each sample, the largest value ($x_{\max}$) and the smallest value ($x_{\min}$). Since R is a random variable, it has a theoretical probability distribution that we could develop if this statistic were of interest. The method of sampling used will affect the probability distribution of any sample statistic.

Sample Mean and Variance

The mean of a set of observations that represent a sample is calculated in the same manner demonstrated in Chapter 1 for the mean of a population [Formula (1.2)]: by summing the product of each value of x times its relative frequency. In this case, we assume that the sample consists of n observations which can be grouped into c different classes, where the frequency of each class is denoted by f_i and the value of x for observations in that group is x_i (for $i = 1, 2, \ldots, c$, where c is the number of classes). The mean of these observations is called the *sample mean*, and is denoted by the symbol $\bar{x}$ (read x-bar):

Sample mean: $$\bar{x} = \frac{1}{n}\sum_{i=1}^{c} x_i f_i. \tag{6.1}$$

The reader may wish to verify the similarity between Formulas (6.1) and (1.2). Also, note that if the frequency of each sample observation is 1, then Formula (6.1) becomes $\bar{x} = (1/n)\sum x_i$. This is comparable to Formula (1.1), which was $\mu = (1/N)\sum x_i$.

Before illustrating the use of Formula (6.1), we present the formula for the variance of a sample. In this case, the sample variance is *not* calculated in the same manner as a population variance, for two reasons. First, we cannot take the sum of squared deviations about μ, for in most sampling problems μ is an unknown. Instead, we will take the sum of squared deviations about $\bar{x}$, or $\sum(x_i - \bar{x})^2$. As was the case for the population variance, each squared deviation must be multiplied by a frequency, so we now have $\sum(x_i - \bar{x})^2 f_i$. The second difference is that, in the present case, we divide the sum of squared deviations not by the number of observations, but by $(n - 1)$. The reason for dividing by $(n - 1)$ is that the resulting measure of variability can be shown to satisfy certain desirable properties for making estimates about the (unknown) population variance. This fact will be explained in more detail in Chapter 7, when we discuss estimation procedures. For now, we merely demonstrate the use of this measure of variability, which is denoted by the symbol s^2.

Sample variance: $$s^2 = \frac{1}{n-1}\sum_{i=1}^{c} (x_i - \bar{x})^2 f_i. \tag{6.2}$$

Recall that we developed a computational formula for variances in Chapter 1 [Formula (1.9)]. The same type of computational formula can be developed for (6.2), where the only difference is that, because our divisor is now $(n - 1)$,

Table 6.1 Number of Federal Grants to Cities with Populations 50,000–200,000

(1) x	(2) f	(3) xf	(4) $(x - \bar{x})$	(5) $(x - \bar{x})^2$	(6) $(x - \bar{x})^2 f$	(7) x^2	(8) $x^2 f$
1	2	2	−1.8	3.24	6.48	1	2
2	3	6	−0.8	0.64	1.92	4	12
3	1	3	0.2	0.04	0.04	9	9
4	3	12	1.2	1.44	4.32	16	48
5	1	5	2.2	4.84	4.84	25	25
	10	28			17.60		96

the old "mean square − square mean" relationship has to be modified slightly, as follows:*

Sample variance (computational form):

$$s^2 = \frac{1}{n-1}\left(\sum_{i=1}^{c} x_i^2 f_i - n\bar{x}^2\right). \tag{6.3}$$

The *sample standard deviation* is denoted by s, and always equals the *square root* of the sample variance.

To demonstrate the use of these formulas, we will consider the case of a United States Senate committee investigating the number of federal grants awarded for local projects (such as HUD and TOPICS Programs) in cities with populations ranging from 50,000 to 200,000. In an attempt to measure the characteristics of the population in this case (all cities of that size in the United States), ten randomly selected cities were surveyed. Table 6.1 shows the results of this survey ($\boldsymbol{x}$ = number of grants during the past year, f is the frequency of each value of $\boldsymbol{x}$, and $c = 5$ classes). The sample mean can be calculated using the information in columns (2) and (3):

$$\bar{x} = \frac{1}{n}\sum_{i=1}^{5} x_i f_i = \frac{1}{10}(28) = 2.8.$$

The average number of grants in the ten cities sampled is thus 2.8.

* In developing (6.3) from (6.2) we might have presented the following intermediate step:

$$s^2 = \frac{1}{n-1}\left(\sum x_i^2 f_i - 2\bar{x}\sum f_i x_i + \bar{x}^2 \sum f_i\right) = \frac{1}{n-1}\left(\sum x_i^2 f_i - 2n\bar{x}^2 + n\bar{x}^2\right)$$

where the terms parentheses represent a simplification of the summation terms in (6.2).

To measure the variance of these data, we can use either Formula (6.2) or (6.3). Using Formula (6.2) and the data in column (6) of Table 6.1, we can calculate s^2 as follows:

$$s^2 = \frac{1}{n-1} \sum_{i=1}^{c} (x_i - \bar{x})^2 f_i = \frac{1}{9}(17.60) = 1.955.$$

If the computational Formula (6.3) had been used, only columns (3) and (8) from Table 6.1 would need to be used:

$$s^2 = \frac{1}{n-1}\left(\sum_{i=1}^{c} x_i^2 f_i - n\bar{x}^2\right) = \frac{1}{9}[96 - 10(2.8)^2] = 1.955.$$

The sample standard deviation is $s = \sqrt{1.955} = 1.398$. Again, one way to check to see if the result $s = 1.398$ is reasonable is by our old rule of thumb. About 68 percent of the sample values should fall within one standard deviation of the mean, $\bar{x} \pm 1s$. In this case, the result appears reasonable, since seven of the ten observations lie in the interval $\bar{x} \pm 1s = 2.8 \pm 1.398 = 1.402$ to 4.198.

Perhaps at this point we should reiterate that one of our objectives in calculating sample means and variances is to be able to make statements about the population mean and variance. Since different samples from the same population may have different means and variances, the only way to determine the true population parameters is to enumerate every item in the population (a census). But a census is generally too costly and time-consuming. Thus, we must be content to *use sample statistics to estimate the population parameters and then to make statements about how reliable or accurate such a sample statistic is in describing the population parameter of interest.*

To establish the reliability or accuracy with which a sample statistic describes a population parameter, one must know how likely it is that specific values of this statistic will occur (1) for every possible value of the population parameter and (2) for every possible sample size. To begin our discussion of the reliability and accuracy of a sample, let us suppose that we can take a large number of random samples, all of size n, and then calculate $\bar{x}$ for each of these samples. These values of $\bar{\mathbf{x}}$ can be put in the form of a frequency distribution. This frequency distribution will have a certain shape, as well as a mean and a variance. Now, if we take *all possible* samples of size n (and the number of samples may be infinite), and determine $\bar{x}$ for each sample, the resulting distribution is the *probability* distribution of all possible values of $\bar{\mathbf{x}}$.

The probability distribution of $\bar{\mathbf{x}}$ is called a *sampling distribution.* We can also calculate a sampling distribution for $\mathbf{s}^2$ by considering all possible values of $\mathbf{s}^2$ from samples of a given size n. Such sampling distributions are necessary for making probability statements about the reliability and accuracy of sample statistics, and will be discussed in detail in the remaining sections of this chapter.

Define: sample design, response bias, nonsampling errors, sampling errors, simple random sampling, systematic sampling, stratified sampling, cluster sampling, two-stage sampling, probabilistic sampling, nonprobabilistic sampling, sample statistic, sample mean, sample variance.

Problems

6.1 Under what conditions is nonprobabilistic sampling more appropriate than probabilistic sampling? Give several examples.

6.2 Distinguish between:

a) systematic sampling and simple random sampling,

b) stratified and cluster sampling,

c) single-stage sampling and multiple-stage sampling,

d) judgment, quota, and convenience sampling.

6.3 In designing a sample survey, what factors are most important in establishing the strata in stratified sampling? The clusters in cluster sampling? How will the cost of sampling affect these decisions?

6.4 A company packages sunflower seeds. Obviously not all the seeds in a given lot will germinate and produce satisfactory flowers. However, the company does not want to package and present for consumer purchase an excessive number of bad seeds. In the long run, about $\frac{3}{4}$ of the sunflower seeds produce while $\frac{1}{4}$ do not. Before packaging a new lot, the company would like to be sure that at least $\frac{3}{4}$ of the new seeds will germinate. One clever(?) student says that they could test a random sample of only 4 seeds, and if 3 of them grew, they could assume that $\frac{3}{4}$ of all the seeds were good.

a) Do you think that the direct relationship between the probability from the sample and the likelihood of good seeds in the entire lot is perfect, as the student suggests? Explain briefly.

b) Suppose exactly $\frac{3}{4}$ of all the seeds were good; what is the probability that exactly 3 out of 4 in a random sample would be good? (Although sampling is without replacement, assume that the number of seeds is very large so that the probability of selecting a good seed, $\frac{3}{4}$, remains the same.)

6.5 Use Table VI to select a random sample of 8 observations from a population of 500 items. Specify where in Table VI you started your process, and which are to be included in the sample.

6.6 The word "random" has been used in several different contexts thus far. Define and distinguish between these uses:

a) random variable

b) random sample

c) observations selected "at random"

6.7 Consider as a single population all undergraduate students in your university.

a) Specify exactly how you would go about taking a simple random sample of 50 students. Would a systematic sampling plan be easier in this situation? Would systematic sampling introduce any biases?

b) Specify exactly how you would go about taking a stratified sample, where the strata are the 4 classes of students (e.g., senior, junior, etc.), and each stratum is sampled in proportion to the number of students in that class.

6.8 Assume that you have been commissioned to design a survey of the age, income, and occupation of the customers who patronize a nationwide chain of stores. Describe how you would proceed with such a study.

6.9 Jerome Kurtz, a former IRS Commissioner, reported in the *Wall Street Journal* that 41% of all U.S. taxpayers used the 1040A short form. You take a random sample of six groups of IRS returns and find that current taxpayers used form 1040A in the following percents: 47, 38, 42, 40, 45, 41. Find the sample mean and standard deviation.

6.10 You are given the following sample of the starting monthly salary for five business majors who recently graduated: \$1100, \$1800, \$1500, \$1450, and \$1600. Find the sample mean and standard deviation.

6.11 Suppose that you are interested in determining the average number of miles/week that a certain group of long-distance runners train. The following random sample resulted: 70, 55, 48, 50, 60, 65, 55, 40, 45, 70.

a) Find the sample mean and variance.

b) Find the sample median, mode, and range.

c) Verify that your value of s is reasonable by using the rule of thumb from Chapter 1.

6.12 Given the following sample distribution for x = number of soft drink machines needing repair per week at a certain university:

x	*Frequency*
1	2
2	7
3	10
4	1

find $\bar{x}$ and s.

6.13 The distribution of the number of defects per square yard of a cotton textile is given as follows:

No. of defects x	*Frequency* f
0	47
1	33
2	14
3	5
4	1
5	0
Total	100

Find the mean and the standard deviation of this distribution.

6.14 Given the following sample distribution by class intervals for the number of hours worked last month by a certain group of business executives;

Class	*Frequency*
190–204	3
205–219	5
220–234	9
235–249	6
250–264	7

a) Find $\bar{x}$ and s and verify that s is reasonable by using the rule of thumb from Chapter 1.

b) Find the mode, median, and range.

6.15 The average weekly wage for all workers in a certain industry is \$220. In a sample of 100 workers, I find $\bar{x}$ = \$235. If the weekly wages of these 100 workers are grouped into classes of 25 each, the following distribution results:

Wages	*f*
138–162	4
163–187	15
188–212	20
213–237	30
238–262	15
263–287	10
288–312	6
Total	100

Find $\bar{x}$ for the grouped data, using class marks, 150, 175, 200, . . . , 300. Why doesn't the mean of the grouped data equal \$235? Why doesn't it equal \$220?

6.16 A manufacturer of razor blades claims that this product will give, on the average, 15 good shaves. Suppose you have five friends who try using one of these razor blades each. The numbers of good shaves reported by your friends are 12, 16, 8, 14, and 10.

a) Find the mean and the standard deviation of this sample.

b) Suggest how you might use this sample evidence to dispute or support the advertiser's claim. Is the sample random?

6.17 A fresh produce distributor has received complaints that his bananas have been arriving spoiled at the retail sale store. He is suspicious of the complaint, since his average delivery time is only 4 days (96 hours), and the bananas are fresh at the time of shipment. He decides to simulate the appropriate conditions and make a test to see how long it takes before the bananas become spoiled. He selects at random a sample of four crates of bananas and measures the number of hours before spoilage occurs. The results for number of hours, x, are given as 106, 102, 104, and 108.

a) Find the mean hours before spoilage.

b) Find the standard deviation for x.

c) On the basis of these measures, do you think that many of the bananas may indeed be arriving spoiled, or are you also suspicious of the complaints? Explain.

d) Find the range of the sample of x values.

e) Give one reason why the answer of part (b) is better than that of part (d) as an estimate of dispersion for the entire population of bananas from which the sample was taken.

6.18 A small town in Southern Indiana recently had the opportunity to receive a Community Action Program (CAP) federal grant for insulation of low-income homes. The city council rejected the grant after a simple random sample of 20 residents indicated 15 against the grant and 5 in favor of it. Comment on this use of a simple random sample.

6.19 Describe a stratified sampling plan that an auditor might formulate for the population described in Problem 1.44.

6.20 An auditor who wants to estimate the aggregate value of an inventory could count and price a subset of the population. This procedure is called

a) attribute sampling

b) acceptance sampling

c) variable sampling

d) discovery sampling

e) systematic sampling.

6.21 If a CPA wants to estimate the percent of credit card customers who had a cash advance in the last month, the plan used should be:

a) attribute sampling

b) discovery sampling

c) acceptance sampling

d) variable sampling

e) none of the above.

6.4 SAMPLING DISTRIBUTION OF $\bar{x}$

*The **sampling distribution of** $\bar{x}$ is the probability distribution of **all possible** values of $\bar{x}$ that could occur when a sample (of size n) is taken from some specified parent population.*

To illustrate the sampling distribution of $\bar{x}$, we will consider a parent population that has only three values, (1, 2, 3), which occur with equal probability. This population could consist of common stocks classified (somewhat arbitrarily) as either $Aaa(=3)$, $Aa(=2)$, or $A(=1)$, each with a probability of

Table 6.2 All Samples of Size $n = 2$ from (1, 2, 3)

Sample	Sample Mean $\bar{x}$	
(1, 1)	1.0	
(1, 2)	1.5	
(2, 1)	1.5	
(1, 3)	2.0	This is the population of all possible $\bar{x}$'s when the sample size is $n = 2$.
(3, 1)	2.0	
(2, 2)	2.0	
(2, 3)	2.5	
(3, 2)	2.5	
(3, 3)	3.0	
Sum	18.0	

$\frac{1}{3}$. The parent population is thus $\boldsymbol{x} = \{1, 2, 3\}$. The mean of this parent population is easily seen to be $\mu = 2.0$; i.e., the average stock is classified as *Aa*. Now, let us assume that you do not know how these stocks are classified, so you decide to take a random sample of size $n = 2$ of the stocks in order to gain information about how they are classified. Your sample of size $n = 2$ will look like one of the nine ordered pairs listed in the lefthand column of Table 6.2.* The righthand column lists the sample mean for each of the nine possible samples.

Each one of the nine samples in column one has the same probability of occurring, $\frac{1}{9}$. This implies that each one of the nine sample means in column two has this same ($\frac{1}{9}$) probability. Now, since only one of the values of $\bar{x}$ in column two equals 1.0, we can write $P(\bar{\boldsymbol{x}} = 1) = \frac{1}{9}$. Similarly, there are two instances when $\bar{x}$ equals 1.5; hence $P(\bar{\boldsymbol{x}} = 1.5) = \frac{2}{9}$. A graph (Fig. 6.1) and a table (Table 6.3) show the probabilities associated with all these possible values of $\bar{\boldsymbol{x}}$.

Thus, Fig. 6.1 and Table 6.3 present the sampling distribution of $\bar{\boldsymbol{x}}$ for $n = 2$, when $\boldsymbol{x} = \{1, 2, 3\}$. It should be emphasized at this point that *the sampling distribution of $\bar{\boldsymbol{x}}$ is itself a population.* As such, we are interested in the parameters of this population of $\bar{\boldsymbol{x}}$'s, especially its mean and variance.

* We assume, until Section 6.7, either that the population is infinitely large, or that sampling is with replacement. However, we note as we proceed that sampling without replacement might change our results.

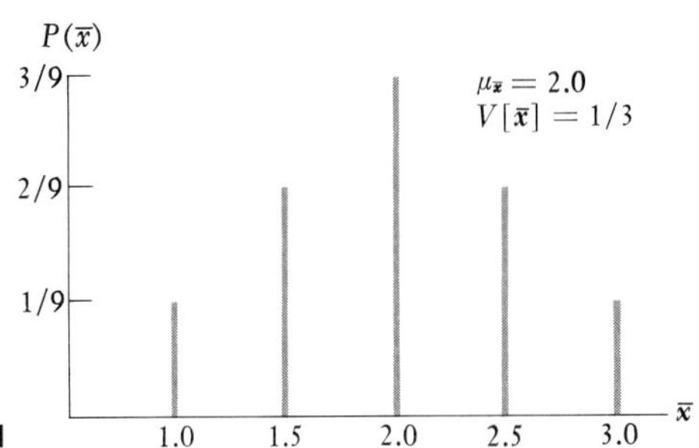

Figure 6.1

Table 6.3

$\bar{x}$	$P(\bar{x})$
1.0	$\frac{1}{9}$
1.5	$\frac{2}{9}$
2.0	$\frac{3}{9}$
2.5	$\frac{2}{9}$
3.0	$\frac{1}{9}$
	1.0

---

*The symbol $\mu_{\bar{x}}$, which traditionally denotes the mean of the sampling distribution of $\bar{x}$, is the mean of all possible sample means.**

Using the notation of expected values, we can thus write $\mu_{\bar{x}} = E[\bar{x}]$. The variance of this population, which is denoted as $V[\bar{x}] = \sigma_{\bar{x}}^2$, will be discussed shortly.

Mean of $\bar{x}$, or $\mu_{\bar{x}}$

Because you plan to take only *one* sample, of size $n = 2$, you could get a sample mean as low as $\bar{x} = 1.0$ or as high as $\bar{x} = 3.0$ (see Table 6.3). A logical question for a researcher to ask in this situation is what would be the *expected* value

* From this point on we will add a subscript to a population parameter or sample statistic whenever it may be unclear which population or sample is being described. For example, s_y refers to the standard deviation of the sample of y-values, while $\mu_{\bar{x}}$ refers to the mean of the population of $\bar{x}$-values. If *no* subscript is used, such as μ or σ or s, these refer to the distribution of x-values under consideration.

of $\bar{\boldsymbol{x}}$ in such a situation, or what is $E[\bar{\boldsymbol{x}}] = \mu_{\bar{x}}$? That is, what is the average of the p.m.f. in Table 6.3? The mean of these values is calculated in the same manner in which we calculated a mean in Chapter 3, except that now the variable is $\bar{\boldsymbol{x}}$, and we denote the mean of this variable as $\mu_{\bar{x}}$.

$$\mu_{\bar{x}} = \sum \bar{x}P(\bar{x}) = (1.0)(\tfrac{1}{9}) + (1.5)(\tfrac{2}{9}) + (2.0)(\tfrac{3}{9}) + 2.5(\tfrac{2}{9}) + (3.0)(\tfrac{1}{9})$$
$$= 2.00.$$

Thus we have shown for this population that $\mu_{\bar{x}} = \mu = 2.00$. The fact that these two means are equal is not a coincidence, as it can be shown that $\mu_{\bar{x}}$ equals μ for *any* parent population and *any* given sample size (and holds for sampling with or without replacement).*

$$\boxed{E[\bar{\boldsymbol{x}}] = \mu_{\bar{x}} = \mu.} \tag{6.4}$$

The Variance of $\bar{x}$

In addition to knowing the mean of the sampling distribution of $\bar{\boldsymbol{x}}$ for a given sample size n, we also need to know its variance.

The variance of the values of $\bar{\boldsymbol{x}}$ is denoted by either $V[\bar{\boldsymbol{x}}]$ or $\sigma^2_{\bar{x}}$.

This variance is defined in the same manner in which we previously defined a variance, namely as the expected value of the squared deviations of the

* Let $\boldsymbol{x}_1, \boldsymbol{x}_2, \ldots, \boldsymbol{x}_n$ represent independent random variables corresponding to the n observations in a sample of size n drawn from a population with mean μ_x (that is, $E[\boldsymbol{x}_i] = \mu$). Now, since $\bar{x} = (1/n)(x_1 + x_2 + \cdots + x_n)$, we can apply the rules of expectation from Section 3.7 as follows:

$$E[\bar{\boldsymbol{x}}] = E\left[\frac{1}{n}(\boldsymbol{x}_1 + \boldsymbol{x}_2 + \cdots + \boldsymbol{x}_n)\right]$$

$$= \frac{1}{n}E[\boldsymbol{x}_1 + \boldsymbol{x}_2 + \cdots + \boldsymbol{x}_n] \qquad \text{By rule 3}$$

$$= \frac{1}{n}(E[\boldsymbol{x}_1] + E[\boldsymbol{x}_2] + \cdots + E[\boldsymbol{x}_n]) \qquad \text{By Formula (3.18)}$$

$$= \frac{1}{n}(\mu + \mu + \cdots + \mu) \qquad \text{Since } E[\boldsymbol{x}_i] = \mu$$

$$= \frac{1}{n}(n\mu).$$

Table 6.4 Calculation of the Variance of a Distribution of Sample Means

$\bar{x}$	$P(\bar{x})$	$(\bar{x} - \mu_{\bar{x}})$	$(\bar{x} - \mu_{\bar{x}})^2$	$(\bar{x} - \mu_{\bar{x}})^2 P(\bar{x})$
1.0	$\frac{1}{9}$	-1.0	1.00	$\frac{1}{9}$
1.5	$\frac{2}{9}$	-0.5	0.25	$\frac{2}{36}$
2.0	$\frac{3}{9}$	0	0	0
2.5	$\frac{2}{9}$	0.5	0.25	$\frac{2}{36}$
3.0	$\frac{1}{9}$	1.0	1.00	$\frac{1}{9}$
Mean 2.0				$V[\bar{x}] = \frac{12}{36} = \frac{1}{3}$

variable ($\bar{x}$ in this case) about its mean (μ_x in this case). Thus,

$$\sigma_{\bar{x}}^2 = V[\bar{x}] = E[(\bar{x} - \mu_{\bar{x}})^2].$$

Suppose that we use the data in Table 6.4 to calculate the variance of the $\bar{x}$'s, or $V[\bar{x}]$. The variance of the $\bar{x}$'s is shown in column 5 to be $\sigma_{\bar{x}}^2 = \frac{1}{3}$; the standard deviation of the $\bar{x}$'s is $\sigma_{\bar{x}} = \sqrt{\frac{1}{3}} = 0.577$. The reader can verify that this standard deviation appears reasonable by noting that $\mu_{\bar{x}} \pm 1\sigma_{\bar{x}} = 2.00 \pm 0.577$ contains 77.8% of the probability distribution of $\bar{x}$, which is not too far away from the rule of thumb of 68%.

Although we calculated the value of $V[\bar{x}]$ directly in this example, in the many problems in which there is a very large (or infinite) number of values of $\bar{x}$ this approach is impractical, if not impossible. Fortunately, one can calculate $V[\bar{x}]$ without going through this process if one knows the variance of the population from which the samples are drawn (i.e., $V[x]$). The reason for this is that the variance of the random variable $\bar{x}$ is related to the variance of the parent population and to the sample size (n) by a very simple formula, which we will present shortly.

On the basis of intuition, it should appear reasonable that the variance of $\bar{x}$ will always be less than the variance of the parent population (except when $n = 1$), because there is less chance that a sample mean will take on an extreme value than there is that a single value of the parent population will take on this value. In order for a sample mean to have an extremely large value, most or all of the sample items would have to be extremely large values. But in such a case, we know, from our study of probability, that the probability of a single extremely large value on a single draw is greater than the probability of n extremely large values on n repeated draws. It would be very unusual in n trials not to draw some middle values or some extremely low values. Such values would balance out the extremely large values and give a less extreme sample mean. Indeed, this intuitive logic is correct. Not only is the variance of $\bar{x}$ always less than or equal to the variance of the parent population, but it can be shown that σ^2 and $\sigma_{\bar{x}}^2$ are very precisely related. *The variance of* $\bar{x}$, *assuming each sample consists of n independent observations, is 1/n times the variance of the parent population.**

$$\textit{Variance of } \bar{x}: \qquad \sigma_{\bar{x}}^2 = \frac{1}{n}\sigma^2. \tag{6.5}$$

When $n = 1$, all samples contain only one observation, and the distributions of x and $\bar{x}$ are identical. That is, $\sigma_{\bar{x}}^2 = \sigma^2/1 = \sigma^2$. As n becomes larger ($n \to \infty$), it is reasonable to expect $\sigma_{\bar{x}}^2$ to become smaller and smaller because the sample means will tend to deviate less and less from the population mean $\mu_{\bar{x}}$. When $n = \infty$ (or for finite populations, when $n = N$), all sample means will equal the population mean and the variance of the $\bar{x}$'s will be zero. To illustrate the relationship described by Formula (6.5), let us return to our stock example involving the population (1, 2, 3). The variance of this population is:

$$\sigma^2 = \frac{1}{N}\sum(x - \mu_{\bar{x}})^2 = \frac{1}{3}[(1-2)^2 + (2-2)^2 + (3-2)^2] = \frac{2}{3}.$$

Since we know that the variance of this population is $\frac{2}{3}$ and the sample size is $n = 2$, we can calculate $V[\bar{x}]$ using Formula (6.5):

$$\sigma_{\bar{x}}^2 = \frac{1}{n}\sigma^2 = \frac{1}{2}\left(\frac{2}{3}\right) = \frac{1}{3}.$$

This value of $V[\bar{x}]$ is exactly the same number we calculated in Table 6.4. Thus, we have verified that Formula (6.5) holds for this particular problem.†

Standard Error of the Mean

It is customary to call the standard deviation of the $\bar{x}$'s (which is the square root of $V[\bar{x}]$ and is denoted as $\sigma_{\bar{x}}$) the *standard error of the mean.* The word "error" in this context obviously refers to sampling error, as $\sigma_{\bar{x}}$ is a measure

* Let $x_1, x_2, \ldots, x_n$ be independent random variables, each having the same variance (that is, $V[x_i] = \sigma^2$):

$$
\begin{aligned}
V[\bar{x}] &= V\left[\frac{1}{n}(x_1 + x_2 + \cdots + x_n)\right] && \text{By definition} \\
&= \left(\frac{1}{n}\right)^2 V[x_1 + x_2 + \cdots + x_n] && \text{By Rule 4 of Section 3.7} \\
&= \left(\frac{1}{n}\right)^2 [V[x_1] + V[x_2] + \cdots + V[x_n]] && \text{By Formula (3.21)} \\
\sigma_{\bar{x}}^2 &= \frac{n}{n^2}(\sigma^2) = \frac{1}{n}\sigma^2 && \text{Since } V[x_i] = \sigma^2.
\end{aligned}
$$

† Formulas (6.5) and (6.6) do not hold when sampling without replacement from a finite population (see Section 6.7).

of the "standard" (or expected) error when the sample mean is used to obtain information or draw conclusions about the unknown population mean.

Standard error of the mean: $$\sigma_{\bar{x}} = \sqrt{\frac{\sigma^2}{n}} = \frac{\sigma}{\sqrt{n}}. \qquad (6.6)$$

In the above example, the standard error of the mean is

$$\sigma_{\bar{x}} = \sqrt{\sigma^2/n} = \sqrt{(\tfrac{2}{3})/2} = \sqrt{\tfrac{1}{3}} = 0.577.$$

As we indicated on page 268, this value of $\sigma_{\bar{x}}$ appears reasonable because 77.8% of the probability distribution of $\bar{x}$ lies within the interval $\mu_{\bar{x}} \pm \sigma_{\bar{x}} = 2.00 \pm 0.577$.

We must emphasize at this point that $\mu_{\bar{x}}$ and $\sigma_{\bar{x}}$ are parameters of the population of all conceivable samples of size n, and these population parameters are *unknown* quantities. In fact, the values of μ, $\mu_{\bar{x}}$, σ, and $\sigma_{\bar{x}}$ are usually *all* unknown quantities, which means that the relationship $\mu_{\bar{x}} = \mu$ and $\sigma_{\bar{x}} = \sigma/\sqrt{n}$ cannot be used to solve for the value of one of these quantities. However, knowledge of the fact that such relationships exist is important in determining how far a sample mean can be expected to deviate from the population mean. The advantage of knowing this information is that we can test *assumptions* about a population by looking at sample results. For example, suppose in our stock market example we had *assumed* (but did not know) that our parent population was $\boldsymbol{x} = \{1, 2, 3\}$. If a single sample of size $n = 2$ from this population yielded $\bar{x} = 1.0$, then we might begin to question our assumption about the population $\boldsymbol{x} = \{1, 2, 3\}$. Our knowledge of the sampling distribution for $\bar{\boldsymbol{x}}$ confirms that $P(\bar{\boldsymbol{x}} = 1.0)$ is only $\frac{1}{9}$ for samples of size 2 from this population—not a frequently occurring event. In Chapter 8 we will formally consider this process of testing assumptions. A sample of (1, 1) might in this case lead us to *incorrectly* reject the assumption that $\boldsymbol{x} = \{1, 2, 3\}$.

Although we have derived the mean and variance of the $\bar{\boldsymbol{x}}$'s, nothing has been said about the *shape* of the sampling distribution of $\bar{\boldsymbol{x}}$. Recall from Chapter 1 that distributions with the same mean and variance may have distinctly different shapes. It is necessary, therefore, to be more specific about the entire distribution of $\bar{\boldsymbol{x}}$'s. To do so, we will first assume that the parent population is normal, and then later drop this assumption.

6.5 SAMPLING DISTRIBUTION OF $\bar{x}$, NORMAL PARENT POPULATION

We already know the mean and variance of the distribution of $\bar{\boldsymbol{x}}$'s, but what is known about its *shape*? It is usually not possible to specify the shape of the $\bar{\boldsymbol{x}}$'s when the parent population is discrete and the sample size is small. However, when the sample is drawn from a parent population (x) which is normally distributed, then the shape of the $\bar{x}$'s can be specified. As you might suspect, in this situation the $\bar{x}$'s are distributed normally. That is,

> *The sampling distribution of $\bar{x}$'s drawn from a normal parent population is a normal distribution.*

Using the fact that the mean of the $\bar{x}$'s is $\mu_{\bar{x}} = \mu$ [Formula (6.4)], and that the variance of the $\bar{x}$'s is $\sigma_{\bar{x}}^2 = \sigma^2/n$, we can now specify that the sampling distribution of $\bar{x}$ is $N(\mu_{\bar{x}}, \sigma_{\bar{x}}^2) = N(\mu, \sigma^2/n)$ whenever the parent population is normal.

To illustrate this sampling distribution of $\bar{x}$, we will suppose that all possible samples of size $n = 20$ are drawn from a normal population which has a mean $\mu = 50$ and a variance $\sigma^2 = 80$; that is, x is $N(50, 80)$. Because all normal distributions are continuous, an infinite number of different samples of size 20 could be drawn. For all of these samples a mean, $\bar{x}$, could be calculated. Since the population mean is $\mu = 50$, the mean of the $\bar{x}$'s is $\mu_{\bar{x}} = 50$. Similarly, since $\sigma^2 = 80$, the variance of the $\bar{x}$'s is $\sigma_{\bar{x}}^2 = \sigma^2/n = \frac{80}{20} = 4$. Finally, because x is normal, $\bar{x}$ will also be normally distributed. All of this information about $\bar{x}$ can be summarized by the following statement: The distribution of $\bar{x}$ is $N(50, 4)$.

This means that 68.3% of the sample means will fall within plus-or-minus one standard deviation of the mean,

$$\mu \pm \sigma_{\bar{x}} = 50 \pm 2 = 48 \quad \text{to} \quad 52;$$

95.4% will fall within plus-or-minus two standard deviations of the mean,

$$\mu \pm 2\sigma_{\bar{x}} = 50 \pm 2(2) = 46 \quad \text{to} \quad 54;$$

and 99.7% of all sample means will fall within plus-or-minus three standard deviations of the mean,

$$\mu \pm 3\sigma_{\bar{x}} = 50 \pm 3(2) = 44 \quad \text{to} \quad 56.$$

Figure 6.2 shows the sampling distribution of $\bar{x}$ for all possible samples of size 20 taken from a population with the distribution $N(50, 80)$.

The following statement summarizes what we now know about the distribution of sample means ($\bar{x}$).

> *If the parent population (x) is normally distributed, with mean μ and variance σ^2, then the distribution of $\bar{x}$ for a given sample size n will be $N(\mu, \sigma^2/n)$.*

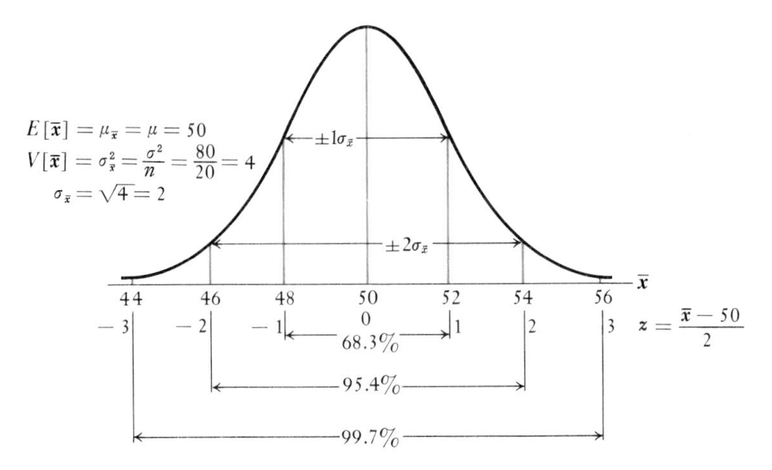

Fig. 6.2 Sampling distribution of $\bar{x}$ for sample of $n = 20$ taken from a population with distribution $N(50, 80)$.

The Standardized Form of the Random Variable $\bar{x}$ (σ Known)*

In Chapter 5 we saw that it is usually easier to work with the standard normal form of a variable than to leave it in its original units. The same type of transformation that was made on the random variable x at that point can now be made on the random variable $\bar{x}$. Recall that in Chapter 5 the variable x was transformed to its standard normal form by subtracting the mean from each value, and then dividing by the standard deviation. The resulting variable, $z = (x - \mu)/\sigma$, was shown to have a mean of zero and a variance of one. Although now we are interested in transforming the variable $\bar{x}$ instead of x, and the standard deviation of this variable is $\sigma/\sqrt{n}$ instead of σ, the transformation is accomplished in exactly the same fashion. One simply must take care always to subtract the mean and divide by the standard deviation corresponding to the variable being standardized. The mean and variance of the resulting variable will always be 0 and 1, respectively. Hence, the random variable

$$z = \frac{\bar{x} - \mu_{\bar{x}}}{\sigma_{\bar{x}}} = \frac{\bar{x} - \mu}{\sigma/\sqrt{n}} \tag{6.7}$$

has a mean of zero and a variance of one. Since we have said that the distribution of the random variable $\bar{x}$ is normal, it follows that the random variable z must

* Whenever $\bar{x}$ is normal, the distribution of $\bar{x}$ is $N(\mu, \sigma/\sqrt{n})$ whether or not σ is known. However, in order to make probability statements about $\bar{x}$, we must standardize this variable, and this standardization requires that we know σ.

also be normally distributed. Thus:

> *When sampling from a normal parent population, the distribution of* $z = (\bar{x} - \mu)/(\sigma/\sqrt{n})$ *will be normal with mean* 0 *and variance* 1. *That is:* $z = (\bar{x} - \mu)/(\sigma/\sqrt{n})$ is $N(0, 1)$.

The standardized normal form of the variable $\bar{x}$ is shown on the z-scale in Fig. 6.2.

The limitations of the preceding discussion should be apparent, for although the normal distribution approximates the probability distribution of many real-world problems, one cannot assume that the parent population is *always* normal. What, for example, will be the shape of the distribution of $\bar{x}$'s when sampling from a highly skewed distribution? We consider this situation in the next section.

6.6 SAMPLING DISTRIBUTION OF $\bar{x}$, POPULATION DISTRIBUTION UNKNOWN, σ KNOWN

When sampling is not from a normal parent population, the size of the sample plays a critical role. When n is small, the shape of the distribution will depend mostly on the shape of the parent population. As n gets large, however, one of the most important theorems in statistical inference states that the shape of the sampling distribution will become more and more like a *normal distribution, no matter what the shape of the parent population.* This theorem, called the central limit theorem, is stated in formal terms below:

> **The central limit theorem:**
>
> *Regardless of the distribution of the parent population (as long as it has a finite mean* μ *and variance* σ^2*), the distribution of the means of random samples will approach a normal distribution (with mean* μ *and variance* σ^2/n*) as the sample size n goes to infinity.*

The difference between this statement and the one above is that no assumption about normality of the parent population needs to be made when n is large.

We will not prove this theorem, but merely show, in Fig. 6.3, graphical evidence of its validity. The first row of diagrams in Fig. 6.3 shows four different parent populations. The next three rows show the sampling distribution of $\bar{x}$ for all possible repeated samples of size $n = 2$, $n = 5$, and $n = 30$, respectively, drawn from the populations shown in the first row. Note in the first column

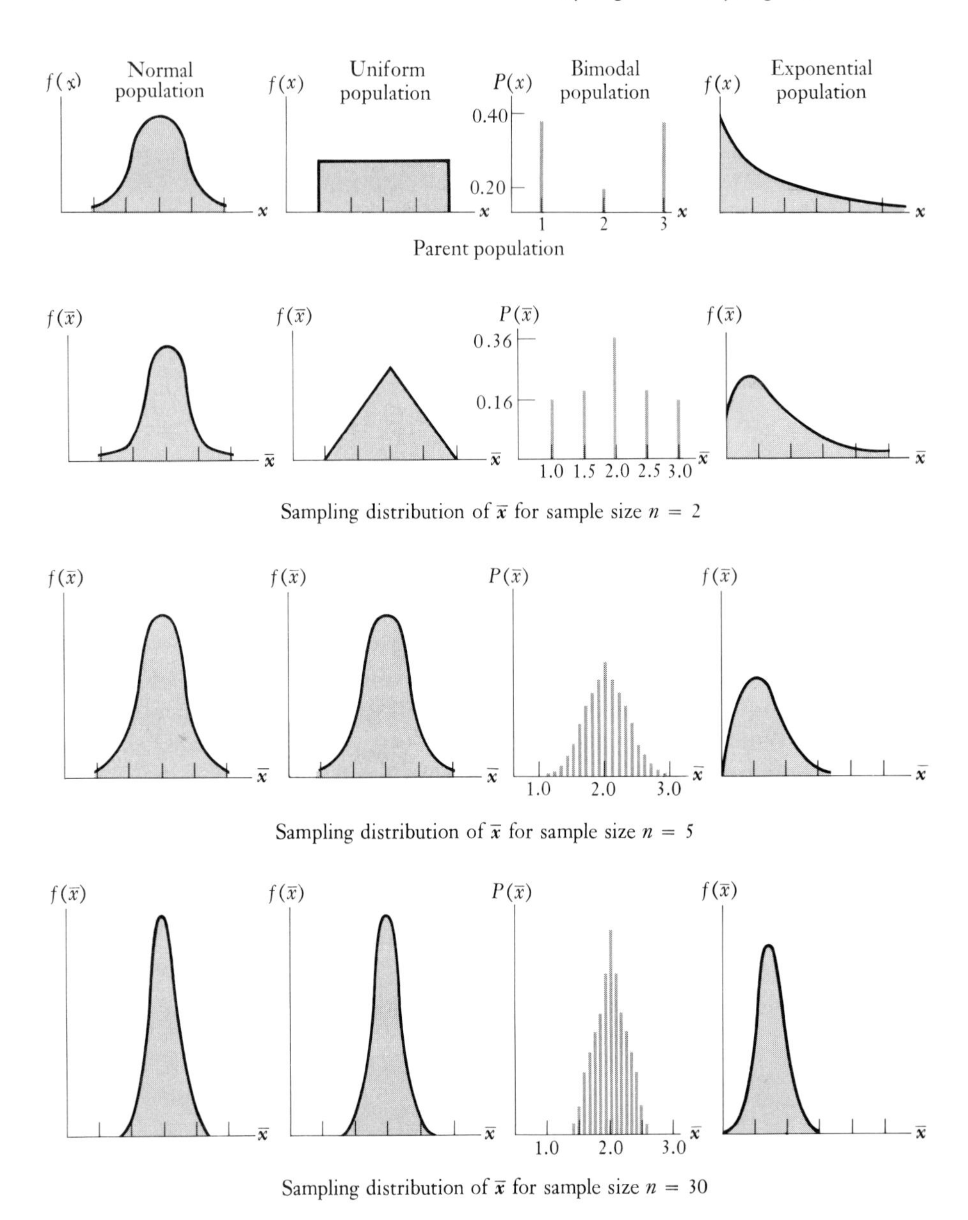

Fig. 6.3 Sampling distribution of $\bar{x}$ for various population distributions when $n = 2$, 5, and 30.

that when the parent population is normal, all the sampling distributions are also normal. Also, note that distributions in the same column have the same mean μ, but their variances decrease as $\sqrt{n}$ increases. This agrees with the central limit theorem.

The second column of figures in Fig. 6.3 represents what is called a *uniform* (or *rectangular*) *distribution.* We see here that the sampling distribution of $\bar{x}$ is already symmetrical when $n = 2$, and it is quite normal in appearance when $n = 5$. Moving to the third column of figures, the parent population is now a bimodal distribution with discrete values of x (the central limit theorem applies whether x is discrete or continuous). Again, by $n = 2$ the distribution is symmetrical, and by $n = 5$ it is quite bell-shaped. The final parent population is the highly skewed exponential distribution. Here we see that for $n = 2$ and $n = 5$ the distribution is still fairly skewed, although it becomes more symmetrical as n increases. When $n = 30$, however, even such a skewed parent population results in a symmetrical, bell-shaped distribution for $\bar{x}$ which, by the central limit theorem, we know to be approximately normally distributed.

In general, just how large n needs to be for the sampling distribution of $\bar{x}$ to be a good approximation to the normal depends, as we saw in Fig. 6.3, on the shape of the parent population. Usually the approximation will be quite good if $n \geq 30$, although the third row of Fig. 6.3 demonstrates that satisfactory results are often obtained when n is much smaller.

Example of the Use of the Central Limit Theorem

Now that the sampling distribution of $\bar{x}$ has been specified, at least for large samples where σ is known, we can demonstrate the use of the central limit theorem. Consider the case of a midwestern telephone company which recently asked for a rate increase for all residential telephones, including a 25-percent increase in student phones. To oppose this increase, a group of students decided to investigate the typical phone costs incurred in their town. The information provided by the telephone company was that, for the city as a whole, the average monthly bill was \$15.30, with a standard deviation of \$4.10. The students, however, were curious as to whether or not dormitory phones incurred similar bills. Since no information was available in this regard, the students decided to take a random sample of $n = 36$, in an attempt to obtain further information.

Now, if the dormitory phone bills come from the same population reported by the telephone company, then we know that the mean of all possible sample means will be $E[\bar{x}] = \mu_{\bar{x}} = \$15.30$, and the standard error of the mean will be $\sigma_{\bar{x}} = \sigma/\sqrt{n} = 4.10/\sqrt{36} = 0.683$. Furthermore, since the sample size is fairly large ($n = 36$), the central limit theorem tells us that the sampling distribution of $\bar{x}$ will be normal; that is,

$$\bar{x} \text{ is } N(\$15.30, 0.683^2).$$

Use of this distribution, shown in Fig. 6.4, allows us to answer a number of different probability questions in the telephone example.

Question 1. Suppose that in the random sample of 36 dormitory residents with phones, we find that the average phone bill is \$14.00. A typical question to ask at this point is: What is the probability that a random sample of $n = 36$

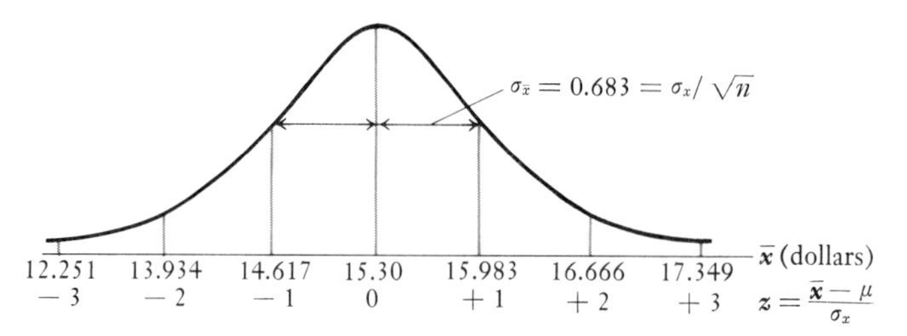

Figure 6.4

will result in an average bill of \$14.00 or less, when $\mu_{\bar{x}} = \$15.30$ and $\sigma_{\bar{x}} = 0.683$? Making use of the standard normal transformation, we know that:

$$P(\bar{x} \leq \$14.00) = P\left(\frac{\bar{x} - \mu_{\bar{x}}}{\sigma_{\bar{x}}} \leq \frac{14.00 - 15.30}{0.683}\right) = P(z \leq -1.90).$$

By the central limit theorem, $\bar{x}$ is approximately normally distributed; hence z is approximately standardized normal and, from Table III,

$$\begin{aligned} P(z \leq -1.90) &= F(-1.90) = 1.0 - F(1.90) \\ &= 1.0 - 0.9713 = 0.0287. \end{aligned}$$

The probability that a sample of 36 gives an average no larger than \$14.00 is thus 0.0287. The reader should sketch this area on Fig. 6.4.

On the basis of the probability value calculated above, does it appear that dormitory phone bills are part of the same population as residential phone bills in general? We leave for the following chapters such questions concerned with drawing conclusions from sample data, but the reader should keep them in mind to retain a feeling of where we are headed.

Question 2. Rather than finding the probability of a specific value of $\bar{x}$, one often wants to determine values a and b, such that the probability is 0.99 (or some other probability) that the sample mean will fall between these values. Recall that we did a similar calculation in our discussion of the normal distribution in Chapter 5. In the present case we know that when $n = 36$, $\bar{x}$ will be approximately normally distributed, with mean $\mu_{\bar{x}} = \$15.30$, and $\sigma_{\bar{x}} = 4.10/\sqrt{36} = 0.683$.

As before, we want the smallest interval including 99 percent, which means that we must exclude half of the remaining probability [or $\frac{1}{2}(0.01) = 0.005$] in each tail of the distribution of $\bar{x}$. To do this, let us first see what values of the standardized normal distribution exclude 0.005 in each tail. From Table III, $F(z) = 0.995$ if $z = 2.575$;* by symmetry, $F(-2.575) = 1.0 - F(2.575) =$

* The reader could determine the value 2.575 by looking for $1 - 0.005 = 0.995$ in the body of Table III. The closest values to $F(0.995)$ are $F(0.9949) = 2.57$ and $F(0.9951) \doteq 2.58$. Linear interpolation between these values yields the approximation $F(0.9950) = 2.575$. A more precise value, taking into account the nonlinear shape of the normal curve, is 2.576.

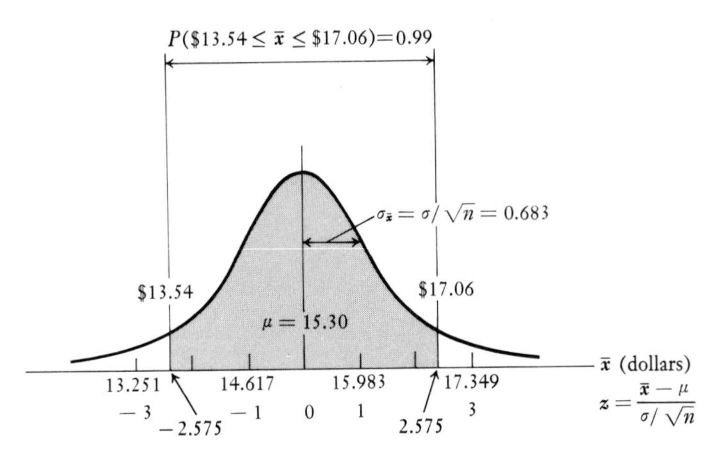

Fig. 6.5 Standardized normal form of the sampling distribution of $\bar{\mathbf{x}}$, sample mean of monthly phone charges, $\mu = \$15.30, \sigma = \$4.10, n = 36$.

0.005. Thus,

$$P(-2.575 \leq z \leq 2.575) = 0.99.$$

Finally, we now have to transform the interval $P(-2.575 \leq z \leq 2.575) = 0.99$ into the original units (dollars) by finding values a and b to satisfy $P(a \leq \bar{x} \leq b) = 0.99$. By standardizing the values of a, $\bar{x}$, and b, we get:

$$P(a \leq \bar{x} \leq b) = P\left(\frac{a - 15.30}{0.683} \leq z \leq \frac{b - 15.30}{0.683}\right).$$

Thus, $(a - 15.30)/0.683 = -2.575$ and $(b - 15.30)/0.683 = 2.575$. Solving these two equations we find $a = \$13.54$ and $b = \$17.06$; this means that the appropriate interval is $P(\$13.54 \leq \bar{x} \leq \$17.06)$. A diagram of the values in this example is shown in Fig. 6.5.

6.7 SAMPLING FROM A FINITE POPULATION WITHOUT REPLACEMENT

In our discussion thus far involving the sampling distribution of $\bar{x}$, we assumed that sampling occurs either from an infinite population or with replacement. But what happens when sampling is not of this type—that is, when samples are drawn from a finite population without replacement? In this case, the value of $\sigma_{\bar{x}}$ will be smaller than $\sigma/\sqrt{n}$. This fact can be demonstrated by referring again to Table 6.2, but now assuming that we sample *without* replacement from the finite population (1, 2, 3). The new set of all samples of size $n = 2$ is given in Table 6.5.

We see from Table 6.5 that $\sigma_{\bar{x}} \cong 0.41$. This standard error is less than $\sigma/\sqrt{n} = \sqrt{2/3}/\sqrt{2} = 0.577$.

Table 6.5 Samples of $n = 2$ Without Replacement from (1, 2, 3)

Sample	Sample Mean ($\bar{x}$)	$(\bar{x} - \mu_{\bar{x}})^2$
(1, 2)	1.5	0.25
(2, 1)	1.5	0.25
(1, 3)	2.0	0
(3, 1)	2.0	0
(2, 3)	2.5	0.25
(3, 2)	2.5	0.25
		1.00

$$\sigma_{\bar{x}}^2 = \frac{1}{6}(1.00) = \frac{1}{6},$$

$$\sigma_{\bar{x}} = \sqrt{\tfrac{1}{6}} \cong 0.41$$

Intuitively, $\sigma_{\bar{x}}$ must be smaller than $\sigma/\sqrt{n}$ when sampling without replacement because σ must approach zero as $n \to N$. For example, if each repeated sample consists of the entire population (i.e., $n = N$), then each sample mean $\bar{x}$ will be identical and equal to the population mean μ; thus, $\sigma_{\bar{x}} = 0$. This is not the case for $\sigma/\sqrt{n}$, because it approaches zero as $n \to$ infinity rather than as $n \to N$.

In order for $\sigma/\sqrt{n} \to 0$ as $n \to N$, we multiply $\sigma/\sqrt{n}$ by an adjustment factor called the *finite population correction factor*. This factor, which is based on the hypergeometric distribution, is defined as follows:

> *Finite population correction factor:*
>
> $$\sqrt{\frac{(N - n)}{(N - 1)}}.$$

Note that this correction factor will always be a number less than 1.0.

To illustrate this correction factor, let's apply it to the value of $\sigma/\sqrt{n}$ in our problem involving the population (1, 2, 3). Recall that $\sigma/\sqrt{n} = 0.577$, $n = 2$, $N = 3$:

$$\sigma_{\bar{x}} = \frac{\sigma}{\sqrt{n}}\sqrt{\frac{N - n}{N - 1}} = 0.577\sqrt{\frac{3 - 2}{2}} \cong 0.41.$$

This result agrees with the value of $\sigma_{\bar{x}}$ calculated in Table 6.5.

When n is small relative to N (say 10% or less), the finite population correction factor is sometimes omitted, and $\sigma/\sqrt{n}$ is used as an approximation to $\sigma_{\bar{x}}$. This approach saves some of the tedious calculations involved with the correction factor when this factor will change $\sigma/\sqrt{n}$ only by a relatively small amount. The example below illustrates how little $\sigma/\sqrt{n}$ changes even when n is slightly larger than 10%.

An important use of the finite correction factor occurs when the standardized z-value in Formula (6.7) is calculated. If we include the correction factor, this standardization becomes:*

$$z\text{-standardization:} \qquad z = \frac{\bar{x} - \mu}{\dfrac{\sigma}{\sqrt{n}}\sqrt{\dfrac{(N-n)}{(N-1)}}}. \qquad (6.8)$$

We illustrate the Formula (6.8) by assuming in our telephone example that the number of dormitory residents with private phones is 300. Since the sampling of 36 residents was without replacement, and a sample of size $n = 36$ equals 12 percent of this population ($\frac{36}{300} = 0.12$), the finite population correction factor should be used.

We correct our z calculation on page 276 as follows:

$$P(\bar{x} \le 14.00)$$

$$= P\left(\frac{\bar{x} - \mu_{\bar{x}}}{(\sigma/\sqrt{n})\sqrt{(N-n)/(N-1)}} \le \frac{14.00 - 15.30}{(4.20/\sqrt{36})\sqrt{(300-36)/(300-1)}}\right)$$

$$= P\left(z \le \frac{-1.30}{0.642}\right).$$

Comparing this result with that on page 276, we see that the denominator has changed from 0.683 to 0.642. Since $-1.30/0.642 = -2.02$, the new probability is

$$P(z \le -2.02) = F(-2.02) = 1 - F(2.02) = 1 - 0.9783 = 0.0217.$$

Define: sampling distribution of $\bar{x}$, standardization of $\bar{x}$, central limit theorem, finite population correction factor.

Problems

6.22 a) Given that the balance due on Sears charge cards last month is a normally distributed random variable with mean $\mu = 50$ and standard deviation $\sigma = 8$, find:

$$P(x \ge 40), \qquad P(x \le 54), \qquad \text{and} \qquad P(44 \le x \le 56).$$

b) If a random sample of size $n = 64$ is drawn from this population, find:

$$P(\bar{x} \le 53), \qquad p(\bar{x} \ge 49), \qquad \text{and} \qquad P(48 \le \bar{x} \le 52).$$

c) Sketch the distribution of x and the distribution of $\bar{x}$.

* When the population size itself is very small, say $N \le 30$, then care must be taken in interpreting this formula, since $\bar{x}$ may not be normally distributed.

6.23 What is a "sampling distribution"? Why is the knowledge of the sampling distribution of a statistic important to statistical inference?

6.24 State the central limit theorem. Why do you think this theorem is so important to statistical inference?

6.25 A telephone company randomly selected 121 long distance calls and found that the average length of these calls was 5 minutes. The population standard deviation is 45 seconds.

a) What is the probability that a sample mean will be large as, or larger than, $\bar{x} = 5$ when the true population mean is $\mu = 4\frac{5}{6}$ minutes? What is the probability that a value will be as small as, or smaller than, $\bar{x} = 5$ when $\mu = 5\frac{1}{5}$?

b) Do your answers to part (a) depend on any assumptions about the distribution of the parent population?

6.26 If the mean of all shoe sizes for 13 year old boys in the United States is 9, and the variance of these sizes is 1, what percent of this population wears a shoe of size 11 or larger? What is $P(\bar{x} > 11)$ if $n = 16$? Assume that the parent population is normally distributed.

6.27 Suppose that a random sample is being drawn from a population of housewives known to have a mean age of 30, with a standard deviation of 3 years. The population is normally distributed.

a) What is the probability that a randomly selected housewife will be over 35 years of age? What is the probability that she will be between 25 and 35?

b) What is the probability, in a sample of 36 housewives, that the mean age will exceed 31? What is the probability that the mean age will be less than 30.5? What is $P(29 \leq \bar{x} \leq 31)$?

c) Does your answer to part (a) of this question depend on the assumption that the parent population is normally distributed? What about your answer to part (b)? Explain.

6.28 a) Funds, Inc., sells bonds maturing in 4, 5, 9, and 10 years. The probability distribution for customers purchasing these bonds is given below:

x	4	5	9	10
$P(x)$	$\frac{1}{2}$	$\frac{1}{6}$	$\frac{1}{6}$	$\frac{1}{6}$

Sketch the probability function and find the mean and the standard deviation.

b) Suppose 115 repeated samples of size $n = 5$ are drawn randomly from this probability distribution. The sample means are calculated and their frequency distribution is given in Table 6.6. Find the mean and the standard deviation of these values. Compare these values to those expected for the distribution of all sample means using samples of size 5.

c) Sketch the distribution of the 115 sample means in a frequency distribution, using class intervals of size 1.0, beginning with 3.5–4.49, 4.5–5.49, etc. Comment on its shape compared to that defined by the central limit theorem.

6.29 Use Table VI to collect a random sample of three observations (with replacement) from the population consisting of the digits 0 through 9. Repeat 4 times.

a) Calculate $\bar{x}$ for each of your five samples of three observations, and then calculate $\bar{\bar{x}}$, the mean of all five sample means.

Table 6.6

$\bar{x}$	Frequency	$\bar{x}$	Frequency
4.0	2	6.4	15
4.2	6	6.6	6
4.4	2	6.8	2
4.6	1	7.0	3
5.0	2	7.2	5
5.2	15	7.4	7
5.4	12	7.6	8
5.6	5	7.8	1
6.0	3	8.2	2
6.2	17	8.4	1
Total			115

b) What is the expected value in the population from which your samples in part (a) were drawn? Is $\bar{\bar{x}}$ reasonably close to $\mu_{\bar{x}}$?

c) Calculate the standard deviation of your five sample means about the grand mean. What is the standard error of the mean in the population from which your samples were drawn? Are the two values reasonably close?

6.30 A study of 100 randomly selected salary offers accepted by students using the Indiana University Placement Service indicated a mean of $\bar{x} = \$15{,}985$. The population in this case consisted of 500 students. Last semester the mean salary offer was $\mu = \$15{,}800$. What is the probability that a sample mean will be as large as \$15,985 if σ is known to be \$1000 and the population mean is \$15,800? Remember to correct for a finite population (assume $N = 500$).

6.31 A cereal company checks the weight of each lot of 400 boxes of breakfast cereal by randomly checking 64 of the boxes. This particular brand is packed in 20-ounce boxes.

a) Suppose a particular random sample of 64 boxes yields a mean weight of 19.95 ounces. How often will the sample mean be this low if $\mu = 20$ and $\sigma = 0.10$? Use the finite population correction factor.

b) What is the difference in the standard error after using the correction factor?

6.32 a) List all possible samples of size $n = 2$ which could be drawn with replacement from the population $\{\mathbf{x} = 4, 5, 9, 10\}$. Draw a graph to illustrate the p.m.f. for the distribution of $\bar{\mathbf{x}}$, assuming each value of $\mathbf{x}$ occurs with probability $\frac{1}{4}$.

b) Use your p.m.f. in part (a) to verify that $E[\bar{\mathbf{x}}] = \mu$ and $V[\bar{\mathbf{x}}] = \sigma^2/n$.

6.33 Sketch the distribution if one takes an infinite number of samples (all of size $n = 100$) from an exponential distribution ($\lambda = 5$).

6.34 A new bathroom soap comes in a jar which is advertised to dispense liquid soap for 300 uses. A consumer agency tests two jars of this soap, and finds 275 and 290 uses. Calculate $\bar{x}$ and s. Comment on this test of a new product.

6.35 a) What is the "finite population correction," and when is it necessary to apply this factor?

b) If $\sigma_x = 50$, $n = 25$, and $N = 100$, what is the corrected standard error of the mean?

6.8 SAMPLING DISTRIBUTION OF $\bar{x}$, NORMAL POPULATION, σ UNKNOWN

In Section 6.5 we discussed the importance in problem solving of using the following standardization:

$$z = \frac{\bar{x} - \mu}{\sigma/\sqrt{n}}.$$

Usually, *our objective in using this type of standardization is to determine the probability of observing some specified value of $\bar{x}$, assuming that the population mean is μ; and then to use this probability in making a decision.* This means we have an assumed value of μ to use in the standardization. But what about the value of σ needed in the denominator? What happens if we do not want to (or cannot) assume a value for σ (i.e., if σ is unknown)? In solving a particular problem where σ is unknown, the sample statistic s can be used in place of σ. That is, our standardization now becomes

$$\frac{\bar{x} - \mu}{s/\sqrt{n}}.$$

The substitution of s for σ is reasonable since we can show* that the expected value of s^2 equals σ^2. That is, $E[s^2] = \sigma^2$.

We have shown previously that, when x is normal, the distribution of $(\bar{x} - \mu)/(\sigma/\sqrt{n})$ is $N(0, 1)$. Unfortunately, when s is substituted for σ, the re-

* To prove that $E[s^2] = \sigma^2$, we use the rules of expectation in Section 3.7.

$$E[s^2] = E\left[\frac{1}{n-1}\sum(x_i - \bar{x})^2\right]$$

$$= \frac{1}{n-1}E[\sum\{(x_i - \mu) - (\bar{x} - \mu)\}^2]$$ By Rule 3 and because $(x - \bar{x}) = (x - \mu) - (\bar{x} - \mu)$

$$= \frac{1}{n-1}\{\sum E[(x_i - \mu)^2] - 2E[n(\bar{x} - \mu)(\bar{x} - \mu)] + \sum E[(\bar{x} - \mu)^2]\}$$ By expansion of square term, by Formula (3.18), and $\sum(x_i - \mu)(\bar{x} - \mu) = n(\bar{x} - \mu)(\bar{x} - \mu)$.

$$= \frac{1}{n-1}(\sum\sigma^2 - 2n\sigma^2/n + \sum\sigma^2/n)$$ Since $\sigma^2 = E[(x - \mu)]^2$ and $\sigma^2/n = E[(\bar{x} - \mu)^2]$

$$= \frac{1}{n-1}(n\sigma^2 - 2\sigma^2 + n\sigma^2/n)$$ Since $\sum_{i=1}^{n}$ (constant) $= n$(constant)

$$= \frac{1}{n-1}\sigma^2(n - 1)$$ Collecting terms,

$$E[s^2] = \frac{n-1}{n-1}\sigma^2 = \sigma^2.$$

If the population is finite, and sampling is *without* replacement, then $E[s^2] \neq \sigma^2$.

sulting distribution is no longer normally distributed, nor is its variance 1.0. Our next task is thus to determine the distribution of the ratio $(\bar{x} - \mu)/(s/\sqrt{n})$. This distribution can be thought of as being generated by the following process:

1. Collect all possible samples of size n from a normal parent population.
2. Calculate $\bar{x}$ and s for each sample.
3. Subtract μ from each value of $\bar{x}$, and then divide this deviation by the appropriate value of $s/\sqrt{n}$. (Remember, s will be different for each sample.)

This process will generate an infinite number of values of the random variable

$$\frac{\bar{x} - \mu}{s/\sqrt{n}}.$$

It is not hard to recognize that the mean of this new distribution still equals zero, since the numerator hasn't changed and it was the numerator that made our original standardization have $E[z] = 0$. The variance of $(\bar{x} - \mu)/(s/\sqrt{n})$ is no longer equal to $V[z] = 1.0$; it is larger than 1.0. This is reasonable when one recognizes that with the ratio $(\bar{x} - \mu)/(s/\sqrt{n})$ one more element of uncertainty (the estimator s) has been added to the standardization. The more uncertainty there is, the more spread out the distribution.

Several additional aspects of the distribution of $\bar{x} - \mu)/(s/\sqrt{n})$ are worth noting. First, we would expect this distribution to be symmetrical, since there is no reason to believe that substituting s for σ will make this distribution skewed either positively or negatively. Secondly, it should be apparent that the variability of this distribution depends on the size of n, for the sample size affects the reliability with which s estimates σ. When n is large, s will be a good approximation to σ; but when n is small, s may not be very close to σ. Hence, the distribution of $(\bar{x} - \mu)/(s/\sqrt{n})$ is a family of distributions in which variability depends on n.

It should be clear from the above discussion that the distribution of $(\bar{x} - \mu_{\bar{x}})/(s/\sqrt{n})$ is not normal, but is more spread out than the normal. The distribution of this statistic is called the "**t**-distribution," and its random variable is denoted as follows:

$$\textit{t-random variable:} \qquad t = \frac{\bar{x} - \mu}{s/\sqrt{n}}. \tag{6.9}$$

The variable t is a continuous random variable. One of the first researchers to work on determining the exact distribution of this random variable was W. S. Gosset, an Irish statistician. However, the Dublin brewery for which Gosset worked did not allow its employees to publish their research; hence,

Gosset wrote under the pen name "Student." In honor of Gosset's research, published in 1908, the **t**-distribution is often referred to as the "Student's **t**-distribution." It is not clear from historical records whether Gosset enjoyed the product of his employer, as do many modern "students."

Student's *t*-Distribution

Since the density function for the **t**-distribution is fairly complex and not of primary importance at this point, we will not present it, but will begin merely by describing the characteristics of this distribution.* As we indicated previously, the **t**-distribution depends on the size of the sample. It is customary to describe the characteristics of the **t**-distribution in terms of the sample size minus one, or $(n - 1)$, as this quantity has special significance.

The value of $(n - 1)$ *is called the number of* ***degrees of freedom*** *(abbreviated d.f.), and represents a measure of the number of independent observations in the sample that can be used to estimate the standard deviation of the parent population.*

For example, when $n = 1$, there is no way to estimate the population standard deviation; hence there are *no* degrees of freedom $(n - 1 = 0)$. There is one degree of freedom in a sample of $n = 2$, since one observation is now "free" to vary away from the other, and the amount it varies determines our estimate of the population standard deviation. Each additional observation adds one more degree of freedom, so that in a sample of size n there are $(n - 1)$ observations "free" to vary, and hence $(n - 1)$ degrees of freedom. The Greek letter ν (nu) is often used to denote degrees of freedom, where $\nu = n - 1$.

A ***t****-distribution is completely described by its one parameter,* $\nu = n - 1 =$ *degrees of freedom. As we stated above, the mean of the* ***t****-distribution is zero,* $E[t] = 0$. *The variance of the* ***t****-distribution, when* $\nu \geq 3$, is $V[t] = \nu/(\nu - 2)$.

Note that this implies that $V[t] \geq 1.0$ for all sample sizes, in contrast to $V[z]$ which is 1.0 no matter what the sample size. For example, when $\nu = 3$ the variance of the **t**-distribution is $3/(3 - 2) = 3.0$. This distribution and the standardized normal are contrasted in Fig. 6.6.

For small sample sizes, the **t**-distribution is seen to be considerably more spread out than the normal. Consider a larger sample size, such as $\nu = 30$;

---

* Mathematically, the random variable **t** is defined as a standardized normal variable z divided by the square root of an independently distributed chi-square variable, which has been divided by its degrees of freedom; that is, $t = z/\sqrt{\chi^2/\nu}$. The chi-square distribution is discussed in Sections 5.6 and 6.9.

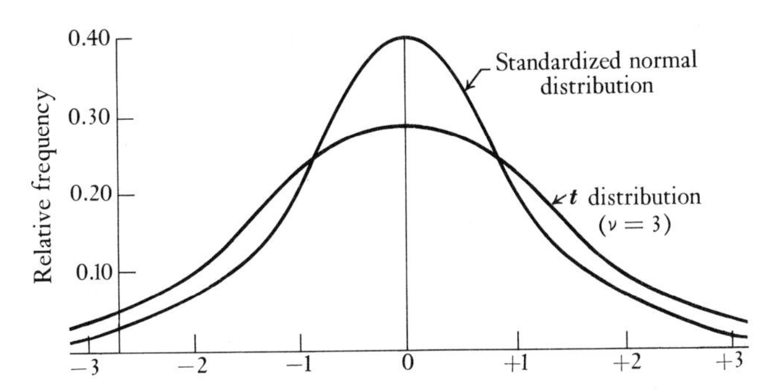

Fig. 6.6 The standardized normal and the t-distributions compared.

then $V[t] = 30/(30 - 2) = 1.07$, which is not much different from $V[z] = 1.0$. As v gets larger and larger, $V[t] \to V[z]$. In the limit, as $n \to \infty$, the t- and z-distributions are identical. Tables of t-values are usually only completely enumerated for $v \leq 30$, because for larger samples the normal gives a very good approximation and is easier to use. For this reason it is customary to speak of the t-distribution as applying to "small sample sizes," *even though this distribution holds for any size n.*

Probability questions involving a t-distributed random variable can be answered by using the t-distribution (shown in Table VII). This table gives the values of t for selected values of $P(t \geq t)$, given across the top of the table, and for degrees of freedom (v), down the left margin. Figure 6.7 shows the values of the t-distribution for $v = 24$ degrees of freedom, taken from Table VII.

Four different values from Table VII are shown in Fig. 6.7:

$$P(t \leq 0.685) = 0.75, \qquad P(t \leq 1.318) = 0.90,$$

and

$$P(t \geq 2.064) = 0.025, \qquad P(t \geq 2.492) = 0.01.$$

Table VII gives probabilities for seven selected t-values for each degree of freedom. More extensive tables are available, and probabilities may be determined mathematically for any t-value.

Fig. 6.7 Various probabilities for the t-distribution for $v = 24$ d.f.

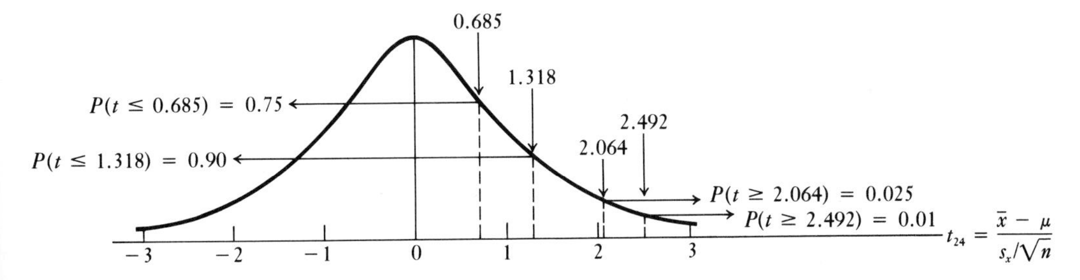

Examples of the *t*-Distribution

As with the normal distribution, the t-distribution is often used to test an assumption about a population mean based on the standardization of an observed sample mean.

> *The* **t**-*distribution is the appropriate statistic for inference on a population mean whenever the parent population is normally distributed and* σ *is unknown.*

To illustrate this type of situation, we will consider the case of a small finance company which has reported to an auditing firm that its outstanding loans are approximately normally distributed with a mean of \$825; the standard deviation is unknown. In an attempt to verify this reported value of $\mu = \$825$, a random sample of 25 accounts was taken. This random sample yields a mean of $\bar{x} = \$780$, with a standard deviation of $s = 105$. The question facing the auditor is how often might one find a sample mean of \$780 or lower when the true mean is \$825? That is, what is $P(\bar{\mathbf{x}} \leq \$780)$? This probability cannot be determined exactly since its value depends on σ, which is unknown. We can approximate it using the t-distribution. To approximate this probability, we must standardize the values in parenthesis. The approximation in this example follows a t-distribution with 24 degrees of freedom because $\mathbf{x}$ is normal, σ is unknown, and $n = 25$. Substituting $\sigma = 105$ and $n = 25$ and using Formula (6.9) yields:

$$P(\bar{\mathbf{x}} \leq \$780) \cong P\left(\frac{\bar{\mathbf{x}} - \mu}{\mathbf{s}/\sqrt{n}} \leq \frac{780 - 825}{105/\sqrt{25}}\right) = P(\mathbf{t} \leq -2.143).$$

Since the t-distribution is symmetrical, the probability we want, $P(\mathbf{t} \leq -2.143)$, is equivalent to $P(\mathbf{t} \geq +2.143)$. Because the number 2.143 does not appear in Table VII for $v = 24$, the exact value of $P(\mathbf{t} \geq 2.143)$ cannot be determined from this table. However, we can determine between which two probabilities the value $P(\mathbf{t} \geq 2.143)$ lies. From Table VII or Fig. 6.7,

$$P(\mathbf{t} \geq 2.064) = 0.025 \qquad \text{and} \qquad P(\mathbf{t} \geq 2.492) = 0.010,$$

which means that $P(\mathbf{t} \geq 2.143)$ lies between 0.025 and 0.010. Thus, we can write

$$0.01 < P(\bar{\mathbf{x}} \leq 780) \cong P(\mathbf{t} \leq -2.143) < 0.025.$$

This result says that a sample mean as low as or lower than \$780 will occur (approximately) between 1% and 2.5% of the time when $\mu = \$825$. Faced with such low probabilities, the auditor might well be concerned with the accuracy of the assumption that $\mu = \$825$.

Suppose that in the above example, instead of the sample results described there, we found the values $\bar{x} = \$842.60$, and $s = 80.0$. For this result we want to determine the probability that $\bar{x}$ is *greater* than or equal to \$842.60 when $\mu = \$825$.

$$P(\bar{x} \geq \$842.60) \cong P\left(\frac{\bar{x} - \mu}{s/\sqrt{n}} \geq \frac{\$842.60 - \$825}{80/\sqrt{25}}\right) = P(t \geq 1.100).$$

From Table VII, $P(t \geq 1.100)$ lies between $P(t \geq 0.685) = 0.25$ and $P(t \geq 1.318) = 0.10$ when $\nu = 24$. Hence,

$$0.25 > P(\bar{x} \geq \$842.60) \cong P(t \geq 1.100) > 0.10.$$

We see from this result that a sample mean of \$842.60 is fairly probable when $\mu = \$825$ and $n = 25$. These values are shown in Fig. 6.8.

As our second example of the use of the t-distribution, let us determine an interval (a, b) such that $P(a \leq t \leq b) = 0.95$, assuming $n - 1 = \nu = 8$ degrees of freedom. As before, the smallest interval is found by putting half the excluded area, $\frac{1}{2}(0.05) = 0.025$, in each tail of the distribution. For example, we want $P(t \geq b) = 0.025$. From Table VII for $\nu = 8$, we see that $P(t \geq 2.306) = 0.025$; hence, $b = 2.306$. Now since the t-distribution is symmetrical, the appropriate value for a is merely the negative of the value of b, or $a = -2.306$. Thus,

$$P(-2.306 \leq t \leq 2.306) = 0.95.$$

Recall, from our previous examples using the standardized normal distribution, that $P(-1.96 \leq z \leq 1.96) = 0.95$. The critical values for z that exclude 0.025 probability in the upper and lower tails are ± 1.96, as opposed to ± 2.306 for the t-distribution with $\nu = 8$. The difference reflects the fact that the t-distribution is more spread out than the z-distribution. Note in Table VII that by *increasing* the value of ν from 8 to 10, then 20, then 60, then 120, and moving down the column for 0.025, the critical values for t *decrease* from

Fig. 6.8 t-distribution values for $\nu = 24$.

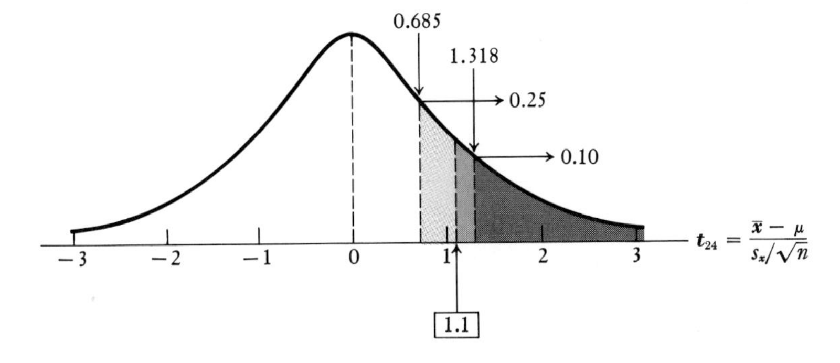

2.306 to 2.228, then 2.086, then 2.000, then 1.98, respectively. For larger values of ν, the spread of the t-distribution closes in to match the spread of the z-distribution. Indeed, when $n = \infty$, the critical value for t exactly equals the value for z (1.96), as shown by the bottom row in Table VII.

Use of the *t*-Distribution When the Population Is Not Normal

It must be emphasized at this point that the t-distribution, as well as the chi-square and F-distributions (discussed in the following sections) assume that samples are drawn from a parent population that is normally distributed. Often there is no way to determine the exact distribution of the parent population. In practical problems involving these distributions, the question therefore arises as to just how critical the assumption is that the parent population be exactly normally distributed. Fortunately, the assumption of normality can be relaxed without significantly changing the sampling distribution of the t-distribution. Because of this fact, the t-distribution is said to be quite "robust," implying that its usefulness holds up under conditions that do not conform exactly to the original assumptions.

We should emphasize again that the t-distribution is appropriate whenever x is normal and σ is unknown, despite the fact that many t-tables do not list values higher than $\nu = 30$. For practical problems this does not cause many difficulties; as the reader will note, the t-values in a given column of Table VII change very little after $\nu = 30$. For example, suppose that a sample of size $n = 400$ results in $\bar{x} = 120$ and $s = 200$, and we would like to calculate $P(\bar{x} \geq 120)$. Assume that σ is unknown and $\mu = 100$.

$$P(\bar{x} \geq 120) \cong P\left(\frac{\bar{x} - \mu}{s/\sqrt{n}} \geq \frac{120 - 100}{200/\sqrt{400}}\right) = P(t \geq 2.00).$$

Although $\nu = 399$ is not listed in Table VII, by looking at $\nu = 120$ versus $\nu = \infty$ we can easily determine that

$$0.025 > P(t \geq 2.00) > 0.01.$$

Thus, if the appropriate ν is not in Table VII we suggest looking at the table entries for ν above and below the one desired.*

Define: t-distribution, degrees of freedom.

---

* Some texts suggest that the normal distribution be used to approximate the t-distribution when $\nu > 30$, since the t- and z-values will then be quite close (the normal value will be slightly smaller than the exact t-value). Because of this procedure, the t-distribution sometimes is referred to *incorrectly* as applying only to "small samples." We prefer to emphasize that the t-distribution is *always* correct whenever σ is unknown.

Problems

6.36 Determine, for each of the following cases, whether the t-distribution or the standardized normal distribution (or neither) is appropriate for answering probability questions relating to sample means.

a) a small sample from a normal population with known standard deviation,

b) a small sample from a nonnormal population with known standard deviation,

c) a small sample from a normal population with unknown standard deviation,

d) a small sample from a nonnormal population with unknown standard deviation,

e) a large sample from a normal population with unknown standard deviation,

f) a large sample from a nonnormal population with unknown standard deviation.

6.37 Suppose that you collect the following sample of four observations, drawn randomly from a normal population, representing the January heating cost for a 3-bedroom house in Champaign, Illinois: 99, 115, 91, 79.

a) What is the probability of obtaining this $\bar{x}$ or one smaller if the population has mean $\mu = 110$ and unknown variance? What is the probability that $\bar{x}$ is this large or larger if the population has mean $\mu = 80$ and unknown variance?

b) What is the probability of obtaining this $\bar{x}$ or one smaller if the population has mean $\mu = 110$ and a standard deviation of $\sigma = 14$? What is the probability that $\bar{x}$ is this small or smaller if the population has mean $\mu = 100$ and $\sigma = 10$?

6.38 a) Describe the difference between the standardized normal distribution and the t-distribution. Under what conditions can each be used?

b) Is $P(z \geq 2.0) = 0.0228$ greater than $P(t \geq 2.0)$ for all sample sizes? How do you explain this fact?

6.39 It is suggested that the average weekly wage of student workers is $120. A random sample of 100 workers yields the distribution in Table 6.7.

a) Calculate $\bar{x}$ and s^2 using class marks 50, 75, 100, . . . , 200.

b) Approximate the probability of obtaining a sample mean this high if $\mu = 120$ and x is normally distributed.

c) Would your result from part (b) lead you to question the supposition that $\mu = 120$?

Table 6.7

Wages	f
38–62	4
63–87	15
88–112	20
113–137	30
138–162	15
163–187	10
188–212	6
Total	100

6.40 Explain why the standardized normal distribution does not provide an accurate description of the sampling distribution of $\bar{x}$ for small samples drawn from a normal population when σ is unknown. What distribution is appropriate in this circumstance?

6.41 Assume an economist estimates that the number of gallons of gasoline used monthly by each automobile in the U.S. is a normally distributed random variable with mean $\mu = 50$ and variance unknown:

a) Suppose that a sample of nine observations yields a sample variance of $s^2 = 36$. Approximate the probability that $\bar{x}$ is larger than 54 if $\mu = 50$. Approximate the probability that $\bar{x}$ is less than 44 if $\mu = 50$. Approximate the probability that $\bar{x}$ lies between 45 and 55.

b) How would your answers to the above problem change if $n = 36$?

6.42 A professional bowler claims her bowling scores can be thought of as normally distributed with mean $\mu = 215$ and unknown variance. In her latest performance, the bowler scores 188, 214, and 204.

a) Calculate $\bar{x}$ and s^2 for this sample.

b) If these three scores represent random samples from a normal population with mean $\mu = 215$, approximate the probability that $\bar{x}$ will be as low as you calculated it to be in part (a).

c) Would you conclude from part (b) that the bowler is "off her game"?

6.43 Approximate the probability of receiving a value of $\bar{x}$ as small as the one in Problem 6.34, assuming $\mu = 300$ hours.

6.44 Solve Problem 6.16(b) using the t-distribution.

6.9 THE SAMPLING DISTRIBUTION OF s^2, NORMAL POPULATION

The only sampling distribution considered thus far has been that of $\bar{x}$, the sample mean. But in many practical problems we need information about the distribution of the sample variance, s^2. That is, we need to investigate the distribution that consists of all possible values of s^2 calculated from samples of size n. The sampling distribution of s^2 is particularly important in problems concerned with the variability in a random sample. For example, the telephone company may be just as interested in the variance in length of calls in a random sample as they are in the mean length. Or a manufacturer of steel beams may want to learn just as much about the variance as the mean of tensile strength of the steel beams.

The same statistician who first worked with the t-distribution, W. S. Gosset, was also one of the first to describe the sampling distribution of s^2. First, note that because s^2 must always be positive, the distribution of s^2 cannot be a normal distribution. Rather, the distribution of s^2 is a unimodal distribution that is *skewed* to the right, and looks like the smooth curve in Fig. 6.9. As

* This section can be omitted without loss in continuity. Although Section 5.6 provides an excellent introduction to this section, we will not presume 5.6 has been studied.

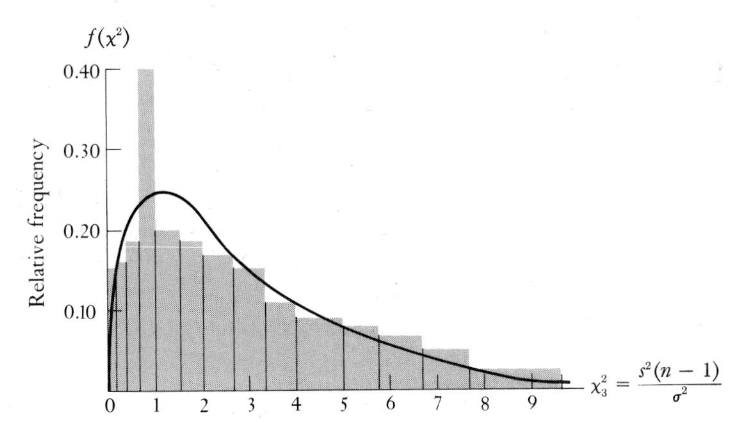

Fig. 6.9 Chi-square approximation to Gosset's data on the heights of criminals.

with the $\boldsymbol{t}$-distribution, sampling is assumed to be from a normal parent population, and the one parameter is the degrees of freedom, ν. A typical problem in analyzing variances is that of determining the probability that the value of $\boldsymbol{s}^2$ will be larger (or smaller) than some observed value, given some assumed value of σ^2. For example, suppose that the variance in diameters of steel ball bearings is specified to be $\sigma^2 = 0.0010$ inches. What is the probability that a random sample of $n = 21$ ball bearings will result in a sample variance at least as large as $s^2 = 0.0016$? That is, what is

$$P(\boldsymbol{s}^2 \geq 0.0016),$$

assuming $\nu = n - 1 = 20$ and $\sigma^2 = 0.0010$?

Unfortunately, we cannot solve problems like this one directly, but must transform them in a way similar to the standardizations for $\bar{\boldsymbol{x}}$. In this case the transformation is accomplished by multiplying $\boldsymbol{s}^2$ by $(n - 1)$, and then dividing the product by σ^2. This new variable is denoted by the symbol $\boldsymbol{\chi}^2$, which is the square of the Greek letter chi. This chi-square variable has one parameter, $\nu = n - 1$, which is its degrees of freedom. Thus,

$$\boldsymbol{\chi}^2_{n-1} = \frac{\nu \boldsymbol{s}^2}{\sigma^2} = \frac{(n - 1)\boldsymbol{s}^2}{\sigma^2}. \tag{6.10}$$

In words, this formula says the following:

> *If $\boldsymbol{s}^2$ is the variance of random samples of size n taken from a normal population having a variance of σ^2, then the variable $(n - 1)\boldsymbol{s}^2/\sigma^2$ has the same distribution as a $\boldsymbol{\chi}^2$-variable with $(n - 1)$ degrees of freedom.*

The subscript on the χ^2 symbol in Formula (6.10) merely serves to remind us of the appropriate degrees of freedom.

Although Gosset was unable to prove Formula (6.10) mathematically, he did demonstrate this relationship in his empirical work. Gosset took the heights of 3000 criminals, calculated the value of σ^2 for these heights, and then grouped these heights into 750 random samples of 4. For each of these 750 samples Gosset, in effect, calculated a value of s^2, multiplied s^2 by $(n - 1) = 3$, and then divided this number by σ^2. The results are plotted in the histogram shown in Fig. 6.9. Note that Gosset's histogram and the chi-square distribution (for $v = n - 1 = 3$) superimposed on it are not in perfect agreement, a fact that Gosset attributed to the particular grouping of heights that he used.

Solving a problem involving s^2 by using Formula (6.10) follows essentially the same process used to solve problems involving $\bar{x}$. For example, to solve the problem mentioned earlier, $P(s^2 \geq 0.0016)$, we transform each value in parenthesis as follows:

$$P(s^2 \geq 0.0016) = P\left[\frac{(n-1)s^2}{\sigma^2} \geq \frac{(20)(0.0016)}{0.0010}\right] = P(\chi^2_{20} \geq 32).$$

The equivalence between $P(s^2 \geq 0.0016)$ and $P(\chi^2_{20} \geq 32)$ is illustrated in Fig. 6.10.

The number of degrees of freedom in a χ^2-distribution determines the shape of $f(\chi^2)$. Since only squared numbers are involved in calculating χ^2, we know that this variable can never assume a value below zero, but it may take on values up to positive infinity. When v is small, the shape of the density function is highly skewed to the right. As v gets larger, however, the χ^2-distri-

Fig. 6.10 Transforming an s^2-value into an equivalent χ^2-value with $v = n - 1 = 20$.

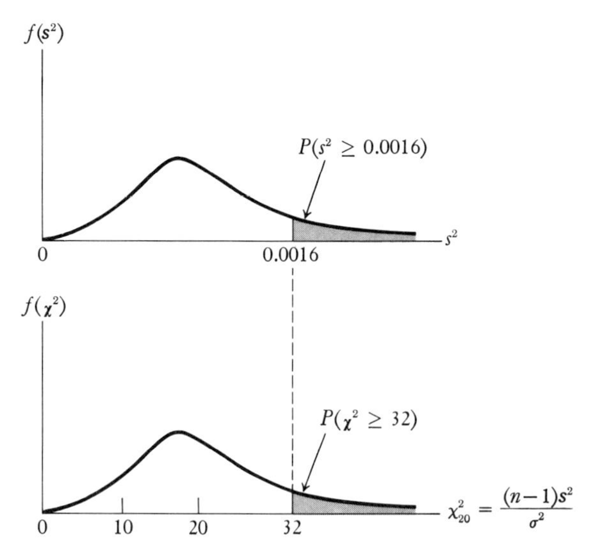

bution becomes more and more symmetrical in appearance, approaching the normal distribution as $n \to \infty$.

The density function for the χ^2-distribution is not of primary importance for our discussion; hence we will not present its formula, but merely concentrate on its characteristics. The mean and the variance of the chi-square distribution are both related to ν as follows:*

$$Mean = E[\chi_\nu^2] = \nu;$$

$$Variance = V[\chi_\nu^2] = E[(\chi_\nu^2 - \nu)^2] = 2\nu.$$

Thus, if we have a chi-square variable involved in a problem using random samples of size $n = 21$, then

$$E[\chi^2] = 20 \qquad \text{and} \qquad V[\chi^2] = 40.$$

Chi-Square Examples

Table V in Appendix B gives values of the cumulative χ^2-distribution for selected values of ν and gives (at the bottom) a formula for the normal approximation to χ^2 which can be used when $\nu > 30$. To illustrate the use of the χ^2-distribution, we will assume that we have taken all possible random samples of size $n = 21$ from some normal parent population. For each of these random samples we then multiply the value of the sample variance (s^2) by $(n - 1)$ and divide the result by the (assumed) population variance (σ^2). When we have finished this hypothetical task (there is an infinite number of such ratios), we will have calculated all possible values of

$$\chi_{20}^2 = \frac{(n-1)s^2}{\sigma^2} = \frac{(20)s^2}{\sigma^2}.$$

The distribution of the statistic given above is the chi-square distribution. We can use Table V in Appendix B to graph a few values of the chi-square distribution for 20 degrees of freedom.

From Fig. 6.11 we see that the ratio $(20)s^2/\sigma^2$ will have a value less than 8.26 only one percent of the time, less than 9.59 two and one-half percent of the time, less than 28.4 ninety percent of the time, and so forth. Let us now use this information to solve our previous ball-bearing problem, which was $P(s^2 \geq$

---

* These relationships are proved below.

$$E[\chi^2] = E\left[\frac{s^2\nu}{\sigma^2}\right] = \frac{\nu}{\sigma^2}E[s^2] \qquad \text{Since } \nu, \sigma^2 \text{ are constants}$$

$$= \frac{\nu}{\sigma^2}\sigma^2 = \nu \qquad \text{Because } E[s^2] = \sigma^2$$

$$V[\chi^2] = V\left[\frac{s^2\nu}{\sigma^2}\right] = \frac{\nu^2}{\sigma^4}V[s^2] \qquad \text{By Section 3.7}$$

$$= \frac{\nu^2}{\sigma^4}(2\sigma^4/\nu) = 2\nu \qquad \text{Since } V[s^2] = 2\sigma^4/\nu$$

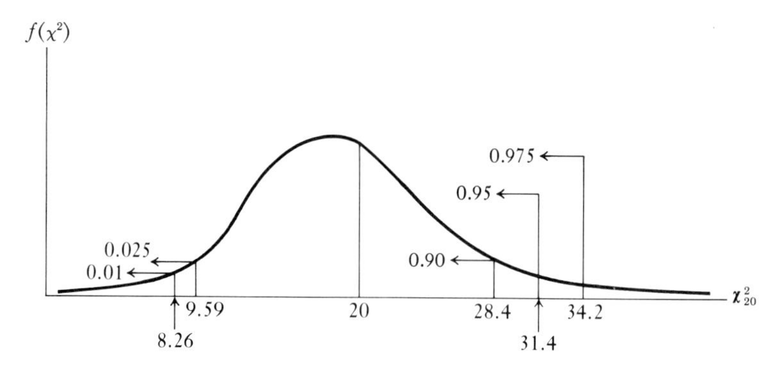

Fig. 6.11 The χ^2-distribution for $\nu = 20$.

0.0016) for $n = 21$. We transform each of the values in parentheses by multiplying by $(n - 1)$, and then divide by the assumed population variance, $\sigma^2 = 0.0010$, as follows:

$$P(\boldsymbol{s}^2 \geq 0.0016) = P\left(\frac{(n - 1)\boldsymbol{s}^2}{\sigma^2} \geq \frac{(20)(0.0016)}{0.0010}\right) = P(\boldsymbol{\chi}^2_{20} \geq 32.0).$$

From Fig. 6.11 we see that $P(\boldsymbol{\chi}^2 \geq 32)$ must be between 0.05 and 0.025; that is,

$$0.05 > P(\boldsymbol{\chi}^2 \geq 32.0) = P(\boldsymbol{s}^2 \geq 0.0016) > 0.025.$$

Since a sample variance as high as 0.0016 will occur relatively infrequently when $\sigma^2 = 0.0010$ (less than five percent of the time), we might question this company's statement that $\sigma^2 = 0.0010$.

6.10 THE *F*-DISTRIBUTION

Another important distribution involving variances is the ***F***-distribution, named in honor of R. A. Fisher, who first studied it in 1924. The ***F***-distribution is particularly useful in problems where one wants to test hypotheses about whether or not two normal populations have the same variance. For example, suppose you are interested in determining whether or not σ_1^2, the variance of some normal population, equals σ_2^2, the variance of a second normal population. Statistically, we can do this by taking a sample from each population, calculating the variance of each sample, and then determining the probability the two sample variances came from populations having equal variances. To determine this probability, let us assume that a random sample from the first population yields a variance of s_1^2, and a random sample from the second population yields a variance of s_2^2. How far the ratio s_1^2/s_2^2 differs from 1.0 can be used to make inferences about whether or not $\sigma_1^2 = \sigma_2^2$. When $\sigma_1^2 = \sigma_2^2$, we would expect the ratio s_1^2/s_2^2 to be close to 1.0. Thus, the more we find that the ratio s_1^2/s_2^2 differs from 1.0, the less confidence we have that $\sigma_1^2 = \sigma_2^2$.

How closely the ratio s_1^2/s_2^2 can be expected to approach 1.0 when $\sigma_1^2 = \sigma_2^2$ depends on the size of the two samples—or, more precisely, on the number of degrees of freedom in each sample. We will denote the number of degrees of freedom in the two samples as v_1 and v_2, and the sample sizes as n_1 and n_2, where $v_1 = n_1 - 1$ and $v_2 = n_2 - 1$. Using this terminology, we can show that:

The ratio of the variances of two independent random samples drawn from normal parent populations having equal variances follows an ***F****-distribution* with $v_1 = n_1 - 1$ and $v_2 = n_2 - 1$ degrees of freedom.*

$$\boldsymbol{F}_{v_1,v_2} = \frac{s_1^2}{s_2^2}. \tag{6.11}$$

The parameters of the $\boldsymbol{F}$-distribution are v_1 and v_2, the degrees of freedom, and we could show that $E[\boldsymbol{F}]$ and $V[\boldsymbol{F}]$ both depend on these parameters. However, we will not have occasion to use either $E[\boldsymbol{F}]$, $V[\boldsymbol{F}]$, or the density function of the $\boldsymbol{F}$-distribution in this book; hence, these values will not be presented.

Figure 6.12 on page 296 shows a typical $\boldsymbol{F}$-distribution; this one represents all possible values of s_1^2/s_2^2 when $v_1 = 10$ and $v_2 = 15$. Note that the $\boldsymbol{F}$-distribution is always positive or zero, and is positively skewed. This is reasonable, since the ratio s_1^2/s_2^2 can never be negative, but can assume *any* positive value. The fact that the $\boldsymbol{F}$-distribution looks something like a chi-square distribution is not coincidental, since these distributions are closely related. Four particular probabilities are indicated in Fig. 6.12, namely $P(\boldsymbol{F} \geq 2.54) = 0.05$, $P(\boldsymbol{F} \geq 3.06) = 0.025$, $P(\boldsymbol{F} \geq 3.80) = 0.01$, and $P(\boldsymbol{F} \leq 0.22) = 0.01$.

To illustrate the use of the $\boldsymbol{F}$-distribution and Formula (6.12), let's assume that two random samples have been drawn from normal populations, and it is desirable to determine the probability that these two samples were drawn from populations having the same variance. Furthermore, we shall assume that the first sample is of size $n_1 = 11$ and the second is of size $n_2 = 16$. The appropriate $\boldsymbol{F}$-distribution for this problem has $v_1 = 11 - 1 = 10$ degrees of freedom, and $v_2 = 16 - 1 = 15$ degrees of freedom, as shown in Fig. 6.12.

---

* The $\boldsymbol{F}$-distribution is usually defined as the ratio of two χ^2 variables, each divided by its degrees of freedom. To illustrate, assume s_1^2 and s_2^2 are from random samples drawn from two normal parent populations, each having the same variance, σ^2. Let the two samples have $v_1 = n - 1$ and $v_2 = n_2 - 1$ degrees of freedom, respectively. We know, from Section 6.9, that the variables $v_1 s_1^2/\sigma^2$ and $v_2 s_2^2/\sigma^2$ will both have a χ^2-distribution. If we take the ratio of these two χ^2 variables, each divided by its degrees of freedom, the result is:

$$\boldsymbol{F} = \frac{\dfrac{v_1 s_1^2/\sigma^2}{v_1}}{\dfrac{v_2 s_2^2/\sigma^2}{v_2}} = \frac{s_1^2/\sigma^2}{s_2^2/\sigma^2} = \frac{s_1^2}{s_2^2}.$$

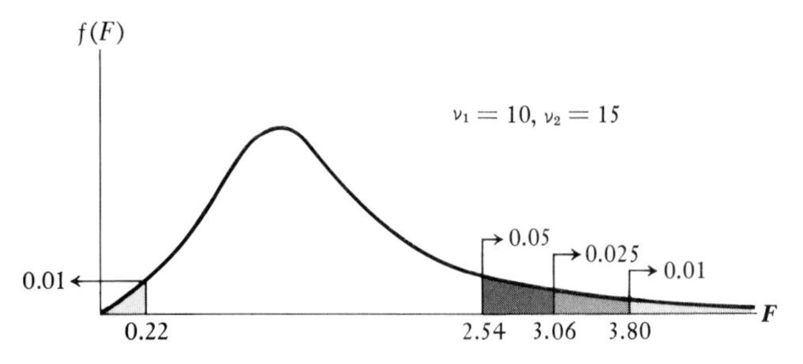

Fig. 6.12 The F-distribution for $\nu_1 = 10$, $\nu_2 = 15$.

Now, let's suppose our first sample of size $n_1 = 11$ yields $s_1^2 = 330$, while the second sample, of size $n_2 = 16$, yields $s_2^2 = 100$. We can calculate the probability that these two samples came from populations having the same variance, by taking the ratio

$$F_{(10, 15)} = \frac{s_1^2}{s_2^2} = \frac{330}{100} = 3.30.$$

From Fig. 6.12 we see that $P(\boldsymbol{F}_{(10, 15)} \geq 3.30)$ is greater than 0.01, but less than 0.025 Hence, we can write $0.025 > P(\boldsymbol{F}_{(10, 15)} \geq 3.30) > 0.01$.

The values for the $\boldsymbol{F}$-distribution shown in Fig. 6.12 were taken from Table VIII in Appendix B. Table VIII (a), (b), and (c) give $\boldsymbol{F}$-values cutting off 0.05, 0.025, and 0.01 in the upper tail of the $\boldsymbol{F}$-distribution.* The reader should verify that when $\nu_1 = 10$ (the numerator) and $\nu_2 = 15$ (the denominator), then parts (a), (b) and (c) of Table VIII yield the values 2.54, 3.06, 3.80, respectively.

In the problem described above, s_1^2 was the larger sample variance; hence the $\boldsymbol{F}$-ratio was a number larger than one and we used the upper value of $\boldsymbol{F}$. When s_1^2 is the smaller sample variance, and thus s_1^2/s_2^2 is a number less than 1.0, it is necessary to have lower-tail values of $\boldsymbol{F}$. Fortunately, lower-tail values are easily found by using the following rules.

A lower-tail value of $\boldsymbol{F}$ *can always be found by reversing the degrees of freedom of the numerator and the denominator, determining the corresponding value in the upper tail of the* $\boldsymbol{F}$*-distribution, and then taking the reciprocal of this number. That is,*

$$\boldsymbol{F}_{(\text{lower value}, \nu_1, \nu_2)} = \frac{1}{\boldsymbol{F}_{(\text{upper value}, \nu_2, \nu_1)}}.$$

* Only upper-tail $\boldsymbol{F}$-values are given because lower-tail values can easily be derived from the fact that if a random variable $\boldsymbol{x}$ has an $\boldsymbol{F}$-distribution with m and n degrees of freedom, then $1/\boldsymbol{x}$ has an $\boldsymbol{F}$-distribution with n and m degrees of freedom.

As an illustration of the calculation of a lower-tail value, recall from Fig. 6.12 that the lower-tail value for 0.01 when $\nu_1 = 10$ and $\nu_2 = 15$ is given as 0.22. This value can be determined by finding first the *upper* value for 0.01 with the degrees of freedom reversed (that is, 15 for the numerator and 10 for the denominator). From Table VIII(c) we see that $F_{(\text{upper value, } 15,\, 10)} = 4.56$. Hence,

$$F_{(\text{lower value, } 10,\, 15)} = \frac{1}{F_{(\text{upper value, } 15,\, 10)}} = \frac{1}{4.56} = 0.22.$$

As Fig. 6.12 illustrates, values of F based on 10 and 15 degrees of freedom will exceed 3.80 one percent of the time, and will be less than 0.22 one percent of the time.

6.11 SUMMARY

Table 6.8 provides a summary of the sampling distributions discussed in this chapter. Figure 6.13 represents a tree diagram that some students have found useful in deciding when to apply each of these distributions.

Define: Chi-square distribution, F-distribution.

Table 6.8 Summary of Sampling Distributions

Random Variable	Situation	Reference Section	Resulting Distribution for Problem Solving	Mean	Variance
$\bar{x}$	Population normal, σ known. sample size n	6.5	$z = \dfrac{\bar{x} - \mu_{\bar{x}}}{\sigma/\sqrt{n}}$	0	1
$\bar{x}$	Population normal, σ inknown, sample size n*	6.8	$t_\nu = \dfrac{\bar{x} - \mu_{\bar{x}}}{s/\sqrt{n}}$	0 with $\nu = n - 1$ d.f.	$\nu/(\nu - 2)$
$\bar{x}$	Population unknown, σ known, $n > 30$	6.6	$z = \dfrac{\bar{x} - \mu}{\sigma/\sqrt{n}}$	0	1
s^2	Population normal, sample size n	6.9	$\chi^2_\nu = \dfrac{(n - 1)s^2}{\sigma^2}$	σ^2 with $\nu = n - 1$ d.f.	$\dfrac{2\sigma^4}{\nu}$
s_1^2, s_2^2	Populations normal, sample sizes of n_1 and n_2	6.10	$F_{\nu_1 \nu_2} = s_1^2/s_2^2$	— with $\nu_1 = n_1 - 1$ and $\nu_2 = n_2 - 1$ d.f.	—

* If $n > 30$, $\bar{x}$ will be approximately normal, and s should be close to σ; hence $(\bar{x} - \mu)(s/n)$ is approximately $N(0, 1)$.

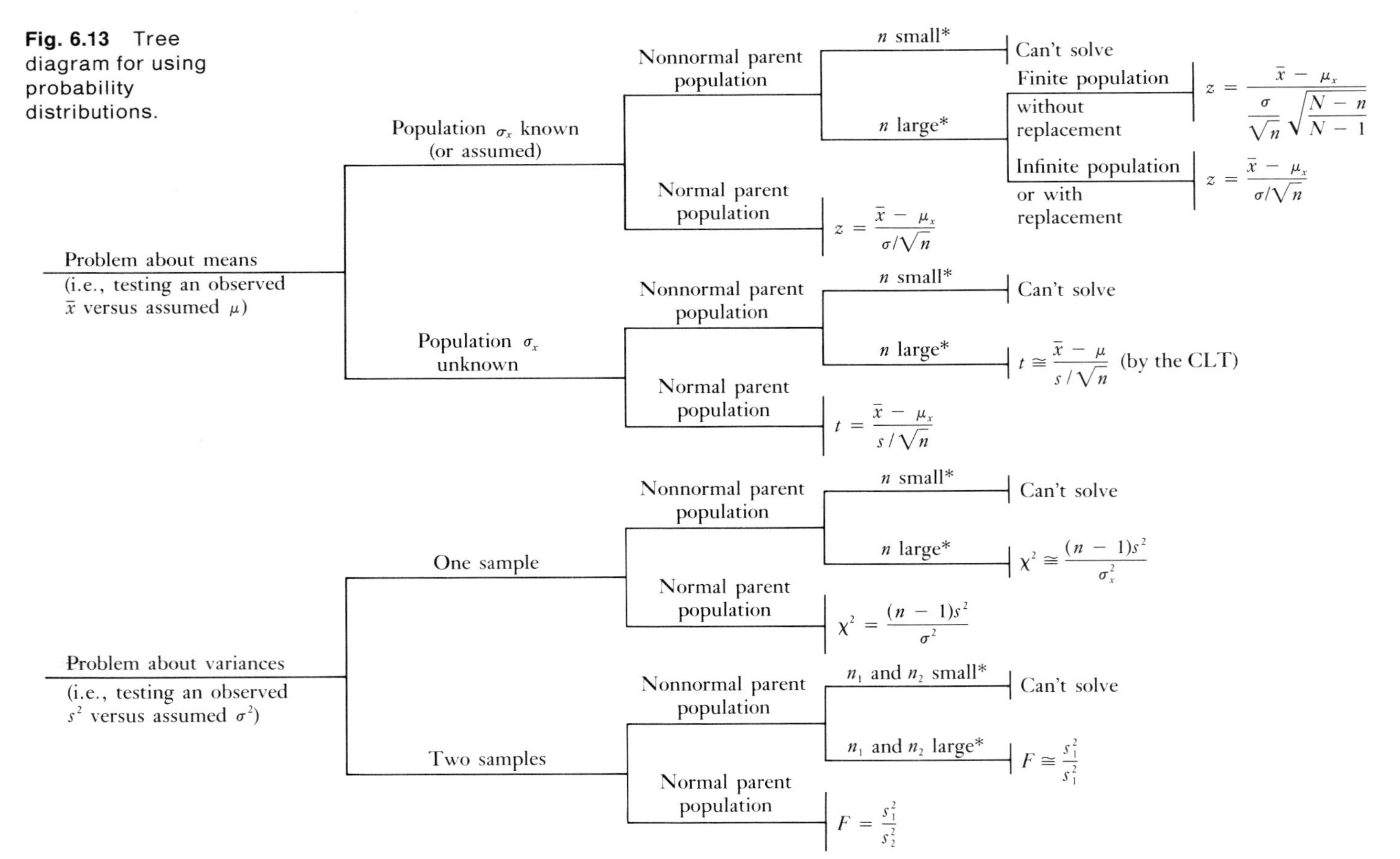

Fig. 6.13 Tree diagram for using probability distributions.

* "Large" and "small" may depend on the accuracy desired and the shape of the parent population. For fairly symmetrical distributions, $n \geq 10$ may be sufficient. In almost all cases, $n \geq 30$ is sufficient for "large."

† Remember, the ***t*** and χ^2 are fairly "robust," meaning that they work well, even if the parent population is not exactly normally distributed but has some normal characteristics (unimodal and not badly skewed).

Problems

6.45 Suppose that you are drawing samples of size $n = 3$ from a normal population with a variance of 8.25. What is the probability that $(n - 1)s^2/\sigma^2$ will exceed 5.99? What is the probability that s^2 will be more than three times as large as σ^2?

6.46 A hamburger chain is concerned with the amount of variability in its "quarter pounder." The amount of meat in these burgers is supposed to have a variance of no more than 0.2 ounces. A random sample of 20 burgers from one chain yields a variance of $s^2 = 0.4$. The parent population is normal.

a) What is the probability that a sample variance will equal or exceed 0.4 if it is assumed that $\sigma^2 = 0.2$?

b) Would you suspect that the meat content of the burgers that this chain is selling varies excessively?

6.47 Calculate the value of s^2 for Problem 6.41(a). How often will s^2 be this low (or lower) if $\sigma^2 = 50$?

6.48 Calculate the value of s^2 for Problem 6.42. How often will s^2 be as large as this value of s^2 if $\sigma^2 = 100$?

6.49 Suppose that a random variable is known to be chi-square-distributed with parameter $v = 24$.

a) What is the mean and the variance of the chi-square distribution for this parameter?

b) What is the probability that the value of χ^2 will exceed 43.0? What is $P(\chi^2 \geq 33.2)$. What is $P(\chi^2 \leq 9.89)$?

c) Use your answers to parts (a) and (b) to draw a rough sketch of the chi-square distribution for $v = 24$.

d) Superimpose on your sketch for part (c) a graph of the normal distribution for $\mu = 24$ and $\sigma^2 = 48$. How closely do the two distributions agree?

6.50 A manufacturer is testing the strength of a new roofing material. While the mean strength is clearly sufficient, the manufacturer is concerned with weakness due to the variability in strength. Past experience with this material suggests that the variance in strength is $\sigma^2 = 1000$ per square inch. If the population is normal and σ^2 is, in fact, equal to 1000, what is the probability a random sample will yield 2050, 2150, 2270, and 2110?

6.51 a) Sketch the F-distribution for $n_1 = 16$ (the numerator) and $n_2 = 13$.

b) Two independent random samples are drawn from normal populations. The first, of size $n_1 = 16$, results in $s_1^2 = 100$. The second, of size $n_2 = 13$, yields $s_2^2 = 250$. What is the probability of receiving these two variances if $\sigma_1^2 = \sigma_2^2$?

6.52 Find the three lower-tail values (0.05, 0.025, 0.01) for the F-distribution involving $v_1 = 10, v_2 = 15$.

6.53 One measure of the risk inherent in a particular stock portfolio is the variance in expected prices. If a random sample of the expected prices of two different portfolios yields $s_1^2 = 4000$ (based on $n_1 = 31$) and $s_2^2 = 7500$ (based on $n_2 = 61$), how likely is it that s_2^2 will be this much larger than s_1^2 if the two portfolios are equally risky and the populations are normally distributed?

6.54 A public-policy researcher is studying the variance in the amount of money requested by certain government agencies this year relative to five years ago. Random samples, of size $n_1 = 21$ and $n_2 = 25$, yielded $s_1^2 = (67{,}223)^2$ and $s_2^2 = (37{,}178)^2$. What is the probability that s_1^2 will be this much larger than s_2^2 if $\sigma_1^2 = \sigma_2^2$ and the parent populations are normally distributed?

6.55 Suppose that an economist is interested in determining the variance of incomes across two communities. Random samples of size $n_1 = 11$ and $n_2 = 10$ result in $s_1^2 = (2000)^2$ and $s_2^2 = (2400)^2$. Assuming that $\sigma_1^2 = \sigma_2^2$ and that the parent populations are normal, what is the probability that s_1^2 will be this much lower than s_2^2?

6.56 The following random sample of the starting salary for MBA finance majors was taken for students working in Chicago ($n_1 = 10$) and New York City ($n_2 = 9$).

	Chicago	**New York City**
1	26,400	27,200
2	29,200	24,100
3	24,900	21,900
4	21,400	28,300
5	26,800	25,500
6	23,100	21,900
7	30,000	26,900
8	19,000	27,100
9	25,500	23,900
10	24,900	
Mean	25,120	25,200

The mean of these two samples is fairly close. Is there enough variability in the samples for us to question the assumption that $\sigma_1^2 = \sigma_2^2$? Assume normal parent populations.

Exercises

6.57 Write a computer program that collects random samples from the following *uniform distribution:*

$$f(x) = \begin{cases} 1.0 & 0 \le x \le 1.0, \\ 0 & \text{otherwise.} \end{cases}$$

a) Write the program to collect 100 random samples of size $n = 25$, and calculate $\bar{x}$ and s^2 for each sample.

b) Sketch the distribution of $\bar{x}$, using some convenient class interval for the x's. Compare the shape of your distribution with that predicted by the central limit theorem.

c) Repeat parts (a) and (b) for $n = 4$.

d) For each sample in part (a), calculate $t = (\bar{x} - 0.5)/(s/\sqrt{n})$, and then sketch these 100 t-values. Does the distribution correspond to that given in Table VII for $v = 24$?

e) For each sample in part (a), calculate $\chi^2 = 24s^2/\sigma^2$, where $\sigma^2 = \frac{1}{12}$. Sketch this distribution. Does this sketch correspond to the distribution of χ^2_{24}?

6.58 Consult a mathematical statistics text and write down the probability density function for the **t**-distribution and the chi-square distribution. How are these two distributions related?

6.59 a) Construct an example to show that $E[s^2] \neq \sigma^2$ when simple random sampling is done without replacement, from a finite population.

b) Assuming a normal parent population, prove that $V[s^2] = 2\sigma^4/\nu$. (*Hint:* Use the fact that $\chi^2 = s^2\nu/\sigma^2$ and $V[\chi^2] = 2\nu$.)

6.60 Research the chi-square distribution and explain what it means to say that this distribution is the sum of the squares of a finite number of independent standardized normal random variables (see Section 5.6).

6.61 Describe how the finite population correction factor and the hypergeometric distribution are related. See if you can justify the fact that the correction factor is $\sqrt{(N - n)/(N - 1)}$.

6.62 The auditor's failure to recognize an error in an amount or an error in an internal-control data-processing procedure is described as a

a) Statistical error

b) Sampling error

c) Standard error of the mean

d) Nonsampling error.

6.63 In connection with his review of charges to the plant maintenance account, Mr. John Wilson, CPA, is undecided as to whether to use probability sampling or judgment sampling. As compared to probability sampling, judgment sampling has the primary disadvantage of

a) Providing no known method for making statistical inferences about the population solely from the results of the sample.

b) Not allowing the auditor to select those accounts which he believes should be selected.

c) Requiring that a complete list of all the population elements be compiled.

d) Not permitting the auditor to know which types of items will be included in the sample before the actual selection is made.

6.64 A CPA's client wishes to determine inventory shrinkage by weighting a sample of inventory items. If a stratified random sample is to be drawn, the strata should be identified in such a way that

a) The overall population is divided into subpopulations of equal size so that each subpopulation can be given equal weight when estimates are made.

b) Each stratum differs as much as possible with respect to expected shrinkage but the shrinkages expected for items within each stratum are as close as possible.

c) The sample mean and standard deviation of each individual stratum will be equal to the means and standard deviations of all other strata.

d) The items in each stratum will follow a normal distribution so that probability theory can be used in making inferences from the sample data.

6.65 In estimating the total value of supplies on repair trucks, Baker Company draws random samples from two equal-sized strata of trucks. The mean value of the inventory stored on the larger trucks (stratum 1) was computed at $1500, with a standard deviation of $250. On the smaller trucks (stratum 2), the mean value of inventory was computed as $500, with a standard deviation of $45. If Baker had drawn an unstratified sample from the entire population of trucks, the expected mean value of inventory per truck would be $1000, and the expected standard deviation would be

a) Exactly $147.50. b) Greater than $250.

c) Less than $45.

d) Between $45 and $250, but not $147.50.

6.66 The required size of a statistical sample is influenced by the variability of the items being sampled. The sample standard deviation, a basic measure of variation, is approximately the

a) Average of the sum of the differences between the individual values and their mean.

b) Square root of the average determined in (a).

c) Average of the sum of the squared differences between the individual values and their mean.

d) Square root of the average determined in (c).

6.67 Write a computer program which demonstrates the central limit theorem. For example, you might consider taking a large number of samples from a Poisson distribution (try $\lambda = 1$). Let the sample size first be $n = 5$, then $n = 25$, then $n = 100$, and show that the resulting distributions of $\bar{x}$ tend to $N(\mu, \sigma^2/n)$ as n gets larger.

GLOSSARY

sample design: a plan specified for obtaining a sample before any data are collected.

sampling and nonsampling errors: sampling errors are those errors that occur because even a perfectly designed sample may not always represent the population exactly. Nonsampling errors are the "human," or avoidable, errors.

random number: a number selected from a population in such a way that it has the same probability of occurring as does every other number in the population.

probabilistic sampling: sample designs that are based primarily on a random selection process. Includes:

a) **simple random sampling:** every item and every group of items has the same probability of being in the sample.

b) **systematic sampling:** selection of every kth item, starting from a random point.

c) **stratified sampling:** selects from layers or strata.

d) **cluster sampling:** selects randomly from groups, or clusters, having similar characteristics.

e) **sequential sampling:** sample items taken not simultaneously but sequentially.

nonprobabilistic sampling: based primarily on a nonrandom selection process. Includes:

a) **quota sampling:** a specified number of values collected.

b) **convenience sampling:** values taken according to what is convenient.

c) **judgment sampling:** primary consideration is the judgment of the person in charge.

sample statistic: a characteristic of a sample.

sample mean: $\bar{x} = \frac{1}{n}(\sum x_i f_i)$

sample variance:

$$s^2 = \frac{1}{n-1}\sum(x_i - \bar{x})^2 f_i$$

$$= \frac{1}{n-1}(\sum x_i^2 f_i - n\bar{x}^2)$$

sampling distribution: the probability distribution of a sample statistic.

expected value of $\bar{\mathbf{x}}$: $E[\bar{\mathbf{x}}] = \mu_{\bar{x}} = \mu$.

variance of $\bar{\mathbf{x}}$: $V[\bar{\mathbf{x}}] = \sigma_{\bar{x}}^2 = \sigma^2/n$.

standard error of the mean: $\sigma_{\bar{x}} = \sigma/\sqrt{n}$.

standardization of $\bar{\mathbf{x}}$: $z = \dfrac{\bar{\mathbf{x}} - \mu}{\sigma/\sqrt{n}}$.

central limit theorem: regardless of the population distribution, as n gets larger, the distribution of $\bar{\mathbf{x}}$ will become more and more like a normal distribution.

finite population correction factor: corrects the standard error of the mean by multiplying by $\sqrt{(N - n)/(N - 1)}$ when sampling from a population of size N without replacement.

degrees of freedom (ν): number of values in a sample which are "free" to vary when calculating a sample statistic. When calculating s, $\nu = n - 1$.

***t*-distribution:** a family of symmetrical p.d.f.'s that depends on the parameter ν (d.f.) with $E[\mathbf{t}] = 0$ and $V[\mathbf{t}] = \nu/(\nu - 2)$ for $\nu \geq 3$. Used particularly in probability questions involving the sample mean when sampling from a normal parent population with unknown variance, $\mathbf{t} = (\bar{\mathbf{x}} - \mu)/(\mathbf{s}/\sqrt{n})$. The $\mathbf{t}$ approaches the normal z as the sample size gets larger.

chi-square (χ^2) distribution: a family of positively skewed p.d.f.'s which depends on the parameter v (degrees of freedom) with $E[\chi^2] = v$, $V[\chi^2] = 2v$. Used in probability questions about a sample variance when sampling from a normal parent population, $\chi^2_{v=n-1} = (n-1)s^2/\sigma^2$.

***F*-distribution:** A family of positively skewed p.d.f.'s which depends on parameters v_1 and v_2. Used to answer probability questions involving the ratio s_1^2/s_2^2 when $\sigma_1^2 = \sigma_2^2$, and sampling is from normal populations.

7

ESTIMATION

"If the polls are so accurate, why are there so many polling companies?"

BILL FOSTER

7.1 INTRODUCTION

In most statistical studies the population parameters are unknown and must be estimated from a sample because it is impossible or impractical (in terms of time or expense) to look at the entire population. Developing methods for estimating as accurately as possible the value of population parameters is thus an important part of statistical analysis. A firm manufacturing electrical components might wish to investigate the average number of defective units in each batch of 1000 items without inspecting each and every component before shipment. The psychologist who wants to determine the mean IQ of all college undergraduates will undoubtedly also have to rely on sample information. In these cases, the value of a sample statistic, such as the sample mean, must be used as an estimate of the population parameter. If the degree of dispersion of defective electrical components from batch to batch or the variability of IQs is of interest, then this parameter also must be estimated from the sample data. Our objective in this chapter, which deals with similar estimation problems, is twofold: first, to present criteria for judging how well a given sample statistic estimates the population parameter; and second, to analyze several of the most popular methods for estimating these parameters.

The random variables used to estimate population parameters are called *estimators*, while specific values of these variables are referred to as *estimates* of the population parameters. The random variables $\bar{x}$ and s^2 are thus estimators of the population parameters μ and σ^2. A specific value of $\bar{x}$, such as $\bar{x} = 120$, is an estimate of μ, just as the specific value $s^2 = 237.1$ is an estimate of σ^2.

It is not necessary that an estimate of a population parameter be one single value; instead the estimate could be a range of values.

Estimates that specify a single value of the population are called ***point estimates,*** *while estimates that specify a range of values are called* ***interval estimates.***

A point estimate for the average IQ of college undergraduates may be 120, implying that our best estimate of the population mean is 120. An interval estimate specifies a range of values, say 115 to 125, indicating that we think the mean IQ for the population lies in this interval.

The choice of an appropriate point estimator in a given circumstance usually depends on how well the estimator satisfies certain criteria. In Section 7.2 we will describe four properties of "good" estimators.

The properties emphasized throughout the remainder of this book are those that have been generally recognized as the most important four properties of a good estimator.

1. *The property of unbiasedness:* On the average, the value of the estimate should equal the population parameter being estimated.
2. *The property of efficiency:* The estimator should have a relatively small variance.
3. *The property of sufficiency:* The estimator should take into consideration as much as possible of the information available from the sample.
4. *The property of consistency:* The estimator should approach the value of the population parameter as the sample size increases.

An estimator is a random variable, since it is the result of a sampling experiment. As a random variable, it has a probability distribution with a specific shape, expected value, and variance. Analysis of these characteristics of the distribution of an estimator permits us to specify desirable properties of the estimator.

7.2 FOUR PROPERTIES OF A "GOOD" ESTIMATOR

Unbiasedness

Normally, it is preferable that the expected value of the estimator exactly equal, or fall close to the true value of the parameter being estimated. If the average value of the estimator does not equal the actual parameter value, the estimator is said to contain a "bias," or to be a "biased estimator." Under ideal conditions, an estimator has a bias of zero, in which case it is said to be "unbiased." This property thus can be stated as follows:

> *An estimator is said to be unbiased if the expected value of the estimator is equal to the true value of the parameter being estimated. That is,*
>
> $E[\text{estimator}] = \text{population parameter}.$

In determining a point estimate for the population mean, it is certainly not difficult to construct examples of a biased estimator. Simply using the largest observation in a sample of size $n > 1$ to estimate μ, and ignoring the rest of the observations, will yield an estimate whose expected value is larger than μ. This is obviously a poor choice for estimating the population mean, especially when there is a much more appealing choice, that of using $\bar{\mathbf{x}}$. The sample mean is the most widely used estimator of all, for one of its major advantages is that it provides an unbiased estimate of μ. The fact that $E[\bar{\mathbf{x}}] = \mu$ was presented in Section 6.4. The parameter other than μ most often estimated is σ^2, the population variance. An unbiased estimator for σ^2 is s^2, since $E[\mathbf{s}^2] = \sigma^2$, as we showed in Section 6.8.*

Although $\mathbf{s}^2$ is an unbiased estimator of σ^2, it is *not* true that s is an unbiased estimator of σ; that is, $E[\mathbf{s}] \neq \sigma$, because the square root of a sum of numbers is not usually equal to the sum of the square roots of those same numbers. To demonstrate this fact, suppose we consider the population discussed in Chapter 6, involving the three elements (1, 2, 3). We saw, in Table 6.3, that there are nine different samples of size $n = 2$, when sampling with replacement. Table 7.1 shows these nine samples, as well as the mean, variance, and standard deviation of each sample (where $s^2 = \sum(x_i - \bar{x})^2/(n - 1)$, and $n = 2$).

The sum of the nine values of $\mathbf{s}^2$ shown in Column (5) is 6.00; this means that the average value of $\mathbf{s}^2$ is $E[\mathbf{s}^2] = \frac{6}{9} = \frac{2}{3}$. Since the population (1, 2, 3) has a variance of $\frac{2}{3}$, we have shown that $E[\mathbf{s}^2] = \sigma^2$.

Now consider the nine values of $\mathbf{s}$ shown in Column (6) of Table 7.1. If $\mathbf{s}$ were an unbiased estimator of σ, the average of these nine values would equal $\sigma = \sqrt{2/3} = 0.817$. The actual average is $5.66/9 = 0.629$; this means that

* Recall that $E[\mathbf{s}^2] \neq \sigma^2$ when sampling from a finite population without replacement.

Table 7.1

(1) (x_1, x_2)	(2) $\bar{x}$	(3) $(x_1 - \bar{x})^2$	(4) $(x_2 - \bar{x})^2$	(5) $s^2 = \frac{1}{n-1}\sum(x_i - \bar{x})^2$	(6) $s = \sqrt{\frac{1}{n-1}\sum(x_i - \bar{x})^2}$
(1, 1)	1.0	0.00	0.00	0.00	0.00
(1, 2)	1.5	0.25	0.25	0.50	0.71
(2, 1)	1.5	0.25	0.25	0.50	0.71
(1, 3)	2.0	1.00	1.00	2.00	1.41
(3, 1)	2.0	1.00	1.00	2.00	1.41
(2, 2)	2.0	0.00	0.00	0.00	0.00
(2, 3)	2.5	0.25	0.25	0.50	0.71
(3, 2)	2.5	0.25	0.25	0.50	0.71
(3, 3)	3.0	0.00	0.00	0.00	0.00
Sum	18.0	3.00	3.00	6.00	5.66

$E[\mathbf{s}] \neq \sigma$. In fact, we see that seven of the nine values of $\mathbf{s}$ *underestimate* the true population standard deviation of $\sigma = 0.817$.* The bias in this case is $0.817 - 0.629 = 0.188$. *Even though* $\mathbf{s}$ *is not an unbiased estimator of* σ, *it is still used as an estimator of* σ *because* $\mathbf{s}^2$ *is an unbiased estimator of* σ^2. There are correction factors which make $\mathbf{s}$ an unbiased estimator of σ. These correction factors depend on the form of the population distribution, and will not be presented here.

As a final example of the property of unbiasedness, consider the problem of estimating p, the population proportion of successes in a binomial distribution. Recall that in Chapter 4 we stated that if a sample yields $\mathbf{x}$ successes in n trials, then the ratio $\mathbf{x}/n$ is an unbiased estimate of p:

$$E[\mathbf{x}/n] = \frac{1}{n} E[\mathbf{x}] = \frac{1}{n}(np) = p.$$

The above result implies that if, in a random sample of 100 voters, 60 people indicate that they intend to vote for Candidate A, then $\frac{60}{100} = 0.60$ is an unbiased estimate of the population proportion of people who would say they intend to vote for Candidate A.

One weakness of the property of unbiasedness lies in the fact that the criterion requires only that the *average* value of the estimator equal the population parameter. It does not require that most, or even *any* of the values of the

* Because the sample standard deviation $\mathbf{s}$ usually *underestimates* σ, it follows that the variable $(\bar{\mathbf{x}} - \mu)/(\mathbf{s}/\sqrt{n})$ will *overestimate* the variable $(\bar{\mathbf{x}} - \mu)/(\sigma/\sqrt{n})$. This is one reason the $\mathbf{t}$-distribution is more spread out than the normal distribution.

estimator be reasonably close to the population parameter, as would seem desirable in a "good" estimator. For this reason, the property of efficiency is important.

Efficiency

For given repeated samples of size *n*, it is desirable that an estimator have values that are close to each other. That is, it would be comforting in estimating an unknown parameter to believe that the value you computed based on a particular random sample would not be much different from the value you or anyone else would compute based on another random sample of the same size. *The property of efficiency implies that the variance of the estimator should be small.* However, having a small variance doesn't make an estimator a good one, unless this estimator is also unbiased. For example, an estimator that always specifies 200 as its estimate of the population parameter will have zero variance. But this estimate will be biased unless the true population parameter happens to equal 200. In other words, a small variance is desirable, but so is unbiasedness.

The property of efficiency of an estimator is defined by comparing its variance to the variance of all other *unbiased* estimators,

---

The most efficient estimator among a group of unbiased estimators is the one with the smallest variance.

The most efficient estimator is also called the *best unbiased* estimator, where "best" implies minimum variance. Figure 7.1 illustrates the distributions of three different estimators (labeled 1, 2, and 3) based on samples of the same size. Of the three distributions, 1 and 2 both have expected values equal to the population parameter; i.e., they are unbiased. The third distribution has a positive bias, since its mean exceeds the population parameter. The variance of the three estimators decreases in size from 1 to 3. However, the third distribution is not the most efficient estimator among this group because it is not unbiased.

Fig. 7.1 Illustration of properties of unbiasedness and efficiency.

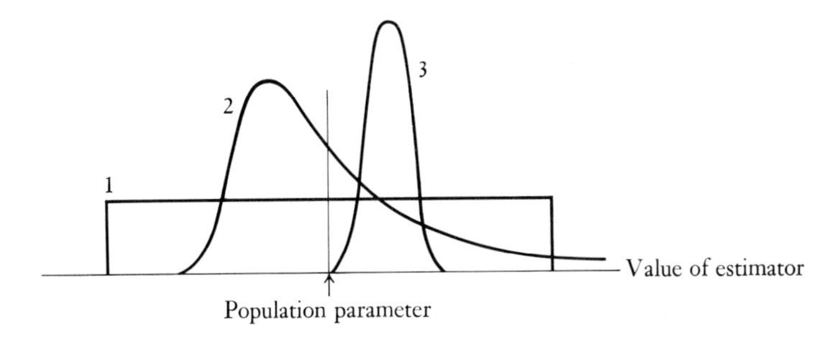

Our definition of efficiency requires that the estimator be *unbiased* and have smaller variance than any other unbiased estimator. Thus, estimator 2 is the most efficient of those illustrated. Whether number 2 is the most efficient of *all* possible unbiased estimators has not been shown and is often difficult to prove.

Relative Efficiency

Since it is generally quite difficult to prove that an estimator is the best among all unbiased ones, the most common approach is to determine the *relative efficiency* of two estimators. Relative efficiency is defined as the ratio of the variances of the two estimators.

> *Relative efficiency:* $\dfrac{\textit{Variance of first estimator}}{\textit{Variance of second estimator}}.$

As an illustration of the use of relative efficiency, consider the sample mean vs. the sample median as estimators of the mean of a normal population. Both estimators are unbiased when we are sampling from a normal population, since the normal is symmetric. From Section 6.4 we know that the variance of $\bar{x}$ equals σ^2/n. It is also possible to find the variance of an estimator using the sample median to estimate the population mean; this variance is $\pi\sigma^2/2n$.* The ratio of these quantities gives their relative efficiency:

$$\frac{V(\text{median})}{V(\bar{x})} = \frac{\pi\sigma^2/2n}{\sigma^2/n} = \frac{\pi}{2} = 1.57.$$

The ratio 1.57 implies that the median is 1.57 times less efficient than the mean in estimating μ. In other words, an estimate based on the median of a sample of 157 observations has the same reliability as an estimate based on the mean of a sample of 100 observations, assuming a normal parent population.

In some problems it is possible to determine precisely the most efficient estimator, that is, the unbiased estimator with the lowest variance. In estimating the mean of a normal population, for example, it can be shown that the variance of any estimator must be greater than or equal to σ^2/n. Since the variance of $\bar{x}$ in this case exactly equals σ^2/n, the sample mean must be the most efficient estimator of μ. In the design of a sample, the most efficient estimator may not always be the best choice because of other factors, such as the time available to collect the sample or the accessibility of the observations. That is, statistical efficiency may have to be sacrificed in order to obtain an estimate in the time allowed; or some other estimator may be less costly to obtain or more meaning-

* For further discussion, see reference [12].

ful and, therefore, may be preferred over the most efficient one. Many high schools, for example, publish statements about a student's academic performance only in terms of the *rank* relative to the rest of the class (e.g., "he ranks tenth in his class"). Trying to determine the most efficient estimator in this case (for example, the average academic performance) may be difficult and such an estimator may not be as meaningful as a measure based on the ranked performance.

Sufficiency

Unbiasedness and efficiency are desirable properties for an estimator, particularly when one is dealing with small samples. Another property of interest is *sufficiency*.

An estimator is said to be sufficient if it uses all the information about the population parameter that the sample can provide.

That is, the sufficient estimator somehow takes into account each of the sample observations, as well as all the information that is provided by these observations. The sample median is not a sufficient estimator because it uses only the *ranking* of the observations and ignores information about the distance between adjacent numerical values. The sufficiency property is of importance in that it is a necessary condition for efficiency.

Consistency

Since the distribution of an estimator will change, in general, as the sample size changes, the properties of estimators for large sample sizes (as $n \to \infty$) become important. Properties of estimators based on limits approached as $n \to \infty$ are called *asymptotic properties*, and these may differ from finite or small-sample properties. The most important of these asymptotic properties is that of *consistency*, which involves the variability of the estimator about the population parameter as the size of n increases:

An estimator is said to be consistent if it yields estimates that converge in probability to the population parameter being estimated as n becomes larger.

To say that an estimator must converge in probability to the parameter being estimated is to imply that, as samples become larger, the expected value of the estimator must approach the true value of the parameter and that the variance of this estimator about the population parameter must become negligible. This requirement is equivalent to specifying that the estimator becomes unbiased and that the variance of the estimator approaches zero as n approaches infinity.

Previously, we showed that $\bar{x}$ is an unbiased estimator of the population mean, and that x/n is an unbiased estimator of the population proportion. It is not difficult to show also that both these estimators are consistent as well as unbiased. We shall prove that x/n is a consistent estimator of p, leaving it to the reader to prove that $\bar{x}$ is also consistent.

$$\begin{aligned} V[x/n] &= \frac{1}{n^2} V[x] && \text{(From Section 3.7, Rule 4)} \\ &= \frac{1}{n^2}(npq) && \text{(Since } x \text{ is binomially distributed)} \\ &= \frac{pq}{n}. \end{aligned}$$

Since the value of pq/n approaches zero as n approaches infinity, and since x/n is an unbiased estimator of p for any sample size, then x/n must be a consistent estimator of p.

While the four properties presented above are certainly all quite desirable, they do not preclude other considerations. We pointed out previously that the inferences (or estimates) made from samples serve as an aid to the process of making decisions, and that samples should be drawn with the objective of minimizing the cost of making an incorrect decision (balanced against the cost of sampling). Since one primary purpose in collecting a sample involves estimating parameters, an estimation procedure should be chosen that will minimize the cost (or loss) of making an incorrect estimate from the sample information. This objective is not necessarily incompatible with any of the above properties of good estimators; in fact, in many cases, when these properties are satisfied the estimator indeed will minimize the cost of making an error.

7.3 ESTIMATING UNKNOWN PARAMETERS

In the 1920's R. A. Fisher developed the method of maximum likelihood as a means of finding estimators that satisfy some (but not necessarily all) of the criteria discussed previously. This method is popular because maximum-likelihood estimators are usually relatively easy to obtain, and are often efficient and approximately normally distributed for large samples. One disadvantage of the method is that maximum-likelihood estimates are not necessarily unbiased for small samples although they are consistent (large sample property).

The maximum-likelihood method estimates the value of a population parameter by selecting the most likely sample space from which a given sample could have been drawn. In other words, the sample space is selected that would yield the observed sample more frequently than any other sample space. The value of the population parameter corresponding to the generation of this sample space is called the maximum-likelihood estimate; the name *maximum likelihood* is derived from this process of selecting the most likely sample space.

As an illustration of the process of determining a maximum-likelihood estimate, consider the problem of estimating the binomial parameter p. Suppose that, in a sample of 5 trials, 3 successes are observed. What sample space (i.e., what binomial population) is most likely to give this particular result; or equivalently, what is the most likely value of p given the observed sample? The most likely population parameter can be determined by calculating the probability of obtaining exactly three successes in five trials for all possible values of the population parameter and selecting that value which yields the highest probability. Table 7.2 examines nine possible values of p, indicating for each value the probability of 3 successes in 5 trials; for example, if $p = \frac{1}{10}$, the appropriate probability is

$$\binom{5}{3}\left(\frac{1}{10}\right)^3\left(\frac{9}{10}\right)^2 = 0.0081.$$

The value of p most likely to yield a sample of 3 successes in 5 trials, as given by Table 7.2, is $p = 0.60$, where the associated probability is 0.3456. The reader will note that this estimate exactly equals the sample proportion $x/n = 3/5 = 0.60$. It is often true that the most likely value for a population parameter is the intuitively appealing one, the corresponding measure of the sample. For

Table 7.2

Value of p	Probability of Three Successes
0.10	0.0081
0.20	0.0512
0.30	0.1323
0.40	0.2304
0.50	0.3125
0.60	0.3456
0.70	0.3087
0.80	0.2048
0.90	0.0729

example, it can be shown that the maximum-likelihood estimator of a population mean is the sample mean. That is, the value of $\bar{x}$ is the most likely value of μ that can be found based on a sample of size n. A proof of this relationship and others involving the derivation of maximum likelihood estimates is given in an appendix to this chapter.

There have been a number of interesting applications of the method of maximum likelihood, one of particular note involving the problem of estimating the size of animal populations from data on recaptured animals. Suppose, for example, that the U.S. Conservation Department wishes to estimate the current number of fish in a lake which was stocked a few years earlier.* Assume 500 fish are caught, marked, and then released. A few days later another catch of 1000 fish is made, and 40 of this new catch are found to have been marked earlier. Based on this information, what is the most likely number of total fish in the lake? If each catch is considered as a random sample, without replacement, of all the fish in the lake, then the hypergeometric distribution (discussed briefly in Chapter 4) can be applied in solving this problem. The probability of catching exactly $x_1 = 40$ marked fish and $x_2 = 960$ unmarked fish, assuming there are n fish in the lake (of which $n_1 = 500$ are marked and $n_2 = n - 500$ are unmarked) is given by Formula 4.4 in Chapter 4:

$$\frac{\binom{n_1}{x_1}\binom{n_2}{x_2}}{\binom{n_1 + n_2}{x_1 + x_2}} = \frac{\binom{500}{40}\binom{n - 500}{960}}{\binom{n}{1000}}$$

By a relatively involved process, the MLE of n can be shown to equal

$$n_1\left(\frac{x_1 + x_2}{x_1}\right),$$

so that the total number of fish most likely to have produced this sample is $500(40 + 960)/40 = 12{,}500$.

Define: Unbiasedness, efficiency, sufficiency, consistency, relative efficiency, maximum-likelihood estimator (MLE).

Problems

7.1 Differentiate between:

a) a point estimate and an interval estimate,

b) unbiasedness and consistency.

* This application is more thoroughly described in W. Feller's text, *An Introduction to Probability Theory and its Applications*, Vol. I, pp. 43–44 (Wiley).

7.2 Elaborate on the four properties of an estimator described in this chapter. Explain why each one is important.

7.3 Given a normal parent population, which of the following estimators are unbiased?

a) mean b) median c) mode d) $\mathbf{x}/n$.

7.4 An accountant selected a random sample of 100 of the commercial accounts in a branch bank. The mean balance among these accounts was found to be $749.13. The accountant then stated that the mean balance must be $749.13 since $\bar{\mathbf{x}}$ is an unbiased estimate of μ. Discuss the reasonableness of this assertion.

7.5 Consider the population $x = \{5, 15\}$.

a) Calculate μ, σ^2, and σ for this population.

b) Make a list of eight possible samples of size $n = 3$, with replacement [For example, (5, 5, 5), (5, 5, 15), (5, 15, 5), etc.]. Calculate $\bar{x}$ for each sample.

c) Show from part (b) that $\bar{\mathbf{x}}$ is an unbiased estimate of μ; that is, $E[\bar{\mathbf{x}}] = \mu$.

d) Calculate s^2 for each sample, and then show that $E[\mathbf{s}^2] = \sigma^2$.

e) Show that the average of the eight values of s is not equal to σ; that is, $E[\mathbf{s}] \neq \sigma$.

f) Calculate the median for each of the eight samples. Is the average of these values equal to μ; that is, is the median an unbiased estimator in this case?

7.6 Calculate the variance of the median values in Problem 7.5(f). Use this variance and the variance of the values of $\bar{x}$ in Problem 7.5(b) to show that $\bar{\mathbf{x}}$ is more efficient than the median as an estimator of μ.

7.7 Repeat Problem 7.5(b), using $n = 2$. Compare the case in which $n = 2$ with that in which $n = 3$, and show that the results support the fact that the mean is a consistent estimator.

7.8 Explain, in your own words, why the properties of efficiency and consistency are both based on the assumption that the estimator is unbiased.

7.9 a) When $n = 4$, the five possible binomial values of x/n are $\frac{0}{4}$, $\frac{1}{4}$, $\frac{2}{4}$, $\frac{3}{4}$ and $\frac{4}{4}$. Calculate $E[\mathbf{x}/n]$ by multiplying each of these five values by the appropriate probabilities in Table I ($n = 4$), using $p = 0.40$. Is $\mathbf{x}/n$ unbiased; that is, is $E[\mathbf{x}/n] = p$?

b) Calculate the variance of the x/n values in part (a). Then calculate the variance of x/n for $p = 0.40$ when $n = 5$. Does the variance decrease from $n = 4$ to $n = 5$, supporting the fact that $\mathbf{x}/n$ is consistent?

7.10 If a random sample of 200 radios in a large production lot yields ten defectives, what is the maximum-likelihood estimate of the proportion of defectives in the entire lot?

7.11 Return to Problem 6.32 of Chapter 6. Calculate the variance for each sample of size $n = 2$ in part (b). Show that $E[\mathbf{s}^2] = \sigma^2$.

7.12 For the data in Table 6.2, Chapter 6,

a) Show that $E[\text{median}] = E[\bar{\mathbf{x}}]$.

b) Show that $V[\text{median}] = V[\bar{\mathbf{x}}]$.

*7.13 Prove that $E[\mathbf{x}/n] = p$ for the binomial distribution.

*7.14 Assume that a sample of two is to be taken from a normal population. Let $\mathbf{x}_1$ represent the first observation, and $\mathbf{x}_2$ the second observation in the sample. Consider the estimator $(2\mathbf{x}_1 + \mathbf{x}_2)/3$.

a) Is this estimator unbiased in estimating μ?

b) Is it sufficient?

c) Find the relative efficiency of this estimator as compared to $(3x_1 + 2x_2)/5$.

*7.15 a) Suppose that x_1 and x_2 are independent random variables having the Poisson probability distribution with parameter λ. Show that the mean of these variables is an unbiased estimator of λ.

b) Compare the efficiency of the mean of x_1 and x_2 with the alternate estimator $(x_1 + 2x_2)/3$.

c) Is the mean of x_1 and x_2 a sufficient estimator?

7.16 Describe, intuitively, what is meant by a maximum-likelihood estimate.

7.17 Show that s^2 is not an unbiased estimator of σ^2 if samples of size $n = 2$ are drawn *without* replacement from the population $\{1, 4, 7, 8\}$.

7.4 INTERVAL ESTIMATION

The particular value chosen as most likely for a population parameter is called a *point estimate.* We know that it would be an exceptional coincidence if this estimate were identical to the population parameter (because of sampling error). Thus, even though the best possible value is used as the point estimate, we should have very little confidence that this value is *exactly* correct. One of the major weaknesses of a point estimate is that it does not permit the expression of any degree of uncertainty about the estimate. The most common way to express uncertainty about an estimate is to define, with a known probability of error, an interval or range of values in which the population parameter is likely to be. This process is known as *interval estimation.*

You will recall that on a number of occasions thus far we have determined values a and b so that $P(a \leq \bar{x} \leq b)$ equals some predetermined value. The values a and b were determined from a knowledge about the parent population and its parameters. The interval (a, b) is called a probability interval for $\bar{x}$. For example, if we calculated $P(a \leq \bar{x} \leq b) = 0.90$, based on a random sample of size n drawn from a population with mean μ, we know the random variable $\bar{x}$ will fall in the probability interval (a, b) 90 percent of the time.

Although it is important to be able to construct probability intervals for $\bar{x}$ based on knowledge of μ, for most practical statistical problems the process must be reversed; i.e., it is μ that is the unknown, and we want to construct a confidence interval for μ based on $\bar{x}$. For example, we may want to develop a method for defining an interval based on $\bar{x}$ such that μ is likely to lie in that interval 90 percent of the times that the method is used—a 90 percent confidence interval. This means that, on the average, 90 such intervals out of every 100 calculated on the basis of means of samples of size n will include the population mean μ.

The use of the future tense in explaining a confidence interval is very important because, once such an interval based on a sample is determined, either the true parameter lies in the interval or it does not. The value of μ cannot be said to have a probability of 0.90 of being within the interval because it is not a random variable, but a constant. If it is in a given interval, then the probability that it is in the interval is 1.0; if not, the probability that it is within the interval is 0.0.

To better distinguish between the two types of intervals, consider the following problem in process control. A manufacturer makes large tile pipes; when the production process is working correctly, the interior diameter of the pipe is normally distributed with mean $\mu = 24''$ and $\sigma = \frac{1}{4}$, or $N(24, 0.25^2)$. At random points in time, a sample of four pipe segments is selected from the production process, to check on the average diameter of all the pipes in the population being produced. The population parameter being investigated here is μ and we know that in this case $\bar{x}$ has a normal distribution, since x is normal. Because $\sigma/\sqrt{n} = 0.25/\sqrt{4} = 0.125$ is the standard deviation of the $\bar{x}$'s, the distribution of $\bar{x}$ is $N(24, 0.125^2)$.

Figure 7.2 represents the two types of intervals discussed above. This figure shows the mean of twenty different samples (twenty dots), each of which was of size $n = 4$. Part (a) of the figure illustrates a probability interval with its center at the population mean, $\mu = 24$ inches, and values $a = 23.755$ and $b = 24.245$, chosen so that 95 percent of the values of $\bar{x}$ will lie between the endpoints $a = 23.755$ and $b = 24.245$. [These endpoints are found, as before, using $z = \pm 1.96$ (Table III) and the process described in Section 6.6.]

In repeated samples of size 4, different values of $\bar{x}$ would be calculated. A 95-percent probability interval means that on the average, 95 out of 100 of these values of $\bar{x}$ should be between a and b. Figure 7.2(a) illustrates twenty different values of $\bar{x}$ calculated from 20 different samples of size 4. Theoretically, 95 percent (or 19) of these should lie within the interval. Actually, 17 of them do and three do not (sample numbers 2, 7, and 16, counting from the bottom). If a large number of the values of $\bar{x}$, or several consecutive ones, should lie outside the interval, it would be a sign that the production process was getting "out of control" and should be stopped and corrected.

Part (b) of Fig. 7.2 illustrates the 95-percent confidence intervals based on these same 20 samples of size 4 each. Since each sample of size 4 can have different values, each sample can have a different sample mean and variance. Each interval depends on these values, so even though the underlying sampling distribution for $\bar{x}$ is the same in each instance, the center and spread of the interval calculated can be different. In Fig. 7.2(b), the center of each interval is at $\bar{x}$; the endpoints depend on s. *The meaning of a 95-percent confidence interval is that the true parameter is expected to be included within such intervals*

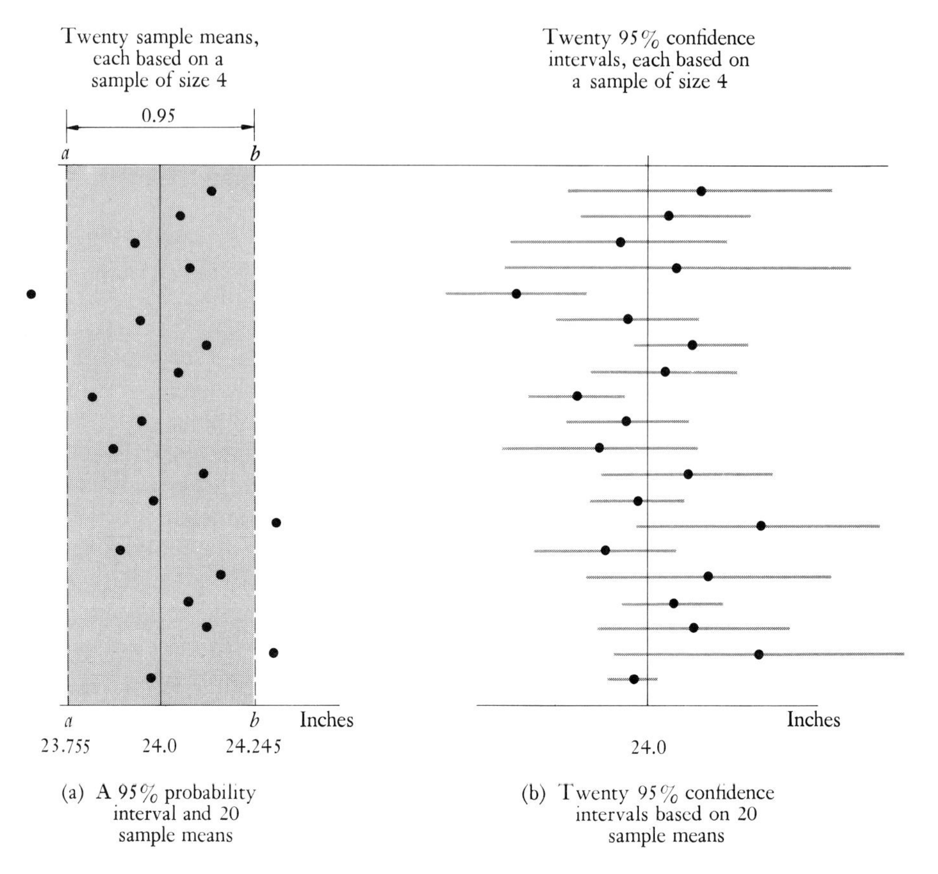

(a) A 95% probability interval and 20 sample means

(b) Twenty 95% confidence intervals based on 20 sample means

Fig. 7.2 Illustration of probability interval and of confidence interval for an unbiased estimator with a normal distribution.

in 95 *out of* 100 *such calculations.* Note in Fig. 7.2(b) that the true population parameter ($\mu = 24$) is included within the interval (the line) in 18 out of 20 samples, and excluded in 2 of the 20 samples (counting from the bottom, the twelfth and sixteenth lines do not include $\mu = 24$).

The Probability of an Error, α

It is more convenient to refer to the probability that a confidence interval will *not* include the parameter than to express the probability that it will. The former probability is usually denoted by α (the Greek letter "alpha"). The value of alpha is often referred to as the probability of making an error, since it indicates the proportion of times that one will be incorrect, or "in error," in assuming that the intervals would contain the population parameter. Since α is defined as the probability that a confidence interval will not contain the population parameter, $(1 - \alpha)$ equals the probability that the population parameter will be in the interval. It is customary to refer to confidence intervals as being of size, "$100(1 - \alpha)\%$". Thus, if α is 0.05, the associated interval is a $100(1 - 0.05)\%$ or a 95-percent confidence interval; if a 90-percent confidence interval is specified, then $\alpha = 1.0 - (\frac{90}{100}) = 0.10$. In our previous example, the process-control problem, α equals 0.05, which means that the probability is 0.05 of making an error in saying that the interval contains the parameter (process is in control) when it in fact does not (the process needs adjustment). In other words, on the average, we would be wrong in only five percent of the cases in which we estimate that the interval includes the population parameter.

There is an obvious trade-off between the value of α and the size of the confidence interval: the lower the value of α, the larger the interval must be.* If one need not be very confident that the population parameter will be within the interval, then a relatively small interval will suffice; if one is to be quite confident that the population parameter will be in the interval to be calculated, a relatively large interval will be necessary. The value of α is often set at 0.05 or 0.01, representing 95- and 99-percent confidence intervals, respectively. This procedure, although widely used, does not necessarily lead to the optimal trade-off between the size of the confidence interval and the risk of making an error.

In general, confidence intervals are constructed on the basis of sample information, so that not only do changes in α affect the size of the interval, but so do changes in n, the sample size. The more observations collected, the more confident one can be about the estimate and, as a result, the smaller the interval needed to assure a given level of confidence. Although it is usually desirable to have as small a confidence interval as possible, the size of the interval must be determined by considering the costs of sampling and how much risk of error one is willing to assume. We shall return several times to this problem of determining the optimal trade-off between the risks of making an error and the sample size. For now we merely caution the reader to be aware that the task of determining an "optimal" trade-off may not be an easy one.

Determining a Confidence Interval

Usually, one of the first steps in constructing a confidence interval is to specify how much confidence one wants to have that the population parameter will fall in the resulting interval, and the size of the sample. In other words, both α and n are usually fixed in advance. It is possible, however, under certain conditions, to consider either α or n as an unknown and to solve for the value of this unknown.

In addition to specifying α in advance, one must also specify what proportion of the probability of making an error is attributable to the fact that the

* We assume here that other factors, such as the sample size, are held constant.

population parameter will sometimes be larger than the upper bound of the confidence interval, and what proportion is attributable to the fact that the population parameter will sometimes be smaller than the lower bound of the confidence interval. As we indicated previously, in determining a probability interval, the *smallest* interval is obtained by dividing the probability to be excluded outside the interval *equally* between the upper and the lower tails of the distribution (assuming a symmetrical distribution). For a confidence interval, however, the decision on how to divide α into these two parts should depend on how serious or costly it is to make errors on the high side, relative to errors on the low side. Since we normally want to avoid the more costly errors, α should be divided so that these errors occur less frequently. Unfortunately, determining the costs of making an error may be quite difficult, so that the common procedure for determining a confidence interval is the same as for probability intervals—that is, to exclude one-half of α, or $\alpha/2$, on the high side and one-half of α on the low side. Thus, if $\alpha = 0.05$, a value of $\alpha/2 = 0.025$ would be the probability that the population parameter will exceed the upper bound of the confidence interval, and $\alpha/2 = 0.025$ would be the probability that the lower bound will exceed the population parameter. (Such a procedure assumes that errors on the high side are equally as serious as errors on the low side.)

In the discussion thus far we have noted that a given confidence interval will depend on the particular sample values received, on α (and the way α is divided), and on the size of the sample. The final factor influencing the boundaries of a particular confidence interval is the sampling distribution of the statistic used to estimate the population parameter. Normally, the procedure for establishing a confidence interval for a population parameter is *first to find a point estimate* of the parameter. One's uncertainty about this point is then determined by finding that interval of values about the point estimate which, according to the sampling distribution of the statistic used, yields the desired degree of confidence. Since different sampling distributions are used for estimating a population mean, the binomial parameter p, and a population variance, we shall describe in separate sections the process of constructing intervals for each of these cases.

7.5 CONFIDENCE INTERVALS FOR μ WITH σ KNOWN

Suppose that we start by constructing a confidence interval for the population parameter μ based on a random sample drawn from a normal parent population *with known standard deviation*. This section is mainly a building block for the sections to follow, for seldom do we know σ, but not know μ. However, there are some circumstances where σ is known from a previous study, or perhaps does not change over time.

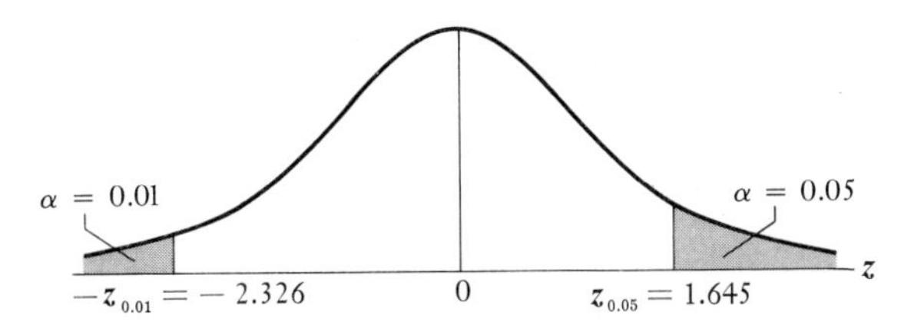

Fig. 7.3 The value of z_α cutting off α percent from the standardized normal distribution.

The natural sample statistic for estimating μ is $\bar{x}$, the sample mean, for reasons we discussed in Section 7.2. Recall that the sampling distribution of $\bar{x}$ has a mean of μ and a standard deviation of $\sigma/\sqrt{n}$. Also recall from Section 6.5 that the variable $z = (\bar{x} - \mu)/(\sigma/\sqrt{n})$ has a standardized normal distribution. Now, suppose we let the symbol z_α represent that value of the standardized normal variable z such that the probability of observing values of z greater than z_α is α. That is,

$$P(z \geq z_\alpha) = \alpha.$$

Similarly, we will let $-z_\alpha$ equal the point such that

$$P(z \leq -z_\alpha) = \alpha.$$

Two such points are illustrated in Fig. 7.3. Their values were derived from Table III in Appendix B.

For many problems in the next several chapters, we will be given a value of α, and then will have to find the z-value cutting off *half* of this value ($\alpha/2$) in each tail of a normal distribution (so that the total area cut off equals α). To distinguish this situation from the one described above, we will let $z_{\alpha/2}$ represent the value such that the probability $P(z \geq z_{\alpha/2}) = \alpha/2$, and let $-z_{\alpha/2}$ equal the point such that $P(z \leq -z_{\alpha/2}) = \alpha/2$. If, for example, $\alpha = 0.05$, then from Table III the value of $z_{\alpha/2}$ satisfying $P(z \geq z_{\alpha/2}) = 0.025$ is seen to be $z_{\alpha/2} = 1.96$. Then the value of $-z_{\alpha/2}$ must be -1.96, since the normal distribution is symmetric. The probability that z falls between the two limits, -1.96 and $+1.96$, is

$$P(-1.96 \leq z \leq 1.96) = 1.0 - 0.05 = 0.95,$$

as shown in Fig. 7.4. In more general terms, the probability that z falls between two limits $-z_{\alpha/2}$ and $+z_{\alpha/2}$ can be written in the following form:

$$P(-z_{\alpha/2} \leq z \leq z_{\alpha/2}) = 1 - \alpha.$$

Note that the interval $-z_{\alpha/2} \leq z \leq z_{\alpha/2}$ is a $100(1 - \alpha)\%$ confidence interval for z. It is also a $100(1 - \alpha)\%$ confidence interval for $(\bar{x} - \mu)/(\sigma/\sqrt{n})$, since $z = (\bar{x} - \mu)/(\sigma/\sqrt{n})$. Unfortunately, this is not the confidence interval we originally set out to derive, for we wanted a $100(1 - \alpha)\%$ confidence interval

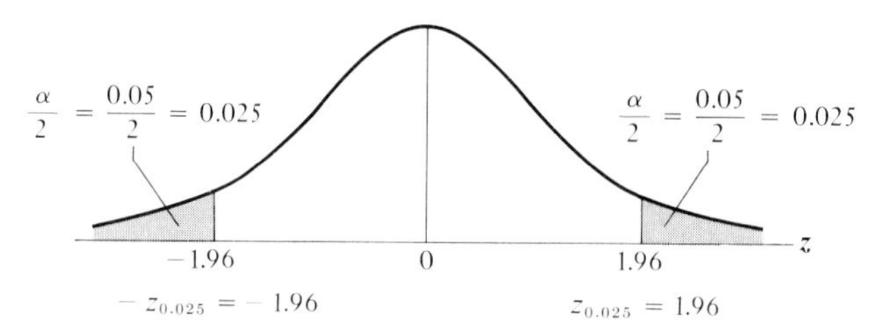

Fig. 7.4 The value of $z_{\alpha/2}$, cutting off a total area of $\alpha = 0.05$ from the standardized normal distribution, leaving an interval including $100(1-\alpha)\% = 95$ percent of the probability.

for μ, not for $(\bar{x}-\mu)/(\sigma/\sqrt{n})$. The difference is not hard to resolve, however, for by rearranging the terms in the confidence interval for $(\bar{x}-\mu)/(\sigma/\sqrt{n})$, it is possible to find an equivalent expression which represents the confidence interval for μ. That is, by rewriting the inequalities in the expression $-z_{\alpha/2} \le z \le z_{\alpha/2}$, we can show that this expression is equivalent to the following inequalities.*

$100(1-\alpha)\%$ confidence interval for μ, where σ is known and the parent population is normal (or $n > 30$):

$$\bar{x} - z_{\alpha/2}\frac{\sigma}{\sqrt{n}} \le \mu \le \bar{x} + z_{\alpha/2}\frac{\sigma}{\sqrt{n}}. \tag{7.1}$$

To illustrate the use of Formula (7.1), consider the problem of estimating the mean monthly rental price of two-bedroom apartments for towns of 100,000 people or less. Suppose we want to construct a 95-percent confidence interval for μ on the basis of a random sample of size $n = 25$. Let us assume for now that the population standard deviation of monthly rental prices is known to be $\sigma = \$20$, and this population is normally distributed. The random sample

* The equivalency is shown as follows:

$-z_{\alpha/2} < z < z_{\alpha/2}$ is equivalent to $-z_{\alpha/2} \le \dfrac{\bar{x}-\mu}{\sigma/\sqrt{n}} \le z_{\alpha/2}$,	By substitution
$-z_{\alpha/2}\dfrac{\sigma}{\sqrt{n}} \le (\bar{x}-\mu) \le z_{\alpha/2}\dfrac{\sigma}{\sqrt{n}}$,	By multiplying each term by $\sigma/\sqrt{n}$
$-\bar{x} - z_{\alpha/2}\dfrac{\sigma}{\sqrt{n}} \le -\mu \le -\bar{x} + z_{\alpha/2}\dfrac{\sigma}{\sqrt{n}}$,	By adding $(-\bar{x})$ to each term
$\bar{x} + z_{\alpha/2}\dfrac{\sigma}{\sqrt{n}} \ge \mu \ge \bar{x} - z_{\alpha/2}\dfrac{\sigma}{\sqrt{n}}$.	By multiplying each term by (-1), thus changing the direction of both inequalities.

of $n = 25$ yields $\bar{x} = \$243$. Now, since we want a 95-percent confidence interval, the appropriate z-values are $z_{\alpha/2} = 1.96$ and $-z_{\alpha/2} = -1.96$. Substituting these two values, as well as $n = 25$, $\sigma = 20$, and $\bar{x} = \$243$, into Formula (7.1) gives the desired 95-percent confidence interval:*

$$\bar{x} - z_{\alpha/2}\frac{\sigma}{\sqrt{n}} \leq \mu \leq \bar{x} + z_{\alpha/2}\frac{\sigma}{\sqrt{n}},$$

$$243 - (1.96)\frac{20}{\sqrt{25}} \leq \mu \leq 243 + (1.96)\frac{20}{\sqrt{25}},$$

$$243 - 7.84 \leq \mu \leq 243 + 7.84,$$

$$235.16 \leq \mu \leq 250.84.$$

We can infer from this analysis, with 95-percent confidence, that the value of μ lies between \$235.16 and \$250.84. That is, on the average, in 95 out of 100 such samples of size $n = 25$, an interval calculated in this manner will include the true population mean μ. We do not know, of course, whether the above interval contains μ.

For the confidence interval calculated above, the probability of making an error in assuming that μ is in the specified interval is $\alpha = 0.05$. If one desires a smaller risk of error, α, then a larger confidence interval must be used. For example, in order to have $\alpha = 0.01$ (that is, a 99-percent confidence interval) the appropriate z-values are found in Table III to be (by interpolating): $z_{\alpha/2} = 2.575$ and $-z_{\alpha/2} = -2.575$ (for $\alpha/2 = 0.005$). The new confidence interval is thus:

$$243 - (2.575)\frac{20}{\sqrt{25}} \leq \mu \leq 243 + (2.575)\frac{20}{\sqrt{25}}$$

$$243 - 10.30 \leq \mu \leq 243 + 10.30$$

$$232.70 \leq \mu \leq 253.30$$

Since this interval is wider than the previous one, we have greater confidence that it includes the population parameter μ. We could further increase our confidence, and decrease the risk of error, α, by extending the interval even more. Of course, there is a limit to the usefulness of the interval when it gets too large. We have almost 100 percent confidence, for example, that the mean apartment rent lies in the interval $200 \leq \mu \leq 300$, but we could have made such a statement without doing any sampling or statistical inference. We sample to get useful information that is more precise. To obtain it, we must be willing to subject our conclusions or inferences to a small controlled level of risk, α.

* For this problem, and the subsequent ones in this chapter, we assume that sampling is from an infinite population, or from a finite population with replacement. When this is not the case, the finite population correction factor must be used—that is, $\sigma/\sqrt{n}$ must be multiplied by $\sqrt{(N - n)}/\sqrt{(N - 1)}$. When $n/N \leq 0.10$, the finite population correction factor may be omitted.

Relax the Assumption of Normality for the Population

The relationship expressed in Formula (7.1) depends on the assumptions that σ is known and that the parent population is normal. When these assumptions are met, Formula (7.1) holds for any sample size, no matter whether n is large or small. But suppose the parent population is *not normal*. In this case, if the sample size n is small, then the distribution of $(\bar{x} - \mu)/(\sigma/\sqrt{n})$ is not normal, and there is no convenient way to determine a confidence interval. On the other hand, when n is large (usually $n > 30$), we know by the central limit theorem that $(\bar{x} - \mu)/(\sigma/\sqrt{n})$ is *approximately* normally distributed; hence the confidence interval specified by Formula (7.1) is still appropriate.

Suppose in our apartment rent example we now drop the assumption of normality and assume the sample size was $n = 64$ rather than 25. Again, if $\sigma = 20$ and $\bar{x} = 243$, then a 95-percent confidence interval is

$$243 - (1.96)\frac{20}{\sqrt{64}} \leq \mu \leq 243 + (1.96)\frac{20}{\sqrt{64}}$$

$$243 - 4.90 \leq \mu \leq 243 + 4.90$$

$$238.10 \leq \mu \leq 247.90$$

7.6 CONFIDENCE INTERVALS FOR μ, WITH σ UNKNOWN

The other assumption we specified at the beginning of the previous section was that the population standard deviation σ is known. This may not be very realistic for many applied problems. When the mean of the population is unknown and must be estimated, it is unlikely that the standard deviation about the unknown mean will be known. Instead, the population standard deviation often must be estimated from the sample standard deviation. Under these circumstances $(\bar{x} - \mu)(s/\sqrt{n})$ has a t-distribution, with $(n - 1)$ degrees of freedom (assuming that the parent population is normal).

The procedure for determining a $100(1 - \alpha)\%$ confidence interval using the t-distribution follows the same pattern as when the normal distribution holds, except that different limits must be used. The limits are found using t-values from Table VII rather than z values from Table III. Let $t_{(\alpha/2, v)}$ represent that value of the t-distribution with $v = n - 1$ degrees of freedom that excludes $\alpha/2$ of the probability in the upper tail. That is, $P(t > t_{(\alpha/2, v)}) = \alpha/2$; and, by symmetry, $P(t < -t_{(\alpha/2, v)}) = \alpha/2$. Then,

$$P(-t_{(\alpha/2, v)} \leq t \leq t_{(\alpha/2, v)}) = 1 - \alpha,$$

or, since $\boldsymbol{t} = (\bar{\boldsymbol{x}} - \mu)/(\boldsymbol{s}/\sqrt{n})$,

$$P(-t_{(\alpha/2,\,\nu)} \leq (\bar{\boldsymbol{x}} - \mu)/(\boldsymbol{s}/\sqrt{n}) \leq t_{(\alpha/2,\,\nu)}) = 1 - \alpha.$$

Solving the inequalities (as before) for μ, it can be shown that,

> $100(1 - \alpha)\%$ *confidence interval for μ, population normal, σ unknown:*
>
> $$\bar{x} - t_{(\alpha/2,\,\nu)}\frac{s}{\sqrt{n}} \leq \mu \leq \bar{x} + t_{(\alpha/2,\,\nu)}\frac{s}{\sqrt{n}}. \qquad (7.2)$$

To illustrate the use of the relationship described by Formula (7.2), suppose that our interest in the rental cost of two-bedroom apartments now concerns the average winter heating expense (per month) in a northeastern city (e.g., Ithaca, N.Y.). In this situation we assume that σ is unknown, and that it is not unreasonable to assume that heating costs are normally distributed. We pick ten apartments at random, and decide that since σ is unknown the $\boldsymbol{t}$-distribution is appropriate [Formula (7.2)]. Our desire is to obtain 99-percent confidence in our interval estimate of the mean heating cost, μ.

Column (1) of Table 7.3 presents the sample selected. Since the sample mean is an integer, $\bar{x} = 70$, we elect to compute the sample variance according to the definition $s^2 = [1/(n - 1)]\sum(x - \bar{x})^2$, and find $s^2 = 65.55$. The sample standard deviation is $s = \sqrt{65.55} = 8.10$, and $s/\sqrt{n} = 8.10/\sqrt{10} = 2.56$.

Table 7.3

	x (score)	$(x - \bar{x})$	$(x - \bar{x})^2$	
	69	-1	1	
	81	11	121	
	67	-3	9	
	80	10	100	
	71	1	1	
	70	0	0	$\bar{x} = (\frac{1}{10})700 = 70$
	78	8	64	
	68	-2	4	$s = \sqrt{\frac{1}{9}(590)} = \sqrt{65.55} = 8.10$
	57	-13	169	
	59	-11	121	$s/\sqrt{n} = 8.10/3.16 = 2.56$
Sum	700	0	590	

From Table VII for the **t**-distribution, we next have to find the critical values for a 99-percent confidence interval, $\pm t_{(\alpha/2,\,v)}$. Using the row for $v = n - 1 = 9$ degrees of freedom, and the column for $\alpha/2 = 0.01/2 = 0.005$, we obtain $t_{(0.005,\,9)} = 3.25$; substituting the appropriate values of $\bar{x}$, s, t, and $\sqrt{n}$ into Formula (7.2) gives

$$\bar{x} - t_{(\alpha/2,\,v)}\left(\frac{s}{\sqrt{n}}\right) \leq \mu \leq \bar{x} + t_{(\alpha/2,\,v)}\left(\frac{s}{\sqrt{n}}\right),$$

$$70 - 3.25\left(\frac{8.10}{3.16}\right) \leq \mu \leq 70 + 3.25\left(\frac{8.10}{3.16}\right),$$

$$70 - 3.25(2.563) \leq \mu \leq 70 + 3.25(2.563),$$

$$70 - 8.33 \leq \mu \leq 70 + 8.33,$$

$$61.67 \leq \mu \leq 78.33.$$

The interval from 61.67 to 78.33 thus represents a 99-percent confidence interval for the mean heating costs for these apartments. We might be disappointed that the interval is so large, and perhaps might have believed we could have guessed that the mean was between 60 and 80. This 99-percent confidence interval is not much different from our nonstatistical guess. However, the statistical work has given us a precise method of inference, a precise point estimate of μ, and exact knowledge about the chance of error in assuming that the true mean does lie in this interval, namely, $\alpha = 0.01$. We can decrease the size of the interval in two ways if we wish—either allow a greater chance of error α, or put more time and effort (cost) into the sampling, to increase n.

7.7 CONFIDENCE INTERVALS FOR THE BINOMIAL PARAMETER *p* USING THE NORMAL APPROXIMATION

The random variable $\boldsymbol{x}/n$ was introduced earlier in this chapter as an estimator of p, the population parameter in a binomial distribution. This statistic, referred to as the *sample proportion*, can be used to determine confidence intervals for populations in applications involving the binomial distribution, such as the proportion of people in a given population who smoke cigarettes, the proportion of voters favoring a certain candidate, or the proportion of defective items produced in a production process.

Recall from Chapter 5 that we showed that when n is large, the number of successes in n independent Bernoulli trials is approximately normally distributed. We approximate the number of successes, $\boldsymbol{x}$, by using the standardized normal variable

$$z = \frac{(\boldsymbol{x} - np)}{\sqrt{npq}},$$

where $np = E[\mathbf{x}]$ and $npq = V[\mathbf{x}]$. Now, if the *number* of successes, $\mathbf{x}$, is normal, the *proportion* of successes in n trials, $\mathbf{x}/n$, must also be normally distributed. Hence, at this point we want to show how the standardized variable $\mathbf{z}$ can be used to approximate the proportion of successes in n trials in precisely the same manner as it approximates the number of successes.

First, we denote our estimator of the population proportion p as $\hat{\mathbf{p}}$, where $\hat{\mathbf{p}} = \mathbf{x}/n$. This estimator has already been shown earlier in this chapter to be unbiased—that is, $E[\hat{\mathbf{p}}] = E[\mathbf{x}/n] = p$. We also showed that the variance of $\hat{\mathbf{p}}$ is $V[\hat{\mathbf{p}}] = pq/n$. Unfortunately, we can't use this variance in constructing a confidence interval because it depends on the unknown parameter p, which we are trying to estimate. The next best thing we can do is use our point estimate of p, $\hat{\mathbf{p}} = \mathbf{x}/n$, in place of p, and $1 - \hat{\mathbf{p}} = 1 - \mathbf{x}/n$ in place of q. Thus, our estimate of the variance is:

$$\textit{Estimated variance:} \quad \frac{\hat{\mathbf{p}}(1 - \hat{\mathbf{p}})}{n}.$$

We can now form a standardized normal variable by taking a normal variable ($\mathbf{x}/n$ in this case), subtracting its mean ($E[\mathbf{x}/n] = p$) and dividing by its (estimated) standard deviation, $\sqrt{\hat{\mathbf{p}}(1 - \hat{\mathbf{p}})/n}$. The appropriate $\mathbf{z}$-variable is:

$$\mathbf{z} = \frac{\hat{\mathbf{p}} - p}{\sqrt{\dfrac{\hat{\mathbf{p}}(1 - \hat{\mathbf{p}})}{n}}}. \tag{7.3}$$

Formula (7.3) can now be used to construct a confidence interval in almost the same fashion as we did for μ in Section 7.5. First, form a $100(1 - \alpha)\%$ confidence interval for $\hat{\mathbf{p}} - p$:

$$-z_{\alpha/2}\sqrt{\frac{\hat{p}(1 - \hat{p})}{n}} \leq \hat{\mathbf{p}} - p \leq z_{\alpha/2}\sqrt{\frac{\hat{p}(1 - \hat{p})}{n}}.$$

Solving this expression for p yields the desired $100(1 - \alpha)\%$ confidence interval.

$100(1 - \alpha)\%$ *confidence interval for* p*:*

$$\hat{p} - z_{\alpha/2}\sqrt{\frac{\hat{p}(1 - \hat{p})}{n}} \leq p \leq \hat{p} + z_{\alpha/2}\sqrt{\frac{\hat{p}(1 - \hat{p})}{n}}. \tag{7.4}$$

To illustrate Formula (7.4), suppose we would like to estimate the proportion of families, from some population, who own two or more cars. A random sample of $n = 144$ families shows that $x = 54$ families have two or more cars. Thus, the best point estimate for the population proportion is the sample proportion $\hat{p} = (x/n) = \frac{54}{144} = 0.375$. Our best estimate of the population variance is

$$\sqrt{\frac{\hat{p}(1-\hat{p})}{n}} = \sqrt{\frac{(0.375)(0.625)}{144}} = 0.040.$$

Suppose we now use these values and Formula 7.4 to construct a 95-percent confidence interval for p (remember, $z_{\alpha/2} = 1.96$ for $\alpha/2 = 0.025$).

$$0.375 - 1.96(0.040) \leq p \leq 0.375 + 1.96(0.040)$$
$$0.375 - 0.078 \leq p \leq 0.375 + 0.078$$
$$0.297 \leq p \leq 0.453.$$

We can thus conclude, with 95-percent confidence, that the population proportion of families who own two or more cars is between 29.7 and 45.3 percent.

7.8 DETERMINING THE SIZE OF *n*

Thus far we have calculated the width of each confidence interval based on the assumption that the sample size, n, is known. In many practical situations, however, the decision-maker does not know what size sample is best. Instead, the decision-maker may prefer to specify the width of the interval desired and use this information to solve for n. The conventional approach to this problem of solving for n in these cases is to ask the decision-maker two questions:

1. The level of confidence desired, that is, the value of $100(1 - \alpha)$; and
2. The *maximum* difference (D) desired between the estimate of the population parameter and the true population parameter—i.e., the value of D is the maximum amount of "error" permitted in estimating the population parameter.

We will consider confidence intervals for both the population mean μ and the population proportion p.

For Statistical Inference on μ

Population Normal, σ Known. First, we consider the problem of determining n when the decision-maker wants a $100(1 - \alpha)\%$ confidence for μ, given that the parent population has a normal distribution with a known standard deviation. In this case, we know the variable

$$z = \frac{\bar{x} - \mu}{\sigma/\sqrt{n}}$$

is $N(0, 1)$. Now if the required level of confidence is $1 - \alpha$, then the above equation results in the following $100(1 - \alpha)\%$ confidence interval for $\bar{x} - \mu$:

$$-z_{\alpha/2}\frac{\sigma}{\sqrt{n}} \leq \bar{x} - \mu \leq z_{\alpha/2}\frac{\sigma}{\sqrt{n}}. \tag{7.5}$$

Since the normal distribution is symmetric, we can concentrate on the righthand inequality, $\bar{x} - \mu \leq z_{\alpha/2}\sigma/\sqrt{n}$. This inequality says that the largest value that $\bar{x} - \mu$ can assume is $z_{\alpha/2}\sigma/\sqrt{n}$. But we also know that our decision-maker says that the largest value for $|\bar{x} - \mu|$ to assume is some amount D. Hence, we can write

$$D = z_{\alpha/2}\frac{\sigma}{\sqrt{n}}.$$

Solving this relationship for n gives the value of n which will assure the decision-maker, with $100(1 - \alpha)\%$ confidence, that $|\bar{x} - \mu|$ will be no larger than D. Solving we get:

Optimal sample size:
$$n = \frac{z_{\alpha/2}^2\sigma^2}{D^2}. \tag{7.6}$$

To illustrate (7.6), suppose that, in the apartment rental problem of Section 7.5, we wanted to find a 95-percent confidence interval for the mean in such a manner that our sample result, $\bar{x}$, and the population mean differ by no more than three dollars—that is, $|\bar{x} - \mu| \leq D = 3$. Assuming, as before, that the parent population is normal, and $\sigma = \$20$, how large should n be to satisfy these conditions?

From Table III, $z_{\alpha/2} = 1.96$ when $\alpha/2 = 0.025$. Substituting this value and $D = 3$, $\sigma = \$20$, into Formula (7.6) yields the appropriate value for n:

$$n = \frac{(1.96)^2(20)^2}{(3)^2} = 170.3.$$

We always round up in this type of problem, to be assured that the sample size is large enough; hence, a random sample of at least 171 students is needed to be assured that, 95 percent of the time, the value of $\bar{x}$ will be within \$3.00 of the true population mean, μ.

Population Not Normal, σ Known. If the population is not assumed to be normal but the standard deviation is known, the same method as above can be used to determine the minimal sample size necessary to satisfy the conditions of confidence and accuracy. By the central limit theorem, we know that the distribution of sample means approaches the normal distribution. Thus, once the necessary sample size is obtained, we can check to see if that size n exceeds 30, and if it does, then we are confident our method of solution was appropriate.

Population Normal, σ Unknown. If the population is normal but the standard deviation is unknown, then the appropriate statistic to use is the t variable, $t = (\bar{x} - \mu)/(s\sqrt{n})$. Again, the maximum difference of $|\bar{x} - \mu| = D$ is obtained from the decision-maker. In this case, we are stuck for a value of s, since s must be calculated from a sample, and we haven't taken a sample yet (the whole purpose is to decide what sample size to take). To make matters even worse, the appropriate t value to use in calculating a $100(1 - \alpha)\%$ confidence interval is $t_{(\alpha/2, v)}$, which again depends on the unknown sample size. To make a long story short, we conclude this paragraph by saying that the solution for n in this case is not a direct process, but can be achieved by a succession of iterative steps which we will not present here.*

For Statistical Inference on p

The size of the sample can be determined in a confidence interval for the binomial parameter p if the maximum difference between $\hat{p}$ and p, $|\hat{p} - p| = D$, is specified in advance. In sampling the number of two-car owners, for instance, it may be desirable to restrict the maximum error to, say, three percentage points, $D = 0.03$. In this case we can *not* use the estimate of the standard deviation used previously in working with a population proportion, $\sqrt{\hat{p}(1-\hat{p})/n}$, since $\hat{p}$ isn't known yet (we haven't taken the sample). If there is a good estimate of p available from other sources (such as a pilot study), then it may be used. If not, then the estimate to be used is $\sqrt{1/4n}$. This estimate is used because it represents the largest (and hence most conservative) value the standard deviation can ever assume, since $\sqrt{pq/n}$ will always be at its maximum when $p = q = \frac{1}{2}$ (try several values to convince yourself). Now, if we substitute $\sqrt{1/4n}$ into the standardized normal variable $z = (\hat{p} - p)/\sqrt{pq/n}$, we obtain

$$z = \frac{(\hat{p} - p)}{\sqrt{1/4n}}.$$

This relationship can be used to derive the following $100(1 - \alpha)\%$ confidence interval for $\hat{p} - p$:

$$-z_{\alpha/2}\sqrt{1/4n} \leq \hat{p} - p \leq z_{\alpha/2}\sqrt{1/4n}.$$

Again, setting $\hat{p} - p$ at the maximum value D we have

$$D = z_{\alpha/2}\sqrt{1/4n} \qquad \text{or} \qquad D^2 = z_{\alpha/2}^2 \frac{1}{4n}.$$

* See reference [1] for an explanation of this process.

Therefore,

$$n = \frac{z_{\alpha/2}^2}{4D^2}. \tag{7.7}$$

Returning to the example problem of estimating "the proportion of families owning two or more cars" within three percentage points with 95-percent confidence, we know $z_{\alpha/2} = 1.96$ (since $\alpha = 0.05$), and $D = 0.03$. Thus,

$$n = \frac{1}{4}\left(\frac{1.96^2}{0.03^2}\right) = 1067.1.$$

We thus conclude that a sample of 1068 people is needed to assure an error of less than three percentage points 95 percent of the time.

7.9 CONFIDENCE INTERVAL FOR σ^2 (OPTIONAL)

Under some circumstances it may be desirable to construct a confidence interval for an estimate of an unknown population variance. As we said before, the telephone company is often interested in the *variability* of the length of telephone conversations, and a contractor who is about to purchase some steel girders will probably be interested in the *variance* of their tensile strengths. A government economist may be just as concerned about the *variability* of taxes paid among individuals as he is about the average tax paid, because the income redistribution effect of taxation is very important. In these cases it may be important to establish limits on just how large or small σ^2 might be; that is, to determine a confidence interval for the population variance.

To construct a $100(1 - \alpha)\%$ confidence interval for σ^2 when sampling from a *normal* population, recall from Section 6.9 that the variable $(n - 1)s^2/\sigma^2$ has a chi-square distribution with $(n - 1)$ degrees of freedom. Denote the point that cuts off $\alpha/2$ of the area of the righthand side of a chi-square distribution with $v = n - 1$ degrees of freedom as $\chi^2_{(\alpha/2, v)}$; that is,

$$P(\chi^2 \geq \chi^2_{(\alpha/2, v)}) = \frac{\alpha}{2}.$$

Since the chi-square distribution is not symmetrical, $-\chi^2_{(\alpha/2, v)}$ does *not* give the appropriate value for cutting off $\alpha/2$ of the lefthand side of this distribution. The point that does give the correct probability is that value of χ^2 that cuts off $1 - \alpha/2$ of the righthand tail, or $\chi^2_{(1-\alpha/2, v)}$. That is,

$$P(\chi^2 \geq \chi^2_{(\alpha/2, v)}) = \frac{\alpha}{2} \quad \text{and} \quad P(\chi^2 \leq \chi^2_{(1-\alpha/2, v)}) = \frac{\alpha}{2}.$$

* This section, which may be omitted without loss in continuity, presupposes that the reader has covered Section 6.9.

The interval between the points $\chi^2_{(\alpha/2,\, v)}$ and $\chi^2_{(1-\alpha/2,\, v)}$ thus contains $1 - \alpha$ probability. It is now possible to define a $100(1 - \alpha)\%$ confidence interval for the variable $(n - 1)s^2/\sigma^2$:

$$\chi^2_{(1-\alpha/2,\, v)} \le \frac{(n - 1)s^2}{\sigma^2} \le \chi^2_{(\alpha/2,\, v)}. \tag{7.8}$$

Again, solving these inequalities for the unknown parameter σ^2 gives

$100(1 - \alpha)\%$ *confidence interval for* σ^2*, with parent population normal:*

$$\frac{(n - 1)s^2}{\chi^2_{(\alpha/2,\, v)}} \le \sigma^2 \le \frac{(n - 1)s^2}{\chi^2_{(1-\alpha/2,\, v)}}. \tag{7.9}$$

It is important to note that, in going from (7.8) to (7.9), the sense of the inequalities has changed. Thus, the larger value of χ^2 now appears in the denominator of the *lower* endpoint for σ^2, while the smaller value of χ^2 is in the denominator of the term that gives the upper endpoint for σ^2.

Consider once again the population of apartment heating costs, and assume that this time we are interested in estimating the *variance* of the population, using the same sample of size $n = 10$ reported in Table 7.2. The best point estimate of σ^2 is s^2, which is computed from this sample to be $s^2 = \frac{1}{9}(590) = 65.55$. To achieve an interval estimate for σ^2 with a known level of confidence and a known risk of error, let us compute a 95-percent confidence interval for σ^2.

Figure 7.5 illustrates the relevant chi-square distribution for $v = n - 1 = 9$ degrees of freedom and the cutoff values for $\alpha/2 = 0.025$ and $1 - \alpha/2 = 0.975$. These values are not equidistant from the mean of the chi-square ($E[\chi^2] = v = 9$) because the χ^2 is a skewed distribution. The cutoff values from Table V in terms of the chi-square distribution for $v = 9$ and $\alpha/2 = 0.025$ and $1 - \alpha/2 = 0.975$ are found to be 19.0 and 2.70, repectively. Thus, $P(2.70 \le \chi^2_9 \le 19.0) = 1 - \alpha = 0.95$.

Substituting the values for $\chi^2_{(0.025,9)} = 19.0$, $\chi^2_{(0.025,9)} = 2.70$, $n = 10$, and $s^2 = 65.55$ into Formula (7.9) gives:

$$\frac{9(65.55)}{19.0} \le \sigma^2 \le \frac{9(65.55)}{2.70}$$

$$31.05 \le \sigma^2 \le 218.50.$$

On the basis of this sample of size 10, one can infer with 95-percent confidence that the population variance lies between 31.05 and 218.50. There is a 2.5 percent chance of error that the true σ^2 may be greater than 218.50 and a 2.5 percent chance of error that it may be less than 31.05.

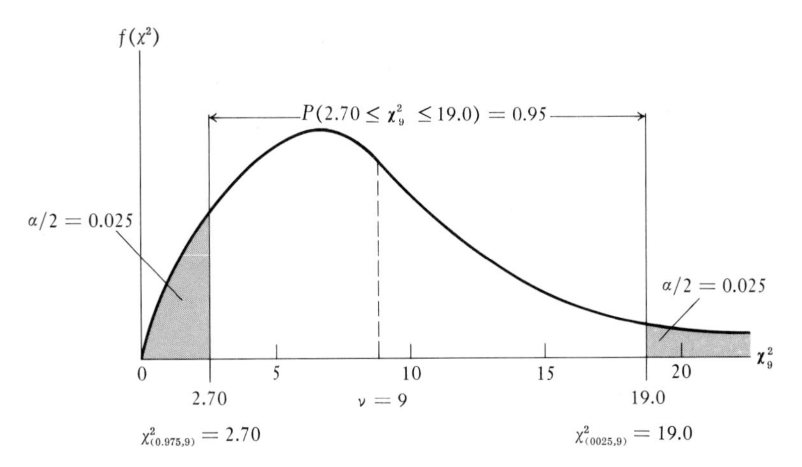

Fig. 7.5 The values of $\chi^2_{(\alpha/2,\,\nu)}$ and $\chi^2_{(1-\alpha/2,\,\nu)}$ cutting off a total area of $\alpha = 0.05$ from the chi-square distribution with $n = 10$, leaving an interval including 95 percent of the probability in the middle.

*7.10 APPENDIX (OPTIONAL)

Deriving a Maximum-Likelihood Estimate (*MLE*)

A maximum-likelihood estimate can be derived in a more rigorous fashion through the use of calculus. Determining a MLE involves finding that value of the population parameter which maximizes the function representing the likelihood of obtaining the sample results. This function is called the *likelihood function* and is denoted by the letter L. In many problems, determining the formula for the likelihood function may not be too difficult, although the notation for describing L may seem formidable. Assume that the random variable $\boldsymbol{x}_1$ represents all possible values of $\boldsymbol{x}$ which could appear first in a series of sample observations, $\boldsymbol{x}_2$ represents all possible values of $\boldsymbol{x}$ which could appear second in a series of sample observations, and so forth, $\boldsymbol{x}_n$ representing all possible values of $\boldsymbol{x}$ which could appear last in a series of n sample observations. Now, let x_1 be an observed value of the variable $\boldsymbol{x}_1$, x_2 be an observed value of $\boldsymbol{x}_2$, and x_n be an observed value of $\boldsymbol{x}_n$. If the parent population is discrete with parameter θ, then the probability that a particular sample $x_1, x_2, \ldots, x_n$ will occur is the likelihood function L, where L is conditional on θ:

$$L = P(\boldsymbol{x}_1 = x_1, \boldsymbol{x}_2 = x_2, \ldots, \boldsymbol{x}_n = x_n \,|\, \theta).$$

If the parent population $\boldsymbol{x}$ is continuous, L equals the joint density function of $x_1, x_2, \ldots, x_n$, which is $f(x_1, x_2, \ldots, x_n \,|\, \theta)$.

Once the function representing L has been determined, the task is to find that value of the population parameter which maximizes this function. Maximizing L can be accomplished in the normal fashion—taking the first derivative with respect to the parameter in question, setting this derivative equal to zero, and then solving for the optimal value of the parameter. In many cases it is easier to maximize the logarithm of L rather than L itself; this transformation will not change the optimal solution, but often will make finding the first derivative much easier.

Perhaps the easiest likelihood function to determine is the one for an experiment generating independent repeated Bernoulli trials. In this case the probability of a sample $x_1, x_2, \ldots, x_n$ depends only on the number of observed successes (and failures), and this probability is given by the binomial distribution. Hence,

$$L = \binom{n}{x} p^x q^{n-x}.$$

Determining the MLE for this particular L is a relatively straightforward differentiation problem for anyone familiar with the process of taking the derivative of a logarithm (see Appendix A).

$$L = \binom{n}{x} p^x q^{n-x};$$

$$\log L = \log\left[\binom{n}{x} p^x q^{n-x}\right]$$

$$= \log\binom{n}{x} + x \log p + (n - x) \log(1 - p);$$

$$\partial \log L/\partial p = x\left(\frac{1}{p}\right) + (n - x)\left[\frac{1}{(1 - p)}\right](-1),$$

$$0 = \frac{x}{p} - \frac{(n - x)}{(1 - p)},$$

$$p = \frac{x}{n}.$$

The sample proportion was previously shown to be unbiased, consistent, and sufficient. The above result proves that, in addition, $\boldsymbol{x}/n$ is the maximum-likelihood estimator of p.

The process described above can be used to find a MLE for the population mean and variance when sampling from a normal population. Assume that n independent observations, $x_1, x_2, \ldots, x_n$, are drawn from a normal population with mean μ and variance σ^2. The probability that these values all occur jointly is $L = f(x_1, x_2, \ldots, x_n)$, and since the n sample values are independent, by the definition of independence presented in Chapter 3 the joint density function can be written as the product of the individual functions, or as

$L = f(x_1)f(x_2)\cdots f(x_n)$. For the normal distribution each one of these individual density functions has the following form:

$$f(x_i) = \frac{1}{\sigma\sqrt{2\pi}} e^{-(1/2)[(x_i-\mu)/\sigma]^2}.$$

Substituting the preceding frequency function for each of the values of i in the formula for L gives the likelihood function for this problem:

$$L = \frac{1}{\sigma\sqrt{2\pi}} \exp\left[-\frac{1}{2}\left(\frac{x_1-\mu}{\sigma}\right)^2\right] \cdot \frac{1}{\sigma\sqrt{2\pi}} \exp\left[-\frac{1}{2}\left(\frac{x_2-\mu}{\sigma}\right)^2\right] \cdots \frac{1}{\sigma\sqrt{2\pi}} \exp\left[-\frac{1}{2}\left(\frac{x_n-\mu}{\sigma}\right)^2\right].$$

The above formula for L can be simplified by combining the common terms:

$$L = \frac{1}{\sigma^n(2\pi)^{n/2}} \exp\left[-\frac{1}{2}\sum_{i=1}^{n}\left(\frac{x_i-\mu}{\sigma}\right)^2\right].$$

Maximizing L with respect to μ yields the MLE for the population mean, while maximizing L with respect to σ^2 yields the MLE for the population variance. Again, it is easier to maximize L by first taking the logarithm of L, finding the first derivative of this logarithm, setting the derivative equal to zero, and solving. Note, in the derivation which follows, that natural logarithms (i.e., logs to the base e, denoted by the symbol ln) are used, rather than common logarithms.

$$\begin{aligned} \ln L &= \ln\left[\frac{1}{\sigma^n(2\pi)^{n/2}} e^{-(1/2)\Sigma[(x_i-\mu)/\sigma]^2}\right] \\ &= \ln\frac{1}{\sigma^n} + \ln\left(\frac{1}{2\pi}\right)^{n/2} - \frac{1}{2}\sum_{i=1}^{n}\left[\frac{(x_i-\mu)}{\sigma}\right]^2 \\ &= -n\ln\sigma - \left(\frac{n}{2}\right)\ln 2\pi - \frac{1}{2\sigma^2}\sum_{i=1}^{n}(x_i-\mu)^2; \end{aligned}$$

$$\frac{\partial \ln L}{\partial\mu} = \frac{1}{\sigma^2}\sum_{i=1}^{n}(x_i-\mu) = \left(\frac{1}{\sigma^2}\right)\left(\sum_{i=1}^{n}x_i - n\mu\right),$$

$$0 = \sum_{i=1}^{n}x_i - n\mu,$$

$$\mu = \sum_{i=1}^{n}x_i/n = \bar{x}.$$

Table 7.4 Summary of Test Statistics and Confidence Intervals for Statistical Inference

Unknown Parameter	Population Characteristics and Other Description	Reference Sections	Test Statistic Involving the Best Sample Estimator	Endpoints for $100(1-\alpha)\%$ Confidence Interval	Reference Formula
Mean μ	Population $N(\mu, \sigma^2)$ or sample size $n \geq 30$, σ *known*	6.5 7.5	$z = \dfrac{(\bar{x}-\mu)}{(\sigma/\sqrt{n})}$	$\bar{x} \pm z_{\alpha/2}(\sigma/\sqrt{n})$	(7.1)
Mean μ	Population $N(\mu, \sigma^2)$, σ *unknown*	6.8 7.6	$t_{v\,\text{d.f.}} = \dfrac{(\bar{x}-\mu)}{(s/\sqrt{n})}$ where $v = n-1$	$\bar{x} \pm t_{(\alpha/2,v)}(s/\sqrt{n})$	(7.2)
Variance σ^2	Population $N(\mu, \sigma^2)$	6.9 7.9	$\chi^2_{v\,\text{d.f.}} = \dfrac{(n-1)s^2}{\sigma^2}$ where $v = n-1$	$\dfrac{(n-1)s^2}{\chi^2_{(\alpha/2,v)}}$ and $\dfrac{(n-1)s^2}{\chi^2_{(1-\alpha/2,v)}}$	(7.9)
Proportion p	Repeated independent trials, $npq \geq 3$, $\hat{p} = x/n$	7.7	$z = (\hat{p}-p)\Big/\sqrt{\dfrac{\hat{p}(1-\hat{p})}{n}}$	$\hat{p} \pm z_{\alpha/2}\sqrt{\dfrac{\hat{p}(1-\hat{p})}{n}}$	(7.4)

This result implies that, given a particular sample $x_1, x_2, \ldots, x_n$, the population mean most likely to produce this sample is $\mu = \bar{x}$. Thus, a value of $\bar{x} = 115$, representing, for example, the mean IQ of a sample of college undergraduates, most likely came from a population with $\mu = 115$.

Instead of differentiating ln L with respect to μ, the first derivative could have been taken with respect to σ^2, yielding a MLE of the population variance. This analysis leads to a rather interesting result—the MLE of σ^2 is *not* an unbiased estimator. Using $\bar{x}$ to estimate μ, it can be shown that the maximum-likelihood estimate of σ^2 is

$$\left(\frac{1}{n}\right) \sum_{i=1}^{n} (x_i - \bar{x})^2,$$

which we know from earlier in this chapter is not an unbiased estimator of σ^2. Thus, if one wants an estimate of the variance of a population which makes the sample appear most likely, then the statistic shown above is the optimal choice; on the other hand, if it seems more desirable to have an estimate which, on the average, equals the true value, then s^2 is the best choice. The cost of making an error in an estimation problem may determine which approach is more appropriate when the sample size is small and the difference between the two is relatively large.

7.11 SUMMARY

Table 7.4 summarizes the test statistics and confidence intervals described in this chapter for various situations of statistical inference.

Define: interval estimation, α error, confidence interval, optimal sample size.

Problems

7.18 a) What factors must be established before a confidence interval can be constructed?

b) What does the Greek letter α represent? How would you suggest the level of α be established?

7.19 a) What is meant by a "$100(1 - \alpha)\%$ confidence interval for μ"?

b) Write down the expression for calculating a $100(1 - \alpha)\%$ confidence interval for μ on the basis of a random sample of size n (large) from a normal population with mean μ and standard deviation σ.

c) Write down the expression for calculating a $100(1 - \alpha)\%$ confidence interval for μ on the basis of a random sample of size n (small) from a normal population with mean μ and unknown standard deviation.

7.20 In one university the average height in a random sample of 256 male students was 70 in., with a standard deviation of 2 in.

a) Construct 95- and 99-percent confidence intervals for the average height of all male students.

b) Repeat part (a), assuming that the sample size had been 100 rather than 256.

7.21 Twenty-five loan applications in a bank were randomly selected for the purpose of determining the average dollar amount requested for each loan.

a) Construct a 95-percent confidence interval for μ assuming that the sample mean was $\bar{x} = \$900$ and the sample standard deviation was \$150. Use the **t**-distribution.

b) Repeat part (a) using the normal distribution rather than the **t**-distribution. Compare your answer to part (a) with the confidence interval in this part.

7.22 A recent survey asked respondents to rate, on a scale from 0 to 100, how good a job they thought the President of the United States had done during the past six months. Assume that the population variance for this survey equals 100. Construct a 95-percent confidence interval for μ, assuming that a random sample of 256 adults yielded a mean score of 61.0. Is it necessary in this case to assume that the parent population is normal?

7.23 A production assembly process is scheduled as a 20-minute operation. A time study based on sixteen randomly selected observation periods (unknown to the employees) shows a sample average of 24.3 minutes. The population standard deviation is six minutes. Find a 90-percent confidence interval for the population mean of this assembly operation, assuming that the population is normally distributed.

7.24 Suppose that you take a survey of the delivery time on a random sample ($n = 25$) of new Boeing 767s from the date of the order, and find the sample mean and variance are 420 days and 25 days, respectively. What are the earliest and latest days a company should expect delivery? Use a 99-percent confidence interval and assume normality.

7.25 Suppose the following nine values represent random observations from a normal parent population: 1, 5, 9, 8, 4, 0, 2, 4, 3. Construct a 99-percent confidence interval for the mean of the parent population.

7.26 A television manufacturer would like to know what proportion of television set owners have color sets. In a sample of 100 randomly selected owners 40 percent were found to own color sets. Construct a 95-percent confidence interval for the population proportion of television owners who have color sets.

7.27 Suppose that Senator Fogbound has engaged a team of public opinion surveyors in an effort to determine the percent of the population who favor his stand on a current issue. The survey company will conduct a random survey of public opinion at a cost of 35 cents per interview. How much will it cost the senator if he insists the sampling error be less than 5 percent 95% of the time and if he has no idea of the percent of the population favoring his stand?

7.28 In order to estimate the percent of all housecleaners who use "Wash Away" detergent, 196 housecleaners were randomly selected and interviewed. If 108 of these housecleaners use this product, what would be a 99-percent confidence interval for the population percent of housecleaners who use "Wash Away"?

7.29 Suppose that the daily volume of stocks exchanged on the NYSE is normally distributed, with $\sigma = 15{,}000$.

a) Based on a sample of $n = 16$, $\bar{x} = 61{,}000$, find a 98-percent confidence interval for μ.

b) How many days will I need to sample in order to have 98-percent confidence that the sample mean will differ from the true mean by no more than 600?

7.30 A population of families has an unknown mean income μ; the standard deviation of these incomes is known to be \$1000. How large a random sample would be needed to determine the mean income if it is desired that the probability of a sampling error of more than \$50 be less than 5 percent?

7.31 A random sample of the IQ of 5 percent of the students in a university resulted in a mean IQ of 120 with a standard deviation of 3. How many students are enrolled at this university if a 99-percent confidence interval for the mean IQ of all students in the university extends three-tenths of a unit to either side of the sample mean? Assume sampling is with replacement.

*7.32 A manufacturer of steel washers periodically samples the washers being produced as a check on the variability of the inside diameter of the washers. A sample of size 20 was checked and found to have a standard deviation of 0.002 in. On the basis of this sample find a 95-percent confidence interval for the true variance.

*7.33 A study of the weather in January in Chicago found that daytime high temperatures are approximately normally distributed. A random sample of $n = 31$ daytime highs resulted in a sample variance of 79.8 degrees. Construct a 95-percent confidence interval for σ^2.

7.34 Construct a 95-percent confidence interval for the variance of the bank loans described in Problem 7.21.

7.35 Suppose that the annual earnings of college graduates are normally distributed, with $\sigma = \$6000$.

a) Based on a sample of $n = 16$ with $\bar{x} = \$28{,}000$, find a 98-percent confidence interval for μ.

b) How many graduates will I need to sample in order to have 98-percent confidence that the sample mean to be calculated will be within six hundred dollars of the true mean?

7.36 A machine that produces ball bearings is stopped periodically so that the diameter of the bearings produced can be checked for accuracy. In this particular case, it is not the mean diameter that is of concern, but the variability of the diameters. Suppose that a sample of size $n = 31$ is taken and the variance of the diameters of the bearings sampled is found to be 0.94 mm.

a) Construct a 95-percent confidence interval for σ^2 assuming that the population is normal.

b) Assume that if this machine is working properly, the variance of the bearings produced will be 0.50 mm. Does this sample indicate that the machine is working improperly? Explain.

7.37 Redo Problem 7.31, assuming that a random sample of 50 percent of all students was tested, and that sampling was without replacement. (Remember to correct for a finite population.)

Exercises

7.38 Suppose that x_1 and x_2 are independent random variables having the Poisson probability distribution with parameter λ. Show that the mean of these variables is an unbiased estimator of λ.

7.39 If $\hat{\theta}$ is an unbiased estimator of the population parameter θ, under what conditions will $\hat{\theta}^2$ be an unbiased estimator of θ^2?

7.40 Which of the properties of a good estimator does $\bar{x}$ have? Prove as many of these as you can.

7.41 At a certain university, there are 53 fraternities and 46 sororities. The average grade-point index of members of this group is 2.55, with a standard deviation of 0.30, and the distribution is quite positively skewed. Suppose that nine of these members are selected at random for a certain project; what is the probability that their average grade-point index is better than 2.75? Can you tell me what values I should expect this average to fall between with probability 0.99?

7.42 The quarterly sales of five different Softee Hamburger drive-in restaurants are 170, 160, 140, 180, and 140 (in thousands of dollars), respectively. You wish to make an estimate of the average sales for this quarter, for all the outlets in the entire chain.

a) On the basis of this sample information, can you make an estimate that will have a 1.00 probability of being correct? Why or why not?

b) Make an interval estimate by using the sample mode plus-or-minus $\frac{1}{3}$ the range.

c) Make an interval estimate by using the mean plus-or-minus the standard deviation of the sample.

d) In probability, which of these estimates is better and why, and how could the better one be made even better?

7.43 Find the amount of bias that results when the statistic $(1/n)\sum(x - \bar{x})^2$ is used to estimate σ^2.

7.44 The Ryder Corporation manufactures a variety of types and sizes of sheet metal. One of their problems has been that of establishing the exact size of the metal plates after they have been cut and stamped, for in some cases it is extremely important that the size of the plate fall within certain tolerance limits. Ryder recently purchased a special gauge that automatically measures the length of the sheet as it passes the gauge. This gauge, however, is subject to random error that has been found to be normally distributed about the true length of the sheet. To compensate for this error, the manufacturers of the gauge designed it so that, instead of producing just one value for the length of the metal passing by, the device actually gives two readings, and these values are independent of each other. Suppose that we let x_1 represent the first estimator of length and x_2 the second.

a) Is $\frac{1}{2}x_1 + \frac{1}{2}x_2$ an unbiased estimator of length?

b) Is $\frac{1}{3}x_1 + \frac{2}{3}x_2$ an unbiased estimator of length?

c) What is the relative efficiency of the estimators in questions (a) and (b)? Which of the two estimators is preferable?

d) Suppose that one of the sheets passing by the gauge is square, and the people at Ryder wish to estimate the area of this sheet. They are unsure whether they should square the two observations first and then average, or average first and then square—i.e., whether they should use

$$\left(\frac{x_1^2 + x_2^2}{2}\right) \quad \text{or} \quad \left(\frac{x_1 + x_2}{2}\right)^2.$$

Which method will provide the better estimator?

7.45 a) Suppose a sample of two items is taken from a population, with observed values x_1 and x_2. Under what conditions will $a_1x_1 + a_2x_2$ be an unbiased estimator of the mean of this population?

b) Assume, in part (a), that a sample of n observations was taken, $x_1, x_2, \ldots, x_n$. Under what conditions will

$$a_1x_1 + a_2x_2 + \cdots + a_nx_n = \sum_{i=1}^{n} a_ix_i$$

be an unbiased estimator of the mean of the population?

c) What is the variance of the unbiased estimator $\sum_{i=1}^{n} a_ix_i$? Under what conditions will this estimator be most efficient?

d) Show that the efficient estimator derived in part (c) is consistent.

7.46 Use a mathematical statistics text to find proof that, for x normally distributed, $(1/n)\sum(x - \bar{x})^2$ is a maximum-likelihood estimate of σ^2.

7.47 If the result obtained from a particular sample will be critical, e.g., the CPA would not be able to render an unqualified opinion (unless every item in the population were examined), which of the following is the most important to the CPA?

a) Size of the population

b) Estimated occurrence rate

c) Specified upper precision limit

d) Specified confidence level.

7.48 From a very large population the auditor selects 400 items at random and finds 16 items in error. The auditor can be 95% confident that the error rate in the population does not exceed:

a) $4\% = \frac{16}{400} \times 100$

b) $5.65\% = 1.65\sqrt{\frac{0.04 \times 0.96}{400}} + 0.04$

c) $5.95\% = 1.96\sqrt{\frac{0.04 \times 0.96}{400}} + 0.04$

d) $6\% = 1.65\sqrt{\frac{0.06 \times 0.94}{400}} + 0.04$

e) $6.4\% = 1.96\sqrt{\frac{0.064 \times 0.936}{400}} + 400$

7.49 Approximately 5% of the 10,000 homogeneous items included in Barletta's finished-goods inventory are believed to be defective. The CPA examining Barletta's financial statements decides to test this estimated 5% defective rate. The CPA learns by sampling without replacement that a sample of 288 items from the inventory will permit specified reliability (confidence level) of 95% and specified precision (confidence interval) of ∓ 0.025. If the specified precision is changed to ∓ 0.05, and the specified reliability remains 95%, the required sample size is:

a) 72 b) 335 c) 436 d) 1543

7.50 An auditor wishes to be 95% confident that the true error rate does not exceed six percent. How large a sample must be taken from a very large population if the auditor estimates an error rate of four percent? Select the closest answer.

a) 150 b) 1250 c) 380 d) 450

7.51 Chalmers asks its Certified Public Accountant to estimate the number of the two thousand 30-day charge accounts that are delinquent. A sample of 100 accounts reveals that twenty are delinquent. Thus, at 95% confidence the best estimate is that:

a) At least 320 are delinquent.

b) At most 560 are delinquent.

c) Between 320 and 480 are delinquent.

d) Between 240 and 560 are delinquent.

e) Not close to one of the above.

*7.52 Suppose that $x_1, x_2, \ldots, x_n$ is a random sample from a Poisson distribution with parameter λ. Find the maximum-likelihood estimate of λ.

*7.53 Find the maximum-likelihood estimate of λ for a random sample of n observations, where $f(x) = \lambda e^{-\lambda x}$, λ is a constant ($\lambda \geq 0$), and $0 \leq x \leq \infty$.

*7.54 Find the maximum-likelihood estimate of θ for a random sample of size n drawn from a population whose density function is $f(x) = 1/\theta$ for $0 \leq x \leq \theta$, and 0 otherwise.

GLOSSARY

point estimate: an estimate of a population parameter that specifies one single value.

interval estimate: an estimate that specifies a range of values.

unbiasedness: an unbiased estimator is an estimator that, on the average, equals the parameter being estimated, i.e., $E[\text{estimator}] = \text{parameter}$.

efficiency: an efficient estimator is an unbiased estimator that has a relatively low variance (when compared to other unbiased estimators).

sufficiency: a sufficient estimator is an estimator that utilizes all the sample information available.

consistency: a consistent estimator is an estimator with variance and bias (if any) that approach zero as n approaches infinity.

maximum likelihood: an estimating procedure that selects the most likely population parameter in view of the sample evidence.

100(1 − α)% confidence interval: an interval estimate that, on the average, will include the population parameter 100(1 − α) times out of 100 that this method is used.

100(1 − α)% C.I. for μ, σ known, normal parent population:

$$\bar{x} \pm z_{\alpha/2}\left(\frac{\sigma}{\sqrt{n}}\right).$$

100(1 − α)% C.I. for μ, σ unknown, normal parent population:

$$\bar{x} \pm t_{\alpha/2,\,v}\left(\frac{s}{\sqrt{n}}\right).$$

100(1 − α)% C.I. for *p*, assuming *n* is "large":

$$\frac{x}{n} \pm z_{\alpha/2}\sqrt{\frac{(x/n)[1 - (x/n)]}{n}}.$$

formula for determining the sample size:

$$n = \frac{z_{\alpha/2}^2\,\sigma^2}{D^2}$$ for a problem relating to means, where σ is known,

or

$$n = \frac{z_{\alpha/2}^2}{4D^2}$$ for a binomial problem where the variance is unknown.

100(1 − α)% C.I. for σ^2:

$$\frac{(n-1)s^2}{\chi_{\alpha/2,\,v}^2} \le \sigma^2 \le \frac{(n-1)s^2}{\chi_{1-\alpha/2,\,v}^2}.$$

8

HYPOTHESIS TESTING

"It is an error to argue in front of your data. You find yourself insensibly twisting them around to fit your theories."

SHERLOCK HOLMES

8.1 INTRODUCTION AND BASIC CONCEPTS

The procedures presented in Chapter 7 describe the process of making both point and interval estimates of population parameters. As we saw then, one advantage of using an interval estimate is that it permits the expression of uncertainty about the true value of the population parameter. Another advantage of using confidence intervals is that they serve to test the validity of *assumed* values of the population parameters. Such assumed values are usually referred to as *statistical hypotheses*. Determining the validity of an assumption of this nature is called the *test of a statistical hypothesis*, or simply *hypothesis testing*.

The major purpose of hypothesis testing is to choose between two competing hypotheses about the value of a population parameter. The process of hypothesis testing brings together many of our previous topics—calculation of sample statistics, random variables, probability distributions, and statistical inference. To illustrate this with a simple example, we will suppose that an engineer in charge of quality control for items produced on an assembly line wishes to choose between the two mutually exclusive and exhaustive hypotheses $p > 0.05$ or $p \leq 0.05$, where p is the proportion of defective items produced. It may be that the production line must be shut down if the proportion of defectives exceeds five percent. Let us assume that the engineer cannot check every item produced because the cost and time delay would be too expensive. Instead, the engineer decides to take a sample and then use the proportion of defectives in the sample (x/n) and a knowledge of sampling distributions (i.e., the sampling distribution of $\boldsymbol{x}/n$) to decide between the two conflicting hypotheses. In the following section the procedure for constructing such a test of hypotheses is presented. Subsequent sections give specific examples of the use of this procedure in some commonly occurring situations.

Types of Hypotheses

In specifying the conflicting hypotheses about the values that a population parameter might assume, it is convenient to distinguish between *simple hypotheses* and *composite hypotheses.* In a simple hypothesis, only one value of the population parameter is specified. If an engineer hypothesizes that the probability of a defective item is $p = 0.10$, this represents a simple hypothesis. A psychologist investigating the hypothesis that the mean IQ of a group is $\mu = 115$ is testing a simple hypothesis. If the exact difference between two population parameters is specified (that is, $\mu_1 - \mu_2 = 0$), this also represents a simple hypothesis. Composite hypotheses, on the other hand, specify not just one value but a *range* of values that the population parameter may assume. The hypotheses $p \leq 0.10$, $\mu \neq 100$, and $\mu_1 - \mu_2 \neq 0$ all represent composite hypotheses because more than one value is specified in each case. As you might suspect, assumptions in the form of simple hypotheses are, in general, easier to test than are composite hypotheses. In the former case we need determine only whether or not the population parameter equals the specified value, while in the latter case it is necessary to determine whether or not the population parameter takes on any one of what may be a very large (or even infinite) number of values.

The two conflicting (i.e., mutually exclusive) hypotheses in a statistical test are normally referred to as the *null hypothesis* and the *alternative hypothesis.* The term "null hypothesis" developed from early work in the theory of hypothesis testing, in which this hypothesis corresponded to a theory about a population parameter that the researcher thought did *not* represent the true value of the parameter (hence the word "null," which means invalid, void, or amounting to nothing). The *alternative hypothesis* generally specified those values of the parameter that the researcher believed did hold true.

Nowadays, it is generally accepted common practice *not* to associate any special meaning to the null or alternative hypotheses, but merely to let these terms represent two different assumptions about the population parameter. We shall see in a moment that, for statistical convenience, it may make a difference which hypothesis is called the null and which is called the alternative. It is *most* convenient always to have the null hypothesis be the hypothesis that contains an equal sign, if either one has an equal sign. The null and alternative hypotheses are distinguished by the use of two different symbols, H_0 representing the null hypothesis and H_a the alternative hypothesis. Thus, a psychologist who wishes to test whether or not a certain class of people has a mean IQ equal to or higher than 100 may establish the following null and alternative hypotheses:

$$H_0: \mu = 100 \quad \text{(Null hypothesis)},$$
$$H_a: \mu > 100 \quad \text{(Alternative hypothesis)}.$$

On the other hand, if the psychologist wishes to test for differences between the mean IQ of two groups, the null hypothesis established may be that the two groups have equal means, with the alternative hypothesis that their means are not equal:

$$H_0: \mu_1 - \mu_2 = 0 \quad \text{(Null hypothesis)},$$
$$H_a: \mu_1 - \mu_2 \neq 0 \quad \text{(Alternative hypothesis)}.$$

The null hypothesis and the alternative hypothesis can both be either simple or composite. The simple null hypothesis $H_0: \mu = 100$, for example, may be tested against a simple alternative hypothesis such as $H_a: \mu = 120$ or $H_a: \mu = 75$, or against a composite hypothesis such as $H_a: \mu \neq 100$, $H_a: \mu > 130$, or $H_a: \mu < 75$. Similarly, the composite null hypothesis $H_0: \mu \leq 100$ may be tested against a simple alternative, such as $H_a: \mu = 120$, or against a composite alternative, such as $H_a: \mu > 100$.

Regardless of the form of the two hypotheses, it is extremely important to remember that the true value of the population parameter under consideration *must* be either in the set specified by H_0 or in the set specified by H_a. By testing $H_0: \mu = 100$ against $H_a: \mu = 120$, for example, one is asserting that the true value of μ equals either 100 or 120, and that *no other values are possible.* One means for assuring that either H_0 or H_a contains the true value of θ is to let these two sets be *complementary*. That is, if the null hypothesis is $H_0: \mu = 100$, then the alternative hypothesis would be $H_a: \mu \neq 100$; or if $H_0: \mu \leq 100$, then $H_a: \mu > 100$. From a statistical point of view, the easiest form to handle is a simple null hypothesis versus a simple alternative hypothesis. Unfortunately, most real-world problems cannot be stated in this form, but instead involve a composite null or a composite alternative hypothesis, or both. If a particular problem cannot be stated as a test between two simple hypotheses, then the next best alternative is to test a simple null hypothesis against a composite alternative. In other words, it is convenient to structure the problem so that the null hypothesis is a simple equality statement.

One- and Two-Sided Tests

If one is fortunate enough to be able to construct the test of hypotheses so that the null hypothesis is simple, then the composite alternative hypothesis may specify one or more values for the population parameter, and these values (or value) may lie entirely above, or entirely below, or on both sides of the value specified by the null hypothesis. A statistical test in which the alternative hypothesis specifies that the population parameter lies entirely above or entirely below the value specified in the null hypothesis is called a *one-sided test*; an alternative hypothesis that specifies that the parameter can lie on either side of the value indicated by H_0 is called a *two-sided test*. Thus, $H_0{:}\mu = 100$ tested against $H_a{:}\mu > 100$ is a one-sided test, since H_a specifies that μ lies on one particular side of 100. The same null hypothesis tested against $H_a{:}\mu \neq 100$ is a two-sided test, since μ can lie on *either* side of 100.

The Form of the Decision Problem

The decision problem that we confront in hypothesis testing is choosing between two mutually exclusive propositions about a population parameter when we are faced with the uncertainty inherent in sampling from a population. The decision-maker has only the sample evidence on which to base the choice of accepting the null hypothesis (which is equivalent to rejecting the alternative hypothesis) or rejecting the null hypothesis (accepting H_a). The standard method of solving this decision problem is, first, to assume that the null hypothesis is true (just as we presume a person's innocence until he or she is proved guilty in a court of law). Then, using the probability theory from Chapters 4, 5, and 6, we can establish the criteria that will be used to decide whether there is sufficient evidence to declare H_0 false. A sample is then taken, the sample evidence is compared to the criteria, and the decision is made whether to accept or reject H_0.

Similar to practice in a court of law, where innocence is maintained as long as any reasonable doubt about guilt remains, in hypothesis testing we reject H_0 only when the chance is small that H_0 is true. With such a procedure, the probability value, upon which we base our conclusion that there is considerable reason to doubt the truth of H_0, is critical. Moreover, since the decision to accept or reject H_0 is based on probabilities and not on certainty, there are chances of error in the decision. Specifically, there are two types of error:

1. One may decide to reject the null hypothesis when this hypothesis is, in fact, true (*Type I error*).
2. One may decide to accept the null hypothesis when this hypothesis is not true (*Type II error*).

There are four possible situations in hypotheses testing, as shown in Fig. 8.1. This figure presents the basic decision problem in hypothesis testing with

		The True Situation May Be:	
		H_0 is True	H_0 is False
Action	Accept H_0	Correct decision	Incorrect decision (Type II error)
	Reject H_0	Incorrect decision (Type I error)	Correct decision

Fig. 8.1 The four possible decision outcomes in hypothesis testing.

reference only to the null hypothesis. A good exercise for the reader would be to construct a similar figure, making reference only to H_a (remember that accepting H_0 implies rejecting H_a, and vice versa).

To illustrate the concept of Type I and II errors, we recall here our psychologist, who established the hypotheses $H_0{:}\mu = 100$ and $H_a{:}\mu > 100$. Now, suppose that the true mean is actually $\mu = 100$. If the psychologist decides (on the basis of some sample evidence) to accept $H_a{:}\mu > 100$, then a Type I error has been made. Let us consider the opposite situation, where the true value of μ exceeds 100. For example, we might assume that $\mu = 105$. A Type II error is made in this case if the psychologist decides to accept the null hypothesis $H_0{:}\mu = 100$.

Another example of Type I and II errors can be drawn from the quality-control problem in Section 4.2. In this problem defective items occur in a production process with one of two probabilities, $p = 0.10$ or $p = 0.25$. Suppose we establish the null hypothesis $H_0{:}p = 0.10$ and let the alternative hypothesis be $H_a{:}p = 0.25$. A Type I error is committed if the decision is to accept the alternative hypothesis $H_a{:}p = 0.25$ when the null hypothesis $H_0{:}p = 0.10$ is true. Similarly, a Type II error is committed if $H_0{:}p = 0.10$ is accepted when $H_a{:}p = 0.25$ is true. We reiterate at this point that the above hypotheses are based on the assumption that the true population value of p is either 0.10 or 0.25. By specifying two simple hypotheses, we are asserting that no other values are possible.

It should be clear that the probabilities of Type I and Type II errors are conditional probabilities. The former depends on the condition that H_0 is true and the latter on the condition that H_a is true. The probability of a Type I error is commonly denoted by the lower-case Greek letter alpha (α) and is called the *level of significance*. That is,

$$\begin{aligned}\alpha &= \text{Level of significance} = P(\text{Type I error}) \\ &= P(\text{reject } H_0 \mid H_0 \text{ is true}).\end{aligned}$$

The level of significance of a statistical test is comparable to the probability of an error, also called α, discussed in Chapter 7. The size of a confidence interval in Chapter 7, $(1 - \alpha)$, is now called a *confidence level*, and represents the complement of P(Type I error).

$$(1 - \alpha) = \text{Confidence level} = 1 - P(\text{Type I error}) = P(\text{accept } H_0 | H_0 \text{ is true}).$$

In constructing a statistical test, we obviously would like to have a small probability of making a Type I error; hence one objective is to construct the test in such a way that *α is small.* This objective, however, ignores the probability of making a Type II error. The probability of making a Type II error—that is, of accepting a false null hypothesis—is usually denoted by the Greek letter beta (β):

$$\beta = P(\text{accept } H_0 | H_0 \text{ is false}).$$

The complement of this probability is

$$1 - \beta = P(\text{reject } H_0 | H_0 \text{ is false}).$$

This probability $(1 - \beta)$ is known as the *power* of a statistical test, since it indicates the ability (or "power") of the test to recognize *correctly* that the null hypothesis is false (and hence, that H_0 should be rejected). Thus, one always wishes that a test will yield a large power (close to one), or equivalently, a low value of β, when H_0 is false. Figure 8.2 presents the same decision problem shown in Fig. 8.1, except that here we identify the *probability* associated with each of the four cells.

Note that the probability of each decision outcome is a *conditional* probability, and that all the elements in each column sum to 1.0, since the events with which they are associated are *complements.* By now it should be apparent that α and β need not add to unity, as these two probabilities are not complementary. Thus, a one-unit change in α does not imply a corresponding one-unit change in β, or vice versa. However, α and β are not independent of each other, nor are they independent of the sample size n. When α is lowered,

Fig. 8.2 The probability of each decision outcome in a hypothesis test.

		The True Situation May Be:	
		H_0 is True	**H_0 is False**
Action	Accept H_0	$1 - \alpha$ (Confidence level)	β (beta)
	Reject H_0	α (alpha)	$1 - \beta$ (Power of the test)
	Sum	1.00	1.00

β normally rises, and vice versa (if n remains unchanged). If n is increased, it is possible for both α and β to decrease, because the sampling error is potentially decreased. Since increasing n usually costs money, the researcher must decide just how much additional money should be spent on increasing the sample size in order to reduce the sizes of α and β. Such analysis, concerned with balancing the costs of increasing the sample size against the costs of Type I and Type II errors, is a fairly complex subject. We will cover this topic briefly in Section 8.9. Until then, we will assume that the sample size is fixed at some predetermined value n.

8.2 THE STANDARD FORMAT OF HYPOTHESIS TESTING

There are many different population parameters, many different potential forms of hypotheses, and many different sample statistics, random variables, and probability distributions that may be involved in testing hypotheses. It is not, therefore, feasible to catalogue all such tests. However, they all follow a similar procedure, which can be learned and then applied to different situations as they arise. This procedure can be summarized by the following five steps:

1. State the null and alternative hypotheses;
2. Determine the appropriate test statistic;
3. Determine the critical region;
4. Compute the value of the test statistic; and
5. Make the statistical decision and interpretation.

We will illustrate this procedure by a simple example that will continue throughout all five steps.

Suppose that it has been asserted that the mean IQ of all the students in a certain university is 130, although certain people claim that the mean must be some number other than 130. For the time being, we will assume that these people do not know whether μ exceeds 130 or is less than 130. Also, we will assume that the IQs of the students in this population are known to be normally distributed with a standard deviation of $\sigma = 5.4$. A random sample of $n = 25$ is proposed to test whether or not the mean is equal to 130. The five steps necessary for performing this test are described below.

Step 1. State the Null and Alternative Hypotheses

In every hypothesis-testing problem the two conflicting hypotheses must be specified clearly. It is usually convenient to formulate the null hypothesis as a simple hypothesis and the alternative hypothesis as a composite hypothesis, although it is not necessary to do so. In any case, the two conflicting hypotheses must be mutually exclusive, and they must be formulated so that the true value of the population parameter is included in either the null or the alternative hypothesis (i.e., it is not permissible for *both* hypotheses to be false).

For our IQ example the parameter in the test is the population mean μ. One hypothesis is that $\mu = 130$, and the other is that $\mu \neq 130$. If we let the simple hypothesis containing the equality statement be H_0, then:

$$\text{Null hypothesis:} \quad H_0: \mu = 130,$$
$$\text{Alternative hypothesis:} \quad H_a: \mu \neq 130.$$

One of these hypotheses must be true, as the values specified by H_a represent the complement of the values specified by H_0.

Step 2. Determine the Appropriate Test Statistic

The second step in testing hypotheses is to determine which of the random variables that we have studied is appropriate to determine whether we should accept or reject H_0. If the parameter being tested is the population mean, then we know that our best estimate of μ is $\bar{x}$. In testing $H_0: \mu = 130$ versus $H_a: \mu \neq 130$, we would want to determine how close $\bar{x}$ is to 130. A value of $\bar{x}$ far below or far above 130 would lead us to accept H_a; conversely, a value slightly below, equal to, or slightly above 130 would lead to acceptance of H_0. In order to decide whether $\bar{x}$ is close enough to 130 to accept H_0, it is necessary to standardize $\bar{\mathbf{x}}$. From the discussion in Chapter 6, we know that if $\bar{\mathbf{x}}$ is normal and σ is known, then the standardization of $\bar{\mathbf{x}}$ is

$$\mathbf{z} = \frac{\bar{\mathbf{x}} - \mu_0}{\sigma/\sqrt{n}}.$$

*The symbol μ_0 in this standardization is the value of μ specified under the null hypothesis.** The random variable $\mathbf{z}$ for this example is called the *test statistic.*

A test statistic is thus a random variable used to determine how close a specific sample result falls to one of the hypotheses being tested. As we will describe in this chapter, most of the distributions studied thus far can be used as test statistics.

A valid test statistic must satisfy three conditions:

1. Its probability function must be known when it is assumed that the null hypothesis is true.

* This is why testing is easier when H_0 is a simple rather than a composite hypothesis. If many values of the population parameter could be true under H_0, then the test statistic is neither easily stated nor easily used.

2. It must contain the parameter being tested.
3. All of its remaining terms must be known or calculable from the sample.

In our example the distribution of z is $N(0, 1)$ when μ is assumed to be $\mu_0 = 130$. Obviously, z contains the parameter μ. The remaining terms are: σ, which is known to be 5.4, $n = 25$, and $\bar{x}$, which can be calculated from the sample.

Step 3. Determine the Critical Region(s)

As we indicated above, certain values of the test statistic lead to acceptance of H_0 while other values lead to the rejection of H_0 (which is equivalent to the acceptance of H_a). In most statistical tests it is important to specify, *before the sample is taken*, exactly which values of the test statistic will lead to rejection of H_0 and which will lead to acceptance of H_0. The former set (leading to rejection of H_0) is called the *critical region*, while the latter set (leading to acceptance of H_0) is called the *acceptance region*. The *critical value* is that point which separates the critical region from the acceptance region. When the alternative hypothesis is two-sided (as in our IQ example), these regions are characterized as shown in the accompanying diagram.

Critical values

Critical region (reject H_0)	Acceptance region (accept H_0)	Critical region (reject H_0)
Large negative z	↑ $z = 0$ z values close to zero	Large positive z

The problem is to determine the *exact* location of the critical values shown in the diagram. These values depend partially on the level of risk of a Type I or Type II error that one is willing to take. For example, the smaller the value of α, the farther outward the critical values in the above diagram will move (thereby making the critical regions smaller). This occurs because when α is small, the decision-maker requires a relatively small critical region (remember, α is the probability that the parameter will be in the critical region when H_0 is true). Figure 8.3 illustrates the effect of reducing the size of α for a given sample size. The size of β affects the critical values in the opposite direction—the smaller the value of β, the more to the *middle* the critical values will move (thereby increasing the size of the critical region). As we indicated previously, the ideal is that both α and β be very small in a given situation. However, for the critical values, the effect of decreasing α is opposite to the effect of decreasing β. It should be clear by now that *decreasing the size of α will increase the size of β*; similarly,

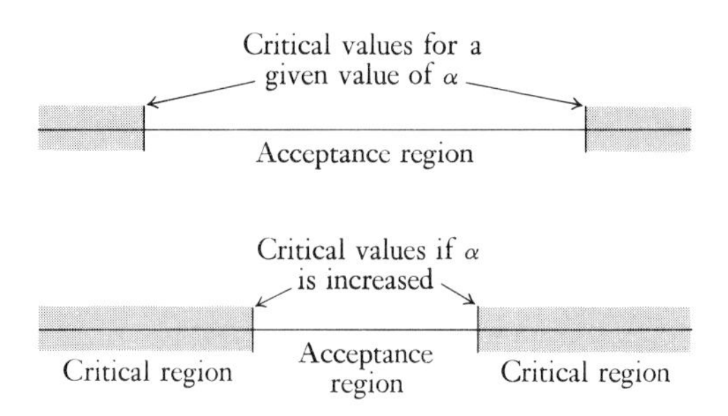

Fig. 8.3 Increasing the size of α will move the critical values toward the middle. Since the acceptance region becomes smaller, the value of β becomes smaller.

if the size of β is decreased, the size of α will increase (assuming that the sample size is fixed).

The trade-off between α and β is most clearly demonstrated with a one-sided test using two simple hypotheses, as shown in Fig. 8.4. Moving the critical value to the right or left in that figure will change both α and β, but not necessarily by the same amount.

The traditional method for selecting a critical region that will lead to the rejection of H_0 is first to establish a value for α and then to choose that critical region which yields the smallest value of β. The rationale behind this procedure is that it is important to establish beforehand the risk that one wants to assume of incorrectly rejecting a true null hypothesis. In other words, the size of the Type I error in this approach is viewed as so much more important than the size of the Type II error that the size of β is considered only after α has been fixed at some predetermined level. The practice of selecting a critical region in this manner stems from the early research on hypothesis testing, in which the null hypothesis usually represented "current opinion" on an issue and the alternative hypothesis represented a viewpoint of the researcher contrary to that commonly accepted. In testing a new drug, such as a cure for cancer, the drug must be assumed to be of no benefit, or even harmful, until it is proved otherwise. The alternative hypothesis is that the drug is indeed beneficial. A

Fig. 8.4 Illustrating the trade-off between α and β.

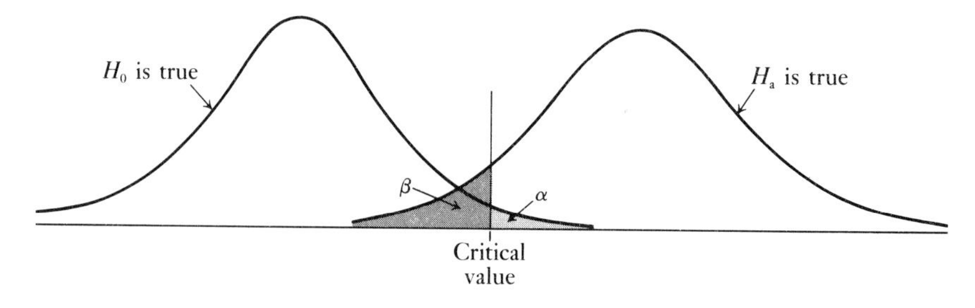

most serious Type I error would be made if a harmful drug (H_0 true) were certified as beneficial.

The value of α, or the level of significance, indicates the importance (i.e., "significance") that a researcher attaches to the consequences associated with incorrectly rejecting H_0. Researchers in the social sciences often use a level of significance of $\alpha = 0.05$, indicating that they are willing to accept a five-percent chance of being wrong when they reject H_0. For most statistical problems α is set, rather arbitrarily, at either 0.05 or 0.01. However, in the medical sciences α is usually set much lower—perhaps as low as 0.005 or 0.001. As we indicated above, medical science has to be *very* concerned about accepting an incorrect hypothesis. Although α usually is never set higher than $\alpha = 0.05$, and values such as $\alpha = 0.025$, $\alpha = 0.01$ and $\alpha = 0.001$ are used frequently, there is no reason why any other value may not be used. If α is set at some predetermined level, then it is extremely important that this value be specified before any data are collected.

In our IQ example, a Type I error occurs when it is concluded that $\mu \neq 130$ when μ is truly 130. Suppose we assume that the researcher uses an $\alpha = 0.05$ level of significance; i.e., this researcher wants to have no more than a five-percent chance of rejecting $H_0\!:\mu = 130$ when this hypothesis is true (a Type I error). This means that the researcher wants the critical region to cut off 5 percent of the appropriate p.d.f. of the test statistic, which from our earlier discussion we know to be the z-distribution. When H_a is two-sided, the optimal critical region will cut off $\alpha/2$ of the area in the upper tail, and $\alpha/2$ of the area in the lower tail. This is the same procedure that we used to construct a $100(1 - \alpha)\%$ confidence interval. If $\alpha = 0.05$, we know from Table III that the values of z that cut off $\alpha/2 = 0.025$ in each tail of the standardized normal distribution are ± 1.96. Figure 8.5 shows the resulting critical regions.

The critical values shown in Fig. 8.5 are expressed in terms of the z-distribution. When constructing critical values, we can express these numbers either as a z-value, or in terms of $\bar{x}$. The advantage of expressing the critical values in terms of $\bar{x}$ is that it is then possible to compare the sample results directly with these values, and to see if the sample falls in the acceptance region

Fig. 8.5 Critical regions for testing $H_0\!:\mu = 130$ against $H_a\!:\mu \neq 130$ with $\alpha = 0.05$, $\sigma = 5.4$, $n = 25$.

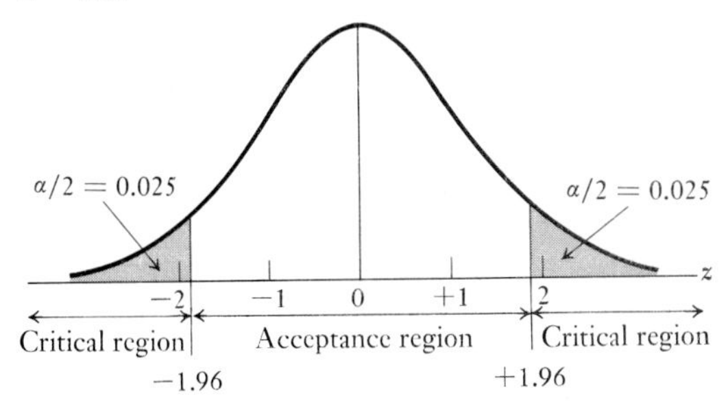

or in the critical regions. Fortunately, it is quite easy to transform a z critical value into its comparable $\bar{x}$ critical value by solving for $\bar{x}$ in the formula that represents the test statistic:

$$z = \frac{\bar{x} - \mu_0}{\sigma/\sqrt{n}}.$$

For our IQ example, the value of μ_0 is 130 (the null hypothesis), $n = 25$, and the known value of σ is 5.4. Substituting these values and $z = 1.96$ into this formula we obtain the $\bar{x}$ critical value for the righthand tail of the distribution:

$$1.96 = \frac{\bar{x} - 130}{5.4/\sqrt{25}}.$$

Solving for $\bar{x}$,

$$\bar{x} = 1.96\left(\frac{5.4}{5}\right) + 130 = 132.12.$$

Similarly, when the critical value is $z = -1.96$, the appropriate critical value of $\bar{x}$ is derived as follows:

$$-1.96 = \frac{\bar{x} - 130}{5.4/\sqrt{25}} \Rightarrow \bar{x} = -1.96\left(\frac{5.4}{5}\right) + 130,$$

$$\bar{x} = 127.88.$$

We could have saved ourselves some work in the above calculation by recognizing that the lower critical value will be the same distance *below* 130 as the upper critical value is *above* 130 (they are both 2.12 units away from 130). These critical values are shown in Fig. 8.6.

From Fig. 8.6, we know that, for random samples of size $n = 25$, 2.5 percent of the time $\bar{x}$ will be less than 127.88, and 2.5 percent of the time $\bar{x}$ will be greater than 132.12, assuming that $H_0\!:\mu = 130$ is true. We must em-

Fig. 8.6 Critical regions for testing $H_0\!:\mu = 130$ against $H_a\!:\mu \neq 130$ with $\alpha = 0.05$, $\sigma = 5.4$, $n = 25$ (in terms of $\bar{\mathbf{x}}$).

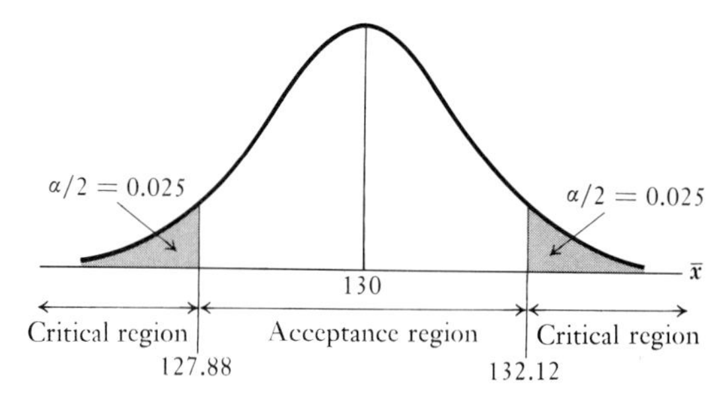

phasize that Figs. 8.5 and 8.6 are *equivalent* ways of presenting the same critical regions. The only difference is that the former is in terms of the test statistic z, while the latter is in terms of the sample mean itself ($\bar{x}$).

One-Sided Tests

For the critical regions calculated above, the alternative hypothesis was two-sided, $H_a:\mu \neq 130$. Now let's change our assumption about this problem, by supposing that it is agreed that the mean IQ cannot exceed 130, but could be lower than 130 (some state legislators seem to feel this way, especially when appropriations for specialized, advanced education programs are under consideration). In this case, the appropriate alternative hypothesis is $H_a:\mu < 130$. *Such an alternative hypothesis is legitimate only if one is certain that the population mean cannot exceed* 130.

The only change caused by a one-sided alternative hypothesis is that we now have a single critical region, rather than the two regions shown in Figs. 8.5 and 8.6. The single critical value is found in much the same way as in the two-sided test. The only difference is that all of the probability associated with the error α is cut off from a single end of the distribution. Whether this is the upper or the lower end depends entirely on whether the alternative hypothesis is a more than ($>$) or a less than ($<$) relationship. In our IQ example, suppose $\alpha = 0.05$ and we want to find the single value which cuts off five percent of the distribution. Note in Fig. 8.7 that we cut off $\alpha = 0.05$ of the *lefthand* portion of the z-distribution, because the values specified by H_a all lie to the *left* of 130 ($\mu < 130$). If the alternative hypothesis had been one-sided on the upper side ($H_a:\mu > 130$), then the critical region would lie entirely in the upper righthand tail.

The value $z = -1.645$ in Fig. 8.7 was derived from Table III in Appendix B by interpolating between $z = -1.64$ [$P(z \leq -1.64) = 0.0505$] and $z = -1.65$

Fig. 8.7 Critical region for testing $H_0:\mu = 130$ against $H_a:\mu < 130$ with $\alpha = 0.05$, $n = 25$, $\sigma = 5.4$.

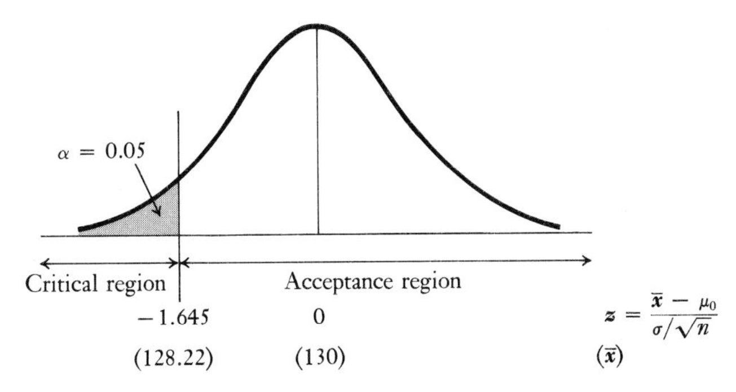

$[P(z \leq -1.65) = 0.0495]$. The $\bar{x}$-value comparable to $z = -1.645$ is determined as follows:

$$z = \frac{\bar{x} - \mu_0}{\sigma/\sqrt{n}},$$

$$-1.645 = \frac{\bar{x} - 130}{5.4/\sqrt{25}},$$

$$\bar{x} = -1.645\left(\frac{5.4}{5}\right) + 130 = 128.22.$$

This critical value of $\bar{x} = 128.22$ is also shown in Fig. 8.7.

Step 4. Compute the Value of the Test Statistic

Now that we have specified the two conflicting hypotheses, determined the appropriate test statistic, and found the critical region(s), we must see whether the sample result falls in the critical region or in the acceptance region. When the critical values are stated in terms of $\bar{x}$, this process is quite simple. For the most part, however, it will be convenient (and it will help the reader to note the similarities between all problems) to use z-values to denote critical values. A value of z calculated on the basis of a sample result is called a *computed z-value*, and is denoted by the symbol z_c.

Computed z-value: $$z_c = \frac{\bar{x} - \mu_0}{\sigma/\sqrt{n}}.$$

To illustrate a computed z-value, we will assume that, in our IQ example, the random sample of $n = 25$ resulted in $\bar{x} = 128$. That is, twenty-five students from this university were selected at random for IQ testing, and the average IQ in this sample was 128. When we substitute this value of $\bar{x}$ into our test statistic, we obtain the following computed z-value (again, assume $\sigma = 5.4$).

$$z_c = \frac{128 - 130}{5.4/\sqrt{25}} = -1.85.$$

We mention here, for later use, that if the t-variable had been the appropriate test statistic, then we would have calculated the value t_c. Similar computed values of the chi-square variable could be denoted χ^2_c.

Step 5. Make the Statistical Decision and Interpretation

> *If the calculated value of the test statistic falls in the critical region, then H_0 is rejected. When the calculated value lies in the acceptance region, then H_0 is accepted.*

This statement holds true for both one- and two-sided tests, and is appropriate when the critical value is expressed either in terms of the sample mean ($\bar{x}$) or in terms of z-values. In any case, it is important for the researcher to summarize or interpret the decision (to reject or accept H_0) *in terms of the original problem*, because the results of statistical tests in business, economics, and the social sciences are often presented to and utilized by people who may not understand statistical terminology.

Let us determine whether or not the sample result $\bar{x} = 128$ (or, equivalently, $z_c = -1.85$) leads to acceptance of $H_0: \mu = 130$ or acceptance of $H_a: \mu < 130$. From Fig. 8.7 it is clear that this sample result falls in the critical region (i.e., below 128.22 for $\bar{x}$, or below -1.645 for z). Thus, we would conclude that the mean is less than 130, although we must admit there is some risk (no more than five percent) that this conclusion is not true.

If our alternative hypothesis had been the two-sided test $H_a: \mu \neq 130$, then the sample result $\bar{x} = 128$ ($z_c = -1.85$) does *not* fall in the critical regions shown in Figs. 8.5 and 8.6. In this case we must conclude that there is not enough evidence to reject the hypothesis that $\mu = 130$. This result emphasizes how important is the selection of the significance level or the type of alternative hypothesis in a statistical test. The more specific one can be in specifying H_a, the more powerful will be the resulting test.

Procedure When H_0 Is Composite

The reader may wonder at this point what happens if the null hypothesis is not simple, but composite. For example, suppose that we wish to test

$$H_0: \mu \geq 130 \qquad \text{versus} \qquad H_a: \mu < 130.$$

For a composite null hypothesis like this, it is impossible to calculate a critical value as we did in Fig. 8.7, because now there is no *one* value for μ specified under the null hypothesis. The common approach to this situation is to *construct the critical value(s) using the most conservative value possible under H_0.* The most conservative value is generally the one closest to the alternative hypothesis. Thus, if the alternative hypothesis is $H_a: \mu < 130$, the closest value specified by the null hypothesis is the value $\mu = 130$. The critical value(s) can now be constructed by treating the null hypothesis as if it were $H_0{:}\mu = 130$. The resulting critical region will look exactly like that shown in Fig. 8.7. Rejection of H_0 in favor of H_a for this extreme value of μ would certainly lead to the same and even stronger conclusions when the other values of μ in the composite null hypothesis H_0 are used.

Beginning students in statistics are often confused in deciding which hypothesis to designate as the null and which the alternative. Theoretically it should make no difference. Because of the wide-spread practice of setting α at some predetermined level, there is an automatic "boost" given to the null hypothesis (because of our reluctance to reject H_0 unless we are convinced it is not true). In this situation, the traditional approach is to designate as H_0 the more "conservative" hypothesis—e.g., the hypothesis which protects the general public, or subscribes to the current theory. Thus, in testing a new theory of relativity, we would let Einstein's theory be H_0. Similarly, in testing a new cure for cancer, we would let H_0 be the assumption that there is no cure for cancer.

These examples have illustrated the general method of hypothesis testing which may be applied to many tests. The most common tests of hypotheses involve μ, the population mean. Tests about μ are usually designed to indicate, on the basis of a *single* sample, which of two hypothesized values of μ should be rejected. In other circumstances, one may be interested in designing a test to indicate, on the basis of a sample from each of *two different* populations, whether or not these *two* samples were drawn from populations having equal means. This breakdown between one- and two-sample tests will be a convenient one for us to follow.

8.3 REPORTING A PROBABILITY (p) VALUE

The five steps outlined above occur in almost all tests of hypotheses, even though the step-by-step details of the process may change from one situation to another. We will illustrate the basic process in a number of examples in this chapter and the following chapter. Before doing so, we should point out that a modification of the steps is often used when it is not possible or desirable to specify the value of α (the level of significancc) *before* taking thc sample. This may happen when the decision-maker is someone other than the person carrying out the research (the statistician). In some cases the decision-maker isn't available to specify an α, or perhaps the statistician is writing a report in which each of the various readers may be thought of as potential decision-makers (all with possibly different α's).

When α is not specified, a common procedure is to determine or "report" a probability value that tells the decision-maker how far away from H_0 an observed sample result (that is, a z_c-value) lies. In a one-sided test involving μ, for example, the p-value is the area in the tail of the test statistic *beyond* the observed sample result z_c (see Fig. 8.8). Thus, for Fig. 8.8, the p-value is $P(z \geq$

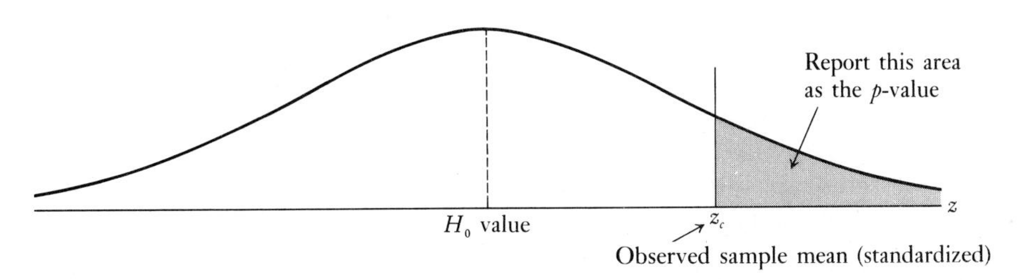

Fig. 8.8 Illustration of the area representing a reported p-value.

z_c). If we had used a one-sided test involving the *left*-handed tail of the z-distribution (contrary to Fig. 8.8), the reported p-value would have been $P(z \leq z_c)$. To illustrate such a calculation, we will apply this procedure to our IQ example, where

$$H_0: \mu = 130 \qquad \text{versus} \qquad H_a: \mu < 130.$$

Recall that the sample result $\bar{x} = 128$ led to a computed z-value of $z_c = -1.85$. *The reported p-value is the probability that the random variable z is less than or equal to z_c.* This probability is (from Table III):

$$P(z \leq z_c) = P(z \leq -1.85) = 1 - 0.9678 = 0.0322.$$

The probability value reported to the decision-maker is thus 0.0322. To a decision-maker who knows little about statistics, we might offer the following statement:

"A sample mean of 128 or lower will occur only 3.22 percent of the time when the true population mean is 130. Since our sample mean was 128, this result casts serious doubt on the validity of the assumption that the mean is, in fact, 130."

A decision-maker who has set α *higher* than 0.0322 should reject H_0 (because then the critical region is larger than 0.0322, and the sample result $\bar{x} = 128$ falls in the critical region). A decision-maker who has set α *lower* than 0.0322 should accept H_0 (because the critical region is smaller than 0.0322, and the sample result $\bar{x} = 128$ falls in the acceptance region). The procedure outlined above is for an alternative hypothesis that is one-sided on the *lower* side of H_0. If H_a is one-sided on the *upper* side of H_0, then the reported p-value is $P(z \geq z_c)$. Otherwise, the procedures described above do not change.

To summarize:

If α exceeds the p-value:	*reject H_0.*
If the p-value exceeds α:	*accept H_0.*

If the alternative hypothesis is no longer one-sided, but rather two-sided, then the process described above must be modified slightly. Suppose, for instance, in the above example that the alternative hypothesis is changed to $H_a\colon \mu \neq 130$. This alternative is *twice* as large as the previous alternative hypothesis (which was $H_a\colon \mu < 130$). Since we now have two critical regions, the probability to be reported must be *twice as large* as in the one-sided case. The p-value for a two-sided alternative must be twice as large as the p-value in the one-sided case, because the p-value in the two-sided case must include both of the one-sided probabilities—that is, $P(z \leq z_c) + P(z \geq z_c)$. Thus, the general rule is:

> *If H_a is two-sided, then the probability value calculated from Appendix B must be doubled.*

Let us now assume that the same sample result, $\bar{x} = 128$, occurs when testing $H_0\colon \mu = 130$ versus $H_a\colon \mu \neq 130$. To find the p-value, we must double the probability $P(z \leq -1.85)$.*

$$2P(z \leq z_c) = 2P(z \leq -1.85) = 2(0.0322) = 0.0644.$$

Thus, we would report that, if the null hypothesis is true, a sample result this far away from the hypothesized value of the parameter (either above or below μ) would occur 6.44 percent of the time. The decision-maker with an α greater than 0.0644 should reject H_0. If α is set at any level of significance below 0.0644, H_0 should be accepted.

Define: Simple, composite, one-sided, two-sided, null and alternative hypotheses; Type I and Type II errors; acceptance and critical regions; α and β probabilities; confidence level; test statistic; critical value; computed z-value; reported p-value.

Problems

8.1 a) Which is more serious, a Type I or a Type II error?

b) Why is it necessary to be concerned with the *probability* of Type I and Type II errors?

c) What is meant by the phrase "the costs of making an incorrect decision"? How are these costs related to the problem of balancing the risks of making an incorrect decision?

* The reader should always make certain that the probability to be doubled is less than 0.50. It would make no sense to double a probability such as $P(z \geq -1.85)$, for doing so would yield a p-value greater than 1.0.

d) How would you go about the process of testing hypotheses in circumstances in which it is impossible to associate, at least directly, any dollar values to the costs of making an incorrect decision? Assume that you are responsible for making an important decision (e.g., you are a doctor who must decide whether or not to operate on a patient).

8.2 A Los Angeles bakery was recently fined \$1200 for selling loaves of bread that were underweight. Assume that the L.A. city attorney has established $H_0: \mu = 24$ oz versus $H_a: \mu < 24$ oz, where 24 oz is the stated weight of each loaf of bread. A sample of 1861 loaves was taken, and σ is known to equal 1.00 oz.

a) Describe, in words, what a Type I and a Type II error would be in this circumstance. What would you guess to be the consequences of each type of error in this situation?

b) Would you accept H_0 or H_a if $\alpha = 0.01$ and the sample mean were $\bar{x} = 23.75$ oz?

c) Do you think $\alpha = 0.01$ is reasonable in this case? Would the decision change if $\alpha = 0.001$?

d) Would you buy bread from this bakery?

8.3 A fast-food chain has been averaging about \$5000 worth of business on a typical Saturday. The manager, concerned that sales are slipping, considers the average sales for four consecutive Saturdays.

a) What null and alternative hypotheses would you establish to determine whether or not sales are slipping (let H_a be one-sided). Describe a Type I and a Type II error in this context.

b) What test statistic should you use if σ is known to equal \$500?

c) Find the critical values for $\bar{x}$ and z, assuming that $\alpha = 0.05$.

d) What decision should be made in regard to the hypotheses if sales are \$4200, \$4400, \$5200, and \$4800 on the four Saturdays? What probability could be reported in this example?

e) What p-value should be reported if H_a is two-sided?

f) Comment on this manager's choice of a "random" sample.

8.4 A manufacturer of steel rods considers that the manufacturing process is working properly if the mean length of the rods is 8.6 in. The standard deviation of these rods always runs about 0.3 in. The manufacturer would like to see if the process is working correctly by taking a random sample of size $n = 36$. There is no indication whether or not the rods may be too short or too long.

a) Establish the null and alternative hypotheses for this problem. What critical values should be used for z and $\bar{x}$ if $\alpha = 0.05$?

b) Assume that the random sample yields an average length of 8.7 in. Would you accept H_0 or H_a? What p-value would you report?

8.5 In a study of family income levels, nine households in a certain rural county are selected at random. The incomes in these households are known to be normally distributed with $\sigma = 2400$. An average income of \$10,000 is reported from this sample survey. Use a statistical test to decide if the null hypothesis, $\mu = \$12{,}000$, can be rejected in favor of H_a, which states that the true average income is less than \$12,000. Use an 0.05 level of significance.

8.6 A manufacturer of rope produces 22,500 ft of rope a day, in lengths of 50 ft each. Each day, 16 of the 50-ft lengths are tested for tensile strength. The variance of tensile strength is always approximately $\sigma^2 = 256$. The mean tensile strength of one sample of 16 observations was 340 pounds. What probability value would you report if the null hypothesis $H_0: \mu = 350$ is tested against $H_a: \mu \neq 350$? How will your answer change if the alternative is $H_a: \mu < 350$? Assume that the tensile strengths are normally distributed.

8.7 A manufacturer claims that its new radial tire will last, on the average, more than 70,000 miles.

a) Should the null hypothesis in this situation be $H_0: \mu \geq 70{,}001$ or $H_0: \mu \leq 70{,}000$? If α is set at 0.05, which hypothesis would the manufacturer prefer? What alternative hypothesis should be established to test the claim? Describe a Type I and Type II error in this situation.

b) Suppose $\sigma = 10{,}000$. What critical region should be used if $\alpha = 0.05$ and $n = 100$? Describe the critical region both in terms of $\bar{x}$ and in terms of z. Assume $\boldsymbol{x}$ is normally distributed.

c) If $\bar{x} = 71{,}250$, would you accept H_0 or H_a using the data in part (b)? What p-value would you report if α is unknown?

8.8 The Parks and Recreation Board is trying to decide whether or not to build new tennis courts in their city. A random sample of 100 people in this city revealed that tennis is played, on the average, 1.2 hours per week during the summer. The population standard deviation is 0.4 hours for all people in the U.S. Test whether this sample indicates that the number of hours that tennis is played in this city differs from the national average of 1.1 hours (use a two-sided alternative). Use $\alpha = 0.01$.

8.9 A statistician is investigating the charge that a minority group of workers is paid less-than-average wages. The population for all comparable workers has a mean wage of \$14,500 with a standard deviation of \$200.

a) Establish the appropriate null and alternative hypotheses. Describe possible consequences of a Type I and Type II error.

b) Find the appropriate critical value if $n = 100$ and $\alpha = 0.05$.

c) Would you accept H_0 or H_a if $\bar{x} = \$14{,}300$?

d) What p-value would you report?

8.10 You are interested in a site for a new restaurant, and a real estate developer claims that, on the average, male resident students at the University eat at least eight meals per week at establishments located more than one mile (beyond walking distance) from their residences. To test this claim, you set $H_a: \mu < 8$, randomly select six students, and record the number of times they ate beyond the "one-mile boundary" in one designated week; the results are 6, 5, 7, 4, 8, and 6. Test, at the 0.01 significance level, whether these results indicate rejection of $H_0: \mu \geq 8$, assuming that $\sigma = 1.5$, and that the population is normally distributed.

8.11 An industrial firm that manufactures small battery-powered toys periodically purchases a large number of flashlight batteries for use in the toys. The policy of this company has been never to accept a shipment of batteries unless it is possible to reject, at the 0.05 level of significance, the hypothesis that the batteries have a mean life of 50 or fewer hours. The standard deviation of the life of all shipments has typically been 3 hours.

a) What null and alternative hypotheses should be established to implement the company policy?

b) Should the company accept a shipment from which a sample of 64 batteries results in a mean life of 50.5 hours?

c) What is the minimum mean life that this company should accept in a sample of 64 batteries?

8.12 One type of aspirin claims that its brand gives faster relief (on the average) from a headache than another brand. Describe the hypothesis that might be used to test this assertion. What are the consequences of a Type I and Type II error in this situation?

8.13 A TV advertisement indicated that the Chrysler Corporation had 50 randomly selected people compare a Horizon car with a Datsun car. What null and alternative hypotheses might Chrysler establish if they wished to see whether their car is preferred over Datsun?

---

*8.4 THE POWER FUNCTION OF A CRITICAL REGION (OPTIONAL)

The examples thus far have involved only simple hypotheses, largely because the problem of calculating and balancing the risks of Type I and Type II errors is considerably more difficult when the null or alternative hypothesis, or both, is *composite*. The difficulty stems from the fact that there is a *different* probability that a given sample falls in the critical region for *each one* of the values specified by a composite hypothesis, so that no one number can express the risk associated with making an incorrect decision.

The different probability values for β that occur when H_a is composite can be presented in a table, graphed, or described by a functional relationship. Often, however, it is more useful to present the values of $(1 - \beta)$. A function describing such probabilities is called a *power function*, since it indicates the ability (or "power") of the test to correctly reject a false null hypothesis. In general, test statistics and critical regions having the highest power are preferred. Although it is beyond the scope of this book to examine the concepts involved in finding a power function for most statistical tests, we must emphasize that the tests presented thus far have made use of these concepts in that we have always selected the *most powerful critical region.* The complexity involved in finding the power of a statistical test again emphasizes the rationale for making the null hypothesis a simple test. If H_0 were a composite hypothesis, then we would also have to use a function to describe all the values of α, or be satisfied with just a single value associated with the most conservative value of μ in H_0.

The power function also indicates the probability that the null hypothesis will be rejected when H_0 is true. The advantage of calculating various power functions for a given statistical test is that by this means it may be possible to eliminate *obviously inferior* critical regions, and perhaps even to decide on an

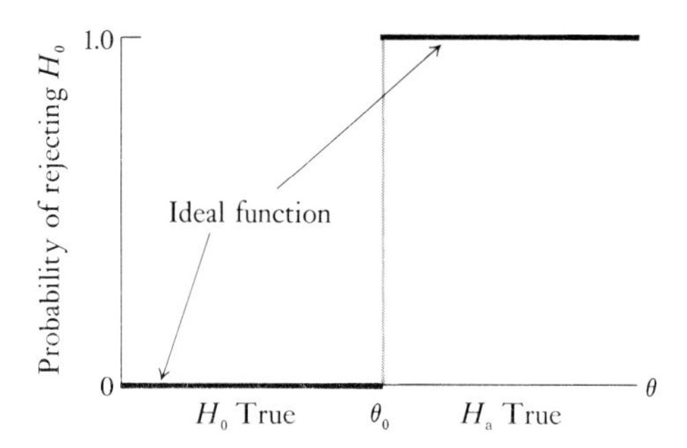

Fig. 8.9 Ideal power function for testing $H_0: \theta < \theta_0$ versus $H_a: \theta < \theta_0$.

optimal critical region. The *ideal power function* is one in which the probability of rejecting H_0 is 1.0 when H_0 is false and 0.0 when H_0 is true. Figure 8.9 shows the ideal power function for a test in which the null hypothesis states that θ is less than some value θ_0, $H_0: \theta < \theta_0$, while the alternative hypothesis states that θ is greater than this value, $H_a: \theta > \theta_0$.

To illustrate the calculation of β [or $(1 - \beta)$] and the trade-offs between α, β, and the sample size, we will consider a firm that manufactures rubber bands. Control over the number of rubber bands placed in each box is kept by sampling and testing hypotheses, rather than by a counting procedure. When the production process is working correctly, the number ($\boldsymbol{x}$) of good bands placed in each box has a mean of $\mu = 1000$, with a standard deviation of 37.5. This variable $\boldsymbol{x}$ is presumed to have an approximately normal distribution; that is, $\boldsymbol{x}$ is $N(1000, 37.5^2)$.

In this case the company wants to test $H_0: \mu = 1000$ against $H_a: \mu \neq 1000$, and the appropriate test statistic is $z = (\bar{x} - \mu_0)/(\sigma/\sqrt{n})$. A Type I error occurs whenever the test results suggest that the process is out of control ($\mu \neq 1000$), when it actually *is* in control ($\mu = 1000$). A Type II error occurs whenever the process is judged to be in control ($\mu = 1000$) and it actually is out of control ($\mu \neq 1000$). In constructing a test of these hypotheses, let us assume that the company periodically selects a random sample of size $n = 9$, and the company policy is to let $\alpha = 0.05$. For these values the boundary points of the acceptance region are

$$\mu_0 \pm z_{\alpha/2} \frac{\sigma}{\sqrt{n}} = 1000 \pm 1.96\left(\frac{37.5}{\sqrt{9}}\right) = 1000 \pm 24.5.$$

This acceptance region is shown in Fig. 8.10.

The question we turn to now is how to calculate β for this problem. Since β is a conditional probability that depends on the value of μ, we will assume that $\mu = 990$. We can now write

$$\beta = P(\text{Accept } H_0: \mu = 1000 \mid \mu = 990).$$

From Fig. 8.10 we see that H_0 is accepted whenever $\bar{x}$ lies between 975.5 and 1024.5. Hence, $\beta = P(975.5 \leq \bar{x} \leq 1024.5 \mid \mu = 990)$. This probability can

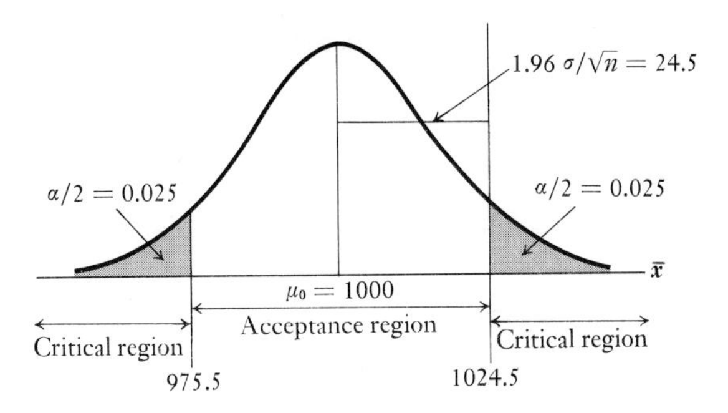

Fig. 8.10 Critical region for test on $H_0{:}\mu = 1000$ against $H_a{:}\mu \neq 1000$, $\sigma = 37.5$, $n = 9$, and $\alpha = 0.05$.

be determined by using the same procedure we learned in Chapter 6 in working with $z = (\bar{x} - \mu)/(\sigma/\sqrt{n})$. First, we transform the problem to standardized normal terms by letting $\mu = 990$, $\sigma = 37.5$ and $\sqrt{n} = \sqrt{9}$, and then we use Table III to find the appropriate probabilities.

$$P(975.5 \leq \bar{x} \leq 1024.5) = P\left(\frac{975.5 - 990}{37.5/\sqrt{9}} \leq \frac{\bar{x} - \mu}{\sigma/\sqrt{n}} \leq \frac{1024.5 - 990}{37.5/\sqrt{9}}\right)$$

$$= P(-1.16 \leq z \leq 2.76) = F(2.76) - F(-1.16)$$

$$= 0.9971 - 0.1230 = 0.8741.$$

Thus, $P(\text{Type II error}) = 0.8741$, as shown in Fig. 8.11. This means that our test procedure is such that when $\mu = 990$, we will *incorrectly* accept $H_0{:}\mu = 1000$ as being true 87.41 percent of the time. The *power* of this test is $1 - \beta = 0.1259$, which means that this test will *correctly* recognize this false null hypothesis 12.59 percent of the time when $\mu = 990$.

Instead of using the value $\mu = 990$ to calculate β, we might have used $\mu = 1010$. These two values of μ are both an equal distance (10 units) away from $H_0{:}\mu = 1000$, so it should not be surprising to learn that the value of β is the same in both cases ($\beta = 0.8741$). Figure 8.12(a) shows the area corresponding to β for $\mu = 1010$, while parts (b) and (c) of this figure show the area for β

Fig. 8.11 Probability of β for the critical region shown in Fig. 8.10 if the true mean is $\mu = 990$.

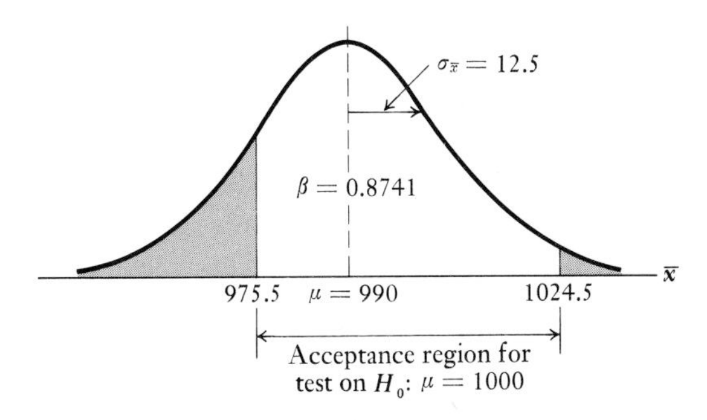

corresponding to $\mu = 970$ and $\mu = 950$, respectively. The calculation of β when $\mu = 970$ is shown below:

$$\begin{aligned} P(\text{Type II error} \mid \mu = 970) &= P(975.5 \leq \bar{x} \leq 1024.5 \mid \mu = 970) \\ &= P\left(\frac{975.5 - 970}{37.5/\sqrt{9}} \leq z \leq \frac{1024.5 - 970}{37.5/\sqrt{9}}\right) \\ &= P(0.44 \leq z \leq 4.36) \\ &= F(4.36) - F(0.44) = 1.000 - 0.670 = 0.3300. \end{aligned}$$

Note that in Fig. 8.12 the size of β decreases as the value of μ gets farther away from $\mu = 1000$. That is, the more incorrect H_0 is, the lower will be the

Fig. 8.12 Values of β for different μ's, $\alpha = 0.05$, $n = 9$, $\sigma = 37.5$.

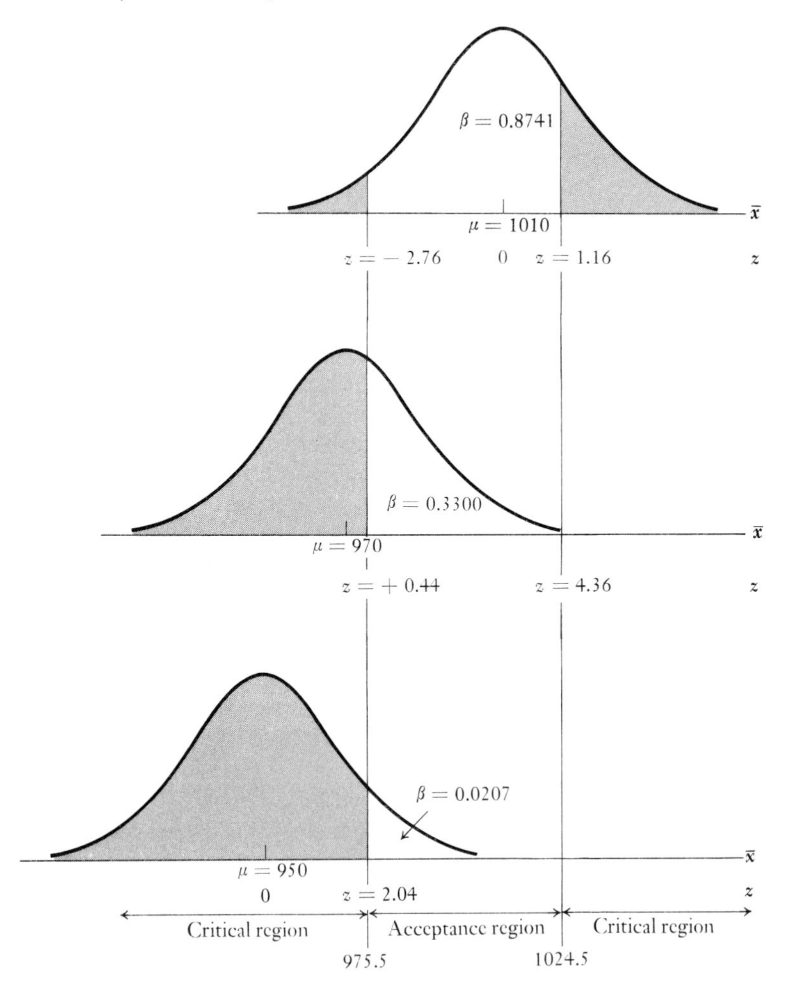

Table 8.1 Value of β and Power for Given True Values of μ

μ	$P(\text{Accept } H_0) = \beta$	$P(\text{Reject } H_0) =$ Power $= 1 - \beta$
950	0.0207	0.9793
970	0.3300	0.6700
980	0.6406	0.3594
990	0.8741	0.1259
1000	$1 - \alpha = 0.95$	$\alpha = 0.05$
1010	0.8741	0.1259
1020	0.6406	0.3594
1030	0.3300	0.6700
1050	0.0207	0.9793

value of β [and the higher $(1 - \beta)$]. This fact is shown in Table 8.1 where we present the value of β and $(1 - \beta)$ for eight different values of μ. The row corresponding to $\mu = 1000$ is placed in a box to emphasize that this is the one case in which H_0 is true; hence β is not defined for $\mu = 1000$. Figure 8.13 is a graph of the power function $(1 - \beta)$.

In comparing the power functions of a number of different tests, we look for tests where the power function rises quickly as the value of μ differs by small amounts from μ_0. The most powerful test would be the one with the steepest ascending power function. In other words, we desire a test such that the probability of recognizing a false null hypothesis increases rapidly, even for rather small differences between the hypothesized value of the parameter and the true value.

The Trade-Offs Between α and β

We have emphasized that when the sample size is fixed, α and β have an inverse relationship. To illustrate this trade-off, we will use the same production-process

Figure 8.13

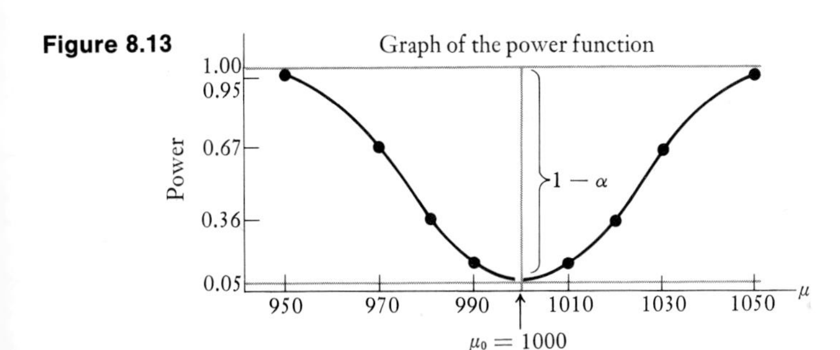

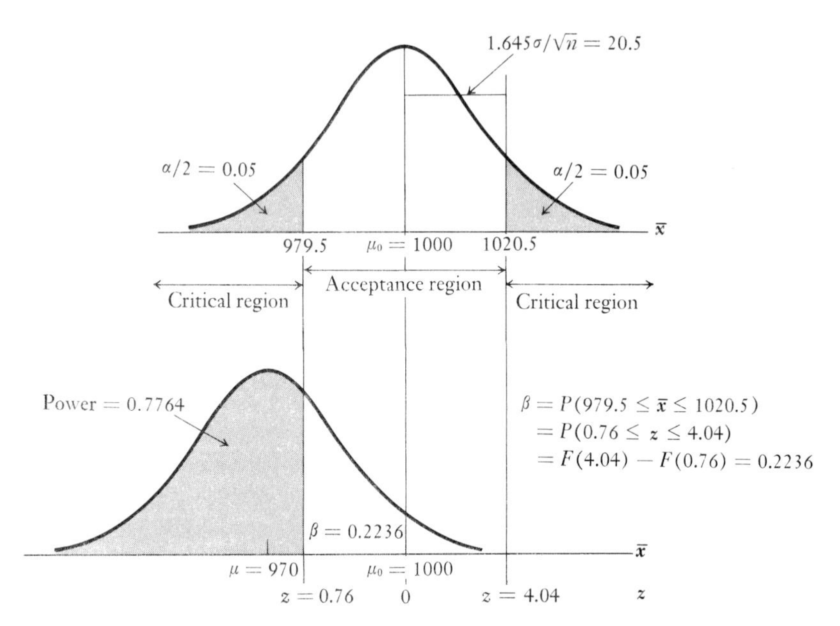

Fig. 8.14 Critical region for $H_0: \mu = 1000$ against $H_a: \mu \neq 1000$ given $\sigma = 37.5$, $n = 9$, $\alpha = 0.10$; and the representation of β given that the true mean is $\mu = 970$.

example described above, but we now will change α from 0.05 to 0.10. Since α has increased, the acceptance region has become smaller.

The effect of the increase of α should be a reduction of β. Figure 8.14 shows the new acceptance region, calculated by letting $z_{\alpha/2} = z_{0.05} = 1.645$. In this situation, the null hypothesis is accepted if $979.5 \leq \bar{x} \leq 1020.5$. The result is that the size of β is reduced from 0.3300 (shown in Table 8.1) to the new value, $\beta = 0.2236$. The power of the test has increased correspondingly. These values are shown in Fig. 8.14.

If we calculated additional values of the power function, we would find all of them larger when $\alpha = 0.10$ than when $\alpha = 0.05$. As we increase α, we narrow the acceptance region, and hence make our test more powerful. Similarly, if we decrease the value of α, then the acceptance region gets larger, β will rise, and the power of the test will drop. Thus, the size of α is related inversely to the size of β and directly to the size of the power, but the trade-off is not one-to-one. In this example, α was increased by 0.05 (0.05 to 0.10), but β decreased by more than twice this amount (0.3300 to 0.2236).

Decreasing α and β by Increasing n

Until now our discussion has been based on the assumption that the size of the sample is fixed in advance. If n is changed, however, the size of both α and β may be changed, because the size of n affects the location of the acceptance and

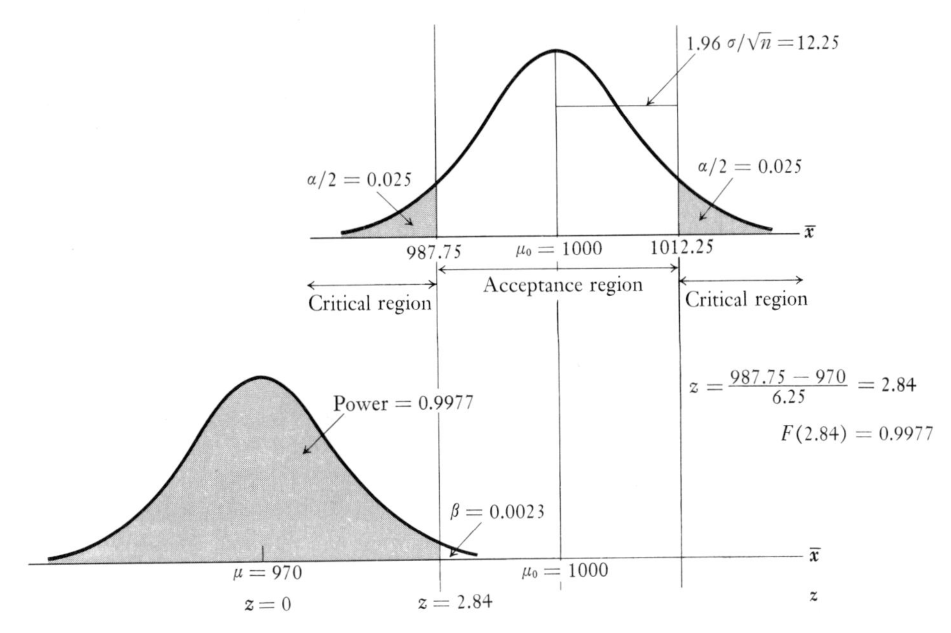

Fig. 8.15 Critical region for the test on $H_0: \mu = 1000$ against $H_a: \mu \neq 1000$, $\sigma = 37.5$, $n = 36$, $\alpha = 0.05$; and the representation of β when the true mean is $\mu = 970$.

critical regions. To illustrate this effect, suppose we return α to its previous level of 0.05, and increase n from 9 to 36. The new critical region for our test, and the determination of β when the true value of μ is 970, are shown in Fig. 8.15.

We see that the increased sample size makes our test more sensitive in distinguishing between H_0 and H_a because the standard error of the mean, $\sigma/\sqrt{n}$, is now half its former value (from $37.5/\sqrt{9} = 12.5$ to $37.5/\sqrt{36} = 6.25$). The null hypothesis will be accepted in this test if $987.75 < \bar{x} \leq 1012.25$. The probability that H_0 will be accepted given that $\mu = 970$ is now only $\beta = 0.0023$. By comparing Figs. 8.14 and 8.15 we see that we have reduced α from 0.10 to 0.05 and reduced β from 0.2236 to 0.0023 merely by increasing n from 9 to 36. Unfortunately, obtaining larger samples is more time-consuming and is often quite costly, so the researcher is faced with the task of balancing the costs of making incorrect decisions against the costs of sampling. We will return to this consideration in Chapter 9.

8.5 ONE-SAMPLE TESTS ON μ

In this section we will discuss additional tests on the mean of a population, μ. In general, there are two types of tests involving μ, one in which σ is assumed to be known, and the other in which σ is assumed to be unknown. In the former case, the z-distribution is the appropriate test statistic. In the latter case the t-distribution is the appropriate test statistic, although the z is used commonly even in these cases, provided that n is large (where "large" generally means over 30). In our examples involving μ in Section 8.2, it was assumed that σ was known; hence we used the z test statistic. In this section, we will present another example in which σ is known, and then move to the situation in which the t-distribution is appropriate.

Another Example in Which σ Is Known

In the following illustration of the testing procedure for μ, we will reconsider the example from Sections 6.6 and 6.7 in which a sample of 36 dormitory residents was taken to judge whether the average monthly phone charge, claimed to be \$15.30, seemed reasonable based on use of the phone by the population of 300 dormitory phone subscribers. The standard deviation for this population was assumed to be $\sigma = \$4.10$. Let us assume that the null hypothesis is to be the simple hypothesis, and that the alternative hypothesis is two-sided:

$$H_0\!:\mu = 15.30, \qquad H_a\!:\mu \neq 15.30.$$

From the central-limit theorem we know that $\bar{x}$ will be approximately normal; hence, we can again use the standardized normal test statistic $z = (\bar{x} - \mu_0)/(\sigma/\sqrt{n})$, where $\mu_0 = \$15.30$, $\sigma = \$4.10$, and $n = 36$. In this case it is not completely accurate to use a standard error of $\sigma/\sqrt{n}$, since we are sampling without replacement from a *finite* population. Rather, the finite population correction factor, $\sqrt{(N - n)/(N - 1)}$, should be used, so that the appropriate test statistic is now

$$z = \frac{(\bar{x} - \mu_0)}{(\sigma/\sqrt{n})\sqrt{(N - n)/(N - 1)}},$$

where $N = 300$, and all other terms are known or can be obtained from the sample.

The next step is to specify the level of α and find the critical regions. However, for this example, we will assume that the researcher does not know the appropriate level of significance set by the decision-maker. Hence, the researcher now proceeds directly to determine the computed value of the test statistic. Assume that a random sample of $n = 36$ yields an average monthly phone bill of $\bar{x} = \$16.90$. The calculated z value is thus:

$$z_c = \frac{\bar{x} - \mu_0}{(\sigma/\sqrt{n})\sqrt{(N - n)/(N - 1)}} = \frac{\$16.90 - 15.30}{(4.10/\sqrt{36})\sqrt{264/299}} = 2.49.$$

The researcher can now report the following p-value:

$$2P(z \geq 2.49) = 2(1 - 0.9936) = 0.0128.$$

Expressed in words that a nonstatistician might understand more readily, this p-value means that only 1.28 percent of the time will one receive a sample mean as far away from \$15.30 as \$16.90, when \$15.30 is the true mean. This result implies that a decision-maker who has an α larger than 0.0128 should reject $H_0: \mu = \$15.30$. If the decision-maker's α is smaller than 0.0128, then he or she should accept the null hypothesis that the true mean is \$15.30.

One-Sample Test When σ Is Unknown

In many circumstances it is unreasonable to assume that σ is known, for when the mean μ is unknown, the population standard deviation often is unknown also (since σ depends on the size of the squared deviations about μ). And when σ is unknown, the z test statistic is no longer exactly correct.

Recall from Chapter 6 that, in order to solve problems involving $\bar{x}$ when σ is unknown, we can use the **t**-distribution if the parent population is normal. In our present context, this means that if μ_0 is the value of μ specified by the null hypothesis, then when σ is unknown and the population is normal the appropriate test statistic for tests on μ is the following **t**-distribution:

$$t_{(n-1)} = \frac{\bar{x} - \mu_0}{s/\sqrt{n}}. \tag{8.1}$$

As an example of a test on μ when σ is unknown, suppose that a promotions expert suggests to an automobile manufacturer (Drof Motor Co.) that it extend its service guarantee from 12,000 miles to 24,000 miles on transmissions, muffler systems, and brakes. This promotions expert says the change would make good advertising copy and be relatively costless to Drof Motors, because such parts seldom require service during this period anyway. The claim is that an average car will run longer than this before the cost of such repairs exceeds \$50.

We will assume that Drof Motors has asked you (as their "expert" on hypothesis testing) to test this claim, by sampling a few car owners and checking their service records, and to give the company advice on extending its guarantee. Now, suppose that you decide to check the service record of 15 randomly selected Drof car-owners from a population of several million. You will let the variable x represent the number of miles driven since purchase when the cumulative service repair cost on the parts under study exceeds \$50. Drof is interested in determining whether the population mean value, μ, equals 24,000 ($H_0: \mu = 24{,}000$), or if μ is less than 24,000 ($H_a: \mu < 24{,}000$). They use a simple null hypothesis since, at this point, they are not concerned with the possibility that μ may exceed 24,000 (technically, however, H_0 is $H_0: \mu \geq 24{,}000$).*

* These hypotheses are convenient and appropriate for Drof Motors. From a consumer's point of view, it might be much more appropriate to see if μ is less than 24,000, with hypotheses such as $H_0: \mu < 24{,}000$ versus $H_a: \mu \geq 24{,}000$.

In this case, a Type I error would occur if it were concluded that $\mu < 24{,}000$ when the null hypothesis was true. In this case, Drof would be afraid to extend the guarantee policy when it *could* do so. A Type II error would occur if $\mu = 24{,}000$ were accepted when $\mu < 24{,}000$; this error might result in excessive service costs if Drof decided to extend the guarantee. Suppose that, in conferring with the owners of Drof Motors, you find that they are willing to assume a risk of 0.025 of incorrectly rejecting $H_0: \mu = 24{,}000$; thus, $\alpha = 0.025$.

Based on past experience, Drof expects the distribution of $\boldsymbol{x}$ to be normal, but they don't know its standard deviation; hence, the appropriate test statistic has a $\boldsymbol{t}$-distribution, $\boldsymbol{t} = (\bar{\boldsymbol{x}} - \mu_0)/(\boldsymbol{s}/\sqrt{n})$. The critical region is that 2.5 percent of the $\boldsymbol{t}$-distribution which lies in the lefthand tail (since H_a is to the "left" of H_0). The boundary point for the acceptance region is found in Table VI, under the heading 0.975 and in the row $v = 14$ (since $n - 1 = 14$), to be $t = 2.145$. Since the $\boldsymbol{t}$-distribution is symmetric, the critical value for a lower one-sided test is -2.145. The critical region is shown in Fig. 8.16. No finite population correction factor is needed here, since the population is very large.

Now we will assume that analysis of the random sample of 15 Drof owners results in the values $\bar{x} = 22{,}500$ and $s = 4000$. The calculated value of t for $(n - 1)$ d.f. is:

$$t_c = \frac{(\bar{x} - \mu_0)}{s/\sqrt{n}} = \frac{22{,}500 - 24{,}000}{4000/\sqrt{15}} = \frac{-1500}{1033} = -1.452.$$

Since the calculated value $t_c = -1.452$ lies in the acceptance region (it is not less than -2.145), we conclude that Drof Motors should accept the claim that $\mu = 24{,}000$. Even though our sample result indicated a value less than 24,000, this value is not far enough below 24,000 to lead to rejection of H_0 at the 0.025 level of significance. This result, however, might be encouraging enough for the promotions manager to test a modified hypothesis on the basis that the guarantee cover fewer parts or fewer miles.

Fig. 8.16 Critical region for $H_0: \mu = 24{,}000$ against $H_a: \mu < 24{,}000$ when σ is unknown and $n = 15$.

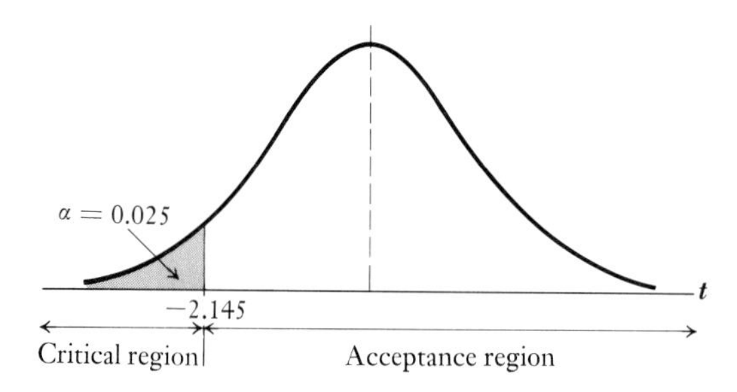

The probability value that would be reported in this example is

$$P(\boldsymbol{t} \leq t_c) = P(\boldsymbol{t} \leq -1.452).$$

Although this probability cannot be determined exactly from Table VII, we can determine from the row for $v = 14$ d.f. that it lies between 0.05 ($t = 1.761$) and 0.10 ($t = 1.345$). Hence, we report the following p-value:

$$0.05 < P(\boldsymbol{t} \leq -1.452) < 0.10.$$

If H_a had been two-sided in this example, the two values 0.10 and 0.05 would have had to be doubled.

Let us extend our Drof Motors example a bit by assuming that the sample size was a considerably larger number, $n = 81$. Suppose that for this larger sample the sample results were:

$$\bar{x} = 23{,}700 \quad \text{and} \quad s = 5000.$$

The calculated value of t in this case is

$$t_c = \frac{23{,}700 - 24{,}000}{5000/\sqrt{81}} = -0.54.$$

We would now like to report a p-value by calculating $P(\boldsymbol{t} \leq -0.54)$ for $n - 1 = 80$ d.f. This calculation poses several problems. First, there is no row in Table VII corresponding to 80 degrees of freedom, so we must move to the closest row, which is 60 d.f. But even for 60 d.f., the value of $t = 0.54$ does not appear among the values listed. From Table VII, row 60, we do know that $0.25 < P(\boldsymbol{t} \leq -0.679)$. Since we also know that $P(\boldsymbol{t} \leq 0) = 0.50$, we can report the following p-value:

$$0.25 < P(\boldsymbol{t} \leq -0.54) < 0.50.$$

As we pointed out in Chapter 6, many statistics texts recommend that the z table be used to approximate the values from the t table when n is "large," which usually means when n is 30 or greater. This process has the advantage of permitting one to report a *single* number for a p-value, rather than the range of probabilities that usually results from the use of t-values in Table VII. The disadvantage of using this process is that beginning students sometimes forget that the $\boldsymbol{t}$-distribution is always the correct test statistic when σ is unknown, and the $\boldsymbol{z}$-distribution is only an approximation (although often a good one).

Finally, the reader may be curious about how to report a p-value when H_a is one-sided and the sample result of the test statistic falls in the *opposite* tail from the critical region. For instance, in our sample of $n = 81$ for Drof Motors, we might have received the sample results

$$\bar{x} = 25{,}000 \quad \text{and} \quad s = 5000.$$

For these results, the calculated t-value is

$$t_c = \frac{25{,}000 - 24{,}000}{5000/\sqrt{81}} = 1.80.$$

Even though this value of t_c is a positive number, we still follow the same procedure outlined above for determining the p-value. That is, we want to find the area to the *left* of the value of t_c, since H_a specifies values smaller than $\mu = 24{,}000$. This p-value is

$$0.95 < P(t \leq 1.80) < 0.975.$$

The probabilities 0.95 and 0.975 come from Table VII, row 60, by using the *complement* of the probability $P(t \geq 1.80)$, which is seen to be

$$0.05 > P(t \geq 1.80) > 0.025.$$

8.6 TWO-SAMPLE TESTS ABOUT μ (σ_1 AND σ_2 KNOWN)

Instead of testing the results of a *single* sample against a hypothesized mean, one may wish to see if *two* different samples came from populations having equal means. This approach is quite common in testing for the effect of different treatments on two groups, where one group is often a "control" group and the other is given a special treatment, such as a new drug, a new approach to learning, or perhaps just a different type of paint. In any case, the objective is to use sample information in an attempt to determine whether the mean of one population differs from the mean of some other population.

As was the case in Section 8.2 an important part of the hypothesis-testing procedure is the specification of the appropriate test statistic. Since the heading of this section indicates that σ_1^2 and σ_2^2 are known, the reader may well guess that the standardized z-distribution will again save the day. It does, but we have to specify a new point estimator (to replace $\bar{x}$) and a new standard error (to replace $\sigma/\sqrt{n}$). As do all standardizations, the z variable in this case takes the following form:

$$z = \frac{\text{point estimator} - \text{hypothesized mean}}{\text{standard error of point estimator}}. \tag{8.2}$$

Assume that we are testing hypotheses about $(\mu_1 - \mu_2)$. It should not surprise you to learn that the best point estimator of $(\mu_1 - \mu_2)$ is $(\bar{x}_1 - \bar{x}_2)$, representing the difference between the mean of the first sample $(\bar{x}_1)$ and the mean of the second sample $(\bar{x}_2)$. The standard error of $(\bar{x}_1 - \bar{x}_2)$ is denoted by the symbol $\sigma_{\bar{x}_1 - \bar{x}_2}$, and can be shown to be*

$$\sigma_{\bar{x}_1 - \bar{x}_2} = \sqrt{\frac{\sigma_1^2}{n_1} + \frac{\sigma_2^2}{n_2}}$$

where n_1 = sample size from population 1 and n_2 = sample size from population 2, and σ_1^2 and σ_2^2 are (known) variances of the two populations. Finally, it can be shown that if $\bar{x}_1$ and $\bar{x}_2$ are normal, then the distribution of $(\bar{x}_1 - \bar{x}_2)$ will also be normal. Thus, we can write the following test statistic, assuming that the null hypothesis is some specified value of $(\mu_1 - \mu_2)$:

Two-sample z-test statistic for testing $\mu_1 - \mu_2$:

$$z = \frac{(\bar{x}_1 - \bar{x}_2) - (\mu_1 - \mu_2)}{\sqrt{(\sigma_1^2/n_1) + (\sigma_2^2/n_2)}}. \tag{8.3}$$

A Simple Example

To illustrate the use of Formula (8.3), we will apply it to test hypotheses about the mean starting salaries of college graduates in two cities, New York City (μ_1) and Chicago (μ_2). Assume that we decide to test

$$H_0: \mu_1 - \mu_2 = 0 \qquad \text{against} \qquad H_a: \mu_1 - \mu_2 \neq 0,$$

using a level of significance of $\alpha = 0.05$. Figure 8.17 (which looks remarkably like Fig. 8.5) shows the critical regions for this test.

Now suppose that a random sample of size $n_1 = 100$ from New York City yields $\bar{x}_1 = \$20{,}250$, while a random sample of size $n_2 = 60$ from Chicago results in $\bar{x}_2 = \$20{,}150$. If we assume that the known variances are $\sigma_1^2 = 40{,}000$ and $\sigma_2^2 = 32{,}400$, then the computed z_c value is

$$z_c = \frac{(20{,}250 - 20{,}150) - (0)}{\sqrt{(40{,}000/100) + (32{,}400/60)}} = 3.26.$$

Since this value of z_c exceeds the critical value of 1.96 shown in Fig. 8.17, the null hypothesis must be rejected in favor of the alternative hypothesis. In

* The variance of the difference, $V[\bar{x}_1 - \bar{x}_2]$ (or $\sigma^2_{\bar{x}_1 - \bar{x}_2}$), can be derived as follows, assuming that the samples are drawn independently:

$$\begin{aligned} V[\bar{x}_1 - \bar{x}_2] &= V[\bar{x}_1] + V[-\bar{x}_2] \\ &= V[\bar{x}_1] + (-1)^2 V[+\bar{x}_2] && \text{(By Formula 3.21)} \\ &= \frac{\sigma_1^2}{n_1} + \frac{\sigma_2^2}{n_2}, && \text{(Since } V[\bar{x}] = \sigma^2/n.\text{)} \end{aligned}$$

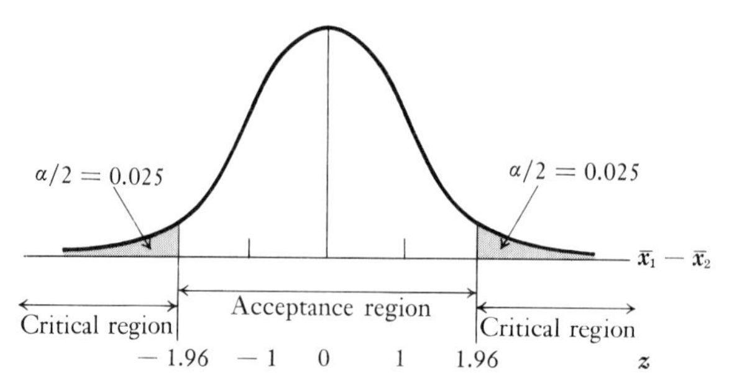

Fig. 8.17 Two-sided test on $\mu_1 - \mu_2$.

other words, the difference between these two means appears to be too large to be attributed entirely to chance. If the researcher in this example wants to report a p-value, then the appropriate probability is

$$2P(z \geq z_c) = 2P(z \geq 3.26) = 2(1 - 0.9994) = 0.0012.$$

The reported probability is thus 0.0012. For any α larger than 0.0012 the decision-maker should reject H_0.

8.7 TEST ON THE DIFFERENCE BETWEEN TWO MEANS (σ_1^2 AND σ_2^2 UNKNOWN, BUT ASSUMED EQUAL)*

The two-sample z-test described in Section 8.6 is appropriate only when both σ_1^2 and σ_2^2 are known. If σ_1^2 and σ_2^2 are unknown, then the t-distribution can be applied under certain circumstances. First, if both parent populations ($\boldsymbol{x}_1$ and $\boldsymbol{x}_2$) are normal, then the $\boldsymbol{t}$-distribution can be used. If $\boldsymbol{x}_1$ and $\boldsymbol{x}_2$ are not normal, but both n_1 and n_2 are reasonably large (say ≥ 15), then again the $\boldsymbol{t}$-distribution can be used. In addition, this two-sample $\boldsymbol{t}$-test assumes that the variances of the two populations are equal ($\sigma_1^2 = \sigma_2^2$), although there is a $\boldsymbol{t}$-test, which we shall describe shortly, that does not require this assumption. Fortunately, the $\boldsymbol{t}$-test is a fairly "robust" distribution (as was pointed out in Section 6.8), so that deviations from the above assumptions may not destroy the usefulness of this approach.

Using the *t*-Distribution

A $\boldsymbol{t}$-test for the difference between two means might seem to have the same format as Formula (8.3) with sample variances s_1^2 and s_2^2 substituted for the

* See reference [16] for a $\boldsymbol{t}$-test, with modified d.f., for the case when σ_1^2 may not equal σ_2^2.

population variances. The change here is due to the stipulation that $\sigma_1^2 = \sigma_2^2$, which means that both s_1^2 and s_2^2 must represent two estimates of the *same* population variance. In other words, because the two populations are assumed to have the same variance, the two sample variances are merely two separate estimates of this population variance. But if s_1^2 and s_2^2 differ, which of these two values should be used to estimate the unknown population variance? The answer is that a *weighted average* of s_1^2 and s_2^2 is the best estimate, where the weights applied are their respective degrees of freedom relative to the total number of degrees of freedom. This weighted average is:

Weighted average of s_1^2 and s_2^2:

$$\frac{(n_1 - 1)}{n_1 + n_2 - 2} s_1^2 + \frac{(n_2 - 1)}{n_1 + n_2 - 2} s_2^2 = \frac{(n_1 - 1)s_1^2 + (n_2 - 1)s_2^2}{n_1 + n_2 - 2},$$

where each sample variance has $(n - 1)$ degrees of freedom, and the total number of degrees of freedom is $(n_1 - 1) + (n_2 - 1) = n_1 + n_2 - 2$. The reader should verify that the two "weights" shown above will always sum to 1.0.

Let us now rewrite Formula (8.3) letting σ^2 denote the two equal variances (that is, $\sigma_1^2 = \sigma_2^2 = \sigma^2$). Factoring σ^2 out of the denominator of (8.3) permits us to rewrite Formula (8.3) as follows:

Formula 8.3, assuming that $\sigma_1^2 = \sigma_2^2 = \sigma^2$:

$$\frac{(\bar{x}_1 - \bar{x}_2) - (\mu_1 - \mu_2)}{\sqrt{\sigma^2[(n_1 + n_2)/n_1 n_2]}}.$$

Substituting our weighted-average formula for σ^2 yields the appropriate ***t***-test statistic:

*Two-sample **t**-test statistic for testing $\mu_1 - \mu_2$, assuming equal variances in the normal parent populations:*

$$\boldsymbol{t}_{n_1 + n_2 - 2} = \frac{(\bar{x}_1 - \bar{x}_2) - (\mu_1 - \mu_2)}{\sqrt{\left(\frac{(n_1 - 1)s_1^2 + (n_2 - 1)s_2^2}{n_1 + n_2 - 2}\right)\left(\frac{n_1 + n_2}{n_1 n_2}\right)}}. \quad (8.4)$$

This ***t***-variable has $(n_1 + n_2 - 2)$ degrees of freedom.

The ***t***-test based on Formula (8.4) can be illustrated by applying it to the previous example on starting salaries of college graduates in New York City and Chicago. Let us now assume that much smaller samples were taken, namely $n_1 = 11$ and $n_2 = 9$, that the population variances are unknown but can be assumed to be equal, and that the parent populations are normal.

Suppose that the decision-maker has set $\alpha = 0.05$, and established:

$$H_0: \mu_1 - \mu_2 = 0 \qquad \text{versus} \qquad H_a: \mu_1 - \mu_2 \neq 0.$$

The results of the two samples are given below:

Sample 1	Sample 2
$n_1 = 11$	$n_2 = 9$
$\bar{x}_1 = 20{,}400$	$\bar{x}_2 = 20{,}300$
$s_1^2 = 50{,}000$	$s_2^2 = 40{,}000$

Substituting these values into Formula (8.4) we obtain the following computed t-value:

$$t_c = \frac{(20{,}400 - 20{,}300) - (0)}{\sqrt{\left(\dfrac{10(50{,}000) + 8(40{,}000)}{11 + 9 - 2}\right)\left(\dfrac{11 + 9}{11 \times 9}\right)}}$$

$$= \frac{100}{95.9} = 1.04.$$

Since this calculated t-value lies between the two critical values for a two-sided test when $\alpha = 0.05$ and $v = 18$ $[t_{0.025,18} = \pm 2.101]$, the decision-maker should accept H_0. The difference of \$100 between $\bar{x}_1$ and $\bar{x}_2$ in these small samples is not large enough to conclude that mean starting salaries differ. The appropriate p-value to be reported has the following form:

$$2P(\boldsymbol{t} \geq t_c) = 2P(\boldsymbol{t} \geq 1.04).$$

This probability cannot be determined exactly, but from Table VII we do know that

$$0.10 > P(\boldsymbol{t} \geq 1.04) > 0.05.$$

Doubling these values we obtain the expression to be reported:

$$0.20 > 2P(\boldsymbol{t} \geq 1.04) > 0.10.$$

Thus, for any α less than 0.10, the decision-maker should accept H_0.

Again, we must point out that many statistics texts recommend using a normal approximation to the $\boldsymbol{t}$-distribution when $n_1 + n_2 - 2$ is relatively large (such as ≥ 30). We have not emphasized that process (although it is quite straightforward),* but rather remind the reader that the $\boldsymbol{t}$-distribution is *always* the correct distribution when σ_1 and σ_2 are unknown. If the exact value of

* If $n_1 + n_2 - 2$ is "large," the t-values will be only slightly larger than the values from the normal distribution; hence the t-values can be approximated by the z-values.

$n_1 + n_2 - 2$ for a particular problem cannot be found in Table VII, use the row with the number of degrees of freedom closest to this value.

The Case of Matched-Pairs Samples

There is another way to test for significant differences between two samples involving small values of n that does *not* use the assumption that the variances of the two populations are equal. In this test it is necessary that the observations in the two samples be collected in the form called *matched pairs.* That is, each observation in the one sample must be paired with an observation in the other sample in such a manner that these observations are somehow "matched" or related, in an attempt to eliminate extraneous factors that are not of interest in the test. In our test for differences in starting salaries, for example, the graduates sampled in the New York area may be considerably older than the graduates sampled in the Chicago area, or they may represent a substantially different mix of undergraduate majors. If such differences are not of interest, then they can be systematically eliminated by selecting a sample in which each person in the New York area is carefully matched—in terms of age, sex, undergraduate major, or any other criterion—with a person in the Chicago area. One of the most common methods of forming matched pairs is to let a subject "serve as his own control," in which case the person is matched with himself at different points in time, or in a "before-and-after" treatment study.

If the observations can be collected in the form of matched pairs, then a **t**-test for differences between the two samples can be constructed on the basis of the *difference score* for each matched pair. This score is calculated by subtracting the score or value associated with the one person or object in each pair from the score of the other person or object in that pair. The **t**-test assumes that these difference scores are normally distributed and independent. If we denote the average difference in scores between the two populations by the capital Greek letter delta Δ, then the hypothesis being tested is $H_0{:}\Delta = k$ where k is the hypothesized average difference ($k = 0$ in a test of significance). If the values from the two matched samples are denoted by x_i and y_i, and the difference score between matched pairs by $D_i = x_i - y_i$, then the average of D_i is our best estimate of Δ. The sample values of D_i can be used in a test similar to a one-sample test on a mean with σ unknown and the population assumed to be normal. The sample standard deviation of the difference scores, s_D, and the sample mean of the difference scores, $\bar{D}$, are used to form:

Test statistic for matched pairs:

$$t_{n-1} = \frac{\bar{D} - \Delta}{s_D/\sqrt{n}}, \tag{8.5}$$

where n is the number of matched pairs in the two samples, and s_D is computed as follows:

$$s_D = \sqrt{\frac{\sum(D_i - \bar{D})^2}{n - 1}} \quad \text{or} \quad \sqrt{\frac{\sum D_i^2 - n\bar{D}^2}{n - 1}}.$$

Suppose, for example, that the observations in Table 8.2 represent the starting salaries for 10 matched pairs from New York City and Chicago. The null and alternative hypotheses to be tested are

$$H_0{:}\Delta = 0 \quad \text{versus} \quad H_a{:}\Delta \neq 0.$$

From Table 8.2 we know that $\bar{D} = \$100$. Using the ten values of $D_i = x_i - y_i$ in the last column of Table 8.2 and Formula (8.5) it is possible to show that $s_D = 240.0$. Substituting these values into Formula (8.5) we obtain the following computed t-value:

$$t_c = \frac{\bar{D} - \Delta}{s_D/\sqrt{n}} = \frac{100 - 0}{240/\sqrt{10}} = 1.32.$$

For $(n - 1) = 9$ degrees of freedom and an α-level of 0.05, the critical values for a two-sided alternative are $t_{0.025,9} = \pm 2.262$. Since t_c falls between these values, the decision should be to accept H_0. In other words, the differences in this sample are not large enough to reject the assumption that starting salaries are equal. If we wish to report a p-value, the appropriate probability is

$$2P(\mathbf{t} \geq t_c) = 2P(\mathbf{t} \geq 1.32).$$

Table 8.2 Data for Matched-Pairs Test

Pair	x_i New York City	y_i Chicago	$D_i = x_i - y_i$ Difference
1	\$20,400	\$20,000	\$400
2	19,800	19,900	−100
3	19,700	20,000	−300
4	20,500	20,400	100
5	20,600	20,600	0
6	20,100	19,900	200
7	20,300	20,400	−100
8	19,900	19,700	200
9	20,400	20,300	100
10	20,700	20,200	500
Sum	202,400	201,400	1000
Mean	20,240	20,140	100

The value of $P(t \geq 1.32)$ is larger than 0.10 and less than 0.25, as shown in Table VII; by doubling these probabilities we get,

$$0.50 > 2P(t \geq 1.32) > 0.20.$$

The statistician now can report that a difference in mean starting salaries as large as \$100 will occur relatively frequently (more than 20% of the time) when the two cities do, in fact, have equal starting salaries.

A matched pairs *t*-test generally involves considerably more time and effort than does the use of Formula (8.4) for testing means. This time and effort is necessary, however, if there are systematic differences that must be eliminated between the two populations.

Define: most powerful critical region, power function, $\sigma_{\bar{x}_1 - \bar{x}_2}$, weighted average of s_1^2 and s_2^2, matched pairs.

Problems

8.14 Explain why, in testing $H_0{:}\mu = 20$ vs. $H_a{:}\mu \neq 20$, we can calculate a single value of α for a given critical region and sample size, but we cannot calculate a single value of β.

8.15 Reconsider Problem 8.2, assuming that the true weight of the loaves of bread in this population is 23.90 oz. What is the value of β if $\alpha = 0.01$? What is the power of the test if $\mu = 23.90$? Using words a nonquantitative decision-maker might understand, interpret the meaning of the values of α and β in this problem.

8.16 Use Problem 8.3(c) with the value of $\alpha = 0.05$.

a) What is the value of β if the true value of μ is \$4900? What is the power of the test?

b) Repeat part (a), assuming that the true values are \$4800, \$4700, \$4600, and \$4500. Construct a graph of the power function, using these values.

8.17 Sketch the power function for Problem 8.8, assuming that the true values of μ are 0.90, 0.95, 1.00, 1.05, and 1.10.

8.18 Return to Problem 8.11, in which the sample size was $n = 64$ and $\alpha = 0.05$.

a) Find β if $\mu = 51$.

b) Show that for the hypothesis and critical region presented in part (a), the values of both α and β will decrease if n is increased to 100.

8.19 Suppose that you are given the following probability density function:

$$f(x) = \begin{cases} 1/\theta & \text{for } 0 \leq x \leq \theta. \\ 0 & \text{otherwise.} \end{cases}$$

a) You decide to test the null hypothesis $H_0{:}\theta = 1$ against the alternative hypothesis $H_a{:}\theta = 2$ by means of a single observation. What is the value of α and β if you select the interval $x \leq 0.5$ as the critical region? Sketch this density function under both the null and alternative hypotheses, and indicate the critical region on this graph.

b) What is the value of α and β if you select $x \leq 0.75$ as the critical region?

c) Which of these two critical regions would be more appropriate if a Type II error is more serious than a Type I error?

8.20 Assuming that, in testing $H_0: \mu = 20$ vs. $H_a: \mu \neq 20$, you decide on the critical region $\bar{x} \leq 15$ and $\bar{x} \geq 25$. Assume $\boldsymbol{x}$ is normally distributed, $\sigma^2 = 25$, and the following four random values are observed: 9, 20, 15, 11.

a) Would you accept or reject H_0?

b) What value of α is assumed here?

c) What probability value would you report?

d) What would be the appropriate critical region for this problem if $\alpha = 0.005$? Would you accept or reject H_0?

8.21 Redo Problem 8.20 assuming that $H_a: \mu < 20$, and the critical region is $\bar{x} \leq 14$.

8.22 A firm that packages deluxe ornamental matches for fireplace use designed a process to place 18 matches in each box. The process was started and allowed to produce 400 boxes. A sample of 16 boxes was then drawn. On the basis of this sample, the number of matches per box averaged 17, while the standard deviation calculated was 2. Would a one-sided test indicate acceptance of the null hypothesis with a mean of 18 if alpha were set at 0.05? Assume that the population is normal.

8.23 Rework Problem 8.10, assuming that σ is unknown and that the population is normal.

8.24 Two types of new cars are tested for gas mileage. One group, consisting of 36 cars, averaged 24 miles per gallon of gas, while the other, consisting of 72 cars, averaged 22.5 miles per gallon.

a) What null and alternative hypotheses would you establish to determine whether or not the mileage differs between the two types of cars?

b) What test statistic is appropriate if $\sigma_1^2 = 1.5$ and $\sigma_2^2 = 2.0$?

c) Construct the appropriate critical values, assuming $\alpha = 0.01$. Would you accept H_0 or H_a?

d) What p-value would you report if α is not specified?

e) Is it necessary in this problem to assume that $\boldsymbol{x}_1$ and $\boldsymbol{x}_2$ are normally distributed? Explain.

8.25 An FBI statistic rates college campuses according to the average number of crimes per month. Assume that this statistic is based on a random sample of 12 different months drawn from many years of data on crimes during the school year.

a) If one major state university (population 1) claims its average number of crimes is less than that of another state university (population 2), what null and alternative hypotheses would you establish?

b) What test statistic is appropriate if $\sigma_1^2 = 400$, $\sigma_2^2 = 800$, and the parent populations are normal?

c) What critical value is appropriate if $\alpha = 0.02$?

d) Assume the $n_1 = n_2 = 12$ sample values used in part (c) yield the following data: $\bar{x}_1 = 370$ and $\bar{x}_2 = 400$. Would you accept H_0 or H_a?

e) A number of large schools have complained (bitterly) about this procedure for reporting crimes. Explain why these complaints may be justified.

8.26 Suppose that an underground newspaper is sold on two college campuses. A random selection of weekly sales figures provides the following data:

Sample	Sample Size	Mean	Standard Deviation
Campus A	10	123	15
Campus B	6	108	$\sqrt{185}$

Based on sales averages, test whether the newspaper sells more at A than at B. Use alpha = 0.05 and a one-sided alternative hypothesis. What probability value would you report? Assume that the parent populations are normally distributed.

8.27 A particular student tells you that you get more ice cream in a 40¢ cone at the Dairy Bar than at the IC-Shoppe. You buy cones at both locations at random times during a seven-week period, and measure the ice-cream content as follows:

	Dairy Bar	IC-Shoppe
Number of cones	8	10
Average (oz.)	7	5.5
Standard deviation (oz.)	1	$\sqrt{1.7}$

Perform an appropriate one-sided test, at the 0.01 level of significance, to determine whether the student's claim is correct. Assume that the parent populations are normal.

8.28 Two sections of a statistics course took the same final examination. A sample of nine scores was randomly drawn from Section A, and a sample of four was randomly drawn from Section B. These scores are arranged in ascending order:

Section A:	65,	68,	72,	75,	82,	85,	87,	91,	95
Section B:	50,	59,	71,	80					

a) What assumptions about the parent population are necessary if we wish to use the t-test to test for significant differences between the means of these sections?

b) At the 0.05 level of significance, can $H_0: \mu_1 - \mu_2 = 0$ be rejected in favor of $H_a: \mu_1 - \mu_2 \neq 0$?

8.29 Two swimmers are recognized throughout the world as among the best in a certain event. In a series of independent practice trials, they achieve the following times in seconds. On the basis of these trials can you detect, with 99 percent confidence, any difference between the performances of these two swimmers? Assume that the times are normally distributed.

Swimmer A	Swimmer B
30.7	31.1
31.2	31.2
31.3	31.4
30.9	31.6

8.30 The monthly sales of birdcages with automatic feeders are recorded in two matched sales districts for five months selected at random. Test at the five-percent significance level to see if sales are significantly higher, on the average, in District B. Assume that the parent populations are normal, and use a matched pairs test.

District A	District B
12	10
16	20
10	16
14	18
8	16

8.31 A student has a new automobile and is trying to determine statistically which of two different gasolines is better. This student measures gas mileage for eight consecutive tankfuls, using Brands A and B alternately. The mileage difference ($B - A$) after every two tankfuls is then computed. The average difference is 2.5 mpg and the standard deviation of the differences is $s_D = 2.0$ mpg. Since the dealer selling Brand A is a friend, this student will switch to Brand B only if 90-percent sure that the Brand B mileage will be at least one mile per gallon better. What should the student do? Assume that the D-values are normally distributed.

8.32 You are considering chartering a boat for some deep-sea fishing. You usually charter with Capt. Mike Ketchum, but you have heard that Capt. Joe Hookum is better at finding the big ones during their feeding hour. You go down to the docks and observe the daily catch from the two different boats. In comparing the six largest fish caught on each boat, you find that the fish from Capt. Joe's boat are, on the average, nine pounds heavier, with a standard deviation (from this average difference) of $s_D = 6$ lbs. Test, with 95-percent confidence, whether or not you should switch from your old salty buddy, Capt. Mike, in order to catch heavier fish. Assume that the differences are normally distributed.

8.33 A county agent experiments with eight acres of land. Half the acreage is treated with fertilizer x, and half is treated with fertilizer y. The average difference in yield between paired acres is ten bushels. The standard deviation of differences is 4. Would you conclude that there is a significant difference of yield between the two differently fertilized tracts? You should have 99-percent confidence, if you do find a difference. Assume that the differences in yields are normally distributed.

8.34 Ten male and female students were matched according to age and driving experience. Their scores on a written test concerned with driving are shown in Table 8.3.

a) Use a matched-pairs **t**-test to determine whether there is a significant difference between the scores of males and females. Assume that $\alpha = 0.05$.

Table 8.3

Pair	Male	Female
1	95	93
2	90	95
3	80	88
4	75	70
5	91	91
6	85	83
7	100	94
8	62	80
9	81	92
10	88	100

b) Use Formula (8.4) to determine whether a significant difference exists between these scores.

c) Is the p-value for part (a) the same as that for part (b)? Explain why or why not.

8.35 An IQ test was given to two different groups of high-school students. The first group resulted in a mean IQ of 112 with a standard deviation of 6, and the second group had an average IQ of 114 with a standard deviation of 4. If the first group was of size $n = 60$ and the second group of size $n = 40$, at what probability value can the null hypothesis $H_0: \mu_1 - \mu_2 = 0$ be rejected in favor of $H_a: \mu_1 - \mu_2 \neq 0$? At what level can H_0 be rejected in favor of $H_a: \mu_1 - \mu_2 < 0$?

8.36 Nine students in each of two statistics classes were carefully matched according to age, sex, grade point average, and SAT scores. The two groups were then taught statistics by two different methods (but with the same instructor). Their scores on the (same) final exam are shown in Table 8.4.

Table 8.4

Pair	Group A	Group B
1	76	83
2	92	91
3	83	72
4	84	93
5	65	75
6	71	87
7	60	75
8	81	79
9	72	74

a) Use a matched-pairs test to accept or reject $H_0: \mu_A - \mu_B = 0$ vs. $H_a: \mu_A - \mu_B < 0$ at $\alpha = 0.05$. What probability value would you report?

b) Repeat part (a), assuming that $H_a: \mu_A - \mu_B \neq 0$.

8.37 Rework Problem 8.6, but this time assume that only 5000 ft of rope are produced each day. (*Hint:* Assume that the population size is now 5000/50 = 100 lengths of rope, and use the finite population correction factor.)

8.38 Under what circumstances is the t-distribution appropriate for testing hypotheses on μ?

8.39 Rework Problem 8.3 assuming that σ is unknown and that the population is normal. Calculate s from the four sample values.

8.40 A city council has received complaints that a local food chain does not provide enough parking spaces. The owner has countered that the average number of cars parked at the store is generally less than 24, the number of parking spaces. The following nine random observations were drawn: 25, 17, 18, 22, 21, 27, 19, 15, 25.

a) Test the null hypothesis $H_0: \mu = 24$ against the alternative hypothesis $H_a: \mu \neq 24$, assuming that $\alpha = 0.05$, and that the population is normal. What probability value would you report?

b) What p-value would you report if $H_a: \mu < 24$?

8.41 A power shovel was designed to remove 31.5 cubic feet of earth per scoop. On a test run, some 25 sample scoops were made; the mean of the samples was 29.3 cubic feet. The standard deviation, as derived from the sample information, was three cubic feet. Test at the 99-percent level of confidence whether the design specifications for this equipment should be revised on the basis of the sample information. Describe the two types of error possible in this test. Assume that the population is normal.

8.42 a) The daily output of a certain department within an industrial plant is presumed to be normally distributed and has a scheduled average of 85 units. Twenty-five days are selected at random, and the output for each day is observed. The average output calculated from this sample is 81 units, and the standard deviation is 9 units. Test with 99-percent confidence whether or not the average output is different from that scheduled.

b) What p-value would you report?

c) Explain the meaning of the beta risk for this test.

8.8 CALCULATING α AND β USING THE BINOMIAL DISTRIBUTION (OPTIONAL)

One objective of this section is to present another example of the process of hypothesis testing. In this case we will use a discrete probability function (the binomial) rather than the continuous function (the z) used in Sections 8.2 and 8.3. A second objective is to illustrate how *both* α and β can be calculated in a problem where the critical region is given, and how changes in the critical region affect the size of α relative to β.

To illustrate the calculation of α and β in a binomial problem, we again return to the production-process example of Chapter 4. In that problem the process needed minor repairs if $p = 0.10$ and major repairs if $p = 0.25$. Now, suppose we let the null hypothesis be $H_0{:}p = 0.10$ and the alternative hypothesis be $H_a{:}p = 0.25$. To decide between these hypotheses we decide to take a sample of $n = 20$ items from the process; furthermore, we decide to accept H_0 if the number of defectives ($\boldsymbol{x}$) is less than four ($\boldsymbol{x} < 4$) and to accept H_a if $\boldsymbol{x} \geq 4$. The value $\boldsymbol{x} = 4$ in this problem is the *critical value*, since it separates those outcomes leading to acceptance of H_0 from those leading to acceptance of H_a.

We now have enough information to calculate α and β. Since α is the probability of accepting $H_a{:}p = 0.25$ when $H_0{:}p = 0.10$ is true, and we accept H_a when $\boldsymbol{x} \geq 4$, $\alpha = P(\boldsymbol{x} \geq 4 \mid p = 0.10)$. Similarly, since β is the probability of accepting H_0 when $H_a{:}p = 0.25$ is true, and H_0 is accepted when $\boldsymbol{x} < 4$, $\beta = P(\boldsymbol{x} < 4 \mid p = 0.25)$. Both these probabilities can be calculated from the binomial values under $n = 20$ in Table I. In fact, we calculated their value in Chapter 4 in our discussion of binomial probabilities. Table 4.2 from Chapter 4, which is reproduced here (as Table 8.5), with the notation of this chapter, gives these probabilities. Thus, for this particular critical value, the probability is $\alpha = 0.1331$ of accepting $H_a{:}p = 0.25$ when $H_0{:}p = 0.10$ is true (a Type I error) and $\beta =$

Table 8.5 (Table 4.2 Reproduced)

Binomial Probabilities for Production Example

No. of Defectives (x)		Decision Rule	If $p = 0.10$: $P(x) = \binom{20}{x}(0.10)^x(0.90)^{20-x}$	If $p = 0.25$: $P(x) = \binom{20}{x}(0.25)^x(0.75)^{20-x}$
0		Minor repairs	0.1216	0.0032
1		Minor repairs	0.2702	0.0211
2		Minor repairs	0.2852	0.0669
3	Critical	Minor repairs	0.1901	0.1339
				$\beta = 0.2251$ (sum of 0–3)
4	Value	Major repairs	0.0898	0.1897
5		Major repairs	0.0319	0.2023
6		Major repairs	0.0089	0.1686
7		Major repairs	0.0020	0.1124
8		Major repairs	0.0004	0.0609
9		Major repairs	0.0001	0.0271
10		Major repairs	0.0000	0.0099
11		Major repairs	0.0000	0.0030
12		Major repairs	0.0000	0.0008
13		Major repairs	0.0000	0.0002
14–20		Major repairs	0.0000	0.0000
			$\alpha = 0.1331$ (sum of 4–20)	
Sum			1.0000	1.0000

0.2251 of accepting $H_0\colon p = 0.10$ when $H_a\colon p = 0.25$ is true (a Type II error). Figure 8.18 is a graph of these probabilities.

We must emphasize that the values of α and β need not add up to one, as these two probabilities are not complementary. They are conditional probabilities based on different conditions. The value of α is conditional on the fact that H_0 is true, while the value of β is conditional on the fact that H_a is true. Thus, a one-unit change in α does not imply a corresponding one-unit change in β, or vice versa. However, since both α and β represent probabilities of events from the same decision problem, they are not independent of each other or of the sample size (n). When α is lowered, β normally rises, and vice versa (if n remains unchanged). The values of α and β also depend on the particular critical value selected (which was $x = 4$ in this case).

It is important to recognize from our production-process example that the value of α and β could not have been calculated if we had not first specified the

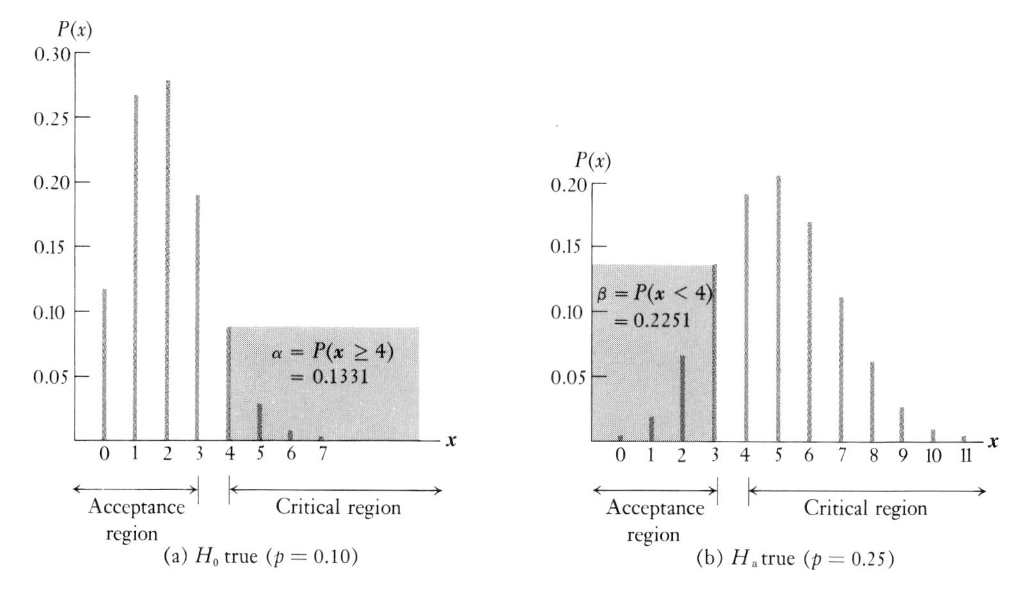

Fig. 8.18 Critical and acceptance regions to test $H_0:p = 0.10$ versus $H_a:p = 0.25$.

size of the sample and a decision rule for accepting or rejecting the null and alternative hypotheses. This will always be the case, for α cannot be calculated until the circumstances are specified under which H_a will be accepted, and β cannot be calculated until the circumstances which lead to the acceptance of H_0 are specified. Different decision rules and sample sizes will lead to different values of α and β. For instance, suppose we change the number of defectives required for a major adjustment from $x = 4$ to $x = 5$, in our production-process example. The value of α and β can again be calculated from Table 8.1, and in this case can be shown to equal

$$\alpha = P(\mathbf{x} \geq 5 | H_0 \text{ true}) = 0.0433$$

and

$$\beta = P(\mathbf{x} < 5 | H_a \text{ true}) = 0.4148.$$

The change in our decision rule has decreased α by 0.0898 (from 0.1331 to 0.0433), while increasing β by 0.1897 (from 0.2251 to 0.4148). The probability of a Type I error has thus decreased, and the probability of a Type II error has increased. Whether or not the trade-off is beneficial in this case, and in general how one goes about balancing the risks associated with a Type I error against the risks associated with a Type II error, are problems we shall discuss in the next section.

To illustrate the effect of sample size on the probability of making Type I and Type II errors, suppose, in the above example, that the sample size could be increased to $n = 50$. Usually a change in sample size means a change in the critical region. Let's rather arbitrarily choose $x \geq 10$ as the critical region for this sample of 50. That is, we will accept $H_a: p = 0.25$ if the number of defectives is greater than or equal to 10, and accept $H_0: p = 0.10$ if $x < 10$. From Table I in Appendix B (under $n = 50$) we can calculate α and β:

$$\alpha = P(\boldsymbol{x} \geq 10 | H_0: p = 0.10) = 0.0245,$$
$$\beta = P(\boldsymbol{x} < 10 | H_a: p = 0.25) = 0.1636.$$

The important point to note is that by increasing the sample size from $n = 20$ to $n = 50$, we have *decreased* the probability of making *both* types of errors. The trade-off in this case is between the increase in cost associated with increasing the sample size, and the reduction in the cost of making an error because α and β are now both smaller.

*Continuous Example for Calculating α and β (Optional)

As another example of the process of determining α and β, recall the probability distribution discussed in Chapter 5 to describe the time between the occurrences of a random variable, where the number of occurrences of this random variable per unit time follows the Poisson distribution. Under these conditions the probability distribution of the time between occurrences is given by the exponential distribution $f(x) = \lambda e^{-\lambda x}$, whose mean is $1/\lambda$. Assume for present purposes that the number of accidents at a busy traffic intersection has been found to follow the Poisson distribution, so that the time between these accidents has the exponential distribution. Over the past 15 years the number of accidents at this intersection has averaged four every 200 days, which means that the average time between accidents has been 50 days. Recently, after a serious accident at this intersection, a new stoplight was installed which, its manufacturers claim, will reduce the number of accidents to one-fourth (i.e., there will be only one accident every 200 days, on the average). To test this assertion, suppose we establish the null hypothesis that the mean time between accidents will not change with the installation of this light ($H_0: 1/\lambda = 50$), and the alternative hypothesis that the mean time between accidents will increase by a factor of four ($H_a: 1/\lambda = 200$).* If we decide to reject H_a if the next accident occurs within 150 days (the noncritical region), and to reject H_0 if the next accident occurs after 150 days (the critical region), what is the probability of a Type I or a Type II error? Letting $\boldsymbol{x}$ equal the number of days until the next accident, the value of α is given by the probability that $\boldsymbol{x}$ exceeds 150 when the null hypothesis holds true. This probability can be calculated by integrating the exponential function $f(x) = \lambda e^{-\lambda x}$ from 150 to infinity. The value of λ in this case must equal 0.02, since H_0 specifies that $1/\lambda = 50$; hence, $\lambda = 0.02$.

* It might, under these circumstances, be more realistic to use a composite alternative hypothesis, such as $H_a: 1/\lambda > 50$. This approach, however, creates problems in calculating the probability of a Type II error which will be discussed shortly.

Under $H_0: 1/\lambda = 50$ (*or* $H_0: \lambda = 0.02$)

$$\alpha = \int_{150}^{\infty} 0.02e^{-0.02x}\,dx = 1 - \int_{0}^{150} 0.02e^{-0.02x}\,dx$$
$$= 1 - \left[-e^{-0.02x}\right]_0^{150} = e^{-3} = 0.0498.$$

The value of β can be calculated in a similar fashion by integrating the exponential function $f(x) = 0.005e^{-0.005x}$ from 0 to 150 (note that λ must equal 0.005 in determining β, since the alternative hypothesis specifies that $1/\lambda = 200$):

Under $H_a: 1/\lambda = 200$ (*or* $H_a: \lambda = 0.005$)

$$\beta = \int_{0}^{150} 0.005e^{-0.005x}\,dx = \left[-e^{-0.005x}\right]_0^{150} = 1 - e^{-0.75} = 0.5277.$$

Thus, by using a critical region of $\boldsymbol{x} > 150$ there is a 4.98 percent chance that the manufacturer's claim will be substantiated when it should not be, and a 52.77 percent chance that the sample will incorrectly indicate the manufacturer's claim to be false. Figure 8.19 shows the probability distribution of $\boldsymbol{x}$ under the null and alternative hypotheses in this example. As before, different critical regions will lead to different values of α and β.

Fig. 8.19 Critical and acceptance regions to test $H_0: \lambda = 0.02$ versus $H_a: \lambda = 0.005$.

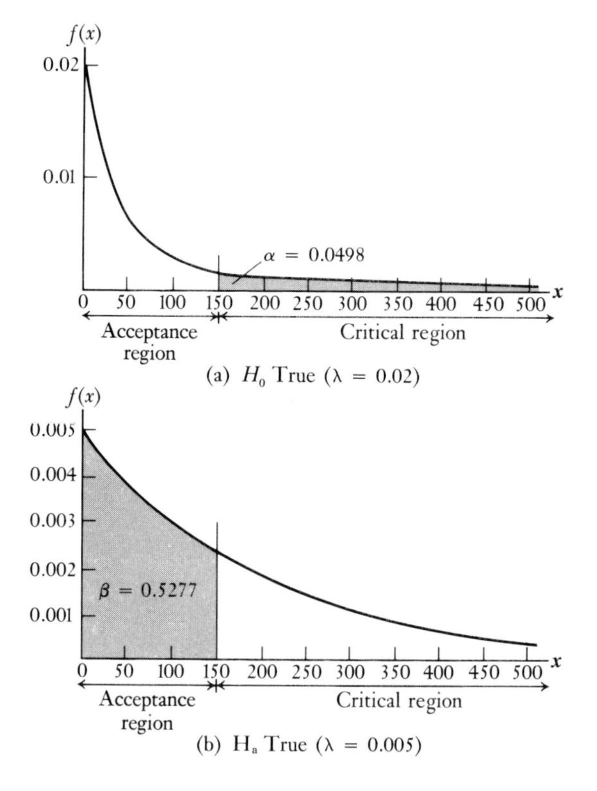

8.9 BALANCING THE RISKS AND COSTS OF MAKING A WRONG DECISION (OPTIONAL)

The practice of establishing α at some predetermined level and then finding the most powerful critical region of size α (i.e., the critical region minimizing the size of β) is widely used in current research. Most modern statisticians, however, for a number of reasons, do not condone the indiscriminate use of this practice. First, as we mentioned earlier, it is no longer considered necessary or even good statistical practice to always establish a null hypothesis representing a theory that the researcher is hoping to disprove. Such a convention may give an artificiality to the research which distorts its true purpose or hinders the effectiveness of the testing procedure. If the null hypothesis no longer represents "current theory" on an issue, then there is little justification for focusing attention primarily on the probability of a Type I error. A much more important objection to the approach of selecting a critical region by first establishing a value of α is that many researchers set α at some arbitrary level, such as $\alpha = 0.05$, or $\alpha = 0.01$, without considering the associated level of β. This practice certainly does not represent an optimal approach to balancing the risks of an incorrect decision, for it ignores the risks associated with a Type II error. While it is not incorrect to set α at some predetermined level, this value should be established by considering the risks of *both* Type I and Type II errors.

We now focus on the important problem of how to choose the "best" critical region to balance off the risks and costs associated with making a Type I error against those associated with making a Type II error. As we pointed out in Chapter 6, the objective of sampling can be stated as one of minimizing the cost of making an incorrect decision (including sampling costs, which will be considered later). Unfortunately, in many circumstances there may be no easy way even to determine what these costs are, much less to try to balance them. In medical research, for example, it may be difficult if not impossible to assess the costs associated with an incorrect decision involving a new drug or surgical technique. But even if these costs could somehow be assessed, how does one go about balancing costs which may include pain, suffering, and even loss of life? On the other hand, if it is possible to identify the relevant costs and to express them in terms of some comparable basis, such as dollars, then we may be able to balance these costs quite explicitly.

An Example of Cost Balancing

The following example is given to describe the process of finding the critical region which minimizes the expected cost of making an incorrect decision. It uses the same quality-control problem of Section 8.8. It should be noted that for the first part of this example we assume the sample size to be fixed at $n = 20$; later, we shall relax this assumption and consider a sample of size $n = 50$. Suppose that all possible costs associated with an incorrect decision (i.e., loss

		Percent defectives	
		$p = 0.10$ H_0 is true	$p = 0.25$ H_0 is false
Decision	Accept H_0 make minor adjustment	Correct decision	\$200
	Reject H_0 make major adjustment	\$500	Correct decision

Fig. 8.20 Costs of making an incorrect decision.

in profit, goodwill, etc.) are those shown in Fig. 8.20. A correct decision is assumed to result in no loss in profit or goodwill.

For each possible critical region we can calculate the expected cost (per sample of 20) which will result if the process is (1) producing 10 percent defectives or (2) producing 25 percent defectives. These expected costs are calculated by multiplying the probability of making each type of error *times* the cost of making that error. We have calculated these values for two critical regions, the one used in Fig. 8.18 of $x \geq 4$, and an alternate one, $x \geq 5$, which gives a smaller α at the expense of a larger β.

a) *Critical region* $x \geq 4$ ($\alpha = 0.1331$ and $\beta = 0.2251$):
   1. $p = 0.10$:
      $$\begin{aligned}\text{Expected cost} &= P(\text{Type I error}) \times \text{Cost of a Type I error}\\ &= 0.1331(\$500) = \$66.55.\end{aligned}$$
   2. $p = 0.25$:
      $$\begin{aligned}\text{Expected cost} &= P(\text{Type II error}) \times \text{Cost of a Type II error}\\ &= 0.2251(\$200) = \$45.02.\end{aligned}$$

b) *Critical region* $x \geq 5$ ($\alpha = 0.0433$ and $\beta = 0.4148$):*
   1. $p = 0.10$:
      $$\begin{aligned}\text{Expected cost} &= P(\text{Type I error}) \times \text{Cost of a Type I error}\\ &= 0.0433(\$500) = \$21.65.\end{aligned}$$
   2. $p = 0.25$:
      $$\begin{aligned}\text{Expected cost} &= P(\text{Type II error}) \times \text{Cost of a Type II error}\\ &= 0.4148(\$200) = \$82.96.\end{aligned}$$

The expected costs given above are *conditional values*, in that each one was calculated by assuming that either $p = 0.10$ or $p = 0.25$. In order to be able to

* The reader can quickly calculate these probabilities of errors from Table 8.5 by moving the decision line down one row and recalculating the sum of terms then included in the brackets. Table I for the binomial with $n = 20$, $p = 0.25$, may, of course, be used directly to get the same values.

determine which of these critical regions is better, we need to know how often the process is expected to be producing 10 percent defectives, relative to the number of times it will be producing 25 percent defectives. Suppose that if an adjustment is required, the probability that it will be a minor adjustment is 0.70, while the probability that a major adjustment is necessary is 0.30. The *total* expected costs associated with each of the two critical regions can now be calculated by taking the product of the expected costs determined above times the probability that each of these costs will be incurred.

a) *Critical region* $x \geq 4$:

$$\begin{aligned} \text{Total expected cost} &= P(\text{Major adj. needed}) \times \text{Exp. cost of an error} \\ &\quad + P(\text{Minor adj. needed}) \times \text{Exp. cost of an error} \\ &= 0.30(\$45.02) + 0.70(\$66.55) \\ &= \$60.09. \end{aligned}$$

b) *Critical region* $x \geq 5$:

$$\begin{aligned} \text{Total expected cost} &= P(\text{Major adj. needed}) \times \text{Exp. cost of an error} \\ &\quad + P(\text{Minor adj. needed}) \times \text{Exp. cost of an error} \\ &= 0.30(\$82.96) + 0.70(\$21.65) \\ &= \$40.04. \end{aligned}$$

Thus, when the process is malfunctioning, the total expected cost for each sample of 20 equals \$60.09 if the critical region is $x \geq 4$ and \$40.04 if the critical region is $x \geq 5$. We leave it as an exercise for the reader to determine that $x \geq 5$ is, in fact, the *optimal* critical region for this problem, with total expected cost smaller than any other critical region.

Changing the Sample Size

Suppose in the above example that the sample size could have been increased to $n = 50$ at a cost of \$10. Our discussion of trade-offs between α, β, and n in Section 8.8 suggests that this increase in sample size can lead to a decrease in both α and β if an appropriate critical region is used. The question is whether or not the decreased probability of making an error is worth the increased sampling costs of \$10. In order to answer this question, let us select a critical region for the new situation (with $n = 50$), and then determine α and β and the total expected cost for this critical region. This total expected cost can then be compared with the preceding optimal of \$40.04.

Since n has increased 2.5 times from 20 to 50, let us arbitrarily try a new boundary value which is 2.5 times the previous old one ($x = 4$); that is, $x = 10$. [We leave it as a good exercise for the reader to determine if a better critical region than $x \geq 10$ could be found—perhaps by using $x \geq 12$ or $x \geq 13$ (which are approximately 2.5 times the previous optimal value of $x = 5$).] The probability of observing a specific number of defectives when $n = 50$ under the two hypotheses, $H_0{:}p = 0.10$ and $H_a{:}p = 0.25$, is shown in Table 8.6. From these values the probabilities of Type I and Type II errors are seen to be $\alpha = 0.0245$ and $\beta = 0.1636$. We thus see that increasing n from 20 to 50 has *reduced* both α and β (see Table 8.6).

Critical region $x \geq 10$:

1. $p = 0.10$:

$$\begin{aligned}\text{Expected cost} &= P(\text{Type I error}) \times \text{Cost of a Type I error}\\ &= 0.0245(\$500) = \$12.25.\end{aligned}$$

2. $p = 0.25$:

$$\begin{aligned}\text{Expected cost} &= P(\text{Type II error}) \times \text{Cost of a Type II error}\\ &= 0.1636(\$200) = \$32.72.\end{aligned}$$

$$\begin{aligned}\text{Total Expected Cost} &= P(\text{Major adj. needed}) \times \text{Exp. cost of an error}\\ &\quad + P(\text{Minor adj. needed}) \times \text{Exp. cost of an error}\\ &= 0.30(\$32.72) + 0.70(\$12.25)\\ &= \$26.58.\end{aligned}$$

Table 8.6 Determining α and β When $n = 50$ and the Critical Region Is $x \geq 10$

x	Decision	If $p = 0.10$, H_0 is True: $P(x) = \binom{50}{x}(0.10)^x(0.90)^{50-x}$		Region		If $p = 0.25$, H_a is True: $P(x) = \binom{50}{x}(0.25)^x(0.75)^{50-x}$
0	Accept H_0, make minor adjustment	0.0052		Noncritical region	$\beta = 0.1636$	0.0000
1		0.0286				0.0000
2		0.0779				0.0001
⋮		⋮				⋮
8		0.0643				0.0463
9		0.0333				0.0721
10	Reject H_0, make major adjustment	0.0152	$\alpha = 0.0245$	Critical region		0.0985
11		0.0061				0.1194
12		0.0022				0.1294
⋮		⋮				⋮
24		0.0000				0.0002
25		0.0000				0.0001
26–50		0.0000				0.0000
Sum		1.0000				1.0000

Thus, if we have to choose between a sample of 20 and critical region $x \geq 5$ (with $\alpha = 0.0433$ and $\beta = 0.4142$), in which the costs will average \$40.04, and a sample of 50 and critical region $x \geq 10$ (with $\alpha = 0.0245$ and $\beta = 0.1636$), in which the costs will average \$26.58 for the incorrect decisions and \$10.00 for the additional observations, it would be better to take the larger sample. It may be, of course, that some other critical region will be even better than $x \geq 10$, or that some other sample size gives a lower expected cost. Given information on the cost of all possible sample sizes, the "optimal" sample size and its associated critical region could be determined for this problem. (Again, parts of this task are left as an exercise for the reader.) In Chapter 9 we shall return to an extended version of this problem and study in more detail the question of sample size.

8.10 TEST ON THE BINOMIAL PARAMETER *p*

The population parameter being examined in a test of hypothesis need not always be the population mean μ. In many situations, the parameter in question is the *proportion* of observations having a certain attribute. When the observations are independent of one another (i.e., they are randomly selected with replacement) and the attribute of interest either occurs or does not occur in each observation, then the appropriate test statistic follows the *binomial distribution.* A test involving an unknown binomial proportion p may be one-sided or two-sided, and the values for p may range from zero to one. For example, suppose the null hypothesis is

$$H_0:p = \tfrac{1}{2}.$$

This hypothesis may be tested against the two-sided alternative $H_a:p \neq \frac{1}{2}$, or perhaps against either of the following one-sided alternative hypotheses: $H_a:p < \frac{1}{2}$ or $H_a:p > \frac{1}{2}$. As we will demonstrate below, hypotheses involving the binomial parameter p can be tested either by using Table I in Appendix B, or by using the normal approximation to the binomial. The extensive example presented in Section 8.9 illustrated the use of Table I for tests involving the binomial parameter p. Another example is presented in this section.

Binomial Tests Using Table I

Recall from Chapter 6 that the best estimator of a binomial proportion p is the sample proportion $\boldsymbol{x}/n$. If the value of n used in making this estimate is one of the values of n presented in Table I in Appendix B (or available from some other source), then we can test hypotheses about p directly (i.e., without using a normal approximation). The only substantive difference between this approach and our previous tests involving μ is that now we are dealing with a discrete rather than a continuous test statistic.

The one difficulty in working with a discrete test statistic comes in Step 3, where it is necessary to define a critical region cutting off an area of size α. Unfortunately, when discrete probability spikes are involved, there may be no critical value(s) cutting off an amount exactly equal to α. For example, suppose that we are searching for a critical value of the variable $\boldsymbol{x}$ that cuts off exactly five percent of the binomial distribution. It may be that one value of $\boldsymbol{x}$ cuts off six percent of the distribution, and the next value of $\boldsymbol{x}$ cuts off four percent. In these cases the usual procedure is to pick the critical region cutting off the *smaller* area, in order to ensure that the probability of a Type I error is no larger than α. This procedure is equivalent to finding the extreme value of $\boldsymbol{x}$ that satisfies the following inequality:

$$P(\boldsymbol{x} \text{ or more successes in } n \text{ trials}) \leq \alpha. \tag{8.6}$$

For a one-sided upper direction test, the *smallest* value of $\boldsymbol{x}$ that satisfies Formula (8.6) will be used. For a one-sided lower direction test, the *largest* value of $\boldsymbol{x}$ such that $P(\boldsymbol{x} \text{ or less}) \leq \alpha$ should be used. For a two-sided test, two values of $\boldsymbol{x}$ would be needed: the smallest value of $\boldsymbol{x}$ such that $P(\boldsymbol{x} \text{ or more}) \leq \alpha/2$ and the largest value of $\boldsymbol{x}$ such that $P(\boldsymbol{x} \text{ or less}) \leq \alpha/2$.

Suppose that to test $H_0{:}p = 0.25$ against $H_a{:}p > 0.25$, a sample of size $n = 20$ is taken and α is assumed to be 0.05. Table I in Appendix B can be used to determine the smallest value of x cutting off probabilities equal to or less than 0.05. If $x = 13$, then 0.0002 is cut off; if $x = 12$, then $0.0008 + 0.0002 = 0.0010$ is cut off. Continuing in this fashion, if $x = 9$, the critical region is of size 0.0410, which is the smallest area less than 0.05. The next smallest value, $x = 8$, determines a critical region of size $0.0410 + 0.0609 = 0.1019$, which is too large. Hence, to assure an α no larger than 0.05, $x \geq 9$ is the appropriate critical region. Any sample that results in 9 or more successes in 20 trials leads to rejection of $H_0{:}p = 0.25$.

Using the Normal Approximation When *n* Is Large

Recall that it is not always convenient to work with binomial tables directly, because the arithmetic may be tedious, or tables may not be available for certain values of p. Fortunately, when the value of n is large and the value of p is not close to either zero or one, then the hypotheses about the binomial parameter p may be tested using the standardized normal distribution as the appropriate test statistic. An often-used rule of thumb is to assume that the standardized normal approximation can be used whenever $npq \geq 3$, where p is the value specified under H_0, $q = 1 - p$, and n is the sample size. Letting p_0

represent the null hypothesis value, the appropriate test statistic is:

Normal approximation to binomial test statistic:

$$z = \frac{(x/n) - p_0}{\sqrt{p_0 q_0/n}}. \tag{8.7}$$

This standardization is exactly like the one described in Section 5.4, Formula (5.7).*

To illustrate the test statistic in Formula (8.7), suppose that during the 1981 air traffic controllers strike President Reagan wanted an advance indication of whether or not the proposed contract would be rejected by a majority vote of the controllers. The hypotheses in this case might be $H_0: p \geq 0.50$ versus $H_a: p < 0.50$ where p is the proportion voting in favor of the proposal. However, recall that composite null hypotheses are somewhat more difficult to work with than simple null hypotheses. Making H_0 simple in this example is quite reasonable, as the concern here is really whether or not the proportion voting in favor of the proposal is less than 50%. Hence, let us make the hypotheses

$$H_0: p = 0.50 \qquad \text{versus} \qquad H_a: p < 0.50.$$

This change in H_0 does not weaken our test; it only means that we are taking the most conservative possible position for H_0.

Let us assume also that α is specified as 0.01, and a random sample of 100 controllers is taken. Since $np_0(1 - p_0) = 100(0.50)(0.50) > 3$, a normal approximation should be quite good. The appropriate normal critical value can be derived now by determining from Table III that $z_{0.01} = 2.33$. Since H_a is to the *left* of H_0, we use $-z = -2.33$ in Formula (8.7) in order to solve for the critical value of x/n.

$$z = \frac{(x/n) - p_0}{\sqrt{p_0 q_0/n}},$$

$$-2.33 = \frac{x/n - 0.50}{\sqrt{(0.50)(0.50)/100}}.$$

Thus,

$$\frac{x}{n} = 0.50 - 2.33\left(\frac{0.50}{10}\right) = 0.3835.$$

* The following test statistic is equivalent to Formula (8.7):

$$z = \frac{x - np_0}{\sqrt{np_0 q_0}}.$$

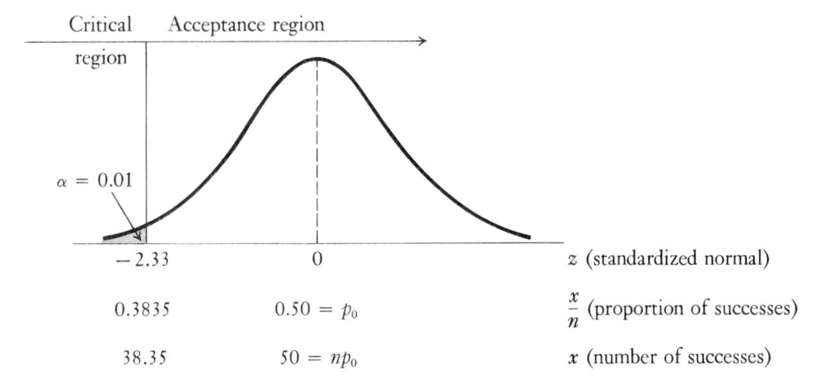

Fig. 8.21 Standardized normal approximation to the binomial test statistic.

The critical value for x/n is 0.3835. If desired, this equation could have been solved for x instead of x/n, since n is known to be 100. In this case the result would have been $x = 38.35$. Both critical values are shown in Fig. 8.21.

The reader may wonder if a correction for continuity is needed in Formula (8.7) before solving the last problem. The answer is no, because we were not using Formula (8.7) to approximate a *specific* binomial probability. Rather, we were using (8.7) to find the appropriate critical value for x/n as well as for x.

Now suppose we change our objective somewhat, and assume a sample has been taken and we wish to find z_c. *When using the normal distribution to approximate a specific binomial probability, the correction for continuity should be used for exactness.* However, if n is large and p is not close to either 0 or 1.0, the change in the results is not great.* In all such problems in this book, the reader may presume that the gain in exactness is negligible and would not change the conclusion of the decision-maker.

In our controllers strike example, let us suppose that 33 of the 100 controllers surveyed indicated that they would vote for the proposed contract, and 67 said that they would vote against it. That is, $x = 33$ and $x/n = 33/100 = 0.33$. This result is clearly in the critical region shown in Fig. 8.21, hence we reject $H_0: p = 0.50$ in favor of $H_a: p < 0.50$.

A calculated value of z for this result is

$$z_c = \frac{(x/n) - p_0}{\sqrt{p_0 q_0/n}} = \frac{0.33 - 0.50}{\sqrt{(0.50)(0.50)/100}} = -3.40.$$

* The appropriate formula to use if the correction for continuity is desired would be:

$$z_c = \frac{(x/n) - p_0 \pm 0.005}{\sqrt{p_0 q_0/n}}$$

where we

1) use $+0.005$ whenever $(x/n) - p_0$ is less than zero, and
2) use -0.005 whenever $(x/n) - p_0$ is greater than zero.

The reported p-value is thus

$$P(z < z_c) = P(z < -3.40) = 1 - 0.9997 = 0.0003.$$

We will leave as an exercise for the reader the verification that the same conclusion results in this problem from using Table I and the approach described earlier. If the correction for continuity had been used, the resulting values would have been $z_c = -3.30$ and p-value $= 0.0012$.

8.11 TEST ON THE DIFFERENCE BETWEEN TWO PROPORTIONS

In Section 8.10 we presented a one-sample test involving the binomial parameter p. The comparable test in this section is designed to distinguish between two population proportions, p_1 and p_2. One illustration of this type of situation is a politician interested in comparing the proportion of people who intend to vote for him or her in one city relative to the proportion in another city. Or perhaps the politician wants to look at how the proportion in a single city changes over time. A similar example is a business firm interested in comparing from one time to another the proportion of units of its product that are defective, or in comparing its proportion of defectives to that of another company.

The usual null hypothesis in testing two population proportions is that p_1 and p_2 are equal: that is,

$$H_0: p_1 - p_2 = 0.$$

The alternative hypothesis can take any of the forms used previously, such as

$$H_a: p_1 - p_2 \neq 0 \qquad \text{or} \qquad H_a: p_1 - p_2 > 0 \qquad \text{or} \qquad H_a: p_1 - p_2 < 0.$$

The test statistic appropriate in this case is the z-distribution, provided that n_1 and n_2 are both sufficiently large. If we let x_1/n_1 and x_2/n_2 represent the sample proportion from the first and second populations, respectively, then the appropriate formula for z_c is*

Calculated z-value for testing $p_1 - p_2 = 0$:

$$z_c = \frac{[(x_1/n_1) - (x_2/n_2)]}{\sqrt{\left(\dfrac{x_1 + x_2}{n_1 + n_2}\right)\left(1 - \dfrac{x_1 + x_2}{n_1 + n_2}\right)\left(\dfrac{n_1 + n_2}{n_1 n_2}\right)}} \qquad (8.8)$$

* Formula (8.8) could be made more precise by adding a correction for continuity. We will assume that n_1 and n_2 are large enough so that this correction changes z_c by a negligible amount.

Table 8.7 Summary of Tests of Hypotheses

Unknown Parameter	Population Description	Reference Section	Test Statistic
Mean μ	Population $N(\mu, \sigma^2)$ or $n \geq 30$, σ known	8.2	$z = \dfrac{\bar{x} - \mu_0}{\sigma/\sqrt{n}}$
Mean μ	Population $N(\mu, \sigma^2)$, σ unknown	8.5	$t_{(n-1)} = \dfrac{\bar{x} - \mu_0}{s/\sqrt{n}}$
Difference $\mu_1 - \mu_2$	Both populations normal, or n_1 and $n_2 \geq 25$, and σ_1^2 and σ_2^2 known	8.6	$z = \dfrac{(\bar{x}_1 - \bar{x}_2) - (\mu_1 - \mu_2)}{\sqrt{(\sigma_1^2/n_1) + (\sigma_2^2/n_2)}}.$
Difference $\mu_1 - \mu_2$	Both populations normal, σ_1^2 and σ_2^2 are unknown, but must be equal	8.7	$t = \dfrac{(\bar{x}_1 - \bar{x}_2) - (\mu_1 - \mu_2)}{\sqrt{\left(\dfrac{[(n_1 - 1)s_1^2 + (n_2 - 1)s_2^2]}{(n_1 + n_2 - 2)}\right)\left(\dfrac{n_1 + n_2}{n_1 n_2}\right)}}.$
Difference $\mu_1 - \mu_2$ (matched pairs)	Both populations normal	8.7	$t = \dfrac{\bar{D} - \Delta}{s_D/\sqrt{n}}.$
Proportion p	Repeated independent trials $npq \geq 3$	8.10 7.7 5.4 4.2	$z = \dfrac{(\boldsymbol{x}/n) - p_0}{\sqrt{p_0 q_0/n}}$
Difference $p_1 - p_2 = 0$	Both n_1 and n_2 are "large."	8.11	$z = \dfrac{[(\boldsymbol{x}_1/n_1) - (\boldsymbol{x}_2/n_2)]}{\sqrt{\left(\dfrac{\boldsymbol{x}_1 + \boldsymbol{x}_2}{n_1 + n_2}\right)\left(1 - \dfrac{\boldsymbol{x}_1 + \boldsymbol{x}_2}{n_1 + n_2}\right)\left(\dfrac{n_1 + n_2}{n_1 n_2}\right)}}$

This test statistic represents an extension of Formula (8.7) to the two-sample situation. To illustrate this test, we will consider some data involving the mandatory use of helmets by motorcycle riders in Indiana. In 1977 the mandatory use of helmets was repealed. A random sample of 400 motorcycle accidents in 1976 indicated "serious" head injuries in 38 of these accidents. For 1978, a random sample of 300 accidents resulted in 49 "serious" head injuries. Since we are interested only in whether p_1 (1976) and p_2 (1978) are equal, or whether $p_1 < p_2$, the relevant hypotheses are:

$$H_0: p_1 - p_2 = 0 \qquad \text{versus} \qquad H_a: p_1 - p_2 < 0.$$

Substituting the appropriate values in Formula (8.8) gives us z_c:

$$z_c = \frac{\left[\left(\frac{38}{400}\right) - \left(\frac{49}{300}\right)\right]}{\sqrt{\left(\dfrac{38 + 49}{400 + 300}\right)\left(1 - \dfrac{38 + 49}{400 + 300}\right)\left(\dfrac{400 + 300}{(400)(300)}\right)}}$$

$$= \frac{(0.095 - 0.163)}{\sqrt{0.00063}} = -2.71.$$

This z_c-value is in the critical region for any α-level above 0.0034; hence we can safely conclude that the 1978 proportion of serious head injuries is significantly higher than the 1976 proportion. One must be careful, however, in assuming that the repealed law "caused" the increase in head injuries. While this change does seem to be the likely cause, there may be other contributing factors (such as all the potholes resulting from the infamous 1977–78 winter). It would appear that this difference between the data from 1976 and those from 1978 is large enough so that it should convince even the most conservative motorcyclist that there is increased safety in wearing a helmet.

Table 8.7 provides a summary of the test statistics in this chapter.

Problems

8.43 In Section 8.9 it was shown that, for the production process under investigation, a sample size of 50 results in a lower expected cost than a sample size of 20 if the additional observations cost only an extra \$10. Suppose that we now have the opportunity to buy 50 more observations (for a total n of 100) for \$15 more (i.e., sampling cost of \$25). Estimate, as best you can, the optimal critical region for this size of sample. Is the total expected cost in this case lower or higher than the cost when $n = 50$?

8.44 One student spokesperson claims that there is more student support for the campus food services this year than last year, when 80% of the students were dissatisfied. A sample of 50 randomly selected students finds that 20 think the food is satisfactory this year; the others express some dissatisfaction. Test the claim that dissatisfaction has decreased, using $\alpha = 0.01$ and Table I. Report a p-value.

8.45 In the discussion in Section 8.9, we calculated the probability of making an incorrect decision concerning adjustments to a production process. The values of α and β were determined for the critical region $x \geq 4$ and for the critical region $x \geq 5$, assuming a sample size of 20.

a) Calculate α and β for this problem for the critical region $x \geq 6$.

b) Calculate the expected cost associated with this critical region when the probability of a defective is 0.10. Do the same thing for the probability of a defective equal to 0.25.

c) Calculate the total expected cost associated with this critical region, assuming that the probability of a minor adjustment is 0.70 and the probability of a major adjustment is 0.30.

d) Is the cost that you determined in part (c) better or worse than the cost calculated in Section 8.9 for the critical region $x \geq 5$? Do you think $x \geq 5$ is the best critical region for this sample size? Explain why, and then try to draw a graph relating the location of the critical region to the total expected costs (let the horizontal axis be the lower bound of the critical regions).

8.46 A nationally known insurance company has been advertising on TV that "9 out of 10 claims are in the return mail two days after receipt." Suppose that you decide to test this assertion by letting $H_0: p = 0.90$ and $H_a: p < 0.90$.

a) What critical region would you establish if $\alpha = 0.05$ and $n = 100$? Use Table I.

b) Would you accept H_0 or H_a if 85 claims were returned within the two days? What p-value would you report?

8.47 a) In Problem 8.13, a test was described comparing the Horizon car with a Datsun involving $n = 50$ randomly selected people. Suppose that we assume that the proportion of the U.S. population preferring the Horizon is actually either $H_0: p = 0.50$ or $H_a: p = 0.75$. Using a critical region of $x \geq 31$, find α and β.

b) Describe the two types of error that could be made and their consequences.

c) If a Type I error is judged to be five times as serious as a Type II error, would the critical region $x \geq 31$ be appropriate? If not, suggest a better one.

8.48 Rework Problem 8.46 using a normal approximation.

8.49 Rework Problem 8.47(a) using a normal approximation.

8.50 A national survey found that one-fifth of all cars on interstate highways are exceeding the 55-mph speed limit. A random sample of 36 cars in Pennsylvania revealed nine exceeding the 55-mph speed limit. Would you conclude that the number of speeding cars in Pennsylvania exceeds the national average?

8.51 The proportion of young egrets that survive to maturity under natural conditions is $\frac{1}{5}$. Because the birds are becoming scarce, a zoo curator traps some of them and attempts to hatch and raise the young in a controlled environment. Out of 64 eggs, the curator is able to raise 16 birds to maturity. Should the procedures of this curator be copied if we wish to increase the population of egrets? Allow for a five-percent chance that you would decide on the expensive controlled procedure when it really is not better.

8.52 Suppose that a supermarket has agreed to advertise through a local newspaper if it can be established that the newspaper's circulation reaches more than 50 percent of the supermarket's customers.

a) What null and alternative hypotheses should be established in this problem in trying to decide, on the basis of a random sample of customers, whether or not the supermarket should advertise in the newspaper?

b) If a sample of size $n = 64$ is collected, and $\alpha = 0.01$, what number (critical value) of supermarket customers need to regularly look at this newspaper's advertisements before a decision should be made to advertise?

8.53 Sixty out of 100 randomly selected shoppers who were classified as having rural backgrounds said they prefer to purchase camera equipment in discount stores. In a comparable study involving 250 shoppers with urban backgrounds, 50 percent said they prefer the discount stores. Use a two-sided test to determine whether these two groups differ in the proportion who prefer discount stores. Use $\alpha = 0.01$.

8.54 Two blackjack players are interested in comparing their ability against common opponents by examining the proportion of times they have won. In a random sample of 200 hands, Player A won 55 percent of the time. Player B won 65 percent of the time in a random sample of 150 hands. Use a one-sided test, and $\alpha = 0.05$, to determine whether Player B can claim to be better at blackjack than Player A.

8.55 A cereal manufacturer will switch to a new TV advertising campaign if the new campaign will increase, by at least 0.10, the proportion of viewers who rate the ad as "highly attractive." In a random sample of 400 viewers, 23 percent rated the current campaign as "highly attractive." After viewing the new campaign, 35 percent of a random sample of 100 viewers rated the new campaign as "highly attractive." Would you recommend the company switch if $\alpha = 0.02$?

8.56 A medical researcher testing the effectiveness of a new drug found that 70 percent of a random sample of 280 patients improved under this drug. In a control group, 140 patients were given a placebo. Fifty percent of these patients improved. Would you conclude that the new drug is more effective than the placebo? Use $\alpha = 0.01$.

Exercises

Note: Some of the tests necessary to solve the following exercises involve sample statistics not specifically described in this chapter (that is, the χ^2 and F). Their properties are described in Chapter 6.

8.57 Suppose that you are presented with an urn containing six balls, some of which are red, the rest of which are white. Let θ represent the number of red balls. You decide to test the null hypothesis $H_0:\theta = 3$ against $H_a:\theta \neq 3$ by drawing two balls from the urn, rejecting H_a if the two balls are the same color, and rejecting H_0 if the two balls are different colors.

a) If sampling occurs with replacement, find the probability of a Type I error. Find the probability of a Type II error, assuming that there are $\theta = 0, 1, 2, \ldots, 6$ red balls.

b) Repeat part (a) assuming that sampling occurs without replacement.

8.58 a) Plot the graph of the power function for parts (a) and (b) of Exercise 8.57.

b) Assume in Exercise 8.57 that you had decided to reject $H_a:\theta \neq 3$ whenever the two balls are different colors, and to reject $H_0:\theta = 3$ whenever the balls are the same color (just the opposite of the above criterion). Plot the power function for this critical region and comment on how it compares with the previous critical region (do this for sampling with and without replacement).

8.59 What assumption is necessary for using the **F**-test to test the null hypotheses $H_0: \sigma_1^2 = \sigma_2^2$?

8.60 In solving Problem 8.34(b) it is necessary to assume that $\sigma_1^2 = \sigma_2^2$. Use an **F**-test to determine whether or not this assumption is reasonable when $\alpha = 0.10$. What p-value would you report if α is unknown?

8.61 The Chamber of Commerce in a small town in Colorado is having a war of words with a small town in Hawaii as to which town enjoys better weather in the summer. Both locations have an average temperature of 85° and both claim to have less variability in temperatures.

a) What null and alternative hypotheses would you establish in this situation? What is the appropriate test statistic?

b) Assume that Senator Proxmire has given you a grant to take a random sample of weather in both locations. You spend two months in Hawaii, and calculate the variance of temperature over these 61 days to be $s_1^2 = 9$. A 31-day stay in Colorado yields $s_2^2 = 16$. Assuming that these are random samples, would you accept H_0 or H_a if $\alpha = 0.10$? What p-value would you report?

c) Comment on how "random" you think these samples really are.

8.62 The telephone company is continually studying the length of phone calls, as well as the variability in lengths. Suppose that the national population variance of calls is $\sigma^2 = 4$ minutes. The telephone company wants to test whether a certain community's calls differ in variability from the national value. The length of calls is assumed to be normally distributed.

a) What null and alternative hypotheses would you establish for the test described above?

b) What critical regions(s) would you use for this test if $\alpha = 0.05$ and $n = 25$? Assume a normal parent population.

c) Would you accept or reject H_0 if the sample of $n = 25$ resulted in $s^2 = 2.5$?

8.63 Basketball players sometimes are rated on how consistent they are at scoring (some are consistently bad). One "expert" has suggested that a good scorer should have a variance in the number of points scored of no more than $\sigma^2 = 25$. Suppose that State U.'s leading scorer made the following number of points in the first five games: 21, 14, 26, 9, and 15. On the basis of this sample (OK, perhaps it is not random) would you accept $H_0: \sigma^2 \leq 25$ or $H_a: \sigma^2 > 25$? [*Hint:* Make H_0 a simple hypothesis.] Use $\alpha = 0.05$. What probability value would you report? Assume that the parent population is normal.

8.64 Suppose that you are convinced that Harvard MBA's have high IQs, but you also believe that there is a higher variability among these people than the national average of $\sigma^2 = 256$.

a) What null and alternative hypothesis would you use to test your assertion?

b) What test statistic is appropriate if all IQs are assumed to be normally distributed?

c) What critical region is appropriate if $\alpha = 0.025$ and $n = 31$?

d) Would you accept H_0 or H_a if the sample result is $s^2 = 350$?

e) What p-value would you report?

8.65 A sample survey firm is contracted by an advertising agency to determine whether or not the average income in a certain large metropolitan area exceeds \$17,500. The agency wants the results of this survey to reject the null hypothesis $H_0: \mu = \$17{,}500$ in favor of $H_a: \mu > \$17{,}500$ at the $\alpha = 0.05$ level of significance when the true mean is as small as \$17,600. If the population standard deviation of incomes in this area is assumed to be \$1000, how large a sample will the survey firm have to take in order to meet the requirements of the advertising agency?

8.66 Suppose that, in Exercise 8.65, the survey firm charges \$5 for each observation it collects for the advertising agency. How much more will it cost the agency to be able to reject $H_0: \mu = \$17{,}500$ at the $\alpha = 0.01$ level rather than at the $\alpha = 0.05$ level?

8.67 It has been estimated that most families in the United States spend approximately 90 percent of their yearly income and save no more than ten percent of their yearly income. Suppose that a random sample of 100 families with high incomes (exceeding \$40,000) shows that sixty percent of these people save more than ten percent of their income.

a) Does this sample support the hypothesis that a majority of families with incomes exceeding \$40,000 save more than 10 percent of their income? What is the null hypothesis in this case? Given these sample results, what is the probability that the null hypothesis is true?

b) Would you conclude from the sample in this problem that families with high incomes will tend to save more than families with more average incomes? Why or why not?

8.68 A public-policy researcher is studying the variance in the amount of money requested by certain government agencies. The alternative hypothesis is that the variance this year (σ_1^2) is larger than the variance five years ago (σ_2^2). Random samples, of size $n_1 = 21$ and $n_2 = 25$, yielded variances of $s_1^2 = (67{,}233)^2$ and $s_2^2 = (37{,}178)^2$. Assume that the two populations are normally distributed. Test, at $\alpha = 0.01$, the hypothesis $H_0: \sigma_1^2 = \sigma_2^2$. What p-value would you report if α is unknown?

8.69 A sample of 10 fibers treated with a standard technique has average strength of 10, with a standard deviation of 3.2. Another sample of 17 fibers treated with a new technique has an average strength of 20 and standard deviation of 3.0. Test at the 0.05 level of significance to see whether the new technique can be expected to produce a population of fibers with average strength *at least* 5 greater than the population of fibers treated with the standard technique. Assume that the populations are normal.

8.70 Go to Table VI in Appendix B and collect two random samples in which each observation is a two-digit number. The first sample should be of size $n_1 = 4$, and the second of size $n_2 = 6$. Calculate $\bar{x}_1$, $\bar{x}_2$, s_1^2, and s_2^2. Assume that you do not know that these samples came from the same population.

a) Use an $\boldsymbol{F}$-test to test $H_0: \sigma_1^2 = \sigma_2^2$ versus $H_a: \sigma_1^2 \neq \sigma_2^2$, setting $\alpha = 0.01$. Are the assumptions necessary for use of the $\boldsymbol{F}$-test met in this case?

b) Use a $\boldsymbol{t}$-test to test $H_0: \mu_1 - \mu_2 = 0$ versus $H_a: \mu_1 - \mu \neq 0$, setting $\alpha = 0.01$. Are the assumptions required for use of the $\boldsymbol{t}$-test met in this case?

8.71 Prove that Formula (8.4) and Formula (8.3) yield identical values of t and z when $n_1 = n_2$, $s_1^2 = \sigma_1^2$, $s_2^2 = \sigma_2^2$.

8.72 Suppose that you decide to test the null hypothesis that the probability of a six on each toss of a single die is 0.17. You intend to toss this die 100 different times, and will reject the null hypothesis if fewer than 10 sixes or more than 24 sixes appear.

a) For the critical region described above, and using Table I, what is the probability of making a Type I error?

b) Use Table III to obtain an approximation to your answer to part (a).

c) What is the probability of a Type I error if the critical region is defined to be the occurrence of 22 or more sixes (a one-tailed test)?

d) If the alternative hypothesis is $H_a: p = 0.25$, and the critical region in part (c) is used, what is the probability of a Type II error?

8.73 (This problem incorporates concepts from Chapters 6, 7, and 8.) Assume you are working for Eastern Airlines, and are concerned with the time it takes for a 727 to fly from Indianapolis to Atlanta. The time has obvious implications for fuel consumption estimates as well as scheduling departures and arrivals. A random sample of 10 flights resulted in the following times: 84, 86, 90, 72, 81, 80, 78, 83, 77, 79. Assume the population is normally distributed. The following questions all refer to this sample.

a) Find an unbiased estimate of μ and σ^2.

b) Construct a 95% confidence interval for μ.

c) You are very concerned with the proportion of times a flight exceeds 85 minutes. Construct a 99% confidence interval for this proportion.

d) Test $H_0: \mu = 85$ vs. $H_a: \mu = 80$ using $\alpha = 0.05$. What p-value would you report if α were unknown?

e) What is β for part (d) using $\alpha = 0.05$, and assuming that $\sigma^2 = 50$?

f) Does it appear reasonable that this random sample was drawn from a population with $\sigma^2 = 50$? Explain.

g) What size sample should have been taken if you had wanted an error in part (b) of no more than 1 minute? Assume $\sigma^2 = 50$.

h) Test $H_0: p = \frac{1}{2}$ vs. $H_a: p < \frac{1}{2}$, where p is the proportion of flights over 85 minutes, using $\alpha = 0.03$.

GLOSSARY

simple hypothesis: a statement that specifies a single value of the population parameter.

composite hypothesis: a statement that specifies more than one value for the population parameter.

null and alternative hypotheses: the two mutually exclusive and exhaustive hypotheses about the feasible values of the population parameter.

one-sided test: all values specified by H_a lie to one side of those given in H_0.

two-sided test: the values of H_a lie on both sides of the values specified in H_0.

Type I error: rejecting H_0 when H_0 is true.

Type II error: accepting H_0 when H_0 is false.

α level of significance: probability of a Type I error.

β: probability of a Type II error.

$(1 - \alpha)$: confidence level.

$(1 - \beta)$: power of a test.

critical region: values of a test statistic that lead to rejection of H_0.

acceptance region: values of a test statistic that lead to acceptance of H_0 (fail to cause rejection of the null hypothesis, based on the sample evidence).

test statistic: the random variable used in a test of hypothesis. It must have a known probability distribution, given that H_0 is true, and contain the parameter being tested; all of its other terms must be either known or calculable from the sample.

computed value: a value of the test statistic, calculated by using a specific sample result.

***p*-value:** the probability, given that H_0 is true, of observing a sample result more extreme than the one observed or calculated.

computed value of z (σ known):

$$z_c = \frac{\bar{x} - \mu_0}{\sigma/\sqrt{n}}.$$

computed value of t (σ unknown):

$$t_c = \frac{\bar{x} - \mu_0}{s/\sqrt{n}}.$$

computed t-value for matched pairs:

$$t_c = \frac{\bar{D} - \Delta}{s_D/\sqrt{n}}.$$

two-sample computed z-value:

$$z_c = \frac{(\bar{x}_1 - \bar{x}_2) - (\mu_1 - \mu_2)}{\sqrt{(\sigma_1^2/n_1) + (\sigma_2^2/n_2)}}.$$

two-sample computed t-value:

$$t_c = \frac{(\bar{x}_1 - \bar{x}_2) - (\mu_1 - \mu_2)}{\sqrt{\left(\frac{[(n_1 - 1)s_1^2 + (n_2 - 1)s_2^2]}{(n_1 + n_2 - 2)}\right)\left(\frac{n_1 + n_2}{n_1 n_2}\right)}}.$$

computed value of z for the normal approximation to a binomial proportion:

$$z_c = \frac{(x/n) - p_0}{\sqrt{p_0 q_0 / n}}.$$

computed z-value for normal approximation to the difference between two proportions:

$$z_c = \frac{[(x_1/n_1) - (x_2/n_2)]}{\sqrt{\left(\frac{x_1 + x_2}{n_1 + n_2}\right)\left(1 - \frac{x_1 + x_2}{n_1 + x_2}\right)\left(\frac{n_1 + n_2}{n_1 n_2}\right)}}.$$

9

STATISTICAL DECISION THEORY

"I was a trembling because I'd got to decide forever betwixt two things, and I knowed it. I studied for a minute, sort of holding my breath, and then say to myself, 'All right, then, I'll go to hell.'"

HUCK FINN

9.1 INTRODUCTION

As noted earlier, one purpose of statistical analysis is to aid in the process of *making decisions under uncertainty*. In Chapters 7 and 8 the process of making inferences based on sample information was discussed. In this chapter we go one step further—from making inferences to making decisions. Life is a constant sequence of decision-making situations. Most decisions are made intuitively without even thinking about them. For example, consider the decision facing you when you want to cross a busy street. You must decide when to attempt to cross the street, keeping in mind how much of a hurry you are in and how heavy the traffic is. The consequence of a "bad" decision here could be quite serious—you could be hit by a car and be killed. Yet you probably do not consciously think of this possibility because you have faced the situation many times in the past. Other decisions may require some thought, but can still be made intuitively. An example might be the choice of a main dish from a menu at a restaurant. You consciously evaluate the various choices, taking into consideration your tastes and the prices, and you make a judgmental decision and give the waiter your order.

Some decisions are not as easy to make as these two examples, however. For instance, consider the decision to purchase common stock. There are many stocks to choose from, and the decision involves the "outlook" for the various stocks, the investor's attitude toward risk, and many other considerations. Complex decisions faced by organizations, such as whether or not to build a new plant or how many units of a given item to produce, are also difficult decisions. It is for decisions such as these that the formal decision-theory procedures to be discussed in this chapter should be useful. Applying the formal models of decision theory may not always be easy, as we shall see, but for important decisions it should prove worthwhile.

As with any statistical procedure, applying decision theory is a modeling problem in which the statistician attempts to build a model that is a reasonable approximation to the real-world decision-making problem of interest but is not so complex that it is difficult to work with. First, it is necessary to carefully identify the problem and to prepare a list of possible decisions, or actions. Next, it is necessary to define the possible events, or states of the world, that seem to have an important bearing on the outcome of the decision. As we shall see, it is desirable to *assign probabilities* to these events, and these probabilities should represent the decision maker's uncertainty concerning the events. Thus, some of the previous material in this book regarding probability will be useful in decision theory. The other key input to problems of decision-making under uncertainty involves the *potential consequences* of the various actions that are being considered. These consequences might be measured in terms of monetary payoffs or losses, or they might involve nonmonetary considerations, as we shall see when we discuss utility.

Once all of the inputs have been determined, the decision-making problem can be solved by using the techniques that will be developed in this chapter. The primary emphasis in this chapter is placed on the mechanics of combining the inputs to determine an optimal decision, but you should keep in mind that the initial modeling of the problem is also very important. Often this modeling procedure is iterative, since the model-building process may lead to the consideration of new actions or events that had previously been ignored.

9.2 CERTAINTY *VS.* UNCERTAINTY

Before we present the theory of making decisions under the condition of uncertainty, it should be useful to discuss the difference between certainty and uncertainty. Formally, a *consequence*, or a *payoff* to the decision-maker, is the result of the interaction of two factors:

1. The *decision*, or the *action*, selected by the decision-maker; and
2. The actual *state of the world which occurs*.

For example, suppose that you are faced with a decision concerning the purchase of common stock. For the sake of simplicity, we shall assume that you intend to invest exactly $1000 in a single common stock and hold the stock for exactly one year, at which time you will sell it at the market price. Furthermore, assume that you are considering only three stocks, A, B, and C, each of which currently sells for $10 a share. Thus, you intend to buy 100 shares of *one* of the three stocks. Your selection of a single stock from the three constitutes your *decision*, or *action*. The prices of the three stocks one year from now constitute the *state of the world*. The combination of your action and the state of the world determines your payoff. Suppose that at the end of one year, the prices of stocks A, B, and C are $15, $5, and $10. The payoffs for the three possible actions are then +$500, −$500, and $0, respectively. For this state of the world, the best decision is to buy stock A, for this results in the highest payoff. If we know the state of the world with certainty, the decision can be made in this manner, and this is called decision-making *under certainty*. In our example, if we know what the prices of the stocks will be in one year, then we simply buy the stock which will give us the maximum payoff—that is, the stock which will increase in value the most during the coming year.

Of course, the assumption that we know what the prices of stocks will be one year hence is not at all realistic. We probably have some ideas regarding the prices, ideas which may be based on our impressions of the economy in general, various industries, and various firms within industries. Our knowledge may be due to a careful study of the stock market, or it may be due to a hot tip from a friend. At any rate, we have knowledge, but not perfect knowledge. As a result, we are no longer operating under the condition of certainty. Instead, we are faced with a problem of decision-making *under uncertainty*. In this example, it is clear that the decision-making problem is considerably more difficult under uncertainty than it is under certainty.

It should be pointed out that in many situations, decision-making under certainty is by no means easy. Often the problem is complex enough to make it very difficult to determine the best action even though all of the relevant factors are known for certain. For example, suppose that a manufacturer must ship a certain product from a number of factories to a number of warehouses. Each factory produces a certain number of units of the product, and each warehouse requires a certain number of units. Furthermore, the cost of shipping from any given factory to any given warehouse depends on the amount shipped, the particular factory, and the particular warehouse. The decision-making problem is this: what shipping pattern minimizes the total transportation costs? That is, what is the least expensive way to transport the product from the factories to the warehouses? In this problem there is no uncertainty; the amounts produced at the various factories, the amounts needed at the various warehouses, and the costs of shipping are all known. Even under certainty, this decision-making problem is clearly not trivial to solve, although it becomes much more complex if uncertainty is introduced. Under certainty, the problem can be solved by a technique known as linear programming. Other decision-making problems under certainty require the use of different types of mathematical optimization procedures that often are classified under the heading of *operations research.* We are concerned with the case of uncertainty rather than certainty, so operations-research procedures are not discussed here, but you should recognize that decision-making under certainty includes many important and by no means mathematically trivial problems.

9.3 CRITERIA FOR DECISION-MAKING UNDER UNCERTAINTY

In the certainty case, the payoffs for each potential action are determined, and the action resulting in the highest payoff is then chosen. Under uncertainty, however, the payoffs cannot be determined for certain simply because the state of the world is not known for certain. For each action, then, there are various possible payoffs corresponding to the various possible states (we shall use the terms "state" and "state of nature" interchangeably with "state of the world"). For example, consider the common-stock example once again, with the assumption that there are only two possible states of nature. In the first (State I), the prices of the stocks A, B, and C at the end of one year are, respectively, \$15, \$5, and \$10. In the second state (State II), the prices are \$8, \$14, and \$12. This results in the payoff table given in Table 9.1, where the payoffs are expressed in dollars. It is not obvious from the payoff table that any one of the three actions is clearly the "best" action. If state of nature I occurs, then A gives the greatest payoff; if state of nature II occurs, then B gives the greatest payoff. An argument can also be made for buying C, for this eliminates the possibility of a negative payoff. This argument is put forth by those supporting the following decision-making criterion: for each action, find the smallest possible payoff, and choose the action for which this smallest possible payoff is largest. In this case the

Table 9.1 Payoff Table for Stock Example

		State of Nature	
		I	II
Actions	Buy A	+500	−200
	Buy B	−500	+400
	Buy C	0	+200

smallest possible payoffs for the three actions are -200, -500, and 0. Clearly zero is the largest of these three numbers, so according to this criterion, the "best" action is to buy stock C. This criterion is called the *maximin* (maximization of minimum gain) criterion.

The maximin criterion can be criticized on two grounds. First, it considers only the smallest payoff for each action and fails to take into account the *largest* payoff. By choosing to buy stock C in the above example, we are avoiding the possibility of a large loss, but at the same time we are giving up a chance for a larger gain. The best we can do with C is to gain \$200, whereas with A it would be possible to gain \$500. A second criticism of the maximin criterion is that it fails to take into consideration the relative likelihoods of the two possible states of nature. These likelihoods can be represented by probabilities, $P(\mathrm{I})$ and $P(\mathrm{II})$, which must sum to one because of the assumption that I and II are the only possible states of nature. If, for example, $P(\mathrm{I}) = 0.90$ and $P(\mathrm{II}) = 0.10$, we might feel that the odds in favor of State I are high enough to warrant the purchase of stock A, which results in a +\$500 payoff if State I occurs.

Before considering the introduction of the probabilities, $P(\mathrm{I})$ and $P(\mathrm{II})$, which represent our uncertainty about the state of nature, let us consider one more possible criterion. This criterion, which we shall label the *maximax* criterion, involves finding the largest possible payoff for each action and choosing the action for which this largest possible payoff is the greatest. In our example, the largest possible payoffs are, respectively, $+500$, $+400$, and $+200$, so the maximax criterion leads to the purchase of stock A. The maximax criterion amounts to trying to attain the highest payoff in the entire payoff table, and as such it is a risk-taking strategy, for it ignores the possibilities of large losses. In the same way, the *maximin* criterion is a risk-avoiding strategy, for it attempts to avoid large losses, even at the sacrifice of possible large gains. Since the maximax criterion considers only the largest entry in each row of the payoff table and does not take $P(\mathrm{I})$ and $P(\mathrm{II})$ into consideration, it can be criticized on the same grounds as the maximin criterion.

In order to make use of the probabilistic nature of the situation, it is necessary to use the concept of *expected value* which was introduced in Chapter 3. For each action, it is possible to compute an expected payoff, and the "best" action would be the one with the highest expected payoff. If payoffs are expressed in terms of money, as in the stock example, we shall call this criterion the EMV (*expected monetary value*) criterion. In our example,

$$\text{EMV (Buy A)} = (\$500)P(\mathrm{I}) + (-\$200)P(\mathrm{II}),$$

$$\text{EMV (Buy B)} = (-\$500)P(\mathrm{I}) + (\$400)P(\mathrm{II}),$$

and

$$\text{EMV (Buy C)} = (\$0)P(\mathrm{I}) + (\$200)P(\mathrm{II}).$$

Suppose $P(\mathrm{I}) = 0.90$ and $P(\mathrm{II}) = 0.10$. Then the three EMV's are, respectively, \$430, $-$\$410, and \$20. In this case, the EMV criterion leads to the purchase of stock A. If, on the other hand, the probabilities are $P(\mathrm{I}) = 0.30$ and $P(\mathrm{II}) = 0.70$, the EMV's would be \$10, \$130, and \$140, and the "best" action would be to buy stock C. With the EMV criterion, then, the payoffs alone do not determine the best action; it is necessary to take into consideration the probabilities representing the uncertainty about the state of nature. The EMV then represents the expected value, or the average payoff, for the different actions.

An alternative way of presenting a decision-making problem involves the use of tree diagrams. The use of tree diagrams is often preferred to the use of payoff tables because the tree diagrams make it easier to see the general structure of decision-making problems, particularly in the case of somewhat complex problems. Tree diagrams are particularly helpful in problems that involve buying additional information, as we shall see in Section 9.5. In Fig. 9.1 a tree diagram for the stock-purchasing example is presented. The point at the lefthand side of the diagram represents the initial position, and the three branches of the tree emanating from that point represent the three actions in the problem. Thus, the lefthand fork is called a decision fork, or an action fork. At the end of each of the three branches representing the possible actions, there is another fork with two branches. These two branches represent the two states of nature, or events, and as a result these forks are called event forks, or chance forks. All forks in a decision tree are either decision forks (involving possible actions) or event forks (involving possible events). This particular tree is quite simple, with only one decision fork and one event fork for each possible action. The numbers on the event branches are the probabilities of the events, and the numbers on the righthand side of the tree diagram are the payoffs. For example, if the decision-maker buys stock A and event I occurs, then the payoff is \$500.

To solve a decision-making problem that is expressed in the form of a decision tree, we start at the righthand side of the diagram and work "backward" to the initial point. Hence, the solution procedure is often called *backward induction.* In working backward, there are only two rules to follow:

1. At chance forks, find expected values (EMV's). Following conventional notation, we will denote chance forks with a circle, ○.

Fig. 9.1 Decision tree for stock example.

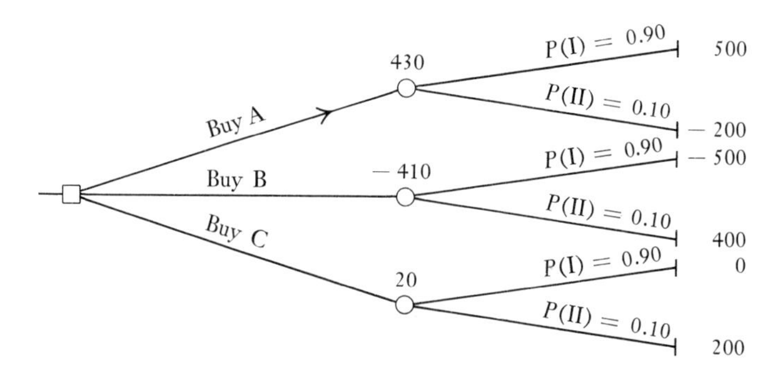

2. At decision forks, choose the action with the highest EMV. Decision forks will be designated with a box symbol, □.

For the tree in Fig. 9.1, EMV's are calculated for each of the three chance forks on the right side of the diagram. These EMV's, which turn out to be \$430, −\$410, and \$20, as calculated previously, are written at the "starting points" of the three chance forks. At the action fork, then, we simply compare these three numbers, note that the largest is \$430, corresponding to the purchase of stock A, and put an arrow on the branch marked "Buy A" to indicate that this is the optimal action. As we shall see later in the chapter, more complicated decision trees require more calculations, but they do not necessitate any procedures different from the procedures used to solve the simple problem represented by this tree diagram.

For another example, consider a contractor who is about to submit a bid for the construction of a new office building. It will cost the contractor \$400,000 to build the proposed building, and he must decide how much to bid. The larger the bid, the more profit the contractor will earn if he wins the bid. However, the contract is awarded to the low bidder, and the contractor knows that some of his competitors will also be submitting bids on this job. After giving the matter serious thought, the contractor decides that the job is not worth his while if he earns less than \$100,000, so the lowest bid he will consider is \$500,000. Furthermore, it is a waste of time to bid more than \$600,000, for he is certain that the low bid will not be above \$600,000. He finally decides to consider bids of \$500,000, \$525,000, \$550,000, and \$575,000. After considering his knowledge about the bidding strategies of his competitors, he decides that the chances of winning the bid are one in two if he bids \$500,000, one in three if he bids \$525,000, one in four if he bids \$550,000, and one in ten if he bids \$575,000. This decision-making problem is represented in Fig. 9.2, and we see that the optimal action is to bid \$500,000. In this problem, the probability of winning the bid goes down

Fig. 9.2 Decision tree for bidding example.

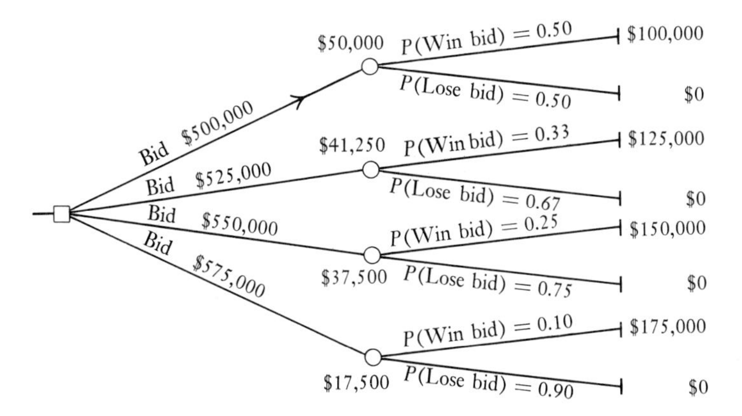

rapidly as the bid increases; had the probability not gone down so rapidly, one of the other bids might have turned out to be optimal.

In Chapter 2 the distinction between frequency probabilities and subjective probabilities was discussed. Note that in both of the examples given in this section, the probabilities were of a subjective nature. In most real-world decision-making problems, the events of interest are not repetitive in nature, so the frequency interpretation of probability is not applicable. As a result, some followers of the frequency interpretation refuse to use probabilities in making decisions, preferring instead to use nonprobabilistic criteria such as the maximin and maximax criteria discussed earlier in this section. Moreover, some statisticians distinguish between what they call "decision-making under risk" (decision-making when objective probabilities are available) and "decision-making under uncertainty" (decision-making when objective probabilities are not available). We do not make this distinction, and we reserve the term "risk" for discussions relating to utility, which will be discussed later in this chapter.

9.4 THE REVISION OF PROBABILITIES

One important input to the decision-making procedure which has not been discussed in any detail is the set of probabilities for the various possible states of nature. How do we arrive at these numbers? In Chapter 2 we discussed two interpretations of probability, the subjective interpretation and the objective interpretation. In this section we shall see how subjective probabilities can be revised on the basis of "objective" sample information. In order to do this, we shall have to recall Bayes' rule from Chapter 2 and restate it in the terminology of decision-making:

Bayes' rule: If $\theta_1, \theta_2, \ldots, \theta_k$ represent the possible states of nature, with probabilities $P(\theta_i)$, $i = 1, \ldots, k$, and x is some other event, then

$$P(\theta_i \mid x) = \frac{P(x \mid \theta_i)P(\theta_i)}{\sum_{j=1}^{k} P(x \mid \theta_j)P(\theta_j)}, \qquad i = 1, \ldots, k.$$

In this form, Bayes' rule provides a means of revising probabilities on the basis of sample evidence. The probabilities $P(\theta_i)$ are the probabilities of the various states of nature *prior to seeing the sample information.* These probabilities may represent a person's judgments (i.e., they may be subjective), or they may represent past sample information (in this sense, they may be regarded as objective). In general, we shall assume that the probabilities are subjective. The event x represents some sample information, and the probabilities $P(\theta_i \mid x)$ are the probabilities of the various states of nature *after seeing the sample information.* We shall call the $P(\theta_i)$ *prior probabilities* and the $P(\theta_i \mid x)$ *posterior probabilities.* The other probabilities appearing in Bayes' rule, which are of the form $P(x|\theta_i)$, will be called *likelihoods*, for they represent the probability, or the likelihood, of the observed sample results *given that* a particular state of nature occurs. The function $P(x|\theta_i)$ is essentially identical to the likelihood function encountered in the appendix to Chapter 7.

The example presented in Chapter 2 concerning a component that is produced by a certain process should help to clarify the preceding concepts. From past data, it is known that, of the components resulting from the production process, about 70 percent have been good and about 30 percent have been defective. (Furthermore, there is no reason to believe that there has been any change in the process that would alter these percentages.) However, management must make a decision as to whether to reject or accept individual components, so some information is desired about the individual components. To provide this information, a testing device is used. The testing device will read positive (+) if it appears that the component is good, and it will read negative (−) if the component appears defective. A large number of components were run through the testing device to check its accuracy, with the following results:

1. Among the good components, 80 percent had a positive reading and 20 percent had a negative reading.
2. Among the defective components, 25 percent had a positive reading and 75 percent had a negative reading.

Thus, the testing device is *more likely* to give a positive reading if the component is good, but the device is by no means perfect.

From the information in the preceding paragraph, Bayes' rule can be used to determine the posterior probabilities that a component is good or defective after the component is tested. Given no information other than the historical data, the prior probabilities are

$$P(\text{Good}) = 0.70 \qquad \text{and} \qquad P(\text{Defective}) = 0.30.$$

If the test yields a positive reading, the likelihoods are

$$P(+|\text{Good}) = 0.80 \qquad \text{and} \qquad P(+|\text{Defective}) = 0.25.$$

On the other hand, if the test yields a negative reading, the likelihoods are

$$P(-|\text{Good}) = 0.20 \qquad \text{and} \qquad P(-|\text{Defective}) = 0.75.$$

Suppose that a particular component is tested and yields a positive reading. Bayes' rule is then applied as follows:

$$P(\text{Good}|+) = \frac{P(+|\text{Good})P(\text{Good})}{P(+|\text{Good})P(\text{Good}) + P(+|\text{Defective})P(\text{Defective})}$$

$$= \frac{(0.80)(0.70)}{(0.80)(0.70) + (0.25)(0.30)} = 0.882$$

and

$$P(\text{Defective} \mid +) = \frac{P(+ \mid \text{Defective})P(\text{Defective})}{P(+ \mid \text{Good})P(\text{Good}) + P(+ \mid \text{Defective})P(\text{Defective})}$$

$$= \frac{(0.25)(0.30)}{(0.80)(0.70) + (0.25)(0.30)} = 0.118.$$

The positive reading increases the probability that the component is good from 0.70 to 0.882, and decreases the probability that the component is defective accordingly. This is intuitively reasonable, since a positive reading is more likely for a good component than for a defective component.

Since the probabilities of the various states of nature in a decision-making problem are supposed to represent the state of uncertainty, we can think of Bayes' rule as a technique for taking into account new information and suitably revising probabilities. The prior probabilities represent the state of uncertainty *prior* to seeing any sample, and the posterior probabilities represent the state of uncertainty *after* seeing a particular sample. The terms *prior* and *posterior*, then, relate only to a particular sample. In the example given above, suppose that the component is tested a second time. Assuming that the two tests are independent and that the testing device essentially "has no memory," the likelihoods given in the previous paragraph are still valid for the second test. The probabilities 0.882 and 0.118, the probabilities posterior to the first sample, now represent the state of uncertainty *prior to the new sample.* Suppose that the second test yields a positive reading, just as the first test did. Applying Bayes' rule a second time,

$$P(\text{Good} \mid +, +)$$

$$= \frac{P(+ \mid \text{Good})P(\text{Good} \mid +)}{P(+ \mid \text{Good})P(\text{Good} \mid +) + P(+ \mid \text{Defective})P(\text{Defective} \mid +)}$$

$$= \frac{(0.80)(0.882)}{(0.80)(0.882) + (0.25)(0.118)} = 0.960$$

and

$$P(\text{Defective} \mid +, +)$$

$$= \frac{P(+ \mid \text{Defective})P(\text{Defective} \mid +)}{P(+ \mid \text{Good})P(\text{Good} \mid +) + P(+ \mid \text{Defective})P(\text{Defective} \mid +)}$$

$$= \frac{(0.25)(0.118)}{(0.80)(0.882) + (0.25)(0.118)} = 0.040.$$

The revised probabilities are now 0.960 and 0.040. The second positive reading has increased the probability that the component is good still more, as would be expected. If more tests were made, the probabilities could be revised again.

One nice feature of Bayes' rule is that if we have a sample of more than one trial, it makes no difference whether the probabilities are revised trial by trial, as in the previous example, or revised just once after the completion of *all* the trials. In this case, starting with prior probabilities of 0.70 and 0.30 and observing two positive readings on two tests, we could apply Bayes' rule as follows:

$$P(\text{Good} \mid +, +)$$
$$= \frac{P(+, + \mid \text{Good})P(\text{Good})}{P(+, + \mid \text{Good})P(\text{Good}) + P(+, + \mid \text{Defective})P(\text{Defective})}$$
$$= \frac{(0.80)^2(0.70)}{(0.80)^2(0.70) + (0.25)^2(0.30)} = 0.960$$

and

$$P(\text{Defective} \mid +, +)$$
$$= \frac{P(+, + \mid \text{Defective})P(\text{Defective})}{P(+, + \mid \text{Good})P(\text{Good}) + P(+, + \mid \text{Defective})P(\text{Defective})}$$
$$= \frac{(0.25)^2(0.30)}{(0.80)^2(0.70) + (0.25)^2(0.30)} = 0.040.$$

This application of Bayes' rule makes use of the fact that the two trials were independent. By independence,

$$P(+, + \mid \text{Good}) = P(+ \mid \text{Good})P(+ \mid \text{Good}) = (0.80)^2$$

and

$$P(+, + \mid \text{Defective}) = P(+ \mid \text{Defective})P(+ \mid \text{Defective}) = (0.25)^2.$$

Note that the resulting probabilities are identical to those obtained by applying Bayes' rule twice, once after each test.

Bayes' rule, then, can be used to revise probabilities as new information is obtained. In this manner we can determine probabilities which represent our current state of uncertainty at the time a decision is to be made. Sometimes no sample information is available, and decisions must be based on prior probabilities which represent the subjective judgments of the decision-maker. In still other cases, no prior information is available, and decisions must be based solely on the sample information. This corresponds to the situation where the prior probabilities for the various states of nature are all equal.

Define: Decision making under uncertainty, actions, state of the world, maximin, maximax, EMV, backward induction, priors, posteriors, likelihoods.

Problems

9.1 Distinguish between making an inference and making a decision.

9.2 Suppose that a friend of yours loves to gamble, and he offers a bet on the toss of a coin. He will pay you $10 if the coin comes up heads, and you will pay him $1 if the coin comes up tails. You can use a coin from your pocket (so as to be sure that your friend is not slipping in a "loaded" coin) and you can do the tossing.

a) What are the two possible actions available to you?

b) What are the two possible events of interest?

c) What are the probabilities for these events?

d) For each combination of an action and an event (for example, you take the bet and the coin comes up heads), what is the consequence, or payoff, to you?

9.3 In Problem 9.2, if you take the bet and the coin comes up tails, does this mean that you made a bad decision? Explain your answer.

9.4 Explain the difference between decision-making under certainty and decision-making under uncertainty.

9.5 It is sometimes said that we are living in an increasingly uncertain world. List some major uncertainties facing managerial decision-makers today.

9.6 Give an example of a decision-making problem under certainty.

9.7 Express Problem 9.2 in the form of a tree diagram. Which action is "optimal" according to the maximin criterion? The maximax criterion? The EMV criterion?

9.8 A firm must decide whether or not to initiate a special advertising campaign for a certain product. The firm feels that a competitor might introduce a competing product, and that if such a new product is introduced, sales of the firm's own product would greatly decrease unless an advertising campaign for the product was in progress. The three actions under consideration are "no advertising," "minor ad campaign," and "major ad campaign." Taking into account the cost of advertising and its anticipated effect on sales, the decision-maker determines the following payoff table (Table 9.2):

a) What should the firm do if the maximin criterion is used?

b) What should the firm do if the maximax criterion is used?

c) On the basis of currently available information concerning the actions of the competing firm, it is decided that the probability is 0.60 that the new product will be introduced. What should the firm do if the EMV criterion is used?

Table 9.2

		State of Nature	
		Competitor Introduces New Product	**Competitor Does Not Introduce New Product**
Action	No advertising	100,000	700,000
	Minor ad campaign	300,000	600,000
	Major ad campaign	400,000	500,000

9.9 A company has $100,000 available to invest. The company can either expand production, invest the money in stocks, or put the money in the bank at a fixed 8% interest. If there is no recession, the company expects to make 14% if production is expanded and 12% if stocks are purchased. If there is a recession, however, expansion will lead to a 6% loss and stocks will provide a 2% loss. The decision will be re-evaluated in one year, so that only the first year's return is of interest now.

a) Draw a tree diagram for this problem.

b) Find the action that is best according to the maximin criterion.

9.10 In Problem 9.9, suppose that the company wants to maximize EMV. If the probability of a recession is 0.2, what should the company do? How large would the probability of a recession have to be before the company would put the money in the bank?

9.11 Discuss the relative merits of the maximin, maximax, and EMV criteria for decision-making.

9.12 During a study of past records, it was found that 10% of the coats manufactured by a particular firm had an imperfection. As a result, an inspector was hired to closely inspect the coats before they leave the factory. Of course, the inspector is not infallible. If a coat has an imperfection, the probability that the inspector will classify it as imperfect is 0.80. If a coat has no imperfection, on the other hand, the probability that the inspector will classify it as imperfect is 0.10.

a) If a coat has been classified as imperfect, what is the probability that it really is imperfect?

b) If a coat has been classified as good, what is the probability that it really is imperfect?

9.13 When someone applies for a loan, a bank gathers information about the applicant and applies a credit rating system to label the applicant either a "good risk," a "moderate risk," or a "bad risk." The bank will lend money to good or moderate risks, but not to bad risks. About 5% of the people borrowing money from the bank fail to repay the loan. Moreover, the bank's records show that 10% of those who failed to repay had been rated as good risks, while 60% of those who *did* repay had been rated as good risks.

a) Without using the credit rating system, what is the probability that an applicant will fail to repay a loan?

b) If the credit rating system labels an applicant a good risk, what is the probability that the applicant will fail to repay?

c) If the credit rating system labels an applicant a moderate risk, what is the probability that the applicant will fail to repay?

9.14 A security analyst feels that daily price changes of a particular security are independent and that the probability that the security's price increases on any given day is either 0.4, 0.5, or 0.6. If it is 0.4, the security is classified as a poor investment; if $p = 0.5$ it is classified as an average investment; and if $p = 0.6$ it is classified as a good investment. The security analyst feels that the security is equally likely to be an average investment or a good investment, and that it is twice as likely to be a good investment as it is to be a bad investment.

a) Find $P(p = 0.4)$, $P(p = 0.5)$, and $P(p = 0.6)$, where p is the probability that the price of the security increases on any given day.

b) The analyst observes the security in question for one day, and on that particular day the price of the security goes up. Use this new information to revise the analyst's distribution of p.

c) The security is observed for a second day and on that day the price goes up again. Once again, revise the analyst's distribution of p.

9.15 In the component-testing example in this section, suppose that a new test is available. This test gives positive readings to 90 percent of the good components and 15 percent of the bad components.

a) If a component receives a positive reading on the new test, what is the probability that it is defective?

b) If a component receives a negative reading on the new test, what is the probability that it is defective?

c) If both tests are run on a component, with positive results from both tests, what is the probability that the component is defective?

9.16 In Problem 9.15, if the old test and the new test cost the same to use, which one would you prefer to use? Explain your answer.

9.5 THE VALUE OF PERFECT INFORMATION

The term "information" has been used in discussing the state of uncertainty facing the decision-maker. The more information he has regarding the states of nature, the better off we would expect him to be. The extreme example is that of perfect information, which corresponds to what has previously been labeled certainty. If the decision-maker has perfect information, he simply chooses the act which results in the largest payoff—there is no uncertainty. In the production-process example, suppose that the decision-maker must decide whether to accept or reject a component. If the component is accepted, it will be marketed. Of course, there is still a chance that it might be defective, and any components that are sold and later found to be defective are replaced immediately with an alternative component that costs more but is guaranteed to be good. The additional cost is borne by the producer (i.e., our decision-maker, in the example) in order to keep from losing business due to unsatisfied customers. If the component is rejected, it is scrapped. This is clearly a problem of decision-making under uncertainty, for the decision-maker is not sure whether the component is good or defective, yet he must decide whether to market it or scrap it. After taking into account the various costs, the decision-maker arrives at the payoff table given in Table 9.3. All payoffs are expressed in dollars.

Prior to testing the component, the decision-maker's prior probabilities are $P(\text{Good}) = 0.70$ and $P(\text{Defective}) = 0.30$, and the EMV's for the two

Table 9.3 Payoff Table for Component Example

		State of the World	
		Component Good	Component Defective
Action	Market component	3	−5
	Scrap component	−1	−1

actions are

$$\text{EMV(Market)} = 3(0.70) + (-5)(0.30) = 0.60$$

and

$$\text{EMV(Scrap)} = -1(0.70) + (-1)(0.30) = -1.00.$$

The optimal decision is to market the component, since this action has the larger of the two EMV's.

What if the decision-maker could obtain perfect information? That is, what if he could find out *for sure* whether the component is good or defective? (In this example, it is possible to conceive of a test that could determine for sure whether a component is good or defective, although such a test might be quite expensive. Usually, however, it is not even *possible* to obtain perfect information in decision-making problems. Nevertheless, it is useful to investigate the value of perfect information and to consider it as an *upper bound* on the value of less-than-perfect information.) With perfect information, the decision-maker could obtain a payoff of \$3 for certain if the component is good and a payoff of −\$1 for certain if the component is defective. But according to the prior probabilities determined from historical data,

$$P(\text{Good}) = 0.70 \quad \text{and} \quad P(\text{Defective}) = 0.30.$$

Thus, *before* obtaining the perfect information (but knowing that he will obtain it), the decision-maker knows that there is a 0.70 chance that he will receive \$3 and a 0.30 chance that he will lose \$1. The corresponding EMV, called the expected payoff under perfect information, is

$$\begin{aligned}\text{EMV(Perfect information)} &= (\text{Highest payoff if good})P(\text{Good}) \\ &\quad + (\text{Highest payoff if defective})P(\text{Defective}) \\ &= 3(0.70) + (-1)(0.30) = 1.80.\end{aligned}$$

Before the decision-maker actually obtains perfect information, then, he can compute the expected payoff under perfect information, and the expected value of perfect information, denoted by EVPI, is

$$\text{EVPI} = \text{EMV}\begin{pmatrix}\text{Perfect} \\ \text{information}\end{pmatrix} - \text{EMV}\begin{pmatrix}\text{Optimal action} \\ \text{under prior information}\end{pmatrix}.$$

Thus, EVPI simply represents how much better off the decision-maker expects to be if he can make his decision with perfect information instead of making his decision on the basis of his current probabilities. For the example, the best EMV under the decision-maker's present state of information is \$0.60, and the difference between EMV(Perfect information) and \$0.60 represents the expected value of perfect information:

$$\text{EVPI} = \$1.80 - \$0.60 = \$1.20.$$

Therefore, if the decision-maker could purchase perfect information, he should be willing to pay up to \$1.20 for it.

The consideration of perfect information is illustrated in decision-tree form in Fig. 9.3. Note that the first fork is a decision fork, with the decision being whether to purchase perfect information before making a final decision regarding the component or to make the final decision *without* obtaining additional information. If the decision is made without additional information, the EMV is \$0.60, whereas if the decision is made with perfect information, the EMV is \$1.80. Comparing the top portion of the tree in Fig. 9.3 with the bottom portion of the tree, it can be seen that with perfect information, the branches

Fig. 9.3 Decision tree for component example.

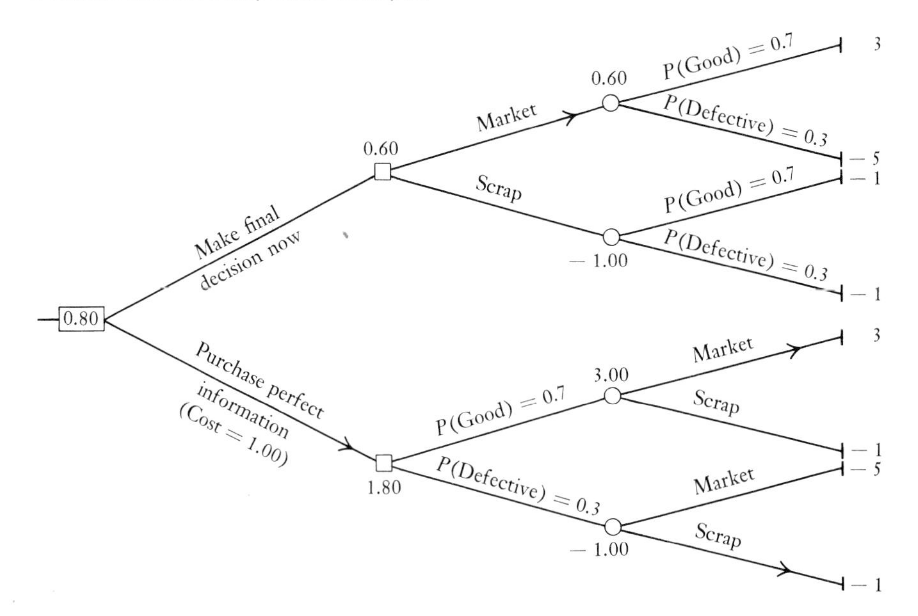

corresponding to whether the item is good or defective come *before* the branches corresponding to the decision whether to market or scrap.

Of course, if the perfect information costs something, its cost must be subtracted from the \$1.80 figure. For example, on the decision tree, the perfect information costs \$1, in which case the *net* EMV with perfect information is \$1.80 – \$1 = \$0.80. This is still better than \$0.60, so the optimal strategy is to purchase the perfect information before making the final decision regarding the component.

Now suppose that the decision-maker observes two tests of the component, with positive readings on both tests. Applying Bayes' rule as in the previous section, the posterior probabilities are 0.960 and 0.040. The EMV's are then

$$\text{EMV}(\text{Market} \mid +, +) = 3(0.96) + (-5)(0.04) = 2.68$$

and

$$\text{EMV}(\text{Scrap} \mid +, +) = -1(0.96) + (-1)(0.04) = -1.00.$$

Using the posterior probabilities, the EMV under perfect information is now

$$3(0.96) + (-1)(0.04) = 2.84.$$

Therefore, the expected value of perfect information is now

$$\$2.84 - \$2.68 = \$0.16.$$

The decision-maker is now willing to pay only up to \$0.16 for perfect information, whereas previously he would have been willing to pay up to \$1.20. It appears that observing the sample information (the results of the two tests) has improved his lot somewhat. After the two positive readings, the probability that the component is good is very high, so it is very unlikely that perfect information will cause him to change his decision. If, on the other hand, he had observed one positive reading and one negative reading, he would not be so certain about whether the component was good or defective, and the expected value of perfect information would not be so low.

9.6 THE VALUE OF SAMPLE INFORMATION

From the expected value of perfect information, the decision-maker knows how much perfect information is worth to him, i.e., how much he should be willing to pay for perfect information. In most real-world decision-making problems, however, perfect information is not available. Even in the component example, it is unlikely that the decision-maker can find out for certain whether a component is good or defective without either destroying the component or conducting a prohibitively expensive examination of the component. The expected value of perfect information is still useful as an upper bound to the amount the decision-maker should be willing to pay for imperfect information, or *sample* information. Given the prior probabilities in the component example, if the expected value of perfect information is \$1.20, the decision-maker surely should not pay any more than that for less than perfect information.

Ideally, of course, it is desirable to compute exactly the expected value of sample information. After a sample is observed, the decision-maker revises his probabilities and recomputes EMV's for the actions under consideration. *Before* the sample is observed, he needs to decide whether or not to take the sample. After all, samples usually have some cost, and the decision-maker wants to know if the sample is expected to be worth the cost involved.

In our example, suppose that the decision-maker has not observed any sample information but that he is contemplating running the test discussed in the previous section. To determine the expected value of sample information, is is necessary to consider all possible sample outcomes and to compute the posterior probabilities and posterior EMV's under each possible sample outcome. We already know that if the test yields a positive reading, the posterior probabilities are

$$P(\text{Good} \mid +) = 0.882 \qquad \text{and} \qquad P(\text{Defective} \mid +) = 0.118.$$

The corresponding expected payoffs are

$$\text{EMV}(\text{Market} \mid +) = 3(0.882) + (-5)(0.118) = 2.06$$

and

$$\text{EMV}(\text{Scrap} \mid +) = -1(0.882) + (-1)(0.118) = -1.00.$$

Using Bayes' rule, the posterior probabilities following a negative reading are

$$P(\text{Good} \mid -) = 0.384 \qquad \text{and} \qquad P(\text{Defective} \mid -) = 0.616,$$

and the expected payoffs are

$$\text{EMV}(\text{Market} \mid -) = -1.93 \qquad \text{and} \qquad \text{EMV}(\text{Scrap} \mid -) = -1.00.$$

Thus, if the test yields a positive reading, the optimal action is to market the component, and the EMV of this optimal action is \$2.06. If the test yields a negative reading, the optimal action is to scrap the component, and the EMV of this optimal action is −\$1.00.

The probabilities for the two possible results of the test, as calculated before the test is run, can be determined as follows:

$$\begin{aligned} P(+) &= P(+ \mid \text{Good})P(\text{Good}) + P(+ \mid \text{Defective})P(\text{Defective}) \\ &= (0.80)(0.70) + (0.25)(0.30) = 0.635 \end{aligned}$$

and

$$\begin{aligned} P(-) &= P(- \mid \text{Good})P(\text{Good}) + P(- \mid \text{Defective})P(\text{Defective}) \\ &= (0.20)(0.70) + (0.75)(0.30) = 0.365. \end{aligned}$$

Therefore, the overall expected payoff *with* the test, as calculated before the test is run, is

$$\text{EMV}\begin{pmatrix}\text{Sample}\\\text{information}\end{pmatrix} = \begin{pmatrix}\text{Highest}\\\text{EMV after }+\end{pmatrix}P(+) + \begin{pmatrix}\text{Highest}\\\text{EMV after }-\end{pmatrix}P(-)$$

$$= 2.06(0.635) + (-1.00)(0.365) = 0.94.$$

Before the sample is actually observed, then, the expected payoff under sample information can be computed, and the expected value of sample information, denoted by EVSI, is

$$\text{EVSI} = \text{EMV}\begin{pmatrix}\text{Sample}\\\text{information}\end{pmatrix} - \text{EMV}\begin{pmatrix}\text{Optimal action}\\\text{under prior information}\end{pmatrix}.$$

EVSI tells the decision-maker how much better off he can expect to be if he chooses to obtain sample information before making his final decision. For the example, the best EMV under the decision-maker's present state of information is \$0.60, and the EVSI is

$$\text{EVSI} = \$0.94 - \$0.60 = \$0.34.$$

Thus, the decision-maker should be willing to pay up to \$0.34 to observe the proposed test.

Suppose that it costs \$0.15 to run the test. This is less than the expected value of the sample information, so the decision-maker should run the test. Formally, a decision as to whether or not to run the test should be based on the expected net gain from sample information, denoted by ENGS:

$$\text{ENGS} = \text{EVSI} - \text{CS},$$

where CS represents the cost of sampling. For the example,

$$\text{EVSI} = \$0.34 \quad \text{and} \quad \text{CS} = \$0.15,$$

so

$$\text{ENGS} = \$0.34 - \$0.15 = \$0.19.$$

Since the ENGS is positive, the test should be run.

The preceding analysis is illustrated in decision-tree form in Fig. 9.4. The lefthand fork represents the decision regarding the running of the test. If the test is not run, the decision to market or scrap is based on the prior probabilities, and this portion of the tree is identical to the top portion of the tree given in Fig. 9.3. If the test is run, the next fork is a chance fork corresponding to the result of the test, a positive reading or a negative reading. If a positive reading is observed, the optimal action is to market and the EMV is \$2.06, whereas if a negative reading is observed, the optimal action is to scrap and the EMV is −\$1.00. The expected value of sample information is \$0.94 − \$0.60 = \$0.34, and since the test costs \$0.15, it costs less than it is worth to the decision-maker.

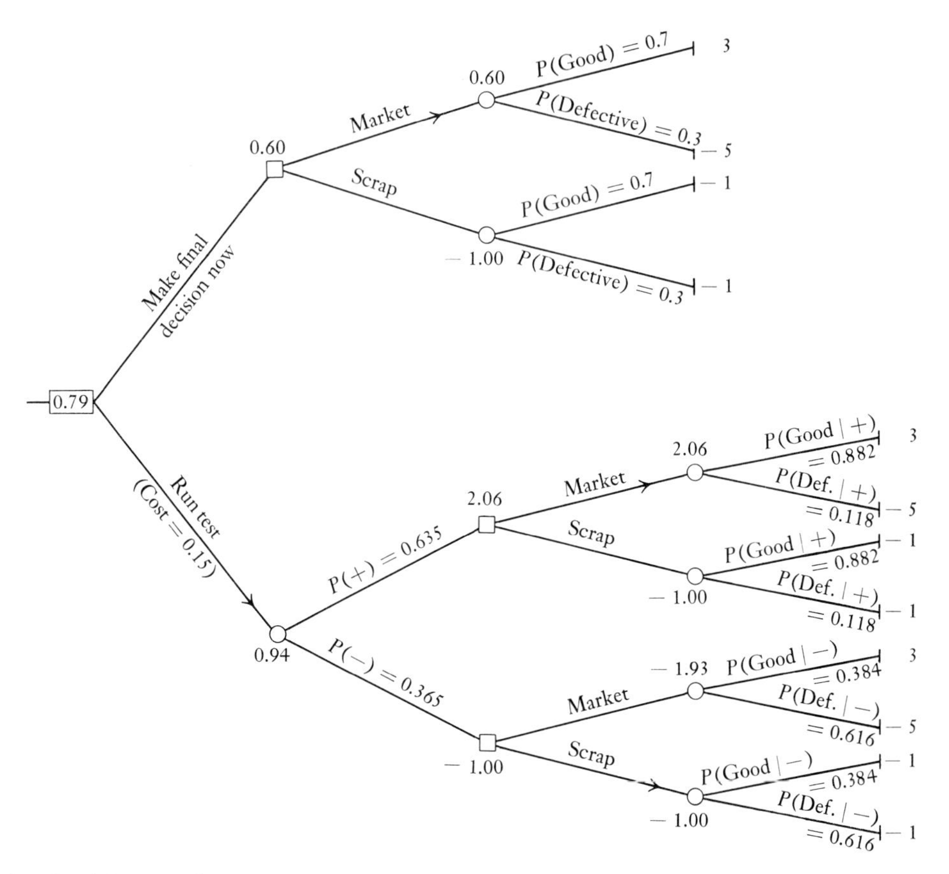

Fig. 9.4 Decision tree for test of component.

Note that even though this tree is much more complex than the simple trees presented in Section 9.3, it can still be solved using only the two simple rules given in that section: Find expected values at chance forks, and choose the action with the largest EMV at decision forks.

9.7 DETERMINING THE OPTIMAL SAMPLE SIZE

In the preceding section, the procedure for determining the EVSI and the ENGS for a particular experiment was presented and illustrated for a situation in which the experiment consisted of a test of a component produced by a certain process. The choice facing the decision-maker was whether or not to conduct the experiment. Often the situation is a bit more complicated, with several potential experiments being considered by the decision-maker. In the example, there might be several different tests that could be performed on the component in an attempt to obtain more information about whether the component is good or defective. In a market-research study, a number of different experiments might be contemplated to obtain more information about consumers' reactions to a new product. In auditing, different ways of obtaining more information about a firm's "books" might be considered. In each of these situations, the choice is not just between running an experiment and not running the experiment; the decision-maker must choose among no experimentation, experiment A, experiment B, experiment C, and so on. If the ENGS can be computed for each experiment, then the optimal experiment is simply the one with the largest ENGS, provided that the ENGS is positive. If the ENGS is negative for all experiments, then it is optimal not to experiment at all.

A situation of particular interest in statistics is that in which the only difference among the experiments that are being considered is a difference in sample size. In the production-process example, for instance, suppose that the testing device discussed in previous sections is the only testing device available to the decision-maker. However, there is nothing to prevent the decision-maker from running the same component through the testing device several times. Thus he wants to choose among the following experiments: do not use the testing device at all; test the component once; test the component twice; and so on. Similarly, a market researcher may have decided on the basic design for a study of consumers' reactions to a new product—his only question now is how many consumers to include in the experiment. The problem here is to find the optimal sample size, and the procedure is simply to compute the EVSI and ENGS for each sample size and to choose the sample size that yields the largest ENGS. (If $n = 0$ yields the largest ENGS, of course, no sample will be taken.)

Returning to the production-process example, suppose that a component can be run through the testing device any number of times, and that the successive tests on the same component can be regarded as independent. In the previous section, the EVSI and ENGS for a sample of size one were computed; now consider a sample of size two. For $n = 2$, there are three possible sample outcomes: two positive readings; one positive reading and one negative reading; and two negative readings. In Section 9.4 it was pointed out that for samples of more than one trial, it is not necessary to revise the probabilities after each trial. Instead, we can simply wait until all trials have been observed and *then* determine the revised probabilities. In that section, we computed the posterior probabilities for a sample of size two with two positive readings,

$$P(\text{Good} \mid +, +) = 0.960 \quad \text{and} \quad P(\text{Defective} \mid +, +) = 0.040.$$

Using these posterior probabilities, the expected payoffs for the two actions are

$$\text{EMV}(\text{Market} \mid +, +) = 3(0.960) + (-5)(0.040) = \$2.68$$

and

$$\text{EMV}(\text{Scrap} \mid +, +) = -1(0.960) + (-1)(0.040) = -1.00.$$

Thus, the optimal decision following two positive readings is to market, and the corresponding EMV is $2.68.

Next, consider a sample with one positive reading and one negative reading. There are two ways to obtain this result (+, − and −, +), so the likelihoods needed for the application of Bayes' rule are

$$P(+, - \text{ or } -, + \mid \text{Good}) = P(+, - \mid \text{Good}) + P(-, + \mid \text{Good})$$
$$= 0.80(0.20) + 0.20(0.80) = 0.320$$

and

$$P(+, - \text{ or } -, + \mid \text{Defective}) = P(+, - \mid \text{Defective}) + P(-, + \mid \text{Defective})$$
$$= 0.25(0.75) + 0.75(0.25) = 0.375.$$

The posterior probabilities are

$$P(\text{Good} \mid +, - \text{ or } -, +)$$
$$= \frac{P(+, - \text{ or } -, + \mid \text{Good})P(\text{Good})}{P(+, - \text{ or } -, + \mid \text{Good})P(\text{Good}) + P(+, - \text{ or } -, + \mid \text{Defective})P(\text{Defective})}$$
$$= \frac{0.320(0.70)}{0.320(0.70) + 0.375(0.30)} = 0.666$$

and

$$P(\text{Defective} \mid +, - \text{ or } -, +)$$
$$= \frac{P(+, - \text{ or } -, + \mid \text{Defective})P(\text{Defective})}{P(+, - \text{ or } -, + \mid \text{Good})P(\text{Good}) + P(+, - \text{ or } -, + \mid \text{Defective})P(\text{Defective})}$$
$$= \frac{0.375(0.30)}{0.320(0.70) + 0.375(0.30)} = 0.334.$$

From these posterior probabilities, the expected payoffs are

$$\text{EMV}(\text{Market} \mid +, - \quad \text{or} \quad -, +) = 3(0.666) + (-5)(0.334) = 0.33$$

and

$$\text{EMV}(\text{Scrap} \mid +, - \quad \text{or} \quad -, +) = -1(0.666) + (-1)(0.334) = -1.00.$$

The optimal decision is to market, and the corresponding EMV is 0.33.

Finally, consider the other possible sample result, two negative readings. Revising the probabilities on the basis of two negative readings yields

$$P(\text{Good} \mid -, -)$$
$$= \frac{P(-, - \mid \text{Good})P(\text{Good})}{P(-, - \mid \text{Good})P(\text{Good}) + P(-, - \mid \text{Defective})P(\text{Defective})}$$
$$= \frac{(0.20)^2(0.70)}{(0.20)^2(0.70) + (0.75)^2(0.30)} = 0.142$$

and

$$P(\text{Defective} \mid -, -)$$
$$= \frac{P(-, - \mid \text{Defective})P(\text{Defective})}{P(-, - \mid \text{Good})P(\text{Good}) + P(-, - \mid \text{Defective})P(\text{Defective})}$$
$$= \frac{(0.75)^2(0.30)}{(0.20)^2(0.70) + (0.75)^2(0.30)} = 0.858,$$

and the expected payoffs are

$$\text{EMV}(\text{Market} \mid -, -) = 3(0.142) + (-5)(0.858) = -3.86$$

and

$$\text{EMV}(\text{Scrap} \mid -, -) = -1(0.142) + (-1)(0.858) = -1.00.$$

The optimal action following two negative readings is to scrap the component, and the EMV is -1.00.

The optimal EMV following each possible sample result has been calculated, and only the probabilities of the three possible sample results are needed to compute EMV (Sample information). These probabilities are

$$P(+, +) = P(+, + \mid \text{Good})P(\text{Good}) + P(+, + \mid \text{Defective})P(\text{Defective})$$
$$= (0.80)^2(0.70) + (0.25)^2(0.30) = 0.46675,$$
$$P(+, - \text{ or } -, +) = P(+, - \text{ or } -, + \mid \text{Good})P(\text{Good})$$
$$+ P(+, - \text{ or } -, + \mid \text{Defective})P(\text{Defective})$$
$$= 0.320(0.70) + 0.375(0.30) = 0.33650,$$

and

$$P(-, -) = P(-, - \mid \text{Good})P(\text{Good}) + P(-, - \mid \text{Defective})P(\text{Defective})$$
$$= (0.20)^2(0.70) + (0.75)^2(0.30) = 0.19675.$$

Therefore, the overall expected payoff with a sample of size two, as calculated before the sample is actually taken, is

$$\text{EMV}(\text{Sample information})$$
$$= (\text{Highest EMV after } +, +)P(+, +)$$
$$+ (\text{Highest EMV after } +, - \quad \text{or} \quad -, +)P(+, - \quad \text{or} \quad -, +)$$
$$+ (\text{Highest EMV after } -, -)P(-, -)$$
$$= 2.68(0.46675) + 0.33(0.33650) + (-1.00)(0.19675)$$
$$= 1.16.$$

The expected value of sample information is

$$\text{EVSI} = \text{EMV}(\text{Sample information})$$
$$- \text{EMV}(\text{Optimal action under prior information})$$
$$= 1.16 - 0.60 = 0.56.$$

Thus, the sample of size two is expected to improve the decision-maker's EMV by \$0.56. If the cost of sampling is \$0.15 per test, as given in the preceding section, the cost of two tests is \$0.30, so the ENGS is

$$\text{ENGS} = \text{EVSI} - \text{CS} = \$0.56 - \$0.30 = \$0.26.$$

Recall, from Section 9.6, that the ENGS for a single test was calculated as \$0.19. Therefore, testing the component twice yields a larger ENGS than testing it only once. What about the possibility of testing it more than twice? We will not take the space here to present the computations for larger sample sizes. (As you might guess the computations become more burdensome as the sample size increases, but that is not a serious problem since it is quite easy in many situations to write a computer program to compute EVSI.) For the production-process example, EVSI, CS, and ENGS for sample sizes up to 6 are given in Table 9.4 and shown graphically in Fig. 9.5. Note that the EVSI increases fairly rapidly at first and then begins to level off. For this example, EVPI = \$1.20, so we know that EVSI can never be greater than \$1.20. The cost of sampling is just a linear function of n, $0.15n$. The ENGS rises at first, but when the incremental gain from a larger sample size levels off and is surpassed by the incremental cost of the larger sample size, the ENGS drops. For the example, the optimal sample size is 3; the decision-maker should test the component three times before deciding whether to market it or scrap it.

The above analysis implies that the decision-maker has more than one decision to make. First of all, he must decide whether or not to purchase sample information. Then, at the point at which he has decided to seek no more sample information, he must make the *terminal decision* between the potential actions available to him. We are primarily concerned with the terminal decision in this chapter. The decision regarding sampling is also important, however. The criterion for this decision is the same as the criterion for the terminal decision—maximize EMV. By comparing the EMV'S for different experiments with each other and with the EMV if *no* experiment is run, the decision regarding sampling can be made. Of course, once a sample has been taken, it is still

Table 9.4 EVSI, CS, and ENGS for Production-Process Example

n	EVSI	CS	ENGS
1	0.34	0.15	0.19
2	0.56	0.30	0.26
3	0.72	0.45	0.27
4	0.83	0.60	0.23
5	0.91	0.75	0.16
6	0.95	0.90	0.05

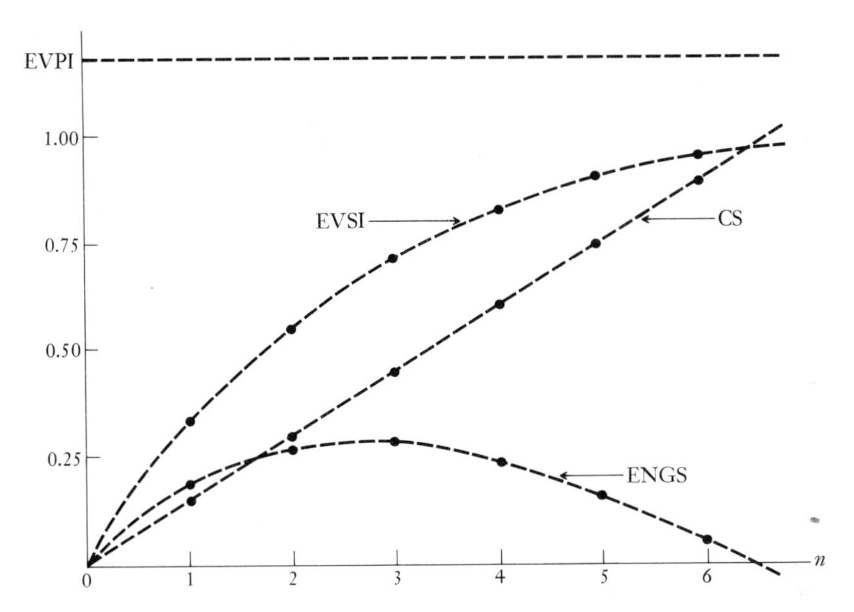

Fig. 9.5 EVSI, CS, and ENGS for production process example.

possible to determine whether or not to seek *more* sample information. It is only after a decision has been made *not* to seek more sample information that a terminal decision is made.

9.8 DECISION-MAKING UNDER UNCERTAINTY: AN EXAMPLE

In this section we attempt to apply the concepts that have been developed in this chapter to a slightly more complicated problem involving information-seeking. Even this problem will seem to be a somewhat simplified representation of the real world, but such simplification is often unavoidable. If every possible factor entering into a problem were included in the formal decision-theory analysis, the analysis would be much too cumbersome. If we can identify the *most important* factors, the formal analysis should be of value even if it is somewhat of a simplification.

A firm is considering the marketing of a new product. For convenience, suppose that the events of interest are simply θ_1 = "new product is a success" and θ_2 = "new product is a failure." The prior probabilities determined by the firm's top management are

$$P(\text{Success}) = 0.3 \qquad \text{and} \qquad P(\text{Failure}) = 0.7.$$

If the product is marketed and is a failure, the firm will suffer a loss of $300,000. On the other hand, if the product is a success, the firm will earn a net profit of

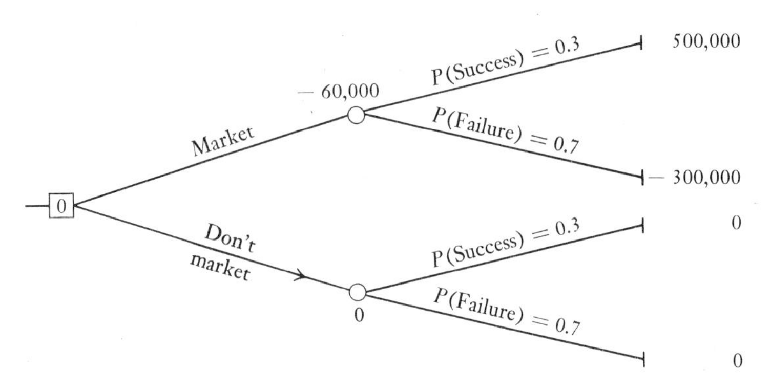

Fig. 9.6 Tree diagram for marketing problem.

\$500,000. If the product is not marketed, the firm will suffer no loss, nor will it earn any profit. This is a relatively simple, straightforward problem with only two actions and two states of nature. It can be represented on a tree diagram, as in Fig. 9.6, and the expected payoffs of the two actions are

$$\text{EMV(Market)} = 500{,}000(0.3) + (-300{,}000)(0.7) = -60{,}000$$

and

$$\text{EMV(Don't Market)} = 0(0.3) + 0(0.7) = 0.$$

It appears, therefore, that on the basis of the current information, as represented by the prior probabilities, $P(\text{Success}) = 0.3$ and $P(\text{Failure}) = 0.7$, the firm should *not* market the product.

Although it looks as though the *optimal decision* is not to market the product, there is a considerable amount of uncertainty concerning the eventual success or failure of the product. Therefore, additional information might be useful to the firm. Perfect information is not available—the only way to tell *for sure* whether the product will succeed is to go ahead and market it. Nevertheless, the expected value of perfect information is easy to calculate and serves as a convenient benchmark, in the sense that it is an upper bound for the expected value of sample information. With perfect information, the firm will market the product and earn \$500,000 if they are assured that the product will be successful; if the perfect information indicates that the product will not succeed in the marketplace, the firm will not pursue the matter and will wind up with a payoff of \$0. *Before* seeing the perfect information, then, the expected payoff under perfect information is

$$\text{EMV(Perfect information)} = 500{,}000(0.3) + 0(0.7) = 150{,}000.$$

Since the optimal action under the prior information is not to market, and this action has an EMV of \$0, the expected value of perfect information (EVPI) is simply

$$\text{EVPI} = \text{EMV(Perfect information)} - \text{EMV}\begin{pmatrix}\text{Optimal action under} \\ \text{prior information}\end{pmatrix}$$

$$= 150{,}000 - 0 = 150{,}000.$$

Thus, the firm should be willing to pay up to $150,000 for perfect information concerning the success or failure of the product.

As we can see from our calculations of EVPI in this problem and from the intuitively reasonable notion that decision-making is much less risky under certainty, a clairvoyant who was "genuine" in the sense of always being able to predict perfectly could earn large sums of money! In the absence of any such clairvoyant, decision-makers have the option to make their decisions without additional information, or to purchase information that is less than perfect. In our example, suppose that the firm is considering the purchase of additional information in the form of a market survey. The market-research department of the firm proposes two possible surveys, labelled A and B. The result of each survey will be an indication of "favorable," "neutral," or "unfavorable." If the product will actually be successful in the marketplace, there is a 0.60 chance that Survey A will give a favorable indication, a 0.30 chance of a neutral indication, and a 0.10 chance of an unfavorable indication. If the product will not be successful, there is a 0.10 chance of favorable indication from Survey A, a 0.20 chance of a neutral indication, and a 0.70 chance of an unfavorable indication. Survey A costs $20,000 to conduct.

Before going on to Survey B, we shall determine the expected value of sample information from Survey A. Recalling that Bayes' rule is of the form

$$\text{Posterior probability} = \frac{\text{(Prior prob.)(Likelihood)}}{\sum\text{(Prior prob.)(Likelihood)}},$$

where the sum is taken over the possible states of nature, we can present the calculations of the posterior probabilities in tabular form. The first column in Table 9.5 gives the possible sample outcomes—favorable (F), neutral (N), and unfavorable (U). The second column gives the two states of nature, success and failure of the product. The third column gives the prior probabilities for θ_1 and θ_2, the states of nature, and the fourth column gives the likelihoods. From the information given above, for example, the likelihood of a favorable indication from Survey A given that the product will be successful is 0.60. The fifth column is the product of the prior probabilities and the likelihoods, and the last column gives the posterior probabilities. In each case, this is the value in the fifth column divided by the sum of all of the elements in the fifth column for a given sample result. This last operation serves to *normalize* the posterior probabilities —to make them sum to one. From the values computed in Table 9.5., we see that the posterior probability that the product will be successful is 18/25 if Survey A yields a favorable indication, 9/23 if Survey A yields a neutral indication, and 3/52 if Survey A yields an unfavorable indication.

Table 9.5 Posterior Probabilities After Survey A

(1) Sample Outcome	(2) θ	(3) Prior Prob.	(4) Likelihood	(5) (3) × (4)	(6) Posterior Prob.
Favorable	θ_1	0.30	0.60	0.18	18/25
	θ_2	0.70	0.10	0.07	7/25
				0.25	
Neutral	θ_1	0.30	0.30	0.09	9/23
	θ_2	0.70	0.20	0.14	14/23
				0.23	
Unfavorable	θ_1	0.30	0.10	0.03	3/52
	θ_2	0.70	0.70	0.49	49/52
				0.52	

Figure 9.7 is a tree diagram for the calculation of the expected value of sample information from Survey A. Observe that if the survey is taken, the branches marked F, N, and U represent the three possible results from the survey. The probabilities on the event branches following these survey results are the posterior probabilities calculated in Table 9.5. Because the portion of the tree following the "Make decision now" branch was presented in Fig. 9.6, we simply put the EMV of this action in Fig. 9.7 instead of reproducing

Fig. 9.7 Tree diagram for EVSI (Survey A).

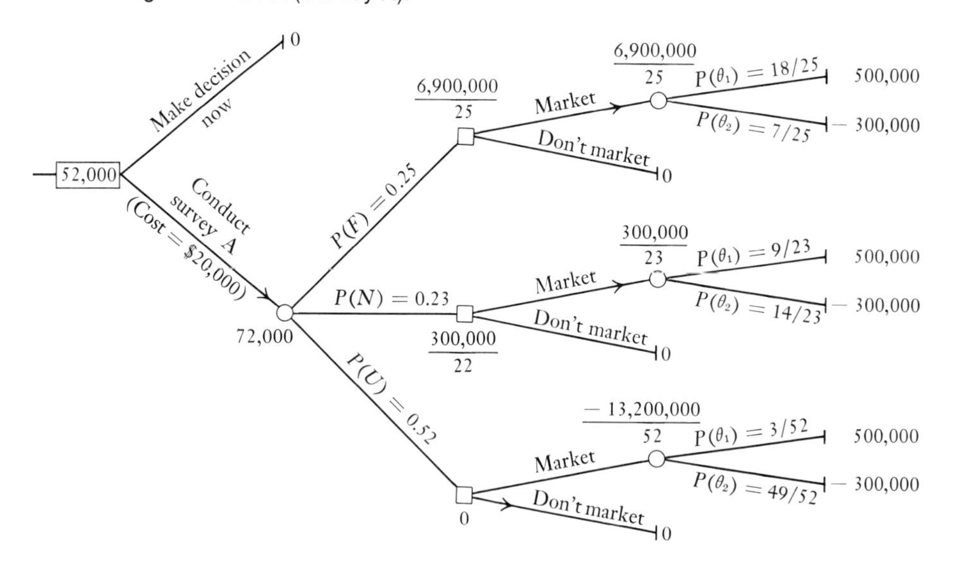

that entire portion of the tree. Looking at the lower portion of the tree once again and working backward from the righthand side of the tree, we see that the expected payoff of marketing the product, if the survey indication is favorable, is

$$500{,}000\left(\frac{18}{25}\right) + (-300{,}000)\left(\frac{7}{25}\right) = \frac{6{,}900{,}000}{25}.$$

If the survey is neutral, we have

$$500{,}000\left(\frac{9}{23}\right) + (-300{,}000)\left(\frac{14}{23}\right) = \frac{300{,}000}{23}$$

and if the survey is unfavorable, we have

$$500{,}000\left(\frac{3}{52}\right) + (-300{,}000)\left(\frac{49}{52}\right) = -\frac{13{,}200{,}000}{52}.$$

We now know what the firm should do after any sample: if the survey indication is favorable or neutral, the firm should market the product (since the the EMV is positive, while the EMV for not marketing is zero); if the survey indication is unfavorable, the firm should not market the product. Incidentally, the EMV's are left in fractional form to simplify the calculations at the next step. For instance, when multiplying 6,900,000/25 by 0.25, the computation is much easier than it would be if the actual division 6,900,000/25 were carried out before multiplying by the 0.25 figure. The probabilities for the three survey outcomes, as determined from Table 9.4, are the sums of the relevant figures in the fifth column of the table—they are 0.25, 0.23, and 0.52. Continuing to work backward, we see that the overall EMV of the "Conduct Survey A" branch is

$$(0.25)\left(\frac{6{,}900{,}000}{25}\right) + (0.23)\left(\frac{300{,}000}{23}\right) + (0.52)(0) = 72{,}000.$$

Since the EMV of the "Make decision now" branch is 0, the expected value of sample information (EVSI) for Survey A is

$$\text{EVSI(Survey A)} = \text{EMV(Survey A)} - \text{EMV}\begin{pmatrix}\text{Optimal action under}\\ \text{prior information}\end{pmatrix}$$

$$= 72{,}000 - 0 = 72{,}000.$$

Therefore, the firm should be willing to pay up to $72,000 for Survey A. But the cost of Survey A is $20,000, so the expected net gain from Survey A is $72,000 − $20,000 = $52,000.

Next, we shall consider Survey B, which costs $30,000. If the product will be successful, there is a 0.80 chance that Survey B will give a favorable indication, a 0.10 chance of a neutral indication, and a 0.10 chance of an unfavorable

Table 9.6 Posterior Probabilities After Survey B

(1) Sample Outcome	(2) θ	(3) Prior Prob.	(4) Likelihood	(5) (3) × (4)	(6) Posterior Prob.
Favorable	θ_1	0.30	0.80	0.24	24/31
	θ_2	0.70	0.10	0.07	7/31
				0.31	
Neutral	θ_1	0.30	0.10	0.03	3/31
	θ_2	0.70	0.40	0.28	28/31
				0.31	
Unfavorable	θ_1	0.30	0.10	0.03	3/38
	θ_2	0.70	0.50	0.35	35/38
				0.38	

indication. If the product will not be successful, there is a 0.10 chance of a favorable indication, a 0.40 chance of a neutral indication, and a 0.50 chance of an unfavorable indication. The calculation of posterior probabilities following each of the possible outcomes from Survey B is presented in Table 9.6, and the tree diagram in Fig. 9.8 enables us to calculate the EVSI for Survey B. Observe that after Survey B is conducted, the optimal decision is to market if a favorable indication is obtained, but not to market otherwise. A neutral indication here leads to not marketing, whereas in Survey A it led to marketing.

Fig. 9.8 Tree diagram for EVSI (Survey B).

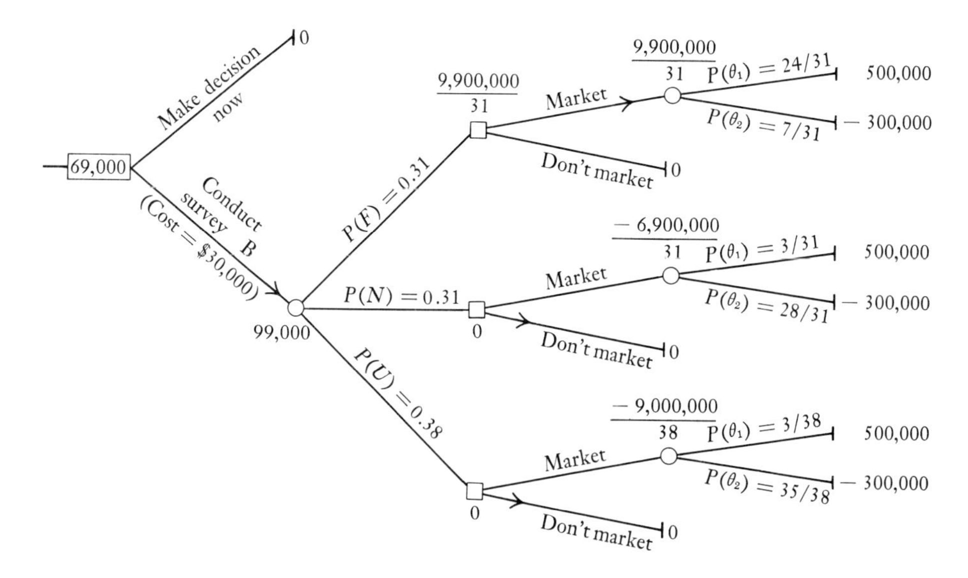

The overall EMV of the "Conduct Survey B" branch is \$99,000, so the expected value of sample information from Survey B is

$$\text{EVSI(Survey B)} = \text{EMV(Survey B)} - \text{EMV}\begin{pmatrix}\text{Optimal action under}\\ \text{prior information}\end{pmatrix}$$

$$= 99{,}000 - 0 = 99{,}000.$$

Therefore, the firm should be willing to pay up to \$99,000 for Survey B. But the cost of Survey B is only \$30,000, so the expected net gain from Survey B is \$99,000 − \$30,000 = \$69,000.

Summarizing the results of the analysis to this point, we see that the EMV of making a marketing decision without additional information is \$0, since the optimal action under the prior probabilities is not to market the product. Perfect information has an expected value of \$150,000, but perfect information is not available to the firm in this situation. Survey A has an expected value of \$72,000 and it costs \$20,000 to conduct, so its net expected value is \$52,000. Survey B has an expected value of \$99,000, which is almost two-thirds of EVPI, and the cost of Survey B is \$30,000, so the net expected gain from Survey B is \$69,000. At this point, it appears that the best strategy for the firm to follow is to conduct Survey B. From Fig. 9.8, the firm should market the product only if Survey B yields a favorable indication.

Thus, we have compared three information-seeking options: Survey A, Survey B, and the option *not* to seek any more information. Of these three options, Survey B is the best choice. Of course, the firm may wish to consider yet other surveys, although it must be kept in mind that Survey B has an EVSI that is just about two-thirds of EVPI. The EVSI of any other survey can be at best only

$$\$150{,}000 - \$99{,}000 = \$51{,}000$$

better than the EVSI of Survey B, and any survey that yields an EVSI that is very close to EVPI might be expected to be quite expensive. One option that can be evaluated quickly is the possibility of conducting *both* Survey A *and* Survey B. Assuming that the two surveys can be considered to be independent, the likelihoods for conducting both surveys are simply the products of the likelihoods of the individual surveys. Since each survey has three possible outcomes, there are 3(3) = 9 possible combinations of outcomes. We will not present the calculations, because this example is already quite lengthy; but it turns out that the EVSI for the combination of both samples is \$102,300. This EVSI is only slightly better than the EVSI of Survey B alone, and the cost is \$20,000 greater than the cost of Survey B alone, so the combination of both surveys is not the best course of action.

Often a sequential information-seeking strategy can be advantageous. In our example, the firm might conduct Survey A and then consider whether or not to conduct Survey B. If Survey A yields relatively conclusive results, it is probably not worthwhile to continue and take Survey B, but if the results are somewhat inconclusive, Survey B might prove valuable. We will not present the calculations, but the expected net gain from this sequential plan is $73,900, which is $4900 greater than the expected net gain from Survey B alone. If Survey A yields a favorable indication, the best strategy is to make a final decision to market the product; if Survey A yields an unfavorable indication, the best strategy is to make a final decision not to market the product. However, if Survey A yields a neutral indication, the best strategy is to conduct Survey B, marketing the product if Survey B is favorable and not marketing it otherwise. Proceeding sequentially rather than deciding in advance to use both surveys has the advantage of avoiding the cost of the second survey when it appears on the basis of the results of the first survey that the second survey is not worthwhile.

This example illustrates the comparison of different sampling plans. In comparing such plans, the plan with the highest net expected gain should be chosen, provided, of course, that this net expected gain is greater than zero. If all net expected gains are less than zero, then the decision-maker should make a final decision without obtaining any additional information.

Define: EVPI, EVSI, ENGS, CS.

Problems

9.17 A store must decide whether or not to stock a new item. The decision depends on the reaction of consumers to the item, and the payoff table (in hundreds of dollars) is shown in Table 9.7. If $P(0.10) = 0.2$, $P(0.20) = 0.3$, $P(0.40) = 0.1$, and $P(0.50) = 0.1$, what decision should be made? For this problem, determine the expected value of perfect information.

Table 9.7

		Proportion of Consumers Purchasing the Item				
		0.10	0.20	0.30	0.40	0.50
Decision	Stock 100	−10	−2	12	22	40
	Stock 50	−4	6	12	16	16
	Don't stock	0	0	0	0	0

9.18 a) Carefully distinguish between the expected *payoff* under perfect information and the expected *value* of perfect information. Calculate both of these values for Problem 9.2.

b) How much should you be willing to pay for perfect information in the stock example in Fig. 9.1 if $P(\text{I}) = 0.30$?

c) In decision-making problems under uncertainty, is perfect information typically available to the decision-maker?

9.19 In Problem 9.9, suppose that the probability of a recession is 0.4. What is your optimal action, and what is the EMV associated with that action? Next, suppose that you could find out for certain whether or not a recession will occur during the coming year. Before obtaining this information, what is your EMV (perfect information)? What is the expected value of the information?

9.20 A corporation considers two levels of investment in a real-estate development, a low participation (A_1) or a high participation (A_2). Two states of nature are deemed possible, a partial success (B_1) or a complete success (B_2). The payoff matrix is estimated to be:

	B_1	B_2
A_1	−200	400
A_2	−500	1000

a) How large does the *a priori* probability of B_1 have to be in order to make action A_1 the better choice?

b) Suppose that the states of nature are presumed initially to occur with probabilities $P(B_1) = 0.4$, $P(B_2) = 0.6$. Then a more careful study is made, which leads to the conclusion that the project will be only a partial success. In previous relevant studies, this same conclusion was obtained in eight out of 10 cases when similar projects were partial successes. Also, this conclusion was obtained in four out of 12 cases when similar projects were *complete* sucesses. Find the revised probabilities of the states of nature, and determine which investment level is appropriate.

9.21 Joe Doakes is considering flying from New York City to Boston in the hope of making an important sale to P. J. Bety, president of NOCO, Inc. If Joe makes this sale he will earn a commission of $1100. Unfortunately, Joe figures there is a 50–50 chance that Bety will be called out of town at the last moment and he will have no chance at a sale. Even if he goes to see Bety, Joe estimates that he has only one chance in five of making the sale. The trip will cost Joe $100, whether or not he gets to see Bety.

a) Draw the tree diagram for Joe, and determine whether Joe should fly to Boston or not.

b) Joe was heard to remark "I'd give my right arm to know if Bety will be in town." How much does he value his right arm?

c) Suppose an information service offers to tell Joe, before he decides to fly, whether or not they think Bety will be in town. The record of this company is such that if they say Bety will be in, the probability that he will, in fact, be in is 0.70, i.e., the posterior probability is $P(\text{in}|\text{say in}) = 0.70$. If they say he will be out, they will be correct 90% of the time; i.e., $P(\text{out}|\text{say out}) = 0.90$. If this service costs $10, and Joe figures that the probability that they will say Bety is in is $\frac{2}{3}$, should he buy the service? Draw the tree diagram. Find ENGS.

9.22 For the component example, find the expected value of sample information for the new test in Problem 9.15. Compare this with the EVSI found in the text for the other test, and discuss any difference in the EVSI figures for the two tests.

9.23 In Problem 9.12, each coat costs $80 to produce (not including inspection costs) and is sold to a distributor for $100. Imperfect coats that are shipped out are returned and repaired at a cost of $30. If a coat is sent back for repairs before being shipped out, the cost is only $10. Any imperfections are found and fixed in the repairing process. The cost of inspection is $1 per coat.

a) If no inspection is made, should each coat be shipped out or put through the repairing process before being sent out?

b) If the inspector classifies a coat as imperfect, should it be shipped out? What is the EMV?

c) If the inspector does not classify a coat as imperfect, should it be shipped out? What is the EMV?

d) What is EMV (sample information) for an inspection?

e) Find the EVSI and ENGS for an inspection.

9.24 In Problem 9.8, suppose that the firm finds out that its competitor is building a new plant. It is likely that the new plant is intended to produce the new product mentioned in that problem, although it could also be used to produce other items. If the competitor really does intend to introduce the new product, the chances are very good (probability 0.8) that a new plant would be built. On the other hand, if the competitor does not intend to introduce the new product, the probability is only 0.3 that a new plant would be built. On the basis of the information that the plant is being built, revise the probabilities concerning the new product and find the optimal decision for the firm by using the revised probabilities.

9.25 In Problem 9.24, suppose that the firm does not yet know whether the competing firm is building a new plant. It may be possible to find out, at some cost, whether the new plant is being built. What is the maximum amount the firm should be expected to pay for such information?

9.26 Suppose that a firm is deciding whether to sell bonds to raise money. A decision tree is constructed, and one of the uncertainties concerns the rate of inflation. Additional information can be obtained: an economist's forecast as to whether the inflation rate will be high, moderate, or low. After performing the necessary calculations, the analyst finds that with a forecast of high inflation, the firm can maximize EMV by selling bonds. However, the same result is obtained with a forecast of moderate inflation or a forecast of low inflation. What can the analyst say about the EVSI associated with the forecast? Would the EVSI for this inflation forecast necessarily be the same for a different firm faced with a different decision-making problem?

9.27 For the component example, suppose that the new test in Problem 9.15 will be used on each item n times. Find EVSI when $n = 1$, when $n = 2$, and when $n = 3$. If the cost of the test is $0.15, like the other test, find the ENGS values for the new test with $n = 1$, $n = 2$, and $n = 3$.

9.28 In Problem 9.17, suppose that sample information is available in the form of a random sample of consumers. For a sample of size *one*,

a) Find the posterior distribution if the one person sampled will purchase the item, and find the posterior EMV's and the optimal action under the posterior distribution.

b) Find the posterior distribution if the one person sampled will not purchase the item, and find the posterior EMV's and the optimal action under the posterior distribution.

c) Calculate the expected value of sample information for this sample of size one.

d) Draw the decision tree.

9.29 In Problem 9.28, suppose that a sample of *two* consumers is being contemplated.

a) List the possible sample outcomes and find the posterior distribution following each possible outcome.

b) Find the optimal action under each of the posterior distributions found in part (a).

c) Find the expected value of sample information for a sample of two consumers.

d) Draw the decision tree.

9.30 a) What is the relationship, if there is any, between the expected value of perfect information and the expected value of sample information?

b) If information is free, is it ever better to obtain sample information instead of perfect information? If information is not free, is sample information ever preferred to perfect information? Explain your answers.

c) For the situation described in Problem 9.17 compare the EVPI (as calculated in that problem), the EVSI for a sample of one consumer (as calculated in Problem 9.28), and the EVSI for a sample of two consumers (as calculated in Problem 9.29). Explain any differences among these three values.

9.31 a) Assume that for the example in Section 9.8, a third survey, Survey C, indicates either favorable or unfavorable (there is no "neutral" indication). The probability of a favorable indication is 0.9 if the product will be successful and 0.3 if the product will not be successful. Find the EVSI for Survey C.

b) Show that the EVSI for the combination of both surveys in the Section 9.8 example is $102,300, as claimed in the text.

9.32 The Dixon Corporation makes picture tubes for a large television manufacturer. Dixon is concerned because approximately 30% of their tubes have been defective. When the television manufacturer encounters a defective tube, Dixon is charged a $20 penalty cost (to pay for repairs and lost time). One way Dixon can avoid this penalty cost is to re-examine and fix each defective tube before shipping. This would cost an extra $7 per tube. Or they can rent a testing device that costs $1 for each tube tested. Since this device is not infallible, its effectiveness was tested by running through it a large number of tubes, some known to be good, and others known to be defective. The results of this study determined the following likelihoods.

		State of Tube	
		Good	**Defective**
Test results	Good	0.75	0.20
	Def	0.25	0.80
		1.00	1.00

Draw the decision tree for Dixon, assuming that they must decide between shipping directly, re-examining each tube, or testing each tube. Calculate ENGS and EVSI.

9.33 The Techno Corporation is considering making either minor or major repairs to a malfunctioning production process. When the process is malfunctioning, the percentage of defective items produced seems to be a constant, with either $p = 0.10$ (indicating minor repairs necessary) or $p = 0.25$ (indicating major repairs necessary). Defective items are produced randomly, and there is no way Techno can tell for sure whether the machine needs minor or major repairs. If minor repairs are made when $p = 0.25$, the probability of a defective is reduced to 0.05. If minor repairs are made when $p = 0.10$, or major repairs made when $p = 0.10$ or $p = 0.25$, then the proportion of defectives is reduced to zero. Techno has recently received an order for 1000 items. This item yields them a profit of $0.50 per unit, except that they have to pay a $2.00 penalty cost for each item found defective. Major repairs to the process cost $100 while minor repairs cost $60. No adjustment can be made to the production process once a run has started. Prior to starting the run, however, Techno can sample items from a "trial" run, at a cost of $1.00 per item. The prior probabilities are $P(\text{major}) = 0.3$, $P(\text{minor}) = 0.7$.

a) Find the optimal action for Techno if they are trying to decide between not sampling at all and sampling one item. Draw the decision tree.

b) Find the optimal action for Techno if they are willing to consider a sample of either one or two items. Draw the decision tree. Calculate ENGS and EVSI.

---

* 9.9 BAYES' RULE FOR NORMAL DISTRIBUTIONS (OPTIONAL)

In Chapters 7 and 8, the importance of the normal distribution in statistics was clearly demonstrated. Most of the confidence intervals and tests of hypotheses discussed in those chapters involved the use of the normal distribution, and these techniques are widely applicable (especially for large samples, due to the Central Limit Theorem). In this chapter we present a method for incorporating prior information as well as sample information in the inferential and decision-making process. Unfortunately, only the discrete case is considered in Section 9.4. The procedures of Section 9.4 can be applied in the continuous case if the continuous distributions of interest are approximated by discrete distributions. Moreover, Bayes' rule is also applicable if the variable of interest is continuous rather than discrete:

Bayes' Rule for Continuous Distributions

If $f(\theta)$ is the prior density function of a parameter θ and $\ell(\text{Sample}|\theta)$ is the likelihood function, then the posterior density function of θ is

$$f(\theta|\text{Sample}) = \frac{f(\theta)\ell(\text{Sample}|\theta)}{\int f(\theta)\ell(\text{Sample}|\theta)\,d\theta}.$$

Note that this is very similar to Bayes' rule for discrete distributions, presented in Section 9.4. In the discrete case, the prior probabilities are multiplied by the appropriate likelihoods and the resulting numbers are divided by their sum, so that they will sum to one. In the continuous case, the prior density function is multiplied by the likelihood function and the resulting function is divided by its integral over all values of θ, so that the posterior density function will integrate to one (recall that a continuous variable must have a density function with a total area of one under the graph of the function).

As previously presented, Bayes' rule is applicable to any choice of a prior distribution and likelihood function. We are particularly interested in the normal distribution, however, so let us consider this special case. Suppose that we are sampling from a normal population with *known* variance σ^2, that we are interested in making inferences about μ (the mean of the population), and that the following two conditions are satisfied:

1. The prior distribution of μ is a normal distribution with mean m and variance v.
2. A sample of size n is observed, with sample mean $\bar{x}$.

Under these conditions, the posterior distribution of μ is a normal distribution with mean

$$M = \frac{(m/v) + (n\bar{x}/\sigma^2)}{(1/v) + (n/\sigma^2)}$$

and variance

$$V = 1/[(1/v) + (n/\sigma^2)].$$

For example, suppose that $\sigma^2 = 144$ and the prior distribution of μ has mean 60 and variance 48 (assuming, of course, that it is a normal distribution). A sample of size 4 is taken, with an observed sample mean of 70. The posterior distribution is then a normal distribution with mean

$$M = \frac{(60/48) + 4(70)/144}{(1/48) + (4/144)} = 65.71$$

and variance

$$V = 1/[(1/48) + (4/144)] = 20.57.$$

The numerical example illustrates a number of interesting features of Bayes' rule as applied to normal distributions. The most important feature is the fact that if the prior distribution is normal and the sample comes from a normal population with known variance, then the posterior distribution is also normal. Remember that if we have a series of samples, Bayes' rule can be applied successively to each sample, thus continually revising the probabilities. In this special case, the posterior distribution is normal, which implies that if another sample is taken, the same procedure can be repeated, using the posterior distribution following the first sample as the prior distribution prior to the second sample (this is possible because all of the conditions given above are still satisfied).

A second point of interest is that the posterior mean lies between the prior mean and the sample mean. This result is intuitively appealing, since the use of Bayes' rule is nothing more than a combination of prior information and sample information. It will always be true that the posterior mean will lie between the prior mean and the sample mean.

A third feature of Bayes' theorem for normal distributions is that the variance of the posterior distribution is smaller than the variance of the prior distribution (in the example, $V = 20.57$ and $v = 48$). This is reasonable, for the new information obtained in the sample should reduce our uncertainty concerning μ and hence reduce the variance, just as an increase in sample size reduces the variance of the sample mean.

In the Bayesian approach to statistics, the posterior distribution should be used instead of the sampling distribution or the likelihood function in making inferences. If we want to estimate μ, for example, we might take M, the mean of the posterior distribution of μ, rather than the estimator $\bar{x}$, which is based solely on sample information. Similarly, confidence intervals and tests of hypotheses should be based on the posterior distribution.

For a slightly more realistic example of Bayes' rule for normal distributions, suppose that an accountant in the credit department of a department store is concerned with μ, the average outstanding balance for the store's charge accounts. It is known from historical data that the standard deviation of outstanding balances is 5 (all values are expressed in dollars). Furthermore, the accountant's judgments about the average balance can be represented by a normal distribution with a mean of $m = 12$ and a variance of $v = 4$. This implies, for example, that he feels that there is approximately a 0.95 probability that μ is between

$$12 - 2\sqrt{4} = 8 \qquad \text{and} \qquad 12 + 2\sqrt{4} = 16.$$

In order to obtain additional information, the accountant randomly selects a sample of 25 accounts, and the average balance on these accounts is 9. Assuming that the population distribution of outstanding balances can be approximated by a normal distribution (this might be questionable in this situation because of a large number of zero balances and a small number of very large balances), the accountant's posterior distribution for μ is a normal distribution with mean

$$M = \frac{(12/4) + 25(9)/25}{(1/4) + (25/25)} = 9.6$$

and variance

$$V = 1/[(1/4) + (25/25)] = 0.8.$$

Thus, after seeing the sample information, the accountant's posterior point estimate for μ is 9.6 and the probability is approximately 0.95 that μ is between

$$9.6 - 2\sqrt{0.8} = 7.81 \quad \text{and} \quad 9.6 + 2\sqrt{0.8} = 11.39.$$

If the accountant is interested in the hypotheses

$$H_0: \mu \geq 10 \quad \text{and} \quad H_a: \mu < 10,$$

he can calculate the posterior probabilities of these hypotheses:

$$P(H_0) = P(\mu \geq 10) = P\left(z \geq \frac{10 - 9.6}{\sqrt{0.8}}\right) = P(z \geq 0.45)$$

$$= 0.3264$$

and

$$P(H_a) = P(\mu < 10) = P\left(z < \frac{10 - 9.6}{\sqrt{0.8}}\right) = P(z < 0.45)$$

$$= 0.6736.$$

Thus, on the basis of the posterior distribution H_a appears to be $0.6736/0.3264 = 2.06$ times as likely as H_0.

These examples illustrate the ease with which Bayes' rule can be applied when the population of interest is normally distributed with unknown mean and known variance, and when the prior distribution of the population mean μ is a normal distribution. In this situation, the normal prior distribution is said to be a *natural conjugate distribution* with respect to the normal population. In working with Bayes' rule for continuous distributions, the analysis is greatly simplified if the prior distribution is a member of the family of distributions that is a natural conjugate family with regard to the population or data-generating process of interest. Otherwise, it is necessary to carry out the integration in the denominator of Bayes' rule for continuous distributions, and this integration can sometimes be quite difficult. Using natural conjugate prior distributions means that the application of Bayes' rule can be expressed in a few simple formulas such as the formulas given above for M and V in the normal case. The combination of a normal prior distribution and a normal population is but one example of a natural conjugate relationship. If we are dealing with a Bernoulli process, for example, as discussed in Chapter 4, the natural conjugate family of prior distributions for p, the Bernoulli parameter, is the family of beta distributions; for the Poisson process, also discussed in Chapter 4, the natural conjugate family for λ, the Poisson parameter, is the family of gamma distributions. We will not discuss these families of distributions here—for a relatively nontechnical discussion of the notion of natural conjugate prior distributions, see Winkler, *An Introduction to Bayesian Inference and Decision.* A more advanced treatise on the subject is Raiffa and Schlaifer, *Applied Statistical Decision Theory.*

9.10 INFERENCE AND DECISION

In a way, statistical inference is closely related to statistical decision theory. The theory of estimation is concerned with deciding on an estimate, and the theory of hypothesis-testing involves a choice between two actions: accepting or rejecting a hypothesis. As we have seen, a formal theory has been developed to determine estimates and to test hypotheses; and this theory, together with probability theory, forms the backbone of the traditional "classical" approach to statistics. Decision theory represents an extension of the classical theory in the sense that it includes the same inputs as the classical theory (likelihoods) and allows other inputs (prior probabilities, payoffs). The decision-theory approach to statistics has, for an obvious reason, been labeled "Bayesian statistics." In this view, the use of prior probabilities and payoffs is extended to estimation and hypothesis-testing procedures. In estimation, for example, the costs of *overestimation* and *underestimation* are introduced into the analysis. In hypothesis-testing, the prior probabilities of each of the hypotheses under consideration are included (e.g., see the example in the preceding section), as well as the costs of Type I and Type II errors.

An example should serve to demonstrate how estimation can be thought of as a decision-making problem. Suppose that an appliance dealer must place an order for television sets and that this is his last chance to order the current model. Next month's order will be for the manufacturer's new model, and any current-model televisions still in stock at that time will be sold by the appliance dealer at a reduced price. The television sets cost the appliance dealer $400 each, and he sells them for $500. Any leftover current models will be sold for $350, however. In a sense, we can think of the number of sets ordered as an estimate of the demand for the sets. Moreover, unlike previous estimation problems we have considered, the costs of overestimating and underestimating can be determined in this problem from the information given above. If too many sets are ordered, the appliance dealer will suffer a loss because he will have to sell leftover sets at a loss of $50 per set. If too few sets are ordered, the appliance dealer will be forgoing a chance to make additional profits of $100 per set. In addition, a complete analysis of this problem might include consideration of the cost of carrying inventory, the cost due to unsatisfied customers who come in to buy television sets and find that there are none in stock, and so on. We ignore these considerations in this example.

Suppose that the appliance dealer is sure that the demand for television sets between now and the time the new models appear will be between six and ten sets. Moreover, on the basis of his past experience with demand near the end of a model year, he feels that the probability distribution of demand is as follows:

θ (Demand)	$P(\theta)$
6	0.10
7	0.25
8	0.30
9	0.25
10	0.10

Clearly, then, he will not order fewer than 6 sets or more than 10 sets. From the information given in the preceding paragraph, a payoff table can be determined for this problem, and the payoff table is given in Table 9.8. For instance, is 8 sets are ordered and the demand is only 7, the appliance dealer makes a profit of \$700 on the 7 sets that are sold but loses \$50 on the one set that is left over and must be sold at a reduced price of \$350. Thus, his overall payoff is \$700 − \$50 = \$650. The expected payoff for the first action, ordering 6 sets, is

$$\begin{aligned}\text{EMV(Order 6)} &= 600(0.10) + 600(0.25) + 600(0.30) \\ &\quad + 600(0.25) + 600(0.10) \\ &= 600.\end{aligned}$$

Table 9.8 Payoff Table for Television Example

		Demand				
		6	7	8	9	10
Action (Number of television sets ordered)	6	600	600	600	600	600
	7	550	700	700	700	700
	8	500	650	800	800	800
	9	450	600	750	900	900
	10	400	550	700	850	1000

Similarly,

$$\text{EMV(Order 7)} = 685,$$
$$\text{EMV(Order 8)} = 732.5,$$
$$\text{EMV(Order 9)} = 735,$$

and

$$\text{EMV(Order 10)} = 700.$$

Hence, the optimal action for the appliance dealer is to order 9 television sets.

Observe, from the probability distribution for demand, that the mean of the distribution is 8 sets. Yet in a decision-theoretic sense, the best estimate of demand is 9 sets. This is because the cost of underestimating demand is $100 per set (the lost profits due to not having enough sets on hand), whereas the cost of overestimating demand is only $50 per set (the loss due to selling leftover sets at a reduced price). Since the cost of overestimation is less than the cost of underestimation, the optimal strategy is to increase the estimate in order to avoid the higher cost of underestimation. If the two costs were equal, then by symmetry the best estimate would be the mean, 8 sets. In this case, the asymmetry in the costs causes the appliance dealer to order one additional set. In general, the optimal estimate in this type of problem is the

$$\frac{100k_u}{(k_u + k_o)}$$

percentile of the probability distribution, where k_u is the cost of underestimation and k_o is the cost of overestimation. In the example, $k_u = 100$ and $k_o = 50$, so we want the $100(100)/(100 + 50) = 67$th percentile. For the given distribution, 9 is the 67th percentile.

In order to illustrate the use of continuous distributions in decision-making problems, consider a modification of the above example. Suppose that the appliance dealer owns a very large chain of appliance stores and that he must place a single order for television sets for the entire chain of stores. The appliance dealer feels that the total demand for television sets in the chain of stores between now and the time the new models appear is approximately normally distributed, with a mean of 140 sets and a standard deviation of 16 sets. All the other details of the example are unchanged. As noted above, in order to maximize EMV, the appliance dealer should order a number of sets equal to the 67th percentile of the distribution of demand. But from the table of normal probabilities, the 67th percentile of a standard normal distribution is 0.44, so the 67th percentile of a normal distribution with mean 140 and standard deviation 16 is

$$140 + 0.44(16) = 147.$$

Thus, 147 sets should be ordered, and in the sense of maximizing EMV, 147 is the optimal estimate of demand.

There is a degree of controversy over this approach to statistics, which utilizes Bayesian procedures to revise probabilities and decision-theoretic procedures to incorporate payoffs or losses into the problem. The main advantage of this approach is that it enables the statistician to include all relevant information, including prior information and information concerning payoffs or losses, in the formal statistical model. Some statisticians object to the inclusion of such information because it is often of a "subjective" nature; objections like this are essentially philosophical rather than practical. From a practical viewpoint, a disadvantage of the Bayesian approach is that it is not always easy to determine the necessary inputs to the formal model. Assessing a prior distribution is not always easy, and information concerning relevant payoffs and losses is often somewhat vague. Nevertheless, for important problems it is worthwhile to consider an analysis of the sort suggested in this chapter, particularly when there is a considerable amount of prior information and the potential payoffs and losses are quite serious.

9.11 UTILITY

Payoffs and Utilities

In this chapter we have used the EMV criterion to make decisions in the face of uncertainty. This criterion states that the action with the highest EMV should be chosen. There are situations in which maximizing EMV is not an appealing strategy, however. In this section we shall attempt to indicate why maximizing EMV does not always seem to be a good approach; and we shall discuss briefly a more general criterion: *maximization of expected utility.*

One weakness of the EMV criterion is that it considers only the expected payoff, or the mean payoff, and not the variability in payoffs. For example, suppose that you were offered the bets given in Table 9.9 concerning a single toss of a fair six-sided die. The EMV's of Bets A and B are as follows (assuming the die is fair):

$$\text{EMV(Bet A)} = (-\$1)(0.50) + (\$1)(0.50) = \$0,$$

and

$$\text{EMV(Bet B)} = (-\$10{,}000)(0.50) + (\$10{,}500)(0.50) = \$250.$$

According to the EMV criterion, you should prefer Bet B to Bet A. Yet if you were forced to choose between the two in an actual betting situation, which bet would you choose? Because of the potential loss of \$10,000 in Bet B, most people would choose Bet A, even though it has a smaller EMV. This is because most of us would be put in deep financial difficulty if faced with a sudden debt of \$10,000. If this is the case, there must be factors involved in this example which are not formally considered by the EMV criterion.

Table 9.9 Payoff Table for Bet on Die

	State of Nature—Face Coming Up on Toss of Die	
	1, 2, or 3	4, 5, or 6
Bet A	−\$1	+\$1
Bet B	−\$10,000	+\$10,500

For a more practical example, consider an investor who has, say, \$10,000 to invest. He is considering three alternatives:

1. A savings account that will yield a fixed amount, 5%;
2. A conservative stock that has a normally distributed return with mean 7% and standard deviation 5%;
3. A speculative stock that has a normally distributed return with mean 10% and standard deviation 15%.

According to the EMV criterion, the investor should look only at the expected returns of the three investments, in which case he will invest all of his money in the investment with the highest return. The three expected returns are 5%, 7%, and 10%, respectively, for the savings account, the conservative stock, and the speculative stock. Therefore, the EMV criterion would have the investor invest everything in the speculative stock. For most investors, however, the risk associated with the investments would be a relevant consideration. The savings account is a riskfree investment, while both of the stocks have an element of risk. The speculative stock, of course, is the riskiest of the three investments. In reality, the investor's decision will probably depend not just on the expected returns, but also on this person's attitude toward risk. To avoid risk at any cost, the investor should put all of the money into the savings account. The investor who is completely indifferent to risk will put all of the money into the speculative stock. Most likely, the investor will attempt to achieve some sort of compromise by *diversifying*. That is, the investor will buy some of the speculative stock in the hope of obtaining a high return; but to reduce the overall risk, the investor will also put some money in the savings account, and may also buy some of the conservative stock. If the EMV criterion is rigidly followed, such diversification is not possible. Thus, the fact that diversification is common in real-world investing suggests that many investors are not willing to act strictly in accordance with the EMV criterion.

The preceding examples point out the fact that the value of a dollar is different for different persons, and that the value of one dollar is not necessarily the same as the value of another dollar for any one particular person. This means that \$1000 is not necessarily 1000 times as desirable to *you* as is a single

Table 9.10 Payoff Table for Bet on Coin

	State of Nature	
	Heads	Tails
Bet C	\$1 million	\$1 million
Bet D	\$100 million	\$0

dollar, even though it *is* exactly 1000 times as valuable in terms of purchasing power. To clarify this somewhat, consider the example in Table 9.10—a choice of two bets concerning the face coming up on a single flip of an unfair coin, with

$$P(\text{Heads}) = 0.90 \qquad \text{and} \qquad P(\text{Tails}) = 0.10.$$

Which bet would you choose? The EMV's of the bets are

$$\begin{aligned}\text{EMV(Bet C)} &= (\$1 \text{ million})(0.90) + (\$1 \text{ million})(0.10) \\ &= \$1 \text{ million,}\end{aligned}$$

and

$$\begin{aligned}\text{EMV(Bet D)} &= (\$100 \text{ million})(0.90) + (\$0)(0.10) \\ &= \$90 \text{ million.}\end{aligned}$$

In spite of the great difference in EMV's, many persons, if given this opportunity on a one-shot basis, would choose Bet C. This is because \$1 million is a very large sum of money, clearly enough to allow a person to lead a quite comfortable life (even at today's prices!). It would be even nicer, of course, to have \$100 million, but most people feel that Bet D is not preferable to Bet C because \$100 million is not so preferable to \$1 million that it is worth a risk of winding up with nothing. In other words, \$100 million is not worth 100 times as much as \$1 million. If it were, then Bet D would be preferred, since the EMV criterion would apply.

A few comments are in order regarding the preceding example. The amounts are much larger than they would be in most decision-making situations. Unless the decision at hand is a very important decision involving a large corporation, we would expect the amounts involved to be much smaller. The large amounts were chosen in order to stress the point, but other examples with smaller amounts could be found which give essentially the same results. Also, the bets are to be offered strictly as a one-shot affair, as we pointed out. If the situation were to be repeated several times, Bet D would in all likelihood be preferred to Bet C, for the probability of winding up with nothing would be greatly reduced (if the choice of bets is repeated just twice, the probability of winding up with nothing if Bet D is chosen twice is just 0.01, as compared with 0.10 for the one-shot affair). If the situation is to be repeated, the problem becomes more complex, involving more actions. For a 5-trial repetition, one action might be (B, B, B, A, A)—in this case, there are 2^5, or 32 actions in all (there are 2 actions per trial, and thus there are $2 \times 2 \times 2 \times 2 \times 2 = 2^5$ actions in 5 trials), and there are 2^5 states of nature.

In these examples, it appears that EMV is not a very good criterion for decision-making under uncertainty. How can we determine a better criterion, one which will take into account the decision-maker's preferences for various consequences or payoffs? The theory of utility prescribes such a criterion: the *maximization of expected utility*, or the EU criterion. In order to understand this criterion, it is first necessary to discuss briefly the concept of *utility*.

Essentially, the theory of utility makes it possible to measure the *relative* value to a decision-maker of the payoffs, or consequences, in a decision problem. First of all, it is necessary to determine what we consider to be the most preferable and least preferable payoffs. These can be assigned utility values of 1 and 0, respectively. The choice of 1 and 0 is arbitrary; we could just as easily have chosen 234 and -101, but 1 and 0 simplify the calculations somewhat. Let us call the most preferable payoff M and the least preferable payoff L. Then suppose there is another payoff, P, the utility of which we would like to determine. We can do this in the following manner. Consider the following betting situation:

Bet I —Receive P for certain,

Bet II—Receive M with probability p,

Receive L with probability $1 - p$.

According to the EU criterion, the bet with the highest expected utility should be selected. But we can calculate the expected utilities, where the utility of a consequence C is denoted by $U(C)$:

$$\begin{aligned} EU(\text{Bet I}) &= U(P). \\ EU(\text{Bet II}) &= p(U(M)) + (1 - p)(U(L)) \\ &= p(1) + (1 - p)(0) = p. \end{aligned}$$

Thus, if $U(P) < p$, Bet II should be chosen; if $U(P) > p$, Bet I should be chosen; and if $U(P) = p$, we are indifferent between the two bets. We shall exploit this last property to determine the utility of P. If we can determine a probability p which makes us *indifferent* between the two bets, then the utility of P is equal to this value, p. In this manner, we can determine the utility of any consequence, or payoff, once the most and least preferable consequences, M and L, have been determined. Note that it is not necessary for the consequences to be stated in terms of dollars, as it is when we are using EMV. Because of this, it is possible to take into consideration both monetary and *non*monetary factors in determining the utility of a consequence. In many business decisions, nonmonetary factors (e.g., factors involving labor, such as working conditions; factors involving prestige, such as the size and design of office buildings; and factors

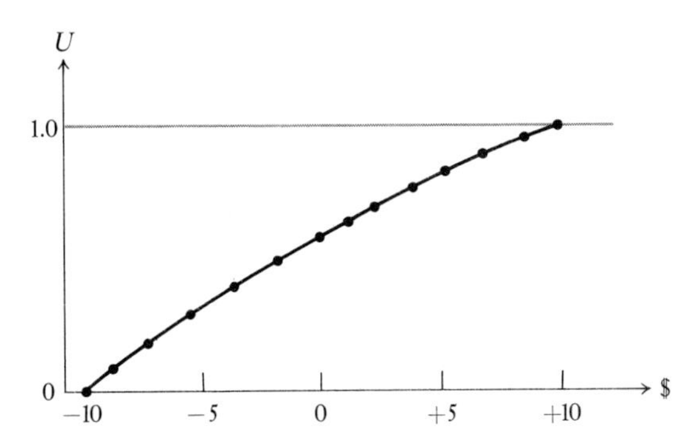

Fig. 9.9 A utility function for money.

involving business ethics) can be quite important, and it is useful to be able to consider these factors in the analysis.

Although nonmonetary factors may be of some importance, it is of interest to determine the relationship between money and utility. Suppose that we are interested in this relationship over a range of monetary values from −\$10 to +\$10. Then $M = +\$10$ and $L = -\$10$. If we use the above analysis, we can determine the utility of any amount between L and M. For example, suppose we decide that we are indifferent between (1) receiving \$5 for certain and (2) receiving M with probability 0.80 and receiving L with probability 0.20. Then U(\$5) = 0.80. If we assess U(\$0), U(\$1), U(−\$1), etc., in a similar manner, we could plot the resulting points on a graph, as shown in Fig. 9.9. After several points are determined, it is possible to draw a rough curve through them. This curve is a *utility function for money*. It is important to note that different persons have different utility functions, and that any single individual may have different utility functions at different points in time. It is not possible to make interpersonal comparisons of utility, however, because of the arbitrary scale (0 to 1) used for utilities.

Types of Utility Functions

Several types, or classes, of utility functions for money can be distinguished, although there are utility functions not falling in any of the classes to be described. In Fig. 9.10 three utility curves are presented. Curve A represents the utility curve of a risk-avoider, Curve B represents the curve of a risk-taker, and Curve C represents the curve of a person who is neither a risk-taker nor a risk-avoider (i.e., a risk-neutral person). To see why these labels apply to the three curves, consider the following bet: win \$10 with probability $\frac{1}{2}$ and lose \$10 with probability $\frac{1}{2}$. This can be thought of as a bet of \$10 on the toss of a fair coin. In terms of EMV, a person should be indifferent about the bet, since it has an EMV of zero. In terms of EU, however, the situation varies with the three curves

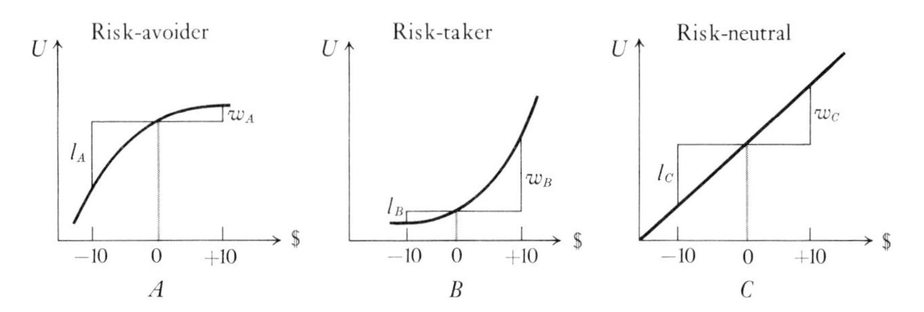

Fig. 9.10 Three types of utility functions.

presented in Fig. 9.10. The gains in utility for the three curves if the bet is won are w_A, w_B, and w_C. The corresponding losses in utility if the bet is lost are l_A, l_B, and l_C.

The expected utility of this bet to the three types of individuals can be calculated by multiplying the probability of winning times the gain in utility from winning, and then subtracting the probability of losing times the loss in utility if the bet is lost. Thus, the expected utility to individuals A, B, and C is:

$$\text{EU}(A) = \tfrac{1}{2}w_A - \tfrac{1}{2}l_A = \tfrac{1}{2}(w_A - l_A),$$
$$\text{EU}(B) = \tfrac{1}{2}w_B - \tfrac{1}{2}l_B = \tfrac{1}{2}(w_B - l_B),$$

and

$$\text{EU}(C) = \tfrac{1}{2}w_C - \tfrac{1}{2}l_C = \tfrac{1}{2}(w_C - l_C).$$

From Curve A we can see that this "fair" bet is unfavorable to the risk-avoider since the amount of utility he can gain, w_A, is less than the amount of utility he can lose, l_A, and as a result $\text{EU}(A) < 0$. The risk-avoider represented by Curve A would thus be expected to *reject* the bet in question since he prefers to avoid the risk of a bet with $\text{EMV} = 0$; in fact, we could find some bets with *positive* EMV's that A would consider unfavorable. The individual represented by Curve B, on the other hand, would be expected to *accept* this bet, as his potential gain in utility, w_B, exceeds the potential loss, l_B, so that $\text{EU}(B) > 0$. In other words, this type of person prefers to take the risk associated with a bet of $\text{EMV} = 0$, and we could, in fact, find some bets with *negative* EMV's that B would consider favorable. And finally, the individual represented by Curve C is indifferent regarding the bet, as this person's utility function for money is linear; that is,

$$w_C = l_C \quad \text{and} \quad \text{EU}(C) = 0.$$

When a person's utility function for money is linear, then choosing the action with the highest EU is equivalent to choosing the action with the highest EMV. To prove this, note that if the curve is linear, it can be written in the form

$$\text{U} = a + b(\text{MV}),$$

where b must be greater than zero since more money is always preferred to less money. But a and b are constants, so, by the laws of expected values,

$$E(\mathrm{U}) = a + bE(\mathrm{MV}).$$

EU is then at a maximum when EMV is at a maximum, so the two criteria are equivalent in this case.

The curves presented in Fig. 9.10 are by no means the only possible forms for utility functions. Several other forms, which we shall not describe, have been proposed; and still other forms, many of which are not mathematically tractable (that is, they cannot be represented by a simple mathematical function), no doubt exist. Most utility curves for individuals probably are similar to Curve A. It is an interesting exercise to select a range of values (say, $-\$100$ to $+\$100$) and attempt to determine your own utility curve.

Maximizing Expected Utility

Once a utility function for money is determined, the EU criterion is easy to apply. Simply take each value in the payoff table for the decision-making problem of interest and use the utility function to convert the payoff to a utility. In this fashion, the entire payoff table can be converted into a utility table. If a tree diagram is used, all of the payoffs at the righthand side of the tree must be converted to utilities *before* the backward-induction procedure is started. It is very important to remember that the conversion from payoffs to utilities must occur *before* any expected values are calculated. To calculate expected payoffs and *then* convert to utilities is an invalid procedure.

For an example of the use of a utility function in a decision-making problem, consider an investor who wants to invest \$10,000. He is considering three investments: a savings account, a conservative stock, and a speculative stock. To simplify the problem, assume that he has decided to invest the entire \$10,000 in a single investment—that is, he has *ruled out* the possibility of diversification. The assessor thinks about his preferences and contemplates various bets in order to asses a utility function, and it turns out that the resulting utility function can be represented quite well by the function

$$\mathrm{U}(M) = 10 - \left(\frac{2000 - M}{500}\right)^2 \quad \text{for } M \le 2000,$$

where M represents dollars. This function, which is graphed in Fig. 9.11, illustrates the use of a mathematical function to represent a decision-maker's utility function. This particular function is a quadratic function; other mathematical functions that are frequently used to represent utility functions are exponential and logarithmic functions (in addition to linear functions, of course). Observe from the graph that the investor's utility function takes on both positive and negative values and that a utility of zero does *not* correspond

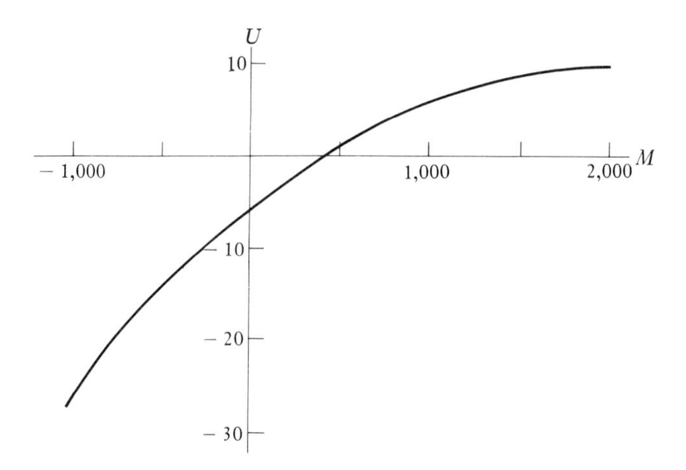

Fig. 9.11 Investor's utility function.

to a dollar value of zero. Earlier in the section we discussed utility functions with values from zero to one, but not all utility functions are so restricted.

The investor feels that his payoff from the investment during the coming year will depend on whether the stock market goes up or down. In either case, the savings account will yield a return of 5%, so his payoff will be \$500 for the year. Furthermore, the investor judges that the conservative stock will yield him a profit of \$1000 if the market goes up but \$0 if the market goes down, and the speculative stock will yield him a profit of \$2000 if the market goes up but a loss of \$1000 if the market goes down. The payoff table, expressed in dollars, is given in Table 9.11. Now suppose that the investor feels that the probability that the market will go up is 0.60. The EMV's are 500 for the bank account, 600 for the conservative stock, and 800 for the speculative stock. Therefore, if his utility function were linear, so that he could use the EMV criterion, it would be optimal for the investor to buy the speculative stock.

Since the investor's utility function is not linear with respect to money, we need to convert the payoff table from dollars into utility values. We can do this

Table 9.11 Payoff Table for Investor

		State of Nature	
		Market Up	**Market Down**
Action (Investment)	Savings account	500	500
	Conservative stock	1000	0
	Speculative stock	2000	−1000

Table 9.12 Utility Table for Investor

		State of Nature	
		Market Up	Market Down
Action (Investment)	Savings account	1	1
	Conservative stock	6	−6
	Speculative stock	10	−26

by reading values from the graph, but since the utility function is expressed as a mathematical function, it is more accurate to simply use this function to determine the utility corresponding to each payoff in Table 9.11. For instance, if the investor chooses the savings account and the market goes up, the payoff is \$500, and the utility corresponding to \$500 is

$$U(\$500) = 10 - \left(\frac{2000 - 500}{500}\right)^2 = 1.$$

Similarly, all of the values in the payoff table can be converted to utilities, and the resulting table of utilities is given in Table 9.12. The expected utilities of the three actions are

$$\text{EU(Savings account)} = 1(0.60) + 1(0.40) = 1,$$

$$\text{EU(Conservative stock)} = 6(0.60) + (-6)(0.40) = 1.2,$$

and

$$\text{EU(Speculative stock)} = 10(0.60) + (-26)(0.40) = -4.4.$$

Thus, the optimal action, according to the EU criterion, is to buy the conservative stock. In this case, the speculative stock was risky enough and the investor was risk-averse enough so that, even though the speculative stock had the highest EMV, the conservative stock had the highest EU (and the savings account was not far behind!). This does not mean that risk-avoiders will avoid *all* risks—in this example, even the conservative stock has a risk of a return of zero, which is much less than the sure return from the savings account.

Important decisions should be based on the EU criterion rather than on the EMV criterion. For simplicity, however, we assumed in much of this chapter that utility was linear with respect to money. This assumption allowed us to use EMV as a decision-making criterion, since maximizing EMV is equivalent to maximizing EU. You should keep in mind, however, that this is a simplifying assumption that may not be realistic in many decision-making situations.

Define: expected utility, risk-avoider, risk-taker, risk-neutral.

Problems

9.34 A production manager is interested in the mean weight of items turned out by a particular process. He feels that the weight is normally distributed with standard deviation 2, and his prior distribution for μ is a normal distribution with mean 110 and variance 0.4. He randomly selects five items from the process and weighs them, with the following results: 108, 109, 107.4, 109.6, 112. What is the production manager's posterior distribution?

9.35 You are attempting to assess a prior distribution for the mean of a process, and you decide that the 25th percentile of your distribution is 160 and the 60th percentile is 180. Assuming that your prior distribution is normal, determine the mean and variance.

9.36 In assessing a distribution for μ, the mean height of a certain population of college students, a physical-education instructor decides that his distribution is normal, the median is 70 inches, and the 20th percentile is 67 inches. He is interested in

$$H_0: \mu \geq 71 \quad \text{versus} \quad H_a: \mu < 71.$$

He takes a random sample of 80 students and observes $M = 71.2$ inches. Assume (from previous studies) that $\sigma = 5.6$ inches is the population standard deviation. Find the posterior odds ratio of H_0 to H_a.

9.37 Since the deregulation of the airline industry, airlines find that their personnel are swamped with calls and that many customers are unable to reach a reservation clerk by telephone without a long holding period. The holding time is felt to be normally distributed with mean μ and variance 4, and the prior distribution of μ is normal with $P(\mu < 3.5) = 0.067$ and $P(\mu > 6) = 0.16$. A call is selected at random and it results in a holding time of 1.2 minutes. Find the posterior distribution of μ.

9.38 In Problem 9.34, find a 95-percent interval estimate for μ and determine $P(\mu > 110)$

a) from the prior distribution,

b) from the posterior distribution.

9.39 A contractor must decide how many houses to build in a new subdivision. All houses built in this subdivision must use solar energy for heating, and the contractor is not sure how strong the demand is for such expensive houses. Each house costs the contractor $80,000 to build and sells for $100,000. However, to build the houses, the contractor must take a loan from the local bank, and the terms of the loan state that any houses unsold at the end of six months will be bought by the bank at $72,000 per house. The contractor's distribution of θ, the number of houses that can be sold within six months, is as follows:

θ	4	5	6	7	8	9	10	11
$P(\theta)$	0.05	0.13	0.22	0.28	0.15	0.09	0.06	0.02

9.40 The manager of a bookstore in a college community must decide how many copies of a particular book to order for the coming semester. Based on the anticipated enrollment of the course in which the book is being used and the number of competing bookstores in town, the manager feels that the demand for the book at his store is normally distributed, with mean 120 and variance 100. The book costs the bookstore

\$10 and retails for \$12. Since the course is being phased out of the curriculum, the book will not be used again, so any unsold copies will be returned to the publisher. The publisher will repurchase the books at \$9 per copy, and the bookstore's cost for postage and handling on returned copies amounts to \$0.25 per copy. How many copies of the book should be ordered?

9.41 The manager of a grocery store must decide how many gallons of milk to order for the coming week. The milk costs the store \$1.50 per gallon and retails for \$1.60, but any milk unsold at the end of the week must be thrown out. The cost due to loss of goodwill when customers try to buy milk and find that it is sold out is judged to be \$0.20 per gallon. From past sales data, it appears that weekly sales of milk are approximately normally distributed with a mean of 820 gallons and a standard deviation of 60 gallons. How much milk should the manager order?

9.42 On April 1, a car dealer in a small town has five new cars on hand and is preparing to place an order for more cars. The additional cars will arrive in a week, and the *next* order of new cars won't arrive until May 1. The dealer feels that the process "generating" car purchases at his dealership can be treated as a Poisson process with an average intensity of ten cars per month. The cars cost the dealer \$3000 and sell for \$3500, and the dealer suffers an opportunity loss of \$200 (due to loss of goodwill and possible loss of a sale) if a customer wants to buy a car but the dealer has none in stock. How many cars should the dealer order?

9.43 Why is the maximization of EMV not always a reasonable criterion for decision-making? What problems does this create for an analyst who is trying to model a decision-making problem using the techniques discussed in this chapter?

9.44 Suppose that you are contemplating drilling an oil well, with the following payoff table (Table 9.13) in terms of thousands of dollars:

Table 9.13

		State of the World	
		Oil	**No Oil**
Action	Drill	100	−40
	Don't drill	0	0

If after consulting a geologist you decide that $P(\text{Oil}) = 0.30$, would you drill or not drill according to the

a) maximin criterion?

b) maximax criterion?

c) EMV criterion?

d) EU criterion, where $U(0) = 0.40$, $U(100) = +1$, and $U(-40) = 0$?

Explain the differences in the results in parts (a) through (d).

9.45 Attempt to determine your own utility function for money in the range from −\$500 to +\$500.

9.46 In Problem 9.8, suppose that $U(M) = 50 - (8 - M)^2$, where M represents the payoff in units of \$100,000. Graph U and discuss the behavior implied by such a utility function. What should the firm do if the objective is to maximize expected utility?

9.47 In Problem 9.9, suppose that the company wants to maximize expected utility. The firm's utility function is

$$U(M) = \begin{cases} M + 5 & \text{if } M \geq 5 \\ 2M & \text{if } M < 5, \end{cases}$$

where M represents the payoff in thousands of dollars. If the probability of a recession is 0.3, what should the company do?

9.48 In Problem 9.47, how large would the probability of a recession have to be before the company would put the money in the bank?

9.49 Compare your answers to Problem 9.48 and Problem 9.10. Explain any differences.

Exercises

9.50 A contractor wants to choose between bidding on the construction of either a dam or an airport. He can bid on only one of these jobs, since his engineers don't have time to analyze both jobs in order to determine an appropriate bid. Preliminary estimates of construction costs from the projects are \$10,000,000 from the airport and \$18,000,000 from the dam. In addition, the out-of-pocket costs for preparing the bids are \$500,000 for the airport and \$1,000,000 for the dam. Moreover, he has a choice of using his standard bidding procedure or trying to bid lower than usual to increase the chances of winning the job. The standard bids would be approximately \$12,500,000 for the airport and \$23,000,000 for the dam, and the lower bids are \$11,500,000 for the airport and \$21,000,000 for the dam. According to the maximin criterion, what should the contractor do? According to the maximax criterion, what should the contractor do?

9.51 In Problem 9.50, the chances of winning the bid are 0.4 for the airport and 0.3 for the dam if standard bids are used, but the chances are 0.7 for the airport and 0.5 for the dam if lower bids are used. What should the contractor do?

9.52 An importer has an option to buy 100,000 tons of scrap iron from a foreign firm for \$5 per ton. The market price in the U.S. for scrap iron is \$8 per ton. The importer must decide immediately whether to purchase the scrap iron, but a license must be obtained later, before the material can be imported. Because of an increasing tendency to protect U.S. business by restricting imports, the importer feels that there is a 50–50 chance that the government will refuse to grant an import license, in which case the contract will be annulled and the importer will have to pay a penalty of \$1 per ton to the foreign government. Should the importer purchase the scrap iron if this person wants to maximize EMV?

9.53 A contractor must decide whether to buy or rent equipment for a job up for bid. Because of lead-time requirements in getting the equipment, he must decide before knowing whether the contract is won. If he buys, a contract would result in a net profit of \$120,000; but failing to win the contract means that the equipment will have to be sold at a \$40,000 loss. By renting, his profit from the contract is only \$50,000 if he wins it, but there will be no loss of money if the job is not won. The probability that he will win the job is 0.4. What should the contractor do?

9.54 An insurance company's records reveal that in a one-year period, approximately 20 percent of drivers under age 21 submit a claim as a result of an accident. Further investigation shows that 30 percent of those submitting claims are students, while 60 percent of those not submitting claims are students. What is the probability that a person randomly chosen from the under-age-21 drivers insured by the company will submit a claim as a result of an accident in the next year?

9.55 In Problem 9.14, suppose that the security analyst observes the security for a total of 15 days, and the price increases on 9 of those days. The analyst could revise his probability distribution each day, but suppose that he decides to wait until the end of the 15-day period. Is there an easy way to determine the relevant likelihoods for the entire 15-day period? (*Hint:* The likelihoods are of the form

$$P\left(\begin{array}{l}\text{Security price goes up on} \\ \text{9 days in a sample of 15 days}\end{array}\middle|\begin{array}{l}\text{Probability of price going} \\ \text{up on any given day is } p\end{array}\right).$$

From Chapter 4, try to find a distribution that can be used to calculate probabilities of this form.) After the 15-day period, what is the analyst's probability that the security is an average investment? Also find the probabilities for "poor investment" and "good investment" after the information from the 15-day sample.

9.56 In Problem 9.52, how much would the importer pay to learn, before deciding about the purchase of the scrap iron, whether the import license would be approved?

9.57 In Problem 9.53, find the EVPI.

9.58 A coin lying on a table is either two-headed or is a fair coin. You are asked to guess whether or not the coin is two-headed. If you guess correctly you win \$2. If you guess incorrectly, you lose \$1. You can't see the coin, but you do have the opportunity of having an impartial observer flip the coin and tell you if it came up heads or tails. This flip costs you \$0.20. Draw a decision tree and use it to determine your optimal action and the value of ENGS and EVSI.

9.59 In Problem 9.58, suppose you can elect to buy either two flips or one flip, at a cost of \$0.20 per flip. Draw the decision tree for this problem, and calculate ENGS. Assume you must decide before the first flip whether or not you want the second flip.

9.60 For Problem 9.59, assume that the cost per flip is \$0.10. Write and run a computer program to determine the optimal sample size.

9.61 a) Redo Problem 9.59, assuming that you can decide whether or not to purchase a second flip *after* viewing the first flip.

b) Redo Problem 9.60 with the same assumption as given above.

9.62 Consider a bookbag filled with 100 poker chips. You know that either 70 of the chips are red and the remainder blue, or that 70 are blue and the remainder red. You must guess whether the bookbag is (70R, 30B) or (70B, 30R). If you guess correctly, you win \$5. If you guess incorrectly, you lose \$3. Your prior probability that the bookbag contains (70R, 30B) is 0.40.

a) If you had to make your guess on the basis of the prior information, what would you guess?

b) If you could purchase perfect information, what is the most that you should be willing to pay for it?

c) If you could purchase sample information in the form of one draw from the bookbag, how much should you be willing to pay for it? Draw the decision tree.

d) If you could purchase sample information in the form of three draws (with replacement) from the bookbag, how much should you be willing to pay for it? Draw the decision tree.

9.63 Do Problem 9.62 with the following payoff table (9.14) (in dollars):

Table 9.14

		State of the World	
		(70R, 30B)	**(70B, 30R)**
Your guess	(70R, 30B)	6	−2
	(70B, 30R)	−6	10

9.64 Write a computer program to calculate EVSI for any size sample from $n = 1$ to $n = 50$ in Problem 9.62(d). What is the optimal sample size if each sample draw costs \$0.25?

9.65 It is felt by a market research group that the total time elapsed between the development of a new product and the time the product reaches the market is normally distributed with mean μ and variance 7.5. The price distribution of μ is a normal distribution with mean 6 and variance 0.5. A sample of 26 new products results in an average "time to market" of 5.8 months. What is the posterior distribution of μ following this sample?

9.66 In Problem 9.65, find a 95-percent interval estimate for μ

a) from the prior distribution,

b) from the sample,

c) from the posterior distribution.

9.67 Envisioning a potentially large new market, a U.S. manufacturer agrees to export food processors to China. The manufacturer will send a shipment of the current year's models to China, and no additional shipments will be made for one year, at which time next year's models will be available. The expected demand for food processors during the first year they are sold in China is normally distributed with mean 12,000 and standard deviation 3000. The net profit for each food processor sold in China is \$10, and any left unsold at the end of the first year will be shipped elsewhere to be sold at a loss of \$15 per unit. How many food processors should be sent in the first year's shipment?

9.68 In Problem 9.53, what is the best strategy if the contractor's utility function is $U(M) = \sqrt{M + 40{,}000}$, where M represents dollars?

9.69 An investor is contemplating an investment having the potential payoffs shown in Table 9.15. He assesses his utility for values from −\$30,000 to \$50,000 in increments of \$10,000 as shown in Table 9.16.

a) Graph the utility function and interpret the shape of the curve.

b) Should he make the investment or not?

Table 9.15

Payoff	Probability
–$30,000	0.2
–$10,000	0.1
$ 5,000	0.1
$20,000	0.4
$50,000	0.2

Table 9.16

M	U(M)	M	U(M)	M	U(M)
–$30,000	0	$ 0	0.60	$30,000	0.90
–$20,000	0.28	$10,000	0.72	$40,000	0.96
–$10,000	0.45	$20,000	0.82	$50,000	1.00

9.70 A very small electronics company produces and sells mini-computers and assorted computer hardware and software. The firm is relatively new, and all of its sales have been domestic. Recently a French firm learned of the company's products, and the firm now has offered to purchase several mini-computers and other items. The proposed contract calls for full payment in French francs at the time of delivery, which is to be no later than six months from the date the contract is offered. The amount is 2,400,000 francs.

This is by far the largest order the company has received in its short lifetime. The six-month deadline can be met, but it will be necessary to put off other orders to do so. The president estimates that if the French order is accepted, the company will have to forgo other sales that would lead to a net profit of about $260,000. The cost of filling the French order, including shipping, is expected to be $200,000.

At the time the payment of 2,400,000 francs is received, the company will need cash, and the francs will have to be converted into dollars immediately. Because of recent fluctuations of the dollar vis-a-vis the franc and other currencies, the president is unsure about how many dollars will be received. After some brief consultations with bankers and economists, the president comes up with the following probability distribution for θ, the exchange rate (in dollars per French franc) six months from now, shown in Table 9.17. Given the current state of the world economy, such a wide range of values of θ is not surprising. Nevertheless, it makes the president very nervous.

One option that is available is to obtain further information about the dollar/franc exchange rate. For $2000, an econometric forecasting firm will forecast whether

Table 9.17

θ	$P(\theta)$
0.18	0.10
0.20	0.20
0.22	0.30
0.24	0.30
0.26	0.10

the dollar is expected to be weak or strong in the next six months. The chances of a forecast of a weak dollar are 0.70 if θ will end up at 0.18 in six months, 0.60 if θ will be 0.20, 0.50 if θ will be 0.22, 0.40 if θ will be 0.24, and 0.30 if θ will be 0.26. The corresponding chances of a forecast of a strong dollar are 0.30 (for $\theta = 0.18$), 0.40, 0.50, 0.60, and 0.70 (for $\theta = 0.26$).

In talking with some friends, the president learns that it might be possible to remove the uncertainty caused by the need to convert francs to dollars in six months. At a cost of $25,000, an insurance policy will guarantee a rate no lower than 0.22 dollars/franc for the changing of 2,400,000 francs six months from now. That is, if $\theta < 0.22$, the insurance company will purchase the francs at a rate of 0.22, while if $\theta \geq 0.22$, the company will change the francs at the market rate, θ. Alternatively, the firm can obtain a "futures" contract, agreeing to deliver 2,400,000 French francs six months from now at a rate of 0.21 dollars per franc.

a) Assuming that the president is risk-neutral, what course of action would you advise?

b) Suppose that the president's utility function for M, payoffs in units of $100,000, is

$$U(M) = 9 - (5 - M)^2.$$

Also, the econometric forecasting firm just went out of business. Now what course of action would you advise?

9.71 The Delaney-Bryce Corporation is a major manufacturer of specialized soap and detergent products. It currently controls 31 subsidiary companies that manufacture disinfecting detergent powder primarily for use by hospitals, linen suppliers, diaper services, and other large institutional laundry facilities. Each of the 31 subsidiaries sells in its own region, and together they serve a large portion of the United States.

Delaney-Bryce established the Ohio Valley Detergent Corporation in Indianapolis last year. Since that time Ohio Valley has captured a larger market share in each of the three quarters its plant has been in operation. The directors expect that the company will have reached maturity sometime in the next three years, and that the rapid growth in sales it has been experiencing during the present start-up period will begin to level off. The warehouse that Ohio Valley has been using for the past year is rapidly becoming inadequate to serve the company's growing sales volume. The directors, knowing that such rapid changes in requirements would occur in the company's first stages, leased the present warehouse facilities for only eighteen months. This lease will expire soon and the directors and officers now wish to negotiate another. Now that Ohio Valley is approaching maturity they wish to acquire warehouse space for a period of three years. Leasing facilities on this long-term basis will save the company money both as a result of lower monthly rent payments and by avoiding the need to regularly renegotiate lease terms.

There are only two warehouse facilities in Indianapolis that the directors of Ohio Valley feel may be adequate. Both contain the necessary equipment and other features that the company's operation requires. The location of each warehouse is also suitable to the directors. But the sizes of the facilities differ, one being 16,500 square feet, the other, 21,000 square feet. The decision must be made, then, as to which of these two different-sized warehouses Ohio Valley Detergent Corporation should lease in order to minimize its expected cost. But before such a decision can be made it will be necessary for the directors to have some reliable prediction of the level of sales the company can expect to maintain over the period covered by the lease. Kenneth Rein, a member of the board, has predicted that over the next three years Ohio Valley will sell about 10.83 million pounds of detergent annually, and that this prediction of sales comes from a normal distribution with a standard deviation of 1.18 million pounds. The 16,500-square-foot warehouse will hold a maximum of 1,835,000 pounds of detergent. Likewise, the 21,000-square-foot warehouse can be used to store at most 2,300,000 pounds.

The company plans to keep on hand at any one time a two-month supply of its product. This means that if Ohio Valley should sell exactly the predicted amount of 10.83 million pounds, it would want always to keep in storage $(\frac{1}{6})(10.83) =$ 1,805,000 pounds of detergent. Note that, for this prediction for two months of sales, the standard deviation is 196,666 pounds, which is one-sixth of the error associated with the prediction of full-year sales (that is, $\frac{1}{6} \times 1{,}180{,}000$).

In addition to this information on warehouse utilization, Rein has given you the following guidelines concerning the costs involved in leasing each of the two warehouses available. As Ohio Valley is most concerned here with avoiding unnecessary expenses, Rein tells you to consider that, for this decision process, the cost will be zero if the company leases the smaller warehouse and, for the duration of the lease, requires no more than the space that the smaller warehouse can provide. The cost is also assumed to be zero if Ohio Valley leases the larger facility and requires over the years more space than the smaller warehouse could have provided. If the company leases the smaller warehouse and sales are at a higher level than can be supplied by this facility, high-cost short-term facilities will have to be leased to supplement the main warehouse. Rein estimates that this added cost, combined with costs of reduced efficiency due to the resulting lack of centralization, will be approximately $500,000 over the entire period of the lease. If the company leases the larger warehouse and sales over the lease years prove to be low enough so that it actually needs only the smaller warehouse, the extra expense will be $325,000 over the life of the lease (lease terms prohibit subleasing of unused space).

Ohio Valley has recently learned that they can purchase a sample survey for $5000. This survey's outcomes will be either "favorable" (meaning large sales) or "unfavorable" (meaning moderate sales). Judging from past records, Ohio Valley estimates that if sales $\leq 1{,}835{,}000$ is the true state of nature, the survey will result in the "unfavorable" outcome about 77% of the time. Conversely, if sales $> 1{,}835{,}000$ is the true state of nature, the survey will result in the "favorable" outcome about 66% of the time. On the basis of this information, draw the decision tree, and calculate EVSI and ENGS.

GLOSSARY

action: a choice, or a decision, available to a decision-maker.

state of the world: one or more events relevant to a decision-making problem.

decision-making under uncertainty: decision-making when the actual state of the world is not known for sure.

consequence: a monetary payoff or other type of outcome to a decision-maker.

maximin criterion: choosing an action to maximize the smallest possible payoff resulting from the action.

maximax criterion: choosing an action to maximize the largest possible payoff resulting from the action.

EMV criterion: choosing an action to maximize expected payoff.

decision tree: a representation of a decision-making problem in terms of a tree diagram.

backward induction: working "backward" through a decision tree to solve a decision-making problem.

Bayes' rule: a formula for revising probabilities when new information is obtained.

prior probabilities: probabilities before seeing some new information.

posterior probabilities: probabilities after seeing some new information.

likelihoods: relative probabilities of some new information, given different states of nature.

EVPI: expected value of perfect information.

EVSI: expected value of sample information.

ENGS: expected net gain from sample information.

CS: cost of sampling.

optimal sample size: the sample size that yields the largest ENGS.

utility: a measure representing a decision-maker's preferences among different consequences.

EU criterion: choosing an action to maximize expected utility.

risk-avoider: a decision-maker who prefers to avoid risk and will not take a gamble with an EMV of zero.

risk-taker: a decision-maker who likes to take risks and is happy to take a gamble with an EMV of zero.

risk-neutral: a decision-maker who is indifferent to risk and will therefore use the EMV criterion.

10

SIMPLE LINEAR REGRESSION

"If a problem has really big numbers in it, the answer is always 'one million.'"

PEPPERMINT PATTY, in *Peanuts*

10.1 INTRODUCTION *

In the past several chapters we have discussed the process of using sample information to make inferences, test hypotheses, or modify beliefs about the characteristics of a population. In this chapter and the next we turn to a related problem, involving two or more variables—making inferences about how changes in one set of variables are related to changes in another set. A description

* Before beginning this chapter the reader may wish to read (or reread) Section 3.9, on bivariate expectations.

of the *nature* of the relationship between two or more variables is called *regression analysis*, while investigation into the *strength* of such relationships is called *correlation analysis*.

Sir Francis Galton, an English expert on heredity in the late 1800's, was one of the first researchers to work with the problem of describing one variable on the basis of one or more other variables. Galton's work centered on the heights of fathers compared to the heights of their sons. He found that there was a tendency toward the mean—exceptionally short fathers tended to have sons of more average height (i.e., taller than their fathers), while just the opposite was true for unusually tall fathers. Galton said that the heights of the sons "regressed" or reverted to the mean, and thus originated the term *regression*. Nowadays the term "regression" much more generally means the description of the nature of the relationship between two or more variables.

Regression analysis is concerned with the problem of describing or estimating the value of one variable, called the *dependent* variable, on the basis of one or more other variables, called *independent* variables. Suppose, for example, that a manager is trying to predict sales for next month (the dependent variable) on the basis of indexes of disposable income, price levels, or any of numerous other independent variables; or perhaps this person is trying to predict the performance of one product under certain conditions of stress or at various temperatures; similarly the decision-maker may be using one or more of a battery of tests in trying to evaluate the ability of prospective employees for new jobs. In these cases, regression analysis is being used in an attempt to *predict* or estimate the value of an unknown dependent variable on the basis of the known value of one or more independent variables. In other cases, regression may be used merely to *describe* the relationship between known values of two or more variables. An economist may use it for this purpose as an aid in understanding the relationship between historical observations over a specified time span, such as the relation of consumption to current and past levels of income and wealth, or the relationship between any one or more of a number of economic indicators and prices, or profits, or sales in a given industry.

No matter whether regression analysis is used for descriptive or predictive purposes, one cannot expect to be able to estimate or describe the *exact* value of sales, or profits, or consumption, or any other dependent variable of this nature. There may be many factors which could cause variations in the dependent variable for a given value of the independent variables, such as fluctuations in the stock market, changes in the weather, a passing fad, or just differences in human ability and motivation. Because of these possible variations, we shall for the most part be interested in determining the *average* relationship between the dependent variable and the independent variables. That is, we will want to be able to estimate the mean value of a dependent variable for any given value of the independent variable. Although regression analysis can involve one or more independent variables, in this chapter we will confine our analysis to the case of *simple* linear regression—i.e., only *one* independent variable. *Multiple* linear regression is presented in Chapter 12.

The Regression Model

For most regression analysis, the average population relationship between the dependent variable (which is usually denoted by the letter **y**) and the independent variable (denoted by the letter x) is assumed to be linear.* A linear function is used because it is mathematically simple, and yet still provides an approximation to the real-world relationship which is sufficient for most practical purposes.

Since we are interested in determining the *mean* value of **y** for a given value of x, we are thus interested in the conditional expectation $E[\mathbf{y}|x]$. Another symbol often used to denote this expectation is $\mu_{\mathbf{y}\cdot x}$, which is read as "the mean of the **y** values for a given x value." By assuming that **y** and x are linearly related, we are saying that all possible conditional means ($E[\mathbf{y}|x] = \mu_{\mathbf{y}\cdot x}$) which might be calculated (one for each possible value of x) must be on a *single straight line.* This line is called the *population regression line.* To specify this line (or any straight line), we need to know its *slope* and *intercept.* Suppose we let α be the **y**-intercept and β be the slope of the line. This line is thus written as follows:†

Population regression line:

$$\mu_{\mathbf{y}\cdot x} = E[\mathbf{y}|x] = \alpha + \beta x. \tag{10.1}$$

To illustrate this population regression line, suppose **y** represents the quantity of beef purchased per month by a household, and x represents the retail price of beef. The value of $\mu_{\mathbf{y}\cdot x}$ is thus the mean quantity of beef purchased per month for some given price of beef. When some exact value of x is specified, it is customary to denote this value as x_i, and to let $\mu_{\mathbf{y}\cdot x_i}$ represent the mean of the y values for the specific value of x. For example, x_i might be a price of beef, such as $x_i = \$2.79/\text{lb}$, and

$$\mu_{\mathbf{y}\cdot x_i} = \mu_{\mathbf{y}\cdot \$2.79} = \alpha + \beta(2.79)$$

would be the mean quantity purchased per month when the price is $2.79/lb. Thus, when speaking of a specific value of x, we merely substitute x_i for x in Formula (10.1).

---

* In this chapter the variable x is not, technically, a *random* variable; hence we will not denote this variable by boldface type. Specific values of x will be denoted by the symbol x_i.

† The meaning of the Greek letters α and β as used in this chapter has no relationship to the meaning of these same letters used in Chapter 8 to describe the probability of Type I and Type II errors.

In addition to estimating the mean value $\mu_{y \cdot x_i}$, we would like to make statements about the *actual* value of $\mathbf{y}$ for a given value x_i. This value is denoted as y_i. For example, we might want to make a statement about the quantity of beef the "J. Doe" family would purchase per month when the price of beef is $x_i = \$2.79$/lb. As we indicated above, the actual value y_i for the given value x_i is usually not equal to $\mu_{y \cdot x_i}$. The difference between y_i and $\mu_{\mathbf{y} \cdot x_i}$ depends upon the accuracy of the regression model in depicting the real-world situation, and by the accuracy with which the variables x and y are measured. It also depends on the predictability of the underlying behavior of the persons, businesses, governments, etc., involved in the model. Any changes in human or institutional behavior could also cause such differences.

The point of the above discussion is that the difference between y_i and $\mu_{\mathbf{y} \cdot x_i}$ is the unpredictable element in regression analysis. For this reason, this difference is usually called the random "error," and denoted by the symbol $\boldsymbol{\varepsilon}_i$. That is,

$$\boldsymbol{\varepsilon}_i = y_i - \mu_{\mathbf{y} \cdot x_i} \qquad \text{or} \qquad y_i = \mu_{\mathbf{y} \cdot x_i} + \boldsymbol{\varepsilon}_i. \tag{10.2}$$

Suppose the price of beef is $x_i = \$2.79$, and the average quantity purchased per month by *all* households at this price is

$$\mu_{\mathbf{y} \cdot \$2.79} = 5 \text{ lbs.}$$

Should the Doe family purchase $y_i = 3.2$ lbs per month, then the error in this case would be

$$\boldsymbol{\varepsilon}_i = 3.2 - 5.0 = -1.8 \text{ lbs.}$$

We can now use Formula (10.2) to describe what is called the "population regression model." This model consists of all the terms which, when added together, sum to y_i. Substituting $y_i = \mu_{\mathbf{y} \cdot x_i} + \boldsymbol{\varepsilon}_i$ into (10.1) we get:

$$\textit{Population regression model:} \qquad y_i = \alpha + \beta x_i + \boldsymbol{\varepsilon}_i. \tag{10.3}$$

An example of this model is shown in Fig. 10.1. When values of x and $\mathbf{y}$ are plotted in this fashion, the diagram is called a *scatter diagram.*

As an illustration of a problem in which regression might be used to estimate an unknown population relationship, consider the task facing the director of admissions in most graduate schools. Because of limited resources, admission can be granted only to a select group of students, usually those predicted to be the most successful in graduate school. The process of deciding which students to admit is a difficult one, in which most directors presumably attempt to relate "success" in graduate school to a large variety of factors such as admission test scores, previous grades, recommendations, age, etc. In general, the only way to determine the nature of such a (population) relationship is on the basis of past data (sample information). If it can be determined that certain of these factors

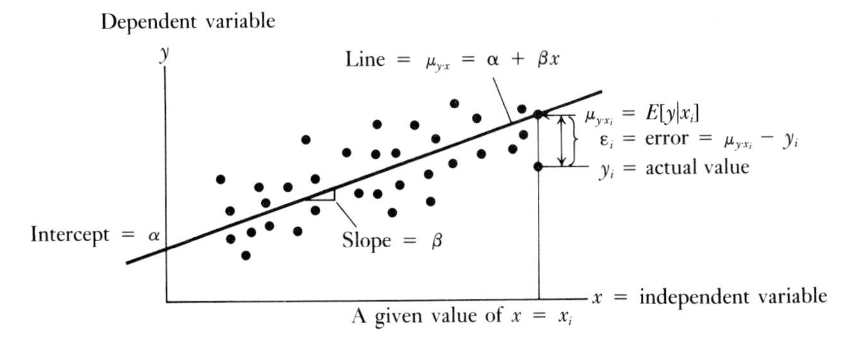

Fig. 10.1 The population regression model.

have been related to some definition of success in the past, then this information may be helpful in separating the potentially successful applicants from those less likely to be successful. Of considerable interest to most admissions directors, for example, is the relationship between an applicant's score on the Graduate Management Admission Test (GMAT) and this person's subsequent performance in graduate school, as measured by grade point average (GPA).

Let us denote by x the Graduate Management Admissions Test scores and let y represent the grade-point averages (GPA). The population in this case might be considered to be all possible candidates for admission during a specified time span, such as 1975 through 1990. Once again it must be pointed out that use of the term "the population relationship" does not imply that there is necessarily an exact relationship, which always holds true, between two variables. The variables GMAT and GPA are obviously not exactly related, as a higher GMAT score doesn't *always* lead to a higher GPA; still, one would expect to find a positive relationship between *mean* GPA's and GMAT scores. It is therefore meaningful to attempt to determine how changes in the independent variable influence the mean value of a dependent variable. The methods of this chapter enable us to estimate the parameters of the population regression model by using sample data.

We can illustrate the characteristics of a population regression model by assuming (for the moment) that the population parameters for this GMAT–GPA example are known quantities. For example, let's assume $\alpha = 0.95$ and $\beta = 0.0039$. This means that the population regression line is

$$\textit{Population regression line:} \quad \mu_{y \cdot x} = 0.95 + 0.0039x,$$

and the population regression model is:

$$\textit{Population regression model:} \quad y_i = 0.95 + 0.0039x_i + \varepsilon_i.$$

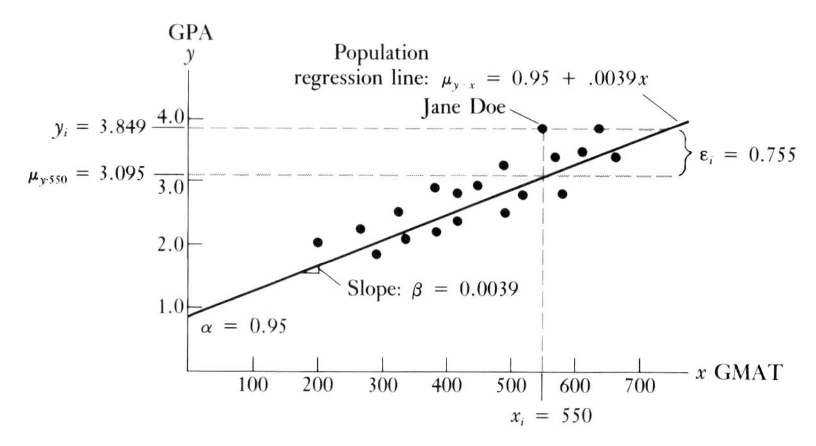

Fig. 10.2 Population regression line for GMAT–GPA example.

This regression model is diagrammed in Fig. 10.2, where each dot represents one observation (i.e., one person) in the population.

In Fig. 10.2 the population mean value for a GMAT of 550 is seen to be

$$E[y|550] = \mu_{y \cdot 550} = 3.095.$$

In other words, the average GPA for all students with GMAT scores of 550 is 3.094. This value is easily derived by substituting $x_i = 550$ into the population regression line, as follows:

$$\mu_{y \cdot 550} = 0.95 + 0.0039(550) = 3.095.$$

Now, let's assume one student in this population, "Jane Doe," had a GMAT of 550, and she earned a GPA of $y_i = 3.849$. For her, the value of the error ε_i is

$$\varepsilon_i = y_i - \mu_{y \cdot 550} = 3.849 - 3.094 = 0.754.$$

Our assumption that α and β are known is, of course, an unrealistic one. Usually α and β can only be estimated on the basis of sample data. For example, we might use as sample data all the GMAT and GPA scores, from 1975 to the present, of students who actually enrolled and graduated. The discussion in the following section introduces the sample regression model as the basis for estimating α and β.

The Sample Regression Model

In regression analysis it is customary to let a be the best estimate of α, b be the best estimate of β, and to let $\hat{y}$ be the resulting estimate of $\mu_{y \cdot x}$. The values a and b are called the *regression coefficients*, and the line relating a, b, and $\hat{y}$ is called the *sample regression line*. This line has the same form as the population regression line (Formula (10.1)):

$$\text{Sample regression line:} \quad \hat{y} = a + bx. \tag{10.4}$$

Just as we did for the population regression line, we can add the subscript i to these variables to indicate specific values. Thus, if x_i is a specific value of x, then $\hat{y}_i = a + bx_i$ is the equation for finding $\hat{y}_i$, which is the best estimate of $\mu_{y \cdot x_i}$ for this value of x. We can also specify a *sample regression model* just as we specified a population regression model. Again we need to define an error term, which in this case is the difference between the predicted value $\hat{y}_i$ and the actual value y_i. This error term is denoted as e, which means that the sample regression error e_i is an estimate of the population error ε_i. The errors e_i are often called "residuals."

$$\text{Residuals:} \quad e_i = y_i - \hat{y}_i \quad \text{or} \quad y_i = \hat{y}_i + e_i.$$

If we substitute the value $y_i = \hat{y}_i + e_i$ into Formula (10.4), we get the sample regression model:

$$\text{Sample regression model:} \quad y_i = a + bx_i + e_i. \tag{10.5}$$

Suppose we illustrate the concept of a sample regression model by assuming that the admissions director in our GMAT–GPA example has selected a random sample of students, one of whom is Jane Doe. Figure 10.3 shows these observations in a scatter diagram; the line drawn through these points is the sample regression line (we will explain later how this line was derived). From

Fig. 10.3 Sample scatter diagram of observed values for **y** and **x**.

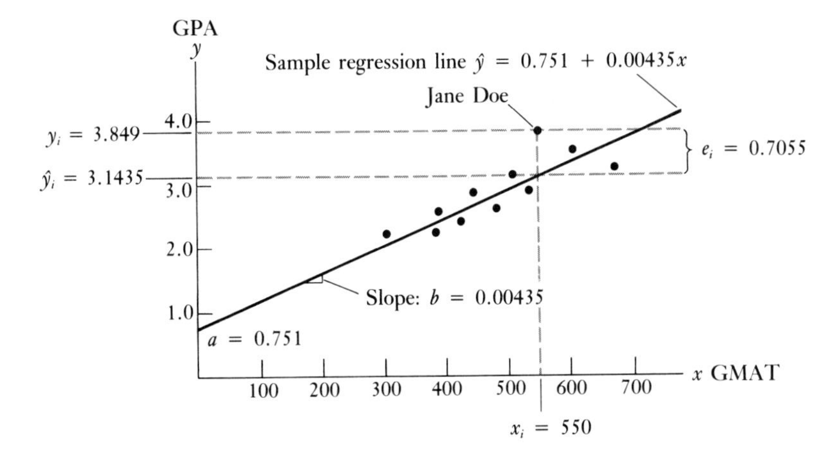

Fig. 10.3 we see that

$$a = 0.751 \qquad \text{and} \qquad b = 0.00435,$$

and the sample regression line is:

Sample regression line: $\hat{y} = 0.751 + 0.00435x.$

Note that, on the basis of our sample regression line, we predict that the mean GPA for all students with a GMAT of 550 will be

$$\hat{y}_i = 0.751 + 0.00435(550) = 3.1435.$$

We can also determine the amount of "error" in predicting Jane Doe's GPA by recalling that her GPA was $y_i = 3.849$. The value of e_i is therefore

$$\begin{aligned} e_i &= y_i - \hat{y}_i \\ &= 3.849 - 3.1435 = 0.7055. \end{aligned}$$

Thus, the value $a = 0.751$ is an estimate of $\alpha = 0.95$; $b = 0.0435$ is an estimate of $\beta = 0.0039$, and for Jane Doe, the error $e_i = 0.7055$ is an estimate of the true error $\varepsilon_i = 0.754$.

Now that we have specified the sample and population regression model, we need a procedure for determining values of a and b which provide the "best" estimates of α and β. The procedure for finding such estimates is called the *method of least squares*.

10.2 ESTIMATING THE VALUES OF α AND β BY LEAST SQUARES

A first step in finding a sample regression line of best fit is to plot the data in a scatter diagram. Such a plot allows us to visually determine whether a straight-line approximation to the data appears reasonable, and to make rough estimates of a and b. Although this approach often yields fairly satisfactory results, there are at least two reasons for having a more systematic approach to finding the "best" straight-line fit to the data. First, different people are likely to find slightly different values for a and b by the freehand drawing method. Secondly, the freehand estimation procedure provides no way of measuring the sampling errors, which are always important in forming confidence intervals or doing tests of hypotheses on population parameters.

What we need is a mathematical procedure for determining the sample regression line that best fits the sample data. The difficulty in establishing such a mathematical procedure to give the line of best fit is in determining the criterion to use in defining "best fit." Perhaps the most reasonable criterion is to find values a and b so that the resulting values of $\hat{y}$ (in the equation $\hat{y} = a + bx$) are as close as possible to the actual values y_i. That is, we want to minimize the values of e_i, where $e_i = y_i - \hat{y}_i$ (which, in Fig. 10.3, is seen to be the vertical distance from $\hat{y}_i$ to y_i).

The approach almost universally adopted to find the line of best fit in regression analysis is to determine the values of a and b which minimize the sum of the *squared errors.* This procedure is known as the *method of least squares.* Since $e_i = y_i - \hat{y}_i$, this method is defined as follows:

Method of least-squares estimation:

$$\text{Minimize} \sum_{i=1}^{n} e_i^2 = \sum_{i=1}^{n} (y_i - \hat{y}_i)^2. \tag{10.6}$$

By minimizing the sum of the squared errors (the vertical deviations), the method of least squares is assuming, in effect, that the values of the independent variable are fixed quantities, known in advance, and that the only random element is the value of the dependent variable. This is why we have defined the population regression line as the conditional distribution of $\mathbf{y}$, *given* x. Suppose that we are testing a new type of automobile engine for the length of time it takes a certain part to wear out, relative to the speed at which the engine is run. The regression relationship in this case might logically be tested by running the engine at a number of predetermined speeds, say 30, 40, 50, and 60 miles per hour, and noting the time to failure. Controlled experiments such as this are common in agricultural research, in studying the effect of certain predetermined levels of various types of fertilizers on crop yields. Even in the GMAT–GPA example, it can be assumed that the GMAT scores of all candidates are known in advance (i.e., before they are admitted) and that GPA is a random variable. In other cases, however, it is not obvious that the independent variable can always be treated as a fixed quantity. In determining the relationship between the height and weight of college freshmen, for instance, both variables should probably be considered as random variables. Fortunately, the assumption that the values of x are fixed quantities (rather than a random variable) is not a crucial one in least-squares regression analysis, so that it is possible to apply the technique even when both $\mathbf{x}$ and $\mathbf{y}$ are random variables.

Determining the least-squares regression line requires finding those values of a and b which minimize $\sum_{i=1}^{n} (y_i - \hat{y}_i)^2$, which is equivalent to minimizing

$$\sum_{i=1}^{n} [y_i - (a + bx_i)]^2,$$

because $\hat{y}_i$ is defined to be equal to $a + bx_i$. For convenience, let

$$G = \sum_{i=1}^{n} (y_i - a - bx_i)^2.$$

Then, in order to minimize G, it is necessary to take the partial derivative of G with respect to the variables of concern, set each of these partials equal to zero, and then solve the resulting equations simultaneously. The reader must take care to understand the fact that the variables in this problem are a and b (not x and y), since it is these two values which are free to vary (x and y are fixed by the data). Hence, it is necessary to set $\partial G/\partial a$ and $\partial G/\partial b$ equal to zero, and then solve for a and b. Solving yields*

$$\sum_{i=1}^{n} y_i = na + b \sum_{i=1}^{n} x_i. \tag{10.7}$$

and

$$\sum_{i=1}^{n} x_i y_i = a \sum_{i=1}^{n} x_i + b \sum_{i=1}^{n} x_i^2. \tag{10.8}$$

Formulas (10.7) and (10.8) are called the *normal equations*. Solving these two equations for a and b results in the line of "best fit" to the sample data, which is the regression line minimizing $\sum_{i=1}^{n} (y_i - \hat{y}_i)^2$. Although it is not difficult to solve Formulas (10.7) and (10.8) simultaneously, the algebra of solving for the regression coefficients is messy enough to justify our decision not to present the process of their solution in this book. Rather, we present

* Solving first for $\partial G/\partial a$ yields

$$\frac{\partial G}{\partial a} = \sum_{i=1}^{n} 2(y_i - a - bx_i)(-1),$$

$$0 = -2 \sum_{i=1}^{n} (y_i - a - bx_i),$$

which is equivalent to (10.7).

The same process is followed for $\partial G/\partial b$:

$$\frac{\partial G}{\partial b} = \sum_{i=1}^{n} 2(y_i - a - bx_i)(-x_i),$$

$$0 = -2 \sum_{i=1}^{n} (y_i - a - bx_i)(x_i),$$

which is equivalent to (10.8).

the value of a and b resulting from the solution of normal equations:

Slope of least-squares line:

$$b = \frac{\frac{1}{n-1}\sum_{i=1}^{n}(x_i - \bar{x})(y_i - \bar{y})}{\frac{1}{n-1}\sum_{i=1}^{n}(x_i - \bar{x})^2}, \tag{10.9}$$

Intercept of least-squares line:

$$a = \bar{y} - b\bar{x}.$$

Perhaps the formula for b in (10.9) will be a bit easier to remember if we point out that its denominator is the *sample variance of* x, since

$$s_x^2 = \frac{1}{n-1}\sum_{i=1}^{n}(x - \bar{x})^2.$$

The numerator of the expression for b is the *sample covariance of* x *and* y. The concept of a *population* covariance was defined in Chapter 3 to be:

$$C[\mathbf{x}, \mathbf{y}] = \frac{1}{N}\sum(x - \mu_x)(y - \mu_y).$$

The symbol s_{xy} is usually used to denote an unbiased estimate of $C[\mathbf{x}, \mathbf{y}]$; that is,

Sample covariance: $\quad s_{xy} = \frac{1}{n-1}\sum(x - \bar{x})(y - \bar{y}).$

Thus

$$b = \frac{s_{xy}}{s_x^2} = \frac{\text{Sample covariance between indep. and dep. variables}}{\text{Sample variance of indep. variable}}. \tag{10.10}$$

Two features of the sample regression line should be noted from Formula (10.9). First, note that if we substitute $a = \bar{y} - bx$ into the equation $\hat{y} = a + bx$, we get $\hat{y} = \bar{y} - b(x - \bar{x})$. This means that, whenever $x = \bar{x}$, then $\hat{y} = \bar{y}$; hence, the regression line always goes through the point $(\bar{x}, \bar{y})$. Secondly, in minimizing $\sum e_i^2$, the least-squares method automatically sets $\sum e_i = 0$. Thus, our line estimating the average relationship between y and x passes through the point of average $(\bar{x}, \bar{y})$, and splits the scatter diagram of observed points so that the positive residuals (underestimates of the true points) always exactly cancel the negative residuals (overestimates of true points). Such a sample regression line therefore correctly (unbiasedly) estimates the population regression line.

Table 10.1 Sample Points for GMAT and GPA and Calculation of Their Means, Variances, and Covariance

	(1) GMAT(x_i)	(2) GPA(y_i)	(3) $(x_i - \bar{x})$	(4) $(y_i - \bar{y})$	(5) $(y_i - \bar{y})(x_i - \bar{x})$	(6) $(x_i - \bar{x})^2$
	480	2.70	−60	−0.40	24	3,600
	490	2.90	−50	−0.20	10	2,500
	510	3.30	−30	+0.20	−6	900
	510	2.90	−30	−0.20	6	900
	530	3.10	−10	0.00	0	100
	550	3.00	+10	−0.10	−1	100
	610	3.20	+70	+0.10	7	4,900
	640	3.70	+100	+0.60	60	10,000
Sum	4320	24.80	0	0	100.00	23,000
Mean	$\bar{x} =$ 540	$\bar{y} =$ 3.10				

To illustrate the technique of finding a least-squares regression line, consider again our GPA–GMAT problem. Suppose we now use the method of least squares to determine a and b for a random sample of eight observations (i.e., eight students).* Columns (1) and (2) in Table 10.1 give the hypothetical data for these students, ordered from the lowest GMAT to the highest score. We will use the rest of the data in Table 10.1 in a moment.

Our first step in analyzing the data in the first two columns of Table 10.1 is to construct a scatter diagram, to see if the assumption of linearity is a reasonable one in this case. Figure 10.4 indicates that it is.

The sums in columns (5) and (6) of Table 10.1 give the information necessary to calculate a and b. Using Formula (10.9), we can determine b as follows:

$$b = \frac{\frac{1}{n-1}\sum_{i=1}^{n}(y_i - \bar{y})(x_i - \bar{x})}{\frac{1}{n-1}\sum_{i=1}^{n}(x_i - \bar{x})^2} = \frac{100/7}{23{,}000/7} = 0.00435.$$

Using this value of b and the means of y and x shown in columns (1) and (2) of Table 10.1, we obtain the following value of a:

$$a = \bar{y} - b\bar{x} = 3.10 - 0.00435(540) = 0.751.$$

* For most practical purposes, a sample size of eight students would not be sufficient. We assume it here only for computational ease.

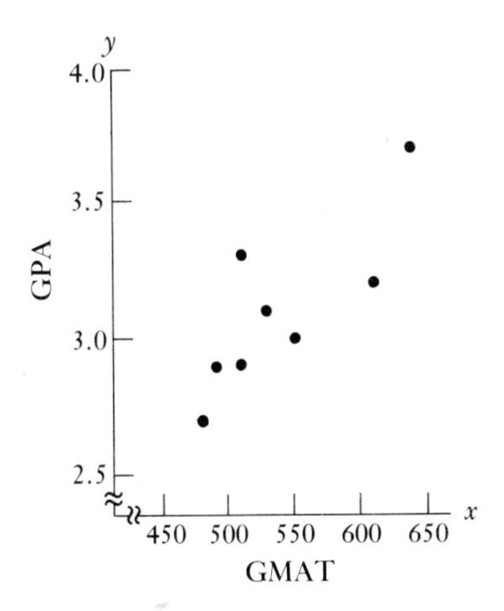

Fig. 10.4 Scatter diagram for GMAT–GPA sample values.

Hence, the least-squares regression line for this example is

$$\hat{y} = 0.751 + 0.00435x. \tag{10.11}$$

Figure 10.5 illustrates this sample regression line. Since the line in Fig. 10.5 was determined by the method of least squares, there is no other line which could be drawn such that the sum of the squared residuals between the points and the line (measured in a vertical direction) could be smaller than for this line. The residuals and the estimated values of y_i for all eight sample points are given in Table 10.2. The sum of the errors in this case (-0.003) differs from zero only because of rounding error.

Fig. 10.5 The least-squares regression line $\hat{y} = 0.751 + 0.00435x$.

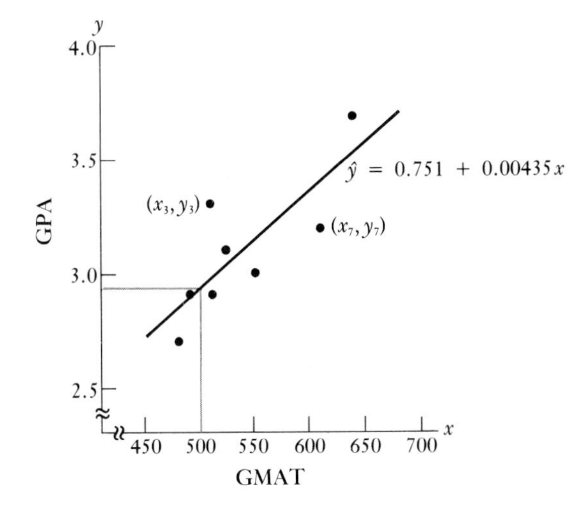

Table 10.2 Observed, Estimated, and Residual Values for the Least-Squares Regression Shown in Fig. 10.5

	x_i	Observed Value (y_i)	Predicted Value ($\hat{y}_i = 0.751 + 0.00435x_i$)	Residual $e_i = (y_i - \hat{y}_i)$
	480	2.70	2.839	−0.139
	490	2.90	2.883	0.017
	510	3.30	2.970	0.330
	510	2.90	2.970	−0.070
	530	3.10	3.057	0.043
	550	3.00	3.144	−0.144
	610	3.20	3.405	−0.205
	640	3.70	3.535	0.165
Sum	4320	24.80	24.80	−0.003

At this time the reader should check on how well the least-squares regression line is understood by verifying that the point of means $(\bar{x}, \bar{y})$ does lie on the line, and by examining the corresponding values of the errors (or "residuals" as they are often called) in Table 10.2. Note from Fig. 10.5 that a positive residual such as $e_3 = 0.33$ means that the point (x_3, y_3) lies above the line and therefore, $\hat{y}_3$ underestimates y_3. On the other hand, a negative residual such as $e_7 = -0.205$ corresponds to an overestimation of y_7 by $\hat{y}_7$, in that the point (x_7, y_7) lies below the regression line.

Further Discussion of the Estimators *a* and *b*

Formula (10.9) provides a convenient method for calculating the regression slope b when the values of x and y are integers. When they are not integers, an alternative formula for b may often prove to be computationally easier. This formula is:

Computational formula for b:

$$b = \frac{n\sum x_i y_i - \sum x_i \sum y_i}{n\sum x_i^2 - (\sum x_i)^2}. \tag{10.12}$$

We will illustrate this formula in terms of our GMAT–GPA example. Before doing so, however, we should point out that, since the value of b depends on the *relative* values of x and y, not on their absolute size, subtracting a constant from each value of x and/or y will not affect the regression slope. The advantage of such subtractions is that the algebra of Formula (10.12) may then become much easier. To illustrate this fact, suppose we subtract some convenient number from each value of x and y in Table 10.1, and then use Formula 10.12 to determine b. We (arbitrarily) select 500 to be subtracted from each value of x, and 3.00 from each value of y. These values are shown in Table 10.3.

Using these sums and (10.12) we can calculate b:

$$b = \frac{8(132.0) - (320)(0.80)}{8(35{,}800) - (320)^2} = 0.00435.$$

This is the same value of b calculated previously. In using this computational formula, one must remember to return to the original data in order to calculate the value of a, for an incorrect a value will result from the coded data.

The least-squares regression line described by Formula (10.11) serves several related purposes. First, the regression coefficients provide point estimates of the population parameters, 0.751 being an estimate of α, and 0.00435 an estimate of β. These point estimates, and the interval estimates which can be derived from them, serve a variety of research needs. Economists, for example, use regression analysis in an attempt to test various assumptions about population characteristics of economic concepts, such as marginal propensity to consume, the price elasticity of demand, and the factor shares of labor and capital in production, and to study the effect of changes in one variable (e.g., a tax change) on one or more other variables (e.g., consumption). Knowing the regression line also enables us to use the values of $\hat{y}$ to estimate the conditional mean of the dependent variable, $\mu_{y \cdot x}$, for specific values of x. Perhaps most

Table 10.3 GMAT-GPA Data Repeated, with 500 Subtracted from Each x Value, and 3.00 from Each y Value

	x	y	xy	x^2	y^2
	−20	−0.30	6.0	400	0.09
	−10	−0.10	1.0	100	0.01
	10	0.30	3.0	100	0.09
	10	−0.10	−1.0	100	0.01
	30	0.10	3.0	900	0.01
	50	0	0	2,500	0
	110	0.20	22.0	12,100	0.04
	140	0.70	98.0	19,600	0.49
Sum	320	0.80	132.0	35,800	0.74

importantly, the regression line permits prediction of the *actual* values of the dependent variable, given a value of the independent variable.

For example, suppose a student has a GMAT score of $x_i = 500$ and we would like to predict the grade-point average, y_i, that this person will earn. The best estimate of y_i, using Formula (10.9), is given by $\hat{y}_i$, which in this case is

$$\hat{y}_i = 0.751 + 0.00435(500) = 2.926.$$

In using a regression line to make predictions about the dependent variable, special care must be taken when the value of the independent variable falls outside the range of past experience (historical data), since it may be that these values cannot be represented by the same equation. The regression equation described above, for example, predicts that the GPA for students with a GMAT of 100 will be

$$\hat{y}_i = 0.751 + 0.00435(100) = 1.186.$$

But since students scoring this low are usually not admitted to graduate school, there is really no way of knowing what GPA they might achieve. Similarly, the equation we derived predicts the GPA for a student with a test score of 800 to be

$$\hat{y}_i = 0.751 + 0.00435(800) = 4.231,$$

which is not even possible. Clearly special care must be taken in this example when attempting to predict outside a range of GMAT scores running from about 400 to 700.

One additional precaution in using regression that should be mentioned here is that of avoiding what is called the "regression fallacy." This fallacy occurs when one attempts to relate the values of a variable at one point in time to the comparable values of that *same* variable at some other point in time. The problem in using regression or correlation in this circumstance arises because of the tendency for unusually high or low values of a random variable to be followed by more average values. Students scoring abnormally high or low on a midterm exam tend to have more average grades on the final. A corporation earning exceptionally high or low profits in one year is likely to have more nearly average profits the next year. This is not to imply that these observations are not related—students *do* tend to make consistently high or low exam scores, and the profits a corporation earns in one period *are* related to those in another period. The random fluctuations in such variables, however, will produce a spurious convergence which will often cause the regression line to have a slope less than would be expected. For example, in Galton's research relating the heights of fathers to the heights of their sons, he found that the regression line had a slope of less than the expected 45° line, as shown in Fig. 10.6. This inclination below the 45° line should not be interpreted as a "regression to mediocrity," but rather the convergence of chance variations to the mean.

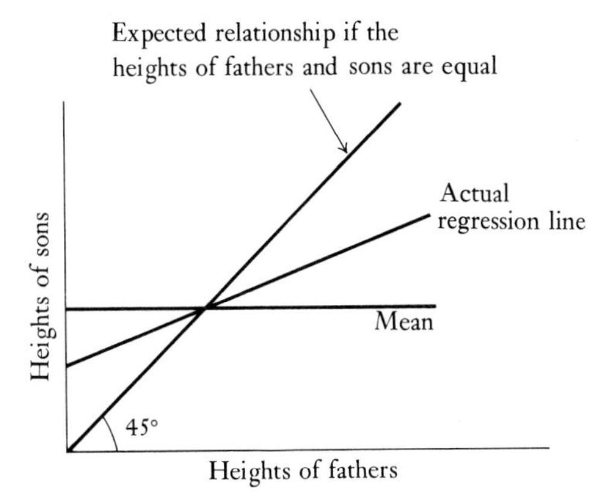

Fig. 10.6 The "regression fallacy."

Several aspects of the method of least squares need to be emphasized at this point. First, this method is just a curve-fitting technique, and as such it requires no assumptions about the distribution of x or $\mathbf{y}$. It is only when probability statements about the parameters of the population are desired that such assumptions are necessary. Secondly, the method of least squares can be adapted to apply to nonlinear populations. Although the formulas necessary for applying the method of least squares to a nonlinear relationship will naturally differ from those used for linear relationships, the objective in fitting the curve is the same in the two cases—to find the line minimizing the sum of the squared residuals. Nonlinear regression models, which are often quite complex, will be discussed briefly in Chapter 12.

When certain assumptions are made about the population, then the resulting estimators a and b can be shown to be unbiased, consistent, and efficient. Furthermore, these assumptions permit us to construct the confidence intervals and test the hypotheses that are so crucial to regression analysis. In other words, the assumptions provide the rationale behind the widespread use of the least-squares approach. In every regression-analysis problem, the researcher must therefore be satisfied that such assumptions are reasonable. The assumptions are concerned with the random variable $\boldsymbol{\varepsilon}$, for this variable describes how well $\mu_{\mathbf{y} \cdot x_i}$ estimates $\mathbf{y}_i$. Making assumptions about the mean and the variance of $\boldsymbol{\varepsilon}$ enables us to make inferences about this variable on the basis of the sample values of e.

10.3 ASSUMPTIONS AND ESTIMATION

Many possible sets of assumptions about the distribution of the variables in the population regression model could be formulated. Five assumptions are commonly used because they yield relatively simple estimators possessing many desirable properties, and they result in test statistics that follow commonly known distributions. In the following section, we describe these five assumptions of simple linear regression.

The Five Assumptions

> *Assumption* 1. *The random variable* ε *is assumed to be statistically independent of the values of* x.

This means that the covariance between the values of the independent variable and the corresponding error terms is zero. This assumption will be true necessarily when x is a *fixed* variable with values known in advance, rather than a *random* variable with values drawn from an underlying sampling distribution.* When x is a fixed variable, it must be independent of the random variable ε since the covariance of a random variable and a constant is always zero. The constant has no variation from its fixed values. When x is a random variable, this assumption may be violated. Suppose that the errors are a *percentage* of the values of x because of measurement error. For example, let y be aggregate consumption and x be national income measured over time. Since these values have increased by a factor of 10 since World War II, it is likely that the size of typical errors of measurement has also increased. Thus, the random variable ε is larger today than it was in 1946. In this case, the probability of observing larger-sized errors increases as the size of the variable x increases, as shown in Fig. 10.7.

Other violations of assumption 1 may occur when an important variable is omitted from the regression analysis. Let us consider the problem of estimating demand for U.S. compact cars where the quantity demanded is y and the price of compacts is x. For simplicity, suppose that we assume that the only variable of importance omitted from this model is the price of the substitute good, import compacts. If the price of U.S. compacts increases, we would expect a lower demand for U.S. compacts, assuming that all other factors are equal. But all other factors are not equal, as demand also depends on the price of imports. If, as the price of U.S. compacts increases, it becomes more and more difficult to predict demand (i.e., the errors increase because we do not know what will happen to the price of imports), then assumption 1 is violated. Figure 10.8 illustrates this situation by a diagram showing *errors* on the y-axis.

Many other examples can be cited in which behavior is fairly exact for moderate values of the independent variables, but erratic for extreme high or

* Both of these interpretations about x are commonly used in regression analysis. With proper handling, the calculational and interpretive results presented in this chapter can be used when x is a random variable as well as when it is a fixed variable.

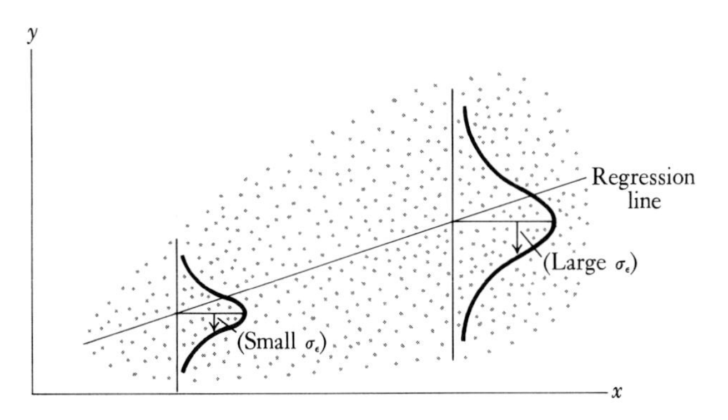

Fig. 10.7 Violation of assumption 1 (and assumption 5).

low values of x. These also correspond to violations of assumption 1. Suppose that y is investment expenditures by firms and x is annual growth in sales. If sales growth is 5–10 percent, firms might make corresponding investment expenditures to replace worn-out equipment and buy some new goods to allow for some expansion of production. This behavioral response would be quite consistent across firms. However, suppose that sales growth for a group of firms was 75–100 percent. More variation in response could be expected. Some might feel very optimistic and make large investment expenditures to double their output. Others might be very cautious and feel that the sales growth was a one-time increase. They might save the extra revenues temporarily and not expand at all. They might not even replace what wears out, and use the current sales figures as motivation to sell out at a good price. Large variations in investment decisions could also be expected in response to a sudden downturn indicated by a negative growth in sales. Thus, the errors ε would be quite large in estimating investment at a high (or low) level of sales growth because the behavior predicted in the model is not systematic under these circumstances. The errors ε and the sales growth x are not independent because the relation fits much better (small ε) in the range of x where investment

Fig. 10.8 Illustration of how larger values of x may lead to larger errors.

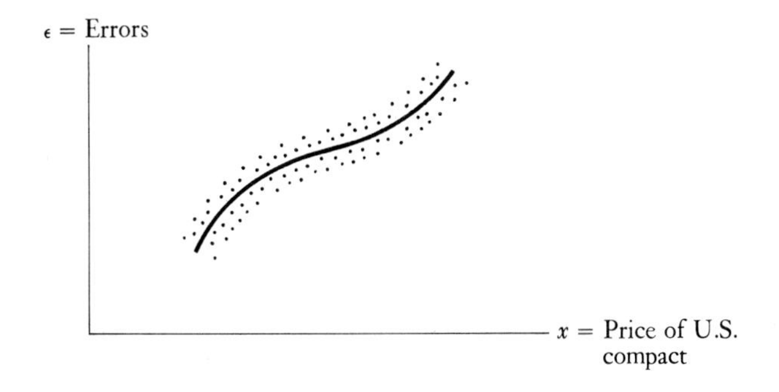

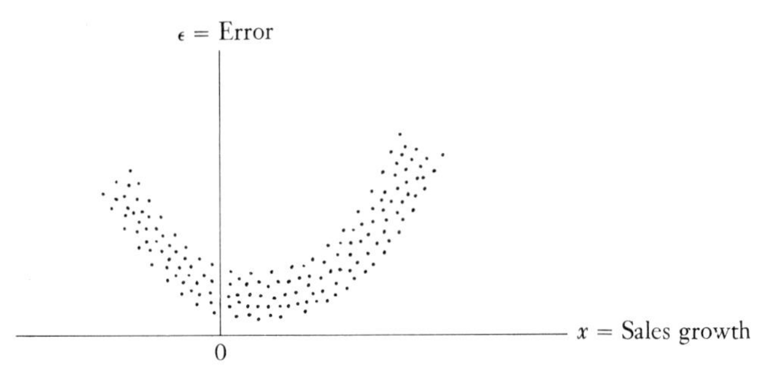

Fig. 10.9 Illustration of errors dependent on sales growth.

behavior is consistent, and fits much worse (large ε) in the ranges of $\boldsymbol{x}$ where firms make widely different decisions on investment expenditures. Figure 10.9 represents this situation.

> *Assumption 2. The random variable ε is assumed to be normally distributed.*

Since ε_i is a composite of many factors (such as errors of measurement, errors in specifying the model, or irregular errors such as economic, political, social, and business fluctuations), it is reasonable to expect that many of these factors tend to offset each other so that large values of ε_i are much less likely than small values of ε_i. Indeed, if many of these factors are unrelated, a form of the central limit theorem guarantees that their joint effect (represented by ε_i) will be normally distributed. Figure 10.10 illustrates the meaning of this

Fig. 10.10 Normally distributed errors ε_i about $\mu_{y \cdot x_i}$ for any value x_i (given that assumptions 3 and 5 also hold).

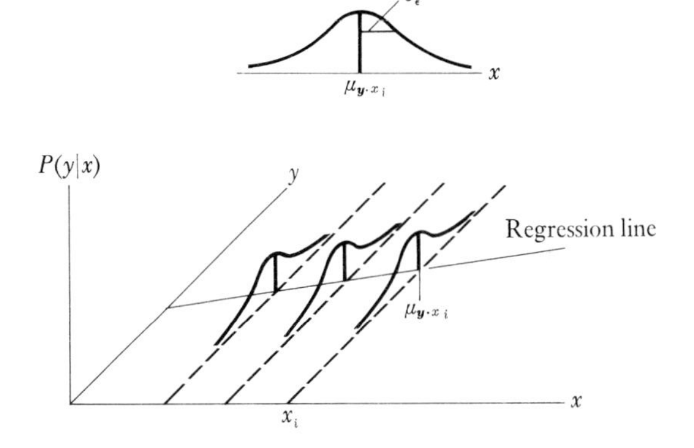

assumption by showing the normal distribution of errors ε_i about the population straight-line relationship between y and x.

Assumption 3. *The variable* ε *is assumed to have a mean of zero*; *that is*,

$E[\varepsilon] = 0$.

This means that, for a given x_i, the differences between y_i and $\mu_{y \cdot x_i}$ are sometimes positive, sometimes negative, but on the average are zero. Thus, the distribution of ε_i about the population regression line $\mu_{y \cdot x}$ (as shown in Fig. 10.10) is always centered at the value $\mu_{y \cdot x_i}$ for any given x_i. Assumption 3 establishes the estimability of α. That is, only when $E[\varepsilon] = 0$ can one get an unbiased estimate of α using the method of least squares.

Assumption 4. *Any two errors*, ε_k *and* ε_j, *are assumed to be statistically independent of each other*; *that is, their covariance is zero*,

$C[\varepsilon_k, \varepsilon_j] = 0$.

This assumption means that the error of one point in the population cannot be related systematically to the error of any other point in the population. In other words, knowledge about the size or sign of one or more errors does not help in predicting the size or sign of any other error. For example, knowing that the error in describing the GPA of one (or more) graduate student is positive does not give you any help in determining whether or not the error for another graduate student will also be positive.

This assumption is violated most commonly when observations are drawn periodically over time (time-series data). A simple graph of the values of ε over time that indicates such a violation is shown in Fig. 10.11. Figure 10.11(a), for example, might represent long-term changes in the prime interest rate, where time is measured in months or years. Figure 10.11(b), on the other hand, could represent fluctuations in the amount of electrical usage in a city, where time is measured in 12-hour periods.

Assumption 5. *The random variables* ε_i *are assumed to have a finite variance* σ_ε^2 *that is constant for all given values of* x_i.

This means that the dispersion or variability of points in the population about the population regression line must be constant. In Fig. 10.10 this constant variance of ε_i is represented by depicting all the normal distributions about $\mu_{y \cdot x_i}$ as having the same standard deviation. No one distribution is more spread out or more peaked than another for a different value of x.

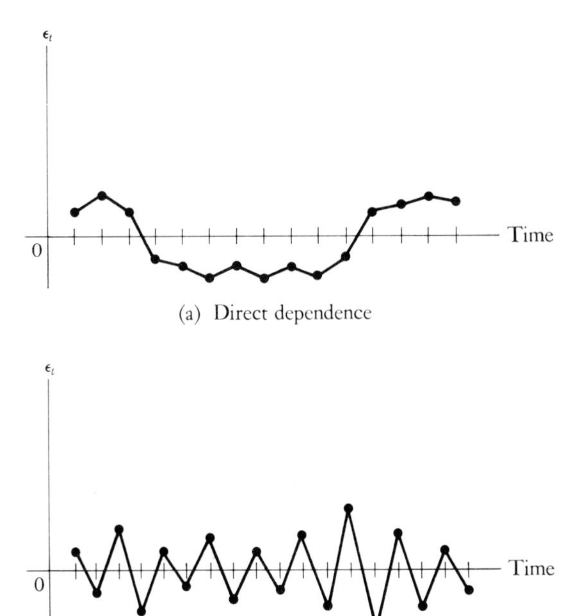

Fig. 10.11 Time-sequence plots of errors ε.

Assumption 5 is violated most often in data measured at the same point of time (cross-sectional data). For example, in a study relating per capita income and state monies allocated for highway repair across the U.S.A., assumption 5 might be violated because of differing state legal codes, climate, or political interests. Similarly, a study relating educational attainment and household income might be affected by differing races or locations (urban, suburban, rural). Figure 10.12 illustrates a case in which the distribution of errors is more spread

Fig. 10.12 Increasing variance of ε as x increases (reproduction of Fig. 10.7).

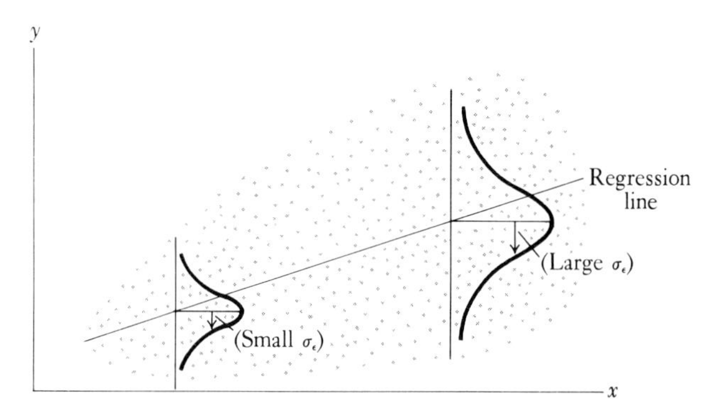

out (large variance of ε) for higher values of x than for lower values of x. Note that Fig. 10.12 is merely a repeat of Fig. 10.7.

We point out here that when assumption 1 is violated because of proportional errors in measuring x, assumption 5 (constant variance) will also be violated. Thus, Fig. 10.7 is also an illustration of the violation of assumption 5, because the distribution of errors is more spread out (i.e., larger variances) for higher values of x than for lower values of x. Usually, assumption 5 is referred to as the assumption of *homoscedasticity*. Populations that do not have a constant variance are called *heteroscedastic*.

Properties of the Regression Coefficients

As we indicated previously, the five assumptions presented above provide the rationale for the widespread use of the least-squares procedure. Given assumption 2, it can be shown that the estimators of α, β, and $\mu_{y \cdot x_i}$ obtained by using the least-squares criterion are *identical* to the estimators that would result using the principle of maximum likelihood estimation. Such estimators have a number of desirable properties (such as consistency).

A second important result related to the least-squares estimators is called the *Gauss–Markov theorem*. This classical result of linear estimation was formulated by the German mathematician and astronomer, Karl F. Gauss (1777–1855), in his early works published in 1807 and 1821. Since these involved applications in physics and planetary motion, they generally remained unknown to social scientists and businessmen until they were restated in a more modern context by A. A. Markov in 1912, in a study of linear processes. In the 1930's the work of Markov was extended and applied directly to least-squares estimation in several ways, and the Gauss–Markov theorem assumed the identity it has today:

> **Gauss–Markov theorem.** *If assumptions* 1, 3, 4, *and* 5 *hold true, then the estimators of* α, β *and* $\mu_{y \cdot x}$ *determined by the least-squares criterion are Best Linear Unbiased Estimators* (i.e., *BLUE*).

In the above context, the term *linear* means that the estimators are straight-line functions of the values of the dependent variable y. They are *unbiased* because their expected value is equal to the population value (given that assumptions 1 and 3 are true). They are *best* in the sense of being *efficient* (if assumptions 4 and 5 are true). That is, the least-squares estimators have a variance smaller than that of any other linear unbiased estimator. Thus, the importance of the Gauss–Markov theorem is that if assumptions 1, 3, 4, and 5 hold, then the least-squares estimators have the desirable properties of unbiasedness and efficiency.

Define: independent and dependent variables, $\mu_{y \cdot x}$, ε_i, intercept (α) and slope (β), population regression model, sample regression model, Gauss–Markov theorem, least-squares method, normal equations, population and sample covariance, regression fallacy.

Problems

10.1 a) Describe what is meant by "the method of least squares." What is a least-squares regression line?

b) What assumptions about the parent population are necessary to fit a least-squares regression line to a set of observations? What assumptions about the parent population are necessary to make interval estimates on the basis of a least-squares regression line?

10.2 Explain why a disturbance term ε is included in the population regression model.

10.3 Explain what is meant by "best" in a Best Linear Unbiased Estimator (BLUE).

10.4 Suppose that y = average test score on college entrance boards and x = hundreds of dollars of expenditures per pupil in the student's respective high school. Given that a regression of y on x gives $\hat{y} = 320 + 50x$, what is the interpretation of the value 50 and what value would you predict if $x = 10$?

10.5 a) Estimate, on the basis of the following data, the weight of a person who is 71 in. tall.

Height (in.)	Weight (lb.)
65	150
70	170
75	160

b) Estimate the height of a person who weighs 153 lbs.

c) Plot the regression lines that you calculated for parts (a) and (b). Why do these lines differ?

10.6 The following data represent the dollar value of sales and advertising for a retail store.

Advertising	Sales
600	5000
400	4000
800	7000
200	3000
500	6000

a) Draw the scatter diagram for these data. Fit a straight line by the free-hand method. Does a linear relationship seem appropriate?

b) Find the least-squares regression line for predicting sales on the basis of advertising.

c) What value for sales would you predict if advertising is \$700? What value would you predict if advertising is zero?

10.7 A study was conducted to determine the relationship (if any) between the number of people in a household (x) and the number of radios in that household (y). The following sample was collected:

People (x):	3	1	5	2	4
Radios (y):	3	2	6	4	5

a) Plot these data to verify that none of the five assumptions is obviously violated.

b) Verify that the least-squares regression line is $\hat{y} = 1.3 + 0.9x$. Show that this line goes through the point $(\bar{x}, \bar{y})$.

*10.8 Prove that a least-squares regression line always passes through the point $(\bar{x}, \bar{y})$.

10.9 A manufacturing firm bases its sales forecast for each year on government estimates for total demand in the industry. The data in Table 10.4 give the government estimate for total demand and this firm's sales for the past ten years. Find the least-squares regression line for estimating sales on the basis of the demand estimate. What sales would you predict when the demand estimate is 300,000?

Table 10.4

Demand Estimate	Sales
200,000	$5,000
220,000	6,000
400,000	12,000
330,000	7,000
210,000	5,000
390,000	10,000
280,000	8,000
140,000	3,000
280,000	7,000
290,000	10,000

10.10 Recalculate the sample regression line in Problem 10.9 by subtracting $200,000 from each government estimate and $8000 from each sales value. Compare this answer with the answer to 10.9.

10.11 The data in Table 10.5 pertain to selling prices and the number of pages of new statistics books. Find the regression equation of y (price in dollars) on x (number of pages in hundreds), using the method of least squares.

Table 10.5

Price	Number of Pages
$15	400
17	600
17	500
15	300
13	400
13	200

10.12 In Section 11.6 of Chapter 11 we will present an example involving the relationship between a stock index (**x**) and the amount of investment over all private business in the U.S. (**y**). Run a computer program to compute the least-squares regression line for the data in Table 11.2. Compare your computer results with the values of a and b given on page 533.

10.4 THE STANDARD ERROR OF THE ESTIMATE

The Components of Variability

Our presentation of the standard error of the estimate will be easier if we first present some of the components of variability in regression analysis.

In regression analysis the difference between y_i and the mean of the **y** values ($\bar{y}$) is often called the *total deviation of* **y**; that is, it represents the total amount by which the ith observation deviates from the mean of all y values. Now it is not too hard to show that this total deviation can be written as the sum of two other deviations, one of which is $(y_i - \hat{y}_i)$, and the other is $(\hat{y}_i - \bar{y})$. Since the deviation $(\hat{y}_i - \bar{y})$ is the part of total deviation which is "explained" (i.e., accounted for by the regression line), this term is called *explained deviation.* On the other hand, the deviation $(y_i - \hat{y}_i)$ is the *error* for the ith sample observation, and since we have no basis for explaining why it occurred, this term is called *unexplained deviation.* That is,

$$\begin{array}{ccccc} \text{Total deviation} & = & \text{Unexplained deviation} & + & \text{Explained deviation} \\ (y - \bar{y}) & = & (y_i - \hat{y}_i) & + & (\hat{y}_i - \bar{y}). \end{array} \tag{10.13}$$

This relationship is illustrated in Fig. 10.13 in the context of our GMAT–GPA example (see Table 10.2).

Because the two parts of total deviation shown in Formula (10.13) are independent, it can be shown* that this same relationship holds when we square each deviation, and sum over all n observations. That is,

$$\sum_{i=1}^{n} (y_i - \bar{y})^2 = \sum_{i=1}^{n} (y_i - \hat{y})^2 + \sum_{i=1}^{n} (\hat{y}_i - \bar{y})^2 . \tag{10.14}$$

---

* To prove this statement, observe that

$$\begin{aligned} \sum(y_i - \bar{y})^2 &= \sum[(y_i - \hat{y}_i) + (\hat{y}_i - \bar{y})]^2 \\ &= \sum[(y_i - \hat{y}_i)^2 + 2(y_i - \hat{y}_i)(\hat{y}_i - \bar{y}) + (\hat{y}_i - \bar{y})^2] \\ &= \sum(y_i - \hat{y})^2 + \sum 2(y_i - \hat{y}_i)(\hat{y}_i - \bar{y}) + \sum(\hat{y}_i - \bar{y})^2 \end{aligned}$$

If the middle term of this last expression equals zero, then we have proved (10.14). By a little manipulation, we can write

$$\sum 2(y_i - \hat{y}_i)(\hat{y}_i - \bar{y}) = 2b\sum(x_i - \bar{x})(y_i - \hat{y}_i)$$

The righthand side was set equal to zero when solving $\partial G/\partial b$ for the least-squares estimate on page 480.

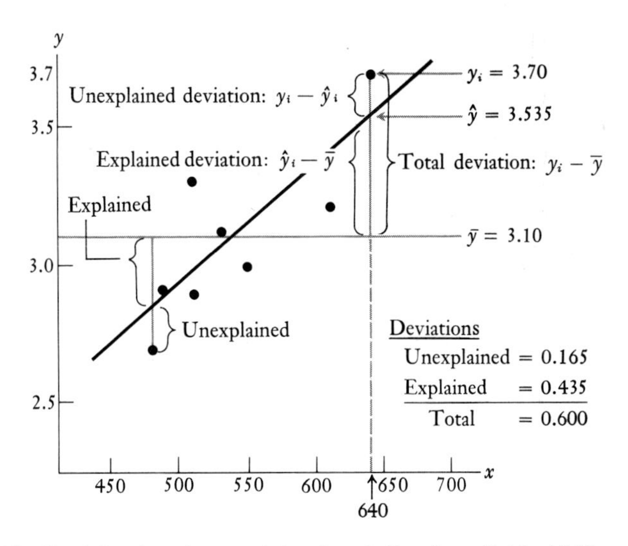

Fig. 10.13 Explained and unexplained variation (see Table 10.2).

The lefthand side of Formula (10.14) is referred to as the *total variation*, or as the *sum of squares total* (which is abbreviated as SST). The first term on the right is the *unexplained variation*, or equivalently, the *sum of squares error* (SSE). You should recognize SSE as $\sum e_i^2$, the term that was minimized in finding the least-squares regression line. The last term in (10.14), which is the sum of squares of the values of $\hat{y}$ about $\bar{y}$, is called the *explained variation*, or the *sum of squares regression* (SSR). Thus, we can rewrite (10.13) as follows:

SST	=	SSE	+	SSR
Total variation	=	Unexplained variation	+	Explained variation

$$\sum_{i=1}^{n}(y_i - \bar{y})^2 = \sum_{i=1}^{n}(y_i - \hat{y})^2 + \sum_{i=1}^{n}(\hat{y}_i - \bar{y})^2. \tag{10.15}$$

The advantage of breaking total variation into these two components is that we can now talk about goodness of fit in terms of the size of SSE. For example, if the line is a perfect fit to the data, then SSE $= 0$. Usually, however, the line is not a perfect fit; hence, SSE $\neq 0$.

We can calculate SST, SSE, and SSR for our GMAT–GPA example from the data in Tables 10.1 and 10.2. First, SST can be derived by squaring the values in column 4 of Table 10.1. These values are shown in column 1 of Table 10.6. The value of SSE is derived by squaring the errors $e_i = (y_i - \hat{y}_i)$, shown in the final column of Table 10.2. These squares (rounded to three

Table 10.6 Calculation of SST, SSE, and SSR

$(y_i - \bar{y})^2$	$(y_i - \hat{y})^2$	$(\hat{y}_i - \bar{y})$	$(\hat{y}_i - \bar{y})^2$
0.16	0.019	2.839–3.10	0.068
0.04	0.000	2.883–3.10	0.047
0.04	0.109	2.970–3.10	0.017
0.04	0.005	2.970–3.10	0.017
0.00	0.002	3.057–3.10	0.002
0.01	0.021	3.144–3.10	0.002
0.01	0.042	3.405–3.10	0.093
0.36	0.027	3.535–3.10	0.189
SST = 0.66	SSE = 0.225		SSR = 0.435

decimals) are shown in column 2 of Table 10.6. Finally, we can calculate SSR $= \sum(\hat{y}_i - \bar{y})^2$ by subtracting $\bar{y} = 3.10$ from each value of $\hat{y}$ in column 3 of Table 10.2, and then squaring these differences. These values are shown in column 4 of Table 10.6.

We thus see that

$$\begin{aligned} \text{SST} &= \text{SSE} + \text{SSR} \\ 0.660 &= 0.225 + 0.435. \end{aligned}$$

In practice there are more efficient ways of calculating SST, SSR, and SSE. For example, SST $= \sum(y_i - \bar{y})^2$ is part of the formula for $V[\mathbf{y}]$, and Chapter 1 gave a computational formula for $V[\mathbf{y}]$. The appropriate form now is:

$$\sum(y_i - \bar{y})^2 = \text{SST} = \sum y_i^2 - \frac{1}{n}\left(\sum y_i\right)^2.$$

A computational formula for SSR is:

$$\text{SSR} = b\sum(x - \bar{x})(y - \bar{y}) = b\left[\sum xy - \frac{1}{n}(\sum x)(\sum y)\right].$$

Once SSR is known, SSE can be derived by letting SSE = SST − SSR.

In Table 10.1 the value of $\sum(x - \bar{x})(y - \bar{y})$ was shown to be 100.0. Since $b = 0.00435$,

$$\text{SSR} = 0.00435(100.0) = 0.435,$$

which agrees with the value given in Table 10.6.

Standard Error of the Estimate

Our measure of the variability about the regression line is the value of SSE divided by its degrees of freedom. Since SSE is the sum of the squared errors, this measure is thus the sample variance of the values of e_i.* The number of degrees of freedom in this measure is $n - 2$, because *two* sample statistics (a and b) must be calculated before the values of $\hat{y}$ can be computed (since $\hat{y} = a + bx$). Hence,

Variance of errors:

$$V[e] = \frac{1}{n-2} \sum_{i=1}^{n} (y_i - \hat{y})^2 = \frac{1}{n-2} (\text{SSE}).$$

Since standard deviations are usually easier to interpret than are variances, the usual measure of goodness of fit for regression analysis is the square root of $V[e]$, which is called the *standard error of the estimate* (denoted by the symbol s_e).

Standard error of estimate:

$$s_e = \sqrt{\frac{1}{n-2} \sum (y_i - \hat{y})^2} = \sqrt{\frac{\text{SSE}}{n-2}}, \tag{10.16}$$

where s_e^2 is an unbiased estimate of σ_ε^2.

To illustrate the calculation of s_e for the GMAT–GPA example, we need to recall from column 2 of Table 10.6 that $\sum e_i^2 = \text{SSE} = 0.225$. Since $n = 8$ for that example, the value of s_e is:

$$s_e = \sqrt{\frac{\text{SSE}}{n-2}} = \sqrt{\frac{0.225}{6}} = 0.1936.$$

The value of s_e can be interpreted in a manner similar to the sample standard deviation of the values of x about $\bar{x}$. That is, given that assumption 2 holds (i.e., the ε_i are normal, with mean of zero), then approximately 68.3 percent of the observations will fall within $\pm 1 s_e$ units of the regression line, 95.4 percent will fall within $\pm 2 s_e$ units of this line, and 99.7 percent will fall within $\pm 3 s_e$ units of it.† Using this information gives one a good indication of the fit of

* Recall from Chapter 6 that an unbiased sample variance is calculated by dividing the sum of squared deviations by the degrees of freedom.

† Technically, this interpretation of s_e should be used only when the sample size is relatively large, as it gives an approximation to the correct interval (see Section 10.6). For present purposes, however, we will ignore this distinction.

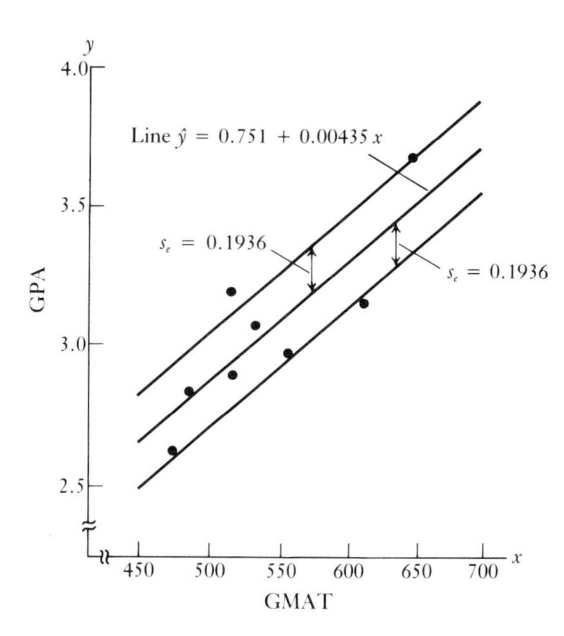

Fig. 10.14 A band of one s_e about the regression line.

the regression line to the sample data. In our example, a range of $\pm 3s_e$ would be $\pm 3(0.1936) = 0.5808$, which means the potential error in estimating GPA on the basis of GMAT usually will be less than 0.5808 grade points. This is a rather large potential error, since the average GPA = 3.10, and the limits on GPA are zero and 4.0; but we must remember that our sample size was very small ($n = 8$). On the other hand, over two-thirds (68.3%) of the actual values of y will fall within $\pm s_e = 0.1936$ grade points of the estimated $\hat{y}$ values. A band of $\pm 1s_e = 0.1936$ about the regression line is illustrated in Fig. 10.14. In this graph we see that six of the eight "errors" are less than 0.1936.

10.5 TEST OF THE SIGNIFICANCE OF THE SAMPLE REGRESSION LINE

We have presented a way of estimating the best regression line fitting the linear relation between $\mathbf{y}$ and x, and we have discussed measures of the strength of the linear relationship. However, we have not given any rules or guidelines to help determine whether knowledge of the independent variable x is useful in predicting the values of y_i. Suppose the population relationship is such that $\beta = 0$. This means that the population regression line must be a horizontal straight line, where $\hat{y} = \bar{y}$. Since $\hat{y}$ is a constant whenever $\beta = 0$, the values of x are of no use in predicting $\mathbf{y}$. If β is not equal to zero, then the values of x

are meaningful in predicting y. Thus, to determine whether or not the estimation of the y values is improved by using the regression line, we need to test the null hypothesis that $\beta = 0$.

Test on the Slope: A *t*-Test

Suppose the null hypothesis is

$$H_0: \beta = 0.$$

Rejecting H_0 in this case means concluding, on the basis of the sample information given, that β does not equal zero, and hence that the regression line improves our estimate of the dependent variable. The alternative hypothesis for this test might take on a number of different forms. For example, we could use the one-sided alternative that the slope is greater than zero ($H_a: \beta > 0$), or the one-sided alternative that the slope is less than zero ($H_a: \beta < 0$). A two-sided alternative would be to hypothesize merely that the slope does not equal zero ($H_a: \beta \neq 0$).

Although the above hypotheses are by far the most common forms used, the hypothesized value need not be zero. To illustrate a problem where β is not assumed to be equal to zero, suppose that it is desirable to test the null hypothesis that the slope of the regression line relating personal income and consumption (i.e., marginal propensity to consume) has not deviated from some historical value, such as $\beta = 0.90$. If we assume that the alternative hypothesis is two-sided, then the appropriate test is $H_0: \beta = 0.90$ against $H_a: \beta \neq 0.90$. The assumed value of β in such a test is usually denoted as β_0; a two-sided alternative would thus be

$$H_0: \beta = \beta_0 \qquad \text{vs.} \qquad H_a: \beta \neq \beta_0.$$

A t-test is used to test the null hypothesis $H_0: \beta = \beta_0$. This test is very similar to the t-test about a population mean, as we are again testing a mean (β), the population is assumed to be normal (the ε_i's), and the population standard deviation is unknown. In the present case, the sample statistic is b (rather than $\bar{x}$), the hypothesized population value is β_0 (rather than μ_0), and the sample standard error is s_b (rather than $s_{\bar{x}}$), where s_b is defined as follows:

Standard error of the regression coefficient b:

$$s_b = s_e \sqrt{\frac{1}{\sum_{i=1}^{n} (x_i - \bar{x})^2}}. \tag{10.17}$$

The value s_b is a measure of the amount of sampling error in the regression coefficient b, just as $s_{\bar{x}}$ was a measure of the sampling error of $\bar{x}$. We can now test the null hypothesis $H_0: \beta = \beta_0$ by subtracting the hypothesized value β_0 from b and dividing by the standard error of the regression coefficient.

$$\textit{Statistic for testing } H_0\!:\beta = \beta_0\!: \qquad t_{(n-2)} = \frac{b - \beta_0}{s_b}. \tag{10.18}$$

This statistic follows a **t**-distribution with $(n - 2)$ degrees of freedom.*

To illustrate the use of Formula (10.18), suppose in our GMAT–GPA example that the null hypothesis $H_0\!:\beta = 0$ is tested against $H_a\!:\beta > 0$ (we use the one-sided test here because it is not reasonable to expect an *inverse* relationship between GMAT and GPA). First we need to calculate s_b by substituting into Formula (10.17) the previously determined values $s_e = 0.1936$ and $\sum(x_i - \bar{x})^2 = 23{,}000$. We obtain:

$$s_b = 0.1936\sqrt{\frac{1}{23{,}000}} = 0.00128.$$

Therefore,

$$t = \frac{b - \beta_0}{s_b} = \frac{0.00435 - 0}{0.00128} = 3.40.$$

For $n - 2 = 6$ degrees of freedom, the probability that **t** is larger than 3.40 falls between 0.01 and 0.005 (see Table VII). Thus, it is highly unlikely that a slope of $b = 0.00435$ will occur by chance when $\beta = 0$; and we can conclude that the regression line does seem to improve our ability to estimate the dependent variable (i.e., we reject H_0).

In addition to being able to test hypotheses about β, it is possible to construct a $100(1 - \alpha)\%$ confidence interval for β. Since the regression coefficient b follows a **t**-distribution with $n - 2$ degrees of freedom and standard deviation s_b, the desired interval is:

$100(1 - \alpha)\%$ *confidence interval for* β:

$$b - t_{(\alpha/2,\,n-2)}s_b \le \beta \le b + t_{(\alpha/2,\,n-2)}s_b. \tag{10.19}$$

A 95-percent confidence interval given that $n = 8$,

$$t_{(\alpha/2,\,n-2)} = t_{(0.025,\,6)} = 2.447,$$

* A similar test statistic could be used to test hypotheses about the population **y**-intercept α_0, based on the sample estimate of a and its standard error (s_a):

$$s_a = \sqrt{s_e^2 \sum_{i=1}^{n} x_i^2 \Big/ \sum_{i=1}^{n} (x_i - \bar{x})^2}.$$

The proper **t**-statistic with $(n - 2)$ degrees of freedom is $t_{(n-2)} = (a - \alpha_0)/s_a$.

and $s_b = 0.00128$ would be:

$$0.00435 - (2.447)(0.00128) \leq \beta \leq 0.00435 + (2.447)(0.00128),$$
$$0.00122 \leq \beta \leq 0.00748.$$

On the basis of our sample of eight students we would thus expect an improvement of between 0.122 and 0.748 points in GPA for each 100-point increase in GMAT. This is rather a wide interval for any precise forecasting of GPA, but again we must point out that the sample size is very small.

Test on the Slope: The *F*-Test

There is still another way of testing the null hypothesis $H_0: \beta = 0$, this one using the measures of unexplained and explained variation. Before the reader bemoans the presentation of one more test to learn, we must point out that the test in this section is particularly important because it can be generalized to problems involving more than just one independent variable (i.e., to the multiple regression case). This use is discussed in Chapter 12.

Recall that SST = SSE + SSR. You may also recall that the degrees of freedom associated with SST is $n - 1$ (since only $\bar{y}$ needs to be calculated before SST can be computed), while the d.f. for SSE is $n - 2$ (both a and b need to be calculated before computing SSE). Since the d.f. for SST must equal the sum of those for SSE and SSR, we see by subtraction that the d.f. for SSR = 1 (because $(n - 1) = (n - 2) + (1)$). Now, a sum of squares divided by its degrees of freedom is called a *mean square*. The two mean squares we will need are *mean square error* (MSE) and *mean square regression* (MSR).

Mean square error: $$\text{MSE} = \frac{\text{SSE}}{n - 2} = s_e^2,$$

Mean square regression: $$\text{MSR} = \frac{\text{SSR}}{1}. \tag{10.20}$$

It is customary to present information about MSE and MSR in what is called an *analysis-of-variance* table, such as Table 10.7.

One word of caution is necessary here: Although the sums of squares and the degrees of freedom are additive, it is *not* true that the mean-square terms are additive. Note, in the analysis of variance table, that no MS term is given in the row labeled "Total." The SST and the degrees of freedom total are given in the table, so that it can be verified that elements in the body of the table do sum to these values.

To illustrate the use of an analysis-of-variance table, suppose we construct such a table for the GMAT–GPA example. Recall, from Section 10.4, that

$$\text{SST} = 0.660 \qquad \text{SSE} = 0.225, \qquad \text{and} \qquad \text{SSR} = 0.435.$$

Table 10.7 Analysis-of-Variance Table for Simple Regression

Source of the Variation	Sum of Squares	Degrees of Freedom	Mean Square
Regression	SSR	1	SSR/1
Error (or residual)	SSE	$n - 2$	SSE/$(n - 2)$
Total	SST	$n - 1$	

Table 10.8 Analysis-of-Variance Table for the Regression of GPA on GMAT

Source	SS	d.f.	Mean Square
Regression	0.435	1	0.435
Error	0.225	6	0.0375
Total	0.660	7	

Hence,

$$\text{MSR} = \frac{\text{SSR}}{1} = \frac{0.435}{1} = 0.435$$

and

$$\text{MSE} = \frac{\text{SSE}}{(n-2)} = \frac{0.225}{6} = 0.0375.$$

Table 10.8 shows the analysis-of-variance table for this example.

Now we return to our new test of $H_0: \beta = 0$. If the value of MSR is high relative to the value of MSE, then a large proportion of the total variability of y is being explained by the regression line, implying that we should reject this null hypothesis. If, however, MSR is small relative to the MSE, then the regression line does not explain much of the variability in the sample values of y, and the null hypothesis would not be rejected. Thus, the null hypothesis $H_0: \beta = 0$ can be tested by using the ratio of the mean squares, MSR/MSE. This ratio can be shown to have an F-distribution with 1 and $(n - 2)$ degrees of freedom:*

$$\textit{Statistic for testing } H_0: \beta = 0: \qquad F_{(1, n-2 \text{ d.f.})} = \frac{\text{MSR}}{\text{MSE}}. \tag{10.21}$$

* Recall that the ratio of two chi-square variables each divided by their d.f. follows the ***F***-distribution. In this case, both SSR and SSE can be shown to be chi-square distributed.

We can now use the data in Table 10.8 to test the hypothesis $H_0: \beta = 0$. If we choose a significance level of $\alpha = 0.05$, the critical value (from Table VIII(a)) is $F_{(0.05;\,1,6\,\text{d.f.})} = 5.99$. The decision rule should thus be to reject H_0 (i.e., accept that the regression line does contribute to the explanation of the variation in $\boldsymbol{y}$) if the calculated value of F based on our sample exceeds 5.99.

Using Formula (10.21) and Table 10.8, we calculate the value of F as

$$F = \frac{\text{MSR}}{\text{MSE}} = \frac{0.435}{0.0375} = 11.6.$$

Since the sample value of $\boldsymbol{F}$ exceeds the critical value of 5.99, we reject the null hypothesis and conclude that the linear relationship does help explain the variation in GPA.

We see from the above discussion that the $\boldsymbol{F}$-test is comparable to the $\boldsymbol{t}$-test for $H_0: \beta = 0$, in that they are both measures of the strength of the relationship in linear regression. In fact, it can be shown that the $\boldsymbol{t}$-test on β and the $\boldsymbol{F}$-test are equivalent tests for the significance of the linear relationship between two variables. The calculated value of $\boldsymbol{F}$ should always equal the square of the calculated value of $\boldsymbol{t}$. In our example,

$$t^2 = (3.40)^2 = 11.56,$$

which differs from $F = 11.6$ only due to rounding errors. The advantage of the $\boldsymbol{F}$-test is that it can be generalized to a test of significance when there is more than one independent variable, while the $\boldsymbol{t}$-test cannot. The $\boldsymbol{t}$-test is more flexible since it can be used for one-sided alternatives while the $\boldsymbol{F}$-test cannot.

10.6 CONSTRUCTING A FORECAST INTERVAL (OPTIONAL)

One of the important uses of the sample regression line is to obtain forecasts of the dependent variable, given some value of the independent variable. The estimated value $\hat{y}_i = a + bx_i$ is the best estimate we can make of both $\mu_{y \cdot x_i}$ (the mean value of $\boldsymbol{y}$, given a value x_i) and of y_i (the actual value of $\boldsymbol{y}$ that corresponds to the given value x_i). Forecasts of both types are frequently desired. Economists may desire to forecast the average or expected level of unemployment, given assumed values of independent variables under policy control. From such forecasts, they might argue which variables should be affected by policy, by how much, and in what direction, so that unemployment can be expected to be reduced toward a certain policy goal, say, 4% unemployed. In other cases, the forecast of the actual value of the dependent variable may be desired, as, for example, in making predictions of the level of unemployment which will occur in the second quarter of the next year, or of the price of General Motors common stock at the end of this year, or of the total yield of this year's corn crop.

Point Estimates of Forecasts

To obtain the best point estimate for forecasts of both the mean value and the actual value of $\mathbf{y}$, the given value of the independent variable (call it x_g) is substituted into the estimating equation to obtain the forecast value (called $\hat{y}_g$):

$$\hat{y}_g = a + bx_g.$$

Thus, $\hat{y}_g$ is an estimate of *both* $\mu_{\mathbf{y}\cdot x_g}$ and y_g.

Suppose that in our GMAT–GPA example we wish to obtain the forecast value for the given GMAT score of $x_g = 500$. Using the estimated regression coefficients $a = 0.751$ and $b = 0.00435$, we obtain

$$\hat{y}_g = a + bx_g = 0.751 + 0.00435(500) = 2.926.$$

Thus, our best estimate for one person who has a GMAT of 500 is $\hat{y}_g = 2.926$. Similarly, our estimate for the *mean* of *all* persons having GMAT's of 500 is also $\hat{y}_g = 2.926$. Although these estimates both equal the same value, we must emphasize that they are interpreted differently. This difference will become important when we investigate the process of making interval estimates. Specifically, we will see that the confidence interval for estimating a single value will always be a lot larger than the confidence interval for estimating the mean value because the former will always have a larger standard error.

Interval Estimates of Forecasts

Recall that an interval estimate uses a point estimate as its starting point, and then uses the standard error of the point estimate to find the endpoints of the interval. From the discussion above we know that the starting point is always the same value, $\hat{y}_g$. And, as we pointed out in Section 10.4, the standard error s_e can be used to form the endpoint of the interval when estimating the actual sample value y_g based on the regression estimate, $\hat{y}_g$. *When making estimates based on values of x not included in the original data, the value of s_e is actually only an approximation to the appropriate standard error.* The appropriate standard error is usually called the *standard error of the forecast*, which we will denote as s_f. We write the formula for s_f below in terms of s_e:

$$s_f = s_e\sqrt{1 + \frac{1}{n} + \frac{(x_g - \bar{x})^2}{\sum(x_i - \bar{x})^2}}. \tag{10.22}$$

Note that s_f will always be larger than s_e since the term under the square root will always be greater than one. Also, note that s_f depends on the particular value of x_g of interest. Finally, we see that if n is large and if the new sample value of x_g is close to $\bar{x}$ (the mean of the previous sample values), then the term under the square root will be close to 1.0; hence, s_e and s_f will be approximately equal. This result should not be too surprising, for we know that the larger the sample, and the less that a given value x_g deviates from $\bar{x}$, the more faith we have in the sampling results and in the subsequent forecast.

We can now use our point estimate, $\hat{y}_g$, and the standard error s_f, to construct a $100(1 - \alpha)\%$ confidence interval for y_g. The appropriate test statistic in this case is the ***t***-distribution with $(n - 2)$ degrees of freedom.

> *Endpoints of a* $100(1 - \alpha)\%$ *forecast interval for* y_g:
>
> $$\hat{y}_g \pm t_{(\alpha/2, n-2)} s_f. \tag{10.23}$$

Suppose we want to construct, on the basis of our sample of $n = 8$, a 95-percent forecast interval for the GPA of an individual student with a GMAT score of 500. From Table VII the value of $t_{(0.25, 6)} = 2.447$, and we know from our previous analysis that

$$\hat{y}_{500} = 2.926, \qquad s_e = 0.1936, \qquad \bar{x} = 540, \qquad \text{and} \qquad \sum(x - \bar{x})^2 = 23{,}000.$$

Substituting these values into Formula (10.23), and using the definition of s_f in Formula (10.22), we get the following endpoints for the forecast interval:

$$\begin{aligned} 2.936 \pm 2.447(0.1936)&\sqrt{1 + (\tfrac{1}{8}) + [(500 - 540)^2/23{,}000]} \\ &= 2.926 \pm 2.477(0.1936)(1.09296) \\ &= 2.926 \pm 0.518 \\ &= 2.408 \quad \text{and} \quad 3.444. \end{aligned}$$

We can thus, assert, with 95% confidence, that the GPA of an individual with a GMAT of 500 will fall between 2.408 and 3.444.

Now we turn to the problem of constructing a forecast interval for $\mu_{y \cdot x_g}$, the *mean* of the y-values. In this case the appropriate standard error is denoted by the symbol $s_{\bar{y}}$, where:

$$s_{\bar{y}} = s_e \sqrt{\frac{1}{n} + \frac{(x_g - \bar{x})^2}{\sum(x_i - \bar{x})^2}}. \tag{10.24}$$

As was the case for s_f, $s_{\bar{y}}$ depends on n, x_g, and s_e. The value of $s_{\bar{y}}$, however, will always be smaller than s_f, since s_f contains one additional positive term under the square-root sign. Again, the appropriate test statistic is the ***t***-distribution, with $(n - 2)$ degrees of freedom. The endpoints of a $100(1 - \alpha)\%$ interval are thus:

> *Endpoints for a* $100(1 - \alpha)\%$ *forecast interval for* $\mu_{y \cdot x_g}$:
>
> $$\hat{y}_g \pm t_{(\alpha/2, n-2)} s_{\bar{y}}. \tag{10.25}$$

Substituting the appropriate values in (10.24), and using Formula (10.25) for $s_{\bar{y}}$, we obtain the following endpoints:

$$2.926 \pm 2.447(0.1936)\sqrt{(\tfrac{1}{8}) + [(500 - 540)^2/23{,}000]}$$
$$= 2.926 \pm 0.212$$
$$= 2.714 \quad \text{and} \quad 3.138.$$

The interval 2.714 to 3.318 thus represents a 95% forecast interval for the mean GPA of all students with GMAT's of 500 ($\mu_{y \cdot 500}$).

Figure 10.15 shows the forecast interval for values of x from $x_g = 450$ to $x_g = 650$ for both $\mu_{y \cdot x_g}$ (the narrow band) and for y_g (the wide band). As we indicated before, the forecast interval is wider for y_g than for $\mu_{y \cdot x_g}$ because predicting the grade-point average for an individual involves a lot more variability than predicting for an average over many such individuals (i.e., s_f is larger than $s_{\bar{y}}$). Note that both bands become increasingly wider for values of x_g further away from $\bar{x}$. This reflects the fact that our least-squares estimating

Fig. 10.15 Ninety-five percent forecast interval (FI) for $\mu_{y \cdot x_g}$ (narrow band) and y_g (wide band).

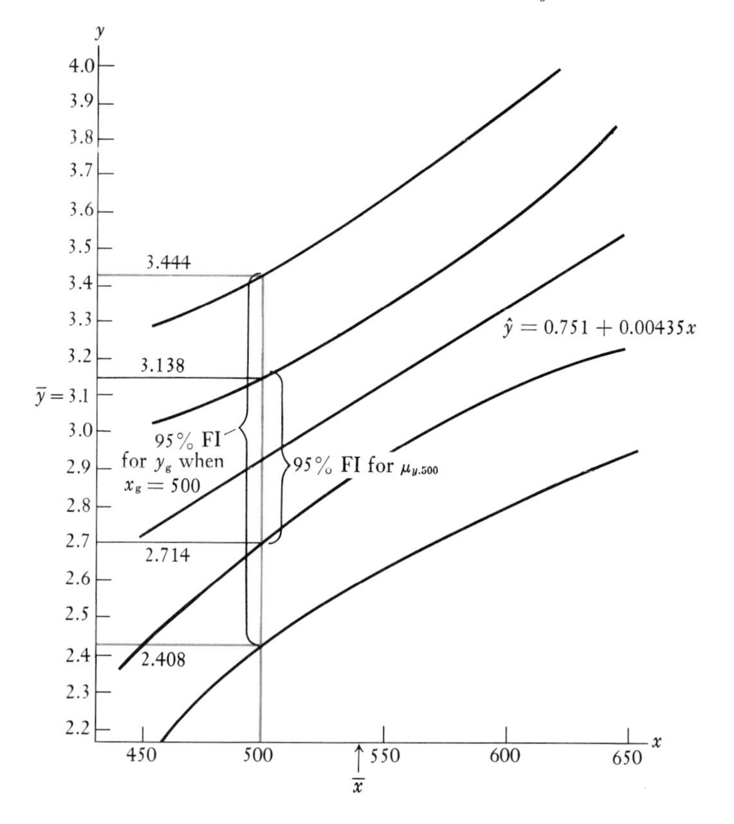

line is most accurate at the point of means $(\bar{x}, \bar{y})$, since it is a representation of the average relationship between $\boldsymbol{x}$ and $\boldsymbol{y}$.

Define: standard error of the estimate, SSE, SSR, SST, s_e, s_b, MSE, MSR, forecast value, s_f, $s_{\bar{y}}$.

Problems

10.13 In a regression problem involving $n = 61$ sample observations, you find $s_{xy} = -720$, $s_x^2 = 150$, $\bar{y} = 27.2$ and $\bar{x} = 14.3$.

a) Find the least-squares regression line.

b) Plot the least-squares regression line between $x = 0$ and $x = 30$.

10.14 a) Find the sample covariance for Problem 10.11. What population parameter does s_{xy} estimate?

b) Find the sample covariance for Problem 10.7.

10.15 Data on the production of spark plugs and average cost were collected and are shown in Table 10.9.

a) Find the regression equation of cost (y) on production (x) using the method of least-squares.

b) Does the scatter diagram for these observations indicate any possible violations of the five assumptions?

c) What is the value of s_{xy} for these data?

d) Find $\hat{y}$ for each value of x, and then calculate $(y - \hat{y})$ for each x-value. Does $\sum(y - \hat{y}) = 0$?

Table 10.9

Average Cost (cents)	Product per month (000's)
3	25
9	20
40	10
15	20
33	15

10.16 Many retail stores earn as much as 50 percent of their yearly sales between Thanksgiving and Christmas. Often, the more warm days there are in this period, the less the volume of sales. The following dollar values (y) represent 5 years of sales data in Philadelphia (in millions of dollars), adjusted for inflation. The x-values represent the number of days in this period with a high temperature of over 55°.

x	5	1	7	6	4
y	121.3	128.0	104.3	107.2	125.5

a) Find the least-squares regression line and estimate sales if $x = 3$.

b) Find SSR, SSE and SST, and test the hypothesis of no linear regression.

c) Form a 95% confidence interval for β_i and for y_i if $x = 3$.

10.17 In a regression problem involving a sample of $n = 100$ observations, $\sum xy = 1321$, $\sum x \sum y = 11{,}017$, $\sum x^2 = 42$, and $(\sum x)^2 = 4018$. If $\bar{y} = 7.2$, find the least-squares regression line.

10.18 Oil consumption in the U.S. has taken on the values between 1960 and 1978 shown in Table 10.10.

a) Use a least-squares regression line to predict consumption for 1981 and 1985.

b) Find a library source with the actual 1981 oil consumption. How close was your 1981 prediction?

Table 10.10

Year	Consumption (millions of barrels a day)	Year	Consumption (millions of barrels a day)
1960	9.7	1970	14.4
1962	10.2	1972	16.0
1964	10.8	1974	16.2
1966	11.9	1976	17.0
1968	13.0	1978	18.0

10.19 One model in an article by J. P. Newhouse in the *Southern Economic Journal* estimated the demand curve of a representative consumer of physician services as

$$\hat{p} = 0.3 + 2.2y,$$

where p = price per visit and y = per capita income, both deflated by the Consumer Price Index. Interpret and then criticize this least-squares model.

10.20 We wish to explore the relationship between family monthly food consumption (y) and family monthly income (x), both measured in hundreds of dollars. We are given information for 100 families as follows:

$n = 100$	$\bar{y} = \$6$	$\bar{x} = \$8$
$\sum xy = 6000$	$\sum y^2 = 4500$	$\sum x^2 = 10{,}000$

a) Find the regression equation $\hat{y} = a + bx$ using the method of least-squares.

b) Using the regression equation, make an estimate for a family with a monthly income of $1500.

10.21 Given the following data for quiz scores (y) and class absences (x) for a certain statistics class:

$\sum y = 1800$	$\sum xy = 4750$	$\sum y^2 = 113{,}000$
$\sum x = 90$	$n = 30$	$\sum x^2 = 400$

a) Find the estimating equation $\hat{y} = a + bx$.

b) Determine a student's expected quiz score if this person had six absences, and comment on the validity of using the estimating equation for subsequent classes or quizzes.

10.22 Define or describe briefly each of the following:

a) standard error of the estimate,

b) standard error of the forecast,

c) standard error of the regression coefficient.

10.23 What is the hypothesis of no linear regression? Describe two methods for testing this hypothesis.

10.24 Assume a least-square regression line was fitted to crop yields from 62 randomly selected acres of land in Iowa. The relationship between yield (y) and the amount of fertilizer applied (x, in 100s of pounds) was $\hat{y} = 16.9 + 1.225x$. SST for this sample was determined to be 594, while SSE was found to be 540.

a) Plot the regression line for values of x between zero and three.

b) If $x = 1$, what is the least-squares estimate of y? What is the least-squares estimate of y if $x = 3$?

c) Find the standard error of the estimate for this sample. Use this answer to construct a 95-percent confidence interval for y_i when $x = 1$ and when $x = 3$. Draw, on your graph for part (a), a band representing a 95-percent confidence interval for $0 \le x \le 3$.

10.25 a) Assume for Problem 10.24 that $\sum(x_i - \bar{x})^2 = 36$. Use this information to find the standard error of the regression coefficient. Test the null hypothesis $H_0\!:\beta = 0$ against $H_a\!:\beta \neq 0$ by means of a **t**-test.

b) Construct a 95-percent confidence interval for the value of β.

c) From the information given in this example, determine MSR and MSE, and use these values to determine the value of $\boldsymbol{F}$ necessary for testing the hypothesis of no linear regression. How is this value of $\boldsymbol{F}$ related to the value of $\boldsymbol{t}$ you calculated in part (a) above? Show that they both lead to rejection of the null hypothesis at the same level of significance.

10.26 Suppose you ask three people the number of airplane trips over 500 miles (y) and the number of automobile trips over 500 miles (x) they took last year, with the following results:

x	0	1	2
y	2	4	3

a) Find the least-squares regression line.

b) Find SSE, SSR, and SST.

c) What percent of the variability in y is explained by the values of x?

d) What is the value of s_{xy}?

10.27 In 1980 the Labor Department published the following chart, representing new factory orders for durable goods.

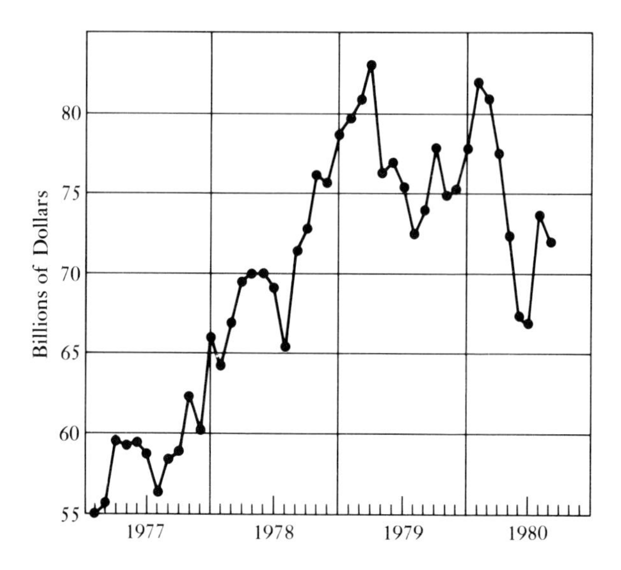

a) Fit a least-squares trend line to these 44 observations (estimate quarterly points as best you can). What value would you predict for the fourth quarter, 1980?

b) Would you feel comfortable estimating new factory orders on the basis of this trend line? Explain.

10.28 The following least-squares analysis resulted from a regression relating the beginning salary of graduating MBA students (y, in thousands) and the number of years of work experience of each student (x).

$$\hat{y} = 25 + 0.96x$$

Source	SS	d.f.
Regression	144	1
Error	342	38

a) What was the sample size in this problem? What is SST?

b) Use an ***F***-test to accept or reject: $H_0:\beta = 0$ vs. $H_a:\beta \neq 0$. What probability value would you report?

c) Calculate the value of s_e, the standard error of the estimate. Use this information to construct a 95% confidence interval for $\hat{y}$ given that $x = 5.0$.

10.29 a) For Problem 10.28, assume that $s_b = 0.24$. Use this information to construct a ***t***-test to accept or reject $H_0:\beta = 0$ vs. $H_a:\beta \neq 0$. Is your answer consistent with your answer for part (b) of Problem 10.28?

b) Construct a 95% confidence interval for β.

c) Assume, for this problem, that when $x_g = 5$, $s_f = 0.32$ (from Formula 10.22). Use this information to construct a 95% confidence interval for y_g.

d) If $s_{\bar{y}} = 0.1965$ when $x_g = 5$ (from Formula 10.24), construct a 95% confidence interval for $\mu_{y \cdot x_g}$.

10.30 a) Test the hypothesis of no linear regression $H_0:\beta = 0$ vs. $H_a:\beta \neq 0$ for the data in Problem 10.6, on page 494.

b) Construct a 95% confidence interval for y assuming that $x = \$700$.

10.31 The least-squares estimating line of the number of motel rooms that are rented (y), based on the number of advance reservations made (x), for a certain Holiday Inn is $\hat{y} = 26 + \frac{3}{4}(x)$, with an average $\bar{y} = 60$.

a) For a particular night, 60 advance reservations are received. Suppose that the manager needs one maid for each nine rooms that need cleaning the following morning. Advise this manager as to the minimum number of maids that should be employed in this particular case, based on your estimate of the number of rooms that will be occupied and will need cleaning.

b) Suppose that, for one day, the number of advance reservations received is 36, and the number of rooms occupied turns out to be 55. Find the total, explained, and unexplained deviations for this case.

10.32 A Peace Corps representative works with five Thai farmers in a cooperative shop rebuilding small (United States surplus) gasoline motors for use in water pumps and on sampans. She recognizes a difference in the workers' individual ability to learn the new job and to do it properly without supervision. After several weeks, she records the following information, where y = average weekly output of correctly rebuilt motors, and x = years of education of each of the five Thai workers.

x	7	5	6	10	4
y	15	7	10	20	8

a) Find the regression equation $\hat{y} = a + bx$.

b) Find SSE and SSR and show that they sum to SST.

10.33 Test the hypothesis of no linear regression for Problem 10.7 (on page 495), using $\alpha = 0.05$. Assume H_a is one-sided.

10.34 For the data in Problem 10.9 (on page 495):

a) Test the hypothesis of no linear regression $H_0:\beta = 0$ vs. $H_a:\beta \neq 0$.

b) Construct an estimate for sales if the demand estimate is 300,000. Use Formula (10.23) to construct a 95% forecast interval for y_g.

10.35 Construct a forecast interval for Problem 10.18 (on page 510), using $x = 1981$ and $x = 1985$ and Formula (10.23). Let $\alpha = 0.10$.

10.36 a) Test the hypothesis of no linear regression, $H_0:\beta = 0$ versus $H_a:\beta > 0$ for Problem 10.20 (on page 510). Use $\alpha = 0.01$.

b) Test $H_0:\beta = 0$ versus $H_a:\beta < 0$ for the data in Problem 10.21. Use $\alpha = 0.05$.

Exercises

10.37 The Ohio Valley Detergent Company has 30 subsidiaries located in the U.S. In an effort to relate the sales of each subsidiary to the surrounding population, a "canned" regression program, which is part of the *Statistical Package for the Social Sciences* (SPSS), was used to regress sales (y) on population (x = pop.). The following output was generated from the 30 subsidiaries (the data for this are given in Problem 12.36).

VARIABLE	B	STD ERROR B	F ---------- SIGNIFICANCE
POP	.53339673	.50520918E-01	111.47006 .000
(CONSTANT)	1386969.3	883403.06	2.4649940 .128

a) Write the least-squares regression equation.

b) What is the value of s_b? Use this value to construct a t-test for $H_0:\beta = 0$ versus $H_a:\beta \neq 0$, using $\alpha = 0.01$.

c) Interpret the value 111.47006 in the last column of the output. What does the .000 below it represent? How are these values related to your answers to part (b)?

d) What value of sales would you predict if population is 19,660,000?

e) Run a computer package on the sales and population data in Problem 12.36 to verify the output given above.

10.38 A leading securities firm, A. Victor and Sons, has decided to bid against E. W. Martin, Inc., on a new $10 million series of first-mortgage bonds. The analysts at Victor have data on past bids by Martin for various bonds. Specifically, they have data on Martin's bid (as a percent of par value) and the Moody rating (Aaa = 1, Aa = 2, A = 3, Baa = 4, Ba = 5) for each of 23 different bonds. Victor decides to use simple linear regressions to relate Martin's bid (y) to the Moody rating (x). They use the "canned" regression program which is part of a set called *Biomedical (BMD) Computer Programs*. The output from this program (whose data are given in Section 12.12) is:

STD. ERROR OF EST. .89386283

ANALYSES OF VARIANCE

	DF	SUM OF SQUARES	MEAN SQUARE	F RATIO
REGRESSION	1	26.305753	26.305753	32.924
RESIDUAL	18	14.381834	.79899076	

VARIABLES IN EQUATION

VARIABLE	COEFFICIENT	STD. ERROR
(CONSTANT	101.53471	)
MOODY RATING	-.81298744	.14169

a) Write the least-squares regression equation.

b) What is the value of s_b? Use this value to construct a t-test for $H_0:\beta = 0$ versus $H_a:\beta \neq 0$, using $\alpha = 0.05$.

c) What bid for Martin would you predict if the Moody rating is Aaa?

d) Run a computer package on the data for Martin's bid and Moody rating in Section 12.12, to verify the output given above.

e) Transform each value of Moody rating by taking $\log_{10}$ (Moody rating), and then run a regression analysis relating this new variable to Martin's bid. Is the new variable a better predictor of Martin's bid? Explain why or why not.

10.39 a) Prove that solving the normal equations does, in fact, lead to Formula (10.9).

b) Prove that Formulas (10.10) and (10.12) are equivalent equations for b.

10.40 Given the following values of x and y:

x	y
1.00	1.0
1.44	2.0
1.96	3.0
3.24	4.0
4.00	5.0
7.84	6.0

a) Plot these six values of x and y on a graph, and then make a freehand estimate of the curve relating the two variables. Would a linear function be appropriate in this circumstance? What type of relationship does y appear to have to the values of x?

b) Transform the variable x into a new variable by taking the square root of x, and then plot this new variable against y. Is this relationship approximately linear?

c) Use the transformation in part (b) to establish a least-squares regression line for the relationship between y and $\sqrt{x}$. Calculate a "list of residuals."

10.41 General Motors continually monitors the relationship between the fuel economy (mpg) of its new cars relative to the gross vehicle weight of each auto. For one division, a test of twelve different models of 8-cylinder cars yielded the weights and mileage results shown in Table 10.11.

a) Calculate the least-squares regression line for these data. What value of mpg would you predict if the weight of a car is 4300 lbs?

b) Test $H_0: \beta = 0$ vs. $H_a: \beta \neq 0$ using $\alpha = 0.01$.

Table 10.11

Weight (in lbs)	mpg	Weight	mpg
3750	20.00	4000	21.00
4000	19.80	4500	19.10
4500	19.30	4750	19.10
4250	19.60	4250	19.40
4250	20.00	4500	19.20
3625	20.20	4000	18.20

10.42 Make up a set of 10 sample values which illustrates a violation of assumption 4. Calculate the least-squares regression line for these data, and then find each value of $(y_i - \hat{y}_i)$. Could you have determined whether the data violate assumption 4 without a scatter diagram merely by looking at these 10 values of $(y_i - \hat{y}_i)$? Explain.

10.43 Find the least-squares estimator for β in the function $y = \beta x^3$, based on n sample observations.

10.44 Derive the normal equations for the least-squares estimation of the function $y = \alpha + \beta x^2$.

10.45 Discuss the following statement: "Cause-and-effect inferences can never be made from regression analysis."

10.46 Determine the least-squares estimate for the following data, assuming that $\hat{y} = ax^b$ (let $\log \hat{y} = \log a + b \log x$). Plot the original data and the least-squares estimate on graph paper, and compute the residuals $e_i = y_i - \hat{y}_i$. Find s_e, using Formula (10.16).

x	1	2	3	4	5
y	1.0	2.1	4.3	8.1	13.0

Determine the least-squares estimate using the line $\hat{y} = a + bx$ and find the residuals. By comparing s_e for both forms of the estimation, determine which model specification provides the best fit.

10.47 The following data represent the growth pattern of a certain plant life, where $\mathbf{x}$ is in months and $\mathbf{y}$ is in inches.

x	1	2	3	4	5	6	7
y	0.80	1.10	1.70	2.60	3.80	5.70	8.50

a) Find the least-squares equation relating $\mathbf{x}$ and $\mathbf{y}$, of the form $\hat{y} = ab^x$. (Take the logarithm of both sides of this equation, letting $\log \hat{y} = \log a + x \log b$.)

b) Plot the original data and your least-squares estimate, and then use this sketch to find the error of prediction for these seven observations.

GLOSSARY

population regression line $[\mu_{\mathbf{y}\cdot x} = E[\mathbf{y}|x] = \alpha + \beta x]$: for the population, intercept (α) plus slope (β) times x equals the mean of the $\mathbf{y}$-values given an x-value.

population regression model $[y_i = \alpha + \beta x_i + \varepsilon_i]$: an error term ($\varepsilon_i$) is added to the regression line to describe the actual value of the ith observation (y_i).

sample regression line $[\hat{y} = a + bx]$: estimate of population regression line.

sample regression model $[\hat{y}_i = a + bx_i + e_i]$: estimate of population regression model.

five assumptions: the population error terms (ε_i) are assumed (1) to be independent of x, (2) to be normally distributed, (3) to have a mean of zero, (4) to be independent of one another, and (5) to have a finite variance (σ_ε^2) that is constant (homoscedasticity assumption).

Gauss–Markov theorem: theorem developed by K. F. Gauss and A. A. Markov which says that the least-squares estimators of the intercept and the slope in the population regression line are the best linear unbiased estimators (BLUE).

least-squares procedure: process of finding regression coefficients a and b by minimizing sum of squared deviations $\sum(y - \hat{y})^2$.

***b*-value:** best estimate of population slope (β); $b = s_{xy}/s_x^2$, where s_{xy} is the sample covariance.

***a*-value:** best estimate of population y-intercept (α); $a = \bar{y} - b\bar{x}$.

SST, SSE, SSR: sum-of-squares total $= \sum(y_i - \bar{y})^2$, sum-of-squares error $= \sum(y_i - \hat{y}_i)^2$, sum-of-squares regression $= \sum(\hat{y}_i - \bar{y})^2$; SST = SSE + SSR.

standard error of the estimate (s_e): standard deviation of sample points about the sample regression line; s_e^2 represents an unbiased estimate of σ_ε^2.

population covariance: $C[x, y] = E[(x - \mu_x)(y - \mu_y)]$. Describes how x and y covary in the population.

hypothesis of no linear regression: $H_0: \beta = 0$.

standard error of b (s_b): indicates the uncertainty of the least-squares estimate of β

$$s_b = s_e \sqrt{\frac{1}{\sum(x - \bar{x})^2}}.$$

***t*-test for $H_0: \beta = \beta_0$:** $t = (b - \beta_0)/s_b$, with $(n - 2)$ d.f.

confidence interval for β: $b - t_{(\alpha/2, n-2)}s_b \leq \beta \leq b + t_{(\alpha/2, n-2)}s_b$.

mean square: a sum of squares divided by its degrees of freedom. For simple regression, mean square error MSE = SSE/$(n - 2)$, and mean square regression MSR = SSR/1.

ANOVA table: analysis-of-variance table.

F-test for $H_0: \beta = 0$: F = MSR/MSE.

forecast value ($\hat{y}_g$): obtained by substituting a given value x_g in the estimating equation $\hat{y}_g = a + bx_g$.

standard error of the forecast (s_f):

$$s_f = s_e \sqrt{1 + \frac{1}{n} + \frac{(x_g - \bar{x})^2}{\sum(x_i - \bar{x})^2}}.$$

forecasted interval for y_g: $\hat{y}_g - t_{(\alpha/2, n-2)}s_f \leq y_g \leq \hat{y}_g + t_{(\alpha/2, n-2)}s_f$.

standard error for mean $\mu_{y \cdot xg}$:

$$s_{\bar{y}} = s_e \sqrt{\frac{1}{n} + \frac{(x_g - \bar{x})^2}{\sum(x_i - \bar{x})^2}}.$$

forecast interval for $\mu_{y \cdot xg}$:

$$\hat{y}_g - t_{(\alpha/2, n-2)}s_{\bar{y} \cdot x} \leq \mu_{\bar{y}} \leq \hat{y}_g + t_{(\alpha/2, n-2)}s_{\bar{y}}.$$

11

SIMPLE LINEAR CORRELATION

"I have a great subject (statistics) to write upon, but feel keenly my literacy incapacity to make it easily intelligible without sacrificing accuracy and thoroughness."

SIR FRANCIS GALTON

11.1 INTRODUCTION

The discussion of linear regression in Chapter 10 concentrated on describing the *nature* of the relationship between two or more variables. We now turn to a very closely related problem, that of determining the *strength* of the linear relationship between these variables. The word "strength" in this context refers to the *degree* of association (or the *closeness of fit*) between variables; for example, how close do two variables come to following an exact *straight-line* relationship, given that a linear function best approximates the population relationship? If the values of two variables form a perfect straight line, then the closeness of the fit, or the strength of the linear relationship, between these two variables is said to be "perfect," because the value of one variable can always be determined from a knowledge of the other. In other words, in determining the strength of the relationship between variables, we are measuring how well the value of one variable can be estimated (or described) on the basis of a knowledge of the other variables.

There are a number of ways to measure strength of association, several of which were presented in Chapter 10. The standard error of the estimate, for instance, provides some information about the strength of the relationship between variables; but this measure is difficult to interpret because it depends so highly on the units used to measure the variables. Testing regression coefficients for significant differences from zero also gives evidence about the closeness of fit between variables; yet this method still does not provide a single indicator measuring the relative strength of association. In this chapter, we shall discuss a measure which does provide such an indicator, the *correlation coefficient.*

Just as Chapter 10 investigated simple linear regression, describing the nature of the relationship between two variables related by a linear function, this chapter investigates *simple linear correlation,* describing the strength of association between two variables assumed to be linearly related. Our objective is to develop a measure which indicates the strength of the tendency for high (or low) values of one variable to be associated with high (or low) values of the other; that is, to formulate a measure of how well two variables, $\mathbf{x}$ and $\mathbf{y}$, "vary together."

11.2 THE SIMPLE LINEAR CORRELATION MODEL

Recall, from Chapter 3, that we presented one measure of fit between two variables $\mathbf{x}$ and $\mathbf{y}$, the *covariance of* $\mathbf{x}$ *and* $\mathbf{y}$, which is

$$C[\mathbf{x}, \mathbf{y}] = E[(\mathbf{x} - \mu_x)(\mathbf{y} - \mu_y)].$$

Although the covariance has many important statistical uses, this measure in general is *not* a good indicator of the relative strength of the relationship between two variables because its magnitude depends so highly on the *units* used to measure the variables. For example, the covariance between two measures of length $\mathbf{x}$ and $\mathbf{y}$ will be much smaller if $\mathbf{x}$ is scaled in feet than if $\mathbf{x}$ is scaled in inches. For this reason it is necessary to "standardize" the covariance of two variables in order to have a good measure of fit. This standardization is accomplished by dividing $C[\mathbf{x}, \mathbf{y}]$ by σ_x and σ_y. The resulting measure is called the *population correlation coefficient*, and is denoted by the Greek letter ρ (rho):

Population correlation coefficient:

$$\rho = \frac{\text{Covariance of } \mathbf{x} \text{ and } \mathbf{y}}{(\text{Std. dev. of } \mathbf{x})(\text{Std. dev. of } \mathbf{y})} = \frac{C[\mathbf{x}, \mathbf{y}]}{\sigma_x \sigma_y}. \tag{11.1}$$

Three values of ρ serve as benchmarks for interpretation of a correlation coefficient. First, let's consider the population where the values of $\mathbf{x}$ and $\mathbf{y}$ all fall on a single straight line that has a positive slope. In this case, which is referred

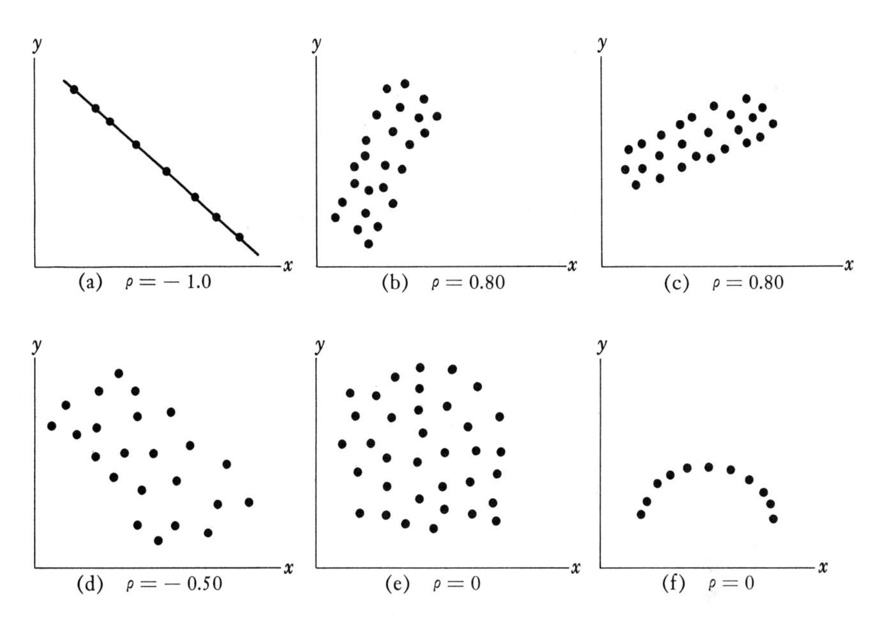

Fig. 11.1 The population correlation coefficient.

to as a "perfect positive linear relationship" between $\boldsymbol{x}$ and $\boldsymbol{y}$, the value of $C[\boldsymbol{x}, \boldsymbol{y}]$ will exactly equal the value of σ_x times σ_y; hence, ρ will equal $+1$.

When the relationship between $\boldsymbol{x}$ and $\boldsymbol{y}$ is a perfect *negative* linear relationship, all values of $\boldsymbol{x}$ and $\boldsymbol{y}$ lie on a straight line with a negative slope. This situation results in a value of $C[\boldsymbol{x}, \boldsymbol{y}]$ which exactly equals $-(\sigma_x)(\sigma_y)$. Thus, in this case, ρ will equal -1.

If $\boldsymbol{x}$ and $\boldsymbol{y}$ are not linearly related (i.e., if the slope of the regression line equals zero), then the value of the correlation coefficient will be zero, since in this case $C[\boldsymbol{x}, \boldsymbol{y}] = 0$, which means that $\rho = C[\boldsymbol{x}, \boldsymbol{y}]/\sigma_x\sigma_y = 0$. Thus, ρ measures the strength of the linear association between $\boldsymbol{x}$ and $\boldsymbol{y}$. Values of ρ close to zero indicate a weak relation; values close to $+1.0$ indicate a strong "positive" correlation, and values close to -1.0 indicate a strong "negative" correlation. Figure 11.1 illustrates some representations of values of ρ for selected scatter diagrams.

Note, from Figs. 11.1(b) and 11.1(c), that two populations that appear quite different can have the same correlation coefficient. Figures 11.1(c) and 11.1(d) show the difference between positive and negative correlation. The last two diagrams show different examples of a population with zero correlation. In Figure 11.1(f), $\boldsymbol{x}$ and $\boldsymbol{y}$ are perfectly related in a *nonlinear* fashion, yet still $\rho = 0$, which emphasizes the fact that ρ measures the strength of the *linear* relationship.

A legitimate question might be raised at this point as to exactly what population characteristic the parameter ρ represents. First, we should point out that the calculation of ρ does not require any assumption about the (joint) distribution of $\boldsymbol{x}$ and $\boldsymbol{y}$. As was the case for regression analysis, however, it is necessary

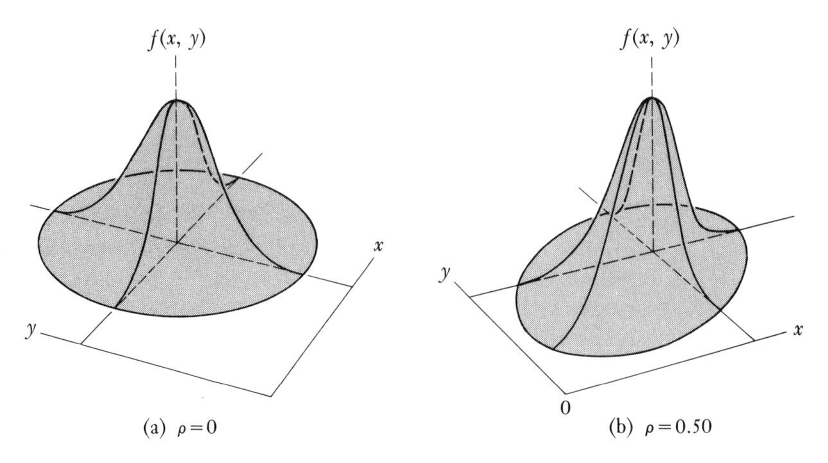

Fig. 11.2 The bivariate normal distribution for $\rho = 0$ and $\rho = 0.50$.

to make assumptions about the population in order to make inferences on the basis of sample data. In this case we need only make a single assumption, namely, that both $\boldsymbol{x}$ and $\boldsymbol{y}$ are normally distributed random variables. Under this assumption, if $\boldsymbol{x}$ and $\boldsymbol{y}$ are independent ($\rho = 0$), their distribution is the perfectly symmetrical bell-shaped curve shown in the first diagram in Fig. 11.2.* If $\boldsymbol{x}$ and $\boldsymbol{y}$ are normal, but are not independent, then the joint distribution of the two variables becomes more and more elongated as ρ moves away from zero. Figure 11.2(b) shows the joint distribution of $\boldsymbol{x}$ and $\boldsymbol{y}$ for $\rho = 0.50$. From these diagrams, it can be seen that ρ measures the degree of symmetry in the joint distribution of $\boldsymbol{x}$ and $\boldsymbol{y}$, perfect symmetry occurring when $\rho = 0$, and the most asymmetrical case occurring when $\rho = +1$ or -1.

We indicated above that, in making inferences from correlation analysis, it is usually necessary to assume that both $\boldsymbol{x}$ and $\boldsymbol{y}$ are normally distributed random variables. As has been the case for many of the measures studied thus far, the fact that the assumption of normality is not exactly met in practical problems does not destroy the usefulness of this technique.

11.3 SAMPLE CORRELATION COEFFICIENT AND COEFFICIENT OF DETERMINATION

The Sample Correlation Coefficient

As in all estimation problems, we use a sample statistic to estimate a population parameter. In this case, the population parameter is ρ; the sample statistic is called the *sample correlation coefficient*, and is denoted by the letter r. The value of r is defined in the same way as ρ, except that we substitute for each population parameter its best estimates based on the sample data. For instance, the best

* The symmetrical shape will be circular only if $\sigma_x = \sigma_y$.

estimate of $C[\mathbf{x}, \mathbf{y}]$ in Formula (11.1) is the sample covariance, which is denoted by the symbol s_{xy}, where

Sample covariance:
$$s_{xy} = \frac{1}{n-1}\sum_{i=1}^{n}(x_i - \bar{x})(y_i - \bar{y}).$$

We already know from Chapter 6 that the best estimate of σ_x^2 is the sample variance $s_x^2 = (1/(n-1))\sum(x_i - \bar{x})^2$, and the best estimate of σ_y^2 is the sample variance $s_y^2 = (1/(n-1))\sum_{i=1}^{n}(y_i - \bar{y})^2$. Substituting these estimates in Formula (11.1) yields the following formula for r.

Sample correlation coefficient:

$$r = \frac{s_{xy}}{s_x s_y} = \frac{\text{Sample covariance of } x \text{ and } y}{(\text{Std. dev. of } x)(\text{Std. dev. of } y)}$$

$$= \frac{\frac{1}{n-1}\sum(x_i - \bar{x})(y_i - \bar{y})}{\sqrt{\frac{1}{n-1}\sum(x_i - \bar{x})^2}\sqrt{\frac{1}{n-1}\sum(y_i - \bar{y})^2}}. \qquad (11.2)$$

This correlation coefficient is often referred to as the Pearson product–moment correlation coefficient, in honor of early research by Karl Pearson.

A sample correlation coefficient is interpreted in the same manner as ρ, except that it measures the strength of the *sample* data (rather than of the *population* values). For example, when $r = \pm 1$, there is a perfect straight-line fit between the sample values of x and y; hence, they are said to have a perfect correlation. If the sample values of x and y have little or no relationship, then r will be close to or equal to zero.

To illustrate the determination of the sample correlation coefficient, once again consider the data for our GMAT–GPA example. The data necessary for calculating r are given in Table 11.1, which is a repeat of Table 10.1. We see that the sample data in this example have a correlation of 0.81, indicating a fairly strong linear relationship. This result agrees with the result of our t-test on the same data in Chapter 10.

Just as we presented a computational formula for calculating b in Chapter 10, so it is possible to modify Formula (11.2) in order to simplify the calculation of r when $\bar{x}$ and $\bar{y}$ are not integers.

Computational formula for r:

$$r = \frac{n\sum x_i y_i - (\sum x_i)(\sum y_i)}{\sqrt{n\sum x_i^2 - (\sum x_i)^2}\sqrt{n\sum y_i^2 - (\sum y_i)^2}}.$$

Table 11.1

	x(GMAT)	y(GPA)	$(x - \bar{x})$	$(y - \bar{y})$	$(x - \bar{x})(y - \bar{y})$	$(x - \bar{x})^2$
	480	2.70	−60	−0.40	24	3,600
	490	2.90	−50	−0.20	10	2,500
	510	3.30	−30	+0.20	−6	900
	510	2.90	−30	−0.20	6	900
	530	3.10	−10	0.00	0	100
	550	3.00	+10	−0.10	−1	100
	610	3.20	+70	+0.10	7	4,900
	640	3.70	+100	+0.60	60	10,000
Sum	4320	24.80	0	0.00	100	23,000
Mean	$\bar{x} =$ 540	$\bar{y} =$ 3.10				

$$r = \frac{\frac{1}{n-1}\sum_{n=1}^{n}(x_i - \bar{x})(y_i - \bar{y})}{\sqrt{\frac{1}{n-1}\sum(x - \bar{x})}\sqrt{\frac{1}{n-1}\sum(y - \bar{y})^2}} = \frac{100/7}{(\sqrt{23{,}000/7})(\sqrt{0.66/7})} = 0.81.$$

We can now show that this formula yields the same value of r for our GMAT–GPA example. Since subtracting a constant from every value of y and x will not change the value of r (just as it didn't change the value of b), we will use the coded data from Table 10.3 (on page 485) to find r:

$$r = \frac{8(132) - (320)(0.80)}{\sqrt{8(35{,}800) - (320)^2}\sqrt{8(0.74) - (0.80)^2}}$$

$$= \frac{800}{\sqrt{184{,}000}\sqrt{5.28}} = 0.81.$$

This value agrees with the r we calculated previously.

The Coefficient of Determination

In most correlation problems, the value of r may be somewhat difficult to interpret. For example, what do we mean when we say that $r = 0.81$ indicates a "fairly strong linear relationship"? What we need is a measure permitting us to interpret, in some relative sense, the strength of fit implied by a particular value of r. Such a measure is given by the square of r (r^2), which is called the *coefficient of determination.*

We can explain the logic behind the interpretation of the coefficient of determination by presenting this measure in terms of SST, SSR, and SSE. Suppose in the relationship SST = SSE + SSR we divide each term by SST, as follows:

$$\frac{\text{SST}}{\text{SST}} = \frac{\text{SSE}}{\text{SST}} + \frac{\text{SSR}}{\text{SST}}.$$

Since SSE is the unexplained variation in y, the ratio SSE/SST is the *proportion* of total variation that is *unexplained* by the regression line. Similarly, the ratio SSR/SST is the *proportion* of total variation that is *explained* by the regression line. This last ratio, SSR/SST, is thus a *relative* measure of the goodness of fit of the sample points to the regression line.

Now, let's explore the relationship between r^2 and SSR/SST. It is not difficult to show that:

$$\frac{\textit{Variation explained}}{\textit{Total variation}} = \textit{Coefficient of determination:}$$

$$\frac{\text{SSR}}{\text{SST}} = r^2.$$

Thus, the coefficient of determination is the square of the correlation coefficient. The advantage of the coefficient of determination is that it is easier to interpret, as we illustrate in the following examples.

Once the value of r^2 is calculated in a regression analysis, we have a measure of goodness of fit for the sample. For example, if $r^2 = 0.70$, this means that 70% of the total variation in the sample $\boldsymbol{y}$ values is explained by the best fitting linear function to the sample values of x and y. Similarly, if $r^2 = 0.50$, then 50 percent of the variation in $\boldsymbol{y}$ is explained by $\boldsymbol{x}$. If the regression line perfectly fits all the sample points, then all residuals will be zero, which means that SSE will be zero; hence, SSR/SST $= r^2$ would equal 1.0. In other words, a perfect straight-line fit will always yield $r^2 = 1$. As the level of fit becomes less accurate, less and less of the variation in $\boldsymbol{y}$ is explained by the relation with $\boldsymbol{x}$ (i.e., SSR decreases), which means that r^2 must decrease. The lowest value of r^2 is 0, which will occur whenever SSR = 0. A value of $r^2 = 0$ thus means that *none* of the sample variation in $\boldsymbol{y}$ is explained by the sample values $\boldsymbol{x}$.

To further illustrate the concept of a coefficient of determination, let's calculate r^2 in our GMAT–GPA example. From page 498 in Chapter 10, we know that SSR = 0.435 and SST = 0.660. Hence,

$$r^2 = \frac{\text{SSR}}{\text{SST}} = \frac{0.435}{0.660} = 0.66.$$

Since we earlier found

$$r = 0.81 \qquad \text{and} \qquad r^2 = (0.81)^2 = 0.66,$$

we see that this definition of r^2 is consistent with our definition of r.* The interpretation of this result is that 66% of the total sample variation in GPA is explained by the values of GMAT. The remaining 34% of the variation in GPA is still unexplained. Probably some other factors omitted from our regression model could help determine some additional percent of the variation. If these other factors could be measured and included as additional independent variables, we would have a *multiple* regression model. Such an extension is considered in Chapter 12.

11.4 THE RELATIONSHIP BETWEEN CORRELATION AND REGRESSION

Simple linear correlation is quite closely related to simple linear regression and can, in fact, be considered as just another way of looking at the problem of describing the relationship between two or more variables. As we have pointed out, the objectives of these two approaches are different, regression describing the *nature* of the relationship between variables, correlation describing the *strength* of this relationship. Although both correlation and regression can be applied to any set of observations without making any assumptions about the parent population, it is necessary to make certain assumptions about the underlying model if we want to construct confidence intervals or test hypotheses. Specifically, in regression analysis it is necessary to assume that the dependent variable is a normally distributed random variable, while in correlation analysis it is usually necessary to assume that *both* the independent and dependent variables are normally distributed random variables. In addition, when certain inferences are desired, it is ncessary to assume, in regression analysis, that the values of the independent variable are not randomly distributed, but are *fixed* quantities, known in advance. Fortunately, for most problems this assumption that the x's are fixed quantities is not crucial, and so it is possible to use the techniques of regression in problems where x, rather than being fixed, is a random variable, and to compare the results of a correlation analysis with the results of a regression analysis on the same data. In the GMAT–GPA example, for instance, if both $\mathbf{x}$ and $\mathbf{y}$ can be considered as random variables, then it is not inappropriate to apply, as we did, the technique of correlation analysis as well as regression analysis. In such cases, these two methods of analysis yield much the same information about the relationship between $\mathbf{x}$ and $\mathbf{y}$.

* The similarity in size between SST and r^2 in this example is purely coincidental.

We can explore the connection between the value of r and the value of the slope b by comparing Formulas (10.10) and (11.2), which are reproduced below:

$$r = \frac{s_{xy}}{s_x s_y} \quad \text{and} \quad b = \frac{s_{xy}}{s_x^2}.$$

Because $s_x^2 = s_x s_y$, by substitution we have

$$r = b\frac{s_x}{s_y}. \tag{11.3}$$

Since s_x and s_y are never negative, a positive correlation must always correspond to a regression line with a positive slope, and a negative r corresponds to a negative slope. That is, the sign of r and b will always be the same. Note also that if $b = 0$, then r must equal 0.

We can confirm that the relationship shown in (11.3) holds, by multiplying the formula for b from Chapter 10, which is

$$b = \frac{\frac{1}{n-1}\sum(\boldsymbol{x} - \bar{x})(\boldsymbol{y} - \bar{y})}{\frac{1}{n-1}\sum(\boldsymbol{x} - \bar{x})^2},$$

by the ratio of s_x/s_y, and noting that the result is equivalent to Formula (11.2). To illustrate the use of (11.3), recall, from Chapter 10, that $b = 0.00435$, $s_x = \sqrt{23{,}000/7}$, and $s_y = \sqrt{0.66/7}$. Hence, from (11.3):

$$r = \frac{0.00435\sqrt{23{,}000/7}}{\sqrt{(0.66)/7}} = 0.81,$$

which agrees with the value of r calculated earlier.

It is important to note, at this point, that the value of the correlation coefficient does not depend on which variable is designated as $\boldsymbol{x}$ and which as $\boldsymbol{y}$. The distinction *is* important in regression analysis, however, for the conditional distribution of $\boldsymbol{y}$, given x, results in a different regression line than the conditional distribution of $\boldsymbol{x}$, given y. Formula (11.3) holds only if the numerator is b multiplied by the standard deviation of the *independent* variable.

A note of caution must be added to anyone attempting to infer cause and effect from correlation or regression analysis, because a high correlation (or a good fit to a regression line) does *not* imply that x is "causing" $\boldsymbol{y}$. It does not even imply that x will provide a good estimate of $\boldsymbol{y}$ in the future or for any other set of sample observations. For example, the weekly Dow–Jones stock index was reported to have an 0.84 correlation with the number of points scored by a New York City basketball team, and liquor consumption in the U.S.A. is supposed to be highly correlated with teachers' salaries. In the latter case, the high correlation undoubtedly results because of the presence of one or more additional influences on both variables, such as increases in the general economic well-being.

The above discussion should not be interpreted to mean that one cannot, or should not, draw inferences or conclusions from regression or correlation analysis, but only that care must be taken in assuming cause and effect. Most graduate schools, for example, believe that admission test scores are one means of estimating academic success; they presumably are basing their opinion not on the fact that higher GMAT scores "cause" higher grades, but rather on the fact that whatever these tests measure (memory, intelligence, vocabulary) *does* have an influence on graduate grades. As another example, economists make frequent use of regression techniques in attempts to determine cause-and-effect relationships, especially those which might prove useful in predicting the future of the economy. These techniques form the basis for the field of econometrics.

11.5 TESTS ON THE CORRELATION COEFFICIENT

Tests for $H_0: \rho = 0$

From the discussion thus far we know that when there is no linear regression (i.e., $\beta = 0$), then the correlation between $\boldsymbol{y}$ and $\boldsymbol{x}$ must be zero (that is, $\rho = 0$). Thus, testing $H_0: \beta = 0$ is exactly equivalent to testing $H_0: \rho = 0$. In Chapter 10, we presented two formulas for testing $H_0: \beta = 0$, one of which was a $\boldsymbol{t}$-test (Formula 10.18) and the other an $\boldsymbol{F}$-test (Formula 10.21). There is a $\boldsymbol{t}$-test that is equivalent to Formula (10.18) but uses the value of r rather than b. The following $\boldsymbol{t}$-distributed random variable, with $(n - 2)$ d.f., can be used:

t-statistic for testing $H_0: \rho = 0$:

$$t_{(n-2)} = \frac{r\sqrt{n - 2}}{\sqrt{1 - r^2}}. \tag{11.4}$$

To illustrate how this ratio can be used to test the hypothesis $H_0: \rho = 0$, we again use our GMAT–GPA example, in which $r = 0.81$. For the same reason that we used the alternative $H_a: \beta > 0$, we now use the alternate hypothesis $H_a: \rho > 0$ (because we expect GMAT and GPA to be positively related). To determine the probability that a value such as 0.81 would occur by chance, given that $\rho = 0$ and $n = 8$, we use Formula (11.4) to obtain:

$$t_6 = \frac{0.81\sqrt{8 - 2}}{\sqrt{1 - 0.81^2}} = 3.40.$$

This is (and must be) *exactly* the same t value obtained when we used (10.18) to test $H_0:\beta = 0$ for the same GMAT–GPA data. As we did then, we reject the null hypothesis that $\boldsymbol{x}$ and $\boldsymbol{y}$ are not related at a level of significance between 0.005 and 0.01 (from Table VII) for $n - 2 = 6$ d.f.

*Test for H_0: $\rho = \rho_0$ (Optional)

Despite the fact that the Pearson product–moment correlation coefficient r represents the most likely value of ρ (i.e., it is a maximum-likelihood estimate), this value of r is not an unbiased estimate of ρ, since $E[\boldsymbol{r}] \neq \rho$. Furthermore, even when both $\boldsymbol{x}$ and $\boldsymbol{y}$ are exactly normally distributed, the sampling distribution of $\boldsymbol{r}$ will not be normal. The sampling distribution of r is obviously not normal when ρ is fairly close to $+1$ or -1, as in these cases the distribution of $\boldsymbol{r}$ must be skewed to the right or the left due to the fact that the sample values of $\boldsymbol{r}$ must lie between -1 and $+1$. This situation is pictured in Fig. 11.3 for $\rho = 0.80$.

The ratio given in Formula (11.4) applies only when the sampling distribution of $\boldsymbol{r}$ is symmetrical; that is, when $\rho = 0$. Fortunately, there is a way to test the hypothesis $H_0:\rho = \rho_0$ for other assumed values of ρ_0. This method, developed by R. A. Fisher, transforms the sampling distribution of $\boldsymbol{r}$ into a variable that is normally distributed for most values of ρ, assuming that the sample size is larger than 10. Suppose we designate this new variable as $\boldsymbol{z}_r$.

Transformation of r to a normally distributed variable:

$$\boldsymbol{z}_r = \tfrac{1}{2} \ln\left(\frac{1 + r}{1 - r}\right). \tag{11.5}$$

The variable $\boldsymbol{z}_r$ is normally distributed with

Mean $\mu_r = \tfrac{1}{2} \ln[(1 + \rho)/(1 - \rho)]$

Fig. 11.3 Sampling distribution of r when $\rho = 0.80$.

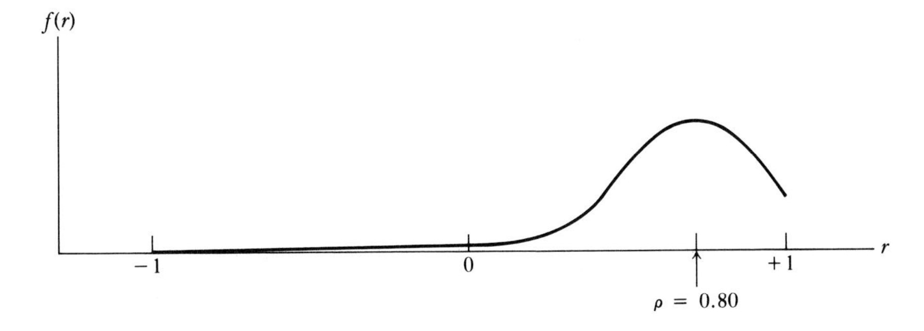

and

$$\text{Standard deviation } \sigma_r = \frac{1}{\sqrt{n-3}}.$$

The value of a sample correlation coefficient can now be tested for a significant difference from the mean, μ_r, for any assumed value of ρ_0, by applying the transformation shown in (11.5) and then calculating the following standardized normal z value:

$$\textit{Standardized variable for testing } H_0: \rho = \rho_0: \qquad z = \frac{z_r - \mu_r}{\sigma_r}. \tag{11.6}$$

To illustrate the use of Formulas (11.5) and (11.6), if the hypothesized value of the population correlation coefficient equals 0.70 ($\rho = 0.70$), then the mean of the variable z_r, calculated from samples drawn from this population, must be:

$$\mu_r = \tfrac{1}{2} \ln\left(\frac{1 + 0.70}{1 - 0.70}\right) = 0.8673.$$

Suppose the sample size in a given problem is $n = 28$. Drawing samples of size $n = 28$ determines a standard deviation equal to

$$\sigma_r = \frac{1}{\sqrt{n-3}} = \frac{1}{\sqrt{25}} = 0.20$$

Now, consider a sample result such as $r = 0.60$, based on a sample of $n = 28$, and coming from an experiment designed to test the following null and alternative hypotheses at the 0.05 level of significance:

$$H_0: \rho = 0.70 \qquad \text{and} \qquad H_a: \rho \neq 0.70.$$

The observed value of z_r is:

$$z_r = \tfrac{1}{2} \ln\left(\frac{1 + 0.60}{1 - 0.60}\right) = 0.6932.$$

To test whether or not the observed result, $z_r = 0.6932$, differs significantly from the mean value $\mu_r = 0.8673$, the standardized z value must be computed:

$$z = \frac{z_r - \mu_r}{\sigma_r} = \frac{0.6932 - 0.8673}{0.20} = -0.871.$$

The probability that z is less than -0.871, $P(z \leq -0.871) = 0.192$, is not low enough to reject H_0 (the critical value for $\alpha = 0.05$ being 1.96); in other words, it must be concluded that a sample result of $r = 0.60$ is not sufficiently unlikely when $\rho = 0.70$ and $n = 28$ to reject this null hypothesis.

A $100(1 - \alpha)\%$ confidence interval for ρ can be constructed on the basis of the information calculated in the example above. To do this, a confidence interval must first be calculated for μ_r, and then the calculated interval for μ_r must be transformed into the equivalent values of ρ by solving for ρ in the relationship $\mu_r = \frac{1}{2}\ln[(1 + \rho)/(1 - \rho)]$. The interval for μ_r is:

$100(1 - \alpha)\%$ *confidence interval for* μ_r:

$$z_r - z_{\alpha/2}\sigma_r \leq \mu_r \leq z_r + z_{\alpha/2}\sigma_r,$$

$$z_r - z_{\alpha/2}\sigma_r \leq \frac{1}{2}\ln\left(\frac{1 + \rho}{1 - \rho}\right) \leq z_r + z_{\alpha/2}\sigma_r.$$

Since solving for ρ in the above relationship becomes rather complex, Table IX in Appendix B gives the values of ρ equivalent to specified values of μ_r. To illustrate the use of this table, we shall compute a 95-percent confidence interval for μ_r based on the sample result given above, $r = 0.60$ for $n = 28$, and then convert this interval into the appropriate confidence interval for ρ:

$$z_r - 1.96\sigma_r \leq \mu_r \leq z_r + 1.96\sigma_r,$$

$$0.6932 - 1.96(0.20) \leq \mu_r \leq 0.6932 + 1.96(0.20),$$

$$0.3012 \leq \mu_r \leq 1.0852.$$

The values of ρ equivalent to $\mu_r = 0.3012$ and $\mu_r = 1.0852$ shown in Table IX are (approximately) $\rho = 0.29$ and $\rho = 0.79$, respectively. The desired 95-percent confidence interval is thus:

$$0.29 \leq \rho \leq 0.70.$$

There are a number of additional significance tests available for use in linear correlation. For example, just as it is possible to test for a significant difference between two (or more) regression coefficients, so can two (or more) correlation coefficients be tested. The methods for these tests are available in a number of statistics textbooks and will not be discussed here.*

11.6 A SAMPLE PROBLEM

Having specified all the concepts and formulas essential for both simple regression and correlation analysis, let us now review them all in an application. One of the important decisions faced by business managers is the amount of *investment* they should make in new plant and equipment and in maintenance and repair of existing capital goods. Economists are very interested in the level of aggregate investment over all private business in the U.S. as this value is an

* Reference [3] is an excellent text in this area.

important factor in determining national income and the potential for growth in an economy.

Suppose a model is specified where the dependent variable y is the level of investment. In such a model, one might want to estimate the amount of investment for a *single* firm by analyzing the relationship between investment and one or more independent variables in a sample of firms; or one may be more interested in estimating the *aggregate* investment for all firms, using as the sample data a number of past time periods. For this example, we choose the latter concept, because the necessary data are easily available without a survey of firms. The Department of Commerce and the Federal Reserve Board, among other agencies, report the key monthly, quarterly, or annual economic aggregate for the U.S. economy.

Any textbook of basic economic principles suggests some theoretical relations between investment and other variables. Generally, it is recognized that the level of investment depends on the availability of funds for investment and on the need for expanding production capacity. Thus any variables which reflect these supply or demand factors might be appropriate in explaining or predicting changes in levels of investment. Such variables as the existing amount of plant and equipment which is depreciating, current and past levels of profits, the interest rate at which funds can be borrowed, indicators of the current general economic conditions, the amount of labor that is unemployed, etc., might be selected as independent variables to help explain variation in investment.

For simplicity, suppose we specify the population regression model to be $y = \alpha + \beta x + \varepsilon$, where x represents a composite price index for 500 common stocks during a given time period, and y represents the amount of aggregate investment *during the following time period.* We might postulate such a relation-

Table 11.2 Date for Estimating the Investment Equation

Observation	Investment y	Stock Index x	Observation	Investment y	Stock Index x
1	62.30	398.4	11	84.30	581.8
2	71.30	452.6	12	85.10	707.1
3	70.30	509.8	13	90.80	776.6
4	68.50	485.4	14	97.90	875.3
5	57.30	445.7	15	108.70	873.4
6	68.80	539.8	16	122.40	943.7
7	72.20	662.8	17	114.00	830.6
8	76.00	620.0	18	123.00	907.5
9	64.30	632.2	19	126.20	905.3
10	77.90	703.0	20	137.00	927.4

Table 11.3 SPSS Output for Investment/Stock Index Example

```
DEPENDENT VARIABLE..      INVEST

MEAN RESPONSE          88.91500          STD. DEV.          24.58678


COEFFICIENTS AND CONFIDENCE INTERVALS.

VARIABLE              B            STD ERROR B          T        95.0 PCT CONFIDENCE INTERVAL

INDEX            .12346865          .12961585E-01   9.5257369    .96237367E-01,   .15069993
CONSTANT         3.8549801


ANALYSIS OF VARIANCE     DF     SUM OF SQUARES       MEAN SQUARE        F        SIGNIFICANCE
REGRESSION                1.       9584.42580        9584.42580     90.73966          .000
RESIDUAL                 18.       1901.25970         105.62554


MULTIPLE R              .91349
R SQUARE                .83447

STD DEVIATION         10.27743
```

ship because we believe that stock market prices are indicative of the general level of business expectations for the future. For this example, we will estimate the population regression line on the basis of 20 quarterly observations. The data, shown in Table 11.2, represents investment measured at an annual rate in billions of dollars.

The canned regression program that is part of the *Statistical Package for the Social Sciences* (*SPSS*) was used to determine the relationship between the dependent variable investment (y = INVEST) and the independent variable stock index (x = STOCK). This analysis is presented in Table 11.3.* The first information given in Table 11.3 provides data about y, namely

$$\bar{y} = 88.915 \qquad \text{and} \qquad s_y = 24.587.$$

Note that we are rounding all values to three decimal places. Table 11.3 next provides us with the regression equation:

$$\hat{y} = 3.855 + 0.123 \text{ INDEX},$$

* Those parts of the SPSS output not covered explicitly in this text have been omitted from Table 11.3. For the interested reader, the omitted terms are defined below, along with their values for the current example.
ADJUSTED R SQUARE (.825): This value of R^2 is adjusted for degrees of freedom. The adjusted R^2 can never exceed the original R^2, and may be negative.
COEFF OF VARIABILITY (11.6 PCT): This measure of relative variability equals $s_e/\bar{y}$.
STD ERROR ($s_a = 9.220$): This is the standard error of the constant a.
BETA (.913): Beta is the standardized regression coefficient. For variable x_i, beta $= b_i s_{x_i}/s_y$. Interpreted as the number of standard deviations change in y caused by a one-standard-deviation change in x_i.
ELASTICITY (.956): The elasticity is $b_i \bar{x}_i/\bar{y}$. Measures the percent change in y caused by a 1% change in x_i.

where the standard error of b is $s_b = 0.0130$. The t-value of 9.526 permits testing the null hypothesis $H_0: \rho = 0$ (or $H_0 = \beta_{\text{INVEST}} = 0$). That same line of output gives a 95% confidence interval for β:

$$0.0962 \leqq \beta \leqq 0.1506$$

The ANOVA table presented next gives further information for testing the strength of the linear relationship.

Source	d.f.	SS	Mean Square	
REG.	1	SSR = 9584.42	MSR = 9584.42	$F = 90.740$
RESIDUAL	18	SSE = 1901.26	MSE = 105.63	

Either the F-value of 90.740 or the t-value presented earlier, $t = 9.526$ (note that $t^2 = F$) can be used to test $H_0: \rho = 0$. For the F-statistic, the program prints out a "significance," which is what we have called a p-value. That is, the printed significance,

$$P(\boldsymbol{F} \geq 90.740) \cong 0.000,$$

says that there is a very small chance of committing the Type I error of concluding that there is some linear correlation (or regression) when this is not the case. That is, if stock prices do reflect business expectations, then this data supports the theory that business firms are more willing to expand their plants and equipment when they foresee "good times" ahead.

Define: correlation coefficient (r), covariance ($C[\boldsymbol{x}, \boldsymbol{y}]$ and s_{xy}), coefficient of determination (r^2).

Problems

11.1 In a simple regression and correlation analysis based on 72 observations, we find $r = 0.8$ and $s_e = 10$.

a) Find the amount of unexplained variation.

b) Find the proportion of unexplained variation to the total variation.

c) Find the total variation of the dependent variable.

11.2 a) In a simple regression, explain the importance of r^2 and s_e, and differentiate between them.

b) A large-scale national survey relating health habits to length of life reported a high *negative* correlation between the number of breakfasts a person skips and the length of life. Would you be willing to infer from this study that eating breakfast regularly will help prolong one's life? Explain.

11.3 Use the data from the following table to compute the regression equation of sales (y) on advertising expenditure (x). Then compute a measure that describes the proportion of the variation of sales which is explained by the regression.

Region	**Sales, y**	**Advertising Expense, x ($10,000)**
A	31	5
B	40	11
C	25	3
D	30	4
E	20	2
F	34	5

11.4 What is the coefficient of correlation between two variables if:

a) One of the variables is constant?

b) The value of one variable always exceeds the value of the other variable by 100?

c) The unexplained variation is twice the explained variation?

11.5 Suppose that the following three values represent observations for the random variables x and y.

x	y
3	4
0	2
3	3

a) Compute the sample correlation coefficient.

b) What is the covariance of x and y? How much of the variability in y is explained by x if the regression line $\hat{y} = a + bx$ is used?

11.6 Run a correlation analysis on the data in Problem 10.41 (page 515), relating MPG and weight for 8-cylinder General Motors cars. What percent of the variability in MPG can be attributed to weight? Test $H_0:\rho = 0$ vs $H_a:\rho > 0$ using $\alpha = 0.05$.

11.7 A study was made of the stopping distance (y) for a 3000-lb car in dry weather at various speeds (x).

y (ft)	270	170	120	60	40
x (mph)	60	50	40	30	20

a) Find the correlation between speed and stopping distance.
b) Test $H_0:\rho = 0$ vs $H_a:\rho > 0$ at $= 0.01$.

11.8 Run a correlation analysis on the data you made up for Problem 10.42 (page 515), using as the two variables e_t and e_{t+1}. Can you reject $H_0:\rho = 0$ using a two-sided alternative if $\alpha = 0.05$?

11.9 What is the difference between regression analysis and correlation analysis? When should each be used? What assumptions about the parent population are necessary for making inferences about ρ?

11.10 Define or describe briefly each of the following:

a) Sample covariance (s_{xy});

b) Coefficient of determination;

c) Pearson product-moment correlation coefficient.

11.11 a) Determine, for the data given in Problem 10.24 (page 511), the value of the correlation coefficient r.

b) Test the null hypothesis $H_0:\rho = 0$ against $H_a:\rho \neq 0$ by means of a t-test. Does this value agree with your answer to Problem 10.25, part (a)?

c) How much of the variability in y is explained by x for these data?

11.12 For the data in Problem 10.6 (page 494),

a) Find the value of r.

b) Test $H_0:\rho = 0$ vs. $H_a:\rho \neq 0$ at $\alpha = 0.05$. What probability value would you report? Does your t-value agree with the answer to the t-value for part (a) of Problem 10.29?

c) Show that $r = b(s_x/s_y)$.

11.13 Suppose that, in analyzing the relationship between 26 observations of two variables, you find that SST $= 120$ and SSR $= 13.2$. The slope of the regression line is positive.

a) What is the value of r for this problem? What percent of the sample variation has been explained?

b) Use a t-test to test $H_0:\rho = 0$ vs. $H_a:\rho \neq 0$ at $\alpha = 0.05$.

c) What probability value would you report?

d) Compute the F-value necessary for testing the null and alternative hypotheses in part (b). Is this value of F consistent with the t value calculated in part (b)? Does it lead to acceptance or rejection of the null hypothesis at the 0.05 level of significance?

11.14 Run a correlation analysis on the data in Problem 10.7 (page 495). Test $H_0:\rho = 0$ vs $H_a:\rho \neq 0$, using $\alpha = 0.01$.

11.15 Given that, for a sample of 17 pairs of observations on y and x, the total variation is 28,416, the SSR is 7104, and the covariation of x and y is $-42{,}624$.

a) Find s_e and r and explain their meaning.

b) Assuming normality, test the hypothesis that there is no correlation between y and x. Use a two-sided alternate hypothesis and let $\alpha = 0.10$.

11.16 a) Find the sample correlation coefficient for the data in Problem 10.15 (page 509). Does the value of the correlation coefficient depend on which variable is called x and which is called y?

b) Show for the data in Problem 10.15 that $r = b(s_x/s_y)$.

11.17 a) Find the sample correlation coefficient for the data given in Problem 10.9.

b) For the above data, test $H_0:\rho = 0$ vs $H_a:\rho \neq 0$ at $\alpha = 0.01$. Follow the directions in Problem 10.10 and recalculate r. Do the transformations affect r?

11.18 For the data in Problem 10.24 (on page 511):

a) Test the null hypothesis $H_0:\rho = 0.50$ vs. $H_a:\rho \neq 0.50$ at $\alpha = 0.05$. What probability value would you report?

b) Construct a 95% confidence interval for ρ.

11.19. For Problem 11.13:

a) Test the null hypothesis $H_0:\rho = 0.05$ vs. $H_a:\rho \neq 0.05$ at $\alpha = 0.05$.

b) What probability value would you report?

11.20 Through a simple regression based on 32 observations, it is found that $r = 0.60$ and $s_e^2 = 100$.

a) Find SST, SSR, and SSE.

b) Do an analysis-of-variance test to determine whether the linear relationship is significant at the 0.01 level.

11.21 The information given below represents part of an *SPSS* output for the data in Problem 11.3, relating sales (y) to advertising expense (x).

a) Interpret this output, giving the regression equation, the values of r and r^2, and test $H_0:\rho = 0$ vs. $H_a:\rho > 0$.

b) What value of sales would you predict if ADEXP = $8000?

```
DEPENDENT VARIABLE          SALES

MEAN RESPONSE          30.000000          STD. DEV.          6.95701

VARIABLE(S) ENTERED ON STEP NUMBER   1..   ADEXP


MULTIPLE R          .90909   ANALYSIS OF VARIANCE  DF     SUM OF SQUARES
R SQUARE            .82645   REGRESSION             1.        200.000000
                             RESIDUAL               4.         42.000000
STD DEVIATION      3.24037
COEFFICIENTS AND CONFIDENCE INTERVALS.

VARIABLE              B              STD ERROR B           T

ADEXP          2.0000000           .45825757          4.3643578
CONSTANT      20.000000
```

11.22 In a regression of y on x based on 11 observations, the value of the coefficient of determination is 0.36.

a) Does this indicate a significant correlation between y and x at the 0.05 significance level if proper normality assumptions are made? Let H_a be two-sided.

b) What is the proportion of variation left unexplained in this regression?

11.23 Suppose that ten observations were obtained in a survey to determine the relationship between an individual's educational level and this person's salary. These observations are shown in Table 11.4 (on page 538).

a) Assuming normal distributions, what is the correlation between years of higher education and income for this sample?

b) Use $\alpha = 0.05$ and your answer to part (a) to test the null hypothesis $H_0:\rho = 0$ against the alternative hypothesis $H_a:\rho > 0$.

Table 11.4

Years of Higher Education	Income
3	$20,000
4	18,000
7	23,000
9	30,000
1	16,000
0	15,000
2	18,000
1	17,000
8	24,000
5	19,000

11.24 The following data represent the dollar value of sales and advertising for a retail store. [This problem is a continuation of Problem 10.6.]

Advertising (in Thousands)	Sales (in Thousands)
$600	$5000
400	4000
800	7000
200	3000
500	6000

a) Draw the scatter diagram for these data. Fit a line by the freehand method. Does the linear approximation seem appropriate?

b) Find the least-squares regression line.

c) What value for sales would you predict if advertising were $700? What value would you predict if advertising were zero?

d) Construct a 95-percent confidence interval for the *mean* value of sales when advertising is $500 (i.e., for $\mu_{y \cdot 500}$).

e) Construct a 95-percent confidence interval for actual values of sales when advertising is $500.

f) Test the null hypothesis that the slope of the regression line is zero against $H_a: \beta \neq 0$ using $\alpha = 0.05$.

g) Find the sample correlation coefficient.

h) Assuming normality, test the null hypothesis $H_0: \rho = 0$ against the alternative $H_a: \rho \neq 0$. At what level of significance can H_0 be rejected?

11.25 Given a regression equation, $\hat{y} = 14 + 6x$, based on 12 observations, test the hypotheses $H_0: \beta = 0$ versus $H_a: \beta > 0$ using a significance level of 0.05. The standard error of b is $s_b = 1.5$.

11.26 In a correlation between corporate net investment and long-term interest rates, using quarterly observations from the third quarter of 1973 through the fourth quarter of 1982, a correla i coefficient of +0.60 is obtained. Determine whether this is a significant positive correlation, using $\alpha = 0.01$, a one-sided alternative hypothesis, and assuming normality.

11.27 Suppose that, in analyzing the relationship between 26 observations of two variables, you find SST equal to 120.0 and SSR equal to 13.2. The slope of the sample regression line is positive. Assume that the populations are normal.

a) What is the coefficient of determination for this problem? What percent of the sample variation has been explained?

b) Use the t-test described in Section 11.5 to test the null hypothesis $H_0\colon \rho = 0$ against $H_a\colon \rho \neq 0$. Can the null hypothesis of no linear correlation be rejected at the 0.05 level of significance?

c) Compute the F-value necessary for testing the null and alternative hypotheses in part (b). Is this value of F consistent with the t-value calculated in part (b)?

11.28 Let x represent income payments in Texas (billions of dollars) and let y represent retail sales of Texas jewelry stores (millions of dollars.) The regression equation is $\hat{y} = 8.505x - 7.41$. The standard error of the estimate is 3.5, and the correlation coefficient is 0.95.

a) For a year in which income payments are \$10.0 billion, what is the best estimate of sales in jewelry stores?

b) What proportion of the variation in the retail jewelry sales is explained by the variation in income payments?

c) This correlation indicates that in Texas higher retail jewelry sales cause higher incomes. Comment.

11.29 For the data in Problem 11.24, find s_x, s_y, and s_{xy}. Show that the value of r you calculated for Problem 11.24 equals $s_{xy}/s_x s_y$.

11.30 Many investors see a strong relationship between the prime rate and the Dow-Jones Industrial Average. The approximate monthly average values for both variables during 1980 are given below.

	Jan	Feb	Mar	Apr	May	June	July	Aug	Sept	Oct	Nov	Dec
Dow-Jones	850	895	778	770	815	870	895	931	935	948	950	930
Prime Rate	15	16	17	19	16	12	11	11	12	14	16	20

Calculate correlation between the Dow-Jones and the prime rate. Test $H_0\colon \rho = 0$ vs $H_a\colon \rho > 0$ at $\alpha = 0.01$.

Exercises

11.31 Calculate r^2 for the data in Problem 10.40 (page 515), and then find r^2 by using $\sqrt{x}$ as the independent variable. Does x or $\sqrt{x}$ provide a better fit to these data?

11.32 Find $C[e_t, e_{t+1}]$ and the correlation between e_t and e_{t+1} for the data you generated for Problem 10.42, on page 515.

11.33 Prove that $r = b(s_x/s_y)$.

11.34 Prove that Formula (11.2) is equivalent to the computational formula for r.

11.35 Prove that $r^2 = \text{SSR/SST}$.

11.36 Run a computer program on the data in Table 11.1 (page 524) to verify that the values of a, b, s_e, and r^2 derived from these data are correct.

GLOSSARY

coefficient of determination: $r^2 = \text{SSR/SST}$ = explained amount of variation in y relative to total to be explained: $0 \leq r^2 \leq 1$.

population covariance: $C[x, y] = E[(x - \mu_x)(y - \mu_y)]$. Describes how x and y covary in the population.

population correlation coefficient: $\rho = C[x, y]/\sigma_x\sigma_y$; a measure for two populations' linear association between x and y: $-1 \leq \rho \leq 1$.

sample covariance (s_{xy}): unbiased estimate of $C[x, y]$,

$$s_{xy} = \frac{1}{n - 1}\sum(x_i - \bar{x})(y_i - \bar{y})$$

sample correlation coefficient: best estimate of ρ; $r = s_{xy}/s_x s_y$.

t-test for $H_0: \rho = 0$: $t = r\sqrt{n - 2}/\sqrt{1 - r^2}$ with $(n - 2)$ d.f.

F-test for $H_0: \rho = 0$: $F = \text{MSR/MSE}$ with $(1, n - 2)$ d.f.

z-test for $H_0: \rho = \rho_0$: $z = \dfrac{z_r - \mu_r}{\sigma_r}$

12

MULTIPLE REGRESSION AND CORRELATION

"There are three types of liars: liars, damn liars, and statisticians."

BENJAMIN DISRAELI

In Chapters 10 and 11 we presented the methods for simple linear regression and correlation. In this chapter, we extend that analysis by presenting the methods for relating a dependent variable to *two or more* independent variables. Multiple regression is presented first, followed by multiple linear correlation.

12.1 MULTIPLE LINEAR REGRESSION: THE POPULATION

As was the case for simple linear regression, we first need to specify certain population relationships. Suppose that there are m independent variables, $x_1, x_2, \ldots, x_m$, where $m \geq 1$, and again let $\mathbf{y}$ be the value of the dependent variable. Also, let

$$E[\mathbf{y} \mid x_1, x_2, \ldots, x_m] = \mu_{y \cdot x_1, x_2, \ldots, x_m}$$

denote the conditional mean of the y values, given specific values of the m independent variables. We now need to specify the relationship between $\mu_{y \cdot x_1, x_2, \ldots, x_m}$, and the values of $x_1, x_2, \ldots, x_m$. This relationship could take on many different forms, most of which would be extremely difficult to handle in a regression analysis. For example, although certain nonlinear relationships can be solved by an extension of the methods of this chapter (see Section 12.7), for the most part we will assume that the variables are all linearly related. By this we mean that the relationship between y and *each* one of the independent variables is linear. Assuming linearity, and letting α equal the y-intercept, β_1 equal the slope of the relationship between y and x_1, β_2 equal the slope between y and x_2, and so forth, the equation relating $\mu_{y \cdot x_1, x_2, \ldots, x_m}$ to the independent variables is called the

Population multiple linear regression equation:

$$\mu_{y \cdot x_1, x_2, \ldots, x_m} = \alpha + \beta_1 x_1 + \beta_2 x_2 + \cdots + \beta_m x_m. \tag{12.1}$$

The coefficients $\beta_1, \beta_2, \ldots, \beta_m$ are called the *partial* regression coefficients, since they indicate the (partial) influence of each independent variable on y, with the influence of all the remaining variables held constant.

To illustrate a regression equation, suppose that the admissions director in our GMAT–GPA example would like to expand the analyses by considering, in addition to Graduate Management Admissions Test scores, undergraduate grades, age, or a variety of other factors which might be useful in predicting the dependent variable (GPA). For example, suppose we designate GMAT as x_1, and undergraduate grade-point average (UGG) as x_2, and assume the mean GPA ($\mu_{y \cdot x_1, x_2}$) is related to these variables as follows:

$$\mu_{y \cdot x_1, x_2} = -1.75 + 0.005x_1 + 0.70x_2. \tag{12.2}$$

The value $\beta_1 = 0.005$ in this case indicates that, after eliminating or taking into account the effect of all other variables (i.e., holding them constant), a one-unit increase in x_1 (GMAT) will increase the mean value of y (GPA) by 0.005 units. Similarly, a one-unit increase in x_2 (undergraduate grades) will increase the mean GPA by 0.70 units, since $\beta_2 = 0.70$. Note that the fact that β_2 is larger than β_1 does not mean that undergraduate grades are more important in determining GPA than are GMAT scores. The value of the partial regression coefficients assumed depends so highly on the choice of units to measure x that it is usually difficult, if not impossible, to make direct comparisons. Had we measured undergraduate grades so that each one-unit change in x was equivalent to a change in grade-point average of one-hundredth of a point, then the value of β_2 would have been 0.007. Figure 12.1 is a graph of the plane represented by the multiple regression equation in Formula (12.2).

As was the case for simple linear regression, we also want to specify the population regression model used to predict the specific values y_i. To do this, we again need to add an error term (ε) to the regression equation, and put the subscript i on each variable (to indicate the relationship represents the ith observation in the population).

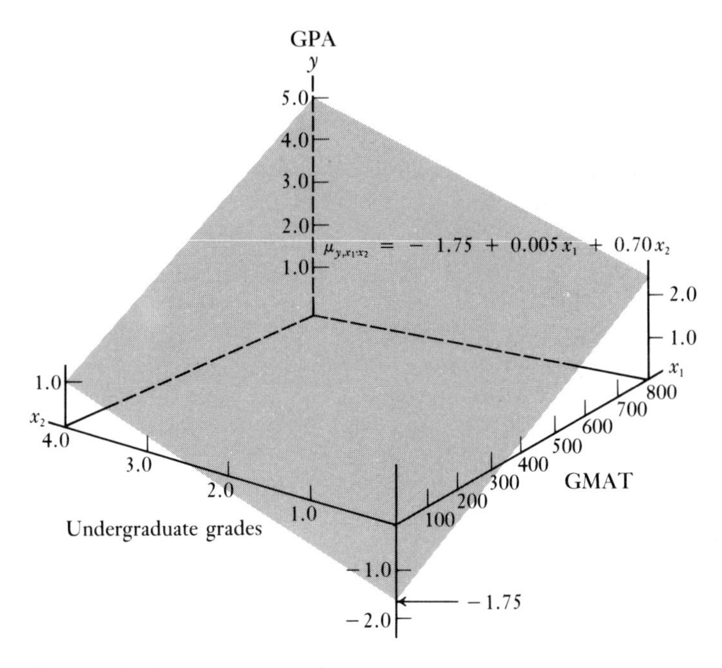

Fig. 12.1 The population regression plane $\mu_{y \cdot x_1 x_2} = -1.75 + 0.005x_1 + 0.70x_2$.

Population regression model:

$$y_i = \alpha + \beta_1 x_{1i} + \beta_2 x_{2i} + \cdots + \beta_m x_{mi} + \varepsilon_i.$$

The process of using sample information to estimate the parameters of a multiple linear regression equation involves the same least-squares technique used in the sample linear regression case.

12.2 MULTIPLE LEAST-SQUARES ESTIMATION

In this section we present the sample regression model, and the least-squares estimate of the population parameters. Suppose we have a sample consisting of n observations for each of the m variables. The problem is to find the sample regression equation which provides the "best fit" to these data, and to use the coefficients of that equation as estimates of the parameters of the population regression equation. This sample regression equation has the following form:

Sample regression equation:

$$\hat{y} = a + b_1 x_1 + b_2 x_2 + \cdots + b_m x_m, \tag{12.3}$$

where the value of $\hat{y}$ is the estimate of the mean value for a given set of values of $x_1, x_2, \ldots, x_m$ (that is, $\hat{y}$ is an estimate of $\mu_{y \cdot x_1, x_2, \ldots, x_m}$), a is the estimate of α, and $b_1, b_2, \ldots, b_m$ are the estimates of the partial regression coefficients $\beta_1, \beta_2, \ldots, \beta_m$.

The sample regression model follows the same pattern; adding the subscript i to each x value, and adding the sample error term e_i, as an estimate of ε_i yields:

Sample regression model:

$$\hat{y}_i = a + b_1 x_{1i} + b_2 x_{2i} + \cdots + b_m x_{mi} + e_i. \tag{12.4}$$

The values of the estimators $a, b_1, b_2, \ldots, b_m$ can be obtained by a direct extension of the least-squares approach presented in Chapter 10. Assuming a sample of n observations (i.e., $i = 1, 2, \ldots, n$), it is again necessary to minimize the sum of the squared deviations of the individual values y_i about the predicted values $\hat{y}_i$:

$$\text{Min } G = \sum_{i=1}^{n} (y_i - \hat{y}_i)^2 = \sum_{i=1}^{n} (y_i - a - b_1 x_{1i} - b_2 x_{2i} - \cdots - b_m x_{mi})^2.$$

The procedure for minimizing G is similar to the simple linear process, except that now partial derivatives must be taken for $(m + 1)$ variables instead of just two; i.e., derivatives need to be taken with respect to a, and with respect to the m variables $b_1, b_2, \ldots, b_m$. Setting these $(m + 1)$ partials equal to zero and solving yields $(m + 1)$ *normal equations.**

The number of normal equations $(m + 1)$ always exceeds the number of independent variables (m) by one. When the normal equations involve only one or two independent variables (i.e., two or three equations), then, solving these equations for a and b_1, or a, b_1 and b_2, is usually relatively simple. For example, when there are two variables and eight observations, as in the

---

* The normal equations are:

$$na + b_1\sum x_{1i} + b_2\sum x_{2i} + \cdots + b_m\sum x_{mi} = \sum y_i. \tag{12.5}$$

$$a\sum x_{1i} + b_1\sum x_{1i}^2 + b_2\sum x_{1i}x_{2i} + \cdots + b_m\sum x_{1i}x_{mi} = \sum x_{1i}y_i, \tag{12.6}$$

$$a\sum x_{2i} + b_1\sum x_{1i}x_{2i} + b_2\sum x_{2i}^2 + \cdots + b_m\sum x_{2i}x_{mi} = \sum x_{2i}y_i, \tag{12.7}$$

$$\vdots$$

$$a\sum x_{mi} + b_1\sum x_{1i}x_{mi} + b_2\sum x_{2i}x_{mi} + \cdots + b_m\sum x_{mi}^2 = \sum x_{mi}y_i. \tag{12.8}$$

These equations are not as forbidding as they may at first appear. Formula (12.5) can be viewed as a generalization of Formula (10.7), while Formulas (12.6) to (12.8) can be viewed as generalizations of Formula (10.8), with different values of the first subscript of x. Thus, Formula (12.6) is similar to (10.8), except that (12.6) sums x_{1i} instead of x_i. Likewise, (12.7) is similar to (10.8) except that it sums x_{2i} instead of x_i. Note that (12.6) can be derived by multiplying both sides of (12.5) by x_{1i} and summing; (12.7) can be derived by multiplying both sides of (12.5) by x_{2i} and summing; and so forth.

GMAT–GPA example, the three normal equations to be solved are:

$$na + b_1 \sum_{i=1}^{8} x_{1i} + b_2 \sum_{i=1}^{8} x_{2i} = \sum_{i=1}^{8} y_i,$$

$$a \sum_{i=1}^{8} x_{1i} + b_1 \sum_{i=1}^{8} x_{1i}^2 + b_2 \sum_{i=1}^{8} x_{1i}x_{2i} = \sum_{i=1}^{8} x_{1i}y_i,$$

$$a \sum_{i=1}^{8} x_{2i} + b_1 \sum_{i=1}^{8} x_{1i}x_{2i} + b_2 \sum_{i=1}^{8} x_{2i}^2 = \sum_{i=1}^{8} x_{2i}y_i.$$

Solving more than three normal equations, or solving when the number of observations is large, requires such tedious arithmetic that computer programs, based on the techniques of matrix algebra, are employed to solve such systems of equations. There are a number of standard computer programs for solving both regression and correlation problems. In Section 12.6 of this chapter we illustrate the solution of the normal equation using matrix algebra.

To illustrate the use of multiple linear regression, suppose that our admissions director has calculated the following least-squares multiple linear regression equation for GPA(y), GMAT(x_1), and undergraduate grade-point average (x_2):

$$\hat{y} = -1.25 + 0.0040x_1 + 0.75x_2. \tag{12.9}$$

This equation provides an estimate of α equal to -1.25, an estimate of β_1 equal to 0.0040, and an estimate of β_2 equal to 0.75. The value of $\hat{\mathbf{y}}$ which results from substituting specific values of x_1 and x_2 into this equation is an estimate of $\mu_{y \cdot x_1, x_2}$. For example suppose we want an estimate of the *mean* GPA for all students with a GMAT score of 500 and an undergraduate grade-point average of 3.10. Substituting $x_1 = 500$ and $x_2 = 3.10$ into the equation given above yields the following estimate:

Estimate of mean GPA, *given* $x_1 = 500$, $x_2 = 3.10$:

$$\hat{y} = -1.25 + 0.0040(500) + 0.75(3.10) = 3.075.$$

Now, suppose we want to predict the GPA of a *single* student, and this student had a GMAT of 500 and an undergraduate grade-point average of UGG $= 3.10$. Our best estimate in this case is the *same* as before:

Estimate of individual GPA, *given* $x_1 = 500$, $x_2 = 3.10$:

$$\hat{y}_i = -1.25 + 0.0040(500) + 0.75(3.10) = 3.075.$$

Even though these two estimates are equal, we must again point out that we have much less confidence in the estimate of a single value (y_i) than we do in the estimate of the mean value ($\hat{y}$). To be more precise about the degree of confidence one has for the estimates in multiple linear regression, we need to again present the assumptions (about the errors ε_i) which are necessary for interpreting the estimators of the $(m + 1)$ parameters (i.e., estimators of α and m values of β).

12.3 ASSUMPTIONS FOR THE MULTIPLE REGRESSION MODEL

Again we must emphasize that the least-squares procedure does not require *any* assumptions about the population, since this procedure is merely a curve-fitting technique. However, in order to be able to test the goodness of fit of a sample regression equation, it is once more necessary to make certain assumptions about the error term (ε) in the population regression model. The first five of these assumptions are parallel to those specified in Section 11.2 for the simple regression model. We repeat them below for the multiple regression case:

Assumption 1. *The error term ε is independent of each of the m independent variables $x_1, x_2, \ldots, x_m$.*

Assumption 2. *The error ε_i for all possible sets of given values of $x_1, x_2, \ldots, x_m$ is normally distributed.*

Assumption 3. *The expected value of the errors is zero for all possible sets of given values $x_1, x_2, \ldots, x_m$.* That is, $E[\varepsilon_i] = 0$ for $i = 1, 2, \ldots, n$.

Assumption 4. *Any two errors ε_k and ε_j are independent; i.e., their covariance is zero,* $C[\varepsilon_k, \varepsilon_j] = 0$ for $k \neq j$.

Assumption 5. *The variance of the errors is finite, and is the same for all possible sets of given values of $x_1, x_2, \ldots, x_m$. That is, $V[\varepsilon_i] = \sigma_\varepsilon^2$ is a constant for $i = 1, 2, \ldots, n$.*

In addition to these five assumptions, two additional conditions are necessary to obtain least-squares estimates in the multiple regression equation.

Condition 1. *None of the independent variables is an exact linear combination of the other independent variables.*

This means that no one variable x_i is an exact multiple of any other independent variable. Further, if $m \geq 2$, this assumption means that no one variable x_i can be written as

$$x_i = a_1x_1 + a_2x_2 + \cdots + a_{i-1}x_{i-1} + a_{i+1}x_{i+1} + \cdots + a_mx_m,$$

where the a's are constants. This assumption is a weak condition, since it requires only that the variables not be *perfectly* related to each other in a linear function. In practice, the independent variables often are partially linearly related to each other, or related to each other in some nonlinear way. Although least-squares estimates can be calculated in these situations, problems sometimes arise in their interpretation.

> **Condition 2.** *The number of observations* (n) *must exceed the number of independent variables* (m) *by at least two* (*that is,* $n \geq m + 2$).

Since, in the multiple regression equation, there are $m + 1$ parameters to be estimated, the number of degrees of freedom is $n - (m + 1)$. This condition merely specifies that there be at least one degree of freedom (that is, $n - (m + 1) \geq 1$). In practice, the sample size must be quite a bit larger than this value; otherwise, any measure of goodness of fit may be more a consequence of having only a small amount of data, rather than of having an accurately specified explanatory model.

12.4 MULTIPLE STANDARD ERROR OF THE ESTIMATE

Just as for simple linear regression, the total variation in a regression problem, $\text{SST} = \sum(y_i - \bar{y})^2$, can be divided into the sum of $\text{SSE} = \sum(y_i - \hat{y}_i)^2$ and $\text{SSR} = \sum(\hat{y}_i - \bar{y})^2$. And once again the standard error of the estimate is defined as

$$s_e = \sqrt{\frac{\text{(Unexplained variation)}}{\text{(Degrees of freedom)}}} = \sqrt{\frac{\text{SSE}}{\text{d.f.}}}$$

In the present case there are $(m + 1)$ parameters to be estimated before the errors can be calculated; hence the d.f. associated with SSE is $n - (m + 1) = n - m - 1$. Thus,

> *Multiple standard error of the estimate:*
>
> $$s_e = \sqrt{\frac{\text{SSE}}{n - m - 1}} = \sqrt{\frac{1}{n - m - 1}\sum_{i=1}^{n} e_i^2}. \tag{12.10}$$

For our GMAT–GPA example, total variation is $\sum(y_i - \bar{y})^2 = 0.6600$ (from Chapter 10). When both x_1 (GMAT) and x_2 (UGG) are included in the analysis, then it can be shown that:*

$$\text{SSE} = \sum(y_1 - \hat{y})^2 = 0.1400$$

and

$$\text{SSR} = \sum(\hat{y}_i - \bar{y})^2 = 0.5200.$$

Note that SSE and SSR sum to SST, as they must.

We can now calculate s_e by remembering that $m = 2$ (two independent variables) and $n = 8$ (observations in the sample).

Standard error of the estimate:

$$s_e = \sqrt{\frac{\text{SSE}}{n - m - 1}} = \sqrt{\frac{0.1400}{8 - 2 - 1}} = 0.1673.$$

As before, we know that when n is large, about 68 percent of all sample points should lie within one standard error of the estimated value $\hat{y}_i$, and about 95 percent should be within two standard errors.

The reader should observe that the standard error with only one independent variable (GMAT) was (from Chapter 10) $s_e = 0.225$, which is larger than the comparable value of 0.1673 when there are two independent variables (GMAT and UGG). In general, s_e may either increase or decrease with the addition of new variables. When new variables are added, SSE can never increase; but since an extra d.f. is lost with each new variable, in some cases s_e will increase.

12.5 TESTS FOR MULTIPLE LINEAR REGRESSION

A variety of test procedures have been developed for the parameters of a multiple regression model. Not all these tests will be discussed here, because their complexity is better handled in a more advanced text. The usual questions in a multiple linear relationship are the overall goodness of fit, as well as the significance of the partial regression coefficients.

The *F*-Test

In multiple linear regression, one of the most commonly held null hypotheses is that there is no linear regression at all in the population—i.e., that all the β values are equal to zero:

$$H_0: \beta_1 = \beta_2 = \cdots = \beta_m = 0.$$

* Since we haven't presented any data for x_2(UGG), we show only the results of these calculations.

Table 12.1

Source	SS	d.f.	Mean Square
Regression	SS Regression	m	SS Regression$/m$ = MSR
Error	SS Error	$n - (m + 1)$	SS Error$/(n - m - 1)$ = MSE
Total	SS Total	$n - 1$	

If this hypothesis is true, then we would expect SSE to be relatively large and SSR to be relatively small. If each of these sums is divided by its degrees of freedom, then a mean square is obtained. The ratio of the mean-square regression (MSR) to the mean-square error (MSE) can be shown to follow an F-distribution (just as it did in Chapter 10).

As we indicated previously, the d.f. associated with SSE is $n - (m + 1)$, because $(m + 1)$ parameters are being estimated (one d.f. is lost for each parameter estimated). The d.f. for SSR is the same as the number of independent variables (m). Note that these d.f. sum to the total degrees of freedom, which must be $(n - 1)$. These values are summarized in the analysis-of-variance table, Table 12.1.

The appropriate statistic to test the null hypothesis is given by the ratio of MSR to MSE, which follows an F-distribution with m and $(n - m - 1)$ d.f.:

Statistic for Testing $H_0: \beta_1 = \beta_2 = \cdots = \beta_m = 0$:

$$F = \frac{\text{SSR}/m}{\text{SSE}/(n - m - 1)} = \frac{\text{MSR}}{\text{MSE}}.$$

We illustrate this test by referring again to the GMAT–GPA example. Since we know that SST = 0.6600, SSR = 0.5200, SSE = 0.1400, and the degrees of freedom are $m = 2$ and $(n - m - 1) = (8 - 2 - 1) = 5$,

$$F = \frac{0.5200/2}{0.1400/5} = 9.29.$$

From Table VIII we see that this value of F will occur between one and two and one-half percent of the time by chance when the null hypothesis of no linear regression is true. Hence, we can probably safely accept the hypothesis that our equation *is* useful in describing the values of y.

Tests on Values of β_i

To determine the significance of an individual coefficient (β_i) in the regression model, a test is used which is similar to that for the slope in simple linear regression. The null hypothesis $H_0:\beta_i = 0$ means that the variable x_i has no linear relationship with y, *holding the effect of the other independent variables constant.* The best linear unbiased estimate of β_i is the sample partial regression coefficient b_i. Under the assumption that the (unknown) error is normally distributed, then the test for this null hypothesis follows the ***t***-distribution with $(n - m - 1)$ d.f.:

Statistic for testing $H_0:\beta_i = 0$: $\quad t = \dfrac{b_i - 0}{s_{b_i}}.$

The value of s_{b_i} is the estimated standard error of the estimate b_i. Calculation of s_{b_i} is quite tedious, but it is always shown in the output of a regression analysis program for a computer. Thus, the determination of t in a practical application is simply done by forming the ratio of the coefficient to its estimated standard error. When the calculated value of t exceeds the critical value of $\boldsymbol{t}$ determined from Table VII, then the null hypothesis of *no significance* can be rejected. It is then concluded that the variable x_i does have an important influence on the dependent variable y, even after accounting for the influence of all other independent variables included in the model.

For our example, the estimated standard errors of the coefficients b_1 and b_2 of the variables x_1 (GMAT) and x_2 (undergraduate grades) are $s_{b_1} = 0.0011$ and $s_{b_2} = 0.216$ respectively. Since $n = 8$ and $m = 2$ in this case, the critical value for a one-sided test on either coefficient (using a significance level of $\alpha = 0.01$) is $t_{(\alpha, n-m-1)} = t_{(0.01, 5)} = 3.365$. Thus, the critical region for a one-sided test when $H_0:\beta_1 = 0$ (or $H_0:\beta_2 = 0$) is all values of $\boldsymbol{t}$ that exceed 3.365. We choose a one-sided test because our *a priori* theoretical propositions were that both x_1 and x_2 were positively related to y.

For the test on β_1,

$$t = \frac{b_1}{s_{b_1}} = \frac{0.0040}{0.0011} = 3.636,$$

and for the test on the significance of β_2,

$$t = \frac{b_2}{s_{b_2}} = \frac{0.750}{0.216} = 3.472.$$

We conclude that both variables, x_1 = GMAT and x_2 = UGG, are significantly related to y = GPA. The variable x_1 is the more influential of the two, since it has the higher t value.

Now that we have found that both regression coefficients provide a good fit for the data, we can proceed to the next logical task, that of determining the best point forecast based on the previous relationships between GPA and GMAT and UGG. In doing so, we must remember that, even if the regression equation has been shown to fit well, and has all very significant coefficients, such results may not hold for *future* data. The relationship for new applications may differ due to some change in the social, political, or economic environment. However, let's be courageous and assume that the past relationship (given in Formula (12.9), on page 545) will hold for students currently applying for admission. If the given values of the independent variables for one such student are $x_1 = 550$ and $x_2 = 3.00$, then the forecast value is:

$$\begin{aligned}\hat{y} &= a + b_1x_1 + b_2x_2 \\ &= -1.25 + 0.0040(550) + 0.75(3.00) \\ &= 3.20.\end{aligned}$$

The estimated GPA for this person is thus 3.20.

It is important to recognize that we have presented only part of the analysis possible in our regression model. We could have constructed a confidence interval for y_i and $\mu_{y \cdot x_1x_2}$, or we might have added additional independent variables, and then performed the same F-test (for the overall relationship) and t-test (for the individual coefficients) for the new data.

Now that we have presented the least-squares approach when there are two or more independent variables, we need to make the comparable extension of simple linear correlation. This extension is begun in Section 12.8. Section 12.6 is a starred (optional) section, which presents the least-squares approach in terms of matrix algebra. Section 12.7 is another starred section which presents a brief discussion of some nonlinear relationships.

12.6 LEAST-SQUARES REGRESSION IN MATRIX FORM (OPTIONAL)

As we indicated earlier, it is possible to perform least-squares analyses much more easily by using matrix algebra. First, suppose we rewrite the sample regression model (12.4) by letting **Y**, **X**, **B**, and **E** be matrices defined as follows:

$$\boldsymbol{Y} = \begin{bmatrix} y_1 \\ y_2 \\ \vdots \\ y_n \end{bmatrix}, \quad \boldsymbol{X} = \begin{bmatrix} 1 & x_{11} & x_{21} & \cdots & x_{m1} \\ 1 & x_{12} & x_{22} & \cdots & x_{m2} \\ \vdots & \vdots & \vdots & & \vdots \\ 1 & x_{1n} & x_{2n} & \cdots & x_{mn} \end{bmatrix}, \quad \boldsymbol{B} = \begin{bmatrix} a \\ b_1 \\ b_2 \\ \vdots \\ b_m \end{bmatrix}, \quad \boldsymbol{E} = \begin{bmatrix} e_1 \\ e_2 \\ \vdots \\ e_n \end{bmatrix}.$$

The sample regression model written in matrix notation is thus:

$$\boldsymbol{Y} = \boldsymbol{XB} + \boldsymbol{E}. \tag{12.11}$$

To find the values $a, b_1, b_2, \ldots, b_m$, the least-squares method minimizes $\sum e_i^2$. In matrix notation, this sum can be written as $\boldsymbol{E}^t\boldsymbol{E}$, where $\boldsymbol{E}^t$ is the *transpose of* **E**. From Formula (12.11), we see that $\boldsymbol{E} = \boldsymbol{Y} - \boldsymbol{XB}$; hence, for the method

* This section may be omitted by readers who have not studied matrix algebra.

of least squares, it is necessary to

$$\textit{Minimize } \mathbf{E}^t\mathbf{E} = (\mathbf{Y} - \mathbf{XB})^t(\mathbf{Y} - \mathbf{XB}).$$

To minimize this expression we must use vector differentiation. Differentiating and setting the first derivative equal to zero yields

$$\mathbf{B} = (\mathbf{X}^t\mathbf{X})^{-1}\mathbf{X}^t\mathbf{Y}, \tag{12.12}$$

where $(\mathbf{X}^t\mathbf{X})^{-1}$ is the *inverse* of the matrix $\mathbf{X}^t\mathbf{X}$. Thus, to solve for the values of the matrix $\mathbf{B}$, it is necessary that the inverse of $\mathbf{X}^t\mathbf{X}$ exist. Such an inverse *will* exist if Condition 2 of our assumptions is met; i.e., if none of the independent variables is an exact linear combination of the other independent variables.

Even with the use of matrix algebra, the solution of regression problems involving two or more independent variables is not an easy task, because finding the inverse $(\mathbf{X}^t\mathbf{X})^{-1}$ may be difficult. Again, we must emphasize that computer programs are usually employed to perform the algebra in regression analyses. To keep our own algebra fairly straightforward, we will illustrate Formula (12.12) by using the GMAT–GPA data, involving just one independent variable (from Table 10.1 in Chapter 10). For this data, $n = 8$, $m = 1$, and

$$\mathbf{Y} = \begin{bmatrix} 2.70 \\ 2.90 \\ 3.30 \\ 2.90 \\ 3.10 \\ 3.00 \\ 3.20 \\ 3.70 \end{bmatrix}, \quad \mathbf{X} = \begin{bmatrix} 1 & 480 \\ 1 & 490 \\ 1 & 510 \\ 1 & 510 \\ 1 & 530 \\ 1 & 550 \\ 1 & 610 \\ 1 & 640 \end{bmatrix}.$$

To solve Formula (12.12), we first find $(\mathbf{X}^t\mathbf{X})$:

$$\underset{\mathbf{X}^t}{\begin{pmatrix} 1 & 1 & 1 & 1 & 1 & 1 & 1 & 1 \\ 480 & 490 & 510 & 510 & 530 & 550 & 610 & 640 \end{pmatrix}} \underset{\mathbf{X}}{\begin{pmatrix} 1 & 480 \\ 1 & 490 \\ 1 & 510 \\ 1 & 510 \\ 1 & 530 \\ 1 & 550 \\ 1 & 610 \\ 1 & 640 \end{pmatrix}} = \underset{(\mathbf{X}^t\mathbf{X})}{\begin{pmatrix} 8 & 4320 \\ 4320 & 2{,}355{,}806 \end{pmatrix}}.$$

The determinant of $(\mathbf{X}^t\mathbf{X})$ is 184,000, and its inverse is

$$(\mathbf{X}^t\mathbf{X})^{-1} = \begin{pmatrix} 12.80326086 & -0.02347826 \\ -0.02347824 & 0.0000434782 \end{pmatrix}.$$

Finally, we need to calculate $\boldsymbol{X}^t\boldsymbol{Y}$:

$$\overset{\boldsymbol{X}^t}{\begin{pmatrix} 1 & 1 & 1 & 1 & 1 & 1 & 1 & 1 \\ 480 & 490 & 510 & 510 & 530 & 550 & 610 & 640 \end{pmatrix}} \overset{\boldsymbol{Y}}{\begin{pmatrix} 2.70 \\ 2.90 \\ 3.30 \\ 2.90 \\ 3.10 \\ 3.00 \\ 3.20 \\ 3.70 \end{pmatrix}} = \overset{\boldsymbol{X}^t\boldsymbol{Y}}{\begin{pmatrix} 24.8 \\ 13.492 \end{pmatrix}}.$$

Multiplying $(\boldsymbol{X}^t\boldsymbol{X})^{-1}$ by $\boldsymbol{X}^t\boldsymbol{Y}$ gives the $\boldsymbol{B}$-matrix:

$$\boldsymbol{B} = \overset{(\boldsymbol{X}^t\boldsymbol{X})^{-1}}{\begin{pmatrix} 12.80326086 & -0.02347826 \\ -0.02347824 & 0.0000434782 \end{pmatrix}} \overset{(\boldsymbol{X}^t\boldsymbol{Y})}{\begin{pmatrix} 24.8 \\ 13.492 \end{pmatrix}} = \overset{\boldsymbol{B}}{\begin{pmatrix} 0.75118540 \\ 0.0043475 \end{pmatrix}}.$$

The values in the $\boldsymbol{B}$ matrix are the least-squares regression coefficients we obtained previously,

$$a = 0.751 \quad \text{and} \quad b = 0.00435.$$

*12.7 NONLINEAR RELATIONSHIPS (OPTIONAL)

In many circumstances it may not be reasonable to assume a linear relationship between the dependent variable and the independent variables. This fact often becomes obvious when constructing a scatter diagram for the sample observations; for this reason such diagrams are very useful. If the relationship between two variables is not linear, then it may be possible to (1) transform the data so that it takes on a linear form, or (2) find a curvilinear function which provides a good fit to the data. (A third method is also used, which employs a number of short straight lines to approximate small portions of a curved line. This method, called *piecewise linear approximation*, will not be discussed here, but can be found in a number of intermediate texts.)

A large number of functions exist for transforming a set of observations into linear form. In some cases, the appropriate transformation can be decided upon after viewing the scatter diagram. Under these circumstances it is often helpful to plot the data on semilogarithmic graph paper, using the log scale for either variable, or on a double-logarithmic graph. If the observations on any one of these graphs follow approximately a straight line, then a logarithmic transformation can be used. For instance, if the relationship is linear when x is plotted on the log scale, then $\hat{y} = a + b \log x$ is the appropriate form; when y is plotted on the log scale, then $\log \hat{y} = a + bx$ is appropriate. When the relationship is linear on a double-logarithmic graph, the appropriate equation is $\log \hat{y} = a + b \log x$. Although a logarithmic transformation on x and/or y is a common form, many other transformations may give the desired linear relationship. To name just a few others, y may have a linear relationship with the *reciprocal* of x, $\hat{y} = a + b/x$, or with the *square* of x, $\hat{y} = a + bx^2$, or perhaps even with a *polynomial* function of x, such as $\hat{y} = a + bx + cx^2$. In this latter case, x^2 can be treated as if it were a new variable, say, $x_2 = x^2$. If we now call the original variable x_1, and let $b = b_1$ and $c = b_2$, then the parabola $\hat{y} = a + bx + cx^2$ can be written as $\hat{y} = a + b_1x_1 + b_2x_2$. The relationship between x and y in this form can be handled by the techniques of multiple linear regression that were described in Section 12.2. In fact, this method of transforming a parabola is, in general, applicable to any higher-order polynomials.

Rather than pick a suitable transformation for x and y on the basis of a scatter diagram, the researcher may often have an *a priori* rationale for selecting one transformation over another or may want to try several "models" and simply determine which one provides the best fit to the sample data (i.e., gives the lowest standard error of the estimate). For example, suppose the relationship between x and y is estimated by an exponential function of the form $\hat{y} = ae^{bx}$. Taking the logarithm to the base e of both sides yields a linear function between $\ln y$ and x:

$$\hat{y} = ae^{bx},$$
$$\ln \hat{y} = \ln(ae^{bx}) = \ln a + bx.$$

The estimates provided by a regression line based on this relationship could be compared with the estimates provided by some other fit to the data, to see which one gives the best fit. Unfortunately, transformations like these may cause problems in regression analysis, especially in the error terms.

Instead of transforming the data into a linear form, it may be possible in some circumstances to directly determine a least-squares estimate of the population parameters for a curvilinear function relating two or more variables. Consider once again the parabolic relationship described above, $\hat{y} = a + bx + cx^2$. To minimize the sum of the squared error, again let $G = \sum_{i=1}^{n} (y_i - \hat{y}_i)^2$, and minimize $G = \sum_{i=1}^{n} (y_i - a - bx_i - cx_i^2)^2$ by taking the partial derivative of G with respect to a, b, and c. The resulting normal equations are:

$$na + b\sum_{i=1}^{n} x_i + c\sum_{i=1}^{n} x_i^2 = \sum_{i=1}^{n} y_i,$$

$$a\sum_{i=1}^{n} x_i + b\sum_{i=1}^{n} x_i^2 + c\sum_{i=1}^{n} x_i^3 = \sum_{i=1}^{n} x_iy_i,$$

$$a\sum_{i=1}^{n} x_i^2 + b\sum_{i=1}^{n} x_i^3 + c\sum_{i=1}^{n} x_i^4 = \sum_{i=1}^{n} x_i^2y_i.$$

Note that these equations have the same form as the normal equations for multiple linear regression (see Eqs. (12.8)). The variable x_1 in Formula (12.8) corresponds to the variable x in the present case, x_2 corresponds to x^2, and the constants b_1 and b_2 correspond to b and c. The above three equations can be solved in the same way as in the multiple regression case to yield the least-squares regression line.

Now that we have presented the analysis for extending simple linear regression ($m = 1$) to multiple regression ($m > 1$), we need to make the comparable extension of simple linear correlation to multiple linear correlation.

Define: $E[y|x_1, x_2, \ldots, x_m]$, hypothesis of no linear regression, partial regression coefficients, normal equations.

Problems

12.1 Discuss the usefulness and value of the extension of regression analysis to include more than one explanatory factor.

12.2 Explain the difference in meaning between the simple regression coefficient in a simple regression analysis and a partial regression coefficient in a multiple regression analysis.

12.3 Describe the population regression equation and the population regression model. Why are their equations different?

12.4 a) Fit a multiple linear regression line to the following four observations:

y	x_1	x_2
7	0	2
8	5	3
3	1	1
6	2	2

b) From your answer to part (a) find the error of prediction for each of the four values of x_1 and x_2.

c) What is the value of "total variation" in this example? What are the values of "explained variation" and "unexplained variation"?

d) Find SSR and SSE, and then compute MSR and MSE. From these answers, determine the value of F for testing the null hypothesis of no linear regression.

12.5 a) Fit a multiple least-squares regression line to the following observations (let y be the dependent variable).

y	x	z
1.0	1.0	5.0
2.0	1.2	3.5
3.0	1.3	3.0
4.0	1.8	4.0
5.0	2.0	1.0
6.0	2.8	2.0

b) What value of y would you estimate if $x = 1.5$, $z = 2.5$?

c) Determine the value of s_e.

12.6 Suppose, in a multiple regression problem, the ANOVA table is as follows:

Source	SS	d.f.
Regression	36	2
Error	64	32

a) How many independent variables are there? What is the sample size?

b) Test the null hypothesis that there is no linear regression. Use $\alpha = 0.05$.

c) Calculate the standard error of the estimate.

12.7 In multiple regression analysis:

a) What measures are used to determine whether the equation fits the data well and may be useful for forecasting?

b) How can you determine which of the explanatory factors included in the model has the most significance in explaining the variation of the dependent variable $\boldsymbol{y}$?

12.8 Given the following results from a multiple regression analysis, using 53 observations (standard errors are in parentheses):

$$\begin{array}{llll}\hat{y} = 6 & + 3x_1 & + 10x_2 & - 4x_3. \\ \quad (1.5) & (2) & (4) & (0.8)\end{array}$$

a) What value of y would you predict if $x_1 = -1$, $x_2 = 3$, and $x_3 = 2$?

b) Calculate the values of the $\boldsymbol{t}$-statistic for one-sided tests of the significance of each individual estimate of the regression coefficients, and find the critical value for such tests if alpha $= 0.05$.

c) Determine which independent variable is most important and which is least important for explaining the variation in the dependent variable.

d) Suppose that $\boldsymbol{x}_1$, $\boldsymbol{x}_2$, and $\boldsymbol{x}_3$ are policy variables that can be manipulated. If x_1 and x_2 are increased by 20 units each while x_3 is increased by 50 units, what is your best estimate of the change in y?

12.9 In a multiple regression based on 40 observations, the following results are obtained:

$$\begin{array}{lllll} & \hat{y} = 10 & + 4x_1 & + 6x_2 & - 2x_3 \\ \text{with standard errors:} & & (1.2) & (5.0) & (0.4)\end{array}$$

a) Explain the meaning of the coefficient for x_2.

b) Using some test statistic, explain which of the independent variables is the most significant.

c) Explain one probable effect of dropping variable $\boldsymbol{x}_2$ from the regression model and reestimating.

d) What value of y would you predict if $x_1 = 4$, $x_2 = 1$, and $x_1 = 2$?

12.10 a) List all the assumptions underlying statistical inference based on a least-squares estimation of a multiple-regression model.

b) Indicate, by a short statement, the essential value of each assumption to the analysis.

c) Suggest one type of situation or type of data for which a particular assumption is likely to be violated.

12.11 Explain how the use of the $\boldsymbol{t}$-distribution differs from the use of the $\boldsymbol{F}$-distribution in testing hypotheses in multiple linear regression.

12.12 A stenographic pool supervisor wishes to use the intermediate sample data provided below to determine a regression equation that can predict the total typing hours, y, for report drafts. She uses as independent variables the number of words in the draft, x_1 (in tens of thousands), and an index x_2 for level of difficulty on a scale of 1 (least difficult) to 5 (most difficult).

$n = 25, \sum y = 200, \sum x_1 = 100, \sum x_2 = 75, \sum x_1 y = 1000,$
$\sum x_2 y = 800, \sum x_1^2 = 600, \sum x_2^2 = 325, \sum y^2 = 3800, \sum x_1 x_2 = 200.$

a) Calculate the coefficients a, b_1, and b_2 for the estimated regression equation.

b) Explain the meaning of the values obtained for b_1 and b_2.

c) Test for the significance of the individual coefficients, using an appropriate one-tailed test and $\alpha = 0.05$.

12.13 During the summer of 1982, 10 State University students were hired by Disney World as entertainers. Before being hired each student was given two different aptitude tests. They were then given three days of training, after which they were rated by a committee. The results are shown in Table 12.2.

Table 12.2

Test I (Score x_1)	Test II (Score x_2)	Rating (y)
74	40	91
59	41	72
83	45	95
76	43	90
69	40	82
88	47	98
71	37	80
69	36	75
61	34	74
70	37	79

a) Calculate a, b_1, and b_2 in the equation $\hat{y} = a + b_1 x_1 + b_2 x_2$.

b) How would you interpret b_1? How would your answer differ if you had estimated the equation $\hat{y} = a + b_1 x_1$?

c) Calculate SSR and SST for part (a).

d) Use your results from part (c) to test the hypothesis of no linear regression.

12.14 a) Solve Problem 12.4 for the regression line relating y and x_1 using matrix algebra.

b) Solve Problem 12.4 for the regression line relating y, x_1, and x_2 using matrix algebra.

12.15 A study investigated the relationship between the average number of hours of sleep and the length of life for ten randomly selected business executives.

Length of Life, y	**Hrs. of Sleep, x**
76.7	6.0
62.9	5.5
70.5	8.5
49.2	4.8
78.4	7.5
81.2	8.5
48.5	5.0
83.4	7.0
65.5	6.3
68.4	9.5

a) Plot the values of x and y, and then find the least-squares line $\hat{y} = a + bx$ for these data. Find s_e^2.

b) Fit the model $\hat{y} = \alpha + \beta_1 x + \beta_2 x^2$ to these data. Find s_e^2.

c) Which model provides the better fit? Explain. Does a lower s_e^2 always mean a better fit? Under what circumstances will the model in part (b) be worse than the model in part (a)?

12.16 In Section 12.7, several approaches to solving nonlinear regression problems were suggested. The problems listed below represent nonlinear problems from Chapter 10.

a) Solve Problem 10.40. b) Solve Problem 10.27.

c) Solve Problem 10.46. d) Solve Problem 10.47.

12.17 The ten observations in Table 12.3, represent the price movement of a certain common stock over a ten-year period.

Table 12.3

x (year)	**y (price)**	**x (year)**	**y (price)**
1	100	6	35
2	120	7	60
3	75	8	75
4	50	9	80
5	40	10	70

a) Sketch the relationship between $\boldsymbol{x}$ and $\boldsymbol{y}$. What type of function does this relationship seem to follow for the given ten years?

b) Use the normal equations derived in Section 12.7 to fit a parabola of the form $\hat{y} = a + bx + cx^2$ to these observations.

c) Determine the error of prediction for these ten observations.

12.18 How will your answers to Problem 12.5 change if the values of $\mathbf{x}$ are changed to $x = 1.0, 1.44, 1.69, 3.24, 4.0, 7.84$? (*Hint:* Notice that $\mathbf{x}$ and $\mathbf{y}$ are no longer linearly related.)

12.8 MULTIPLE LINEAR CORRELATION

Coefficient of Multiple Correlation

Multiple linear correlation bears the same relationship to simple linear correlation as multiple linear regression does to simple linear regression; that is, it represents an *extension of the techniques* for handling the relationship between only two variables to the set of methods for handling the relationship between more than two variables. In multiple linear correlation, the objective is to estimate the *strength* of the relationship between a variable $\mathbf{y}$ and a group of m other variables $x_1, x_2, \ldots, x_m$. The measure usually used for this purpose is an index of the strength of the relationship in the sample data, which we call the *coefficient of multiple correlation* and denote by the symbol $R_{y \cdot x_1, x_2, \ldots, x_m}$. This measure can be interpreted in a manner similar to r, as a multiple linear correlation coefficient represents the simple linear correlation coefficient between the sample values y_i and estimates of these values provided by the multiple regression equation. The value of a multiple coefficient R lies between zero and one, since it is not possible to indicate the sign of *each* of the regression coefficients which relate $\mathbf{y}$ to the variables $x_1, x_2, \ldots, x_m$ by a *single* plus or minus sign. Since $R_{y \cdot x_1, x_2, \ldots, x_m}$ is merely a generalization of the formula for r on page 525, it is defined in precisely the same manner:

Coefficient of multiple correlation:

$$R_{y \cdot x_1, x_2, \ldots, x_m} = \sqrt{\frac{\text{SSR}}{\text{SST}}}. \tag{12.13}$$

As before, the square of a correlation coefficient is much easier to interpret than the coefficient itself. The square of a coefficient of multiple correlation is called the *multiple coefficient of determination*, and is defined as follows:

Multiple coefficient of determination:

$$R^2_{y \cdot x_1, x_2, \ldots, x_m} = \frac{\text{SSR}}{\text{SST}}. \tag{12.14}$$

We see from Formula (12.14) that R^2 is interpreted exactly like r^2, in that it measures the proportion of total variation which is explained by the regression line.

To illustrate the use of Formulas (12.13) and (12.14), recall that in our GMAT–GPA example, SST = 0.66, and with x_1 and x_2 in the analyses, that

$$\text{SSE} = 0.1400 \qquad \text{and} \qquad \text{SSR} = 0.5200.$$

Given these values, we can calculate the coefficient of multiple correlation between y_i and $\hat{y}$, which is $R_{y \cdot x_1, x_2}$, as follows:

$$R_{y \cdot x_1, x_2} = \sqrt{\frac{\text{SSR}}{\text{SST}}} = \sqrt{\frac{0.5200}{0.6600}} = 0.89.$$

This result indicates that 79 percent of the variability of GPA is explained by differences in GMAT scores and differences in undergraduate grades, since the square of the coefficient of multiple correlation is $R^2_{y \cdot x_1, x_2} = 0.89^2 = 0.79$. Because 66 percent of this variability was originally explained using GMAT scores alone ($r^2 = 0.81^2 = 0.66$), the addition of undergraduate grades to the analysis explains an additional 13 percent of the variability in GPA. The fact that GMAT explains 66 percent of the variability and undergraduate grades only an additional 13 percent does not imply that GMAT scores are better estimators of GPA than undergraduate grades, but merely the fact that GMAT scores were analyzed before we even talked about using undergraduate grades. If undergraduate grades had been the first variable examined, this variable could have explained either more or less than 66 percent of the variability in $\mathbf{y}$.

It is only when two variables are independent that the additional amount of variability explained is not affected by the order in which they enter the analysis. If the two variables are correlated, then adding one of these variables to the analysis will automatically explain some of the variability in $\mathbf{y}$ which could have been explained by the other variable. Adding additional variables to a regression analysis can never *decrease* SSR; hence, R must always either increase or stay the same as more independent variables are included in the model.

The value of R measures the degree of association between the dependent variable and *all* of the variables $x_1, x_2, \ldots, x_m$ *taken together*. One may, however, be more interested in the degree of association between $\mathbf{y}$ and *one* of the variables $x_1, x_2, \ldots, x_m$ with *the effect of all the other variables removed*, or equivalently, *held constant*. The measure of strength of association in this case is called a *partial correlation coefficient*.

Partial Correlation Coefficient

Partial correlation analysis measures the strength of the relationship between $\mathbf{y}$ and one independent variable, in such a way that variations in the other independent variables are taken into account. Thus, a partial correlation coefficient is analogous to a partial regression coefficient, in that all other factors are "held constant." Simple correlation, on the other hand, ignores the effect of all other variables, even though these variables might be quite closely related to the dependent variable, or to one another.

Partial correlation measures the strength of the relationship between **y** and a *single independent variable* by considering the *relative* amount that the unexplained variance is reduced by including this variable in the regression equation. For instance, in our GMAT–GPA example, we might want to calculate the partial correlation between **y** and x_2, where the linear effect of x_1 is held constant (i.e., eliminated). This partial correlation is denoted by the symbol $r_{y, x_2 \cdot x_1}$, where the variables before the dot indicate those variables whose correlation is being measured (**y** and x_2), and the variable(s) after the dot (x_1) indicate those whose influence is being held constant.

As before, the *square* of a correlation coefficient is usually easier to interpret than the coefficient itself. In the case of a partial regression coefficient, this square is called a *partial coefficient of determination.* A partial regression coefficient of determination measures the proportion of the unexplained variation in **y** that is *additionally* explained by the variable *not* being held constant.

Using this definition, we can now interpret $r^2_{y, x_2 \cdot x_1}$ as follows:

$$r^2_{y, x_2 \cdot x_1} = \frac{\begin{pmatrix}\text{Extra variation in } \mathbf{y} \text{ explained} \\ \text{by the additional influence of } x_2\end{pmatrix}}{\text{Variation in } \mathbf{y} \text{ unexplained by } x_1 \text{ alone}}.$$

To illustrate the calculation of $r^2_{y, x_2 \cdot x_1}$ recall that in our GMAT–GPA example the total variation in **y** (GPA) is $\sum(y_i - \bar{y})^2 = 0.66$. Based on the *simple* relationship between **y** and x_1 (GMAT), the amount of variation in **y** unexplained was SSE $= 0.225$. When x_2 was added to the analysis, SSE decreased to 0.1400. This means that the extra variation explained by x_2 is $0.225 - 0.1400 = 0.085$. These values are shown in Fig. 12.2.

Fig. 12.2 The elements of variation in determining $r^2_{y, x_2 \cdot x_1}$.

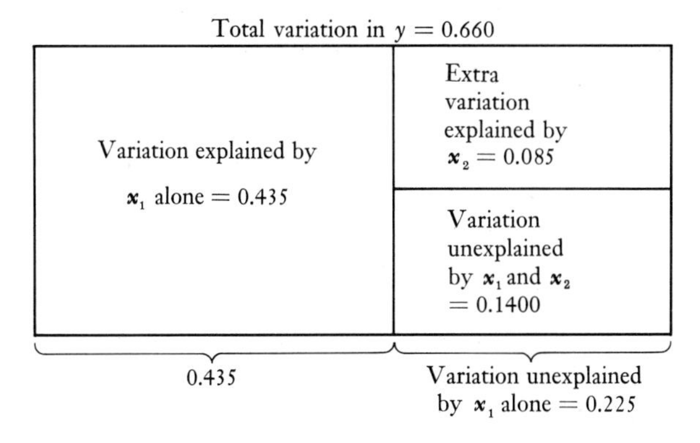

Using the values in Fig. 12.2, we can now calculate $r^2_{y, x_2 \cdot x_1}$:

$$r^2_{y, x_2 \cdot x_1} = \frac{\text{Extra explained variation with } x_2}{\text{Unexplained variation with } x_1 \text{ alone}}$$

$$= \frac{0.085}{0.225} = 0.378.$$

In other words, 37.8% of the unexplained variability in $\mathbf{y}$ (after considering x_1) is now explained by the addition of x_2 (undergraduate grades). The square root of this value gives the partial correlation coefficient $r_{y, x_2 \cdot x_1} = \sqrt{0.378} = 0.614$ between GMAT and undergraduate grades, holding GMAT scores constant.*

The value $r_{y, x_2 \cdot x_1}$ represents the partial correlation coefficient for x_2, holding x_1 constant. We could just as easily calculate the partial correlation coefficient for x_1, holding x_2 constant, if the value of SSE (x_2) is known. If we were interested in the effect of adding a third variable, x_3, to the analysis, holding x_1 and x_2 constant, the appropriate partial correlation coefficient would be:

$$r_{y, x_3 \cdot x_1, x_2} = \frac{\text{Extra variation in } \mathbf{y} \text{ explained by additional influence of } x_3}{\text{Variation in } \mathbf{y} \text{ unexplained by } x_1 \text{ and } x_2 \text{ alone}}$$

Notice the similarities in the definition of a partial correlation coefficient and the definition of a multiple correlation coefficient (in Formula (12.13) on page 559); i.e. a partial correlation coefficient measures reductions in the *unexplained* variation of $\mathbf{y}$, while a multiple correlation coefficient measures reductions in the *total* variation of $\mathbf{y}$.

12.9 MULTICOLLINEARITY

We return now to a consideration of special problems that arise when one of the conditions specified in Section 12.3 is violated. This section deals with the violation or near violation of Condition 1, which specifies that none of the independent variables can be an exact linear combination of the other independent variables. If the independent variables, $\mathbf{x}_1, \mathbf{x}_2, \ldots, \mathbf{x}_m$ are perfectly linearly related to each other, then they are linearly *dependent*. In this case, no estimates of the partial regression coefficients can be obtained, since the normal equations will not be solvable; that is, the method of least squares breaks down and no estimates can be calculated. Perfect dependence seldom

* In this case we would have to know, in advance, that the sign of $r_{y, x_2 \cdot x_1}$ is positive rather than negative. Or, we could define $r_{y, x_2 \cdot x_1}$ as the ordinary correlation of y^* and x_2^*, where the * symbol represents residuals from linear regression on x_1.

occurs in practice, because most investigators are careful not to include in the regression model two or more explanatory variables that represent the same influence on the dependent variable y. Indeed, even if an investigator did accidentally include two or more such variables, it is unlikely that the *sample* observations representing measures of these variables would be perfectly related, because some slight errors of measurement and sampling are almost inevitable.

Sometimes, however, special problems do occur, when two or more of the independent variables are strongly (but not perfectly) related to one another. This situation is known as *multicollinearity*. When multicollinearity occurs it is possible to calculate least-squares estimates, but difficulty arises in the interpretation of the strength of the effect of each variable.

Detection of a Multicollinearity Problem

From the above discussion we see that a high correlation between any pair of explanatory variables x_i and x_j may be used to help identify multicollinearity. It is possible, however, for all independent variables to have relatively small *mutual* correlations and yet to have some multicollinearity among three or more of them. Sometimes it is possible to detect these higher-order associations by using a multiple correlation coefficient that deals only with the explanatory variables. Suppose that we use the symbol R_j to denote the multiple correlation coefficient of variable x_j with all the other $(m - 1)$ independent variables, $x_1, x_2, \ldots, x_{j-1}, x_{j+1}, \ldots, x_m$. Such a measure could be determined for each of the independent variables. Generally, if one or more of these values, $R_1, R_2, \ldots, R_j, \ldots, R_m$, is approximately the same size as the multiple correlation coefficient $R_{y \cdot x_1 \cdots x_m}$, then multicollinearity is a problem. In other words, if the strength of the association among any of the independent variables is approximately as great as the strength of their combined linear association with the dependent variable, then the amount of overlapping influence may be substantial enough to make the interpretation of the separate influences difficult and imprecise.

To illustrate, we present a model with four independent variables,

$$\mathbf{y} = \alpha + \beta_1 \mathbf{x}_1 + \beta_2 \mathbf{x}_2 + \beta_3 \mathbf{x}_3 + \beta_4 \mathbf{x}_4 + \boldsymbol{\varepsilon}.$$

The multiple correlation coefficient for this model is

$$R_{y \cdot x_1 x_2 x_3 x_4} = 0.90.$$

To check for multicollinearity, one would first calculate the six simple correlations between pairs of independent variables

$$r_{x_1 x_2}, \quad r_{x_1 x_3}, \quad r_{x_1 x_4}, \quad r_{x_2 x_3}, \quad r_{x_2 x_4}, \quad r_{x_3 x_4}.$$

If one of these is close to unity, then imprecise estimation will result. The next step would be to calculate the multiple correlation coefficients of each independent variable with the other three: that is,

$$R_{x_1 \cdot x_2x_3x_4}, \qquad R_{x_2 \cdot x_1x_3x_4}, \qquad R_{x_3 \cdot x_1x_2x_4}, \qquad \text{and} \qquad R_{x_4 \cdot x_1x_2x_3}.$$

If any of these are as large as

$$R_{y \cdot x_1x_2x_3x_4} = 0.90,$$

then the problem of multicollinearity may be substantial. There is really no statistical method for testing whether these values indicate high multicollinearity or not, since this is not a problem of statistical inference about the population, but merely a property of the sample observations.

Effects of Multicollinearity

When multicollinearity occurs, the least-squares estimates are still unbiased and efficient. The problem is that the estimated standard error of the coefficient (say, s_{b_i} for the coefficient b_i) tends to be inflated. This standard error tends to be larger than it would be in the absence of multicollinearity because the estimates are very sensitive to any changes in the sample observations or in the model specification. In other words, including or excluding a particular variable or certain observations may greatly change the estimated partial coefficient. When s_{b_i} is larger than it should be, then the t-value for testing the significance of β_i is smaller than it should be. Thus, one is likely to conclude that a variable $\mathbf{x}_i$ is not important in the relationship when it really is.

If one is interested primarily in the forecasts of y_i, or $\mu_{y \cdot x_1, x_2, \ldots, x_m}$, rather than in the significance of the separate coefficients $b_1, b_2, \ldots, b_m$, then multicollinearity may not be a problem. Suppose that the combined fit for the regression equation is very good. If the observed linear relationships among all the independent variables can be expected to remain true for some new observations, then the regression model should also give a close fit for the new sample values even if multicollinearity is present.

Correction of Multicollinearity

When multicollinearity in a regression model is severe and more precise estimates of the coefficients are desired, one common procedure is to select the independent variable *most seriously involved* in the multicollinearity and remove it from the model. The difficulty with this approach is that the model now may not correctly represent the population relationship, and all estimated coefficients would contain a *specification bias*. It would be better to try to replace the multicollinear variable with another that is less collinear but may still measure the same theoretical construct. For example, if the theoretical variable "business expectations" is measured by a stock price index that is highly collinear with retained earnings, then it may be possible to replace the stock index with some other measure, perhaps an index of business expectations obtained by surveying executives in the 500 largest corporations. In this way, the multicollinearity may be reduced while the theoretical base for the model is still retained.

12.10 VIOLATION OF ASSUMPTION 4 OR 5

Recall that assumption 5 states that each ε_i has the same variance ($V[\varepsilon_i] = \sigma_\varepsilon^2$), and assumption 4 states that the covariance between any two disturbance variables ε_k and ε_j is zero,

$$C[\varepsilon_k, \varepsilon_j] = 0.$$

We mentioned that these two assumptions are crucial in obtaining simple least-squares estimates of the regression coefficients that are *efficient*. This means that these estimators have a smaller variance than any other linear unbiased estimator that might be devised. If one or both of these assumptions is violated, then the estimator calculated by the method of least squares would not have the smallest variance; some *other* estimator that uses more information would be the efficient one. This loss in efficiency occurs whenever either of two problems is encountered: *heteroscedasticity* or *autocorrelation*. These terms are defined below.

Heteroscedasticity occurs when assumption 5 *is violated.*
It means that the variance of the disturbances ε_i is not constant, but changing.

Autocorrelation occurs when assumption 4 *is violated.*
It means that there is a correlation between the error terms.

The effect of either of these problems is a least-squares estimate of the regression coefficient for which the standard error of the coefficient is not minimized. Thus, tests of hypotheses or confidence intervals based on this property will not be correct.

Detection of Heteroscedasticity

To detect a situation where the variance of the errors is not constant, it is often useful to plot each $\hat{y}$-value against its corresponding residual $(y_i - \hat{y})$. For example, the V-shaped slope of the boundary lines for the scatter of points in Fig. 12.3(a) suggests an increasing variance of the residuals as the value $\hat{y}$ increases. Such a plot may indicate that the fit of the model is not uniform and that the disturbances may not have a constant variance. A changing variance could also be indicated if the boundary lines approximated an inverted V or if they were close together at some points and wider apart at others, as, for

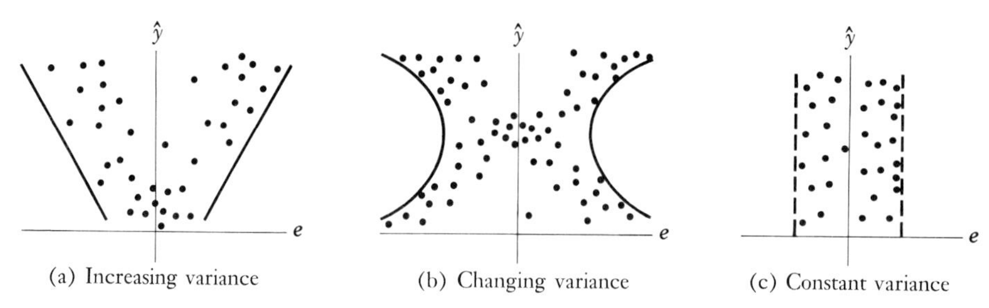

Fig. 12.3 Plotting residuals against $\hat{y}$ to detect heteroscedasticity.

example, in Fig. 12.3(b). Assumption 5 of constant variance does *not* seem to be violated if the boundary lines are approximately parallel, as in Fig. 12.3(c). If the variance of ε is suspected to be related to the size of a particular independent variable, such as x_j = time, then we may plot e_i against the specific x-variable rather than against $\hat{y}$.

Frequently, the assumption of constant variance is not seriously violated when using economic or business data *measured over time* unless some significant structural change occurred to affect the observations, such as a new law, a war, a revolution, or some natural disaster. More often, the problem of heteroscedasticity arises when cross-sectional data, *at a given point in time*, is used, such as employment or production data across firms, or tax and revenue data across states. In these cases, the disturbances may not have constant variances because of differing factors related to the size or the legal code of the different cross-sectional entities. For example, large corporations have different structures and operate under different tax laws than do small business firms. Thus, one would expect a specified model to represent one of these types better than the other. The variance of disturbances for the one type that it fits best will be smaller than the variance of disturbances for observations of the other type.

Detection of Serial Correlation or Autocorrelation*

Violations of assumption 4 tend to occur most frequently when the observations for the variables are taken at periodic intervals *over time*. If some underlying factors not specified in the model exert an influence on the fit of the model over several time periods, then the disturbances tend to be correlated to each other. Consider a change in corporate tax laws that might affect both the amount of investment (due to investment tax credit or depreciation write-offs) and the amount of retained earnings (due to taxes on profits). This legal factor may not

* The terms *serial correlation* and *autocorrelation* often are used to refer to the same phenomenon, namely, the correlation of a variable with a lagged version of itself. Correlation of residuals from cross-sectional data, when it occurs, is called *spatial correlation*.

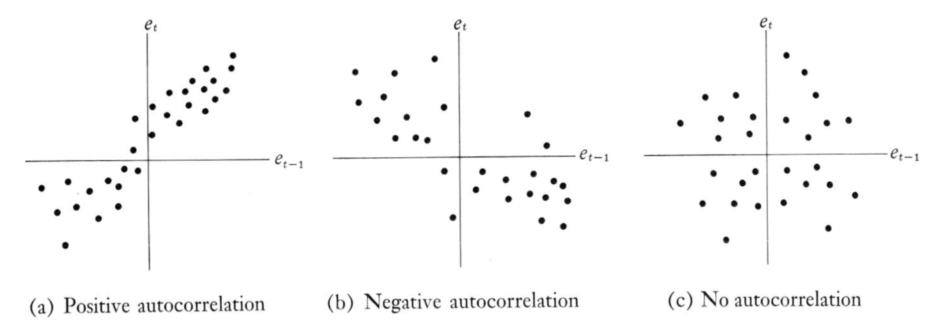

Fig. 12.4 Plots of successive residuals.

be represented in the model, but its effect may be seen in the errors of the regression equation. For example, the average relationship estimated may give values too high before the tax-law change and too low afterwards. The residuals would then tend to be all negative for observations taken before the tax change and all positive for observations made after the change. Thus, the residuals would not be occurring at random but would systematically be related to each other, i.e., autocorrelated.

The most frequently considered form of autocorrelation is the linear association of successive residuals. If we denote a residual at time period t by e_t and the previous residual by e_{t-1}, then first-order autocorrelation refers to the simple linear correlation of e_t with e_{t-1} over the entire set of observations, $t = 2, 3, 4, \ldots, n$. A measure of this correlation is given by the correlation coefficient between these two variables, $r_{e_t e_{t-1}}$. A geometric representation can be obtained by making a scatter diagram of the points corresponding to each pair (e_t, e_{t-1}). Figure 12.4 illustrates three cases: positive autocorrelation (a); negative autocorrelation (b); and no autocorrelation (c). When the points are predominantly in the positive quadrants (a), this means that successive residuals tend to have the same sign. If most points lie in the negative quadrants (b), then successive residuals tend to have opposite signs. If the scatter of points is spread over all quadrants (c), successive residuals tend to be independent, or not correlated.

Although autocorrelation clearly exists in both Fig. 12.4(a) and (b), it is generally difficult to determine whether autocorrelation is present merely by using a scatter diagram. Hence, we need a test statistic to determine whether or not to accept the null hypothesis of independence (no autocorrelation) among successive error terms. The test used most often for this purpose is called the Durbin–Watson test.

The Durbin–Watson Test

The Durbin–Watson (D–W) test is designed to test the null hypothesis that there is no first-order autocorrelation among the error terms. The alternative hypothesis is that first-order autocorrelation does exist. To be more precise, let us suppose that the relationship between the population errors ε_t and ε_{t-1} can be expressed as follows:

$$\varepsilon_t = \rho\varepsilon_{t-1} + v_t,$$

where ρ is the population autocorrelation coefficient between ε_t and ε_{t-1}. The v_t are the error terms and they satisfy assumptions 1–5 as applied to this model. When $\rho = 0$, no autocorrelation exists; the further ρ is away from zero toward $+1.0$ or -1.0, the greater the autocorrelation.

The test statistic for autocorrelation is the value of d determined by the following equation:

Durbin–Watson statistic

$$d = \frac{\sum_{t=2}^{n} (e_t - e_{t-1})^2}{\sum_{t=1}^{n} e_t^2}. \tag{12.15}$$

The value of $\boldsymbol{d}$ will always fall between zero and four—i.e., $0 \leq d \leq 4$. If no autocorrelation exists, the value of $\boldsymbol{d}$ is expected to be the number two. When positive autocorrelation is present, the difference between e_t and e_{t-1} tends to be relatively small [see Fig. 12.4(a)], and d tends toward zero. With negative autocorrelation [see Fig. 12.4(b)], the difference between e_t and e_{t-1} tends to be relatively large and d tends toward the value 4.0. Thus, the null hypothesis of no autocorrelation will be rejected for relatively low values of $\boldsymbol{d}$, which indicate positive autocorrelation, as well as for relatively large values of $\boldsymbol{d}$, which indicate negative autocorrelation.

The test statistic $\boldsymbol{d}$ was developed in 1950 by two Englishmen, a statistician and a social scientist, for use in tests of hypotheses about the existence of first-order autocorrelation. The contribution of Durbin and Watson was to determinc thc distribution of the test statistic $\boldsymbol{d}$ and to develop a table of critical values for testing the null hypothesis, $H_0: \rho = 0$: no autocorrelation exists. The form of this table (Table X in Appendix B) includes two critical values, a d-upper (d_U) and a d-lower (d_L) for specified sample sizes n, for the number of independent variables in the regression model, $m = 1, 2, 3, 4$, or 5, and for two significance levels, $\alpha = 0.01$ or 0.05. The testing process that they proposed is slightly different from the ones we have used thus far, since the $\boldsymbol{d}$-statistic leads to one of *four* conclusions, instead of the usual two. The possible conclusions are: no autocorrelation, positive autocorrelation, negative autocorrelation, or "don't know." The reason for this inconclusive region in the test result is that the

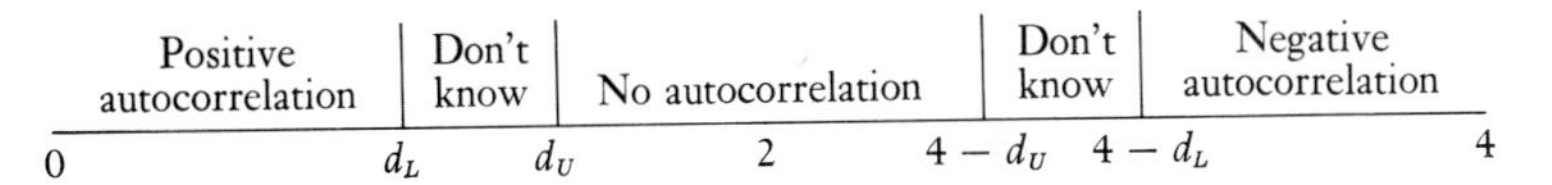

Fig. 12.5 Alternate conclusions based on the Durbin–Watson ***d***-statistic.

distribution of the test statistic ***d*** depends to some extent on the particular characteristics of the interrelationships among the independent variables in each particular problem, and no generalizations of these characteristics can be found that unambiguously restrict the ***d***-distribution. Figure 12.5 illustrates the possible values of the ***d***-statistic and the conclusions associated with them.

The following examples will show the reader how to use Table X and Fig. 12.5 to detect the problem of autocorrelation.

Example 1 Suppose that we suspect the presence of positive autocorrelation in a particular regression problem involving $m = 3$ independent variables, $n = 45$ observations, and a calculated d-value, from Formula (12.15), of $d_c = 1.31$. Using the portion of Table X for $\alpha = 0.05$, we find $d_L = 1.38$ and $d_U = 1.67$. Since our computed value d_c is below d_L, we conclude that positive autocorrelation is a problem. We have in our model a significant violation of assumption 4.

Example 2 Suppose that we suspect the presence of negative autocorrelation in a particular regression problem where $m = 4$ independent variables, $n = 70$, $d_c = 2.94$, and $\alpha = 0.01$. Using the portion of Table X for $\alpha = 0.01$, we find $d_L = 1.34$ and $d_U = 1.58$. Since we are testing for negative autocorrelation and Table X only gives values for the test on positive autocorrelation, we must transform the d-statistic, using the property of symmetry. We find that the transformed critical values are $4 - d_L = 2.66$ and $4 - d_U = 2.42$. Since our computed value $d_c = 2.94$ lies above $4 - d_L$, we conclude that a significant problem of negative autocorrelation exists.

Most commonly, one does not know, before calculating d from Formula (12.15) or before looking at a plot of successive residuals as in Figure 12.4, whether to be suspicious of positive or negative autocorrelation (although positive correlation occurs much more frequently in business and economic applications of regression analysis). In this case, one really is using a two-sided test and should realize that *the proper significance level is found by doubling the values of* 0.05 *or* 0.01 *given in Table X*.

Finally, recent research on the distribution of ***d*** and on alternate new statistics for detecting autocorrelation indicate that the inconclusive region in the D–W test may usually be reduced in the direction of d_U. That is, a single critical value separating the no-autocorrelation region from the positive-autocorrelation region seems closer to d_U than to d_L (and $4 - d_U$ is the better value in the negative autocorrelation test). Therefore, current users of the ***d***-statistic would use the d_U- and d_L-values as indexes of the severity of the problem, replacing the original conclusions in Fig. 12.5 as follows:

$d < d_L$:	serious problem of positive autocorrelation that requires correction.
$d_L < d < d_U$:	weaker problem of positive autocorrelation for which a correction is probably worthwhile.
$d_U < d < 4 - d_U$:	no problem of autocorrelation worth correcting.
$4 - d_U < d < 4 - d_L$:	weaker problem of negative autocorrelation for which correction is probably worthwhile.
$4 - d_L < d$:	serious problem of negative autocorrelation that requires correction.

Example 3 Suppose that we wish to test for autocorrelation in a particular regression problem with $m = 2$ independent variables, $n = 100$, and $d_c = 1.60$. Using the table for $\alpha = 0.01$, we find $d_c > d_U = 1.58$. Thus, we accept the condition of no autocorrelation (along with the possibility of some unknown Type II error—that we are accepting this when it is false). Using the table for $\alpha = 0.05$, we find that $d_c < d_L = 1.63$. In this case, we would conclude that positive autocorrelation is a problem (accepting a potential Type I error of 0.10, since we must double the α-value in a two-sided test for which we did not know *a priori* whether the computed d_c would lie above or below the value 2.0). In this example the test conclusion is critically affected by the choice of the significance level $\alpha = 0.01$ or 0.05.

Correction of Heteroscedasticity or Autocorrelation

As we indicated previously, when either assumption 4 or assumption 5 is violated, the ordinary least-squares (OLS) estimators that we have described are not efficient. Another estimating procedure exists which also gives linear unbiased estimates and in which the variance of the estimators is smaller than that of OLS estimators. This procedure corrects for violations of assumption 4 or 5 by using a more complete estimating procedure, called *generalized least squares* (GLS), which explicitly uses information about the variances and co-variances of the error terms ε_i in the calculation. This information is usually determined from an analysis of the residuals e_i obtained from a first OLS estimation of the regression model. The purpose in using this information is to generate a new model situation in which the new error terms are free of the violations of assumptions 4 and 5.

For example, if the problem is a first-order positive autocorrelation of the errors, then a method is needed whereby the data or the model can be transformed so that the revised error terms are free of autocorrelation. If the problem is heteroscedasticity, then this information about the way the variance increases or decreases (as depicted in Fig. 12.3(a) or (b)) should be used to weight the original observations before determining the least-squares estimates. In general, those observations for which the variance of the errors is smallest should be the more reliable, and therefore should be weighted more heavily. Since these methods are more complex, they will not be detailed here.*

12.11 DUMMY VARIABLES IN REGRESSION ANALYSIS

Thus far, the variables we have used in regression problems have been "quantitative variables," which means that they represent variables that are either measured or counted. In some types of problems it is desirable to use another type of variable called a *qualitative variable*, which merely indicates whether or not an object belongs to a particular category or possesses a particular quality. For example, in a regression analysis in which the dependent variable is the consumption expenditures of families in the United States, one may be interested not only in relating $\boldsymbol{y}$ to family income ($\boldsymbol{x}_1$) but also to whether or not the family lives in an urban or rural community ($\boldsymbol{x}_2$). The variable $\boldsymbol{x}_1$ is a quantitative variable (it measures income) while the variable $\boldsymbol{x}_2$ is a qualitative variable (it indicates whether or not the family is classified as rural or urban). A variable such as $\boldsymbol{x}_2$ often is called a "dummy variable." To construct such a variable, we might let $x_2 = 0$ if the family lives in a rural community, and let $x_2 = 1$ if they live in an urban community.

The introduction of a dummy variable does not change the multiple regression process described thus far. That is, all computations are made in the same way as for a regression analysis involving only quantitative variables. One characteristic of the addition of a dummy variable, $\boldsymbol{x}_2$, is that we know that its value in the regression equation is either zero or one. Hence, we can write two regression equations, one using $x_2 = 0$, and the other using $x_2 = 1$. To illustrate this, we present the following least-squares regression line where $x_2 =$ dummy variable:

$$\hat{y} = a + b_1x_1 + b_2x_2.$$

If $x_2 = 1$, then substituting $x_2 = 1$ into this equation yields

$$E[\boldsymbol{y}|x_2 = 1] = a + b_1x_1 + b_2(1) = (a + b_2) + b_1x_1.$$

Since $x_2 = 1$ indicates an urban family, this equation represents the regression line for urban families. If we substitute $x_2 = 0$ in the original regression line, we get the regression line for rural families.

$$E[\boldsymbol{y}|x_2 = 0] = a + b_1x_1 + b_2(0) = a + b_1x_1.$$

* See James L. Murphy, *Introductory Econometrics*, Chapter 13, for a more complete discussion of the methods of regaining efficient estimators in these cases.

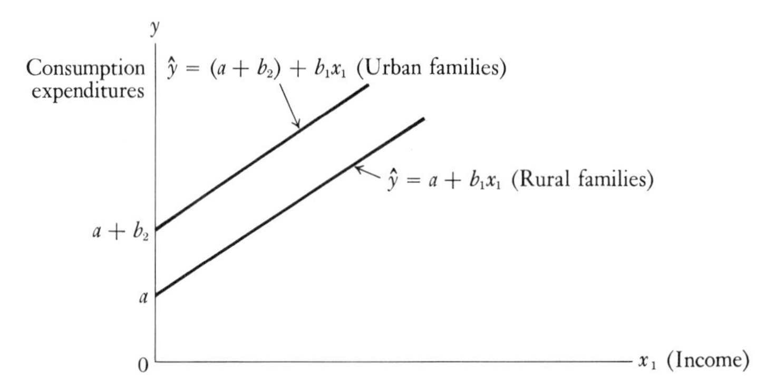

Fig. 12.6 Two regression lines resulting from the use of a dummy variable.

Thus, we have derived two regression lines from the original regression model, as shown in Fig. 12.6.

In this example the distinction between rural and urban families shifts the regression intercept from a to $(a + b_2)$. (Our graph shows a positive value for b_2.) Note that the slope of both straight lines is the same (b_1).

If one wishes to allow for a different slope in the relation between consumption and income for rural versus urban families, the specification of the model that includes the dummy variable would be different. We would then specify the regression model as

$$\hat{y} = a + (b + cx_2)x_1,$$

where x_2 = dummy variable. Then, if $x_2 = 1$ for an urban family, the line is

$$E[y|x_2 = 1] = a + (b + c)x_1 \qquad \text{with a slope of } (b + c).$$

When $x_2 = 0$ for a rural family, the line is

$$E[y|x_2 = 0] = a + bx_1 \qquad \text{with a slope of } b.$$

A positive value of c would indicate that urban families have a higher propensity to consume out of extra income received than do rural families, who would

Fig. 12.7 Two regression lines resulting from the use of a dummy slope variable.

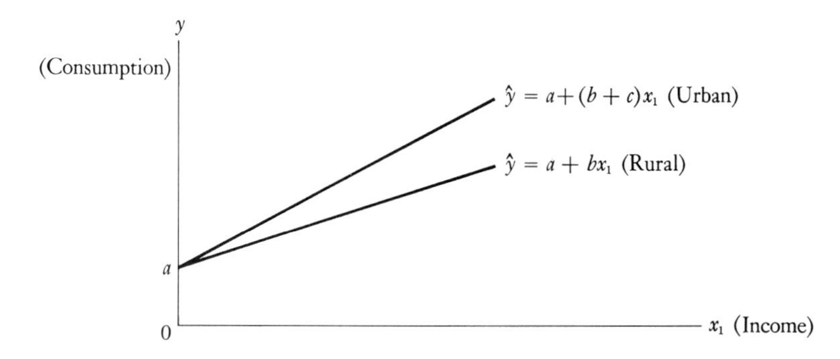

have a higher propensity to save extra income. A negative value of c would give evidence of the converse. In both cases, the intercept is the same value a. Figure 12.7 illustrates the situation of a dummy slope variable. The reader should recognize that dummy variables may be used to allow for both different intercepts and different slopes in the same model.

An Example Using a Single Dummy Variable

To illustrate the use of dummy variables, we will present two examples: one involving a single dummy variable, and a second example involving three dummy variables. In each example dummy variables will be used for intercepts, since the algebra is simpler. Dummy variables used for different slopes would be similar. The data for these examples are presented in Table 12.4.

Table 12.4 presents quarterly data, measured at an annual rate, on retail sales (in millions of dollars) as the dependent variable (y) and gross national product (in billions of dollars) between 1972 and mid 1975 as x_1. Now, suppose that we want to investigate the effect on retail sales of a huge oil price increase initiated by the OPEC countries in October, 1973. To do this, we introduce the dummy variable x_2, and let $x_2 = 1$ for any period after the price

Table 12.4

	Quarter	Retail Sales, y	GNP, x_1	Price Increase Variable, x_2	Quarter Variable x_3	x_4	x_5
1972	1	978	1112.5	0	1	0	0
	2	1123	1143.0	0	0	1	0
	3	1125	1169.3	0	0	0	1
	4	1260	1204.7	0	0	0	0
1973	1	1121	1248.9	0	1	0	0
	2	1275	1277.9	0	0	1	0
	3	1257	1308.9	0	0	0	1
	4	1381	1344.0	1	0	0	0
1974	1	1172	1358.8	1	1	0	0
	2	1368	1383.8	1	0	1	0
	3	1382	1416.3	1	0	0	1
	4	1454	1430.9	1	0	0	0
1975	1	1260	1416.6	1	1	0	0
	2	1462	1440.9	1	0	1	0

Source: *Survey of Current Business.*

increase, and $x_2 = 0$ before the price increase. A least-squares regression using these variables yields (the standard errors are in parentheses):

$$\hat{y} = -242.11 + \underset{(0.4122)}{1.2346x_1} - \underset{(89.61)}{42.456x_2}.$$

By substituting $x_2 = 0$ into this equation, we get the regression equation for the period before the price increase,

$$\hat{y} = -242.11 + 1.2346x_1 - 42.456(0)$$
$$= -242.11 + 1.2346x_1.$$

Substituting $x_2 = 1$ yields the equation for the period after the price increase.

$$\hat{y} = -242.11 + 1.2346x_1 - 42.456(1)$$
$$= -284.566 + 1.2346x_1.$$

The short-run impact of the oil price increase, as represented by these data, was to dampen retail sales by about 42 million dollars per year. Perhaps a shift in sales from large expensive cars to cheaper gas-saving cars accounts for much of this.

As is the case with all regression variables, we are interested in whether or not the coefficient of a dummy variable differs significantly from zero. In the present example, we might have hypothesized that the sign of the coefficient b_2 will be negative due to a shift from "gas-guzzlers" to "gas-sippers"; i.e., we might be testing

$$H_0: \beta_2 = 0 \qquad \text{versus} \qquad H_a: \beta < 0.$$

The calculated value t_c for these hypotheses is

$$t_c = \frac{b_2}{s_{b_2}} = -\frac{42.456}{89.61} = -0.474.$$

Since the value $t_c = -0.474$ is not in the critical region for $(n - m - 1) = (14 - 2 - 1) = 11$ d.f. for any reasonable α, we must accept the null hypothesis $H_0: \beta_2 = 0$. Therefore, our conclusion must be that the price increase did not significantly affect retail sales. The size of the apparent change in sales is not uncommon even during periods of no change in oil prices.

An Example Using Multiple Dummy Variables

Again, we will use Table 12.4 to illustrate how two or more dummy variables can be used simultaneously. Suppose that we decide that there may be consistent differences in retail sales related to the quarters of the year (winter, spring, summer, fall). To include the quarter of the year as a variable in the problem we must add *three* dummy variables to the analysis, as shown in the final three columns of Table 12.4. If $x_3 = 1$ (and $x_4 = x_5 = 0$) this indicates the first (or winter) quarter; when $x_4 = 1$ (and $x_3 = x_5 = 0$), this indicates the second

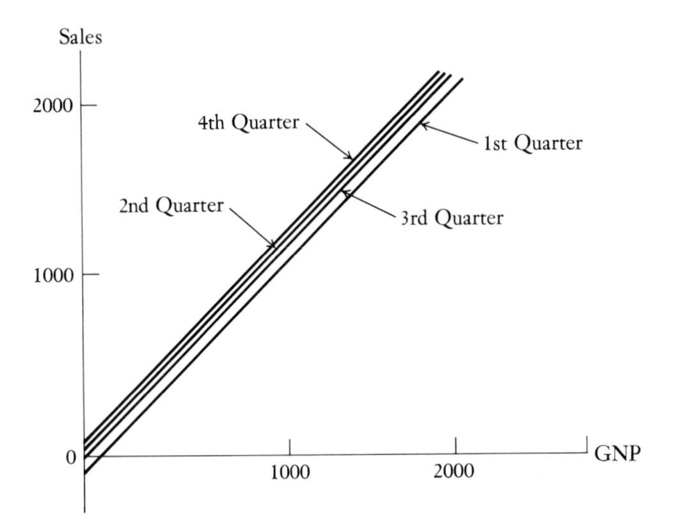

Figure 12.8

(or spring) quarter; for the third (or summer) quarter, $x_5 = 1$ (and $x_3 = x_4 = 0$). Note that we only need *three* dummy variables to indicate *four* quarters, for the fourth quarter (fall) is designated by the absence of a 1 in the x_3, x_4, or x_5 columns (i.e., $x_3 = x_4 = x_5 = 0$). In general, if the model includes an intercept term a, then the number of dummy variables needed to specify h different categories or levels is $h - 1$. Thus, to indicate four different quarters, three dummy variables are required. Also, to indicate two different categories (such as rural versus urban, or before OPEC price change versus after), only one dummy variable is needed.

The least-squares regression equation for the data in Table 12.4 (omitting variable $\mathbf{x}_2$) is given below (with standard errors in parentheses).

$$\begin{aligned}\hat{y} = 76.023 + &\underset{(0.0503)}{0.972x_1} - \underset{(15.614)}{191.115x_3} - \underset{(15.487)}{43.295x_4} - \underset{(16.598)}{83.770x_5}.\end{aligned}$$

When the appropriate values of x_3, x_4, and x_5 are substituted, the above equation yields the four different regression lines shown in Fig. 12.8. For example, the equation for the first quarter is derived by setting $x_3 = 1$, $x_4 = 0$, and $x_5 = 0$:

$$\begin{aligned}\hat{y} &= 76.023 + 0.972x_1 - 191.115(1) - 43.295(0) - 83.770(0)\\ &= -115.092 + 0.972x_1.\end{aligned}$$

The values of $\hat{y}$ for the other three quarters are:

$$\begin{aligned}\hat{y} &= 32.728 + 0.972x_1 \text{ (second quarter)},\\ \hat{y} &= -6.747 + 0.972x_1 \text{ (third quarter)},\\ \hat{y} &= 76.023 + 0.972x_1 \text{ (fourth quarter)}.\end{aligned}$$

Testing the Joint Effect of a Subgroup of Variables

From Fig. 12.8 we see that the regression lines do appear to differ from quarter to quarter. At this point we could use a ***t***-test to determine whether each of the coefficients b_3, b_4, and b_5 is significantly different from zero. Perhaps a more meaningful test, however, would be to test all the explanatory contribution of all three dummy variables simultaneously, as follows:

H_0: the joint effect of the dummy variables is not significant

versus

H_a: the dummy variables as a group do contribute to the explanation of the variation in y.

A test statistic useful in this circumstance is derived by comparing the amount of unexplained variation (SSE) when all the variables are in the equation (GNP and the three dummy variables) with the amount of unexplained variation when the dummy variables are not included, labeled as SSE_S for the sum of squares of the errors in the "*s*horter" or "*s*maller" form of the model (including only GNP). If there are j dummy variables, the following ***F***-test is appropriate.

$$\boldsymbol{F}_{(j,\, n-m-1)} = \frac{(\text{SSE}_S - \text{SSE})/j}{\text{SSE}/(n - m - 1)}. \qquad (12.16)$$

For our retail data, $\text{SSE}_S = 7595.58$, and $\text{SSE} = 369.144$, $j = 3$, and $(n - m - 1) = (14 - 4 - 1) = 9$. Substituting these values into (12.16) yields

$$F_c = \frac{[7595.58 - 369.144]/3}{369.144/9} = \frac{7408.8}{41.016} = 58.729.$$

Since this value of F_c exceeds the critical value shown in Table VIII(c) of $F(3, 9) = 6.99$, we can reject the null hypothesis that seasonal factors do not affect retail sales, and conclude that the quarterly dummy variables are useful in explaining variation in retail sales.

As a final note, we add that Formula (12.16) is useful in any situation where one is considering the inclusion of one or more additional variables (not necessarily dummy variables) in a multiple regression analysis. One way to do this is to fit a regression equation twice to the same data, once with all variables included in the analysis, and once excluding all variables with coefficients that are hypothesized to equal zero. The results of the original run will yield SSE, while the second run will yield SSE_S. Then, Formula (12.16) can be used with j = number of independent variables in the subgroup excluded, and m = total number of independent variables.

12.12 MULTIPLE LINEAR REGRESSION: AN EXAMPLE

Suppose that a leading securities firm, A. Victor and Sons, has decided to compete in the sealed bidding on a new $10 million series of first-mortgage bonds. These bonds, which must be purchased as a single unit, are to be awarded to the highest bidder. Bids are to be in terms of a percent of the par value of the issue. Moody's Investors Service rates the bonds as Aa, the second highest rating.

In the process of determining a bid price, the analysts at Victor learn that there will be only a single competitor bidding against them for the bonds, the Martin Company. Furthermore, they find in their files a complete record of the twenty previous occasions in which Victor and Martin were both bidding

Table 12.5

Martin's Bid (% of Par) y	Par Value (Millions) x_1	Number of Bidders* x_2	Moody Rating x_3
97.682	13.0	2	A
98.424	6.0	5	A
101.435	9.0	5	Aa
102.266	5.5	7	Aaa
97.067	7.0	3	Baa
97.397	9.5	2	Ba
99.481	17.0	2	Aa
99.613	12.5	5	A
96.901	13.5	2	Ba
100.152	12.5	3	Aaa
98.797	13.0	4	Baa
100.796	7.5	6	Aa
98.750	7.5	2	Aa
97.991	12.0	3	Ba
100.007	14.0	4	Aaa
98.615	11.5	6	Ba
100.225	15.0	2	Aa
98.388	8.5	6	Baa
98.937	14.5	7	A
100.617	9.5	5	Aaa

* Including Victor and Martin.

on the same issue. These records show Martin's bid for the issue, the par value of the issue, the number of bidders participating, and the Moody rating. After examining this information, shown in Table 12.5 on page 577, the analysts at Victor decide to use linear regression analysis in an attempt to better assess what Martin might bid for the current issue.

Before calculating a linear regression equation for the data in Table 12.5, we first plot the relationship between the dependent variable (Martin's bid) and each of the three independent variables (par value, number of bidders, and the Moody rating) in order to visually check whether or not the relationship in these three cases is approximately linear. Before doing so it is necessary to attach a numerical value to the Moody ratings. Suppose we give the highest rating, Aaa, a value of 1, the second highest rating, Aa, a value of 2, A a value of 3, Baa a value of 4, and the lowest rating, Ba, a value of 5. The scatter diagrams in Fig. 12.9 show the relationship between Martin's bid price and each of the three independent variables. Note that a linear approximation is not unreasonable for the first two relationships, but that Martin's bid and the Moody rating follow a curvilinear pattern rather than a linear one. Although it may not be apparent from the scatter diagram, it is possible to show (by some pro-

Fig. 12.9 Scatter diagrams.

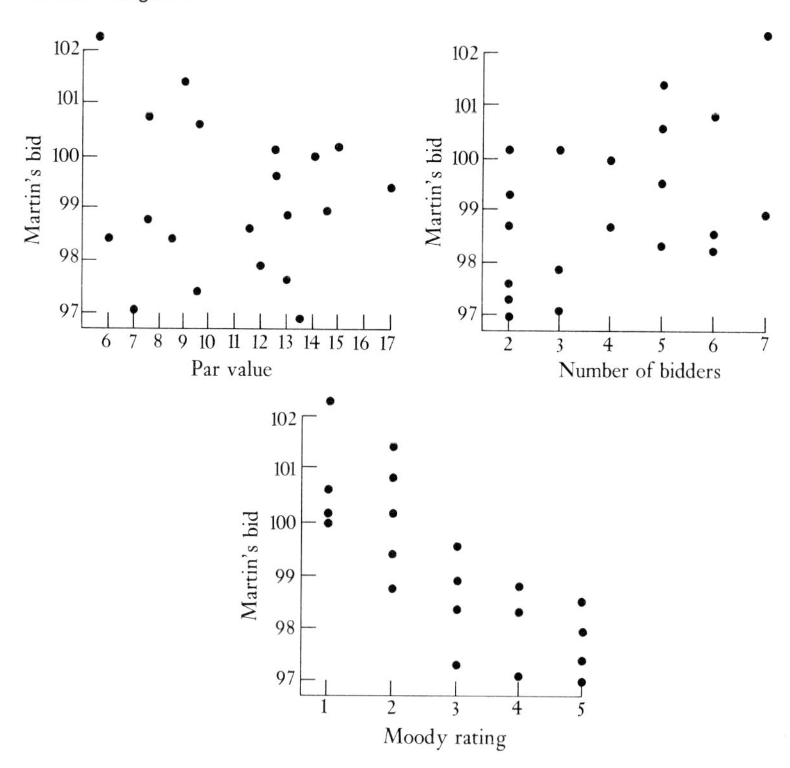

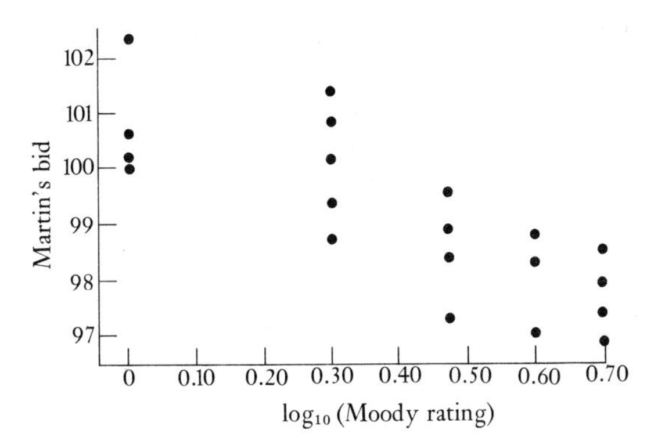

Fig. 12.10 Scatter diagram of Martin's bid and $\log_{10}$ (Moody rating).

cedures not discussed here) that Martin's bid price is related to *percent* changes in the Moody rating rather than absolute changes measured on a scale from 1 to 5. Fortunately, this type of curvilinear relationship can be transformed into a linear one by taking the logarithm of each value of the independent variable: that is, let the rating Aaa $= \log_{10}1 = 0$, Aa $= \log_{10}2 = 0.301$, A $= \log_{10}3 = 0.477$, Baa $= \log_{10}4 = 0.602$, and Ba $= \log_{10}5 = 0.699$. A scatter diagram of Martin's bid price plotted against $\log_{10}$ (Moody rating) is seen in Fig. 12.10 to approximate the desired linear form.

Using the "Canned" Program BMD02R

Determining the multiple linear regression equation for the data in Table 12.5 by hand calculation would be a tedious task. Fortunately, there are a number of the so-called "canned" computer programs available for calculating a variety of statistical measures, including correlation and regression. Perhaps the best known set of programs is the *Biomedical* (*BMD*) *Computer Programs*, written by the Health Services Computing Facility, School of Medicine, University of California, Los Angeles. One of these programs, the stepwise multiple regression program BMD02R, was used to analyze the data in Table 12.5.

BMD02R is a stepwise multiple regression program. As this name implies, the program computes the multiple regression equation in steps. The first step is to compute the *simple* regression line between the dependent variable and that particular independent variable that will reduce SSE to its smallest value. In other words, the program chooses, from the set of all independent variables specified, the one explaining the most variation in y, and then calculates the regression equation between y and this variable. In each succeeding step, one additional variable is added to the equation, the variable added always being the one which results in the greatest reduction in SSE. Although BMD02R is classified as a regression program, its output also provides many of the correlational measures described in this chapter. We present this output in a modified form in Figs. 12.11 through 12.14. Figure 12.11 shows the first output of BMD02R, namely, and identification of the number of cases (n), the number of variables (including the independent variable), the mean and the standard deviation of each variable, and the simple correlation coefficient (r) between each of these variables.

The correlation matrix gives the simple correlation coefficient between each pair of variables, the diagonal elements all being equal to one because a variable is always perfectly correlated with itself. Note that the highest correlation is between Martin's bid and $\log_{10}$ (Moody rating), equaling -0.799. From the definition of a correlation coefficient given by Formula (12.13), the variable explaining the most variability in y will be that independent variable

Fig. 12.11 BMD02R output.

BMD02R – STEPWISE REGRESSION –
HEALTH SCIENCES COMPUTING FACILITY, UCLA

PROBLEM CODE	BMD02R
NUMBER OF CASES	20
NUMBER OF ORIGINAL VARIABLES	4
NUMBER OF VARIABLES ADDED	0
TOTAL NUMBER OF VARIABLES	4
NUMBER OF SUB-PROBLEMS	1

VARIABLE	MEAN	STANDARD DEVIATION
MARTIN'S BID	99.17704	1.46334
PAR VALUE	10.92500	3.27380
NO. OF BIDDERS	4.05000	1.79106
LOG_{10} (MOODY RATING)	0.40078	0.25055

COVARIANCE MATRIX

VARIABLE	MARTIN'S BID	PAR VALUE	NO. OF BIDDERS	LOG_{10} (M.R.)
MARTIN'S BID	2.141	-0.638	1.265	-0.293
PAR VALUE		10.718	-2.180	0.047
NO. OF BIDDERS			3.208	-0.072
LOG_{10} (M.R.)				0.063

CORRELATION MATRIX

VARIABLE	MARTIN'S BID	PAR VALUE	NO. OF BIDDERS	LOG_{10} (M.R.)
MARTIN'S BID	1.000	-0.133	0.483	-0.799
PAR VALUE		1.000	-0.372	0.057
NO. OF BIDDERS			1.000	-0.160
LOG_{10} (M.R.)				1.000

```
STEP NUMBER     1
VARIABLE ENTERED    LOG10 (MOODY RATING)

MULTIPLE R                    0.7988
STD. ERROR OF EST.            0.9044

ANALYSIS OF VARIANCE
                          DF    SUM OF SQUARES    MEAN SQUARE    F RATIO
        REGRESSION         1            25.963         25.963     31.742
        RESIDUAL          18            14.723          0.818

          VARIABLES IN EQUATION                    VARIABLES NOT IN EQUATION

       VARIABLE      COEFFICIENT  STD. ERROR      VARIABLE             PARTIAL CORR.

        (CONSTANT       101.04691)
   LOG10 (MOODY RATING) -4.66551    0.82810          PAR VALUE              -0.14526
                                                NO. OF BIDDERS              0.59717
```

Fig. 12.12 BMD02R output, step number 1.

most highly correlated with y. In a stepwise procedure this variable should thus be the first one entered in the regression equation, and we see in the output given in Fig. 12.12 that it is.

The term labeled "constant" represents the value of a. Thus, $a = 101.04691$. Following the constant term, the program prints out the value of the regression coefficient, which is $b = -4.66551$. Since y = Martin's bid price and $x_3 = \log_{10}$ (Moody rating),

$$\hat{y} = 101.04691 - 4.66551x_3.$$

The remaining values in the computer output provide additional information about the variables, the regression coefficients, and the fit of the regression equation to the sample observations. The analysis-of-variance table, for example, indicates that the ratio MSR/MSE is $F = 31.742$. Since the critical value of F with 1,18 d.f. is 8.29 for $\alpha = 0.01$, we can reject the null hypothesis of no linear regression. Also given in the output is the standard error of the estimate, $s_e = 0.9044$. The standard error of the regression coefficient is $s_{b_3} = 0.82810$.

The partial correlations shown in Fig. 12.12 can be used to indicate which of the two variables *not* in the equation would explain more variability when added to the analysis. For instance, the partial correlation associated with par value (x_1) is $r_{y,x_1 \cdot x_3} = -0.14526$. The square of this value, $r^2_{y,x_1 \cdot x_3} = 0.021$, indicates that the additional variation explained by adding x_1 would be 2.1 percent. Similarly, in Fig. 12.12 we see that $r_{y,x_2 \cdot x_3} = 0.59717$, which means that $(0.59717)^2 = 0.3566$, or 35.66 percent extra variation is explained by x_2 (number of bidders). Since x_2 explains more additional variation than x_1, x_2 is added in Step 2 of BMD02R.

Step Two in BMD02R

The variable which reduces SSE the most is x_2 = number of bidders; hence, this variable is added next. (See Fig. 12.13 on page 582.)

```
STEP NUMBER    2
VARIABLE ENTERED   NO. OF BIDDERS

MULTIPLE R                   0.8759
STD. ERROR OF EST.           0.7456

ANALYSIS OF VARIANCE
                    DF     SUM OF SQUARES    MEAN SQUARE     F RATIO
        REGRESSION   2           31.214         15.607        28.009
        RESIDUAL    17            9.473          0.557

           VARIABLES IN EQUATION                       VARIABLES NOT IN EQUATION

       VARIABLE       COEFFICIENT    STD. ERROR          VARIABLE    PARTIAL CORR.

        (CONSTANT           99.70602)
 NO. OF BIDDERS              0.29735    0.09687           PAR VALUE      0.09988
 LOG10 (MOODY RATING)       -4.32464    0.69245
```

Fig. 12.13 BMD02R output, step number 2.

We can now see that adding the variable *number of bidders* to the regression equation increases the multiple correlation coefficient to

$$R = 0.8759,$$

which means that about 77% ($R^2 = 0.8759^2$) of the variability in Martin's previous bid prices has now been explained. The standard error of the estimate has been reduced to 0.7465, and the multiple regression equation is

$$\hat{y} = 99.706 + 0.297x_2 - 4.325x_3, \tag{12.17}$$

where x_2 = number of bidders and $x_3 = \log_{10}$ (Moody rating).

At this point the question is whether or not x_1 (par value) should be entered into the regression equation. Adding it will reduce the unexplained variation by less than 1 percent (0.09988^2), and there is serious doubt whether β_1 differs from zero. The question of when to stop adding variables is comparable to the choice of an α-level in hypothesis testing, and as such it will depend on the risk the researcher is willing to incur of including a variable in the regression equation which has a β-value of zero. Recall that adding an additional variable can never *increase* the amount of unexplained variation in a given problem but may decrease SSE by such a small amount that when dividing this sum by the number of degrees of freedom in order to compute the mean square error, the one extra degree of freedom lost may result in an *increase* of the standard error of the estimate. Such a result occurs in the present case, when adding the variable par value, as shown in the final output from BMD02R for this example (Fig. 12.14). Note that the standard error of b_1,

$$s_{b_1} = 0.058,$$

is larger than the value of the coefficient itself, as $b_1 = 0.023$. Thus, even a 68-percent confidence interval will include the value $\beta_1 = 0$, and one cannot have much confidence that the value of β_1 differs from zero.

To complete this example, let's assume that Victor has decided to use the

```
STEP NUMBER     3
VARIABLE ENTERED   PAR VALUE

MULTIPLE R                     0.8772
STD. ERROR OF EST.             0.7656

ANALYSIS OF VARIANCE
                          DF     SUM OF SQUARES     MEAN SQUARE     F RATIO
          REGRESSION       3             31.308          10.436      17.805
          RESIDUAL        16              9.378           0.586

                  VARIABLES IN EQUATION

          VARIABLE          COEFFICIENT   STD. ERROR

              (CONSTANT       99.38828)
    PAR VALUE                  0.02321     0.05779
    NO. OF BIDDERS             0.31314     0.10685
    LOG10 (MOODY RATING)      -4.32394     0.71019
```

Fig. 12.14 BMD02R output, step number 3.

output from Step 2 of BMD02R, since x_1 (par value) does not appear to aid their predictive ability. From Formula (12.17) we know that the appropriate regression equation is

$$\hat{y} = 99.706 + 0.297x_2 - 4.325x_3,$$

where x_2 = number of bidders and $x_3 = \log_{10}$ (Moody rating). Since, in the present case, there are two bidders and the Moody rating is Aa, we substitute $x_2 = 2$ and $x_3 = \log_{10}2 = 0.301$ into this equation to get the *expected* bid for Martin:

$$\hat{y} = 99.706 + 0.297(2) - 4.325(0.301) = 98.998.$$

From this result and the output of Step 2 of BMD02R, Victor can now construct a probability distribution for the bid they expect from Martin. The mean of this distribution is $\hat{y} = 98.998$, and its standard deviation is (from Fig. 12.13) $s_e = 0.7456$, as shown in Fig. 12.15. If we knew the profit Victor expects to make for each possible bid, we could use the probability distribution in Fig. 12.15 to find the bid that would maximize expected profits. This task is left as an exercise for the reader (Problem 12.45).

Figure 12.15

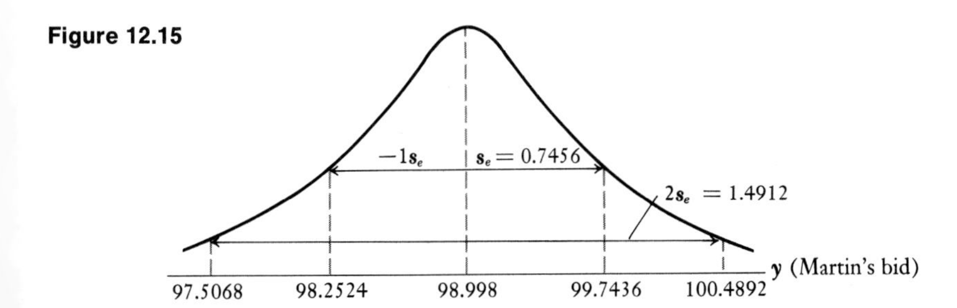

It is possible in using BMD02R to have the computer print the error of prediction, $y_i - \hat{y}_i$, for each set of observations in the sample. These errors, usually given in a "List of residuals," can be examined to check on the assumptions of linearity and homoscedasticity. In fact, most regression programs allow for the option of having the computer plot a scatter diagram showing the relationship between residual values and each of the independent variables. This process of constructing scatter diagrams in regression analysis is an important one, for it is all too easy, especially when using the computer, to compute a regression line and not check the assumptions of linearity, independence, uniform variance, and, for small samples, normality.

Define: multicollinearity, autocorrelation, heteroscedasticity, Durbin–Watson test, ordinary least squares (OLS), generalized least squares (GLS), dummy variables, stepwise multiple regression.

Problems

12.19 Explain the meaning of multicollinearity, and specify one of its effects that you think is important.

12.20 Calculate the residuals for the regression equation in Fig. 12.13 (see Table 12.5). Assume that these observations represent a variable measured over time. Calculate the value of the Durbin–Watson statistic d and interpret the result. Let $\alpha = 0.02$, $m = 2$, and use a two-sided alternative hypothesis.

12.21 A study was made of 53 countries to determine whether the formal character of a country's political constitution has a systematic impact on decentralization of public revenues. The following regression results were obtained:

$$\hat{y} = \underset{(12.1)}{96} - \underset{(-1.3)}{1.21x_1} - \underset{(-2.3)}{0.004x_2} - \underset{(-5.5)}{0.6x_3} - \underset{(-4.7)}{15.9x_4},$$

$$R^2 = 0.65$$

where

y is central government share of total public revenues, expressed as a percentage,
x_1 is the natural logarithm of population size in thousands,
x_2 is per-capita income in 1965 U.S. dollars,
x_3 is Social Security contribution as a percentage of total public revenue,
x_4 is 1 for countries with a federal constitution and 0 otherwise.

The values in parentheses are t-ratios of the regression estimates to their standard errors.

a) Interpret the meaning of the coefficient of x_4 in descriptive terms.

b) Using a one-percent level of significance, test the hypothesis that the formal character of a country's constitution affects the proportion of public revenue that the central government obtains against the hypothesis that it doesn't.

c) Using a five-percent level of significance and the statistic

$$F = (R^2/m)/[(1 - R^2)/(n - m - 1)]$$

which is equal algebraically to

$$F = (\text{SSR}/m)/[\text{SSE}/(n - m - 1)],$$

test the null hypothesis of no linear relationship.

d) If the computer program used for this problem gave a Durbin–Watson statistic of 1.7, what interpretation would you give to this information?

12.22 Given the following information on a linear multiple-regression model:

y = average yield in bushels of corn per acre on an Iowa farm;

x_1 = amount of summer rainfall, District 3 weather station, Iowa;

x_2 = average daily use in machine hours of tractors on the farm;

x_4 = amount of fertilizer, type XS80, used per acre.

The sample includes observations for ten crop years.

Results:

$$\hat{y} = \underset{(10)}{16} + \underset{(25)}{75x_1} + \underset{(4)}{6x_2} + \underset{(8)}{48x_3}$$

Regression equation
Standard errors of regression coefficients

$n = 10$, $s_e = 20$ bushels, $s_y = 40$ bushels,

$r^2_{y,x_1 \cdot x_2x_3} = 0.60.$

Answer the following questions:

a) What are the degrees of freedom for t-distributed test statistics for regression?

b) Explain which variable appears to be the most important in explaining the variation of yield.

c) From the regression results, is it proper to argue that more machine hours of tractor use causes more yield, or that more yield requires more machine hours of tractor use? Explain.

d) Find the coefficient of multiple correlation, $R^2_{y \cdot x_1x_2x_3}$.

e) Account for the different values of

$$R^2_{y \cdot x_1x_2x_3} \quad \text{and} \quad r^2_{y,x_1 \cdot x_2x_3}$$

by explaining the different meanings of the two coefficients.

12.23 Using 12 observations, a model $y = \alpha + \beta_1x_1 + \beta_2x_2 + \beta_3x_3 + \beta_4x_4 + \varepsilon$ is estimated by the method of least squares. Here SST = 400, SSE = 170, and the amount of variation explained jointly by x_1, x_2, and x_3 is 200.

a) Find $r^2_{y,x_4 \cdot x_1x_2x_3}$ and explain what it means.

b) Do an ANOVA test with $\alpha = 0.05$ to determine whether this linear relationship is significant.

12.24 Consider a linear-regression model, $y = \alpha + \beta_1x_1 + \beta_2x_2 + \varepsilon$, where

y = learning by grade 12, as measured by an academic test score composite, with mean 300 and standard deviation 150, for the entire population of 12th-graders;

x_1 = school expenditures per pupil during 3 years of high school (in hundreds of dollars);

x_2 = an index of socioeconomic status of the individual, with mean 10 and standard deviation 2, for the entire population of 12th-graders.

Based on a sample of 25 twelfth-grade-level individuals who were arrested on drug possession charges, the following results are obtained. Analyze, interpret, and explain these data in the way you think most appropriate and meaningful.

Variable	Mean	Standard Deviation
y	306.67	175.98
x_1	12.58	9.31
x_2	11.17	8.95

Correlations:

$r_{yx_1} = 0.83;$ $r_{yx_2} = 0.35;$ $r_{x_1x_2} = 0.10.$

Coefficient	Estimate	Standard Error	t-Value
a	10.16	11.90	0.85
b_1	17.60	0.62	28.30
b_2	4.30	2.90	1.48

Multiple $R = 0.92;$ $s_e = 3.015.$

Analysis of Variance	SS	d.f.	Mean Square
Regression	1090	2	545.00
Residual	200	22	9.09

12.25 In a multiple regression analysis of changes in annual average U.S. interest rates ($\boldsymbol{y}$) on three explanatory variables ($\boldsymbol{x}_1$, $\boldsymbol{x}_2$, and $\boldsymbol{x}_3$), the following results are found:

$\sum(y - \bar{y})^2 = 600,$ $\sum e^2 = 150,$

and the variation explained by x_1 and x_2 is 350.

a) Find the multiple coefficient of determination, and explain its meaning.

b) Find $r^2_{y,\,x_3 \cdot x_1x_2}$, and interpret its meaning.

12.26 Suppose the variation in y is 500 units and the model

$\hat{y} = a + b_1x_1 + b_2x_2$

leaves 240 units unexplained (based on 15 observations). Extending the model to include variable $\boldsymbol{x}_3$ explains 80 more units of variation in y. Find $R^2_{y \cdot x_1x_2x_3}$ and $r^2_{y,\,x_3 \cdot x_1x_2}$.

12.27 Suppose that in a multiple regression of $\boldsymbol{y}$ on variables $\boldsymbol{x}_1$, $\boldsymbol{x}_2$, and $\boldsymbol{x}_3$ we obtain SST = 1000, SSE = 200, and the variation explained by only x_2 and x_3 is 400. Find $R^2_{y \cdot x_1x_2x_3}$ and $r^2_{y,\,x_1 \cdot x_2x_3}$.

12.28 Discuss whether each of the following statements is true or false.

a) If SSE = SST, then $b = 0$.

b) If $R_{y \cdot x_1x_2x_3} = 1$, then $r_{y,\,x_1 \cdot x_2x_3} = 0$.

c) If $R_{y \cdot x_1x_2x_3} = R_{y \cdot x_1x_2}$, then $r_{y,\,x_3 \cdot x_1x_2} = 0$.

d) $R^2_{y \cdot x_1x_2x_3} \geq R^2_{y \cdot x_1x_2}$.

e) $r^2_{y,\,x_1} + r^2_{y,\,x_2} = R^2_{y \cdot x_1x_2}$.

12.29 In a multiple regression of y on the variables x, z, and w, the total variation is 200, the residual variation is 20, and the variation explained by only variables z and w is 120 units.

a) Find $R^2_{y \cdot x,z,w}$.

b) Find $r^2_{y,x \cdot zw}$.

12.30 Given a multiple regression model, $\hat{y} = a + b_1x_1 + b_2x_2 + b_3x_3$ based on 24 observations on each variable. Suppose that the total variation, SST, equals 300, the unexplained variation is 60, and the amount of variation explained by variables x_1 and x_2 together is 160.

a) Calculate the value of the multiple coefficient of determination and interpret its meaning.

b) Prepare a diagram similar to Fig. 12.2 to explain the meaning and value of $r^2_{y,x_3 \cdot x_1x_2}$.

c) Complete an analysis-of-variance table to test the significance of the linear relation.

12.31 In a multiple regression problem, the data in Table 12.6, are used.

Table 12.6

	y	x_1	x_2	x_1^2	x_2^2	x_1y	x_2y	x_1x_2	y^2
	9	3	2	9	4	27	18	6	81
	10	4	3	16	9	40	30	12	100
	2	1	2	1	4	2	4	2	4
	9	2	3	4	9	18	27	6	81
	20	5	5	25	25	100	100	25	400
$\sum$	50	15	15	55	51	187	179	51	666
Mean	10	3	3						

a) The normal equations for two independent variables are

$$\sum y = na + b_1\sum x_1 + b_2\sum x_2,$$
$$\sum x_1y = a\sum x_1 + b_1\sum x_1^2 + b_2\sum x_1x_2,$$
$$\sum x_2y = a\sum x_2 + b_1\sum x_1x_2 + b_2\sum x_2^2.$$

Find a, b_1, and b_2.

b) Find SSE, SSR, and SST. Use this information to calculate s_e and R^2.

c) If $r^2_{y,x_1} = 136.9/166$, and $r^2_{y,x_2} = 140.167/166$, find $r^2_{y,x_1 \cdot x_2}$ and $r^2_{y,x_2 \cdot x_1}$.

d) Test the null hypothesis of no linear correlation $H_0{:}\rho = 0$ versus $H_a{:}\rho > 0$. Use $\alpha = 0.05$.

e) Set up the ANOVA table for this problem.

12.32 A time-series study has been done based on annual data for the period 1950–1975 on the demand for money as a function of the current interest rate, last period's interest rate, and the change in the interest rate from year to year. Last period's interest rate reflects habit and inertia, and the change in interest rates reflects expectations of change based on recent changes. The regression model is:

$$y_t = \beta_0 + \beta_1x_{1t} + \beta_2x_{2t} + \beta_3x_{3t} + \varepsilon$$

where

y_t is the money supply at time t,

x_{1t} is the interest rate at time t,

x_{2t} is the interest rate at time $t - 1$,

x_{3t} is the change in the interest rate from last period, $x_{1t} - x_{2t}$.

The Dubin–Watson statistic is 1.1, $R^2 = 0.95$, and $n = 26$.

a) Using a five-percent level of significance, test the hypothesis of no positive autocorrelation against the alternative of positive autocorrelation.

b) Based on your answer to (a), either test the hypothesis that the regression does not explain the demand for money against the alternative that it does, or explain why such a test is inappropriate in this case.

12.33 Given the following multiple-regression results (Tables 12.7–12.9), where $n = 12$,

Table 12.7

Variable	Mean	Standard Deviation	
y	306.67	174.98	$r_{yx_1} = 0.9348$
x_1	12.58	9.31	$r_{yx_2} = 0.3501$
x_2	11.17	8.95	$r_{x_1x_2} = 0.0096$

Table 12.8

Coefficient	Estimator	Standard Deviation	
b_1	17.61	0.623	$R_{y \cdot x_1x_2} = 0.995$
b_2	6.70	0.647	$s_e = 19.220$
a	10.17	11.972	

Table 12.9

Observation	Observed y	Residual
1	650	−1.63
2	80	11.99
3	120	−18.46
4	180	+32.34
5	360	17.79
6	140	−21.07
7	450	7.11
8	550	11.43
9	280	−14.37
10	300	−16.30
11	350	−17.43
12	220	

a) Which independent variable is most important in determining y?

b) Find the residual for the twelfth observation if the observed values for x_1 and x_2 are 8 and 9, respectively.

c) What is the number of degrees of freedom for the F-test on the multiple linear association represented by this estimated model?

d) What percentage of the total variation in y has been explained by this regression?

e) Using a plot, comment on the validity of the assumption of homoscedasticity in this model.

f) Using a plot, comment on the validity of the assumption of no autocorrelation in this model.

g) Do you think multicollinearity is a problem in this estimation? Explain your answer.

h) If new observations on x_1 and x_2 are obtained, a corresponding value y can be predicted by using the regression equation. Discuss the accuracy of such a prediction if x_1 and x_2 are 20 and 30, respectively, as compared to the prediction of $\hat{y}$ if x_1 and x_2 are 10 and 12, respectively.

12.34 Let y = individual income for persons selected from among full-time workers in manufacturing, and let x = age. A simple regression model is written $y = \alpha + \beta x + \varepsilon$.

a) Suppose that I wish to include in the model the factor of sex. Specify and interpret the dummy variable to be included in order to differentiate levels of income depending on sex as well as on age.

b) Suppose that now I wish to allow for different impacts of age on income, depending on whether or not the person completed college. Specify a dummy slope variable to be added to the model, and explain the separate regression lines that can be derived.

c) Make a sketch with income and age on the axes, and illustrate the different potential regression lines you have implied in parts (a) and (b).

d) Finally, add a set of dummy variables to the model to allow for different income levels for five different areas of the country: northeast, midwest, southeast, southwest, and west. In your model specify the difference in the intercepts for females in the northeast compared to males in the west.

12.35 Consider the following BMD02R computer output.

Step 2

Source	SS	d.f.	MS	F-ratio
Regression	100	2	50	10.0
Residual	325	65	5	

(Constant 17.321)

Variables in Equation			Variables not in Equation	
Variable	**Coefficient**	**Std. Error**	**Variable**	**Partial Corr**
x_4	−2.317	0.157	x_1	−0.2314
x_2	4.539	1.128	x_3	0.1788

a) Write down the least-squares equation.

b) Determine the standard error of the estimate.

c) Find the value of multiple R.

d) Test the $H_0:\beta_i = 0$ vs. $H_a:\beta_i \neq 0$ for both β_2 and β_4.

e) Test $H_0:\beta_2 = \beta_4 = 0$ vs. $H_a:\beta_2$ or $\beta_4 \neq 0$.

f) Which variable will enter next in the stepwise procedure? How much will SSE be reduced by this new variable?

12.36 In Exercise 10.37 Kenneth Rein of the Ohio Valley Detergent Corporation predicted sales for the next three years to average 10.83 million pounds of detergent annually,

Table 12.10

	Population	Unemployment Rate	Advertising Expense	Competition	Sales
1.	7,500,000	5.1	59,000	0	5,170,000
2.	8,710,000	6.3	62,500	1	5,780,000
3.	10,000,000	4.7	61,000	1	4,840,000
4.	7,450,000	5.4	61,000	1	6,000,000
5.	8,670,000	5.4	6,100	1	6,000,000
6.	11,000,000	7.2	12,500	1	6,120,000
7.	13,180,000	5.8	35,800	1	6,400,000
8.	13,810,000	5.8	59,900	1	7,100,000
9.	14,430,000	6.2	57,200	2	8,500,000
10.	10,000,000	5.5	35,800	1	7,500,000
11.	13,210,000	6.8	27,900	1	9,300,000
12.	17,100,000	6.2	24,100	2	8,800,000
13.	15,120,000	6.3	27,700	2	9,960,000
14.	18,700,000	5.0	24,000	3	9,830,000
15.	20,200,000	5.5	57,200	3	10,120,000
16.	15,000,000	5.8	44,300	3	10,700,000
17.	17,600,000	7.1	49,200	4	10,450,000
18.	19,800,000	7.5	23,000	4	11,320,000
19.	14,400,000	8.2	62,700	2	11,870,000
20.	20,350,000	7.8	55,800	2	11,910,000
21.	18,900,000	6.2	50,000	3	12,600,000
22.	21,600,000	7.1	47,600	4	12,600,000
23.	25,250,000	4.0	43,500	4	14,240,000
24.	27,500,000	4.2	55,900	5	14,410,000
25.	21,000,000	7.0	51,200	4	13,730,000
26.	19,700,000	6.4	76,600	3	13,730,000
27.	24,150,000	5.0	63,000	3	13,800,000
28.	17,650,000	8.5	68,100	4	14,920,000
29.	22,300,000	7.1	74,400	5	15,280,000
30.	24,000,000	8.0	70,100	5	14,410,000

with a standard deviation of 1.18 million pounds. These estimates were based on a regression analysis using a random sample of 30 observations Rein considered to be comparable to the Indianapolis situation. He used four independent variables: population, unemployment rate, advertising expense, and the number of competitors. Rein estimates the value of these independent variables in Indianapolis for the next three years to be

19,660,000, 6.4, $35,000, and 2, respectively.

a) Use a computer program to verify Rein's estimates on the basis of the data given in Table 12.10.

b) Test the significance of the coefficient for each independent variable using $\alpha = 0.01$.

12.37 The data in Problem 12.36 was analyzed using an interactive multiple regression package. In the output presented below, OVDC stands for Ohio Valley Detergent Company, and the variables x_1, x_2, x_3, x_4 and y are in the same order (left to right) as presented in Table 12.10. For this analysis, each population value in Table 12.10 was divided by 1000.

```
COMMAND? REGRESSION

   TYPE? MULTIPLE

   DEPENDENT VARIABLE FILE NAME? OVDC:M5(5)

   INDEPENDENT VARIABLES FILE NAME(S)? OVDC:M5(1-4)

DEPENDENT VARIABLE:
   Y  =OVDC:M(5)

INDEPENDENT VARIABLES:
   X1 =OVDC:M(1)          X2 =OVDC:M(2)          X3 =OVDC:M(3)
   X4 =OVDC:M(4)

    TOTAL PERIODS USED =  30


                  DF      CUM SS         MEAN SS
     REGRESSION    4    2.84350E+08    7.10876E+07
     RESIDUAL     25    3.48149E+07    1.39260E+06
     TOTAL        29    3.19165E+08    1.10057E+07


      R-SQ    ADJ R-SQ     COR COEFF   RES STDERR     OVERALL-F
     .89092    .87833       .94388    1180.08300     51.04678 (DF= 4, 25)


     VAR    REGR COEFF   UPPER LIMIT   LOWER LIMIT    STD ERROR    PAR-F/T-SQ
     X1        0.3646       0.5442        0.1850        0.0872      17.48624
     X2      540.3555     957.3904      123.3206      202.4441       7.12440
     X3        0.0220       0.0476       -0.0036        0.0124       3.12650
     X4      612.2122    1319.3110      -94.8870      343.2520       3.18109
CONSTANT   -1790.7930


     DURBIN-WATSON STATISTIC =      1.96659
     APPROX SERIAL CORRELATION      0.01600
```

a) Write the regression equation and verify Rein's prediction in Problem 12.36.

b) Interpret each part of the computer output.

12.38 Consider the following correlation matrix:

	y	x_1	x_2	x_3
y	1.00	0.30	0.50	0.20
x_1		1.00	0.95	0.05
x_2			1.00	0.01
x_3				1.00

a) In a stepwise multiple regression procedure, which variable would enter first? Why?

b) Which variable would you guess would enter second in a stepwise procedure? Which value is larger, $r^2_{y,x_3 \cdot x_2}$ or $r^2_{y,x_1 \cdot x_2}$?

c) If $r^2_{y,x_3 \cdot x_2} = 0.40$, explain what this number means.

12.39 Suppose you are trying to decide how many independent variables to include in a multiple regression analysis.

a) Can adding additional variables ever decrease R?

b) Can adding additional variables ever decrease the value of F?

c) Can adding additional variables ever increase the value of $\sum(y - \hat{y})^2$? Will s_e ever decrease? Explain your answers.

12.40 Give an argument that explains why any one of the standard assumptions for regression analysis would be violated, in each of the following situations, for a simple model of the form $\mathbf{y} = \beta\mathbf{x} + \boldsymbol{\varepsilon}$. Also, suggest how the violation of this assumption would affect the properties of the ordinary least-squares (OLS) estimators.

a) $\mathbf{y}$ measures wealth of an individual and $\mathbf{x}$ measures this person's age; $V[\mathbf{x}]$ increases with age.

b) Observations on $\mathbf{y}$ and $\mathbf{x}$ are daily stock averages and volume of trading, respectively.

12.41 Explain the meaning of each of the following assumptions in which $\boldsymbol{\varepsilon}$ is a random term in a linear-regression model. Give an example of some specified model that might violate each assumption, and explain why.

a) $C[\boldsymbol{\varepsilon}_i, \boldsymbol{\varepsilon}_j] = 0$ for $i \neq j$;

b) $V[\boldsymbol{\varepsilon}_i] = \sigma_i^2$ for all i.

12.42 The application of OLS treats all observations as equally important. State one situation in which this may be an inappropriate procedure, and explain the general principle or method of a better procedure.

Consider a simple model, $y_i = \alpha + \beta x_i + \varepsilon_i$, for which it is known that $\varepsilon_t = 0.3\varepsilon_{t-1} + v_t$ where the v_t are normally and independently distributed with constant variance and mean zero.

a) Explain the problem of using ordinary least squares in this situation.

b) Construct the appropriate expression that will minimize the sum of squares, if the variables are transformed to correct for the problem. [*Hint:* Let the new variables be $y_t^* = y_t - 0.3y_{t-1}$ and $x_t^* = x_t - 0.3x_{t-1}$.]

12.43 a) If $C[e_t, e_{t+1}] \neq 0$, what problems does this present for a linear regression analysis?

b) If a population is not homoscedastic, what problems does this present for a linear regression analysis?

12.44 Consider the following values of y, x, and z, as shown in Table 12.11. Find the least-squares multiple regression equation $\hat{y} = a + bx + cz$. What difficulties are encountered, and why do they occur?

Table 12.11

y	x	z
0.5	1.00	5.0
1.5	1.50	4.5
3.0	2.00	4.0
4.5	4.00	2.0
5.5	4.50	1.5
6.0	5.50	0

12.45 Assume, for the example in Section 12.12, that Victor and Sons have determined that they can sell the bonds at an average price of 100.50% of par value, and that their total cost (fixed and variable) will be 0.25% of par value. Use these data and Fig. 12.15 to determine the optimal bid that Victor should make in order to maximize net profit.

12.46 Suppose we want to add a second variable to the analysis of investment presented in Section 11.6 of Chapter 11. This new variable is x_2 = retained earning of the firms, as shown in Table 12.12.

Table 12.12

Retained Earnings of U.S. Firms (x_2) in Billions of Dollars					
Observation	x_2	**Observation**	x_2	**Observation**	x_2
1	16.2	8	14.3	15	26.1
2	17.4	9	10.9	16	29.0
3	14.8	10	16.0	17	24.6
4	14.6	11	16.2	18	27.8
5	8.2	12	16.4	19	23.3
6	14.9	13	20.4	20	21.6
7	15.1	14	20.5		

a) Run a computer program to find the least-squares regression equation relating y (investment) to both x_i (stock prices) and x_2 (retained earnings).

b) Test the hypothesis: H_0:no linear regression and also test $H_0:\beta_i = 0$ for both β_1 and β_2.

GLOSSARY

standard error of the estimate (s_e):

$$s_e = \sqrt{\mathrm{SSE}/(n - m - 1)}.$$

coefficient of determination (R^2):

$$R^2 = \mathrm{SSR}/\mathrm{SST}.$$

partial correlation coefficient:

$$r^2_{y, x_2 \cdot x_1} = \frac{\text{extra amount explained by adding } x_2}{\text{amount unexplained by } x_1 \text{ alone}}.$$

multicollinearity: a condition in which two or more of the independent variables are strongly (but not perfectly) related to one another in a linear relationship.

heteroscedasticity: the variance of the error terms is not constant.

autocorrelation: there is a correlation between the error terms.

Durbin–Watson test: a test to detect autocorrelation.

dummy variable: a variable that reflects qualitative data by assigning either zero or one as the value of the variable.

hypothesis of no linear regression (H_0):

$$H_0: \beta_1 = \beta_2 = \cdots = \beta_m = 0$$

OLS: Ordinary least squares.

GLS: Generalized least squares.

13

TIME SERIES AND INDEX NUMBERS

"Statistics are the heart of democracy."

SIMEON STRUNSKY

13.1 INTRODUCTION TO TIME SERIES

In Chapter 11 we studied methods for describing the nature of the relationship between two variables. In this chapter we turn to what is, in a sense, a subset of this type problem, in which the independent variable under investigation is time.

Recording observations of a variable that is a function of time results in a set of numbers called a *time series*. Most data in business and economic publications take the form of a time series—e.g., the monthly sales receipts in a retail store, the annual Gross National Product (GNP) of the U.S., and indexes of consumer and wholesale prices, to name just a few. The analysis of time-series data in such circumstances usually focuses on two types of problems:

1. Attempting to estimate the factors (or components, as they are called) that produce the pattern in the series; and
2. Using these estimates in forecasting the future behavior of the series.

In this book we shall concentrate our attention mainly on estimating the components of the time series, rather than studying the forecasting implications provided by these estimates. Our approach will emphasize economic time series because of their central importance in the planning function performed by many businesses and government agencies.

Components of a Time Series

In general, the fluctuations in an economic time series are assumed to result from four different components: trend (T), seasonal variation (S), cyclical variation (C), and irregular or random variation (I). *Trend* is the long-term movement in a time series. (GNP, for example, has grown at a rate of approximately three to four percent a year over the past 20 years. The tendency toward a decreasing work week and increasing price levels over the past several decades also illustrates long-term movements or trends.) *Seasonal variation*

Fig. 13.1 The four components of a time series.

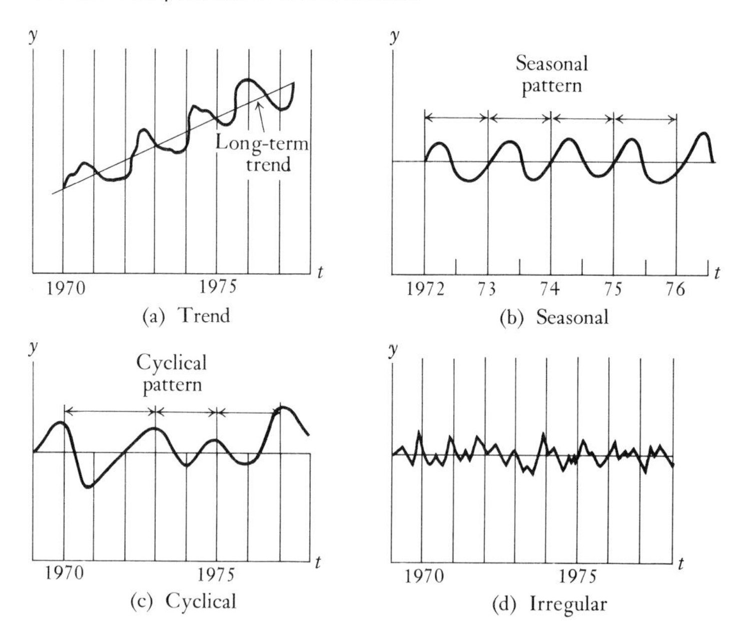

represents fluctuations that repeat themselves within a fixed period of one year. Many economic series have seasonal highs or lows due to changes in the supply of certain factors (youth employment in the summer, food harvests in the fall), or in demand for certain factors (ski equipment in the winter, toys before Christmas, gardening supplies in the spring), or in marketing factors (seasonal changes in clothing, new car models in the fall). The *cyclical* component of a time series represents a pattern repeated over time periods of differing length, usually longer than one year. (Business cycles, with their stages of prosperity, recession, and recovery are important examples of such cyclical movements.)

The movements in a time series generated by trend, seasonal and cyclical variation are assumed to be based on systematic causes; that is, these movements do not occur merely by chance, but reflect factors that have a more or less regular influence. Exactly the opposite holds true for random or *irregular variation*, which is, by definition, fluctuation that is unpredictable, or takes place at various points in time, by chance or randomly. (Floods, strikes, and fads illustrate the irregular component of a time series.) Figure 13.1 presents a graphical view of the four components of a time series. In each case the dependent variable $\mathbf{y}$ is expressed as a function of the independent variable t (time).

The Time-Series Model

Now that we have specified the components of a time series, the problem becomes one of estimating each of these components for a given series. That is, we want to be able to estimate what portions of the value of $\mathbf{y}$ for any given year are attributable respectively to trend, seasonal factors, cyclical factors, or random or irregular variation. In order to separate these values, we must make some assumptions about how these components are related in the population under investigation; these assumptions are referred to as the *time-series model.*

Time-series models usually fall into one of two categories, depending on whether their components are expressed as sums or products. The first of these, called the *additive model*, assumes that the value of $\mathbf{y}$ equals the *sum* of the four components, or that $\mathbf{y} = T + S + C + I$. By assuming that the components of a time series are additive we are, in effect, assuming that these components are independent of one another. Thus, for example, trend can affect neither seasonal nor cyclical variation, nor can these components affect trend. The other major type of relationship between the components expresses $\mathbf{y}$ in the form $\mathbf{y} = T \times S \times C \times I$ and is called the *multiplicative model.* A model of this form assumes that the four components are related to one another, yet still allows for the components to result from different basic causes. Note that one can transform a multiplicative model into a linear (i.e., additive) model by taking the logarithm of both sides,

$$\log y = \log T + \log S + \log C + \log I.$$

Other models exist in addition to the additive and multiplicative ones, although these models usually take the form of combinations of additive and multiplicative elements, such as

$$y = S + T \times C \times I,$$

or

$$y = C + T \times S \times I.$$

For the purpose of estimating each of the components of a time series, models normally treat S, C, and I as deviations from the trend. In other words, trend is usually estimated first, and the variation that can be attributed to trend is eliminated from the values of y. The variation remaining in y must be due to either seasonal, cyclical, or irregular factors. Each of these components of the time series can be isolated by other statistical techniques. While some of these methods are excessively tedious, the basic concepts are similar to the simple methods to be discussed in this chapter. Once all the components of a time series are estimated, then forecasts of the value of the time series at some future point in time can be made by estimating first the value of the trend component at that point, and then modifying this trend value by an adjustment that takes into account the seasonal and cyclical components.

Some Purposes and Problems of Forecasts

The primary purpose of forecasting future values of a time series is to facilitate planning. A business manager wants to forecast the trend in sales in order to make long-range planning decisions about investment in more plant capacity and new equipment. Or there may be a need to forecast cyclical movements in order to take advantage of lower interest-rate periods to conduct a bond sale, or of a period of high investor expectations to release a new issue of common stocks. Seasonal variation is important for short-run planning of inventories and employment levels, as the manager needs to be prepared for periods of high demand for certain products and services. As consumers, we recognize these seasonal components also, and plan to make purchases at times of special sales when prices are lower. Of course, the marketing experts use the seasonal components to plan advertising campaigns to entice buyers to purchase during the high season when the product or service is most desirable. We are sure that any salesman would prefer to handle ski equipment in the fall and winter, and air conditioners in the spring and summer, rather than vice versa.

The most obvious problem inherent in forecasting future values of a time series is the potential size of the (unpredictable) irregular component. This component is a random variable whose size depends upon a large number of independent factors that affect the economic variable y. Psychological and sociological variables of individual and group behavior, as well as other economic and business considerations, can affect the forecast. In addition, the irregular component may consist of one single occurrence in the time period being forecast, such as a flood, a major political event, or an energy crisis. Neither the size nor the direction of the irregular component can be predicted.

There is also an important problem in trying to forecast the more regular parts of the time series, the trend, cyclical, and seasonal components. In forecasting, it is necessary to assume that no change occurs in the fundamental causes underlying these regular patterns. For this reason, forecasts far beyond the range of presently observable values of the time series are always very dubious. In most cases this means that the forecaster must not try to predict very far into the future, for fear of being greatly embarrassed.

13.2 LINEAR TREND

The first step in analyzing a time series is usually estimation of the trend component T. As was true in regression analysis the first decision usually must be whether or not the trend can be assumed to be linear. Several linear methods will be discussed briefly, before we discuss the methods for estimating trend when the relationship is not linear.

The linear trend is written $y = a + bx$, where x is a measure of time. Often a graph of the time-series data indicates quite well whether or not this linear relationship provides a good approximation to the long-term movement of the series. If a linear relationship is appropriate, then there are several methods for roughly approximating T. A fairly accurate approximation can often be obtained by merely drawing the line that, by free-hand estimation, seems to represent best the long-run movement of the points. A more systematic approach is to use what is called the *method of semi-averages*. In this approach the data is divided into two equal parts, one representing the values associated with the first half of the years under investigation, and the second representing the remaining years. The average value of the independent (time) and dependent variables is calculated for each of these parts, and then the points representing these averages are connected by a straight line representing the trend line. For data containing an odd number of observations, the middle observation may be left out when dividing the data into two equal parts. Similarly, it may be advisable to eliminate one or two observations when calculating the mean value of y, if these observations are clearly atypical of the rest of the series and if their inclusion will disturb the whole trend line.

The method of least squares, developed in Chapter 10, represents the most popular method of fitting a trend line to time-series data. Recall that, in order to determine the values of a and b in a least-squares analysis, it is necessary to solve the following two normal equations:

$$\sum_{i=1}^{n} y_i = na + b \sum_{i=1}^{n} x_i,$$

$$\sum_{i=1}^{n} x_i y_i = a \sum_{i=1}^{n} x_i + b \sum_{i=1}^{n} x_i^2. \tag{13.1}$$

We will describe the use of these equations shortly.

Scaling and Interpreting the Time Variable

In any time-series analysis it is important to define carefully the units of the time variable x, and to scale this variable so that it is easy to manipulate. Since time is continuous, it really makes little difference if the units used to express time are years, months, weeks, days, or any other desirable period. Also, because the point in time that is selected as $x = 0$ has no influence on the analysis, any point in time can be assigned this value. For whatever period is used, however, it usually is convenient to let $x = 0$ represent the *middle* of that period. For example, if x is in months, then $x = 0$ would be the middle of one of the months; similarly, if x is in periods six months long (half-years), then $x = 0$ would represent the middle of this period (i.e., after 3 months). A simple coding rule used often in assigning the values of x is:

a) If the number of time periods is odd, let x be in the same units as the observed time periods (years, months, etc.) and assign $x = 0$ to that time period falling in the exact center. Then let time periods before $x = 0$ be denoted by $\ldots, -3, -2, -1$, and future time periods be denoted by $+1, +2, +3, \ldots$
b) If the number of time periods is even, let x be in units one-half as large as the observed time periods (half-years, and half-months, etc.), and assign $x = 0$ to the midpoint in time between the two middle observations. Then denote time periods before $x = 0$ as $\ldots, -5, -3, -1$, and denote future periods as $+1, +3, +5, \ldots$

This coding procedure has the advantage that the mean of the x-values will always be $\bar{x} = 0$, and also that the middle of each time period is represented by an integer value of x. This means that the solution of the least-squares trend equation is computationally easier, and the interpretation and use of the

Table 13.1 Monthly Observations of Employment in a Firm

y Employment	Time Period	Value of x
16	March, 1981	−5
17	April, 1981	−3
18	May, 1981	−1
20	June, 1981	+1
18	July, 1981	+3
24	August, 1981	+5

trend equation is simpler. We will illustrate both of these two types of scaling. First, suppose that the time-series data we have available concerns the variable y = employment in a firm, measured over the six-month period shown in Table 13.1. Since the number of time periods is even, we use coding procedure (b).

The units of x in Table 13.1 are expressed in units of half-months, with $x = 0$ corresponding to the date, midnight May 31, 1982. The values of x now represent the number of half-months before or after the end of May, 1982. For example, the value $x = 5$ represents August 15, 1982, which is 5 half-months after May 31st. Note that the sum of the x-values is zero, which means that $\bar{x} = 0$. To forecast a trend value for November, 1982, the value $x = 11$ would be substituted into the trend equation. To forecast a trend value at the *end* of the year (December 31, 1982), the value $x = 14$ would be used.

To take another example, consider the sales data (y) in Table 13.2. This data uses x measured in years because the number of time periods is odd, with the origin $x = 0$ corresponding to July 1, 1982. The value $x = -1$ thus corresponds to July, 1981, and $x = 2$ is July 1, 1986. Again note that $\sum x_i = 0$, and $\bar{x} = 0$. We will use the data in the remaining columns of Table 13.2 shortly.

Table 13.2 Sales (in Thousands of Dollars)

Year	x	y	x^2	xy
1979	−2	$2	4	−4
1980	−1	6	1	−6
1981	0	10	0	0
1982	1	13	1	13
1983	2	16	4	32
Sum	0	47	10	35

Using Simple Regression to Find the Trend Line

By using the coding rules outlined above, the solution of the normal equations in Formula (13.1) for time series is relatively simple. Since $\sum_{i=1}^{n} x_i = 0$ under these coding rules, if we substitute $\sum x = 0$ into the two normal equations, the result is the following two "reduced" equations.

Reduced equations:

$$\sum_{i=1}^{n} y_i = na \qquad \text{or} \qquad a = \frac{\sum_{i=1}^{n} y_i}{n} = \bar{y},$$

$$\sum_{i=1}^{n} x_i y_i = b \sum_{i=1}^{n} x_i^2 \qquad \text{or} \qquad b = \frac{\sum_{i=1}^{n} x_i y_i}{\sum_{i=1}^{n} x_i^2}. \tag{13.2}$$

Let us now substitute the appropriate values from Table 13.2 into these reduced equations in order to find the least-squares regression line for the sales data. The value of a equals $47/5 = 9.4$, while the value of b is $35/10 = 3.5$. Hence, the line of best fit is

$$\hat{y} = 9.4 + 3.5\,x$$

with x in units of a year; $x = 0$ is July 1, 1981, $x = 2$ is July 1, 1983. An estimate of the value of sales for 1986 ($x = 5$) based on trend factors only, would thus be

$$\begin{aligned} \hat{y} &= 9.4 + 3.5(5) \\ &= 26.9. \end{aligned}$$

The trend line based on the data of Table 13.2 is shown in Fig. 13.2. A good exercise for the reader would be to determine and graph the trend line for the data in Table 13.1.

The x-values shown in Table 13.2 and represented by the regression line $\hat{y} = 9.4 + 3.5x$ are based on yearly data. Later in this chapter, when we discuss seasonal variation, it will be necessary to switch a trend line based on yearly data to a trend line based on quarterly, monthly, or even weekly time periods. In some cases one may wish to present data using annual magnitudes in quarterly periods, or quarterly magnitudes in quarterly periods, etc. For each of these situations there will usually be different values plotted at a different point in time for the same data set. Consider, for example, Fig. 13.3, where the upper line represents the trend line for the yearly data in Table 13.2.

The lower line in Fig. 13.3 represents the same data shown in quarterly magnitudes. That is, the values represented by circles are one-fourth the size of the values represented by squares. Note that these circle values are plotted

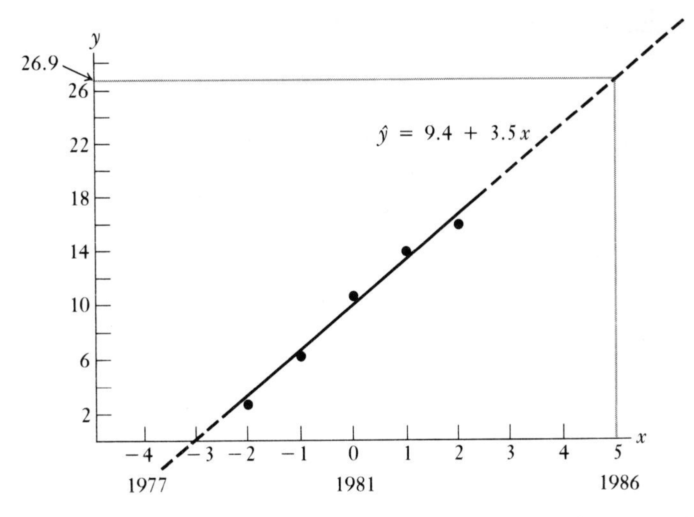

Fig. 13.2 Trend line for data of Table 13.2. Forecast value of sales for mid-1986 = 26.9.

Fig. 13.3 Annual (□) versus quarterly (○) magnitudes used to plot a trend line.

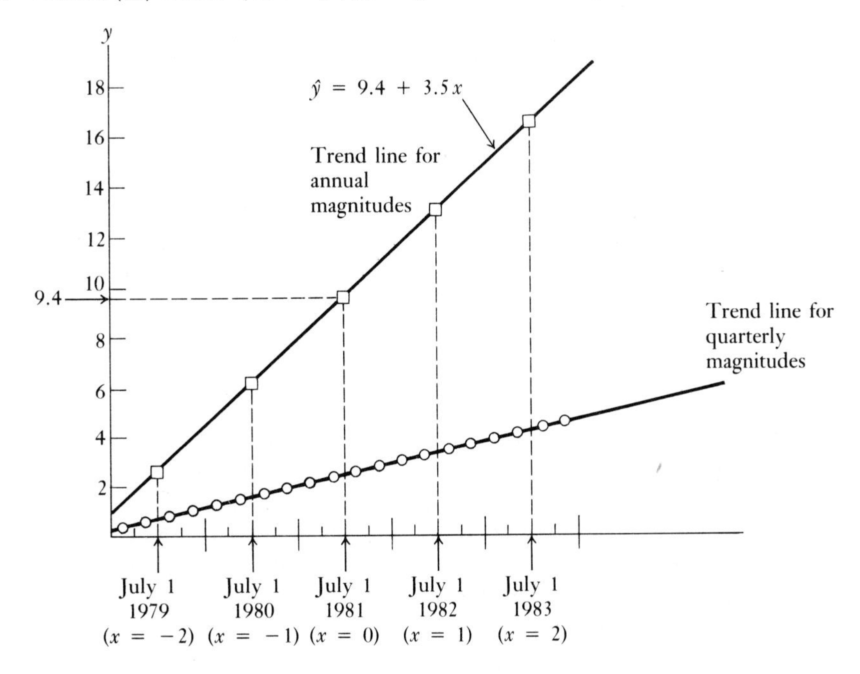

at the middle of each quarter (e.g., Feb. 15, May 15, Aug. 15, and Nov. 15), and thus no quarterly value falls at the same point in time as do the yearly values (July 1).

In reducing annual magnitudes to quarterly magnitudes for the trend equation $\hat{y} = a + bx$, where x is in years, it is necessary to divide both a and b by 4. This gives quarterly magnitudes, but the values of x are still measured in years ($x = 1$ is 1 year), and the origin remains at July 1. To obtain quarterly magnitudes with x measured in quarters, we need to further divide its coefficient (b) by 4. This yields

$$\hat{y} = \frac{a}{4} + \frac{b}{16}x,$$

where $x = 0$ is July 1, 1981 and $x = 1$ is October 1, 1981. Unfortunately, this equation is still centered on July 1, which is not the middle of one of our quarters. Suppose that we decide to shift the origin in Fig. 13.3 one-half quarter ($1\frac{1}{2}$ months) to the right, from July 1, 1981 to August 15, 1981. To do this we need to add $\frac{1}{2}$ to x in the equation above. The equation is now

$$\hat{y} = \left(\frac{a}{4}\right) + \left(\frac{b}{16}\right)\left(x + \frac{1}{2}\right) = \left(\frac{a}{4}\right) + \left(\frac{b}{32}\right) + \left(\frac{b}{16}\right)x,$$

where $x = 0$ is August 15, 1981 and $x = 1$ is November 15, 1981. For our example, $a = 9.4$ and $b = 3.5$. Hence, the new equation is

$$\begin{aligned}\hat{y} &= \frac{9.4}{4} + \frac{3.5}{32} + \left(\frac{3.5}{16}\right)x \\ &= 2.46 + 0.219x.\end{aligned}$$

The reader should verify that this equation does, indeed, represent the lower trend line in Fig. 13.3, where $x = 0$ is August 15, 1981.

In some situations trend values are reported quarterly, but annual magnitudes are given. GNP is often presented in this manner. For this case the trend equation would be 4 times as large as presented above. Applying this concept to our sales example we obtain

$$\hat{y} = 4(2.46 + 0.219x) = 9.84 + 0.876x.$$

13.3 NONLINEAR TRENDS

The problem of fitting a trend line to a nonlinear time series is essentially the same problem we mentioned in Chapter 11 concerning nonlinear regression—that of finding an equation that best describes the relationship between an independent variable (time, in this case) and the dependent variable (the time-series values). As is true in fitting a regression line, it is not sufficient merely to find an equation that provides a good fit to the data; it also is necessary to find a model that is justifiable in terms of the underlying economic nature of the series. In estimating the trend in a time series, there are a number of nonlinear equations that can be justified under a wide variety of circumstances.

Exponential Curve

Time series are often used to describe data that increase or decrease at a constant proportion over time, such as population growth, the sales of a new product, or the spread of a highly communicable disease. Data taking this form can be approximated by an equation referred to as the *exponential curve:*

$$\textit{Exponential curve:} \qquad y = ab^x. \tag{13.3}$$

The form of the exponential curve depends on the values of a and b. If b is between zero and one, then the value of y will decrease as x increases. When b is larger than one, y will increase as x increases. The value of a gives the y-intercept of the curve, as shown in Fig. 13.4.

Note that by taking the logarithm of both sides of Formula (13.3), we can transform the exponential curve into a linear relationship,

$$\log y = \log(ab^x) = \log a + x \log b.$$

Our model is now linear, and the least-squares approach can be used to find the line of best fit. To illustrate, consider the sales data given in Table 13.3 and graphed in Fig. 13.5. Again, the values of x have been specified so that $\sum_{i=1}^{n} x_i = 0$. From Fig. 13.5 we see that sales from 1979 to 1983 were not linear, and that an equation of the type shown in Fig. 13.4(b) would not be unreasonable. To use least-squares regression analysis to find the values of a and b in the equation $y = ab^x$, it is necessary to substitute log a for a, log b for b, and log y for y in the normal equations. Since $\sum_{i=1}^{n} x_i = 0$, the appro-

Fig. 13.4 The exponential curve $y = ab^x$.

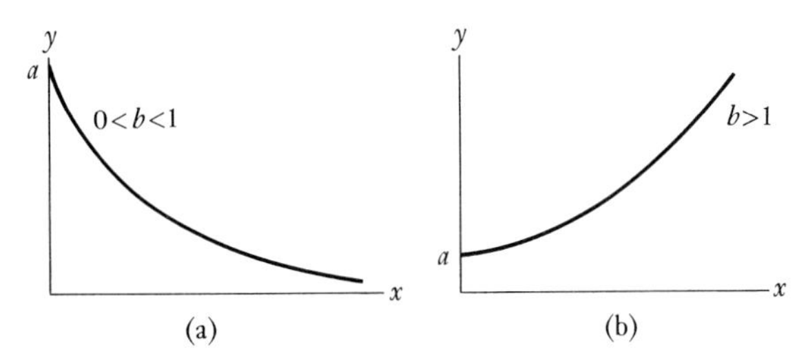

Table 13.3 Sales

Year	x	y
1979	-2	$1
1980	-1	3
1981	0	6
1982	1	14
1983	2	41

priate equations are thus:

$$\log a = \frac{\sum_{i=1}^{n} \log y_i}{n}, \qquad \log b = \frac{\sum_{i=1}^{n} x_i \log y_i}{\sum_{i=1}^{n} x_i^2}. \tag{13.4}$$

To solve these equations we need to transform the data in Table 13.3 by finding the logarithm of y and finding $x \log_{10} y$ (given in Table 13.4). We can now solve for a and b. Substituting the values calculated in Table 13.4 into Formula (13.4) yields:

$$\log a = 0.8028, \qquad \log b = 0.3895.$$

Taking the antilog of these values yields the lest-squares estimates $a = 6.35$ and $b = 2.45$. Thus, the exponential trend equation is

$$\hat{y} = (6.35)(2.45)^x.$$

Using this equation to forecast sales for 1984 ($x = 3$) yields $(6.35)(2.45)^3 = 93.34$. The fit provided by this equation for all values of x from -2 to $+2$ is shown in Fig. 13.5.

Figure 13.5

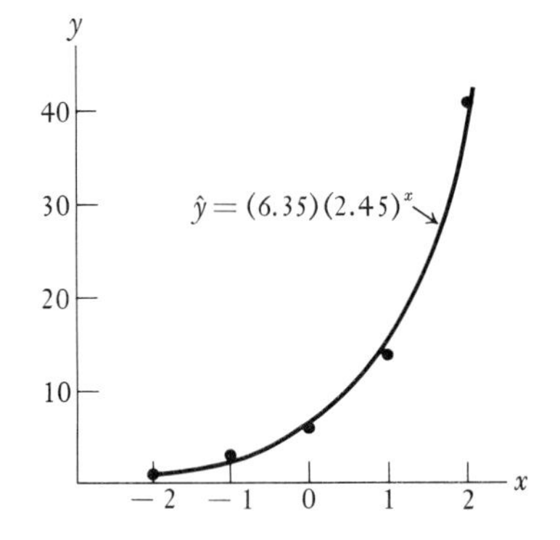

Table 13.4 Sales (Thousands of Dollars)

x	y	$\log_{10} y$	$x \log_{10} y$	x^2
−2	1	0.000	0.000	4
−1	3	0.477	−0.477	1
0	6	0.778	0.000	0
1	14	1.146	1.146	1
2	41	1.613	3.226	4
Sum		4.014	3.895	10

Modified Exponential Curve

In a number of circumstances it is desirable to allow for more flexibility in deciding on the position of the trend line than is provided by the exponential curve, without altering the basic form of this curve. One way to accomplish this objective is to modify the exponential curve by adding a constant to the equation. Suppose we add the constant c. The resulting equation is called the *modified exponential curve:*

$$\textit{Modified exponential curve:} \quad y = c + ab^x. \tag{13.5}$$

The modified exponential, like the exponential itself, can assume many different forms depending on the values of a, b, and in this case c. Although the addition of the constant c merely serves to shift the exponential curve up or down by a constant amount, such a shift is convenient in describing a time series with values that approach an upper or lower limit, as shown in Fig. 13.6. Changes in trend equations from annual magnitudes to monthly, or from monthly to quarterly, can be done following the same logical processes.

A special problem arises when data are analyzed for a product or service that is produced for only a fraction of a year in each year. For example, sugar beet processing plants operate for about four months starting in September

Fig. 13.6 The modified exponential curve $y = c + ab^x$.

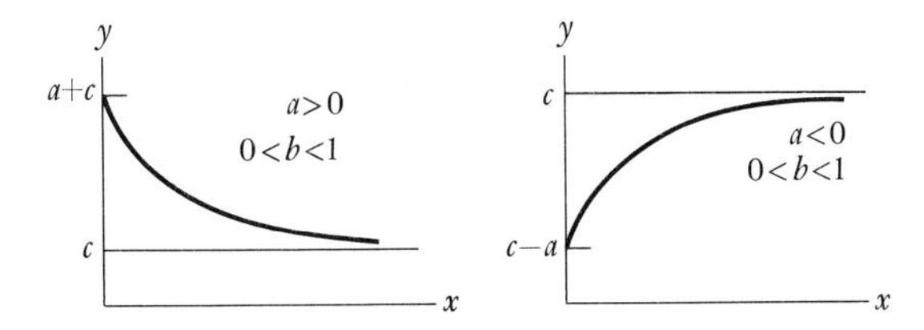

Table 13.5 Sales

Year	x	y
1979	-2	\$3
1980	-1	22
1981	0	39
1982	$+1$	46
1983	$+2$	51

of each year, and all sugar refining processes shut down completely when the last beets are processed around the beginning of the following calendar year. When a trend line is being plotted for monthly production of sugar from sugar beets, the x-values must be expressed in months, but there will be only four values of x in each year.

Finding the best fit to a modified exponential curve is not as easy as in the case for the exponential curve, as there is no simple transformation that makes the equation linear, and finding the best least-squares fit directly is a fairly difficult task. There is one relatively straightforward method if one is willing to use just three of the observations to determine the equation for estimating trend. With three observations of x and y we can form three equations in three unknowns, and then solve these equations simultaneously. To illustrate, consider the sales data given in Table 13.5 and graphed in Fig. 13.7.

Suppose that we choose the three observations representing $x = -2, x = 0$, and $x = 2$, and substitute these values into Formula (13.5). The result is the

Figure 13.7

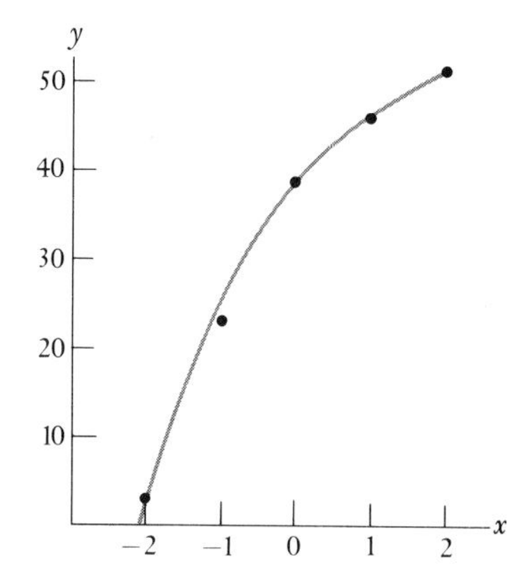

following three equations in three unknowns:

$$3 = c + ab^{-2},$$
$$39 = c + ab^{0},$$
$$51 = c + ab^{2}.$$

Solving these three equations simultaneously we obtain $a = -18.1$, $b = 0.58$, and $c = 57.1$. Thus, our modified exponential curve is

$$\hat{y} = 57.1 - 18.1(0.58)^{x}.$$

If we want to forecast y for 1981 ($x = 3$) on the basis of these data, then the appropriate predicted value is

$$\hat{y} = 57.1 - 18.1(0.58)^{3} = 53.46.$$

Figure 13.7 shows the fit of this equation for values of x from -2 to $+2$.

The procedure just described for fitting a trend line to a set of observations is called the "three-point method." When more than just a small number of observations is involved, it is customary to apply this method by dividing the relevant years into three equal periods. The mean value of y for each of these periods is used for each of the three points, rather than the value of y from a single year. In either case, the three-point method should be viewed as providing only an approximation to the more precise fitting obtainable by the method of nonlinear least squares (which is not presented here).

Logistic Curve

There are a number of additional curves used to estimate trend in a time series, two of which are suitable for mention here. The first is known as a *logistic curve*, and is defined by the following equation:

$$\textit{Logistic curve:} \qquad y = \frac{1}{c + ab^{x}}. \tag{13.6}$$

Note that the logistic curve is just the reciprocal of the modified exponential curve. Its rate of growth or decline is relatively rapid at first, but slows down in the later stages of the series, as shown in Fig. 13.8. Bacterial growth or sales increases for a new company sometimes exhibit such a pattern over time. Solving for the values of a, b, and c in a logistic curve presents the same problem as solving for these values in the modified exponential curve. The same approach described previously, using three of the time-series values to establish three equations in three unknowns, can be used to determine a logistic curve by letting $1/\mathbf{y}$ be the dependent variable, rather than $\mathbf{y}$. The resulting equation, $1/\mathbf{y} = c + ab^{x}$, has the same form as the modified exponential.

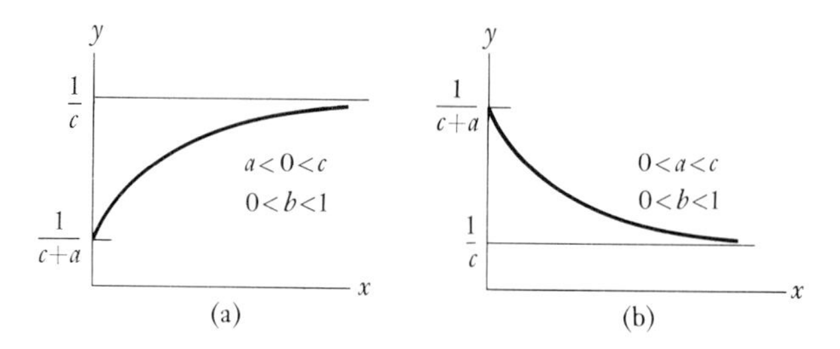

Fig. 13.8 The logistic curve $y = 1/(c + ab^x)$.

Gompertz Curve

The final trend equation we shall discuss is called the *Gompertz curve*, named after Benjamin Gompertz, who used the curve in the early 1800's in work concerning mortality tables. This curve is similar to the exponential curve, except that the constant a is raised to the b^x power instead of to the x power:

$$\textit{Gompertz curve:} \qquad y = ca^{b^x}. \tag{13.7}$$

When the value of b in a Gompertz curve is between zero and one, the power to which a is raised will approach zero as x increases; hence the value of y will become closer and closer to c, as shown in part (a) of Fig. 13.9. When the value of b is greater than one, the curve will either increase without bound (when $a > 1$) or will approach zero (when $0 < a < 1$) as x gets larger and larger, as shown in part (b) of Fig. 13.9. The Gompertz curve can be fitted to time-series data in essentially the same fashion described for the modified exponential curve, by taking the logarithm of both sides of Formula (13.7), as follows:

$$\log y = \log c + (b^x)\log a. \tag{13.8}$$

Formula (13.8) is a special form of the modified exponential curve, with unknowns $\log a$, b, and $\log c$, instead of a, b, and c.

Fig. 13.9 The Gompertz curve $y = ca^{b^x}$.

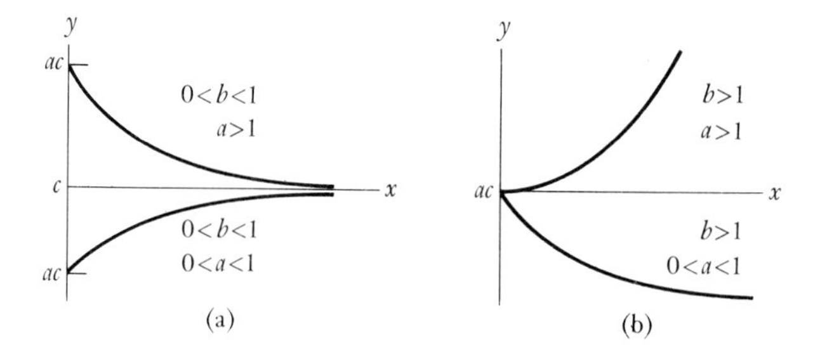

The problem of deciding which of the many available nonlinear equations to use in estimating trend in a given circumstance is often a difficult one. It is not unusual, in fitting the commonly used mathematical curves to a set of time-series observations, to find that two or more equations provide approximately the same closeness of fit, so that for the purpose of merely describing the data there is little to choose between these curves. The problem arises when such curves are used to predict future values of the time series (i.e., extrapolations into the future), for the curves tend to diverge rather quickly, and unacceptably large divergences may result from even a small extrapolation. Thus, as we pointed out earlier in this chapter, it is not sufficient to merely find a trend line providing a "good fit" to the data. It is necessary to fit a curve that can be justified by a general assessment of the underlying nature of the series. In other cases, no curve may fit well because there may be no single persistent relationship between the dependent variable and time.

13.4 MOVING AVERAGES TO SMOOTH A TIME SERIES

In the discussion thus far, we have assumed that a trend line can be calculated without first removing, or at least minimizing, the effect of seasonal, cyclical, and random movements in the series. In some circumstances, however, it may be easier to estimate trend if the effects of these fluctuations are removed from the data. Methods for removing these effects are usually referred to as smoothing techniques.

The Method of Moving Averages

The most commonly used method of smoothing data is called the *method of moving averages*. A moving average is actually a series of averages, where each average is the mean value of the time series over a fixed interval of time, and where all possible averages of this time length are included in the analysis. A 12-month moving average, for example, must include the mean value for each 12-month period in the series. These averages represent a new series that has been smoothed to eliminate fluctuations that occur within a 12-month period. Since a seasonal pattern will repeat itself every 12 months, a 12-month moving average will eliminate fluctuations caused by the seasons of the year. Note that a 12-month moving average also will smooth any other fluctuations with a pattern that repeats itself over an interval of less than a year's duration, such as daily or weekly patterns. Similarly, a five-year moving average could be used to eliminate a pattern or cycle that repeats itself every five years. Such a moving average would also smooth seasonal patterns, as well as daily and weekly fluctuations.

Table 13.6 Profits (Thousands of Dollars)

Year	y	Centered M.A.	Year	y	Centered M.A.
1967	2		1975	13	11.4
1968	4		1976	14	12.6
1969	5	5.2	1977	11	14.0
1970	7	6.0	1978	14	15.4
1971	8	6.8	1979	18	17.2
1972	6	8.0	1980	20	
1973	8	9.2	1981	23	
1974	11	10.4			

To illustrate the process of calculating a moving average, we will use this approach to smooth the data in Table 13.6. These data appear to fluctuate in a cyclical pattern that repeats itself every five years. In calculating a five-year moving average, we first must determine the mean value for the initial five years, 1967 through 1971. This mean value, which equals $(\frac{1}{5})(2 + 4 + 5 + 7 + 8) = 5.2$, is "centered" on (i.e., placed in the middle of) the five years being averaged, 1969. Similarly, the next moving average, 6.0, is computed from the values corresponding to 1968 through 1972 and centered at the year 1970. This process continues until the last observation is included; thc last moving average is 17.2, and this value is centered at the year 1979. The moving average values, when compared to the original values, tend to reduce the variation and represent a smoothed version of the time series in which the cyclical component has

Fig. 13.10 Values of annual profits and five-period moving average values (–○–) from Table 13.6.

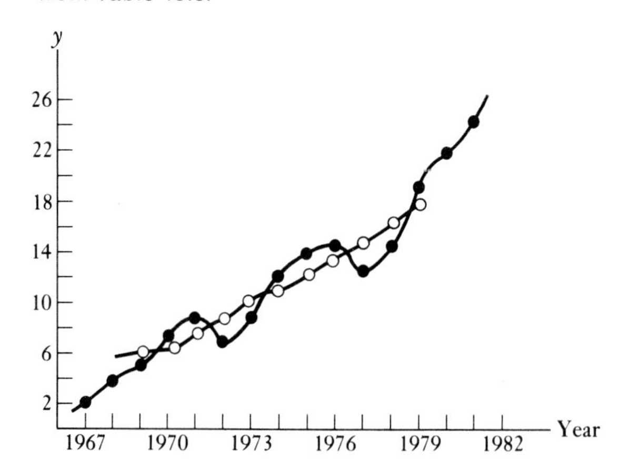

been reduced. If the observation units were one quarter of a year or less, the moving average technique would smooth out seasonal variations also. The comparison between the original time series and its representation by the moving average values is shown in Fig. 13.10.

This moving-average procedure gives equal weights to all five observations in calculating each mean value. Sometimes, one may wish to increase the relative importance of one or more observations by using a *weighted moving average*. To calculate a weighted moving average, each observation being averaged is given a weight that reflects its relative importance. In a five-year weighted moving average, for example, the weights might be 1, 2, 3, 2, 1, based on the assumption that the middle value in a series of observations should have the largest weight, and the first and fifth observation should have the smallest weights. Using this weighting system on the data in Table 13.6 we obtain a weighted moving average centered at 1969 equal to

$$\frac{1(2) + 2(4) + 3(5) + 2(7) + 1(8)}{1 + 2 + 3 + 2 + 1} = 5.22.$$

The second weighted moving average would be centered at 1970 and equal to

$$\frac{1(4) + 2(5) + 3(7) + 2(8) + 1(6)}{1 + 2 + 3 + 2 + 1} = 5.67.$$

By continuing this process we could derive the entire series for the weights 1, 2, 3, 2, 1. Had we used some other weighting system, an entirely different moving average would be determined. In general, the weights used depend on the degree to which the analyst wishes to emphasize particular values.

The Use of Moving Averages for Smoothing

It is important to note that by smoothing a time series we have not solved the problem of estimating the trend component for that series. Smoothing the series merely serves to eliminate some of the variability not attributable to trend, in the hope that the trend component then can be more easily identified. One method for identifying the appropriate trend curve (e.g., exponential, Gompertz, logistic) is to look at certain characteristics of a curve's moving average. In their book *Mathematical Trend Curves*, for example, Gregg, Hossell, and Richardson advise using a transformation of each curve's slope to decide among the common trend equations, since dispersions in the projection of these curves will arise largely because of differences in the rate of change of the curve. They suggest finding, for each curve considered, the transformation of the slope that yields a straight line when the transformed values are plotted against time. The estimated slopes of the time series, for each year in the period under consideration, are then subjected to all such transformations. If one of these transformations yields values that approximate a straight line, then the curve corresponding to this transformation may be expected to be a satisfactory predictor.

The following transformations yield straight lines for the curves discussed in Section 13.3.

Trend Curve	Transformation
Exponential	Slope/moving average
Modified exponential	Logarithm of the slope
Gompertz	Logarithm of {slope/moving average}
Logistic	Logarithm of {slope/(moving average)^2}

The reader is referred to the book mentioned above for a further discussion and examples of the process of fitting trend curves to time-series data.

Although the use of a moving average may help to identify trend, moving average methods have several weaknesses as a smoothing device. First of all, this method is often only partially successful in removing all the effects of S, C, and I from $\mathbf{y}$ when determing T, because C and I usually do not have regular fluctuations. In addition, the moving average method tends to introduce spurious cyclical movements into the data being smoothed, so that an analyst who tries to remove cyclical effects from his data may, in fact, introduce a non-existent cycle.

A final problem with the use of the moving average is thc arbitrary choice of its length, denoted by h, the number of consecutive values used in the averages. The larger the value of h, the more the moving average smooths out the original data, but the greater is $(h - 1)$, the number of total observations at the beginning and end of the data for which no moving average value can be determined. Using the five-period moving average, we see in Table 13.6 that a total of four observations are lost in the moving-average series. The size of h must be large enough for smoothing purposes but small enough to retain sufficient observations. Also, to smooth cycles the value of h should be selected to coincide with the length of the cycle (or an integer multiple of its length). If the cycle length varies, some complex moving-average techniques that use different values of h may be most appropriate.

13.5 ESTIMATION OF SEASONAL AND CYCLICAL COMPONENTS

Until now we have been concerned with estimating only trend. Often, however, it may be just as (or even more) important to be able to estimate the *seasonal* component in a time series. From a planning point of view, for example, it is often necessary for business managers to take into consideration fluctuations other than trend when financing future operations, purchasing materials or merchandise, establishing employment practices, etc. Many of these fluctuations are of a seasonal nature, caused by the weather, as in the case of the increase in swim-wear sales during the summer months, or by social customs, as in the case of the increase in retail sales during the Christmas season.

Seasonal Index

The seasonal component of a time series is usually expressed by a number, called a seasonal index number, which expresses the value of the seasonal fluctuation in each month as a percent of the trend value expected for that month. A value of 104 thus means that, because of seasonal factors, the time-series value is expected to be four percent above the trend value. If the index number is 93, then the time-series value is expected to be seven percent less than the trend value.

Because of seasonal variation, special care must be taken in forecasting sales for any given month, or in comparing sales between months. Such forecasts or comparisons are usually made by using a seasonal index to *deseasonalize* the data. Since the base value of a seasonal index is always 100, the value of the seasonal index for each month represents how much above or below the overall monthly average that particular month usually falls. To illustrate this type of index, we will assume that bicycle sales for a given manufacturer have averaged 50,000 units a month for the past five years. However, the average sales in each month are not always equal to 50,000. In February, for example, sales over the past five years have averaged 30,000 units. Similarly, May has averaged 50,000 units, and December, 80,000 units. Since February has averaged only 60% of the overall average sales of 50,000, the seasonal index for February is $S_{\text{Feb.}} = 60.0$. The indexes for May and December would be $S_{\text{May}} = 100.0$ and $S_{\text{Dec.}} = 160$.

In deseasonalizing a time-series value, one first divides by the seasonal index, and then multiples by 100. That is,

$$\textit{Seasonally adjusted value} = \frac{\text{Original value} \times 100}{\text{Seasonal index}}. \tag{13.9}$$

We can illustrate the use of this formula to adjust values in our bicycle problem by assuming that the actual sales for February and December in a given year were 32,000 and 84,000, respectively. Adjusting these values we obtain:

$$\textit{Adjusted February sales} = \frac{32{,}000 \times 100}{60} = 53{,}000;$$

$$\textit{Adjusted December sales} = \frac{84{,}000 \times 100}{160} = 52{,}500.$$

Note how this seasonal adjustment makes the comparison of sales between the two months very easy. Here we see that February sales were slightly better than December sales on a seasonally adjusted basis. Because the average seasonal index for these two months exceeds 100 $[(60 + 160)/2 = 110]$, their total adjusted values (105,500) must be less than the total sales that actually occurred (116,000). If a seasonal index has been correctly formulated, the adjusted total for twelve months will always equal the actual total over all twelve months.

Ratio-to-Trend Method

In order to determine seasonal index numbers, it is necessary to estimate the seasonal component of a time series. Suppose that a given time series can be represented by the multiplicative model $y = T \times S \times C \times I$. One method for estimating the seasonal component in this model is called the *ratio-to-trend method*, which estimates S by removing trend from the series, but does not attempt to remove cyclical and irregular variation. Assume that a value of T has been calculated for each monthly value of y in the series. Trend can be eliminated by dividing y by these monthly trend values, $y/T = S \times C \times I$; the remaining fluctuations are assumed to represent, primarily, seasonal variation. We will illustrate the ratio-to-trend method using the data in Table 13.7, which represent the monthly sales in a department store over the three-year period from 1979 to 1981 (Tables 13.7, 13.8, and 13.9 appear on page 617).

The trend component of these observations can be estimated by fitting a least-squares regression line to the data. Rather than fit a line to all 24 points, we can estimate the trend line by using just the three monthly averages shown in the last column of Table 13.7 as follows:

Year	x	y	x^2	xy
1979	−1	73.0	1	−73.0
1980	0	86.8	0	0.0
1981	1	99.5	1	99.5
Sum	0	259.3	2	26.5

Since $\sum x_i = 0$, we can use the reduced form of the normal equations presented in Formula (13.2),

$$a = \sum_{i=1}^{n} y_i/n \quad \text{and} \quad b = \sum_{i=1}^{n} x_i y_i \Big/ \sum_{i=1}^{n} x_i^2.$$

Substituting the appropriate values into these equations we obtain a value of a equal to $259.3/3 = 86.4$ and b equal to $26.5/2 = 13.3$. These values determine

Table 13.7 Monthly Sales

Year	Jan.	Feb.	Mar.	Apr.	May	June	July	Aug.	Sept.	Oct.	Nov.	Dec.	Monthly Average
1979	67	71	72	73	71	70	67	64	66	74	82	99	73.0
1980	80	86	89	89	84	83	79	76	77	87	98	114	86.8
1981	92	98	101	103	96	96	89	87	90	99	112	131	99.5

Table 13.8 Monthly Trend

Year	Jan.	Feb.	Mar.	Apr.	May	June	July	Aug.	Sept.	Oct.	Nov.	Dec.
1979	67.2	68.3	69.4	70.5	71.6	72.7	73.8	74.9	76.0	77.1	78.2	79.3
1980	80.4	81.5	82.6	83.7	84.8	85.9	87.0	88.1	89.2	90.3	91.4	92.5
1981	93.6	94.7	95.8	96.9	98.0	99.1	100.2	101.3	107.4	103.5	104.6	105.7

Table 13.9 Seasonal Index Numbers

Year	Jan.	Feb.	Mar.	Apr.	May	June	July	Aug.	Sept.	Oct.	Nov.	Dec.
1979	99.7	103.9	103.7	103.5	99.1	96.2	90.7	85.4	86.8	95.9	104.8	124.8
1980	99.5	105.5	107.7	106.3	99.0	96.6	90.8	86.2	86.3	96.3	107.2	123.2
1981	98.2	103.4	105.4	106.2	97.9	96.8	88.8	85.8	87.8	95.6	107.0	123.9
Average	99.1	104.3	105.6	105.3	98.7	96.5	90.1	85.8	87.0	95.9	106.3	124.0

the following trend line:

$$\hat{y} = 86.4 + 13.3x, \tag{13.10}$$

with x in units of one year and $x = 0$ at mid-1980. Since we want to estimate monthly trend changes, this equation will be slightly easier to work with if x is expressed in months, so we need merely to divide the slope of the line by 12.* The resulting equation,

$$\hat{y} = 86.4 + 1.108x, \tag{13.11}$$

indicates that y will increase 1.108 each month due to trend, accumulating an annual increase of 13.3 as in Formula (13.10). When $x = 0$ the value of $\hat{y}$ is 86.4 for mid-1980. The difficulty with this value is that the value of $\hat{y}$ on July 1,

* To change from one trend equation to an equivalent one with different units for x, simply multiply the slope by the ratio of the size of the new unit for x to the size of the original unit for x. Since one year has 12 months, the appropriate ratio here is $\frac{1}{12}$.

1980 is really not representative of the entire month of July. A more representative date for July would be July 15th, or $\frac{1}{2}$ month later. The value of $\hat{y}$ estimated by the trend line for July 15 can be determined by substituting $x = \frac{1}{2}$ into Eq (13.11), so that $86.4 + 1.108(\frac{1}{2}) = 87.0$. To find the trend value for preceding months, approximately 1.1 is subtracted successively from 87.0. These values are shown in Table 13.8.

We have now determined a value of T for each month in the series. The next step is to divide each month's sales value shown in Table 13.7 by the corresponding monthly trend value given in Table 13.8. The results of this division are shown in the first three rows of Table 13.9. Since there are three different index numbers for each month in Table 13.9, we combine these values by taking their average, as shown in the bottom row of Table 13.9. The values shown in the bottom row of Table 13.9 are the monthly seasonal index values, and may be used to deseasonalize observations, as discussed before. The deseasonalized data can be more accurately examined for trend and cycles than the original data, since any confusion of seasonal movements is eliminated.

A final step that is often necessary in the construction of seasonal indexes is called *leveling the index*. The need for leveling may arise from the construction of seasonal indexes by either the ratio-to-moving-average method (which will be discussed in the next section), or the ratio-to-trend method. For example, in the ratio-to-trend method of constructing the index, raw seasonal index values are obtained by "averaging" the ratio-to-trend values for each quarter or month as the case may be. This averaging process is done without a restriction as to what the average of these "averages" must be over the entire year. However, the average of these "averages," or new seasonal indexes, must be 100 if the index is to alter only the pattern but not the level of the raw data series when deseasonalizing the series. For example, if the average over the year of the raw seasonal indexes were 90, the use of the raw seasonal indexes would raise the average level of the deseasonalized series by $[(\frac{100}{90}) - 1]$ or about 11 percent above the average level of the raw data series. Clearly, this is an unintended result of deasonalizing the series. The seasonal index in Table 13.9 averaged 100; hence, leveling was not necessary in this case.

In order to level an index, one finds the average of the raw seasonal indexes over the year and divides that average value into each raw seasonal index or "average" for each month or quarter. The effect of this computation is to increase or decrease all values of the raw seasonals in the same proportion and to make the leveled index average exactly 100 over the year. Use of the leveled index to deseasonalize a series alters the pattern within the year and tends to reduce monthly distortions in the level of the index.

Perhaps Fig. 13.11 will help summarize some of the material in this section. The lower portion of this figure shows the original sales data from Table 13.7 and the trend component of this data (Table 13.8). The seasonal index at the top of the figure was derived by dividing each of the trend values into the sales data, and then averaging, for each month, over the three years 1979, 1980, 1981.

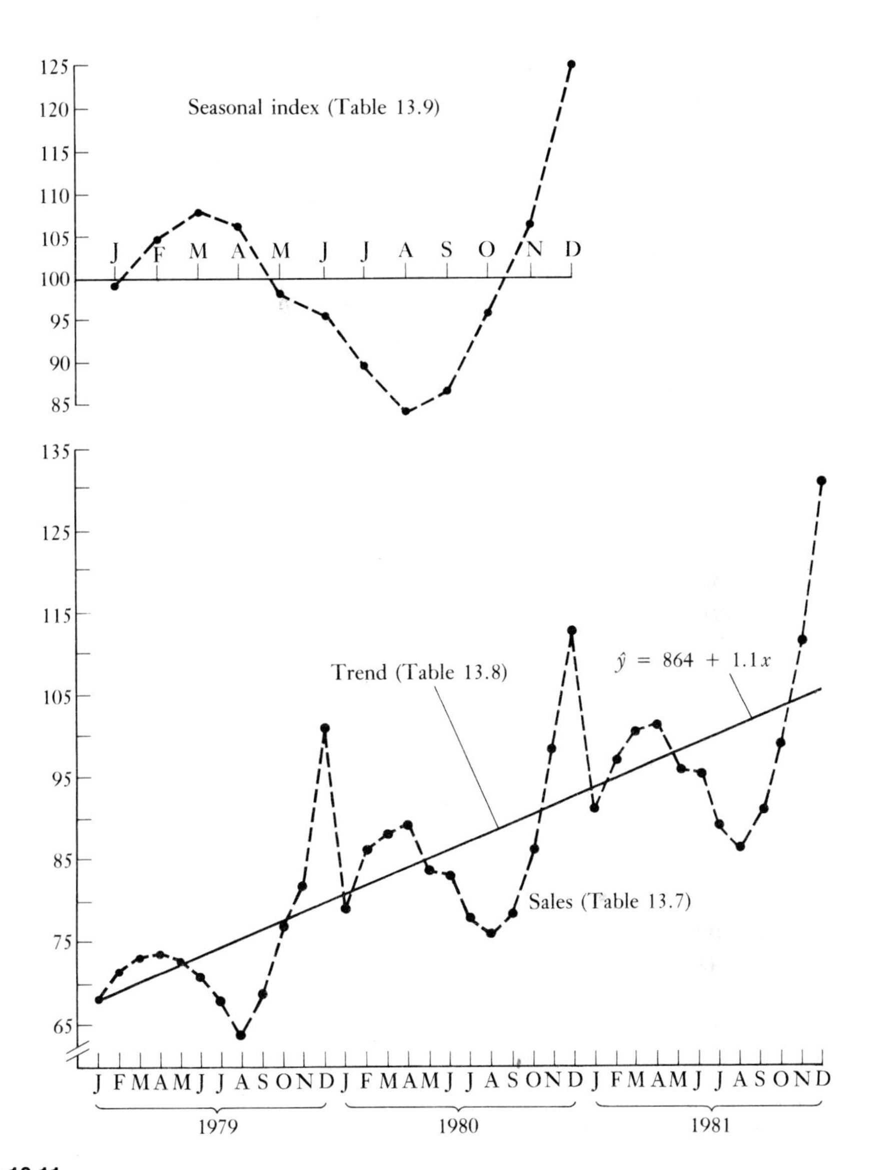

Figure 13.11

Ratio-to-Moving-Average Method

The ratio-to-trend procedure we have just described produces a seasonal index that still contains C and I. Although the process is considerably more cumbersome, there is a way to calculate an index, called the *ratio-to-moving-average method*, in which these fluctuations have been removed. The first step in this approach is to smooth the data by using a 12-month moving average. If we disregard the irregular component, the causes and occurrences of which are unknown, the smoothed series will contain fluctuations attributable only to T and C, since a 12-month moving average will remove seasonal variations. In terms of the multiplicative model we have, in effect, divided $\boldsymbol{y}$ by S, so that the new series is $\boldsymbol{y}/S = T \times C$. If the *original* time-series values, $\boldsymbol{y}$, are now divided by this newly calculated series containing only T and C, the result is $y/(T \times C) = (T \times S \times C)/(T \times C) = S$. Thus, by a rather roundabout route, we have isolated S and determined a seasonal index in which the variations attributable to T and C have been removed. To illustrate how this process works, we again use the the data of Table 13.7.

First, a 12-month moving average is taken over all months in the series. This moving average is shown in columns 3, 6, and 9 of Table 13.10. Now, to associate the moving-average values with the fifteenth rather than the first day of each month, these values are "centered" by taking the average of each two adjacent months. The centered moving-average values are shown in columns 4, 7, and 10 and represent $T \times C$ and some parts of I.

We have now eliminated S and some remaining parts of I from the series. To find a seasonal index we must divide the original time-series values in Table 13.7 by the 12-month centered moving-average values in Table 13.10. The result of this division, shown in Table 13.11, is two index numbers for each month of the year.

Table 13.10

(1) Month	(2) 1979 Sales	(3) 12-Month M.A.	(4) Centered 12-Month M.A.	(5) 1980 Sales	(6) 12-Month M.A.	(7) Centered 12-Month M.A.	(8) 1981 Sales	(9) 12-Month M.A.	(10) Centered 12-Month M.A.
Jan.	67			80		80.8	92		93.6
					81.3			94.0	
Feb.	71			86		81.8	98		94.5
					82.3			94.9	
Mar.	72			89		82.8	101		95.5
					83.3			96.0	
Apr.	73			89		83.8	103		96.5
					84.3			97.0	
May	71			84		85.0	96		97.6
					85.7			98.2	
June	70			83		86.3	96		98.9
		73.0			86.9			99.6	
July	67		73.6	79		87.4	89		
		74.1			87.9				
Aug.	64		74.7	76		88.4	87		
		75.3			88.9				
Sept.	66		76.1	77		89.4	90		
		76.8			89.9				
Oct.	74		77.5	87		90.6	99		
		78.1			91.2				
Nov.	82		78.7	98		91.7	112		
		79.2			92.1				
Dec.	99		79.8	114		92.7	131		
		80.3			93.2				

Table 13.11 Seasonal Index Numbers

Year	Jan.	Feb.	Mar.	Apr.	May	June	July	Aug.	Sept.	Oct.	Nov.	Dec.
1979							91.0	85.7	86.9	95.5	104.1	124.0
1980	99.0	105.1	107.1	106.1	98.8	96.2	90.4	86.0	86.2	96.0	106.7	122.9
1981	98.4	103.9	105.9	106.8	98.5	97.0						
Aver.	98.7	104.5	106.5	106.5	98.7	96.6	90.7	85.9	86.6	95.8	105.4	123.5

Since these index numbers may differ, some irregular component variation I is still included. By averaging the month $S \times I$ estimates for each calendar month, some more of the irregular component is eliminated. In a long time series with at least five values of $S \times I$ for each month, this averaging virtually isolates the pure seasonal index component S, with T, C, and I removed. The averages of these two values are the seasonal indexes found in the bottom row of Table 13.11, which are quite similar to those in Table 13.9, where the ratio-to-trend method was used.

Dividing a time series by the appropriate seasonal index values, such as those shown in Table 13.11, "seasonally adjusts" a time series so that the changes in the series do not reflect merely seasonal changes. Quarterly GNP figures, for example, are usually adjusted to remove seasonal fluctuations. We must point out, as we did at the end of the discussion on the ratio-to-trend method, that "leveling the index" may be necessary to ensure that the seasonal index has an average of 100.

Determining the Cyclical Variation

In some analyses, it is desirable to isolate the cyclical component of a time series so that turning points and peaks and troughs may be studied. If a stable cyclical pattern, such as a 3.4-year business cycle or a 20-year housing construction cycle could be isolated, it would greatly aid economists in determining the underlying causes and in forecasting future movements. Quite a wide variety of complicated methods have been developed for studying the cyclical component of a time series. However, we can use a roundabout method again, to find C in a simple way.

Given a time series $y = T \times C \times S \times I$, the seasonal component S can be determined by the ratio-to-moving-average method. The values of y can then be divided by the seasonal index S to obtain $y \times 100/S = T \times C \times I$. Next, we can eliminate the trend component from $T \times C \times I$ by a least-squares fit, using the most appropriate linear or nonlinear curve. Dividing $T \times C \times I$ by the values of $\hat{y}$ obtained from this least-squares analysis leaves us with only the components $C \times I$. To eliminate I, a weighted moving average of three or five periods can be determined, using the $C \times I$ data. If a longer moving average were used, the cycles might be smoothed out too much and partially eliminated also. The best type of weights to use are large weights for the center values and and much smaller weights for the values farther away from the center. If the weights are constructed so that their sum is equal to one, then no division by the sum of the weights is necessary. For example in a five-period moving average, the weights might be -0.1, 0.3, 0.6, 0.3, and -0.1. The advantage of this extra computation using weights is that it produces a smoother curve for the cycle component C, while it is also more sensitive to the original fluctuations because it preserves the amplitude of the cycles more faithfully. Using the odd-period moving average is convenient for centering the resulting values (which constitute the component C). For example, based on the monthly data for our sales problem, the cyclical component for April would be:

$$C_{\text{April}} = -0.1(C \times I)_{\text{Feb.}} + 0.3(C \times I)_{\text{March}} + 0.6(C \times I)_{\text{April}} + 0.3(C \times I)_{\text{May}} - 0.1(C \times I)_{\text{June}}.$$

Values of C for other months may be found in a similar way.

*13.6 AN EXPONENTIAL SMOOTHING MODEL (OPTIONAL)

Assume that a forecast is to be made of the sales of a certain product at time $t + 1$ (e.g., month 4) on the basis of information available at time t (month 3). To make this forecast, the exponential method uses two components, both based on the logic that the best prediction of what will happen in time period $t + 1$ is a function of information available at time period t. These two components are (1) the sales *estimated* to take place in time period t, and (2) the sales which *actually* occurred during time period t. We use both actual and estimated values because either one alone might be a poor basis for prediction of future sales. For example, actual sales might be unusually high or low in one month because of some random factor, such as a strike or bad weather. Estimated sales in time period t might also be a poor predictor of what will happen in time period $t + 1$ if the conditions which led to such an estimate no longer hold true in estimating sales for period $t + 1$. The name "exponential smoothing" is derived from the fact that in this procedure the weights given to past data decrease exponentially with the age of the data.

In exponential smoothing, actual sales in the period being estimated, $t + 1$, are considered a random variable. That is, many different values could take place, and there exists a probability distribution describing the relative frequency of these events. If S_{t+1} represents this random variable, then $\bar{S}_{t+1}$ stands for its expected value or mean. Unfortunately, $\bar{S}_{t+1}$ cannot be determined directly since there is usually no way of determining the probability distribution of S_{t+1}. We must therefore estimate the value of $\bar{S}_{t+1}$. Suppose we let $\hat{\bar{S}}_{t+1}$ represent our estimate of the expected value of S_{t+1}—that is, it is our forecast of sales for time period $t + 1$. The components necessary for making the estimate $\hat{\bar{S}}_{t+1}$ are the actual sales for time period t, S_t, and the mean sales which were estimated (in time period $t - 1$) to occur in time period t, $\hat{\bar{S}}_t$. These two components are weighted in determining $\hat{\bar{S}}_{t+1}$ by a smoothing constant, $a(0 \leq a \leq 1)$, as follows:

$$\textit{Estimated mean sales in period } t+1: \qquad \hat{\bar{S}}_{t+1} = aS_t + (1 - a)\hat{\bar{S}}_t. \qquad (13.12)$$

The smoothing constant a in Formula (13.12) determines the relative importance attached to S_t in estimating $\bar{S}_{t+1}$ as compared to the importance of $\bar{S}_t$. An a value greater than $\frac{1}{2}$ gives more weight to actual sales data, while an a value of less than $\frac{1}{2}$ will give more weight to the estimated sales data. It should be noted that the difficult choice of the value of the smoothing constant a is usually based on an analysis of reliable historical data. For the purpose of illustrating the method of exponential smoothing, suppose we assume that $a = \frac{3}{4}$. In the example which follows estimated mean sales are thus determined by using the relationship:

$$\hat{\bar{S}}_{t+1} = \frac{3}{4} S_t + \frac{1}{4} \hat{\bar{S}}_t.$$

Time (t)	Estimated Sales ($\hat{\bar{S}}_t$)	Actual Sales (S_t)	Estimated Sales ($\hat{\bar{S}}_{t+1}$)
1	$\hat{\bar{S}}_1 = 100$	$S_1 = 116$	$\hat{\bar{S}}_2 = 112$
2	$\hat{\bar{S}}_2 = 112$	$S_2 = 128$	$\hat{\bar{S}}_3 = 124$
3	$\hat{\bar{S}}_3 = 124$	$S_3 = 120$	$\hat{\bar{S}}_4 = 121$

Formula (13.12) can be expanded to show that the weights given to past data decrease exponentially with the age of the data. First note the following:

$$\begin{aligned}
\hat{\bar{S}}_t &= aS_{t-1} + (1 - a)\hat{\bar{S}}_{t-1}, \\
\hat{\bar{S}}_{t-1} &= aS_{t-2} + (1 - a)\hat{\bar{S}}_{t-2}, \\
&\;\;\vdots \qquad\qquad \vdots \\
\hat{\bar{S}}_{t-n} &= aS_{t-n-1} + (1 - a)\hat{\bar{S}}_{t-n-1}.
\end{aligned} \qquad (13.13)$$

Now by repeated use of Formula (13.13), expansion of (13.12) yields:

$$\begin{aligned}
\hat{\bar{S}}_{t+1} &= aS_t + (1 - a)[aS_{t-1} + (1 - a)\hat{\bar{S}}_{t-1}] \\
&= aS_t + a(1 - a)S_{t-1} + (1 - a)^2\hat{\bar{S}}_{t-1} \\
&= aS_t + a(1 - a)S_{t-1} + (1 - a)^2[aS_{t-2} + (1 - a)\hat{\bar{S}}_{t-2}] \\
&= aS_t + a(1 - a)S_{t-1} + a(1 - a)^2S_{t-2} + (1 - a)^3\hat{\bar{S}}_{t-2} \\
&\;\;\vdots \\
&= aS_t + a(1 - a)S_{t-1} + a(1 - a)^2S_{t-2} + \cdots \\
&\quad + a(1 - a)^{n+1}S_{t-n-1} + (1 - a)^{n+2}\hat{\bar{S}}_{t-n-1}.
\end{aligned} \qquad (13.14)$$

From Formula (13.14) it can be seen that the weights assigned to past data are a, $a(1 - a)$, $a(1 - a)^2, \ldots$, so that these weights do, in fact, decrease exponentially. It can be shown that the sum of these weights equals one by using the following fact about the sum of a geometric series:

$$\sum_{i=0}^{n} x^i \cong \frac{1}{1 - x} \qquad \text{(for } n \text{ large and } 0 < x < 1\text{)}.$$

If we let $1 - a = x$, then from the above equation

$$(1 - a)^i \cong \frac{1}{1 - (1 - a)} = \frac{1}{a}. \tag{13.15}$$

Multiplying both sides of (13.15) by a gives

$$a \sum_{i=0}^{n} (1 - a)^i \cong a(1/a) = 1.$$

Since $a\sum_{i=0}^{n} (1 - a)^i$ equals the sum of the weights in (13.14), we have shown that these weights sum to one.

Define: Trend (T), seasonal variation (S), cyclical variation (C), irregular variation (I), additive model, multiplicative model, method of semi-averages, reduced normal equations, exponential curve, modified exponential curve, logistic curve, Gompertz curve, method of moving averages, weighted moving average, seasonal index, ratio-to-trend method, ratio-to-moving-average method, leveling an index, exponential smoothing.

Problems

Note. Additional problems covering the material in Section 13.3 are found in the Exercises at the end of this chapter.

13.1 What is a time series? Why is it necessary to distinguish time-series analysis from regression analysis?

13.2 Personal consumption expenditures (y) for a given population during the years 1979–1983 were as shown in Table 13.12.

Table 13.12

y	Years
370	1979
390	1980
395	1981
410	1982
435	1983

a) Fit a straight-line trend line for y by the method of least squares. Be sure to indicate the origin.

b) Give the amount of total deviation, explained deviation, and unexplained deviation for the 1982 observation.

13.3 Suppose that the trend line for average annual wages of members of a local electricians' union from 1960–1980 is $\hat{y} = 16{,}500 + 300x$, with origin at mid-1969 and x in units of one year.

a) What is the average annual increase in wages over this period?

b) Suppose a local nonunion electrician made $19,000 in 1980. How much did this person's income differ from the trend value in 1980 for union members?

13.4 A study is being made to determine the growth of annual honey production (y) over time (x) measured in years before and after 1981. The following data is given ($x = 0$ is 1980):

x	−2	−1	0	1	2
y	6	9	12	11	15

a) Find the regression equation of y on x.

b) Find the estimated value of honey production for 1983.

13.5 The value of building permits in a certain town over an eight-year period is given in Table 13.13.

Table 13.13

x (years)	Value (thousands)	x (years)	Value (thousands)
1975	$300	1979	$310
1976	150	1980	490
1977	210	1981	380
1978	400	1982	400

a) Plot the above data and make a freehand estimate of the trend line.

b) Use the method of semi-averages to construct a trend line.

c) Use the method of least squares to determine a trend line.

d) Compare your estimates of the trend line for parts (a), (b), and (c). What value would you estimate for 1983 under each of these methods?

e) Find the least-squares regression line for quarterly magnitudes, with x measured in years. Shift the origin one-half quarter to the right.

f) Find the trend line, using annual values reported quarterly.

13.6 The values shown in Table 13.14, represent sales data for the years 1977–82.

a) Plot the values and then estimate a linear trend line by the freehand method.

b) Use the method of semi-averages to estimate a trend line.

c) Construct a least-squares regression line to estimate trend.

Table 13.14

Year	Sales (units)
1977	18,000
1978	19,000
1979	23,000
1980	24,000
1981	26,000
1982	28,000

d) Find the least-squares regression line for quarterly magnitudes, with x measured in years. Shift the origin one-half unit to the left.

e) Find the trend line, using annual values reported quarterly.

13.7 The sales of a new product had the growth pattern shown in Table 13.15.

a) Fit a Gompertz to these data.

b) Fit a logistic curve.

c) Graph both of the curves derived above, and then indicate which model appears to give the better fit.

Table 13.15

x (years)	Sales (thousands)	x (years)	Sales (thousands)
1	1.6	4	4.6
2	2.7	5	5.1
3	3.9	6	5.4

13.8 Describe one advantage and one disadvantage of using a moving-average smoothing method in time-series analysis.

13.9 Find a three-year moving average for the data shown in Table 13.16.

Table 13.16

Year	Sales (thousands)
1974	$18
1975	20
1976	22
1977	19
1978	21
1979	24
1980	21
1981	23
1982	27

13.10 Given the amounts of money shown in Table 13.17, (to the nearest thousand dollars) gambled and lost each year by a certain businessman during his vacations in Las Vegas: Find the four-period moving average for the amounts lost.

Table 13.17

Year	Amount	Year	Amount
1973	2	1978	23
1974	7	1979	29
1975	5	1980	68
1976	16	1981	2
1977	12		

13.11 a) Compute the trend equation for the following series (Table 13.18) by the method of least squares.

b) Compare the trend value for 1980 with the three-period moving average value for 1980.

Table 13.18

Year	Tons	Year	Tons
1975	30	1979	51
1976	44	1980	68
1977	50	1981	65
1978	42		

13.12 a) State two reasons for estimating the seasonal component of a time series.

b) State two reasons for estimating the trend equation for a time series.

c) State two disadvantages to the use of a (simple) moving average as a fit of trend.

13.13 Assume that the trend line (annual total equation) for suits is $\hat{y} = 3600 + 480x$ with origin at October, 1980. The seasonal index for April is 80. Estimate the seasonally adjusted output for April, 1984.

13.14 The operating season for a certain tomato cannery is from June to October. Suppose that the observations shown in Table 13.19 represent monthly sales, in thousands, for this cannery from 1981 to 1983.

Table 13.19

Year	June	July	August	September	October
1981	75	86	102	105	90
1982	83	89	110	115	92
1983	84	95	113	118	89

a) Use the ratio-to-trend method to find a seasonal index for sales for the months June to October.

b) Find a seasonal index using the ratio-to-moving-average method, using a five-month moving average.

c) Assume that the multiplicative model $y = T \times S \times C \times I$ holds, and that there is no irregular variation. Decompose the index for August of 1982 into the component parts T, S, and C, using your answer to part (a).

d) Repeat the above process, using your answer to part (b).

13.15 Explain how to find the cyclical component of a time series. Explain the difference between cyclical and seasonal variations in a time series.

13.16 The Business Research Department of the Carolina Corporation forecasts sales for next year of $12 million, based on a trend projection. It is expected that no sharp cyclical fluctuations will occur during the year, that the effect of trend *within* the year will be negligible, and that the past pattern of quarterly seasonal variation will continue. The pattern is as follows:

Quarter	1st	2nd	3rd	4th
Seasonal Index	130	90	75	105

Prepare a forecast of quarterly sales from the above information for the first and second quarters of next year.

13.17 In finding the cyclical component of a time series, explain how to obtain C once the CI component has been isolated.

13.18 a) The unadjusted sales index of Company C-B is 102 for January, 1982. The seasonally adjusted index for the same month is 133. Find the seasonal index for January.

b) The annual sales for 1984 are forecast to be $240,000. The seasonal index for March is computed to be 90. Give a reasonable forecast of the sales for March, 1984.

13.19 Construct, using the data in Problem 13.6, an estimate of sales for 1983 using the exponential smoothing model. Use only the data from 1979–82 and give twice as much weight to 1982 sales data compared to past sales data.

13.20 In using the exponential smoothing model to estimate future sales, both present and past sales are incorporated into the estimate. The value of the constant a determines just how much weight is given present sales versus past sales.

a) What value of a should be used if it is desired that exactly 34.39 percent of the total weight be given to sales from the current period plus sales from the three periods preceding the current period?

b) What should the value of a be if 59.04 percent of the weight must come from these same four periods?

13.21 The index of seasonal variation for cement production for selected months is 60 for January, 70 for March, 122 for August, and 100 for November.

a) In which of these months is cement production the greatest?

b) In which of these months is the production most typical of the monthly average?

c) Production in a certain region increased from 6,060,000 barrels in January, 1980, to 11,590,000 barrels in August, 1980. Determine the percentage change in cement production, allowing for seasonal variation.

13.22 Assume that you have been given the following quarterly seasonal index.

Quarter	1st	2nd	3rd	4th
Seasonal Index	90	115	95	108

Note that these seasonal values do not sum to 100. Prepare a new seasonal index by leveling these values so that the sum equals 100.

13.23 Consider the monthly data given in Table 13.20, representing female employment levels for the two-year period 1979–80 (in millions).

a) Plot the above data, letting $x = 0$ represent January 1, 1980.

b) Use the method of least squares to determine a trend line.

c) Construct a seasonal index using the ratio-to-trend method.

d) Find a seasonal index using the ratio-to-moving-average method, based on a five-month moving average.

Table 13.20

	Year	
Month	**1979**	**1980**
Jan	22.1	22.9
Feb	21.9	22.3
Mar	22.6	23.0
Apr	23.4	23.9
May	24.5	25.0
June	26.0	26.5
July	26.0	26.6
Aug	25.9	25.7
Sept	25.7	25.9
Oct	25.4	25.7
Nov	24.6	24.1
Dec	24.1	24.6

13.7 INDEX NUMBERS

Earlier in this chapter, index numbers were used to express the seasonal components of a time series. Our objective at that point and in the following discussion is to develop a measure that summarizes the characteristics of large masses of data. An index number accomplishes this purpose by relating the values of a dependent variable, such as sales or price, to an independent variable, such as time, income, or occupation. Although index numbers are used in many areas of the behavioral and social sciences, their main application involves describing business and economic activity, such as changes in price, production, wages, and employment, over a period of time.

Uses of Index Numbers

The primary purposes of an index number are to provide a value useful for comparing magnitudes of aggregates of related variables to each other, and to measure the changes in these magnitudes over time. Consequently, many different index numbers have been developed for special uses. Let us briefly mention some of their common uses.

First, index numbers are useful summary measures for policy guides. The Federal Reserve Board may use index numbers on interest rates, employment, or consumer credit as inputs to discussion on appropriate open-market transactions. Presidential advisors may use price and income indexes to set wage and price guidelines.

Second, many index numbers have been developed as indicators of business conditions, including the *Forbes*, *Fortune*, or *Business Week* indexes. The National Bureau of Economic Research, the Bureau of Labor Statistics, and the Commerce Department all publish various index numbers for this same purpose.

Also, some of these indexes are used commonly for comparing changes among different sectors of the economy. Growth in the agricultural or mining sector might be compared to growth in the manufacturing sector. Levels of state and local government expenditures might be analyzed and compared among regions by using index numbers.

A fourth use of certain special indexes, such as wage, productivity, and cost-of-living indexes, is in wage contracts and labor-management bargaining. Management often likes to tie wage increases to productivity increases, while labor unions like to relate the need for wage increases to cost-of-living increases. Similar relations among indexes are used in escalator clauses to adjust insurance coverage, change retirement and social security benefits, etc.

Finally, a fifth and very common use of an index number is as a deflator. Certain measures of economic activity are divided by price or cost indexes in order to obtain the *real* or constant dollar value of these measures. That is, an adjustment is made for the changing value of the dollar, so that more meaningful comparisons can be made over time.

Using an Index as a Deflator

The use of an index as a deflator is so common that it merits illustration. Suppose that a person's income increases from $10,000 to $15,000 over a given period, while the consumer price index (C.P.I.) increases from 100 to 130. The nominal increase in income of \$5000 is offset by general inflation indicated by the 30-percent increase in the C.P.I. If this index is assumed to be relevant to the consumer purchases of such a person, it could be used to "deflate" the income increase and to find the "real" increase in income. Since the base value of any index number is 100, the real income is obtained by the rule:

$$\text{Real income} = \frac{(\text{Nominal income}) \times 100}{\text{Consumer price index}}.$$

The real income at the beginning and at the end of the period being studied, respectively, is:

$$\frac{10{,}000 \times 100}{100} = \$10{,}000 \qquad \text{(Beginning real income)},$$

$$\frac{15{,}000 \times 100}{130} = \$11{,}538 \qquad \text{(End real income)}.$$

With the extra \$5000 income, this person could buy only an extra \$1538 worth of goods and services because prices increased. Real income is usually called the "purchasing power" of the money income. In comparing incomes, wages, or rents of individuals, or gross national product, or personal income per capita of different countries the use of an appropriate deflator is common practice. In this way, the true standard of living is more easily recognized.

As another example, suppose that a state government allocated \$25 million for highway construction in 1975, when costs of labor, materials, land, equipment, etc., were measured by a construction cost index to be 125 (relative to a base period equal to 100, in 1972). Suppose that in 1982 some highway projects receive funding of \$30 million. However, the index of costs in 1982 has risen to 180. We find the real value of the available money by deflating it in terms of constant base-year costs in each case, using the cost index:

$$\text{Real value in 1975} = \frac{\$25 \text{ million} \times 100}{125} = \$20 \text{ million};$$

$$\text{Real value in 1982} = \frac{\$30 \text{ million} \times 100}{180} = \$16.67 \text{ million}.$$

The 1982 allocation will build only 83 percent as much as the lower 1975 allocation would have permitted seven years earlier.

Constructing an Index

Having discussed the nature and uses of index numbers, we turn in the next section to a discussion of constructing a price index. While this is only one type of index, it shares many problems of construction with indexes of all other types. The reader should be ready to recognize the following usual considerations.

Each index number must have a base period for which the value of the index is 100.0. The selection of a typical base year is somewhat arbitrary but very important. If a particularly exceptional period were selected in which the values of the variables were quite extreme, then all other values of the index would be affected.

A second arbitrary choice involves the selection of the type of index to use (simple, relative, weighted, etc.) and of the appropriate weights to use in combining the values of different items included in the aggregate measure. Incidentally, the choice of items to include is often very debatable. An attempt should be made to include the most common and representative items. An index for new car prices in the United States may not include prices for all possible cars that could be purchased, but it surely must include the most popular cars of the largest auto-makers. Similarly, an index of United States wages per work-hour in manufacturing must include a sample of representative types of workers across a representative group of industries selected to be representative of all regions of the country. The problems inherent in such selections are numerous. The final result may be some representative index number that does not suit anyone in particular.

The most compelling practical problem in constructing an index is obtaining sufficient and accurate data for all the items included. To obtain a consumer price index, one might sample consumers to find out how much they paid and what quantity they bought of a large list of products, or one might sample storekeepers to find out their sales and prices on all the different items. Once a set of quantities and prices is obtained, the calculation of the index is essentially a problem involving weighted averages.

Effectively, an index number is simply the ratio of two numbers expressed as a percentage. For example, consider the data in Table 13.21. Six values of the civilian labor force for selected years are shown in column 2. If the year 1954 is

Table 13.21 Civilian Labor Force Index

(1) Year	(2) Civilian Labor Force (Millions)	(3) Index with 1954 = 100
1950	63	98.4
1952	63	98.4
1954	64	100.0
1956	68	106.3
1958	71	110.9
1960	72	112.5

chosen as the base year, then the index number for year i, as shown in column 3, is obtained as follows:

$$(\text{Index number})_i = \frac{y_i \times 100}{y_{\text{Base}}}.$$

For example, the index number for 1960 is $(72 \times 100)/64 = 112.5$. This indicates a 12.5-percent increase in 1960 relative to the base year 1954.

13.8 PRICE INDEX NUMBERS

Since prices play a major role in every economy, there has been great interest in developing appropriate price indexes. In this section, several different types of price index are defined, and some of the procedures for determining their value are outlined. Most of these use the simple *price relative* as a cornerstone. It is an index number for prices of a single item similar to the index number for civilian employment in Table 13.21. A price relative is the ratio of the price of a certain commodity in a given period (p_n) to the price of that same commodity in some base period (p_0). If we let I_n represent the index number for a given year relative to the base year, then $I_n = (p_n/p_0) \times 100$. Following this convention, the index for the base year is 100, since $I_0 = (p_0/p_0) \times 100$. In general, a price relative shows the percentage increase or decrease in the price of a commodity from the base period to a given period.

To illustrate the construction of a price relative, we will apply the concept to the price of eggs for the years from 1977 to 1979. Assume that the price of of eggs for these three years was 60, 68, and 80 cents a dozen, respectively. With 1977 as the base period or year (1977 = 100), the price relatives for the three years are:

$$\text{1977 price relative} = \left(\frac{p_0}{p_0}\right) \times 100 = \frac{60}{60} \times 100 = 100;$$

$$\text{1978 price relative} = \left(\frac{p_1}{p_0}\right) \times 100 = \frac{68}{60} \times 100 = 113;$$

$$\text{1979 price relative} = \left(\frac{p_2}{p_0}\right) \times 100 = \frac{80}{60} \times 100 = 133.$$

Thus the price of eggs increased 13 percent from 1977 to 1978 and 33 percent from 1977 to 1979.

Simple Indexes

Rather than describe changes in a single commodity, one may wish to describe changes in the general price level. To do this we combine a representative

Table 13.22

Commodity	Quantity Units	Prices 1977 (p_0)	1980 (p_3)	Price Relative $\left(\frac{p_3}{p_0}\right)$
Milk	1 qt	$0.60	$0.75	1.25
Bread	1 loaf	0.60	0.80	1.33
Cheese	4 oz	0.40	0.45	1.13
Margarine	1 lb	0.55	0.65	1.18
Oranges	1 doz	1.80	1.95	1.08
Total		$3.95	$4.60	5.97

number of commodities, sometimes called a *market basket*, into an aggregate price index. The prices of these commodities in a given year are compared to the prices of the same market basket in a base year. One such index, called the *simple aggregate price index*, can be expressed as follows:

Simple aggregate price index:
$$I_n = \frac{\sum p_n}{\sum p_0} \times 100.$$

Suppose that we use the data in Table 13.22 to compute a simple aggregate price index for 1980, based on 1977 prices. The sum of the prices in 1980 (p_3) is $4.60, while the sum of the prices in 1977 (p_0) is $3.95; the ratio of these two numbers times 100 gives the simple aggregate price index

$$I_3 = \frac{\sum p_3}{\sum p_0} \times 100 = \frac{4.60}{3.95} \times 100 = 116.5.$$

The simple aggregate price index has two main disadvantages. First, it fails to consider the relative importance of the commodities; secondly, the use of different units to measure each commodity affects the price index. The second disadvantage can be overcome merely by changing to a *simple average of price relatives* index. As the name suggests, this index equals the average of all price relatives. Since $\sum(p_n/p_0)$ equals the sum of these price relatives and N equals the total number of commodities, then:

Simple average of price relatives:
$$I_n = \frac{\sum(p_n/p_0)}{N} \times 100.$$

A simple average of price relatives index constructed for the data given in Table 13.22 can be determined by dividing the sum total of the price relative, 5.97, by the number of commodities, 5 (and then multiplying by 100):

$$I_3 = \frac{5.97}{5} \times 100 = 119.4.$$

Weighted Price Indexes

Weighted index numbers overcome the first disadvantage of the simple index. They permit consideration of the relative importance of the commodities in the market basket, which is usually measured in terms of the total amount of money spent on each commodity during a year. The product of the price of a commodity and the quantity consumed in a given year represents the total amount of money consumers spend on that particular commodity. If q_i represents the quantity consumed in year i, then $p_0 q_0$ equals the total amount spent on a particular commodity in the base year, and $p_n q_n$ represents the amount spent some other year. A *weighted average of price relatives* index weights the various price relatives in a market basket by the total amount spent on that commodity. This index may take on two different forms, depending on whether base-year weights ($p_0 q_0$) or the weights of some given year ($p_n q_n$) are used in constructing the index.

Weighted average of price relatives:

$$\text{Base-year weights:} \quad I_n = \frac{\sum\left(\frac{p_n}{p_0} p_0 q_0\right)}{\sum p_0 q_0} \times 100;$$

$$\text{Given-year weights:} \quad I_n = \frac{\sum\left(\frac{p_n}{p_0} p_n q_n\right)}{\sum p_n q_n} \times 100. \tag{13.16}$$

A weighted average of price relatives index can be constructed from the data in Table 13.22 if the quantities purchased for each commodity are known. Suppose that the monthly purchases of a typical consumer are those shown in Table 13.23. The weighted price relative for milk, for instance, using base-year weights of $p_0 q_0 = \$18.00$, would be

$$\frac{p_3}{p_0} p_0 q_0 = (1.25)(\$18.00) = \$22.50.$$

The weighted price relative values given in Table 13.24 are calculated in the same fashion.

Table 13.23

	Quantities					
Commodity	1977 (q_0)	1980 (q_3)	p_0q_0	p_3q_0	p_0q_3	p_3q_3
Milk	30	35	\$18.00	\$22.50	\$21.00	\$26.25
Bread	25	20	15.00	20.00	12.00	16.00
Cheese	20	30	8.00	9.00	12.00	13.50
Margarine	10	5	5.50	6.50	2.75	3.25
Oranges	15	20	27.00	29.25	36.00	39.00
Total			\$73.50	\$87.25	\$83.75	\$98.00

Table 13.24

Commodity	Base-Year Weights $\frac{p_3}{p_0}p_0q_0$	Given-Year Weights $\frac{p_3}{p_0}p_3q_3$
Milk	\$22.50	\$32.81
Bread	20.00	21.33
Cheese	9.00	15.25
Margarine	6.50	3.84
Oranges	29.25	42.25
Total	\$87.25	\$115.48

We can now use this information to calculate the two types of weighted average of price relatives. Using base-year weights p_0q_0:

$$\frac{\sum[(p_3/p_0)p_0q_0]}{\sum p_0q_0} \times 100 = \frac{\$87.25}{\$73.50} \times 100 = 118.7.$$

Using given-year weights p_3q_3:

$$\frac{\sum[(p_3/p_0)p_3q_3]}{\sum p_3q_3} \times 100 = \frac{\$115.48}{\$98.00} \times 100 = 117.8.$$

In this example, the two methods give results that happen to be quite close together.

A popular means of constructing a weighted price index does not employ weighted price relatives but rather weights each price directly by multiplying it by the quantity of that commodity consumed in either the base year or some other year. The value $p_n q_0$, for example, represents the price of a commodity in year n weighted by the quantity of the commodity consumed in the base year. The total theoretical value of the commodities in year n is thus $\sum p_n q_0$. If we now take the ratio of this value to the actual value of these goods in the base year, $\sum p_0 q_0$, the resulting index is called a *Laspeyres price index:*

$$\text{Laspeyres price index:} \qquad I_n = \frac{\sum p_n q_0}{\sum p_0 q_0} \times 100. \tag{13.17}$$

If, instead of q_0, the quantity q_n consumed in the given year n is used to weight prices, the resulting index is called a *Paasche* price index:

$$\text{Paasche price index:} \qquad I_n = \frac{\sum q_n q_n}{\sum p_0 q_n} \times 100. \tag{13.18}$$

Note that the Laspeyres price index is equivalent to the weighted average of price relatives index using base-year weights, since

$$\frac{\sum[(p_n/p_0)p_0 q_0]}{\sum p_0 q_0} \times 100 = \frac{\sum p_n q_0}{\sum p_0 q_0} \times 100.$$

Similarly, the Paasche price index is like the weighted average of price relative index using given year weights, as we can see by comparing Formulas (13.16) and (13.18). The Laspeyres and Paasche price indexes can be constructed from the data in Table 13.23 by using the information in the last four columns. The result obtained is about the same as for the weighted average of price relatives indexes:

$$\text{Laspeyres price index:} \qquad \frac{\sum p_3 q_0}{\sum p_0 q_0} \times 100 = \frac{\$87.25}{\$73.50} \times 100 = 118.7,$$

$$\text{Paasche price index:} \qquad \frac{\sum p_3 q_3}{\sum p_0 q_3} = \frac{\$98.00}{\$83.75} \times 100 = 117.0.$$

Since construction of the Paasche index requires determination of new weights each year, while the Laspeyres index does not, the Laspeyres index is used much more often. Furthermore, the Laspeyres index has the added advantage that indexes obtained from this formula may be compared from year to year, while indexes obtained by using the Paasche formula can be compared easily only with the base year. Both indexes, however, tend to reflect a slight bias when reporting price changes. Under the usual conditions of a downward sloping demand schedule, when people tend to purchase more of low-priced items and less of high-priced items, the numerator of the Laspeyres index will be somewhat higher than it should be, resulting in an overestimation of price increases. At the same time, the numerator of the Paasche index will tend to be lower than it should be, resulting in an underestimation of price increases. These biases are only slightly evident in the results of the preceding examples, wherein the Laspeyres index is 118.7 while the Paasche index is 117.0.

13.9 ECONOMIC INDEXES AND THEIR LIMITATIONS

We have presented just a brief introduction to the topic of index numbers. In practice, the process of constructing and updating an index can become quite involved. Before leaving the subject, we shall describe a number of the more widely used indexes and some of their limitations.

The Consumer Price Index (C.P.I.) and the Producer Price Index (formerly the Wholesale Price Index), both published by the Department of Labor, Bureau of Labor Statistics, are the two best-known price indexes. These two indexes represent excellent examples of attempts to summarize large masses of data in a single price index. The C.P.I. is based primarily on 265 different items in 68 separate expenditure classes (foods, fuels, apparel, etc.) which are priced throughout each month in 85 representative urban areas. The C.P.I. is a modified Laspeyres index with 1967 as the base year. It represents average changes in prices paid by consumers in retail markets and represents about 80 percent of the total noninstitutional civilian population of the United States. Half of these are wage earners and clerical workers, but also included are salaried workers, self-employed, retired, and unemployed persons. It does not include farm families. It is compiled from a survey of such persons in 85 urban areas chosen to represent all towns with a population over 2500. Persons are selected based on a probability sample from the 1980 Census of Population. The graph of the Consumer Price Index for the years 1973 through 1980 is shown in Fig. 13.12. The Producer Price Index involves about 10,000 price quotations for that Tuesday during the week that contains the thirteenth day of the month. These prices are gathered by mail questionnaire for approximately 2800 commodities. This index gives an indication of prices received in primary markets (not retail) of the United States by producers in all stages of processing. It includes price movements of goods in manufacturing, agriculture, forestry, fishing, mining, gas, electricity, and public utilities, and for finished goods, intermediate goods, and raw materials.

There are two major economic quantity indexes: (1) the *Index of Industrial Production* (I.I.P.), published by the Federal Reserve Board, and (2) the Export and Import Indexes, published by the U.S. Department of Commerce. The Export and Import Indexes calculate quantity changes in both exports and

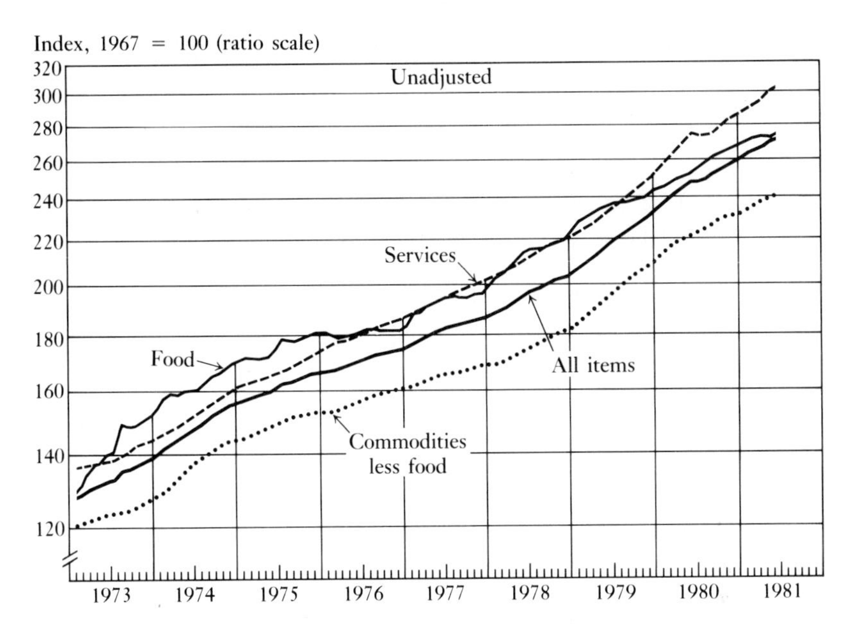

Fig. 13.12 Consumer price indexes: commodities and services.

imports. The I.I.P., which is a *weighted average of relative quantity* index, measures changes in manufacturing output.

The limitations of economic indexes follow directly from the problems suggested earlier in the construction of an index. Since the useful and meaningful application of an index often depends on what year is selected as the base, it is important to use as a base a period with "normal" or "average" economic activity. Sometimes the base-year value can be the average price or quantity of several years rather than a single year. Exactly what items are included in the index also has an important bearing on the validity of the index. In the case of the C.P.I., every few years the Bureau of Labor Statistics undertakes a detailed study of the consumer "market basket" in order to determine which old items to replace and which new items to add.

Particular items selected in the market basket must conform to detailed specifications (variety, brand, size, special features, etc.) that are basically the same across the country, although the final selections of items reported are store-specific and depend on local sales information. The reporting stores (including about 2300 food stores, 18,000 rental units, and 22,000 others in many categories) are determined from the Point-of-Purchase Survey of 23,000 families in the United States. These families are asked for information on the name, location, and amount spent in retail stores for many different categories of goods and services. The weights used in the index are based on information from the Consumer Expenditure Survey, most recently conducted in 1973 in 216 different urban areas of the country. In the case of the Producers Price

Index, there is a semi-annual change in the commodities sampled. For example, in January 1978, 83 items were dropped and 71 new items were added.

Since, over a period of time, the quality of products can change, some items fall out of use completely, and others appear that had no earlier counterpart, it is difficult, if not impossible, to compare the prices of many of today's goods with those of an earlier period. For example, it is unrealistic to compare the price of an old icebox with the price of today's modern refrigerator, and the price of a television set cannot be compared with the price of an article that was not on the market fifty years ago.

This inability to compare goods, and consequently prices, over time has led to charges that price indexes exaggerate the real increase in prices, and that the C.P.I., in particular, exaggerates the change in the cost of living because the indexes do not take into account the improvement in the quality of goods. Even though the Bureau of Labor Statistics attempts to factor out all the quality changes that can be detected in the items used in computing the C.P.I., it is extremely difficult to measure the worth of these increases to consumers. The economist Lloyd G. Reynolds illustrates this problem quite well in the following situation: Give a family a 1955 Sears Roebuck Catalog and a 1980 Sears Catalog and $1000, and allow them to make up an order list from either the 1955 or the 1980 catalog, but not from both. Most families would probably use the 1980 catalog, which must mean that they consider the higher 1980 prices more than offset by new and improved quality products. This implies that there was no real increase in consumer prices between 1955 and 1980, even though the C.P.I. indicated a sizeable increase.

Both the consumer and the producer price indexes are available in seasonally adjusted or unadjusted form. The seasonally adjusted indexes reveal more clearly underlying cyclical trends and are most useful in regression analysis involving other quantities that are seasonally adjusted. Unadjusted data is of primary interest to users who need information on the actual dollar value of transactions, such as marketing specialists, purchasing agents, and commodity traders. Unadjusted indexes are also used generally to escalate contracts such as purchase agreements and real-estate leases. The Bureau of Labor Statistics revises the seasonal adjustment factors for these indexes annually.

Because there are so many difficulties in constructing meaningful price indexes, they should be used only as guides and indicators of price movements, and should not be quoted as indisputable facts. Also, one should remember that they represent prices to some "average" person with "average" tastes and preferences. Whether or not they are relevant to a particular person or group must always be questioned before they are applied.

Define: index number, deflator, real income, price relative, weighted price index, Laspeyres index, Paasche index, base year, consumer and producer price indexes.

Problems

13.24 Go to a library or to your local newspaper and find the current value of the *Consumer Price Index* (C.P.I.) and the *Producer Price Index* (P.P.I.).

a) Determine how the value of these indexes has changed over the past year, as well as over the past ten years. How has the composition of the items included in these indexes changed over the past ten years?

b) Find as many examples as you can from the news media illustrating current uses of these indexes, (e.g., labor-management negotiations, inflation reports, etc.).

13.25 Suppose that a budget request for a new university building was \$3.5 million in 1977, and a similar request is \$4.5 million in 1982. Also, an index of building and construction costs (assumed to be applicable to this situation) with base year 1979 = 100, is equal to 95 in 1977 and 140 in 1982. Compare the relative real value of the building that could have resulted from the two budget requests.

13.26 The Bureau of Labor Statistics consumer price index for services increased from 107 to 115 during a certain period.

a) If a consumer budgeted \$20 per month for services at the beginning of this period, how much should be budgeted for the same level of services at the end of the period?

b) Suppose a person's salary increased from \$500 to \$552 per month during this period. What is the real change experienced in purchasing power for services?

13.27 In the past year, the GNP of a nation has risen from \$200 billion to \$212 billion, which represents a 6% growth rate. However, the price index used to calculate the national product has also risen from 125 to 130. What is the real growth rate of this nation's GNP after accounting for inflation?

13.28 If an index of money wages of workers (1967 = 100) was 250 in 1982 and an appropriate index of living costs (1967 = 100) was 200 in the same year, what has been the percent increase from 1967 to 1982 in:

a) money wages of workers?

b) real wages of workers?

13.29 Suppose that the average beginning salary for business graduates has increased from \$17,000 to \$20,000 over a period of years when the consumer price index increased from 100 to 120. What is the change in the "real" income?

13.30 Suggest two important practical difficulties in determining a Laspeyres price index.

13.31 Explain how a Laspeyres price index differs from a simple arithmetic average of price relatives.

13.32 Suppose that the market-basket values in Table 13.25 on page 642 were observed in 1955 and 1956:

Construct the following:

a) price relatives for each commodity,

b) a simple aggregate price index,

c) a simple average of price relatives index,

Table 13.25

	1955		1956	
Item	Price	Quantity	Price	Quantity
Milk	15¢ qt.	25 qt.	15¢ qt.	20 qt.
Eggs	50¢ doz.	10 doz.	45¢ doz.	15 doz.
Bread	15¢ loaf	30 loaves	20¢ loaf	35 loaves

d) a weighted average of price relatives index, using both base-year and given-year weights,

e) a Laspeyres and a Paasche price index.

13.33 Given the data in Table 13.26

a) Construct a Laspeyres price index using 1979 as the base year;

b) Construct a Paasche price index for these years.

Table 13.26

Year	Price A	Quantity A	Price B	Quantity B
1979	25 cents	10	10 cents	30
1980	30 cents	20	20 cents	30
1981	30 cents	40	25 cents	20
1982	40 cents	45	20 cents	25

13.34 Suppose that price and quantity data (in relevant units) for sales of major appliances are collected in order to construct a consumer durable-goods price index. From the data, in Table 13.27, determine a Laspeyres price index for 1982 relative to the base period, 1977, and interpret its meaning.

Table 13.27

	1977		1982	
Item	p	q	p	q
Range	1.5	8	2	12
Refrigerator	3	10	3.8	12
Air conditioner	1	6	2	15

13.35 Given the following data on monthly purchases by a certain family, as shown in Table 13.28 on page 643.

a) Compute a weighted aggregate Laspeyres price index, using 1950 as the base year.

b) Assume that your index applied in general for all foods. Also assume that a certain family spends $1200 for food in 1950 and $1750 in 1960. In which year would they have had more to eat for their money?

Table 13.28

	1950		1960	
Commodity	p	q	p	q
Eggs (doz.)	0.40	10	0.60	9
Sugar (lbs.)	0.10	35	0.20	30
Butter (lbs.)	0.50	5	0.40	6

13.36 The data on student expenses in Table 13.29 was dug up by the campus newspaper for an editorial on the "good old days."

Table 13.29

Item	Lunch	Gasoline	Movie
1960			
Price (cents)	30	25	50
Quantity	6	1	4
1964			
Price (cents)	50	30	90
Quantity	6	2	2
1966			
Price (cents)	60	35	100
Quantity	6	3	3

a) Compute the Laspeyres price index for the three years, using 1960 as the base.

b) Assuming that these data are representative of all commodities and all students, ascertain, by deflation, the real level of living, relative to 1960, of students who spent \$500 per semester in 1960, \$600 in 1964, and \$700 in 1966.

Exercises

13.37 a) What are the four components of a time series? Give an example of a time series in which you would expect all four components to arise, and explain why you would anticipate them.

b) Can you think of any type of time series that has only two or three of the four components? Are there any other components that might be present in a time series? Give specific examples.

13.38 Describe the additive and the multiplicative models used in time-series analysis. Which of these models do you think is more realistic for most economic time series? Explain why. Give examples where each model would be appropriate.

13.39 Assume that the sales volume in a certain industry can be described by the multiplicative model $y = T \times S \times C \times I$. One month last year, the trend estimate of sales was 44,000 units and actual sales were 55,000 units. If we assume that the seasonal index was $S = 95$, and the index for cyclical movement was $C = 119$, what index value must be associated with the irregular movement, I?

13.40 The estimate of trend accounted for $180,000 of a department store's sales last October. Assuming a multiplicative model, no irregular variation, and a cyclical index of 110, find the seasonal index for this month. Actual sales were $210,000.

13.41 Given a trend equation, $\hat{y} = 37.50 + 6x$ with origin at July 1, 1977 and x in units of one year, where y is truck sales in thousands, convert the equation to monthly units with origin at March 1, 1979.

13.42 Given the following trend equation: $\hat{y} = 137.50 + 8x$. Origin is July 1, 1977; x is expressed in units of one year; y is fruit sales in thousands. Find the monthly trend value for February 15, 1979.

13.43 A producer of one-man portable helicopters had sales over the five years, 1978–1982, of 1, 2, 3, 5, and 9 units respectively.

a) Find the trend line for sales, letting $x = 0$ represent 1980.

b) Find the standard error of estimate for this trend line, and explain its meaning.

c) What is your estimate for sales in 1984?

d) For the second observation (year 1979), give the total deviation, the explained deviation, and the unexplained deviation.

13.44 a) Work Problem 10.47, part (a).

b) Fit a modified exponential curve to the data in (a).

c) Does your answer to part (a) or part (b) provide a better fit to the data? Explain why one model is better than the other.

13.45 a) Fit an exponential curve of the form $y = ab^x$ to the data in Table 13.30 on population growth.

Table 13.30

Year	Population
1978	100
1979	400
1980	700
1981	1600
1982	3900

b) Plot this time series and the curve you calculated above on a graph. What value would you predict for population in 1984?

c) Fit a modified exponential curve to these data.

d) Plot on a graph both the exponential approximation and the modified exponential approximation to these data. Which gives the better approximation? Explain why one model is better than the other.

13.46 A company has determined a seasonal index and a trend line for their monthly sales. The seasonal index for December is 140. The trend line for monthly sales is $\hat{y} =$

163,250 + 4520x, with x in units of months, and origin at mid-April, 1981. Forecast company sales for December, 1983.

13.47 Given the data for sales of boating supplies shown in Table 13.31.

a) Give one representation of the trend equation for sales of boating supplies.

b) Find seasonally adjusted sales for the four quarters of 1981.

c) Assuming no change in the trend or seasonal pattern and assuming that other factors remain constant, forecast second-quarter sales for 1983.

Table 13.31

1981 Quarter	Actual Sales	Trend Values	Seasonal Index
1	100,000	90,000	80
2	150,000	95,000	130
3	120,000	100,000	110
4	110,000	105,000	80

13.48 The secular trend of sales (Table 13.32) for the Jones Department Store is accurately described by the equation $\hat{y} = 120{,}000 + 1000x$, where x represents a period of one month and has a value of zero in December, 1980. The seasonal indexes for the company's sales are:

a) Ignoring cyclical and random influences, forecast sales for February, 1982, May, 1985, and December, 1983.

b) What factors could cause these estimates to be incorrect?

c) What may be done to compensate for inaccuracies as they become apparent?

Table 13.32

1980											
J	F	M	A	M	J	J	A	S	O	N	D
100	80	90	120	115	95	75	70	90	95	120	150

13.49 The following data are the means of ratios of original data to the 12-month moving averages for the sales of a retail store. The data shown in Table 13.33 on page 646 cover a period of ten years.

a) Compute the index of seasonal variation from these ratios for the months of March, April, and August. Be sure to adjust so that the indexes average 100.

b) Suppose that the total sales for next year are estimated in December of this year at 120 million. What would be the best estimate of sales for April and August?

13.50 In the library, find a 10-year time series on annual household income for the United States, a region, your state, or for some occupation, race, or sex subgroup. Find a time series on consumer prices that is relevant for this group of households. Calculate the adjusted annual household income, corrected for price changes, and interpret your results in terms of growth of real income over the period.

Table 13.33

Month	Modified Means of Ratios
Jan.	56
Feb.	60
Mar.	100
Apr.	110
May	105
June	102
July	80
Aug.	72
Sept.	88
Oct.	105
Nov.	120
Dec.	145
Total	1143

13.51 Given the following information in Table 13.34.

a) For the data above, find the estimating line of import price on the quantity imported.

b) Find the correlation coefficient for this data.

c) Find the percentage of variation in import price that is explained by the variation in the quantity imported. What is the name given to this measure?

d) Suppose that the trend line for import price is $\hat{y} = 2.8 + 0.5x$, where $x = 0$ for 1971. Is the import price for these years above or below the trend estimates?

e) Do you think it would be a good idea to remove seasonal variation from this data, as well as trend, before studying the cyclical relatives? Give a reason.

f) Find the values of the three-year moving average of quantity imported for the years 1973 and 1976.

Table 13.34

Year	Import Price y	Quantity Imported, x
1971	2	6
1972	3	5
1973	6	4
1974	5	5
1975	4	7
1976	3	10
1977	5	9
1978	7	7
1979	8	8
1980	7	9

13.52 Using the library, find the price and quantity sold of shares of ten selected stocks on the last day of the previous four years. Using one year as the base period, construct your own:

a) simple price index of the stocks;

b) Laspeyres price index of these stocks.

c) Find the Dow-Jones stock index at year-end for these same years and, using the same base period, construct an index of the Dow-Jones index to compare with your index in part (b).

d) Explain how your price index of stocks differs in construction and meaning from a typical stock index.

GLOSSARY

time series: periodic observations of a variable that is a function of time.

trend (T): long-term movement in a time series.

seasonal variations (S): fluctuations repeated over a period of one year.

cyclical variation (C): pattern repeated over periods of varying lengths, usually longer than one year.

irregular variation (I): unpredictable or random fluctuations.

additive model: $y = T + S + C + I$.

multiplicative model: $y = T \times S \times C \times I$.

method of semi-averages: division of data into two parts with a trend line connecting the average value of each of the two groups.

reduced equations: normal equations from regression, simplified by choosing x so that $\sum x_i = 0$.

exponential curve: $y = ab^x$.

modified exponential curve: $y = c + ab^x$.

logistic curve: $y = 1/(c + ab^x)$.

Gompertz curve: $y = ca^{b^x}$.

method of moving averages: a series of averages in which each average is the mean value over an interval of time of fixed length centered at the midpoint of each interval.

weighted moving averages: a moving average in which each observation is given a weight that reflects its relative importance.

seasonal index: a number that expresses the value of the seasonal fluctuation during a period of time.

ratio-to-trend method: method for estimating S by dividing both sides of $y = T \times S \times C \times I$ by T. Does not eliminate $C \times I$.

ratio-to-moving average: a method for estimating S by first isolating $T \times C$, and then dividing $T \times S \times C$ by $T \times C$.

index number: a number that relates the values of a particular variable to its value in a selected base period.

price relative: the ratio of the price of a certain commodity in a given period to the price of that commodity in some base year.

market basket: the group of commodities selected to represent the typical purchases of an "average" person.

Laspeyres price index: a price index constructed by weighting each price by the quantity consumed in the base year.

Paasche price index: a price index constructed for year n by weighting each price by the quantity consumed in year n.

14

ANALYSIS OF VARIANCE

"A single death is a tragedy,
a million deaths is a statistic."

JOSEPH STALIN

14.1 INTRODUCTION

In Chapter 8, tests of hypotheses concerning the means of two populations were discussed. Of particular interest was the test of $H_0:\mu_1 - \mu_2 = 0$ (which is equivalent to $H_0:\mu_1 = \mu_2$) against the alternative hypothesis

$$H_a:\mu_1 - \mu_2 \neq 0$$

(which is equivalent to $H_a:\mu_1 \neq \mu_2$). In other words, we take samples from each of the populations and use these samples to see if there are any significant differences between the population means. This type of test is widely applied: we may want to compare the output of two production processes, the gasoline mileage of two automobiles, the academic performance of two groups of students, and so on. In many situations we may be interested in more than two populations: there may be three production processes of interest, five automobiles, or four groups of students. It is a computationally burdensome and statistically improper practice to run the test for the difference between two means on all possible pairs of populations. If there are 15 populations, for example, there are $\binom{15}{2} = 105$ possible pairs. If each pair is tested at the 0.10 significance level, say, then we would expect to get approximately 10 "significant" results purely by chance, making it difficult to interpret the results. What is needed is a method for *simultaneously* investigating the differences among the means of several populations, and the body of statistical procedures which is called "analysis of variance" is such a method.

The analysis of variance is a method of estimating how much of the total variation in a set of data can be attributed to certain assignable causes of variation and how much can be attributed to chance. In the simplest model, the total variation in a set of J samples (from J populations) is divided into two parts: (1) the variation *between* the J samples, and (2) the variation *within* the J samples. The latter variance is attributable only to chance, while the former variance is attributable in part to chance and in part to any differences which may exist among the means of the J populations. By comparing these variances, we can investigate for differences among the J means.

14.2 THE SIMPLE ONE-FACTOR MODEL

Suppose that we are interested in comparing the means of J different populations, and that we therefore have J samples, one from each population. Furthermore, suppose that the sample size of the jth sample is n_j ($j = 1, \ldots, J$). Now we are ready to state a model for the composition of any observed value y_{ij}, corresponding to the ith observation in the jth sample. The model used is a simple linear model:

$$y_{ij} = \mu + \tau_j + \varepsilon_{ij}, \qquad \text{with } j = 1, 2, \ldots, J, \quad i = 1, 2, \ldots, n_j. \tag{14.1}$$

According to this model, an observation is the sum of three components:

1. the grand mean μ of the "combined" populations,
2. a treatment effect τ_j associated with the particular population from which the observation is taken,
3. a random-error term ε_{ij}.

The treatment effects can be thought of as deviations from the grand mean. In general, we assume that the proportional representation of the various treatment groups in the experiment is the same as the proportional representation of the groups in the combined population. This condition can be expressed in the form

$$\sum_{j=1}^{J} n_j \tau_j = 0.$$

The random-error terms reflect variability within each of the J populations. Although it is not necessary to introduce distributional assumptions at this point (such assumptions will be necessary in the next section), the random-error terms are generally taken to be independent and normally distributed, with mean zero and variance σ^2. Note that there is a similarity between the simple one-factor analysis-of-variance model given by Formula (14.1) and the linear regression model presented in Chapter 10; we shall discuss this similarity later in this chapter.

An alternative way to write the one-factor model is

$$y_{ij} = \mu_j + \varepsilon_{ij}, \tag{14.2}$$

where μ_j is the mean of the jth population. This is equivalent to 14.1, for the mean μ_j is simply the sum of the grand mean μ and the treatment effect τ_j. Recall that we want to compare the means of the J populations, so that the following hypothesis is of interest:

$$H_0: \mu_1 = \mu_2 = \cdots = \mu_J.$$

But if the J means are equal, each mean μ_j is equal to the grand mean μ, and then there are no "treatment effects," so this hypothesis can be rewritten as follows:

$$H_0: \tau_1 = \tau_2 = \cdots = \tau_J = 0.$$

To illustrate the model, consider a simple situation in which three drugs are being tested in order to investigate their effects on an individual's pulse rate. Six individuals participate in the study involving the drugs, and they are randomly assigned to the three treatments (i.e., the three drugs) such that two individuals are in each treatment group. Thus, $J = 3$ and $n_1 = n_2 = n_3 = 2$. For each subject, the difference between the pulse rate ten minutes after taking the drug and the pulse rate just before taking the drug is measured. Suppose that the grand mean is 20 and that the treatment effects and error terms are all zero. Then each subject's pulse rate would increase by exactly 20, and our data would look like Table 14.1.

Table 14.1

Drug 1	Drug 2	Drug 3
20	20	20
20	20	20

Table 14.2

Drug 1	Drug 2	Drug 3
20 + 4 = 24	20 − 6 = 14	20 + 2 = 22
20 + 4 = 24	20 − 6 = 14	20 + 2 = 22

Now, in addition, suppose that the treatment effects are nonzero:

$$\tau_1 = +4,$$
$$\tau_2 = -6,$$

and

$$\tau_3 = +2.$$

The first drug tends to increase the pulse rate 4 units more than the grand mean, the second drug tends to increase the pulse rate 6 units less than the grand mean, and the third drug tends to increase the pulse rate 2 units more than the grand mean. The data are then as shown in Table 14.2.

Observe that there are now differences *between* the three samples, or treatments, due to the nonzero treatment effects. *Within* each sample, however, the observations are identical, because the random-error terms are all zero. According to the model presented in this section, the only variability within samples is attributable to chance (random error). It is highly unrealistic to assume no random variability, so let us add some nonzero error terms to the data of the drug example and see what the data might look like (Table 14.3).

The increase in pulse rate for a particular individual now depends on the grand mean, on the specific treatment effect associated with the particular drug, and on random variation. There are not only differences *between* samples (attributable both to differences in treatment effects and to random error), but there are also differences *within* each sample (attributable solely to random error).

The above example should serve to explain the model which is represented by Formulas (14.1) or (14.2). Of course, the model assumes that the grand mean, the treatment effect, and the random error term corresponding to any

Table 14.3

Drug 1	Drug 2	Drug 3
20 + 4 + 3 = 27	20 − 6 + 1 = 15	20 + 2 − 2 = 20
20 + 4 − 2 = 22	20 − 6 − 3 = 11	20 + 2 + 3 = 25

observation are additive. Although the model is reasonably simple, it should be adequate to describe numerous situations in which there are several samples.

Naturally, in applications we shall not be given the values of μ, τ_j, and ε_{ij} and asked to construct the raw data. Instead, the data will be given and the objective will be to estimate treatment effects and the grand mean and to test for differences among the J population means (i.e., test to see if any of the estimated treatment effects is significantly different from zero). To estimate μ, we simply take the grand sample mean,

$$\bar{y} = \frac{\sum_{j=1}^{J} \sum_{i=1}^{n_j} y_{ij}}{\sum_{j=1}^{J} n_j}.$$

The sample means are used to estimate the corresponding population means. Thus, an estimator of μ_j is

$$\bar{y}_j = \frac{\sum_{i=1}^{n_j} y_{ij}}{n_j}.$$

To estimate the treatment effects, then, the estimated grand mean is subtracted from each of the estimated population means:

$$\hat{\tau}_j = \bar{y}_j - \bar{y}.$$

(Recall that it is assumed that the random error terms have zero mean and are independent.)

For example, suppose that a production manager is interested in comparing the output of three machines. He conducts an experiment, running each machine under identical conditions (or as nearly identical as possible) and recording the output per minute. This is done for five minutes for Machine 1, ten minutes for Machine 2, and six minutes for Machine 3, so that $n_1 = 5$, $n_2 = 10$, and $n_3 = 6$. The data are given in Table 14.4. The estimated treatment effects are thus

$$\hat{\tau}_1 = \bar{y}_1 - \bar{y} = 8.4 - 8.2 = 0.2,$$

$$\hat{\tau}_2 = \bar{y}_2 - \bar{y} = 6.7 - 8.2 = -1.5,$$

and

$$\hat{\tau}_3 = \bar{y}_3 - \bar{y} = 10.7 - 8.2 = 2.5.$$

Define: Analysis of variance, treatment effect (τ_j), grand mean, random error (ε_{ij}).

Table 14.4

Machine 1	Machine 2	Machine 3	
10	6	11	
6	7	8	
8	9	13	
12	4	10	
6	6	10	
	10	12	
	5		
	6		
	8		
	6		
$\sum_{i=1}^{5} y_{i1} = 42$	$\sum_{i=1}^{10} y_{i2} = 67$	$\sum_{i=1}^{6} y_{i3} = 64$	$\sum_{j=1}^{3}\sum_{i=1}^{n_j} y_{ij} = 173$
$\bar{y}_1 = 8.4$	$\bar{y}_2 = 6.7$	$\bar{y}_3 = 10.7$	$\bar{y} = 8.2$

Problems

14.1 In order to compare the average tread life of three brands of radial tires, random samples of six tires of each brand are selected. A machine which simulates road conditions is used to find the tread life (in thousands of miles) of each tire. If all three brands have identical distributions of tread life, would you expect the 18 tires to last exactly the same number of miles? To what would you attribute differences in tread life among the 18 tires?

14.2 In the experiment in Problem 14.1 suppose that we observe six tires with tread lives of 36 (thousand miles), six with tread lives of 40, and six with tread lives of 42. Consider the following two cases:

Case 1: For each brand, we see two tires with tread lives of 36, two with 40, and two with 42.

Case 2: The six Brand A tires have tread lives of 36, the six Brand B tires have tread lives of 42, and the six Brand C tires have tread lives of 40.

For these two extreme cases, compare the variation between the three brands with the variation within the three brands.

14.3 A firm wishes to compare four programs for training workers to perform a certain manual task. Twenty new employees are randomly assigned to the training programs, with five in each program. At the end of the training period, a test is conducted to see how quickly trainees can perform the task. The number of times the task is performed per minute is recorded for each trainee, with the following results:

Program 1:	9	12	14	11	13
Program 2:	10	6	9	9	10
Program 3:	12	14	11	13	11
Program 4:	9	8	11	7	8

Estimate the treatment effects for the four programs.

14.4 In Problem 14.1, the observed tread lives are as follows:

Brand A:	38	44	37	39	41	37
Brand B:	41	48	46	42	47	46
Brand C:	40	35	36	32	39	37

Estimate the population mean tread lives and the treatment effects for the three brands.

14.5 A number of company presidents were sampled at random from the United States, and the annual income of each president sampled was recorded, as was the region of the U.S. in which the company's headquarters were located. The incomes (in thousands of dollars) shown in Table 14.5 were observed in the sample. Find estimates for the mean income of company presidents in the U.S. and the mean income of company presidents in each of the six regions. Also, assuming that the sample sizes accurately reflect the corresponding population sizes (e.g., there are twice as many companies with headquarters in the Northeast as in the Northwest), estimate the treatment effects associated with the different regions.

Table 14.5

Southeast	Southwest	Northeast	Northwest	Midwest	Far West
84	89	98	118	93	120
117	128	116	103	86	131
110	84	91	91	138	91
91	123	117	86	131	96
114	82	95	114	122	100
89	127	114		103	95
89	87	111		112	
113	91	95			
86	124	92			
	130	109			

14.3 THE *F*-TEST IN THE ANALYSIS OF VARIANCE

In the previous section we presented the simple one-factor model for the analysis of variance and determined estimates of the grand mean, sample means, and treatment effects. We mentioned that the hypothesis of interest for comparing the means of the J populations is

$$H_0: \tau_1 = \tau_2 = \cdots = \tau_J = 0.$$

In this section we shall discuss a test of this hypothesis against the alternative hypothesis that one or more of the treatment effects is nonzero.

The sample variance of the entire set of data (all J samples pooled together) can be written in the following form:

$$s^2_{\text{Total}} = \frac{\sum_{j=1}^{J} \sum_{i=1}^{n_j} (y_{ij} - \bar{y})^2}{\sum_{j=1}^{J} n_j - 1}. \tag{14.3}$$

This is the usual form for a sample variance; the numerator is the sum of the squared deviations about the grand sample mean, and the denominator is the total sample size minus one. It is convenient to consider the numerator and denominator of Formula 14.3 separately. The numerator is a sum of squares (SS), and will be called SS Total. The denominator can be thought of as the number of degrees of freedom associated with s^2_{Total}.

In order to conduct an analysis of variance, it is necessary to break s^2_{Total} into two parts, the first representing the variability within the samples (treatments) and the second representing the variability between the sample (treatment) means. To find the sum of squares *within* the jth sample, simply calculate $\sum_{i=1}^{n_j} (y_{ij} - \bar{y}_j)^2$. Summing this over all samples gives us

$$\text{SS Within} = \sum_{j=1}^{J} \sum_{i=1}^{n_j} (y_{ij} - \bar{y}_j)^2.$$

There are $n_j - 1$ degrees of freedom associated with each individual sample, so the total number of degrees of freedom "within" is

$$\sum_{j=1}^{J} (n_j - 1) = \sum_{j=1}^{J} n_j - J.$$

The sum of squares between the sample means is

$$\text{SS Between} = \sum_{j=1}^{J} n_j(\bar{y}_j - \bar{y})^2,$$

and, since there are J samples, there are $J - 1$ degrees of freedom associated with SS Between.

The partition of total variance into two sources is such that the sum of d.f. between and d.f. within equals the total degrees of freedom, for

$$(J - 1) + \left(\sum_{j=1}^{J} n_j - J\right) = \sum_{j=1}^{J} n_j - 1.$$

In addition to the fact that the number of degrees of freedom *between* samples and the number of degrees of freedom *within* samples sum to the total number of degrees of freedom, it can be shown that the sum of SS Between and SS Within equals SS Total.* That is,

$$\text{SS Total} \quad = \quad \text{SS Within} \quad + \quad \text{SS Between},$$

or

$$\sum_{j=1}^{J}\sum_{i=1}^{n_j}(y_{ij}-\bar{y})^2 = \sum_{j=1}^{J}\sum_{i=1}^{n_j}(y_{ij}-\bar{y}_j)^2 + \sum_{j=1}^{J} n_j(\bar{y}_j-\bar{y})^2.$$

It is convenient to arrange the preceding terms as shown in Table 14.6, in an analysis-of-variance table comparable to that used in Chapter 10. Note that the total variation has been separated into two sources, between samples and within samples. For each of these sources, the sum of squares and degrees of freedom can be calculated, and these are used to determine the mean square between (MS Between) and mean square within (MS Within). Each MS is determined by dividing the appropriate SS by the degrees of freedom. Note, incidentally, that when $J = 2$, MS Within is identical to the pooled estimate of σ^2 given in the two-sample case in Chapter 8.

---

* To prove this statement, observe that

$$\begin{aligned}\text{SS Total} &= \sum_j\sum_i(y_{ij}-\bar{y})^2 = \sum_j\sum_i[(y_{ij}-\bar{y}_j)+(\bar{y}_j-\bar{y})]^2\\ &= \sum_j\sum_i(y_{ij}-\bar{y}_j)^2 + \sum_j\sum_i(\bar{y}_j-\bar{y})^2 + 2\sum_j\sum_i(y_{ij}-\bar{y}_j)(\bar{y}_j-\bar{y}).\end{aligned}$$

But for any j, $(\bar{y}_j - \bar{y})$ is the same for all values of i, so that the last term can be written as follows:

$$2\sum_j\sum_i(y_{ij}-\bar{y}_j)(\bar{y}_j-\bar{y}) = 2\sum_j(\bar{y}_j-\bar{y})\sum_i(y_{ij}-\bar{y}_j).$$

Now, for any j,

$$\sum_i(y_{ij}-\bar{y}_j) = \sum_i y_{ij} - \sum_i \bar{y}_j = \sum_i y_{ij} - n_j y_j = 0.$$

Therefore,

$$2\sum_j\sum_i(y_{ij}-\bar{y}_j)(\bar{y}_j-\bar{y}) = 0,$$

and

$$\begin{aligned}\text{SS Total} &= \sum_j\sum_i(y_{ij}-\bar{y}_j)^2 + \sum_j\sum_i(\bar{y}_j-\bar{y})^2\\ &= \text{SS Within} + \sum_j\sum_i(\bar{y}_j-\bar{y})^2.\end{aligned}$$

Finally, for any j, $(\bar{y}_j - \bar{y})^2$ is a constant with respect to the index of summation i, so that

$$\sum_j\sum_i(\bar{y}_j-\bar{y})^2 = \sum_j n_j(\bar{y}_j-\bar{y})^2 = \text{SS Between},$$

and thus

$$\text{SS Total} = \text{SS Within} + \text{SS Between}.$$

Table 14.6 Analysis of Variance with Unequal Sample Sizes

Source of Variation	SS	d.f.	MS = SS/d.f.
Between samples (treatments)	$\sum_{j=1}^{J} n_j(\bar{y}_j - \bar{y})^2$	$J - 1$	SS Between/$J - 1$
Within samples	$\sum_{j=1}^{J}\sum_{i=1}^{n_j} (y_{ij} - \bar{y}_j)^2$	$\sum_{j=1}^{J} n_j - J$	SS Within$\Big/\sum_{j=1}^{J} n_j - J$
Total	$\sum_{j=1}^{J}\sum_{i=1}^{n_j} (y_{ij} - \bar{y})^2$	$\sum_{j=1}^{J} n_j - 1$	

Frequently the sample sizes for the various samples, or treatments, are equal. In this case, we can let n denote the sample size for each sample (i.e., $n_j = n$ for $j = 1, \ldots, J$). Using equal sample sizes has the advantage of minimizing adverse effects of violations of the assumption that the J populations have the same variance. The analysis-of-variance table for the special case of equal sample sizes is presented in Table 14.7.

In order to test the hypothesis that the treatment effects are all zero, it is necessary to make some distributional assumptions. Thus, we will invoke the assumptions mentioned in the previous section concerning the random-error terms:

1. For each sample (for each $j = 1, 2, \ldots, J$), the random-error terms ε_{ij} are normally distributed with mean zero and variance σ^2, and the variance σ^2 is the same for all samples.
2. The random-error terms are independent.

In addition, of course, it is assumed that the linear model given by (14.1) is applicable. That is, it is assumed that the three factors affecting y_{ij} (the grand

Table 14.7 Anova with Equal Sample Sizes

Source of Variation	SS	d.f.	MS = SS/d.f.
Between samples (treatments)	$n\sum_{j=1}^{J} (\bar{y}_j - \bar{y})^2$	$J - 1$	SS Between /$J - 1$
Within samples	$\sum_{j=1}^{J}\sum_{i=1}^{n} (y_{ij} - \bar{y}_j)^2$	$J(n - 1)$	SS Within/$J(n - 1)$
Total	$\sum_{j=1}^{J}\sum_{i=1}^{n} (y_{ij} - \bar{y})^2$	$Jn - 1$	

mean, the treatment effect, and a random-error term) behave in an additive fashion. Under these assumptions, it can be shown that the expectations of the mean squares are

$$E(\text{MS Between}) = \sigma^2 + \frac{\sum_{j=1}^{J} n_j \tau_j^2}{J - 1} \tag{14.4}$$

and

$$E(\text{MS Within}) = \sigma^2. \tag{14.5}$$

Observe that under the null hypothesis of zero treatment effects, $E(\text{MS Between}) = E(\text{MS Within}) = \sigma^2$. Furthermore,

$$\frac{(J - 1)(\text{MS Between})}{\sigma^2}$$

has a chi-square distribution with $J - 1$ degrees of freedom, and

$$\frac{\left(\sum_{j=1}^{J} n_j - J\right)(\text{MS Within})}{\sigma^2}$$

has a chi-square distribution with $\sum_{j=1}^{J} n_j - J$ degrees of freedom. That these two variables have chi-square distributions follows from the fact that if we have a sample from a normally distributed population, vS^2/σ^2 has a chi-square distribution with v degrees of freedom. Also, the two variables are independent.

In Chapter 6 we pointed out that the ratio of two independent χ^2 variables divided by their degrees of freedom has an **F**-distribution. Taking the ratio of the two chi-square variables given above (each divided by its d.f.) gives the following result:

$$\boldsymbol{F} = \frac{\text{MS Between}}{\text{MS Within}}.$$

Thus, the ratio of MS Between to MS Within has an **F**-distribution with $J - 1$ and $\sum_{j=1}^{J} n_j - J$ degrees of freedom, respectively, provided that the null hypothesis of zero treatment effects is true. This **F**-statistic can be used to test the hypothesis that the treatment effects are all zero. From (14.4), any nonzero treatment effects will tend to increase MS Between, and thus the null hypothesis should be rejected only for large values of **F**; i.e., the righthand tail of the **F**-distribution should be used.

Before presenting some computational forms for the one-factor model and working an example, it should be useful to review the procedure and the rationale for the analysis of variance. The basic idea is to determine whether all of the variation in a set of data is attributable to random error (chance) or whether some of the variation is attributable to chance and some is attributable to differences in the means of the J populations of interest. First, the sample variance for the entire set of data is computed and is seen to be composed of two parts: the numerator, which is a sum of squares, and the denominator, which is the degrees of freedom. The total sum of squares can be partitioned into SS Between and SS Within, and the total degrees of freedom can be partitioned into d.f. Between and d.f. Within. By dividing each sum of squares by the respective d.f., MS Between and MS Within are determined; these represent the sample variability between the different samples and the sample variability within all of the samples, respectively. But the variability *within* the samples must be due to random error alone, according to the assumptions of the one-factor model. The variability *between* the samples, on the other hand, may be attributable both to chance and to any differences in the J population means. Thus, if MS Between is significantly greater than MS Within (as measured by the F-test), then the null hypothesis of zero treatment effects must be rejected. Thus, the analysis of variance is a procedure for a *simultaneous* comparison of a number of means (as opposed, for example, to a series of paired comparisons of the possible pairs of means which can be chosen from the entire set of means of interest).

14.4 COMPUTATIONAL FORMS FOR THE ONE-FACTOR MODEL

The formulas given in the analysis-of-variance table in the previous section, which may be helpful in gaining an understanding of the analysis-of-variance procedure, are computationally difficult to apply. The following computational formulas are much easier to use when determining sums of squares:

$$\text{C.F. (correction factor)} = \frac{\left(\sum_{i}^{n_j}\sum_{j}^{J} y_{ij}\right)^2}{\sum_{j}^{J} n_j},$$

$$\text{SS Total} = \sum_{i}^{n}\sum_{j}^{J} y_{ij}^2 - \text{C.F.},$$

$$\text{SS Between} = \sum_{j}^{J}\left[\frac{\left(\sum_{i}^{n_j} y_{ij}\right)^2}{n_j}\right] - \text{C.F.},$$

and

$$\text{SS Within} = \sum_{i}^{n_j}\sum_{j}^{J} y_{ij}^2 - \sum_{j}^{J}\left[\frac{\left(\sum_{i}^{n_j} y_{ij}\right)^2}{n_j}\right].$$

In practice, SS Within is calculated simply by subtracting SS Between from SS Total. While these formulas may appear more complex than the theoretical formulas presented in the previous section, they are in fact much easier to apply.

For an example illustrating the use of analysis of variance, suppose that a marketing manager is interested in the effect of different types of packaging on the sales of a particular new item. Three different types of packaging have been suggested for the item. The manager draws a random sample of 60 stores from the population of stores that would stock the item. The three different types of packaging are each used at 20 of the stores, with the stores assigned randomly to the three "treatments." The sales of the item at each of the stores are carefully recorded for a period of one month, with the results given in Table 14.8.

Table 14.8 Sales (by type of packaging)

Type A	Type B	Type C	
52	28	15	
48	35	14	
43	34	23	
50	32	21	
43	34	14	
44	27	20	
46	31	21	
46	27	16	
43	29	20	
49	25	14	
38	43	23	
42	34	25	
42	33	18	
35	42	26	
33	41	18	
38	37	26	
39	37	20	
34	40	19	
33	36	22	
34	35	17	
$\sum_{i=1}^{20} y_{i1} = 832$	$\sum_{i=1}^{20} y_{i2} = 680$	$\sum_{i=1}^{20} y_{i3} = 392$	$\sum_{j=1}^{3}\sum_{i=1}^{20} y_{ij} = 1904$

For example, the first store in Group A (the stores with Type A packaging) sold 52 items during the month of the experiment, the second store sold 48 items, and so on down to the last store in Group C, which sold 17 items. Here we have three samples ($J = 3$), with each sample size equal to 20 ($n_1 = n_2 = n_3 = 20$). Using the computational forms, we get

$$\text{C.F.} = \frac{(1904)^2}{60} = 60{,}420.3,$$

$$\begin{aligned}\text{SS Total} &= (52)^2 + (48)^2 + \cdots + (22)^2 + (17)^2 - 60{,}420.3 \\ &= 66{,}872 - 60{,}420.3 = 6451.7,\end{aligned}$$

$$\begin{aligned}\text{SS Between} &= \frac{(832)^2}{20} + \frac{(680)^2}{20} + \frac{(392)^2}{20} - 60{,}420.3 \\ &= 65{,}414.4 - 60{,}420.3 = 4994.1,\end{aligned}$$

and

$$\text{SS Within} = \text{SS Total} - \text{SS Between} = 6451.7 - 4994.1 = 1457.6.$$

The analysis of variance is summarized in Table 14.9.

For the test of the hypothesis that the treatment effects are all zero,

$$F = \frac{2497.1}{25.6} = 97.5,$$

with 2 and 57 degrees of freedom. This is an extremely large F-value, significant far beyond the 0.01 level. The marketing manager can feel very confident in asserting that the type of packaging does have some effect on the sales of the item in question.

The marketing manager may want to estimate the effect of each type of packaging. The estimated grand mean is

$$\bar{y} = \frac{1904}{60} = 31.7,$$

and the three sample means are

$$\bar{y}_1 = \frac{832}{20} = 41.6,$$

$$\bar{y}_2 = \frac{680}{20} = 34.0,$$

and

$$\bar{y}_3 = \frac{392}{20} = 19.6.$$

Table 14.9

Source	SS	d.f.	MS
Between	4994.1	3 − 1 = 2	2497.1
Within	1457.6	60 − 3 = 57	25.6
Total	6451.7	60 − 1 = 59	

Thus, the estimated treatment effects of Types A, B, and C are, respectively,

$$\hat{\tau}_1 = \bar{y}_1 - \bar{y} = 41.6 - 31.7 = 9.9,$$
$$\hat{\tau}_2 = \bar{y}_2 - \bar{y} = 34.0 - 31.7 = 2.3,$$

and

$$\hat{\tau}_3 = \bar{y}_3 - \bar{y} = 19.6 - 31.7 = -12.1.$$

These estimates indicate that Type A packaging produces the greatest sales. If the marketing manager must decide on a single type of packaging for the product, Type A would surely be his best bet on the basis of the experimental results. Incidentally, this suggests a situation in which the results of the F-test in the analysis of variance may be of absolutely no importance to the statistician. In the packaging example, suppose that the marketing manager must choose a single type of packaging and that the costs of the three types are identical. In this case, he should choose the type of packaging with the greatest estimated treatment effect, regardless of whether the difference between the treatment effects is statistically "significant" according to the F-test. Even if the differences are very small and "insignificant," as long as he must choose one type of packaging, he might as well choose the one which looks the best in light of the experiment. Of course, if the statistician is not primarily interested in a decision-making problem, but just wants to make inferences about differences in the treatment effects, then the F-test is of interest.

Observe that this example is a special case of the simple one-factor experimental design, the case with the same number of observations in each treatment group. To see how the analysis is conducted with *unequal* sample sizes, take the last observation in Sample B and assume that it came from Sample A instead. Thus, $n_1 = 21$, $n_2 = 19$, and $n_3 = 20$. SS Total is unchanged, but

$$\text{SS Between} = \frac{(867)^2}{21} + \frac{(645)^2}{19} + \frac{(392)^2}{20} - 60{,}420.3 = 4954,$$

and

$$\text{SS Within} = 6452 - 4954 = 1498.$$

Table 14.10

Source	SS	d.f.	MS
Between	4954	2	2477
Within	1498	57	26.3
Total	6452	59	

The summary table is given in Table 14.10. As would be expected, the shifting of one observation from Group B to Group A does not modify the basic result that the type of packaging has an effect on sales. However, it does serve to demonstrate the calculations in one-factor analysis of variance with unequal group sizes.

14.5 MULTIPLE COMPARISONS

The F-test developed in Section 14.3 enables us to test the hypothesis that the treatment effects are all zero, i.e., that there are no differences among the means of the J treatments, or populations, that are being investigated. If the F-test leads to rejection of the null hypothesis at a given significance level, this is an indication that there *are* differences among the J population means. Such a result does not, however, provide any information regarding differences between *pairs* of populations chosen from the J populations. A significant value of F indicates that at least one of these pairwise differences is significant, but it does not indicate which differences are significant and which are not significant.

Sometimes the difference between two particular treatment effects is of interest. For instance, in the example of the previous section, the marketing manager might want to investigate $\tau_1 - \tau_2$, the difference in the effects of Type A packaging and Type B packaging. A $100(1 - \alpha)\%$ confidence interval for this difference is given by

$$(\hat{\tau}_1 - \hat{\tau}_2) \pm t_{\alpha/2,\, \Sigma n_j - J}\sqrt{\text{MS Within}\left(\frac{1}{n_1} + \frac{1}{n_2}\right)}.$$

Here, MS Within is being used to estimate σ^2 (from Formula (14.5), E(MS Within) $= \sigma^2$). For example, a 95-percent confidence interval for $\tau_1 - \tau_2$ is

$$(9.9 - 2.3) \pm 2.00\sqrt{25.6(2/20)},$$

or

$$(4.4,\ 10.8).$$

Since this confidence interval does not "cover" zero (i.e., zero is not included in the interval), $\tau_1 - \tau_2$ is said to be significantly different from zero at the $1 - 0.95 = 0.05$ level of significance.

It is possible, of course, to compute such an interval estimate for each possible pair of treatments. Unfortunately, this leads to the same difficulty noted in Section 14.1 regarding the use of tests for the difference between two means on all possible pairs of populations. Techniques are available, however, for simultaneously determining confidence intervals for all possible differences. One such technique, developed by Henry Scheffe, is called the S-method. The S-method can be used to simultaneously handle all possible *contrasts* involving the treatment effects, where a contrast is a linear combination of the form $\sum_{j=1}^{J} c_j\tau_j$, with $\sum_{j=1}^{J} c_j = 0$. We are interested in a special set of contrasts: differences between pairs of treatment effects. Note that the difference $\tau_1 - \tau_2$, for instance, is simply a contrast with $c_1 = 1$, $c_2 = -1$, and $c_3 = c_4 = \cdots = c_J = 0$.

The differences $\tau_k - \tau_l$ and $\tau_l - \tau_k$ are equivalent since one is simply the negative of the other. Thus, there are $J(J - 1)/2$ possible pairwise comparisons from a set of J populations. Using the S-method, the probability is at least α that all the $J(J - 1)/2$ differences $\tau_k - \tau_l$ $(k \neq l)$ *simultaneously* satisfy

$$(\hat{\tau}_k - \hat{\tau}_l) \pm \sqrt{(J - 1)F_{(\alpha,\, J-1,\, \Sigma n_j - J)}(\text{MS Within})\left(\frac{1}{n_k} + \frac{1}{n_l}\right)}.$$

Here $\hat{\tau}_k$ and $\hat{\tau}_l$ are simply the estimated treatment effects of population k and population l; $F_{(\alpha,\, J-1,\, \Sigma n_j - J)}$ is the critical value of the F-distribution, which can be found in Table VIII for $\alpha = 0.05$, $\alpha = 0.025$, and $\alpha = 0.01$; and MS Within is used once again to estimate σ^2. For the marketing example, the following three intervals hold simultaneously, with a 95-percent level of confidence:

$$(9.9 - 2.3) \pm \sqrt{2(3.15)(25.6)(2/20)}, \quad \text{or} \quad (3.6, 11.6) \text{ for } \tau_1 - \tau_2;$$

$$(9.9 + 12.1) \pm \sqrt{2(3.15)(25.6)(2/20)} \quad \text{or} \quad (18.0, 26.0) \text{ for } \tau_1 - \tau_3;$$

and

$$(2.3 + 12.1) \pm \sqrt{2(3.15)(25.6)(2/20)} \quad \text{or} \quad (10.4, 18.4) \text{ for } \tau_2 - \tau_3.$$

In general, an interval computed via the S-method for a particular difference is wider than the corresponding interval computed as though that difference were the only difference of interest. A 95-percent confidence interval for $\tau_1 - \tau_2$ was computed as (4.4, 10.8), whereas the corresponding interval computed from the S-method is (3.6, 11.6). For the former interval, of course, the "95-percent confidence" refers to that interval alone, whereas in the S-method, it refers to an entire set of intervals.

Since we are simultaneously comparing pairs of treatment effects, or populations, in the S-method, such a procedure is often called a *multiple-comparisons* procedure. In fact, the confidence intervals can be used to simultaneously test all possible differences. If a given interval does not include zero, then that difference is said to be significant. The **F**-test simply tells us whether there are any significant differences among the populations or not. Methods of multiple comparisons enable us to identify the particular differences that are significant at any given level of significance. In some instances, we are simply interested in the presence or absence of differences among the populations, and the

F-test provides the information of interest. In other cases, we are concerned with pairwise comparisons, in which case a multiple-comparisons procedure should be used following an F-test with significant results.

Define: SS Within, SS Between, SS Total, MS Between, MS Within, C.F., contrasts, multiple comparisons.

Problems

14.6 For the data given in Problem 14.3, find the total sum of squares, the sum of squares between programs, and the sum of squares within programs. Construct an analysis-of-variance table for this experiment. Test the hypothesis of equality among the means of the four programs, using $\alpha = 0.05$.

14.7 Construct an analysis-of-variance table showing sums of squares, degrees of freedom, and mean squares for the data given in Problem 14.4. Test the hypothesis of zero treatment effects at the $\alpha = 0.01$ level.

14.8 For the example concerning the output of three machines with the data given in Table 14.4, find SS total, SS between machines, SS within machines, MS between machines, and MS within machines. At the $\alpha = 0.05$ level, can the production manager reject the hypothesis of equal mean output for the three machines?

14.9 Compute SS total, SS between, and SS within for the two cases in Problem 14.2. Compare the two cases. Can you see why these are called "extreme" cases?

14.10 In the F-test in the one-factor model for the analysis of variance, why is the null hypothesis of equal means rejected only for large values of F? Discuss the sources of variation contributing to the mean square terms in the numerator and denominator of F.

14.11 See if you can prove, without consulting the text, that

SS Total = SS Between + SS Within.

14.12 Do Problem 14.6 using the computational formulas presented in this section.

14.13 An experiment concerning the output per hour of four machines gave the results shown in Table 14.11. Construct an analysis-of-variance table and test the hypothesis of equality among the four population means, using $\alpha = 0.01$.

Table 14.11

A	B	C	D
160	134	104	86
155	139	175	71
170	144	96	112
175	150	83	110
152	156	89	87
167	159	79	100
180	170	84	105
154	133	86	93
141	128	83	65

14.14 Conduct an analysis-of-variance for the data in Problem 14.5 using $\alpha = 0.05$ for the F-test. Can you conclude that there are regional differences in incomes of company presidents?

14.15 For the data in Table 14.4, use the computational formulas to find the relevant sums of squares.

14.16 Show that the computational formulas given in this section for SS Total, SS Between, and SS Within are equivalent to the formulas given for these sums of squares in Section 14.3.

14.17 Use the S-method to find a set of interval estimates for all possible pairwise differences between programs in Problem 14.3 with a confidence level of 95 percent.

14.18 Carry out the multiple-comparisons procedure at the 99-percent confidence level for the data in Table 14.4. At this confidence level, are any of the pairwise differences between machines significant?

14.19 In Problem 14.13 estimate the treatment effects and use the S-method to find simultaneous confidence intervals at the 95-percent confidence level for the six possible pairwise differences between machines.

14.20 Instead of using the multiple-comparisons procedure in Problem 14.18, compute interval estimates separately for each pairwise difference at the 99-percent confidence level. Compare the resulting intervals with those found in Problem 14.18 and explain why they are not identical.

14.6 THE TWO-FACTOR MODEL

In the one-factor model, we worked with a set of J samples, or treatments, and attempted to investigate possible differences between them. In some situations, the statistician may want to investigate simultaneously *two* sets of treatments. In the marketing example, suppose that in addition to the type of packaging, a second factor of interest is advertising. For this factor, assume that there are two treatments (or two levels of the factor), advertising and no advertising. Interest is now focused on two distinct experimental factors, the type of packaging and the advertising, either or both of which possibly influence sales. A random sample of sixty stores is selected, with ten assigned randomly to each of the *six* possible treatment *combinations*. That is, ten stores have Type A packaging and no advertising; ten have Type A and advertising; ten have Type B and no advertising; and so on. This experiment represents an instance where two different sets of experimental treatments are completely crossed: each category or level of one factor (packaging) occurs with each level of the other factor (advertising). Since there are three levels of packaging and two levels of advertising, there are six distinct sample groups, each with a particular combination of the two factors. Furthermore, this experiment is said to be *balanced*, since each of these six groups has the same sample size.

In this experiment, the marketing manager is interested in three questions:

1. Are there systematic effects due to type of packaging alone (irrespective of advertising)?
2. Are there systematic effects due to advertising alone (irrespective of type of packaging)?
3. Are there systematic effects due neither to type of packaging alone, nor to advertising alone, but attributable only to the *combination* of a particular type of packaging with a particular level of advertising?

Note that the experiment could be viewed as two separate experiments carried out on the same set of stores: (a) there are three groups of twenty stores each, differing in type of packaging; and (b) there are two groups of thirty stores each, differing in level of advertising. The third question cannot, however, be answered by the comparison of types of packaging alone or by the comparison of levels of advertising alone, but is a question of *interaction*, the unique effects of combinations of treatments. This is an important feature of the two-factor analysis-of-variance model: we shall be able to examine *main effects* of the separate experimental variables or factors just as in the one-factor model (a "main effect" is the variability due to different levels of a *single* variable) as well as *interaction effects*, differences apparently due only to the unique *combinations* of treatments.

In order to present the two-factor model formally, suppose that there are J levels of the first factor and K levels of the second factor, and that there are n observations for each combination of one level from each factor. If y_{ijk} denotes the ith observation in the group with level j on the first factor and level k on the second factor, then the model may be written as follows:

$$y_{ijk} = \mu + \tau_j + \lambda_k + (\tau\lambda)_{jk} + \varepsilon_{ijk},$$

with $i = 1, \ldots, n, j = 1, \ldots, J$, and $k = 1, \ldots, K$. The observation is thought of as the sum of five components: the grand mean μ over the entire population; a treatment effect τ_j associated with the particular level (the jth level) of the first factor; a treatment effect λ_k associated with the particular level (the kth level) of the second factor; an interaction effect $(\tau\lambda)_{jk}$ associated with the particular combination of the jth level of the first factor and the kth level of the second factor; and finally, a random error term ε_{ijk}. It is assumed that the sum of the treatment effects for the first factor is zero, that the sum of the treatment effects for the second factor is zero, and that for any given level of either of the two factors, the sum of the interaction effects across the other factor is zero. Furthermore, in order to make inferences concerning the parameters of the model, the random-error terms are assumed to be independent and normally distributed, with means of zero and identical variances. Observe that just as there is a similarity between the simple one-factor model and the bivariate linear regression model, there is also a similarity between the above model and the multiple regression model discussed in Chapter 12.

To illustrate the two-factor model, suppose, in the marketing example, that the grand mean is 30 and that all other terms are zero; each observation in the experiment should equal 30. Alternatively, suppose that the treatment effects for type of packaging are +4, −3, and −1, respectively. Then the observations will look like this (only one observation is presented for each group, since all observations within a group will be equal because the random-error term is zero):

	Type of Packaging		
	A	**B**	**C**
Advertising	30 + 4 = 34	30 − 3 = 27	30 − 1 = 29
No advertising	30 + 4 = 34	30 − 3 = 27	30 − 1 = 29

On the other hand, it might turn out that nonzero effects exist only for advertising (not for type of packaging), with the effects of *advertising* and *no advertising* being +3 and −3:

	Type of Packaging		
	A	**B**	**C**
Advertising	30 + 3 = 33	30 + 3 = 33	30 + 3 = 33
No advertising	30 − 3 = 27	30 − 3 = 27	30 − 3 = 27

Now suppose that there are *both* packaging and advertising effects:

	Type of Packaging		
	A	**B**	**C**
Advertising	30 + 4 + 3 = 37	30 − 3 + 3 = 30	30 − 1 + 3 = 32
No advertising	30 + 4 − 3 = 31	30 − 3 − 3 = 24	30 − 1 − 3 = 26

Table 14.12

	Type of Packaging		
	A	B	C
Advertising	33	37	29
	36	33	34
	40	39	36
No advertising	30	12	34
	34	19	29
	34	18	33

Observe that the linear model specifies that the effect of a combination of levels of the two factors is the sum of the individual effects and an interaction effect (which has been assumed to be zero). Finally, let us add interaction effects:

	Type of Packaging		
	A	B	C
Advertising	$30+4+3-2=35$	$30-3+3+7=37$	$30-1+3-5=27$
No advertising	$30+4-3+2=33$	$30-3-3-7=17$	$30-1-3+5=31$

The effect associated with a combination of treatments is no longer the simple sum of the two individual treatment effects, because there are nonzero interaction effects. Of course, in an actual experiment there will also be nonzero random-error terms, and the observations within a particular group will no longer be identical. For example, assuming a sample size three for each combination of factors, the data might look something like Table 14.12.

14.7 INFERENCES IN THE TWO-FACTOR MODEL

Just as in the one-factor model, the basic idea in the two-factor model is to determine how much of the total variation in the data can be attributed to chance and how much can be attributed to certain "effects." In the two-factor model, three effects which are of interest are those associated with the first factor (Factor τ), those associated with the second factor (Factor λ), and interaction effects. The total sum of squares is partitioned into four components:

$$\text{SS Columns (Factor } \tau) = \sum_j Kn(\bar{y}_j - \bar{y})^2, \tag{14.6}$$

$$\text{SS Rows (Factor } \lambda) = \sum_k Jn(\bar{y}_k - \bar{y})^2, \tag{14.7}$$

$$\text{SS Error} = \sum_i \sum_j \sum_k (y_{ijk} - \bar{y}_{jk})^2, \tag{14.8}$$

and

$$\text{SS Interaction} = \sum_j \sum_k n(\bar{y}_{jk} - \bar{y}_j - \bar{y}_k + \bar{y})^2. \tag{14.9}$$

The number of observations in treatment combination jk (level j of A, level k of B) is n, the total number of observations for level j of τ is Kn, and the total number of observations for level k of λ is Jn. The sample mean for the n observations in treatment combination jk is denoted by $\bar{y}_{jk}$, the sample mean for the Kn observations in factor level j of τ is $\bar{y}_j$, the sample mean for the Jn observations in factor level k of λ is $\bar{y}_k$, and the grand sample mean is $\bar{y}$. The sum of the four SS terms above is simply

$$\text{SS Total} = \sum_i \sum_j \sum_k (y_{ijk} - \bar{y})^2. \tag{14.10}$$

Using Formulas (14.6) through (14.10), the analysis-of-variance table for the two-factor model is given in Table 14.13.

The assumptions in the two-factor model are similar to those in the one-factor model:

1. For each treatment combination jk, the random error terms ε_{ijk} are normally distributed with mean zero and variance σ^2, and the variance σ^2 is the same for all treatment combinations.
2. The random error terms are independent.

Table 14.13

Source	SS	d.f.	MS
Columns (Factor τ)	SS Columns	$J-1$	SS Columns/$(J-1)$
Rows (Factor λ)	SS Rows	$K-1$	SS Rows/$(K-1)$
Interaction	SS Interaction	$(J-1)(K-1)$	SS Interaction/$(J-1)(K-1)$
Error	SS Error	$JK(n-1)$	SS Error/$JK(n-1)$
Total	SS Total	$JKn-1$	

Under these assumptions, the expectations of the mean squares can be calculated:

$$E(\text{MS Columns}) = \sigma^2 + \frac{Kn \sum_j \tau_j^2}{J - 1},$$

$$E(\text{MS Rows}) = \sigma^2 + \frac{Jn \sum_k \lambda_k^2}{K - 1},$$

$$E(\text{MS Interaction}) = \sigma^2 + \frac{n \sum_j \sum_k (\tau\lambda)_{jk}^2}{(J - 1)(K - 1)},$$

and

$$E(\text{MS Error}) = \sigma^2.$$

There are several tests of interest in this model. First, consider the hypothesis that the effects of the J levels of Factor τ are all zero:

$$H_0: \tau_1 = \tau_2 = \cdots = \tau_J = 0,$$
$$H_a: \text{at least one } \tau_j \neq 0.$$

If H_0 is true, $E(\text{MS Columns}) = \sigma^2$, and, by the same line of reasoning used to develop the $\boldsymbol{F}$-test in the one-factor model,

$$F = \frac{\text{MS Columns}}{\text{MS Error}}$$

has an $\boldsymbol{F}$-distribution with $J - 1$ and $JK(n - 1)$ degrees of freedom. This $\boldsymbol{F}$-statistic can be used to test H_0; the null hypothesis will be rejected for large values of $\boldsymbol{F}$.

A second hypothesis of interest is the hypothesis that the effects of the K levels of Factor λ are all zero:

$$H_0: \lambda_1 = \lambda_2 = \cdots = \lambda_K = 0,$$
$$H_a: \text{at least one } \lambda_k \neq 0.$$

Under the null hypothesis,

$$E(\text{MS Rows}) = \sigma^2,$$

and

$$F = \frac{\text{MS Rows}}{\text{MS Error}}$$

has an F-distribution with $K - 1$ and $JK(n - 1)$ degrees of freedom. Finally, we might also be interested in the hypothesis that all interaction effects are zero:

$$H_0:(\tau\lambda)_{jk} = 0 \quad \text{for all } j = 1, \ldots, J \quad \text{and} \quad k = 1, \ldots, K,$$
$$H_a:(\tau\lambda)_{jk} \neq 0 \quad \text{for at least one combination } (j, k).$$

The statistic used to test this hypothesis is

$$F = \frac{\text{MS Interaction}}{\text{MS Error}},$$

which has an F-distribution with $(j - 1)(k - 1)$ and $jk(n - 1)$ degrees of freedom.

14.8 COMPUTATIONAL FORMS FOR THE TWO-FACTOR MODEL

The formulas for determining SS which were presented in the previous section are very difficult to apply. The following computational formulas are much more convenient:

$$\text{C.F. (correction factor)} = \frac{\left(\sum_i \sum_j \sum_k y_{ijk}\right)^2}{JKn},$$

$$\text{SS Total} = \sum_i \sum_j \sum_k y_{ijk}^2 - \text{C.F.},$$

$$\text{SS Columns} = \frac{\sum_j \left(\sum_i \sum_k y_{ijk}\right)^2}{Kn} - \text{C.F.},$$

$$\text{SS Rows} = \frac{\sum_k \left(\sum_i \sum_j y_{ijk}\right)^2}{Jn} - \text{C.F.},$$

$$\text{SS Error} = \sum_i \sum_j \sum_k y_{ijk}^2 - \frac{\sum_j \sum_k \left(\sum_i y_{ijk}\right)^2}{n},$$

and

$$\text{SS Interaction} = \text{SS Total} - \text{SS Rows} - \text{SS Columns} - \text{SS Error}.$$

Table 14.14

	Type of Packaging			
	A	B	C	
Advertising	52	28	15	
	48	35	14	
	43	34	23	
	50	32	21	
	43	34	14	
	44	27	20	
	46	31	21	
	46	27	16	
	43	29	20	
	49	25	14	
	$\sum_i y_{i11} = 464$	$\sum_i y_{i21} = 302$	$\sum_i y_{i31} = 178$	$\sum_j \sum_i y_{ij1} = 944$
No advertising	38	43	23	
	42	34	25	
	42	33	18	
	35	42	26	
	33	41	18	
	38	35	26	
	39	37	20	
	34	37	19	
	33	40	22	
	34	36	17	
	$\sum_i y_{i12} = 368$	$\sum_i y_{i22} = 378$	$\sum_i y_{i32} = 214$	$\sum_j \sum_i y_{ij2} = 960$
	$\sum_k \sum_i y_{i1k} = 832$	$\sum_k \sum_i y_{i2k} = 680$	$\sum_k \sum_i y_{i3k} = 392$	$\sum_k \sum_j \sum_i y_{ijk} = 1904$

To illustrate the use of these formulas, suppose that the experiment involving the type of packaging and advertising (which was discussed in Section 14.4) were actually carried out, with the results shown in Table 14.14.

Using the computational formulas, we get

$$\text{C.F.} = \frac{(1904)^2}{60},$$

$$\text{SS Total} = (52)^2 + (48)^2 + \cdots + (22)^2 + (17)^2 - \text{C.F.} = 6451.7,$$

$$\text{SS Columns} = \frac{(832)^2}{20} + \frac{(680)^2}{20} + \frac{(392)^2}{20} - \text{C.F.} = 4994.1,$$

$$\text{SS Rows} = \frac{(944)^2}{30} + \frac{(960)^2}{30} - \text{C.F.} = 4.2,$$

$$\text{SS Error} = (52)^2 + (48)^2 + \cdots + (22)^2 + (17)^2 - \frac{(464)^2 + (302)^2 + \cdots + (214)^2}{10} = 643.2,$$

and

$$\text{SS Interaction} = 6451.7 - 4994.1 - 4.2 - 643.2 = 810.2.$$

The analysis-of-variance table is given in Table 14.15.

For the hypothesis that there are no interaction effects, we have

$$F = \frac{\text{MS Interaction}}{\text{MS Error}} = \frac{405.1}{11.9} = 34.0,$$

with 2 and 54 d.f., which is much greater than the critical value corresponding to $\alpha = 0.05$. Thus, the marketing manager may conclude with considerable confidence that there are nonzero interaction effects. For the tests concerning

Table 14.15

Source	SS	d.f.	MS
Columns (packaging)	4994.1	2	2497.05
Rows (advertising)	4.2	1	4.2
Interaction	810.2	2	405.1
Error	643.2	54	11.9
Total	6451.7	59	

column effects (packaging) and row effects (advertising), the statistics are

$$F = \frac{\text{MS Columns}}{\text{MS Error}} = \frac{2497.05}{11.9} = 209.8$$

with 2 and 54 d.f. and

$$\text{F} = \frac{\text{MS Rows}}{\text{MS Error}} = \frac{4.2}{11.9} = 0.35$$

with 1 and 54 d.f. The former statistic is highly significant, whereas the latter is clearly not significant. The following assertions can be made:

1. There is apparently little or no effect of advertising *alone* on sales.
2. The type of packaging alone *does* seem to affect sales.
3. There is apparently an interaction between advertising and type of packaging, meaning that the magnitude and the direction of the effects of type of packaging differ for the two different advertising levels.

In short, the type of packaging makes a difference in sales, but the kind and extent of the difference depends on the level of advertising.

It should be mentioned that as a result of the fact that the interaction effect was found to be statistically significant, the outcomes of the other two F-tests were of less importance than would be the case if the interaction effect had not been found significant. The statistician is usually interested in whether a particular factor (e.g., advertising) has an effect on the variable of primary interest (e.g., sales), although he may not care whether the factor has a significant effect by itself or only through the interaction term. In our example, advertising apparently does have an effect on sales through its interaction with type of packaging, so the marketing manager will have to consider advertising in making decisions on the basis of the given data. However, had the interaction term not been significant, the other two tests would be of much greater interest.

Although the linear analysis-of-variance model is believed to be quite widely applicable, there are admittedly many situations in which the assumptions concerning the error terms are not satisfied. Therefore it is of some interest to investigate the "robustness" of the analysis-of-variance procedure to violations of these assumptions. The least important of the assumptions is that of normality; non-normality of the error terms usually has little effect on the results, particularly if the sample size is reasonably large. The assumption that the error terms all have the same variance is somewhat more important, especially when the various samples or groups have unequal sample sizes. If it is suspected that the variance might not be constant, it is a good idea to make the sample sizes equal. Finally, violations of the last assumption (independence of the error terms) are quite serious and may affect the results of the analysis quite severely.

Given the preceding results, the marketing manager is interested in estimating the packaging effects and the interaction effects. The effects of packaging can be estimated by subtracting the grand sample mean from the respective column means:

$$\hat{\tau}_1 = 41.6 - 31.7 = 9.9,$$
$$\hat{\tau}_2 = 34.0 - 31.7 = 2.3,$$

and

$$\hat{\tau}_3 = 19.6 - 31.7 = -12.1.$$

To estimate interaction effects, subtract the appropriate row and column sample means from the grand sample mean and add the sample mean for the particular combination of interest. For example,

$$\text{Est. } (\tau\lambda)_{32} = \frac{1904}{60} - \frac{392}{20} - \frac{960}{30} + \frac{214}{10} = 1.5.$$

All six interaction effects can be estimated in this manner. The rationale behind this procedure is quite simple: from the sample mean for the combination of interest, we would like to subtract the row and column effects and the grand mean. In subtracting the appropriate row and column means, we are subtracting the row and column effects and at the same time subtracting the grand mean twice. Therefore, we have to add back the grand mean. In other words, we have

$$\text{Est. } (\tau\lambda)_{jk} = \bar{y}_{jk} - \bar{y}_j - \bar{y}_k + \bar{y}.$$

But

$$\hat{\tau}_j = \bar{y}_j - \bar{y},$$

so

$$\bar{y}_j = \bar{y} + \hat{\tau}_j.$$

Similarly,

$$\bar{y}_k = \bar{y} + \hat{\lambda}_k.$$

Therefore, we have

$$\begin{aligned}\text{Est. } (\tau\lambda)_{jk} &= \bar{y}_{jk} - \bar{y} - \hat{\tau}_j - \bar{y} - \hat{\lambda}_k + \bar{y} \\ &= \bar{y}_{jk} - \hat{\tau}_j - \hat{\lambda}_k - \bar{y},\end{aligned}$$

which is exactly what we wanted in the first place.

Define: Main effects, interaction effect, two-factor model.

Problems

14.21 Construct examples of two-factor experiments in which

a) row effects only are present;

b) row and column effects, but no interaction effects, are present;

c) row, column, and interaction effects are all present.

14.22 In the marketing example in Section 14.6, suppose that the row effects for advertising are zero. Does this mean that advertising has no impact upon sales?

14.23 In an analysis-of-variance, an attempt is made to estimate how much of the total variation in the data can be attributed to certain assignable causes of variation and how much can be attributed to chance. If row, column, and interaction effects are all present in a two-factor experiment, how can the variation due to chance be estimated?

14.24 The interaction effect for the combination of advertising and type A packaging in the marketing example in Section 14.6 is -2, while the interaction effect for the combination of advertising and type B packaging is $+7$. How can you interpret these figures?

14.25 In a study of the amount of air pollution in a city, the two factors of interest are the day of the week (Monday, Tuesday, Wednesday, Thursday, Friday) and the time of day (morning, afternoon). The study reveals nonzero effects for each factor individually and a nonzero interaction effect. Explain what each of these effects represents.

14.26 For the data in Table 14.12,

a) find the grand sample mean, the sample mean for each row, the sample mean for each column, and the sample mean for each of the six treatment combinations;

b) compute SS Columns, SS Rows, SS Interaction, SS Error, and SS Total;

c) determine the mean squares for columns, rows, interaction, and error;

d) construct an analysis-of-variance table;

e) conduct F-tests with $\alpha = 0.05$ for nonzero column effects, row effects, and interaction effects.

14.27 A consumer research firm wants to compare three brands of radial tires in terms of tread life over different road surfaces. Random samples of four tires of each brand

Table 14.16

		Brand		
		A	B	C
Road Surface	Asphalt	36 39 39 38	42 40 39 42	32 36 35 34
	Concrete	38 40 41 40	42 45 48 47	37 33 33 34
	Gravel	34 32 34 35	34 34 30 31	36 35 35 33

are selected for each of three surfaces (asphalt, concrete, gravel). A machine that can simulate road conditions for each of the road surfaces is used to find the tread life (in thousands of miles) of each tire. The data are shown in Table 14.16. Construct an analysis-of-variance table and conduct **F**-tests with $\alpha = 0.05$ for the presence of nonzero brand effects, road surface effects, and interaction effects.

14.28 Determine estimates for the six interaction effects in the marketing example in Section 14.8.

14.29 Use the computational formulas presented in Section 14.8 to do Problem 14.27 and find estimates for any effects that are significant at the 0.05 level.

14.30 A manufacturer frequently sends small packages to a customer in another city via air freight, and in many cases it is important for a package to reach the customer as soon as possible. Three different firms offer air freight service, including pickup and delivery, on a 24-hour basis. The head of the manufacturer's shipping department would like to know if the firms differ in speed of service and if the time of day makes any difference. An experiment is designed to investigate these issues. Packages are sent at random times, and the air freight firm used for each package is also randomly chosen. The customer records the time that each package arrives so that the time elapsed during shipment can be determined. These times are rounded to the nearest hour; the experimental results for a total of 54 packages are shown in Table 14.17. Analyze these data by constructing an analysis-of-variance table, conducting **F**-tests, and determining estimates for the treatment effects.

Table 14.17

		Firm		
		Speedy Air Freight	**ABC Shipping**	**Eagle Air Freight**
	Morning	8, 6, 6, 12, 7, 8	11, 11, 9, 10, 8, 11	7, 4, 6, 4, 9, 7
Time	**Afternoon**	7, 10, 8, 11, 9, 11	10, 13, 10, 12, 11, 10	10, 8, 6, 5, 8, 6
	Night	13, 11, 14, 11, 15, 12	12, 16, 15, 15, 10, 17	8, 11, 9, 9, 9, 10

14.31 A consumer research group is interested in how the price of the leading brand of aspirin varies, if it varies at all, across different areas of a metropolitan area and across different types of retail outlets. The area is divided into four regions: center city, lakefront, west side, and north suburbs. Three types of retail outlets are considered: drugstores, discount stores, and grocery stores. For each region and each type of store, four stores are chosen at random and the price of a large bottle of aspirin is recorded (in dollars). The prices observed are shown in Table 14.18. Carry out an analysis of variance, including appropriate tests and estimates, and provide an interpretation of the results of the analysis.

Table 14.18

	Region			
	Center City	Lakefront	West Side	North Suburbs
Drugstore	2.46 2.63 2.52 2.42	2.85 2.61 2.73 2.68	2.44 2.29 2.48 2.37	2.65 2.73 2.51 2.60
Discount Store	2.27 2.39 2.24 2.30	2.34 2.30 2.43 2.28	2.35 2.19 2.28 2.37	2.36 2.27 2.43 2.30
Grocery Store	2.81 2.68 2.63 2.72	2.70 2.64 2.79 2.62	2.68 2.76 2.59 2.65	2.78 2.74 2.95 2.84

14.9 EXPERIMENTAL DESIGN

The discussion of sampling theory in Chapter 6 demonstrated that there is often more to taking a sample than just determining the sample size, n. First, there is the problem of carefully defining the population of interest. Once this is done, various sampling plans, such as simple random sampling, stratified sampling, and so on, must be considered. Once a sampling plan is chosen, the actual items to be sampled can be determined. This procedure falls under the general heading of "sample design," and the overall objective in designing a sample is to get the greatest possible precision for a given cost, or to attain some given level of precision in the least costly manner.

In sampling theory, it is presumed that the sample consists of observations on an *existing* population; that is, some members of the population are observed and the desired information about these members is recorded. This contrasts with the idea of a *controlled experiment*, which is widely encountered in scientific research. In a controlled experiment, instead of merely observing an existing population, the statistician attempts to control, or to modify, certain factors of interest, and he then observes the effect of these modifications on the results of the experiment. Rather than observing the proportion of defective items in a sample from a production process, the experimental statistician might try a modification of the process and observe the proportion of defectives, both with and without the modification. The statistician must do this carefully, however, to ensure that all factors other than the one being varied remain constant. In terms of the production example, he should attempt to make the conditions under which the process is run *with* and *without* the modification as similar as possible. If he uses the modification during the night shift but not during the day shift, then any differences which appear may be caused by the different sets of personnel rather than by the modification. Unfortunately, it is often quite difficult, if not impossible, to hold such factors constant in actual problems. In the marketing example used to illustrate the analysis of variance, for instance, the various stores in the experiment will obviously differ in some respects (location, number and type of customers, etc.). In business and the social sciences it is not as easy to control extraneous factors as it is in the physical sciences.

The marketing example illustrated the idea of experimental design. First, the factors of interest were determined (type of packaging and advertising), each with a certain number of "levels" (three types of packaging, two levels of advertising). The total sample size was chosen and the 60 experimental units (stores) were selected. The next problem was the allocation of stores to the six possible treatment combinations. This was done randomly, thus invoking the principle of randomization, which is an important principle in experimental statistics. The use of randomization permits the experimenter to make inferential statements from the observed data.

This example should give you an idea of what experimental design is all about. Of course, the example was purposely made simple in order to illustrate a particular analysis-of-variance model, the two-factor model with interaction. A textbook example such as this may give the impression that an experiment is "forced" to fit a certain model. In practice, however, various models are generally considered, and the model that seems to best fit the situation at hand is chosen. Experimental design proceeds from the problem situation to the model, not vice versa. In the marketing example, the marketing manager may decide that factors such as "type of store" may cause a great deal of variation and that it would be desirable to somehow reduce this variation. A generalization of the one-factor model called a *randomized blocks* model allows the experimenter to allow for some of this extraneous variation separately instead of including it in the estimated random variation. For example, if there are five different types of stores, three stores of each type could be selected and the three types of packaging could be randomly assigned to these three stores. The types of stores are the "blocks," and the experimental assignment of treatments to stores within each block is random; hence the term "randomized blocks." Of course, a similar extension could be used in the two-factor situation with combinations of type of packaging and advertising–no advertising assigned randomly to stores, within each type of store.

In some instances, of course, more than two factors are of interest. In general, experimental designs involving several factors are called *factorial* designs and, conceptually at least, there is no limitation on the number of factors that can be considered. As the number of factors increases, the experimental design becomes more complex and the analysis more difficult. If there are four factors and each has four levels (i.e., four possible treatments within each factor), then there are 4^4 or 256 combinations of factor levels. If the design is complete in the sense that all combinations are included and if the samples sizes are identical for all combinations, then the overall sample size will be a multiple of 256. By carefully considering the situation at hand, it may be possible to reduce the number of combinations. For instance, some interaction effects, particularly higher-order interaction effects, might be eliminated from consideration *a priori* by the experimenter. Alternatively, various models are available that allow the experimenter to reduce the overall sample size. For example, in a design known as a *Latin square* design, only one level of a third factor is used with each possible combination of levels of the first two factors. Two other types of designs that lead to reductions in the overall sample size are *incomplete block* designs and *fractional factorial* designs.

As we have pointed out, there is a difference between "sample design" and "experimental design," and there are also some similarities. In both cases, the statistician is attempting to obtain as much information as he can for the smallest possible cost. He would like his results to be as precise as possible. Of course, he can increase precision by increasing sample size, but this may be quite costly in terms of both time and money. Another way to obtain more precise results is to design the experiment or sampling plan more carefully. For instance, it is often possible to increase the precision of estimates obtained from a sample without increasing the sample size by adopting, say, a stratified sampling plan instead of simple random sampling. The same idea holds in experimental design. Designs such as those mentioned in the previous paragraph may enable an experimenter to greatly reduce the number of experimental units needed to make inferences in certain situations. Complex experimental designs such as these are beyond the scope of this book; the student who is interested in pursuing this subject further should see references [1], [11], and [13].

14.10 REGRESSION AND ANALYSIS OF VARIANCE: THE GENERAL LINEAR MODEL

At various points in this chapter we have noted the similarity between the linear analysis-of-variance model and the linear regression model. The bivariate linear regression model is similar to the simple one-factor analysis-of-variance model, and the multiple regression model is similar to a two-(or more-) factor analysis-of-variance model without the interaction terms. Even the assumptions concerning the random-error terms are the same in the two models. The discussion of regression analysis in Chapters 10 and 12 included terms like "sum of squares." Therefore, such terms should not have been new to you in this chapter.

Suppose that, in a simple one-factor model, the J treatment categories correspond to J values of the independent variable, x, and that we want to investigate the linear regression of a second variable, $\mathbf{y}$, on x. The linear regression model can be written as follows:

$$\mathbf{y}_{ij} = \alpha + \beta x_j + \boldsymbol{\varepsilon}_{ij}.$$

There are J levels of the factor $\mathbf{x}$, and a sample of size n_j is taken in the jth level. On the basis of the entire sample (which is of size $\sum_j n_j$), the parameters α and β are estimated by using the least-squares criterion, and the estimated linear regression is

$$y_{ij} = a + bx_j + e_{ij},$$

or

$$y_{ij} = \hat{y}_j + e_{ij}.$$

Here y_{ij} is the observed value of $\mathbf{y}$ and $\hat{y}_j$ is the value predicted by the estimated linear regression.

Using the above model, the deviation of an observed value y_{ij} from the grand mean $\hat{y}$ can be thought of as the sum of two parts:

$$y_{ij} - \bar{y} = (y_{ij} - \hat{y}_j) + (\hat{y}_j - \bar{y}).$$

The first term on the righthand side of this equation is simply the deviation of the particular observation from the predicted value for its group or treatment level. The second term is the deviation of the predicted value itself from the grand mean.

By an argument like that used earlier in this chapter and in Chapter 10, it can be shown that over all observations in all groups, the total sum of squares can be partitioned into SS Error and SS Linear Regression, which reflect, respectively, random error and the "linear regression effect." The formula for SS Total is

$$\text{SS Total} = \sum_j \sum_i (y_{ij} - \bar{y})^2,$$

and for SS Linear Regression,

$$\text{SS Linear Regression} = \sum_j n_j(\hat{y}_j - \bar{y})^2.$$

The analysis-of-variance table is given in Table 14.19.

In practice, the usual computational formula used to determine SS Total is

$$\text{SS Total} = \sum_j \sum_i y_{ij}^2 - \frac{\left(\sum_j \sum_i y_{ij}\right)^2}{\sum_j n_j}.$$

Table 14.19

Source	SS	d.f.	MS
Linear regression	SS Linear Regression	1	SS Linear Regression/1
Error	SS Error	$\sum_j n_j - 2$	SS Error $\Big/ \left(\sum_j n_j - 2\right)$
Total	SS Total	$\sum_j n_j - 1$	

The sums of squares associated with the linear regression can be expressed in terms of the sample correlation coefficient and the sample variance of y:

$$\text{SS Linear Regression} = \left(\sum_j n_j\right) r^2 s_y^2.$$

Finally, SS Error is found by subtracting SS Linear Regression from SS Total. The statistic used to test the hypothesis that there is no linear regression effect is

$$F = \frac{\text{MS Linear Regression}}{\text{MS Error}},$$

which has an F-distribution with 1 and $\sum_j n_j - 1$ degrees of freedom, respectively.

This section demonstrates the fact that a linear regression model can be thought of as a simple analysis-of-variance model. Both linear regression analysis and the analysis of variance fall under the heading of the *general linear model.* In multivariate statistical problems the general linear model is of great value, both because it is often a realistic model and because nonlinear models are very difficult to work with when there are many variables.

Define: Experimental design, controlled experiment, randomized block, factorial design, Latin square design, general linear model.

Problems

14.32 Try to explain in your own words exactly what the term "experimental design" encompasses.

14.33 Carefully explain the difference between the terms "sample design" and "experimental design."

14.33 Discuss the role of randomization in experimental statistics.

14.35 In Problem 14.3, 20 new employees are randomly assigned to the training programs. Would it affect the experiment if some other assignment procedure were used? For example, what if the first five employees to be hired were put in the first program, the next five in the second program, and so on? What if the personnel director decided which employees should be assigned to each program?

14.36 In Problem 10.18, a linear regression is used to predict yearly oil consumption in the U.S. as a function of the year. The data were analyzed with a "canned" regression program which is part of a set called *Interactive Data Analysis* (IDA) programs. The analysis-of-variance table for this example is provided in the following outputs from the computer program:

SOURCE	SS	DF	MS	F
REGRESSION	8.05121E+01	1	8.05121E+01	501.68
RESIDUALS	1.28388E+00	8	1.60485E-01	
TOTAL	8.17960E+01	9	9.08844E+00	

Note that this program uses the term "residuals" where we have used the term "errors"; these terms are interchangeable. With $\alpha = 0.05$, test the hypothesis that there is no linear regression effect against the alternative that there is such an effect.

14.37 For the data in Problem 10.20, find SS Total add SS Linear Regression and complete the analysis-of-variance table for the regression.

14.38 A production manager knows that a particular task can be completed faster if more workers are used. To investigate the relationship between the time required to finish the task and the number of workers assigned to the task, an experiment is conducted. The experimental results are as follows:

Number of Workers:	1	2	3	4
Time Required to Finish Task (in minutes):	15, 18, 10, 14, 16	12, 10, 15, 14, 12	12, 8, 8, 9, 9	6, 4, 4, 7, 5

The total sample size is 20, with a sample of five for each choice of number of workers. Find the analysis-of-variance table and conduct an F-test with $\alpha = 0.05$ for the presence of a linear regression effect.

14.39 Construct an analysis-of-variance table for the regression of crop yield on amount of fertilizer in Problem 10.24.

Exercises

14.40 Discuss the general rationale behind the analysis-of-variance. Since the hypotheses of interest involve means, why does the analysis-of-variance focus on variances?

14.41 Explain the advantage, if any, of a comparison of J means ($J > 2$) by an analysis of variance and an F-test over the practice of carrying out a t-test separately for each pair of means.

14.42 Suppose that four randomly selected groups of five observations each were used in an experiment. Furthermore, imagine that the overall sample mean was 60. What would the data be like if the F-test resulted in $F = 0$? What would the data be like if

$F \to \infty$? If the hypothesis of equality of the means for the four groups were true, how large should we expect F to be?

14.43 Students registering for a review course in accounting in preparation for a CPA examination were randomly divided into three sections of twelve students each. In the first section, the students spent all their time in class working various types of accounting problems. The second section emphasized accounting concepts and was taught primarily in a lecture format. The third section represented a mixture of lectures concerning concepts and practice with actual problems. At the end of the course, the 36 students took the CPA examination, and their scores were as follows:

Section 1: 72, 58, 81, 70, 68, 60
74, 56, 64, 67, 66, 71

Section 2: 52, 78, 64, 69, 73, 55
56, 62, 58, 67, 61, 70

Section 3: 68, 76, 82, 80, 91, 84
77, 71, 79, 85, 82, 66

a) Perform an analysis-of-variance on these data and discuss the results.

b) Determine a set of interval estimates for all possible pairwise differences, using a 95-percent confidence coefficient. From these interval estimates, which differences are significant?

14.44 An experiment was designed to investigate differences in the life of lightbulbs from three different manufacturers. Five lightbulbs were randomly selected from each of three large shipments of bulbs from the different manufacturers. For each bulb, the number of hours until failure was recorded:

Manufacturer I: 120, 90, 105, 100, 125
Manufacturer II: 100, 130, 125, 140, 120
Manufacturer III: 110, 75, 100, 90, 100

a) Construct an analysis-of-variance table and test the null hypothesis of no treatment effects at the 0.05 level of significance.

b) Determine an estimate of σ^2, the error variance.

c) Use the S-method for multiple comparisons to investigate the differences between I and II, between II and III, and between I and III, in terms of the life of their lightbulbs.

14.45 There are many small firms in a particular industry. Some of the firms are unionized under Union A, some are unionized under Union B, and others are not unionized. A random sample of firms reveals the following average pay increases (in percentage terms) over the past year:

Union A: 12, 16, 13, 10, 8, 14, 17, 11
Union B: 8, 13, 7, 9, 11, 6, 8, 8
Nonunion: 10, 12, 8, 11, 7, 9, 10, 8

Carry out an analysis-of-variance, including an **F**-test for equality of means and a set of interval estimates for pairwise differences in means with a 95-percent confidence coefficient. What can you conclude about the impact of the two unions and of the lack of a union on pay increases?

14.46 A new type of pill which, when taken daily, is intended to reduce the chance of catching a cold has been developed in the research laboratory of a drug manufacturer. In order to test the drug, an experiment is designed. Forty-five people are randomly divided into three groups of 15 each. Each experimental subject takes one pill per day, with one group receiving the new pill, one group receiving Vitamin C, and one group receiving a placebo (a pill containing no medication or vitamins). The subjects do not know which of the pills they are receiving. The experiment is conducted for six months, and the number of days with colds is recorded for each subject, with the following results:

New Pill: 4, 15, 8, 6, 9, 8, 18, 0, 12, 6, 7, 10, 11, 2, 6
Vitamin C: 13, 7, 2, 0, 11, 8, 5, 3, 10, 9, 8, 8, 4, 7, 1
Placebo: 9, 12, 17, 13, 5, 19, 6, 8, 10, 11, 2, 14, 15, 1, 12

a) Estimate the treatment effects.

b) Are these treatment effects significantly different from zero at the $\alpha = 0.01$ level of significance?

c) Interpret the results of the experiment.

d) The experimenter could just skip the placebo with the third group and tell the subjects which groups they are in. Would that make any difference?

14.47 One year ago, a contractor built and sold houses with three different types of heating in a new subdivision. The three heating systems utilize gas, electricity, and solar energy, respectively. The contractor is planning an addition to the subdivision, and he must decide what heating systems to use in the new houses that are to be built. The impact of the heating system on the initial cost of the house is an important consideration, as is the anticipated cost of heating the house. To obtain more information on operating costs, the contractor finds the heating costs for the past winter for the houses sold last year. These heating costs, in hundreds of dollars, are as follows:

Gas Heat	Electric Heat	Solar Heat
6.1	8.2	5.2
7.2	7.5	6.1
5.4	8.7	5.7
5.8	7.4	5.4
6.3	7.8	6.2
6.7	9.2	
7.4		
6.4		

a) Estimate the population means and the treatment effects for the three heating systems.

b) Construct an analysis-of-variance table for this example.

c) Are the costs different for the three heating systems at the 0.05 level of significance?

14.48 The director of a medical insurance plan is concerned about the length of time patients spend in the hospital. From records of patients with a certain type of illness, six male patients and six female patients are chosen at random from each of three hospitals. The number of days spent in the hospital by each patient is given in

Table 14.20

	Hospital A		Hospital B		Hospital C	
Male Patients	29	36	14	5	22	25
	35	33	8	7	20	30
	28	38	10	16	23	32
Female Patients	25	35	3	5	18	7
	31	32	8	9	15	11
	26	34	4	6	8	10

Table 14.20. Carry out an analysis of variance, testing for and estimating hospital effects; sex effects, and interaction effects.

14.49 In Problem 14.3, suppose that the firm was interested in the effect of age as well as the effect of the training program. New employees are divided into "below 30" and "30 or older" age groups, and four from each age group are randomly assigned to each program. After the program, the number of times the task is performed per minute is recorded for each trainee, with the results shown in Table 14.21. At the $\alpha = 0.01$ level, test for nonzero program effects, age effects, and interaction effects.

Table 14.21

		Program 1	Program 2	Program 3	Program 4
Age	Below 30	11, 13, 10, 14	10, 0, 7, 9	12, 10, 13, 10	6, 11, 8, 8
	30 or over	10, 9, 12, 10	9, 11, 6, 8	8, 11, 10, 11	9, 7, 7, 6

14.50 To compare the air pollution in different cities and at different times of day, air samples are taken at four randomly selected points within each city. Each sample is analyzed, and an air pollution index is used to indicate the overall degree of pollution in the sample (a larger number means more pollution). Results are shown in Table 14.22. Carry out an analysis of variance, with appropriate tests and estimates.

Table 14.22

	City 1	City 2	City 3
9 a.m.	14, 20, 28, 22	27, 31, 18, 25	6, 12 13, 10
12:30 p.m.	21, 25 19, 32	22, 26, 28, 24	12, 15, 11, 9
4 a.m.	20, 34 22, 24	19, 27 23, 29	28, 20 25, 23

14.51 In Problem 12.37, a linear regression with sales as the dependent variable and population as the independent variable yields the following output from an *Interactive Data*

Analysis (IDA) program:

SOURCE	SS	DF	MS	F
REGRESSION	2.55090E+08	1	2.55090E+08	111.47
RESIDUALS	6.40758E+07	28	2.28842E+06	
TOTAL	3.19166E+08	29	1.10057E+07	

Does this indicate a significant linear relationship at the $\alpha = 0.05$ level?

14.52 In Problem 12.37, the computer program used in Problem 14.51 was used for a linear regression predicting sales as a function of unemployment rate, with the following output:

SOURCE	SS	DF	MS	F
REGRESSION	3.11670E+07	1	3.11670E+07	3.03
RESIDUALS	2.87999E+08	28	1.02857E+07	
TOTAL	3.19166E+08	29	1.10057E+07	

Is there a significant linear relationship at the $\alpha = 0.05$ level?

14.53 In Problem 12.37, the following analysis of variance was generated by computer for the multiple regression predicting sales as a function of the other four variables (population, unemployment rate, advertising expense, competition):

SOURCE	SS	DF	MS	F
REGRESSION	2.84351E+08	4	7.10877E+07	51.05
RESIDUALS	3.48153E+07	25	1.39261E+06	
TOTAL	3.19166E+08	29	1.10057E+07	

Is there a significant multiple linear regression effect at the $\alpha = 0.05$ level?

14.54 A division of a large multinational food products corporation produces bakery goods, including frozen cakes which are sold under the "Pastry Shop" label in supermarkets throughout the United States, Canada, and in some foreign countries as well. A new cake is being developed by the division, and there is some disagreement about the sweetening agent to be used in the cake. Top management feels that sugar should be used because any other sweetener will hurt the reputation of the "Pastry Shop" label. However, cost estimates indicate that corn syrup would be considerably cheaper, and members of the marketing research group point out that a switch from sugar to corn syrup in another of the division's products met with a favorable response in test marketing and did not lead to a reduction in sales when implemented on a regular basis. The director of the research laboratory, however, thinks that a switch from sugar to corn syrup is not a good idea because an artificial sweetener would be even cheaper and, with recent developments, tastes more like sugar than does corn syrup. The proponents of sugar and corn syrup are united in their stand against artificial sweeteners, pointing out that the government has banned certain artificial sweeteners from time to time and that the public is becoming increasingly wary of foods with artificial ingredients. The laboratory director disagrees, blaming poor taste for lack of acceptance of previous artificial sweeteners. The new artificial sweetener has been well received in taste tests and has been approved for use by the government.

In order to gain more information about consumer response to the new cake, a taste-testing experiment is conducted. A random sample of consumers is taken, and each person in the sample is asked to taste various items and to rate each one on a scale from 1 to 5, with 1 representing "tastes awful" and 5 representing "tastes great". One of the items is the new cake, and other items are included to gather information about them as well as to prevent the subjects from focusing just on the new cake. The results for the new cake in terms of the frequencies of the different ratings are shown in Table 14.23. From these figures, the average rating is 2.83 with sugar, 2.55 with corn syrup, and 2.39 with artificial sweetener. Are the differences among these sample means large enough to be attributable to something other than chance (use $\alpha = 0.05$)?

Table 14.23

Rating	Sugar	Corn Syrup	Artificial Sweetener
1	14	16	19
2	30	35	42
3	26	32	22
4	19	12	15
5	11	5	2

In the experiment, sugar comes out on top. But the average rating for the new cake with sugar is still disappointing. Some of the company's current products were also included in the experiment, and they all received average ratings in the neighborhood of 3.5. Top management's first response to the taste test is to scrap the plans for the new cake altogether. When someone points out that a similar cake marketed by a competitor is very successful, the president decides that further investigation is warranted before dropping the project. The investigation reveals that the argument about the sweetener distracted attention from another change. To cut costs, vegetable oil had been substituted for butter, which is used in all of the company's other cakes. The president becomes very upset at this news and demands a replication of the experiment with butter used in place of vegetable oil. The results are as shown in Table 14.24.

Table 14.24

Rating	Sugar	Corn Syrup	Artificial Sweetener
1	3	4	10
2	12	10	23
3	31	28	36
4	32	37	23
5	22	21	8

Analyze the experimental data (treat the two sets of data as a single experiment). Indicate what you think should be done about the new cake, and justify your recommendations.

14.55 The management of "Nightlife" magazine is studying the sensitivity of sales to subscription price. In one part of this study twenty-seven subscribers were asked to indicate the maximum amount they would pay for a one-year subscription. These 27 people were classified according to their level of education and according to their degree of extroversion. The results of this study are shown below.

		Level of Extroversion		
		Ambivert	Extrovert	Introvert
Highest Level of Education	Secondary School	$12, 13, 10	14, 17, 13	8, 5, 7
	Some College	10, 14, 15	15, 16, 19	6, 7, 9
	College Grad	15, 14, 11	18, 14, 17	9, 7, 6

An SPSS program for a two-way analysis of variance was used to analyze these data. What conclusions can "Nightlife" draw from the SPSS output given below?

```
          EDUC       HIGHEST EDUCATION LEVEL  REACHED
          EXT        DEGREE OF EXTROVERSION
* * * * * * * * * * * * * * * * * * * * * * * * * * * * * * * * * * * * * * * * * * * *
                                      SUM OF                      MEAN              SIGNIF
SOURCE OF VARIATION                  SQUARES        DF          SQUARE        F       OF F
MAIN EFFECTS                         365.556         4          91.389   24.431       .001
   EDUC                               10.667         2           5.333    1.426       .266
   EXT                               354.889         2         177.444   47.436       .001
2-WAY INTERACTIONS
   EDUC      EXT                       1.778         4            .444     .119       .974

EXPLAINED                            367.333         8          45.917   12.275       .001
RESIDUAL                              67.333        18           3.741
TOTAL                                434.667        26          16.718
     27 CASES WERE PROCESSED.
```

GLOSSARY

analysis of variance: a method for investigating simultaneously the differences among the means of several populations.

grand mean: the overall mean of the populations of interest.

treatment effect: an effect associated with the particular population from which an observation is taken.

random-error term: a term reflecting variability within a population.

sum of squares: a sum of squared deviations about a sample mean.

SS Within: the sum of squares within the samples.

SS Between: the sum of squares between the sample means.

SS Total: the total sum of squares in an experiment.

mean square: a sum of squares divided by the associated degrees of freedom.

MS Within: the mean square within the samples.

MS Between: the mean square between the sample means.

F **= MS Between/MS Within:** a statistic used to test the hypothesis of equal population means.

analysis-of-variance table: a table showing the sources of variation, with sums of squares, degrees of freedom, and mean squares for an experiment.

C.F.: a correction factor used in computational formulas for the analysis of variance.

multiple comparisons: the simultaneous comparison of various combinations of treatment effects.

***S*-method:** a method for making multiple comparisons.

contrast: a linear combination of treatment effects.

two-factor model: a model of a situation in which populations are distinguished on the basis of two factors.

main effect: an effect associated with only a single factor.

interaction effect: an effect associated with a combination of factors.

experimental design: the planning of an experiment.

controlled experiment: an experiment where certain factors are under the control of the experimenter.

randomization: the use of random selection to allocate experimental units to treatment groups.

balanced experiment: an experiment with equal sample sizes for all possible treatment combinations.

randomized blocks: a model for reducing random variation by assigning treatments within "blocks."

factorial design: an experimental design involving several factors.

Latin square: an experimental design that leads to a reduction in sample size by systematically omitting certain combinations of treatments.

general linear model: a model relating an observation to one or more variables or factors in a linear fashion.

15

NONPARAMETRIC STATISTICS

"It ain't so much the things we don't know that get us in trouble. It's the things we know that ain't so."

ARTEMUS WARD

15.1 INTRODUCTION

Most of the statistical tests considered thus far have specified certain properties of the parent population which must hold before these tests can be used. A t-test, for example, requires that the observations come from a normal population; and if this test is used in testing for differences between means, the two populations must have equal variances. The same type of assumption is necessary in the analysis-of-variance tests presented in Chapter 14, and we assumed a bivariate normal distribution in making probability statements about correlation coefficients. Although most of these tests are quite "robust," in the sense that the tests are still useful when the assumptions about the parent population are not exactly fulfilled, there are still many circumstances when the researcher cannot or does not want make such assumptions. The statistical methods appropriate in these circumstances are called *nonparametric tests* because they do not depend on any assumptions about the parameters of the parent population.

In addition to not requiring assumptions about the parameters of the parent population, most nonparametric tests do not require a level of measurement as strong as that necessary for parametric tests. By "measurement" we mean the process of assigning numbers to objects or observations, the level of measurement being a function of the rules under which the numbers are assigned. The problem of measurement is so important to a discussion of nonparametric statistics that we begin this chapter by first studying the most common levels of measurement.

15.2 MEASUREMENT

The measurement of quantifiable information usually takes place on one of four levels, depending on the strength of the underlying scaling procedure used. The four major levels of measurement are represented by nominal, ordinal, interval, and ratio scales.

The weakest type of measurement is given by a *nominal scale*, which merely sorts objects into categories according to some distinguishing characteristic and gives each category a "name" (hence, *nominal*). Since classification on a nominal scale does not depend on the label or symbol assigned to each category, these symbols may be interchanged without affecting the information given by the scale. Classifying automobiles by makes constitutes a nominal scale, as does distinguishing Republican from Democratic voters, or apples from oranges. No quantitative characteristics may differentiate these objects; if it is necessary to permit quantitative distinctions, a measurement *stronger than nominal* must be assumed. In most nominal measurement, one is concerned with the number (or frequency) of observations falling in each of the categories.

An *ordinal scale* offers the next highest level of measurement, one expressing the *relationship of order*. Objects in an ordinal scale are characterized by relative rank, so that a typical relationship may be "higher," "greater," or "preferred to"; only the relations "greater than," "less than," or "equal to" have meaning in ordinal measurement. When a football team is "ranked" nationally, for example, such a measurement implies an ordinal scale if it is impossible (or meaningless) to say how *much* better or worse this team is compared to others. Most subjective attributes of objects or persons (e.g., flavor, beauty, honesty) are difficult (if not impossible) to consider on a scale higher than the ordinal. Distinguishing service personel by rank (e.g., captain, major) is another example of ordinal measurement.

A third type of scale is given by *interval measurement*, sometimes called *cardinal measurement*. Measurement on an interval scale assumes an exact knowledge of the quantitative difference between objects being scaled. That is, it must be possible to assign a number to each object in such a manner that the relative difference between them is reflected by the difference in the numbers. Any size of unit may be used in this type of measurement, as long as a one-unit change on the scale always reflects the same change in the object being scaled. The choice of a zero point (origin) for the data also can be made arbitrarily. Temperature measured on either a Celsius or a Fahrenheit scale represents interval measurement, as the choice of origin and unit for these scales is arbitrary. Temperature measured on an absolute scale, however, does not represent interval measurement, as this scale has a natural origin (the zero point is that at which all molecular motion ceases). As another example, most IQ measures represent interval scales, since there is no natural origin (zero intelligence?), and the choice of a unit can be made arbitrarily. The name "interval measurement" is used because this type of scale is concerned primarily with the distance *between* objects, that is, the "interval" between them.

The strongest type of measurement is represented by *ratio scales*, or scales which have all the properties of an interval scale *plus* a natural origin—only the unit of measurement is arbitrary. Fixing the origin (the zero point) permits comparisons not only of the intervals between objects, but of the absolute value of the numbers assigned to these objects. Hence, in this type of scale, "ratios" have meaning, and statements can be made to the effect that "x is twice the value of y." Weight, length, and mass are all measured by using a ratio scale. Distance, whether in terms of kilometers, miles, or feet, is an example of a ratio measurement, since all of these scales have a common origin (the zero point, representing "no distance").

15.3 PARAMETRIC *VS.* NONPARAMETRIC TESTS

In addition to assuming some knowledge about the characteristics of the parent population (e.g., normality), parametric statistical methods require measurement equivalent to at least an interval scale. That is, to find the means and variances necessary for these tests, one must be able to assume that it is meaningful to compare intervals. It makes no sense to add, subtract, divide, or multiply ordinal-scale values because the numbers on an ordinal scale have no meaning except to indicate rank order. There is no way to find, for example, the average between a captain and a major in terms of military rank.

To avoid the parametric assumptions normally required for tests based on interval or ratio scales, most nonparametric tests assume only nominal or ordinal data. That is, such tests ignore any properties of a given scale except ordinality. This means that if the data are, in fact, measurable on an interval scale, nonparametric tests waste (by ignoring) this knowledge about intervals. By wasting data such tests gain the advantage of not having to make parametric assumptions, but sacrifice power in terms of using all available information to reject a false null hypothesis. Nonparametric tests, for example, typically involve medians rather than means, because determining a mean requires interval data while determining a median requires only ordinal data.

We have already studied one type of nonparametric test in Chapters 4 and 8, when applications of the binomial distribution were presented. As we indicated then, the binomial distribution can be applied in an experiment in which all outcomes fall into one of two categories, and where there is a constant probability that an observation will fall into these two categories on each of a series of independent trials. The number of successes and failures resulting from Bernoulli trials represent measurement on only a nominal scale because these numbers reflect not the quantitative characteristics of any variable, but only the frequency of observations falling into two arbitrarily defined categories. In addition, the assumption of independence in these trials does not represent a parametric assumption. The binomial distribution is also useful in several types of nonparametric tests, as we will discuss in the following section.

In the sections to follow, we have divided the discussion of nonparametric techniques into the following headings:

1. Tests equivalent to the t-test for independent samples;
2. Tests equivalent to the t-test for matched pairs;
3. Goodness-of-fit tests;
4. The chi-square test for independence; and
5. Correlation measures for ranked data.

We must point out that this grouping is rather arbitrary, as the tests involved could be (and often are) classified under many different headings. Also, we have presented only a few of the numerous nonparametric tests available. The reader interested in a more detailed description of the techniques in this area is referred to references [14] and [15].

15.4 TESTS EQUIVALENT TO THE t-TEST FOR INDEPENDENT SAMPLES

Recall that in Chapter 8 we used the t-distribution to test for the difference between the means of two independently drawn samples (Section 8.7). There are a number of nonparametric equivalents to this test that can be used for data weaker than interval scaling, or when the researcher wishes to avoid the assumptions of the t-distribution. We will present two tests which can be used in such circumstances, the *Mann–Whitney* U*-test*, and the *Wald–Wolfowitz runs test*.

Mann–Whitney *U*-test

The Mann–Whitney U-test is one of the most powerful nonparametric tests, and is a useful alternative to the two-sample t-test described in Section 8.7. This test is designed to determine whether or not two samples were drawn from the same population. Thus, the null hypothesis is that the two populations are identical, and the alternative hypothesis is that they are not the same.

The first step in the Mann–Whitney U-test is to consider all the scores representing the two samples as a single set of observations, and to rank this entire group from the lowest to the highest score. If the null hypothesis that the two samples were drawn from the same population is true, then the observations from the two samples will be fairly well scattered throughout this ranking of both groups. If the two samples do not come from identical populations, then the observations of one sample will tend to be bunched together, either at the low end of the rankings or at the high end of the rankings. Such patterns can be detected by calculating a value of U, which is the statistic for the Mann–Whitney test. The statistic U for the Mann–Whitney test is calculated by counting the number of times the scores from one sample precede each score in the other sample. If the count is quite large or quite small relative to the value expected under the null hypothesis, then the two samples may not be randomly interspersed, but one set of observations may have come from a different population than the other.

Suppose we are interested in determining whether the final examination scores resulting from one method of teaching statistics are different from those resulting from another method. A sample of two groups (Groups A and B) is taken, with the results shown in Table 15.1.

Table 15.1

Group A	Group B
55	65
59	77
61	80
64	80
64	84
70	86
73	88
75	91
76	91
82	93
83	
95	

Table 15.2

Score	55	59	61	64	64	65	70	73	75	76	77
Section	A	A	A	A	A	B	A	A	A	A	B
Rank	1	2	3	4	5	6	7	8	9	10	11
Score	80	80	82	83	84	86	88	91	91	93	95
Section	B	B	A	A	B	B	B	B	B	B	A
Rank	12	13	14	15	16	17	18	19	20	21	22

To calculate the value of the statistic $\boldsymbol{U}$ for these data, we need to determine the total number of times an observation in the A Group precedes each score in the B Group. Suppose we call this number T_A. The format for calculating T_A is given in Table 15.2, where the scores from the A and B Groups are arranged in ascending order in a single set of observations.

To calculate T_A, focus on each B score and count the number of A scores lower than this score. For example, the B score of 65 has 5 scores preceding it. Similarly, there are nine A scores preceding each of the B scores 77, 80, 80. Finally, 11 scores precede each of the last six B scores. The value of T_A is thus

$$T_A = 5 + 3(9) + 6(11) = 98.$$

To complete the Mann–Whitney Test we also need to calculate T_B, which is the total number of times a score in the B Group precedes each score in the A Group. A good exercise for the reader would be to verify that

$$T_B = 4(1) + 2(4) + 10 = 22.$$

The value of the Mann–Whitney statistic is defined to be the *minimum* of the two values, T_A, T_B. Defining it this way means that the more similar the two samples are, the higher will be the value of $\boldsymbol{U}$. Hence, we will reject H_0 when $\boldsymbol{U}$ is small. Since the value of $\boldsymbol{U}$ depends only on the ranks of the scores in the two groups, it is possible to determine the probability of various values of $\boldsymbol{U}$. For small samples these probabilities have been tabled and are available in several sources.* For instance, in our present example, the value of the Mann–Whitney statistic is $\boldsymbol{U} = \min\{98,22\} = 22$. When comparing a sample of size $n_A = 12$ (Group A) with a sample of $n_B = 10$ (Group B), the critical region for $\alpha = 0.02$ is $\boldsymbol{U} \leq 24$ (that is, $P(\boldsymbol{U} < 24) = 0.02$). Because our observed value of $\boldsymbol{U}$ was 22,

* See *Annals of Mathematical Statistics*, **18**, pp. 52–54, or Siegel, *Nonparametric Statistics*, pp. 271–277.

this means we can reject the null hypothesis that the two samples do not differ at the $\alpha = 0.02$ level of significance.

For large sample sizes the above procedure for calculating T_A and T_B can become quite tedious. Fortunately however, these two values can be determined quite easily by using the combined ranking of all $n_A + n_B$ observations. If r_A = the sum of the ranks of the scores from the A Group, and r_B the sum of the ranks from the B Group, then

$$T_A = n_A n_B + \frac{n_A(n_A + 1)}{2} - r_A,$$

and

$$T_B = n_A n_B - T_A.$$

We should point out that these formulas can be used for *any* values of n_A and n_B (not just large samples), and that it makes no difference which sample is labeled A and which one is labeled B. In our present example, the sums of the ranks of the A Group are seen in Table 15.2 to be

$$r_A = 1 + 2 + 3 + 4 + 5 + 7 + 8 + 9 + 10 + 14 + 15 + 22 = 100.$$

This means that

$$T_A = (12)(10) + \frac{(12)(13)}{2} - 100 = 98,$$

and

$$T_B = (12)(10) - 98 = 22.$$

Tables of the critical values of $\boldsymbol{U}$ are available for sample sizes up to about 20. When n_A or n_B is larger than 20, and the two sample sizes are not too different in size, then the sampling distribution of $\boldsymbol{U}$ can be approximated with the following normal distribution:

Standardization of $\boldsymbol{U}$ *for Mann–Whitney test:* $$z = \frac{\boldsymbol{U} - E[\boldsymbol{U}]}{\sigma_U}.$$

where

$$E[\boldsymbol{U}] = \frac{n_A n_B}{2}$$

and

$$\sigma_U^2 = \frac{n_A n_B(n_A + n_B + 1)}{12}.$$

When ties occur in the Mann–Whitney test, they are usually treated by assigning the *average* of the ranks of those observations which are tied. Suppose, for example, that the observations corresponding to ranks two and three are identical. In this case, both observations would be given a rank of 2.5, which is the average of the ranks 2.0 and 3.0.

The Wald–Wolfowitz Runs Test

Another test which can be used in place of the ***t***-test for independent samples is the *Wald–Wolfowitz runs test.* Although this test is not as powerful as the Mann–Whitney ***U***-test, it is useful in some situations where that test may not be appropriate. The null hypothesis in this test is the same as in the Mann–Whitney test—namely, that the two samples were drawn from the same population. Again, the alternative hypothesis is that the two populations differ in some respect. To test for differences between two samples, the observations from both samples are placed in a single group, and then ranked (just as they were in the Mann–Whitney test). The number of runs in this ranking can now be counted, where a run is a sequence of the ranked observations all of which come from the same sample. An indication of whether or not the two samples come from the same population is given by the total number of runs. A large number of runs will occur when the ranks corresponding to the two samples are fairly randomly intermixed, and hence in this case it is reasonable to accept H_0, that they came from the same population. A small number of runs will occur whenever there is some systematic difference between the two samples. For example, if the ranks corresponding to one sample are consistently lower than those in the other sample, this suggests that the central location of the two samples differs. Similarly, if one sample has a smaller *spread* than the other sample, then the ranks corresponding to this first sample will bunch in the center of the array, resulting in a low number of runs. Tables are available to determine whether a given total number of runs is large enough to reject H_0 (see Table XI in Appendix B).

To illustrate the Wald–Wolfowitz runs test, consider again the problem of determining whether the two sets of examination scores shown in Table 15.1 differ significantly. As before, we arrange all 22 of the observations in order from the lowest to the highest (see Table 15.2). The number of runs in this array is shown below to be seven.

Scores 55 59 61 64 64 65 70 73 75 76 77 80 80 82 83 84 86 88 91 91 93 95

Group A A A A A | B | A A A A | B B B | A A | B B B B B B | A

Seven runs 1 2 3 4 5 6 7

From Table XI we see that the critical region for $\alpha = 0.05$, when $n_A = 12$ and $n_B = 10$, is $r \leq 7$, where $r =$ the number of runs. Since our sample result of seven runs falls in this critical region, we can reject the null hypothesis that these two samples of final grades were drawn from the same population.

Table XI presents critical values only for small values of n_A and n_B. When both sample sizes are fairly large (e.g., greater than 19), then the following normal approximation can be used:

Standardization of r *for runs test:*
$$z = \frac{r - E[r]}{\sigma_r},$$

where

$$E[r] = \frac{2n_A n_B}{n_A + n_B} + 1$$

and

$$\sigma_r^2 = \frac{2n_A n_B(2n_A n_B - n_A - n_B)}{(n_A + n_B)^2(n_A + n_B - 1)}.$$

When ties occur in this test, the usual procedure is to assign ranks so as to make the number of runs as large as possible (i.e., ranks are assigned in a manner least favorable to rejecting H_0).

15.5 TESTS EQUIVALENT TO THE *t*-TEST FOR MATCHED PAIRS

The tests presented in this section represent the nonparametric alternatives to the matched-pairs *t*-test discussed in Section 8.7. Again, we must point out that these tests are not the only ones available for this purpose, nor is our way of classifying them the only way they can be grouped. Two tests are presented here, the *sign test*, and the *Wilcoxon test.*

Sign Test

A *sign test* is designed to determine whether significant differences exist between two samples which are related in such a manner that each observation from one sample can be matched with a specific observation from the other sample. For example, one may wish to study the behavior of identical twins under two "treatments," the "before and after" effect of a certain drug, or the attitudes of husbands in contrast to the attitudes of their wives. In the sign test, ordinal data is assumed, so that it is meaningful to rank the observations from one sample (e.g., the husbands) only as higher than, equal to, or lower than their corresponding value in the other sample (the wife group). An easy way to record which sample has the higher value for each matched pair (husband versus wife) is to give each of these pairs a "sign," either a plus (+) sign representing the fact that the first sample has the higher value, or a minus (−) sign representing the fact that the second sample has the higher value. The null hypothesis is usually that the two samples were drawn from populations with the same median, so that the probability of a plus sign (p) or a minus sign (q) for each matched pair is $p = q = \frac{1}{2}$. This hypothesis can be tested, by using the binomial distribution, as illustrated in the following example.

Suppose that an IQ test is given to eleven randomly selected pairs consisting of one brother and one sister from the same family, to test the null hypothesis that this sample was drawn from a population in which the median IQ of a brother and his sister do not differ. The alternative hypothesis is that either the brother or the sister has a higher IQ ($H_a: p \neq \frac{1}{2}$). Table 15.3 gives the data appropriate for use in the sign test.

Note that the only relevant fact about these matched scores is whether the sister's score is higher or lower than the brother's score—the IQ scores themselves cannot be used unless interval measurement is assumed. The binomial distribution can now be used to test the null hypothesis that the probability that the brother's score will exceed the sister's score (or vice versa) equals $\frac{1}{2}$. The probability of receiving 9 or more +'s in a sample of 11, when $p = q = \frac{1}{2}$ (the null hypothesis) can be found (in a binomial table which contains $n = 11$) to be 0.0327. That is,

$$P(x \geq 9) = \sum_{x=9}^{11} \binom{11}{x} \left(\frac{1}{2}\right)^x \left(\frac{1}{2}\right)^{11-x} = 0.0327.$$

Since we did not predict whether the brothers or the sisters were expected to have the higher IQ's, the appropriate binomial test is a two-sided test, and the

Table 15.3

Pair	Sister's Score	Brother's Score	Sign
1	129	115	+
2	111	108	+
3	117	123	−
4	120	104	+
5	116	110	+
6	101	98	+
7	107	106	+
8	127	119	+
9	105	95	+
10	123	130	−
11	113	101	+

above probability must be doubled, to account for the fact that there are two critical regions. The null hypothesis can thus be rejected at α-levels of 0.0654 or higher. If the alternative hypothesis for this problem had been a one-sided test (e.g., if it had been predicted that the sister would score higher), then H_0 could rejected at the 0.0327 level, or higher.

As in the binomial test, the null hypothesis in the sign test need not specify that $p = q = \frac{1}{2}$. Suppose a random sample of 10 people have been interviewed and asked to rate each of two products (A and B) on a scale from 0 to 100. We might have hypothesized that the probability of A being preferred to B is not $\frac{1}{2}$, but some other value, say $\frac{3}{4}$ (that is, $P(\text{A} > \text{B}) = \frac{3}{4}$), the alternative to this hypothesis being that $P(\text{A} > \text{B}) < \frac{3}{4}$. The sample results of the ten interviews are given in Table 15.4.

Under the null hypothesis, the probability of five (or fewer) +'s is

$$P(\mathbf{x} \leq 5) = \sum_{x=0}^{5} \binom{10}{x} \left(\frac{3}{4}\right)^x \left(\frac{1}{4}\right)^{5-x} = 0.078.$$

It is not possible, on the basis of this sample, to reject the null hypothesis $H_0: p = \frac{3}{4}$ at levels of significance less than 0.078.

One of the problems which might occur in a sign test is that there may be one (or more) matched pairs in which the two scores are identical (a tie). In the case of a tie, the usual procedure is to drop that matched pair from the analysis, and simply work with the smaller sample size. Note that the sign test does not require knowledge about the *magnitude* of the difference between the matched pairs, as does the **t**-test. If these differences *are* known, then a more powerful nonparametric test is the Wilcoxon test.

Table 15.4

Consumer	Product A Score	Product B Score	Sign
1	75	58	+
2	85	92	−
3	61	69	−
4	55	50	+
5	82	71	+
6	88	84	+
7	45	78	−
8	90	79	+
9	63	69	−
10	71	80	−

Table 15.5

Couple	Sister's Score	Brother's Score	Difference	Rank of Difference	Signed Rank
1	129	115	14	10	+10
2	111	108	3	2.5	+2.5
3	117	123	−5	4	−4
4	120	104	16	11	+11
5	116	110	6	5	+5
6	101	98	3	2.5	+2.5
7	107	106	1	1	+1
8	127	119	8	7	+7
9	105	95	10	8	+8
10	123	130	−7	6	−6
11	113	101	12	9	+9

The Wilcoxon Test

The Wilcoxon test has the same null hypothesis as the sign test—namely, that the median difference between two populations equals zero. The test in this case, however, takes into account the magnitude of the difference between each matched pair. These magnitudes are first ranked according to their *absolute* value. Then each of these ranks is given either a (+) sign or a (−) sign, depending on whether sample A was larger than B (the plus sign) or sample B larger than A (the minus sign). Now, if the null hypothesis is true, we would expect the sum of those ranks with + signs to be about equal to the sum of those ranks with minus signs. If the two sums differ by very much, we would infer that the two populations are not identical.

Suppose we let T_+ equal the sum of the positive ranks and T_- equal the sum of the *absolute value* of the negative ranks. The Wilcoxon test is based on the statistic T, which is defined to be the *minimum* of T_+ and T_-; that is, $T = \min\{T_+, T_-\}$. Critical values of T for small samples are given in Table XII in Appendix B.

To illustrate the Wilcoxon test, consider again the data in Table 15.3 representing IQ's for eleven brothers and sisters. Table 15.5 shows the difference scores and ranks necessary for calculating T.

Note, in the final column of Table 15.5, that we assigned the average of the two ranks to the two difference scores which are tied (a difference of 3). If some difference score had been 0, this matched pair would be dropped from the analysis. For the present data, the value of T_+ is

$$T_+ = 10 + 2.5 + 11 + 5 + 2.5 + 1 + 7 + 8 + 9 = 56,$$

and

$$T_- = 4 + 6 = 10.$$

Hence, $T = \min\{56, 10\} = 10$. From Table XII the critical region when $\alpha = 0.05$ for $n = 11$ matched pairs is $T \leq 11$ (assuming a two-sided alternative); that is, we reject H_0 whenever T is eleven or smaller. Since our calculated value is $T = 10$, we reject the null hypothesis of no difference between the two populations.

When the number of matched pairs (n) is not small (for example, $n > 8$), then the distribution of T can be shown to be approximately normally distributed:

*Standardization of **T** for Wilcoxon test:* $z = \dfrac{T - E[T]}{\sigma_T}$,

where

$$E[T] = \frac{n(n+1)}{4},$$

and

$$\sigma_T^2 = \frac{n(n+1)(2n+1)}{24}.$$

Define: Nominal, ordinal, interval, ratio measurement, nonparametric tests.

Problems

15.1 Distinguish between parametric and nonparametric statistical tests. Under what circumstances is each type of test most appropriate? Give several specific examples of problems where a nonparametric test would be more appropriate than a parametric test.

15.2 Briefly describe and distinguish between the four levels of measurement—nominal, ordinal, interval, and ratio—giving several examples of each type of scale.

15.3 Identify each of the following numbers as representing measurement on either nominal, ordinal, interval, or ratio scales:

a) the numbers designating years (for example, 1982, 1983, etc.)

b) the numbers on football players' jerseys,

c) the numbers representing golf scores in a tournament, and

d) social security numbers.

15.4 Two groups of students were given a written test on driving skills. There were ten students in the first group (A) and eight in the second (B). Their scores are shown in Table 15.6.

Table 15.6

A:	25, 30, 42, 44, 58, 59, 75, 79, 87, 90
B:	45, 49, 62, 63, 68, 69, 69, 71

a) Use these data to calculate the statistic U for the Mann–Whitney U-test. What null hypothesis is being tested in this case? Can H_0 be rejected if the critical value for $n_1 = 10$, $n_2 = 8$ is 17 at the 0.05 level of significance (a two-tailed test)?

b) Use the runs test to determine whether the null hypothesis that the samples came from the same population can be rejected.

15.5 A study was conducted using the Mann–Whitney U-test on the number of colds caught by a group of smokers (Group A) vs. a group of nonsmokers (Group B). In this case $n_A = 40$, $n_B = 40$, and the calculated value of U was $U = 300$, with the smoking group having the larger number of colds. Use the normal approximation to determine whether or not these samples can be considered to be drawn from the same population.

15.6 When the Wald–Wolfowitz runs test was applied to the data of Problem 15.5, the result was a total of 30 runs. Use a normal approximation to test the hypothesis that the samples can be considered to be drawn from the same population.

15.7 The ten students in group A in Problem 15.4 took a road test for driving skills as well as a written test. Their scores on both tests are shown in Table 15.7. Use a sign test to test the null hypothesis at $\alpha = 0.05$ that a student's score on the road test is not different from his score on the written test. At what level of significance can H_0 be rejected for a two-sided test? At what level can H_0 be rejected if it was hypothesized that the road test scores would be higher than the written test scores?

Table 15.7

Student:	1, 2, 3, 4, 5, 6, 7, 8, 9, 10
Written:	25, 30, 42, 44, 58, 59, 75, 79, 87, 90
Road:	38, 36, 50, 45, 30, 78, 76, 85, 65, 76

15.8 a) Use the Wilcoxon test on the data in Problem 15.7. Use Table XII in Appendix B to reject or accept H_0, assuming that H_a is a two-sided alternative.

b) Use the normal approximation to the Wilcoxon test to test the same hypothesis as in part (b). Do your answers agree?

15.9 Two groups of overweight men were matched in pairs according to age, weight, occupation, and a number of other criteria; one-half of these men were put on one weight-reducing program, the other half on another program. The weight losses are as shown in Table 15.8.

Table 15.8

Pair	First Plan	Second Plan
1	25 lb	15 lb
2	39 lb	22 lb
3	21 lb	30 lb
4	48 lb	12 lb
5	8 lb	0 lb

Use a sign test to determine whether the weight losses under the two plans differ at $\alpha = 0.05$. At what level of significance can the null hypothesis be rejected?

15.10 Senator Fogbound has taken a random sample of 20 of his constituents in order to decide whether a majority of the people favor the legislation he is proposing.

a) At what level of significance can he reject the null hypothesis $H_0: p = \frac{1}{2}$ in favor of $H_a: p \neq \frac{1}{2}$ if 15 of the 20 respondents indicate that they favor his legislation? At what level of significance can the null hypothesis be rejected if $H_a: p > \frac{1}{2}$?

b) What critical region would you establish to test the null hypothesis $H_0: p = 0.65$ against the alternative $H_a: p > 0.65$ at the 0.01 level of significance? What would the critical region be for testing this same null hypothesis against $H_a: p \neq 0.65$ at the 0.05 level of significance?

15.11 In one U.S. state a law was recently passed requiring all motorcyclists to wear protective helmets. Before this law motorcyclists were fatally injured in 5 percent of all motorcycle accidents reported to the state police. During the year after the law was passed there was only one fatal injury in the 100 accidents reported to the police. Is this sample sufficient to reject the null hypothesis $H_0: p = 0.05$ in favor of $H_a: p < 0.05$?

15.12 A company advertising on national TV claims that eight of ten doctors prefer its product over the leading rival product. In a random sample of 100 doctors interviewed by the Consumer Protection Agency, 72 rated this product higher than the other product. On this basis would you reject $H_0: p = 0.80$ in favor of $H_a: p < 0.80$ using a sign test? Would you reject H_0 if $\alpha = 0.01$? Would you reject H_0 if $\alpha = 0.05$? Use both Table I and the normal approximation.

15.13 Four hundred randomly selected registered voters were asked, six months ago, to rate the "effectiveness" of the President of the United States. Last month, these same people were again asked to make the same evaluation. Sixty percent of the second ratings were lower than the first ratings and forty percent were higher. Use a sign test and $\alpha = 0.05$ to determine whether the President's rating has significantly declined. (*Hint.* Use $H_0: p = 0.50$, a two-sided alternative, and a normal approximation.)

15.14 Suppose that ten applicants for graduate programs are rated by two members of the graduate faculty, on a scale from 1 to 10 (with "ten" indicating excellence). Test whether the ratings (at $\alpha = 0.05$) are significantly different between the two professors. Use a sign test. Table follows on page 708.

	Rating of Graduate Applicants									
Professor A	8	4	2	6	6	9	4	5	7	5
Professor B	6	5	4	7	8	10	7	6	6	8

15.6 GOODNESS-OF-FIT TESTS

The tests presented thus far represent nonparametric procedures designed to see how closely two *sample* probability distributions correspond to one another in order to test the hypothesis that they came from the same *population* probability distribution. In many statistical problems the researcher is interested in a closely related problem, that of determining how closely an observed (sample) probability distribution fits some theoretical probability distribution. In this section, we will present two tests designed for this purpose, called goodness-of-fit tests.

We have already studied one type of goodness-of-fit test in our applications of the binomial distribution. The binomial distribution is a goodness-of-fit test, in the sense that it compares the frequency of sample observations in two categories with the frequency of observations expected under the null hypothesis. The binomial distribution is used to determine how "good" the fit is: for close fits we accept H_0; otherwise H_0 is rejected. In addition to being able to use the binomial test on problems of this nature, it is possible to use a statistical test based on the chi-square distribution. This test, called the *chi-square test*, has the advantage (over the binomial) of being generalizable to problems involving more than just the two nominal categories used in a binomial test. When there are just two categories, however, the binomial test is preferred because it is more powerful.

The Chi-Square Test

The chi-square variable is used in this situation to test how closely a set of observed frequencies corresponds to a given set of expected frequencies. The expected frequencies can be thought of as the average number of values expected to fall in each category, based on some theoretical probability distribution. For example, one probability distribution which is often useful is to assume that the expected frequencies in the various categories will all be equal. The observed frequencies can be thought of as a *sample* of values from some probability distribution. The chi-square variable can be used to test whether the observed and expected frequencies are close enough so we can conclude they came from the same probability distribution. For this reason the test is called a "goodness-of-fit" test.

Suppose we assume there are c categories ($c > 1$) and the expected frequency in each of these categories is denoted as $E_1, E_2, \ldots, E_c$, or equivalently, E_i ($i = 1, 2, \ldots, c$). Similarly, the c observed frequencies will be denoted as $O_1, O_2, \ldots, O_c$, or O_i ($i = 1, 2, \ldots, c$). To test the goodness of fit of the observed frequencies (O_i) to the expected frequencies (E_i), we use the following statistic, which can be shown to be approximately a chi-square variable with $(c - 1)$ degrees of freedom.*

Test statistic for goodness of fit: $$\chi^2_{(c-1\text{ d.f.})} = \sum_{i=1}^{c} \frac{(O_i - E_i)^2}{E_i}. \qquad (15.1)$$

Formula (15.1) measures the goodness of fit between the values of O_i and E_i as follows: When the fit is good (that is, when O_i and E_i are generally close), then the numerator of (15.1) will be relatively small; hence the value of χ^2 will be low. Conversely, if O_i and E_i are not close, then the numerator of (15.1) will be relatively large, and the value of χ^2 will also be large. Thus, the critical region for the test statistic given by (15.1) will always be in the *upper* tail of the χ^2 distribution because we want to reject the null hypothesis whenever the difference between E_i and O_i is relatively large. For example, suppose in a particular problem involving 16 categories, the fit between the 16 values of O_i and E_i from Formula (15.1) yields $\chi^2 = 30.0$. From Table V in the row corresponding to $c - 1 = 15$. d.f., we find that $P(\chi^2 \geq 25) = 0.05$. Thus, at the $\alpha = 0.05$ level of significance, we can reject the null hypothesis that the observed values came from the same distribution as the expected values.

To illustrate the use of the chi-square test, suppose that an automobile dealer, in trying to arrange vacations for his salesmen, decides to test the (null) hypothesis that his sales of new cars were equally distributed over the first six months of last year. His expected frequency distribution thus specifies that $E_1 = E_2 = \cdots = E_6$. The alternative hypothesis is that sales were *not* equally distributed over the six months. If we assume that he sold 150 new cars in this period, the expected frequency under the null hypothesis would be 25 cars sold in each month. The expected and observed sales are given in Table 15.9.

The null and alternative hypotheses are:

$H_0: E_1 = E_2 = \cdots = E_5 = 25;$

H_a: The frequencies are not all equal.

The chi-square statistic for this example has $c - 1 = 5$ degrees of freedom. If we let $\alpha = 0.025$, then the appropriate critical region, shown in Fig. 15.1,

* Technically, the chi-square distribution is only an approximation to the distribution of the test statistic (15.1).

Table 15.9 Monthly New Car Sales

	Months						
	Jan.	Feb.	Mar.	Apr.	May	June	Total
Expected sales (E_i)	25	25	25	25	25	25	150
Observed sales (O_i)	27	18	15	24	36	30	150

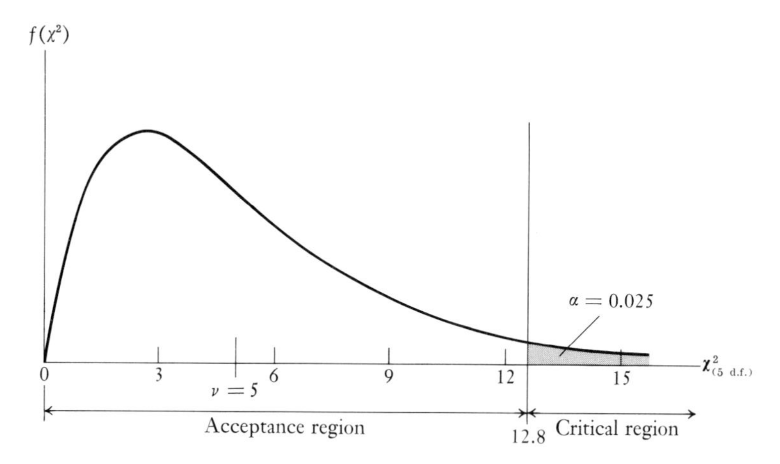

Fig. 15.1 Chi-square critical region for $\nu = 5$, $\alpha = 0.025$.

is derived from the row $\nu = 5$ in Table V. From this figure we see that the null hypothesis will be rejected in favor of the alternate hypothesis of unequal frequencies in monthly sales if the calculated value of χ^2 exceeds 12.8.

The value of χ^2 can now be calculated as follows:

$$\chi^2 = \frac{(27-25)^2}{25} + \frac{(18-25)^2}{25} + \frac{(15-25)^2}{25} + \frac{(24-25)^2}{25} + \frac{(36-25)^2}{25} + \frac{(30-25)^2}{25}$$
$$= 12.0.$$

Thus, at the significance level of $\alpha = 0.025$, we *fail to reject* the null hypothesis that all monthly sales are equal (i.e., that *deviations from sales equal to* 25 are due to random occurrences). However, from Table V, note that for $c - 1 = 5$ degrees of freedom and $\alpha = 0.05$, the critical value for the test would be 11.1. Therefore, if $\alpha = 0.05$ had been selected as the level of significance for rejecting H_0, then the null hypothesis *would* have been rejected. In this case, the choice of the α-level is crucial to the decision. The costs of having too many or too few salesmen on hand in a given month should be given more consideration before setting α at an arbitrary level. Also, a larger number of past months may be sampled, if the sales pattern is presumed not to have changed. For example, the sum of sales for the past three years in the month of January, and February, and March, etc., might be used to test for equal monthly sales

In using the chi-square distribution in these circumstances, one must be careful not to use categories having small expected frequencies. The rule of thumb in chi-square tests is generally that the expected frequency should be at least *five;* recent research, however, has indicated that an expected value of *one or more* in each category is usually sufficient. The easiest way to increase the expected frequencies in a chi-square test is by collapsing two or more adjacent categories, as we illustrate in the following example.

Consider again the supermarket example from Chapter 4 (Section 4.5). The observed and expected frequencies shown in Table 15.10 are derived from Table 4.3, except that now we have collapsed the categories 9, 10, 11, 12, 13 into a single category labeled ≥ 9. This ensures that all expected frequencies are ≥ 1. For example, from the Poisson probabilities in Table 4.4, we know that the probability of 0 arrivals is 0.0224. Over 100 observations, the theoretical or expected frequency would be 100(0.0224) = 2.24. Similarly, from Table 4.4 the probability $P(\mathbf{x} \geq 9) = 0.0159$; hence, the expected frequency for the category ≥ 9 is 100(0.0154) = 1.590. All such theoretical frequencies are shown in Table 15.10.

Table 15.10

Arrivals	Observed Frequency (O_i)	Expected Frequency (E_i)	$(O_i - E_i)^2/E_i$
0	1	2.240	0.6864
1	8	8.500	0.0294
2	19	16.150	0.5029
3	23	20.460	0.3153
4	17	19.440	0.3063
5	15	14.770	0.0036
6	8	9.360	0.1976
7	3	5.080	0.8517
8	3	2.410	0.0695
≥ 9	3	1.590	1.2503
	100	100.000	4.2130

From the last column in Table 15.10, we see that:

$$\chi^2 = \frac{\sum(O_i - E_i)^2}{E_i} = 4.2130.$$

For this problem, there are $c - 1 = 9$ d.f.; the critical value from Table V for $\alpha = 0.05$ is thus 16.9. Since our observed value 4.2130 is less than 16.9, we conclude that the sample probability distribution is a close fit to the theoretical distribution; i.e., we cannot reject H_0 (that these distributions are the same).

The Kolmogorov–Smirnov Test

We showed above how the χ^2 test can be used to measure goodness-of-fit when the data is in nominal form (i.e., categories). When the data is in at least *ordinal* form, then the *Kolmogorov–Smirnov* test can be used. This test has the advantage over the χ^2 test in that it is generally more powerful, it is easier to compute, and it does not require a minimum expected frequency in each cell.

The Kolmogorov–Smirnov test involves a comparison between the theoretical and sample *cumulative* frequency distributions. To make this comparison, the data is put into classes (or categories) which have been arrayed from the lowest class to the highest class. Suppose we use the symbol F_i to denote the cumulative relative frequency for each category of the theoretical distribution, and S_i to denote the comparable value for the sample frequency. The Kolmogorov–Smirnov test is based on the maximum value of the absolute difference between F_i and S_i. If we denote the statistic for this test as $\boldsymbol{D}$, then:

Statistic for Kolmogorov–Smirnov test: $$\boldsymbol{D} = \operatorname*{Max}_i |F_i - S_i|.$$

The decision to reject the null hypothesis (that the sample and theoretical distributions are equal) is based on the value of $\boldsymbol{D}$: the larger $\boldsymbol{D}$ is, the more confidence we have that H_0 is false. Note that this is a one-tail test, since the value of $\boldsymbol{D}$ is always positive, and we reject H_0 for large values of $\boldsymbol{D}$. Table XIII in Appendix B gives the critical values of $\boldsymbol{D}$ for various probability values.

To illustrate the Kolmogorov–Smirnov test, let's once again consider the supermarket data from Chapter 4. In this case, we repeat the data from Table 4.4, which shows the relative frequencies for both the observed and theoretical (Poisson) distribution. Table 15.11 repeats these values and includes the cumulative values F_i and S_i, and each difference $|F_i - S_i|$.

From the last column in Table 15.11, we see that the maximum value of $|F_i - S_i|$ is $\boldsymbol{D} = 0.0365$. Since the sample size in the example is $n = 100$, the last column of Table XIII in Appendix B gives the appropriate critical values of $\boldsymbol{D}$: for $\alpha = 0.01$, $1.63/\sqrt{n} = 1.63/\sqrt{100} = 0.163$. Because $D = 0.0365$ is less than this critical value, we do not reject H_0. As we found before, the agreement

Table 15.11

Arrivals	Observed Relative Frequency	Cumulative Observed Frequency (S_i)	Poisson values (Theoretical Relative Frequency)	Cumulative Relative Frequency (F_i)	$\|F_i - S_i\|$
0	0.010	0.0100	0.0224	0.0224	0.0124
1	0.080	0.0900	0.0850	0.1074	0.0174
2	0.190	0.2800	0.1615	0.2689	0.0089
3	0.230	0.5100	0.2046	0.4735	0.0365
4	0.170	0.6800	0.1944	0.6679	0.0121
5	0.150	0.8300	0.1477	0.8156	0.0144
6	0.080	0.9100	0.0936	0.9092	0.0008
7	0.030	0.9400	0.0508	0.9600	0.0200
8	0.030	0.9700	0.0241	0.9841	0.0141
9	0.020	0.9900	0.0102	0.9943	0.0043
10	0.010	1.0000	0.0039	0.9982	0.0018
11	0.000	1.0000	0.0013	0.9995	0.0005
12	0.000	1.0000	0.0004	0.9999	0.0001
13	0.000	1.0000	0.0001	1.0000	0.0000

between the *observed* and *theoretical* values is sufficiently close for us to believe they come from the same distribution.

Note that the Kolmogorov–Smirnov test uses the *ordinal* nature of the classes, while the χ^2 test makes use of only the *nominal* properties of this data. We should also point out that the Kolmorgorov–Smirnov test can also be used for testing the goodness of fit between two *sample* cumulative relative frequency distributions. In this case, the procedure is exactly the same as before, except that the statistic D has a different distribution (which we shall not present).

15.7 CHI-SQUARE TEST FOR INDEPENDENCE

In the chi-square test of Section 15.6, a set of *observed* values, classified into c categories according to a *single* attribute, were tested for goodness of fit against a set of *expected* values. At this time, we extend that analysis by assuming that more than one attribute is under investigation, and we want to determine whether or not these attributes are independent. For example, instead of investigating car sales relative to the single attribute "months of the year," we might wish to construct a test to determine if the attributes "car model" (such as sedans *vs.* hardtops) and the attribute "months of the year" are *independent* in their effect on sales. Similarly, in our supermarket example, we might have been interested in determining whether or not the pattern of arrivals per minute is independent of the attribute "days of the week."

In examples of this type, a direct extension of the chi-square test of Section 15.6 can be used to test the null hypothesis that the attributes under investigation are independent. For such problems there is generally no "theory" available to use in determining the expected frequency for each category. However, we can use the observed data to calculate expected frequencies, under the assumption that the null hypothesis (of independence) is true. For example, suppose in our car-sales example that 20 percent of the total number of sales fall in the month of June. If the null hypothesis is true, namely, that car model (sedan, hardtop) and months of the year are independent, then we would expect about 20 percent of all sedan sales to fall in June, and 20 percent of all hardtop sales to fall in June. The same relationship should hold for all six months under study. If the alternative hypothesis is true (i.e., these attributes are not independent), then we would expect to find differences in the proportion of sedan and hardtop sales across the six months.

To continue the above example, let's assume that out of the 150 cars sold over the six-month period, 50 were sedans and 100 were hardtops. Since June sales represented 20 percent of this total, if H_0 is true, the dealer would thus "expect" 20 percent of the 50 sedan sales to occur in June, which would be $0.20(50) = 10$ cars. Similarly, he would "expect" 20 percent of the 100 hardtop sales to occur in June, or $0.20(100) = 20$ cars. Note that he expects to sell 10 sedans plus 20 hardtops, for a total of 30 cars, which is exactly 20 percent of the 150 cars sold. Also, notice that he expects in June to sell hardtops and sedans in a ratio of 2 to 1. This ratio exactly agrees with his total sales ratio of 100 hardtops to 50 sedans (or 2 to 1). The entire set of expected frequencies (E_{ij}) and observed frequencies (O_{ij}) for this problem are given in Table 15.12. In each cell the expected frequency is in the upper left corner and the observed frequency is in the lower right corner, thus:

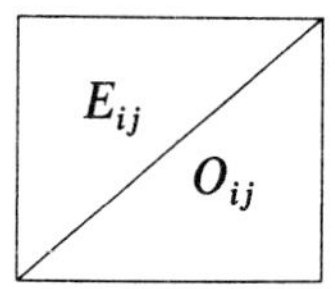

We perform the chi-square test for this type of problem exactly as we did before, except that we now sum over all cells in *two* rows rather than just one.

$$\chi^2 = \sum_{i=1}^{2}\sum_{j=1}^{6}\frac{(O_{ij} - E_{ij})^2}{E_{ij}}$$

$$= \frac{(3 - 9)^2}{9} + \frac{(3 - 6)^2}{6} + \cdots + \frac{(18 - 20)^2}{20}$$

$$= 14.15.$$

Table 15.12 Observed and Expected Sales of Automobiles

	Month						
	Jan.	**Feb.**	**Mar.**	**Apr.**	**May**	**June**	**Total**
Sedans	9 / 3	6 / 3	5 / 4	8 / 12	12 / 16	10 / 12	50
Hardtops	18 / 24	12 / 15	10 / 11	16 / 12	24 / 20	20 / 18	100
Totals	27	18	15	24	36	30	150

The number of degrees of freedom for this problem is 5, for if the marginal totals in Table 15.12 are considered fixed, then only 5 of the 12 cells are free to vary at any one time. The probability $P(\chi_5^2 > 14.15)$ for $v = 5$ degrees of freedom lies between 0.025 and 0.01 (see Table V). Thus, for any level of significance higher than 0.025, we can reject the null hypothesis that the attributes "car model" and "months of the year" are independent. That is, we conclude that the proportion of sedans to hardtops *does* vary from month to month.

Generalizing the Chi-Square Test

The chi-square test illustrated above can be generalized to include problems involving any number of categories for each attribute. Let's designate the two attributes as A and B, where attribute A is assumed to have r categories ($r > 1$) and attribute B is assumed to have c categories ($c > 1$). Furthermore, label the total number of observations in the problem n. A representation of these n observations in matrix form is shown in Fig. 15.2, where O_{ij} equals the observed frequency in the ith row and the jth column. A matrix in this form is called a *contingency table.*

The dots in the column and row totals in the matrix indicate that these numbers represent the sum of a particular set of values. For example, the number $O_{.1}$ represents the sum of all the observed values in the first column, while $O_{1.}$ represents the sum of all the observed frequencies in the first row. The symbol $O_{..}$ represents the sum over all rows and columns, hence $O_{..}$ must equal n, the total number of observations.

Calculating the expected frequency E_{ij} for each cell in a contingency table involves multiplying the *proportion* of the total number of observations falling

Attribute B (columns); Attribute A (rows)

	1	2	3	$\cdots$	j	$\cdots$	c	Totals
1	O_{11}	O_{12}	O_{13}	$\cdots$	O_{1j}	$\cdots$	O_{1c}	$O_{1\cdot}$
2	O_{21}	O_{22}	O_{23}	$\cdots$	O_{2j}	$\cdots$	O_{2c}	$O_{2\cdot}$
3	O_{31}	O_{32}	O_{33}	$\cdots$	O_{3j}	$\cdots$	O_{3c}	$O_{3\cdot}$
$\vdots$	$\vdots$	$\vdots$	$\vdots$		$\vdots$		$\vdots$	$\vdots$
i	O_{i1}	O_{i2}	O_{i3}	$\cdots$	O_{ij}	$\cdots$	O_{ic}	$O_{i\cdot}$
$\vdots$	$\vdots$	$\vdots$	$\vdots$		$\vdots$		$\vdots$	$\vdots$
r	O_{r1}	O_{r2}	O_{r3}	$\cdots$	O_{rj}	$\cdots$	O_{rc}	$O_{r\cdot}$
Totals	$O_{\cdot 1}$	$O_{\cdot 2}$	$O_{\cdot 3}$	$\cdots$	$O_{\cdot j}$	$\cdots$	$O_{\cdot c}$	$O_{\cdot\cdot} = n$

Fig. 15.2 Chi-square contingency table.

in the jth category for Attribute B (which is $O_{\cdot j}/n$) times the *number* of observations falling in the ith category of Attribute A (which is $O_{i\cdot}$).*

Expected frequency in the ith *row*, jth *column:*

$$E_{ij} = \left(\frac{O_{\cdot j}}{n}\right)(O_{i\cdot}) = \frac{O_{i\cdot}O_{\cdot j}}{n}. \tag{15.2}$$

We remind the reader of our rule of thumb that the expected frequency in each cell must be at least 1.0. As soon as the expected frequency is determined for each of the cells in the contingency table (using Formula (15.2)), the calculated value of χ^2 can be determined by the following formula:

Chi-square statistic for independence:

$$\chi^2_{(r-1)(c-1)} = \sum_{i=1}^{r}\sum_{j=1}^{c} \frac{(O_{ij} - E_{ij})^2}{E_{ij}}. \tag{15.3}$$

* This expectation is a direct result of the relationship presented in Chapter 2, which says that two discrete events A_i, B_j are independent if and only if

$$P(A_i \cap B_j) = P(A_i)P(B_j).$$

Since our estimates of $P(A_i)$ and $P(B_j)$ are $O_{i\cdot}/n$ and $O_{\cdot j}/n$, respectively, the product $(O_{i\cdot}/n)(O_{\cdot j}/n)$ is our estimate of the joint probability. Multiplying this product by the total number of observations (n) gives Formula 15.2.

Table 15.13 Frequencies for Stock Problem

		Industry				
		I	II	III	IV	Total
Stock Prices	High	13.8 / 15	10.4 / 8	8.7 / 10	12.1 / 12	45
	Med.	18.5 / 20	13.9 / 16	11.5 / 12	16.1 / 12	60
	Low	7.7 / 5	5.7 / 6	4.8 / 3	6.8 / 11	25
	Total	40	30	25	35	130

The number of degrees of freedom for this χ^2 statistic can be determined by noting that in calculating the expected frequency for each cell we must assume that the marginal totals ($O_{i\cdot}$ and $O_{\cdot j}$) are fixed quantities. This means that one degree of freedom is lost for each row and each column, so that the total number of degrees of freedom is $(r - 1)(c - 1)$.

Sample Problem To illustrate the use of Formula (15.3), consider the problem of trying to determine whether the prices of certain stocks on the New York Stock Exchange are independent of the industry to which they belong. Assume four categories of industries are investigated (labeled I, II, III, and IV), and that stock prices in these industries are classified in one of three categories ("high-priced," "middle-priced," or "low-priced"). The data from such an analysis might look like the values shown in Table 15.13, where the *expected* values are in the upper left of each cell, and the *observed* values in the lower right, as before:

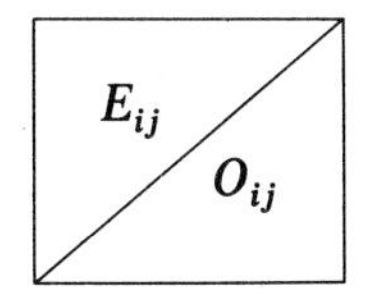

To illustrate the calculation of expected frequencies, note that the expected frequency of high-priced stocks in Industry I is found by multiplying the proportion of stock in Industry I to the total number of observations (which is $\frac{40}{130}$) by the number of observations in the high-priced category (45). This product is $(\frac{40}{130})(45) = 13.8$, which is shown in the first cell. Similarly, the expected frequency for high-priced stocks in Industry II is $(\frac{30}{130})(45) = 10.4$. Note that for each row and column the sum of the expected frequencies must be the same as the sum of the observed frequencies. The number of degrees of freedom for this problem is $(r - 1)(c - 1) = (2)(3) = 6$, and the calculated value of χ^2 is:

$$\chi_6^2 = \frac{(15 - 13.8)^2}{13.8} + \frac{(8 - 10.4)^2}{10.4} + \cdots + \frac{(11 - 6.8)^2}{6.8}$$

$$= 6.264.$$

To be significant at the 0.05 level, the value of χ^2 has to be greater than 12.6 for six degrees of freedom (see Table V). Since the computed value $\chi^2 = 6.264$ is less than this value, the null hypothesis cannot be rejected at the 0.05 level of significance. We thus conclude that the price of stocks is *independent of the industry* associated with that stock.

15.8 NONPARAMETRIC MEASURES OF CORRELATION

Contingency Coefficient

The Pearson product–moment method of correlation described in Chapter 11 assumes that the variable under consideration can be measured on an interval scale. But as we have pointed out in this chapter, interval measurement may be inappropriate or even impossible in a variety of circumstances. If only nominal data are available, then the value of χ^2 can be used to provide a measure of the degree of association between two variables. When there is no association between two variables, then the observed frequency in each cell of a chi-square table should closely correspond to the expected frequency in that cell, because expected frequencies are calculated under the assumption (the null hypothesis) that the two variables are not related. The higher the degree of association between the variables, the larger will be the discrepancy between the observed and expected cell frequencies, and hence the larger the value of χ^2. It is convenient, in defining a nonparametric measure of correlation, to have a statistic which equals zero when there is no association between the variables and one which approaches 1.0 as the amount of association increases. One such statistic, which is called the *contingency coefficient* and is denoted by the letter C, is defined as follows:

Contingency coefficient: $$C = \sqrt{\frac{\chi^2}{(n + \chi^2)}}.$$

In Section 15.6 we calculated the value of χ^2 for testing the association between months of the year and observed sales of new cars, and found it to be 12.0 for $n = 150$. The contingency coefficient for this example is thus:

$$C = \sqrt{12.0/(150 + 12.0)} = 0.27.$$

Determining whether a contingency coefficient significantly differs from zero is equivalent to the χ^2 test for the difference between the observed and expected frequencies. We saw, in Section 15.6, that a value of $\chi^2 = 12.00$ permits rejection of the null hypothesis at a level of significance smaller than 0.05 and larger than 0.025. Thus, the null hypothesis that the contingency coefficient for this example equals zero can be rejected at a probability between 0.05 and 0.025.

The contingency coefficient is not appropriate when ordinal data is available since in that case we need a method of *rank correlation.* There are two important methods available for correlating ordinal data, *Spearman's rank-correlation* method and *Kendall's rank-correlation* approach, both of which will be presented in this section.

Spearman's Rho

Research published by C. Spearman in 1904 led to development of what is perhaps the most widely used nonparametric measure of correlation. This measure, usually denoted either by the letter r or by the Greek word rho, has thus become known as Spearman's rho, or r_S. Spearman's rank correlation coefficient r_S is the ordinary correlation coefficient using ranks as the original data. A perfect positive correlation ($r_S = +1$) means that the two samples rank each object identically, while a perfect negative correlation ($r_S = -1$) means that the ranks of the two samples have an *exactly inverse* relationship. Values of r_S between -1 and $+1$ denote less than perfect correlation. To measure correlation, Spearman's test squares the difference between the rank of an object in one sample and its rank in the second sample. If this squared difference is denoted as d_i^2 for the ith pair of observations, then the sum of these squared differences over a set of n pairs of observations is $\sum_{i=1}^{n} d_i^2$. The value of $\sum_{i=1}^{n} d_i^2$ is used as a measure of the distance between the ranks in the two samples:*

Spearman's rank-correlation coefficient: $$r_S = 1 - \frac{6 \sum_{i=1}^{n} d_i^2}{n^3 - n}. \qquad (15.4)$$

As an example of the use of Spearman's rho, suppose that we calculate the correlation between the brother and sister scores given in Table 15.3. These

* Formula (15.4) is equivalent to Formula (11.2) applied to ranks (except when there are ties).

Table 15.14 IQ Scores

Pair	Sister	Rank	Brother	Rank	d_i	d_i^2
1	129	11	115	8	3	9
2	111	4	108	6	−2	4
3	117	7	123	10	−3	9
4	120	8	104	4	4	16
5	116	6	110	7	−1	1
6	101	1	98	2	−1	1
7	107	3	106	5	−2	4
8	127	10	119	9	1	1
9	105	2	95	1	1	1
10	123	9	130	11	−2	4
11	113	5	101	3	2	4
					Sum	54

$$r_S = 1 - \frac{6\sum_{i=1}^{n} d_i^2}{n^3 - n} = 1 - \frac{6(54)}{11^3 - 11} = 0.755.$$

scores, their ranks, and their differences are reproduced in Table 15.14. The rank correlation between the two samples is 0.755.

In order to test the significance of a given value of r_S, it is ncessary to determine the probability that a given value of r_S will occur under the null hypothesis. This probability depends on the number of permutations of the two variables that give rise to the particular value of r_S. Tables are available which give the critical values of r_S for small values of n. For example, the $r_S = 0.755$ value is significant at the 0.01 level when $n = 11$. When n is large ($n \geq 10$), the significance of an obtained value of r_S under the null hypothesis can be determined from the following $\boldsymbol{t}$-variable:*

Test statistic for Spearman's rho: $$\boldsymbol{t} = r_S\sqrt{\frac{n-2}{1-r_S^2}}.$$

* Note that this t-statistic is directly comparable to the t-statistic used for r in Chapter 11 (Formula 11.4).

Table 15.15

Team	AP Rank	UPI Rank	d_i	d_i^2
A	1	1	0	0
B	2	4	−2	4
C	3	2	1	1
D	4	5	−1	1
E	5	3	2	4
			Sum	10

$$r_S = 1 - \frac{6(10)}{5^3 - 5} = 1 - \frac{60}{120} = 0.50$$

This statistic can be shown to follow the ***t***-distribution with $(n - 2)$ degrees of freedom. To illustrate, suppose we had used this formula for the brother-sister IQ problem. In this case $r_S = 0.755$, $n = 11$, and the computed value of $\boldsymbol{t}$ is

$$t = 0.755\sqrt{(11 - 2)/(1 - 0.755^2)} = 3.454.$$

From the ***t*** table in Appendix B, this value is seen to be significant at the 0.005 level of significance. (Note that we have assumed that a one-tailed test is appropriate in this example. That is, the alternative hypothesis is $H_a\!:p > 0$. If a two-tailed test had been used (i.e., $H_a\!:p \neq 0$), then the significance level would be $2(0.005) = 0.01$ rather than 0.005.)

As another example of the use of Formula (15.4) suppose that we calculate the rank correlation between the nation's top five football teams for a given year, as reflected in polls by the Associated Press (AP) and the United Press International (UPI). Assume that the teams in the top five are identical (teams A, B, C, D, and E), but that they occur in a different order in the two polls; the poll results are shown in Table 15.15. The rank correlation between the AP and UPI ratings is 0.50.

Kendall's Tau

An alternative method for determining a rank correlation coefficient is to calculate Kendall's correlation coefficient. This statistic, developed by the statistician M. G. Kendall, is denoted by the Greek letter τ (tau), and called Kendall's tau. Although Kendall's tau is suitable for determining the rank correlation of the same type of data for which Spearman's rho is useful, the two methods use different techniques for determining this correlation, so their values will not normally be the same. Spearman's rho is perhaps more widely used, but Kendall's tau has the advantage of being generalizable to a partial correlation coefficient.

The rank correlation coefficient τ is determined by first calculating an index which indicates how the ranks of one set of observations, *taken two at a time*, differ from the ranks of the other set of observations. The easiest way to determine the value of this index is to arrange the two sets of rankings so that one of them, say the first sample, is in ascending order, from the lowest score (rank) to the highest score (rank). The other set, representing the second sample, will will not be in ascending order unless the ranks of the two samples agree perfectly. Now, consider all possible combinations of the n ranks in this second sample, taken two at a time (i.e., all pairs); assign a value of $+1$ to each pair in which the two ranks are in the same (ascending) *order* as they are in the first sample, and a value of -1 to each pair in which the two ranks are not in the same (ascending) order as they are in the first sample. The sum of these $+1$ and -1 values is an indication of how well the second set of rankings agrees with the first set. Since there are $\binom{n}{2}$ combinations of n objects taken two at a time, this sum (or index) can assume any value between $+\binom{n}{2}$ and $-\binom{n}{2}$. Kendall's tau is defined as the ratio of the computed value of this index to the maximum value it can assume [which is $\binom{n}{2}$]:

Kendall's rank-correlation coefficient: $$\tau = \frac{\text{Computed index}}{\text{Maximum index}}.$$

Note that when there is perfect positive correlation, τ will equal $+1$, since the computed index and the maximum index will both equal $\binom{n}{2}$; if there is a perfect negative correlation, the computed index will equal $-\binom{n}{2}$ and τ will equal -1.

To illustrate the calculation of Kendall's tau, consider once again the football rankings given in Table 15.15. We must now look at each pair of observations in the UPI column to see if that pair has the same rank order as it does in the AP column. For example, if we compare Team A with Team B, we see their rankings in the UPI column are 1 and 4. Since Team A is ranked ahead of B in both polls, we assign a $+1$ to this pair. Similarly, the pairs A–C, A–D, and A–E all have the same order in both polls, so we assign a $+1$ to each of these pairs. If, however, we look at the pair B–C (where the UPI ranks are 4 and 2) we see these ranks are in *reverse* order, so the value -1 is assigned to this pair. The remaining assignments for this problem are shown in Table 15.16.

We see from Table 15.16 that out of the $\binom{n}{2} = \binom{5}{2} = 10$ different pairs, seven of the UPI ranks are in the same order as in the AP poll (and hence are assigned a $+1$ value), and three are in the reverse order (a -1 value). The numerator of Kendall's tau (the computed index) is thus $7 - 3 = 4$, and

$$\tau = \frac{\text{Computed index}}{\text{Maximum index}} = \frac{4}{10} = 0.40.$$

Table 15.16

Pair	Value	Pair	Value	Pair	Value
A vs B	+1	B vs C	−1	C vs E	+1
A vs C	+1	B vs D	+1	D vs E	−1
A vs D	+1	B vs E	−1		
A vs E	+1	C vs D	+1		

As another illustration of the calculation of Kendall's tau, suppose we calculate τ for the data in Table 15.14 on brother-sister IQ's. This table is repeated in Table 15.17, with the *sister's* scores put in ascending order.

In this problem, there are $\binom{n}{2} = \binom{11}{2} = 55$ paired comparisons. A good exercise for the reader would be to verify that 42 out of the 55 comparisons in the brother column are in the same order as the sister column (+1) and 13 are in the reverse order (−1). The calculated index for Kendall's tau thus is $42 - 13 = 29$, and τ is:

$$\tau = \frac{29}{55} = 0.527.$$

Note that the value of Kendall's tau in the two examples we have considered (0.40 and 0.527) is considerably less than the comparable value of Spearman's rho (0.50 and 0.755). Both coefficients, however, utilize the same amount of information about the association between two variables, and for a given set of observations both will reject the null hypothesis that two variables are unrelated

Table 15.17

Pair	Sister	Rank	Brother	Rank
6	101	1	98	2
9	105	2	95	1
7	107	3	106	5
2	111	4	108	6
11	113	5	101	3
5	116	6	110	7
3	117	7	123	10
4	120	8	104	4
10	123	9	130	11
8	127	10	119	9
1	129	11	115	8

in the population at the same level of significance. For small samples, tables are available for determining the probability of a given value of τ under the null hypothesis. Since τ and r_S must both reject the null hypothesis at the same level of significance, $\tau = 0.527$ for $n = 11$ must be significant at the 0.01 level as was $r_S = 0.755$. When $n \geq 10$, the statistic τ may be considered to be normally distributed with mean $\mu_\tau = 0$, and standard deviation:

$$s_\tau = \sqrt{\frac{2(2n + 5)}{9n(n - 1)}}.$$

The distribution of τ can thus be transformed into the following z-variable:

Standardization of μ_τ *for Kendall's tau:* $$z = \frac{\tau - \mu_\tau}{s_\tau}.$$

In our example, the sample of $n = 11$ resulted in a value of Kendall's tau of $\tau = 0.527$. The standard deviation of a sample of this size is

$$s_\tau = \sqrt{2(2 \times 11 + 5)/9 \times 11(11 - 1)} = 0.234.$$

The standardized normal value is thus

$$z = \frac{0.527}{0.234} = 2.252.$$

Since $P(z > 2.252) = 0.0122$, we can reject H_0 when $\alpha \geq 0.0122$ for a one-sided test, and when $\alpha \geq 2(0.0122) = 0.0244$ for a two-sided test.

Define: Goodness-of-fit tests, observed and expected frequencies, contingency coefficient, Spearman's rho (r_S), Kendall's tau (τ).

Problems

15.15 Identify the level of measurement required for the following tests:

a) the binomial test,

b) the χ^2-test,

c) the runs test,

d) the sign test,

e) the Mann–Whitney U-test,

f) Kendall's tau and Spearman's rho,

g) the contingency coefficient,

h) Kolmogorov–Smirnov test.

15.16 Universities are often criticized for having hired less than a proportional number of members of minority groups for faculty positions. Suppose that a national survey indicates that of all earned doctorates, 20% belong to females, 10% belong to male minorities, and 70% belong to white males. A random sample of 1000 faculty positions yields the following breakdown:

White Males	Women	Minority Males
800	150	50

a) Calculate the expected frequency for each cell, using the population proportions 70%, 20%, and 10%. Use the χ^2 test to determine whether the observed and expected frequencies are close enough for the null hypothesis to be accepted. Use $\alpha = 0.01$.

b) What p-value would you report?

15.17 In one of his classical experiments on heredity, Gregor Mendel observed the color of the plants bred from a purple-flowered and a white-flowered hybrid. Out of 929 plants observed, 705 had a purple flower, and 224 had a white flower.

a) Test the hypothesis that the probability of observing a purple-flowered plant is $\frac{3}{4}$, using the chi-square test and letting $\alpha = 0.05$.

b) Repeat the test using a test statistic with a binomial distribution.

15.18 Recent medical research indicates that more females who are lefthanded were born in November–December than in any other two-month period. Consider the following data, (Table 15.18) representing a sample of 2000 females who are lefthanded. Use $\alpha = 0.01$ and a chi-square test to determine whether this sample indicates a difference across the six two-month intervals in the number of lefthanders.

Table 15.18

Month of Birth	Number	Month of Birth	Number
Jan–Feb	317	July–Aug	278
Mar–Apr	289	Sept–Oct	321
May–June	301	Nov–Dec	494
		Total	2000

15.19 A cafeteria proposes to serve four main entrees. For planning purposes, the manager expects that the proportions of each that will be selected by his customers will be:

Selection	Hot Dogs and Chili	Roast Beef	Steak	Fish
Proportion	0.20	0.50	0.20	0.10

Of the first 50 customers 15 select hot dogs and chili, 20 select roast beef, 5 steak, and 10 select fish. The manager wonders whether to revise his preparation schedule, or whether this deviation from his expectations is merely chance variation that should balance out overall. Make an appropriate test, at the 0.01 level of significance, on which to base your advice to this manager.

15.20 Return to Problem 4.42, and once again determine the Poisson approximation to the observed frequencies for deaths from the kick of a horse in the Prussian Army. Use the χ^2-test to determine whether the observed frequencies and the expected frequencies under the Poisson distribution differ significantly. At what level of significance can H_0 be rejected?

15.21 Use the Kolmogorov–Smirnov test to measure the goodness of fit for the data in Problem 4.42. Is H_0 accepted or rejected?

15.22 A study was recently completed to determine whether or not *the order* in which a person's name appears on a political ballot has any influence on that person's chances of being elected. Ninety-two elections were observed, each having four candidates, and the results are shown in Table 15.19.

Table 15.19

	Position on Ballot				
	1	**2**	**3**	**4**	**Total**
No. of wins	29	21	17	25	92

a) Use a Kolmogorov–Smirnov test to examine the relationship between these data and the distribution one would expect if the probability of winning were equal for all four positions. Is H_0 accepted or rejected?

b) Use a χ^2 test to accomplish the same purpose as part (a).

15.23 The records at a large metropolitan hospital (Table 15.20), listed the following births during the three shifts of the day during a two-week period.

Table 15.20

	7 A.M. –3 P.M.	3 P.M. –11 P.M.	11 P.M. –7A.M.
Males	15	5	10
Females	5	10	15

a) Determine the expected frequency in each cell. What null hypothesis is being tested with a χ^2-test?

b) How many degrees of freedom are there? At what level of significance can H_0 be rejected?

15.24 What is meant by the phrase "goodness-of-fit test"? How is the χ^2 distribution used to make this test?

15.25 Fifty terminally ill patients in a hospital were randomly assigned to either a control group or an experimental group. The experimental was given a new drug being tested for this particular illness. The control group was kept on the medicine they had been receiving. The number of deaths in these two groups during the next year is given in Table 15.21.

Table 15.21

	Living	Deaths
Experimental	9	16
Control	1	24

a) What null hypothesis should be tested here? What level of α would you recommend?

b) At what level of significance can H_0 be rejected?

15.26 Calculate the contingency coefficient for the data in Problem 15.23.

15.27 Use the data in Problem 15.17 to find the value of C, the contingency coefficient.

15.28 Use the data in Problem 15.7 to:

a) Compute Spearman's rho, and test this value to determine whether the hypothesis of no correlation can be rejected.

b) Compute Kendall's tau and again determine the level of significance at which H_0 can be rejected.

15.29 In listing the top five restaurants in New York City, two guides list the same places, but in different order. Denoting the five restaurants as A, B, C, D, and E, the ratings are as shown in Table 15.22. Find Spearman's rho and Kendall's tau.

Table 15.22

Rank	First Guide	Second Guide
1	A	B
2	B	C
3	C	A
4	D	E
5	E	D

15.30 Suppose that the national average for the driving test in Problem 15.4 is 60. Test the hypothesis that the frequency of students scoring above and below the national average is not the same for these two groups (use a χ^2 test).

Exercises

15.31 Discuss the similarities and differences between the χ^2 (goodness-of-fit) test (Section 15.6) and the χ^2 test for Independence (Section 15.7). Are they two different tests, or merely different versions of the same test?

15.32 Show that Spearman's rho is the rank-order equivalent of the correlation coefficient presented in Chapter 11. (*Hint:* See Siegel, *Nonparametric Statistics*, McGraw-Hill, p. 203.)

15.33 There is a test called a one-sample runs test which can be used to test the hypothesis that the observations in a sample are random. For example, during one stretch of 23 days last April, the weather in Boston was classified as either "sunny" or "cloudy." The following pattern was observed, where S = sunny and C = cloudy:

S S C C S S S C C C S S S S S C C C C S S S S.

Suggest how a runs test similar to the Wald–Wolfowitz test could be used to test for randomness here. What kinds of patterns would suggest nonrandomness? Would H_0 be rejected only when the number of runs is small?

GLOSSARY

parametric versus nonparametric tests: parametric tests generally require some assumptions, such as the normality of the parent population, and a measurement level of at least an interval scale. Nonparametric tests require no such assumptions.

levels of measurement:

nominal—categorize data by "names" only;
ordinal—scale has the property of order;
interval—scale has order plus a constant interval;
ratio—scale has order, a constant interval, plus a unique zero point (making ratio statements meaningful).

Mann–Whitney *U*-Test: a nonparametric test to determine whether or not two samples were drawn from the same population.

Wald–Wolfowitz Runs Test: a nonparametric test to determine whether or not two samples were drawn from the same population.

The Wilcoxon Test: a nonparametric test to determine if the median difference between two populations equals zero.

sign test: an ordinal test involving paired samples and information on whether one object's rating is larger (+) or smaller (−) than the rating of its paired object.

χ^2-test for independence: test to determine whether two attributes are independent by comparison of observed frequencies relative to expected frequencies.

contingency table: a table listing the frequency of observations across two or more attributes.

contingency coefficient: a measure of the goodness-of-fit for a contingency table.

Kolmorgov–Smirnov Test: a goodness-of-fit test when the data are in at least ordinal form.

Spearman's rho (r_S): a nonparametric correlation coefficient based on ranks.

Kendall's tau (τ): a nonparametric correlation coefficient based on ranks.

SELECTED BIBLIOGRAPHY

1. Cochran, W. G., and G. M. Cox, *Experimental Designs* (2nd ed). New York: Wiley, 1957.
2. Chatterjee, S., and B. Price, *Regression Analysis by Example*. New York: Wiley, 1977.
3. Draper, N. R., and H. Smith, *Applied Regression Analysis*. New York: Wiley, 1966.
4. Durbin, J., and G. S. Watson, "Testing for Serial Correlation in Least Squares Regression II," *Biometrika*, **38**, pp. 159–178.
5. Feller, W. *Probability and Its Applications*, Vol. 2 (2nd ed). New York: Wiley, 1957.
6. Hamburg, M., *Basic Statistics* (2nd ed). New York: Harcourt Brace Jovanovich, 1979.
7. Harnett, D. L., and J. L. Murphy, *Introductory Statistical Analysis* (2nd ed). Reading, Mass.: Addison-Wesley, 1980.
8. Huff, D., *How to Lie with Statistics*. New York: Norton, 1954.
9. Hoel, P. G., *Introduction to Mathematical Statistics* (4th ed). New York: Wiley, 1971.

10. Johnson, J., *Econometric Methods* (2nd ed). New York: McGraw Hill, 1972.
11. Mendenhall, W., *An Introduction to Linear Models and the Design and Analysis of Experiments.* Belmont, Cal.: Wadsworth, 1968.
12. Mood, A. M., F. A. Graybill, and D. C. Boes, *Introduction to the Theory of Statistics* (3rd ed). New York: McGraw-Hill, 1974.
13. Neter, J., and W. Wasserman, *Applied Linear Statistical Models*, Homewood, Ill.: Irwin, 1974.
14. Runyon, R. P., *Nonparametric Statistics: A Contemporary Approach.* Reading, Mass.: Addison-Wesley, 1977.
15. Siegel, S., *Nonparametric Statistics.* New York: McGraw-Hill, 1956.
16. Snedecor, G. W., and W. G. Cochran, *Statistical Methods* (6th ed). Ames, Iowa: The Iowa State University Press, 1967.
17. Winkler, R. L., *An Introduction to Bayesian Inference and Decision.* New York: Holt, Rinehart and Winston, 1972.
18. Winkler, R. L., and W. L. Hays, *Statistics: Probability, Inference, and Decision* (2nd ed). New York: Holt, Rinehart and Winston, 1975.

APPENDIX A

SUBSCRIPTS, SUMMATIONS, VARIABLES AND FUNCTIONS, CALCULUS REVIEW

"Education is . . . hanging around until you've caught on."

ROBERT FROST

A.1 SUBSCRIPTS AND SUMMATIONS

Throughout this book we use certain symbols to distinguish between the numbers in a set of data, and to indicate the sum of such numbers. For example, we may wish to distinguish between the monthly sales of a certain business, and then sum these monthly sales to get the yearly sales. To do this, suppose that we let the symbol x denote the monthly sales of this firm. Furthermore, we will add a subscript to this symbol to denote which month is being represented. Thus, x_1 = sales in first month, x_2 = sales in second month, and so forth, with x_{12} = sales in the twelfth month. That is, if sales in the sixth month were 120 units, then we would write $x_6 = 120$. The notation x_i thus stands for "sales in the ith month," where i can be any number from 1 to 12;

that is, $i = 1, 2, \ldots, 12$. The dots in this last expression are used to indicate "and so on."

Now, assume that we want to sum the sales for all 12 months in a year, which is

$$x_1 + x_2 + \cdots + x_{12}.$$

Another way of writing this sum is to use the Greek letter $\sum$ (capital sigma). This symbol is read as "take the sum of." At the bottom of this $\sum$ sign we usually place the first value of i which is to be included in the sum. The last value of i to be summed is usually placed at the top of the sum sign. Thus,

$$\sum_{i=1}^{12} x_i$$

is read as "sum the values of x_i starting from $i = 1$ and ending with $i = 12$." That is,

$$\sum_{i=1}^{12} x_i = x_1 + x_2 + \cdots + x_{12}.$$

Similarly, suppose that we want the sum of only the last seven months in the year. This sum is written as follows:

$$\sum_{i=6}^{12} x_i = x_6 + x_7 + \cdots + x_{12}.$$

In statistics we often will not know in advance what the final value in a summation will be. For example, we know that we want to sum a set of sales values, but we do not know how many values there are to be summed. To designate this situation, we will let the symbol n represent the last number in the sum (where n can be any integer value, such as $1, 2, 3, \ldots$). The notation

$$\sum_{i=1}^{n} x_i = x_1 + x_2 + \cdots + x_n$$

is thus read as "the sum of n numbers, where the first number is x_1, the second is x_2, and the last is x_n." In summing monthly sales over a year, we would thus let $n = 12$, so that $\sum_{i=1}^{n} x_i = \sum_{i=1}^{12} x_i$.

Perhaps we should mention that, in some chapters in this book, we have sometimes omitted the limits of summation, and simply written $\sum x_i$. This notation should be interpreted to mean "sum all relevant values of x_i." In these instances we have made sure that the reader always knows what the relevant values of x_i are. Also, we might point out that the choice of symbols in designating a sum of numbers is often quite arbitrary. For example, we might have used the letter y to denote monthly sales (instead of x), and used the letter j as a subscript (instead of i). In this case $\sum_{j=1}^{12} y_j$ would denote the sum of the twelve monthly values.

Double Summations

In a number of chapters in this book we have found it convenient to use *two* subscripts instead of just one. In these instances the first subscript indicates one characteristic under study, and the second subscript some other characteristic. For example, suppose that we let x_{ij} = sales in the ith month by the jth salesman. The notation $x_{6,2} = 15$ would indicate that in the sixth month ($i = 6$), salesman number 2 ($j = 2$) sold 15 units. Using the same procedure as described above, we can denote the total sales over 12 months by the jth salesman as the sum of x_{1j} (sales in the 1st month by the jth salesman) plus $x_{2j}, \ldots,$ plus $x_{12,j}$ (sales in the 12th month by the jth salesman). That is,

Total sales by salesman j:

$$\sum_{i=1}^{12} x_{ij} = x_{1j} + x_{2j} + \cdots + x_{12,j}.$$

Another example of a similar type of sum is the sum of sales in the ith month (where i is some number between 1 and 12) over all the salesmen in the company. If we let m = total number of salesmen, then this sum is x_{i1} (sales in month i by salesman # 1)plus $x_{i2}, \ldots,$ plus x_{im} (sales in month i by salesman m). That is,

Total sales in month i:

$$\sum_{j=1}^{m} x_{ij} = x_{i1} + x_{i2} + \cdots + x_{im}.$$

Finally, we might wish to sum over all months ($i = 1, 2, \ldots, 12$) and all salesmen ($j = 1, 2, \ldots, m$). This sum could be written as:

Total sales over all months and all salesmen:

$$\sum_{\text{All } j} \sum_{\text{All } i} x_{ij} = \left\{ \begin{array}{llll} x_{11} & + x_{12} & + \cdots & + x_{1m} \\ + x_{21} & + x_{22} & + \cdots & + x_{2m} \\ \vdots & & & \\ + x_{12,1} & + x_{12,2} & + \cdots & + x_{12,m} \end{array} \right\}$$

A.2 VARIABLES AND FUNCTIONS

Variables

Variables and the relationship between variables represent an important part of statistics. Hence, it is important that we define these concepts carefully.

A variable is a quantity that may assume any one of a set of values. For example, we might describe the worth of a common stock by the variable "current worth on the stock market." The values of this variable are the different prices the stock can assume. Or, we might be interested in describing how well a specific brand of alkaline battery works by defining the variable "the length of time before failure when in constant use." The values of this variable are the various times it might take before the battery fails.

Variables are often classified according to whether their values are *discrete* or *continuous*. The values of a discrete variable are individually distinct; i.e., they are separable from one another. The price of a common stock, for instance, represents a discrete variable because the prices a stock can assume are all separate values, distinguishable from one another. The following examples also represent discrete variables:

1. the number of defectives in a production lot,
2. the amount of advertising expenditure a certain company plans for next year,
3. the amount of federal income tax owed by an individual.

Most discrete variables represent some quantity that can be "counted."

The values of a *continuous* variable are not separable from one another; each value is immediately adjacent to and indistinguishable from the next. Quantities that are *measured* are usually continuous variables; for example, measures of time, weight, length, and area typically represent continuous variables. Thus, in our earlier example, the time it takes a battery to fail represents a continuous random variable. There are always an infinite number of values of a continuous variable.* The following variables also are continuous:

1. the percentage increase in the consumer price index last month,
2. the amount of gasoline available in the U.S.A. next year,
3. the quality of the air in Los Angeles yesterday, measured in a way so as to include all numbers from 0 to 100.

One of the practical difficulties with continuous variables is that the devices used to measure such variables usually are read only in a discrete manner. For example, the variable "amount of gasoline needed to fill a car" is clearly a continuous random variable, since this amount may be *any* value between zero and the capacity of the gas tank. From a *practical* point of view, however, this variable is discrete because most gas pumps cannot be read (at least accurately) beyond a few decimal points (usually $\frac{1}{10}$ of a gallon). *For most statistical analysis it makes little difference if we treat such variables as discrete or continuous, although a continuous variable is often easier to manipulate than is a discrete variable with many different values.*

* The number of values of a discrete variable may be either finite or infinite.

Functions

If a unique value of some variable y is associated with every possible value of another variable x, then the variable y is said to be "a function of" the variable x. To illustrate a functional relationship, we will let x represent the number of gallons of gasoline you purchase at a service station, and let y be the amount of money you must pay for this gasoline. In this case y is a function of x [written $y = f(x)$] because the exact (unique) amount (y) you will be charged for every possible gasoline purchase (x) is known (assuming the price doesn't change before you get there).

There are three commonly used methods for describing a functional relationship: (1) a table, (2) a graph, and (3) an equation. As we will illustrate below, the first two of these methods work well for discrete functions, while the latter two work well for continuous functions.

Discrete functions. By a discrete function, we mean the function in any situation where x is a discrete variable. If x is discrete, then y must be discrete as well. To illustrate a discrete function, we propose that the Environmental Protection Agency (EPA) is testing a new car to determine its gas mileage (y) at various speeds (x). This car was tested at $x = 10$, 20, 30, 40, and 50 miles per hour (MPH). The miles per gallon (MPG) at these speeds were $y = 21.6$, 26.1, 27.8, 25.3, and 19.5 respectively. This information is shown in Table A.1.

Note that the variable x is discrete, since all possible values of this variable are distinguishable from one another—i.e., they are individually distinct. It is important to understand the symbolic notation in writing functions. For example, the notation $f(10) = 21.6$ means that when $x = 10$, the value of $f(x)$ is $y = 21.6$. Similarly, $f(50) = 19.5$ means that $y = 19.5$ when $x = 50$. Figure A.1 is a graph of the function relating MPH (x) and MPG (y).

We must resist the temptation to connect the points in Fig. A.1 with a line, since such a line might incorrectly lead a viewer to assume that the function is defined for speeds other than $x = 10$, 20, 30, 40, and 50 MPH. It may be possible to define a function that relates additional values of x to y, but in this example the function is defined only for five x-values. When the number of

Table A.1

Miles Per Hour (x)	Miles Per Gallon [$y = f(x)$]
$x = 10$	$y = f(10) = 21.6$
$x = 20$	$y = f(20) = 26.1$
$x = 30$	$y = f(30) = 27.8$
$x = 40$	$y = f(40) = 25.3$
$x = 50$	$y = f(50) = 19.5$

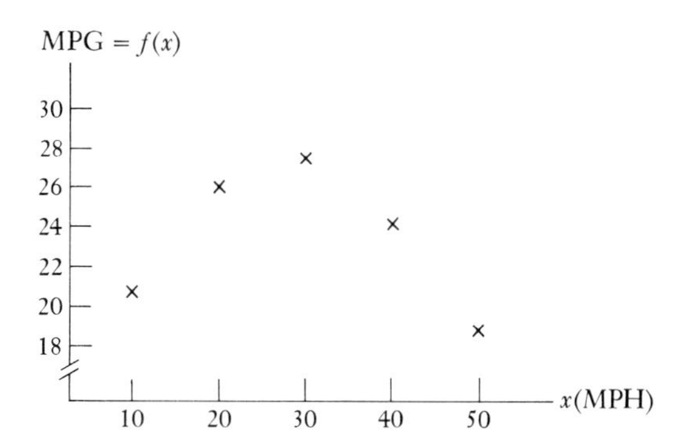

Fig. A.1 Graph of the function relating x = MPH and y = MPG.

x-values is large, a formula is often useful to describe a functional relationship. Chapter 4 gives several examples of such formulas.

Continuous Functions. When the random variable x is continuous, then the functional relationship between x and y usually must be expressed as a formula, or in a graph. Consider a simple example, where the variable x is temperature measured on the Fahrenheit scale, and the variable y is temperature measured on the Celsius scale. For converting values from the Fahrenheit scale (x) to the Celsius scale (y), the following functional relationship is used:*

$$y = \frac{5}{9}x - \frac{160}{9}.$$

This relationship represents a function because a unique value of y is specified for each value of x. The function is continuous, since the values of x are indistinguishable. Now, suppose that we want to graph this function. First, we recognize that it is a straight line, since the exponent of the variable x is 1. To graph a straight line we need only two points. The easiest two points to take are usually the one where $x = 0$ and the one where $y = 0$. When $x = 0$,

$$y = \frac{5}{9}(0) - \frac{160}{9} = -17.78.$$

Similarly, when $y = 0$ we can solve for x as follows:

$$0 = \frac{5}{9}x - \frac{160}{9},$$

$$\frac{5}{9}x = \frac{160}{9},$$

* We could have written this formula as

$$f(x) = \frac{5}{9}(x - 32).$$

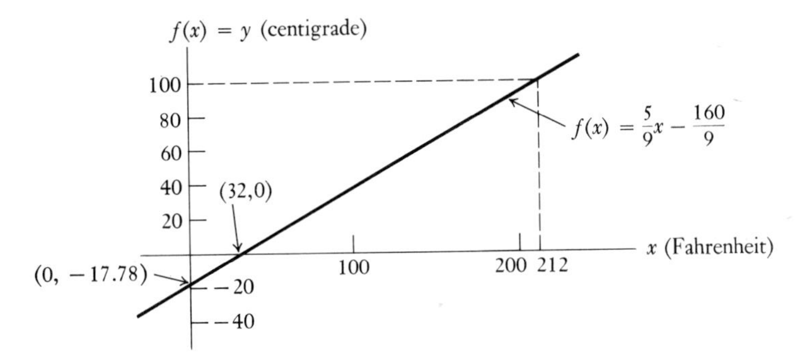

Fig. A.2 A graph of $y = f(x) = \frac{5}{9}x - \frac{160}{9}$.

so

$$x = \frac{160}{9} \cdot \frac{9}{5} = \frac{160}{5} = 32.$$

We now have two points on our function, (0, −17.78) and (32, 0). This function is graphed, in Fig. A.2, by connecting these two points.

From either the straight line in Fig. A.2, or the function itself, we can solve for any additional point. For example, most of us are familiar with the boiling point of water at 212 degrees Fahrenheit, $x = 212$. The comparable value of y is $f(212)$:

$$y = f(212) = \frac{5}{9}(212) - \frac{160}{9},$$

or

$$y = f(212) = 100,$$

which is the Celsius temperature for the boiling point of water. We could have solved for any one of the infinite number of different y-values in a similar manner.

A.3 CALCULUS REVIEW

Introduction

The study of calculus is usually divided into two distinct, though related, parts: differential calculus and integral calculus. Differential calculus is concerned primarily with the slope of a function or curve at any given point on that function; that is, the rate of change. Once the slope of a curve is known, it is usually possible to determine certain important characteristics of the curve, such as where it reaches a maximum or minimum. In statistical analysis, integral calculus is usually associated with the process of finding the area under a curve, i.e., between a curve and the x-axis. As is the case for differential calculus, integral calculus provides information extremely useful for describing certain characteristics of a function or a curve.

Differential Calculus

The slope of a function, or curve, representing the relationship between an independent variable x and a dependent variable y is merely "the rate of change of y divided by the rate of change of x." The slope of a line thus describes how a change in x affects the variable y. If a positive change in x results in a positive change in y, the slope will have a positive sign; when a positive increment in x produces a *negative* change in y, the slope is negative. For different values of x, the slope may vary.

For a *straight* line the difference between *any* two distinct points x_1 and x_2 represents the change in x, and the corresponding difference, $y_1 - y_2$, the change in y. Thus the rate of change of y with respect to x is

$$\frac{(y_1 - y_2)}{(x_1 - x_2)},$$

which represents the slope of a straight line. This value is constant for all points on the line. A horizontal straight line has a slope of zero, while the slope of a vertical straight line is undefined.

For a curve other than a straight line, the process becomes more difficult, since the rate of change of y with respect to changes in x is no longer constant, but changes as x changes. Therefore, the slope must be evaluated at each point on the curve by finding the appropriate rate of change of y with respect to x. To do this, assume that the point $\boldsymbol{a}$ represents any arbitrary point on the curve. Call the x and y values corresponding to the point $\boldsymbol{a}$,

$$x_0 \quad \text{and} \quad f(x_0),$$

where $y = f(x)$. The rate of change of y with respect to x at the point $\boldsymbol{a}$ is equal to the slope of the line *tangent to the curve at that point*. Differential calculus provides a relatively straightforward method for the derivation of the slope (or equation) of this tangent line. Figure A.3 sets the stage for the development to follow.

To find the slope at $\boldsymbol{a}$, calculus proceeds by finding successively better and better approximations to the slope of the tangent line T, which is the slope of concern. These approximations are represented by the line from $\boldsymbol{a}$ to $\boldsymbol{b}$, where the x-coordinate of the point $\boldsymbol{b}$ is located a small amount, Δx, to the right of x. The slope of the straight line $\boldsymbol{ab}$ equals

$$\frac{\Delta y}{\Delta x} = \frac{f(x_0 + \Delta x) - f(x_0)}{\Delta x},$$

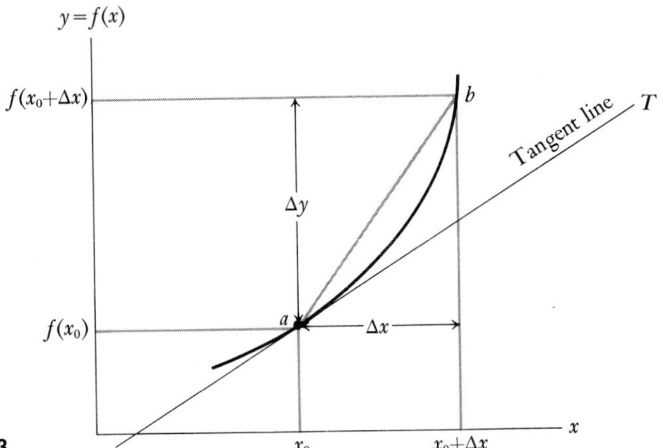

Figure A.3

since $\Delta y = f(x_0 + \Delta x) - f(x_0)$. As Δx is made smaller and smaller, approaching zero ($\Delta x \to 0$), the slope of the line ***ab*** approaches the slope of the tangent line. At the limit, these slopes are equal. Thus

$$\lim_{\Delta x \to 0} \frac{\Delta y}{\Delta x} = \lim_{\Delta x \to 0} \frac{f(x_0 + \Delta x) - f(x_0)}{\Delta x} = \text{Slope at point } \boldsymbol{a}.$$

The usual notation representing slope is dy/dx (or $df(x)/dx$), which is the *first* derivative (rate of change *or* slope) of y with respect to x. Higher derivatives just represent the process of taking the derivative of a derivative, i.e., the rate of change *of the slope*. The second derivative therefore represents the derivative of the slope of a function. Sometimes prime marks are used to designate derivatives: $f'(x)$ thus stands for the first derivative, $f''(x)$ the second derivative, and so forth.

Maximum or minimum points of a function can often be determined by finding the values of x for which the function has a slope of zero. These values can be found by setting the first derivative equal to zero and solving for x. When the first derivative equals zero, the value of the second derivative describes whether the function is a maximum or a minimum. A negative second derivative indicates a relative maximum, while a positive second derivative indicates a relative minimum. If the second derivative equals *zero*, the curve has an inflection point at the value of x.

Partial differentiation is the process of finding the rate of change (slope) between two variables whenever the dependent variable (y) is a function of more than one variable. For partial differentiation, all variables except y and the dependent variable of interest are treated as constants and differentiated accordingly. The symbol $\partial y/\partial x$ [or $\partial f(x)/\partial x$] represents the partial derivative of y with respect to x.

It should be noted that the *letters* used to denote the variables in differentiation or integration problems are not significant—i.e., the relationship could just as easily be $r = f(t)$ with the derivative dr/dt, or $x = f(y)$ and dx/dy, etc. Throughout most of this book, however, we shall assume y to be the dependent variable, and $y = f(x_1, x_2, \ldots, x_n)$, where $x_1, x_2, \ldots, x_n$ are the independent variables.

Differentiation Formulas

There are many rules to aid the process of finding derivatives. We shall present only a few of the most widely used rules, stressing those with particular applications in this book.

1. $y = k; \quad \dfrac{dy}{dx} = 0$

   The derivative of a constant (horizontal straight line) is zero.

2. $y = ax + b; \quad \dfrac{dy}{dx} = a$

   The derivative of any straight line, $y = ax + b$, is a constant.

3. $y = f(x) \pm g(x); \quad \dfrac{dy}{dx} = \dfrac{df(x)}{dx} \pm \dfrac{dg(x)}{dx}$

   The derivative of the sum (difference) of two functions is the sum (difference) of the derivatives.

4. $y = f(x)g(x); \quad \dfrac{dy}{dx} = f(x)\dfrac{dg(x)}{dx} + g(x)\dfrac{df(x)}{dx}$

   The derivative of a product of two functions is the first function times the derivative of the second plus the second function times the derivative of the first.

5. $y = e^{f(x)}; \quad \dfrac{dy}{dx} = e^{f(x)}\dfrac{d[f(x)]}{dx}$

   The derivative of the constant e, to a power which is a function of x, $f(x)$, is $e^{f(x)}$ times the derivative of the exponent.

6. $y = [f(x)]^n; \quad \dfrac{dy}{dx} = n[f(x)]^{n-1}\dfrac{df(x)}{dx}$

   The derivative of a function raised to a power, n, is n times the function raised to the $(n - 1)$ power times the derivative of the function.

7. $y = \log_e x; \quad \dfrac{dy}{dx} = \dfrac{1}{x}$

   The derivative of the natural logarithm of x is $1/x$.

Exercises

A.1. Find $\frac{dy}{dx}$ for each of the following functions:

a) $y = x^3$ [*Ans.*: $3x^2$]

b) $y = 20 + 5x^2$ [*Ans.*: $10x$]

c) $y = 5e^{-2x}$ [*Ans.*: $-10e^{-2x}$]

d) $y = (x^2 - 1)^2$ [*Ans.*: $4x(x^2 - 1)$]

e) $y = 2 \log_e x$ [*Ans.*: $\frac{2}{x}$]

f) $y = x \log_e x$ [*Ans.*: $1 + \log_e x$]

g) $y = x^2 e^{2x^2}$ [*Ans.*: $2xe^{2x^2} + 4x^3 e^{2x^2}$]

A.2. Find $\frac{\partial G}{\partial a}$ and $\frac{\partial G}{\partial b}$ for each of the following functions:

a) $G = 5a^2 + 6bx + 7b^2$ [*Ans.*: $\frac{\partial G}{\partial a} = 10a$

$\frac{\partial G}{\partial b} = 6x + 14b$]

b) $G = \sum(y - a - bx)^2$ [*Ans.*: $\frac{\partial G}{\partial a} = \sum(y - a - bx)(-2)$

$\frac{\partial G}{\partial b} = \sum(y - a - bx)(-2x)$]

A.3. Find the maximum and minimum for the following functions:

a) $y = 2x^3 + 3x^2 - 12x + 2$ [*Ans.*: $x = -2$ is a max

$x = +1$ is a min]

b) $y = (x^2 - 1)^2$ [*Ans.*: $x = 0$ is a max

$x = \pm 1$ are mins]

Integral Calculus

Finding the area under a curve is usually relatively simple when straight lines alone are involved. Nonlinear functions complicate the problem by often requiring the application of the methods of integral calculus for exact solution. As was the case for differential calculus, integration proceeds by successive approximations until, at the limit, the approximating function is identical to the value sought. Figure A.4 shows the graphical representation of an approximation for the area under the curve $y = f(x)$, between a and b, given by the sum of the areas of the 5 rectangles. This sum is equal to

$$\sum_{i=1}^{5} f(x_i)\Delta x,$$

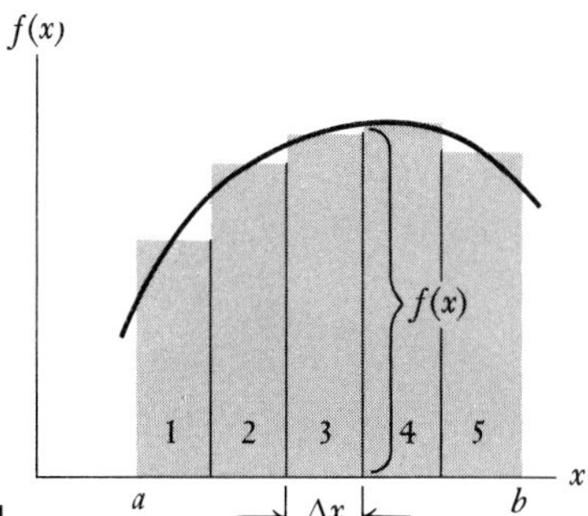

Figure A.4

where Δx is the width of each interval and $f(x_i)$ is the height. Large values of Δx obviously do a poor job of approximating the area in question. As Δx becomes smaller and smaller, approaching zero ($\Delta x \to 0$), the approximation improves until, at the limit, the exact area can be determined. Thus,

$$\lim_{x \to 0} \sum f(x_i)\Delta x = \begin{pmatrix} \text{Area between } a \text{ and } b \\ \text{under the curve } y = f(x) \end{pmatrix},$$

and is written $\int_a^b f(x)\,dx$ (the term dx denotes the independent variable of interest). A fundamental theorem of calculus states that

$$\int_a^b f(x)\,dx = \int_{-\infty}^b f(x)\,dx - \int_{-\infty}^a f(x)\,dx = F(b) - F(a),$$

where $F(a)$ and $F(b)$ are proper integrals evaluated over the interval $-\infty$ to a and b, respectively. That is, if we take the area under the curve from minus infinity up to the point b, and subtract from it the area from minus infinity up to the point a ($a \leq b$), then we are left with the area between the points a and b. The notation $[F(x)]_a^b$ represents this evaluation, namely, $F(b) - F(a)$.

Integration Formulas

1. $\int_a^b dx = \Big[x\Big]_a^b = b - a$

2. $\int_a^b kf(x)\,dx = k\int_a^b f(x)\,dx$

3. $\int_a^b [f(x) \pm g(x)]\,dx = \int_a^b f(x)\,dx \pm \int_a^b g(x)\,dx$

4. $\int_a^b x^n\,dx = \left[\dfrac{x^{n+1}}{n+1}\right]_a^b = \dfrac{b^{n+1}}{n+1} - \dfrac{a^{n+1}}{n+1}$

5. $\int_a^b e^{g(x)}\dfrac{d[g(x)]}{dx}\,dx = \Big[e^{g(x)}\Big]_a^b = e^{g(b)} - e^{g(a)}$

6. $\int_a^b \dfrac{dx}{x} = \Big[\log_e x\Big]_a^b = \log_e b - \log_e a$

7. $\int_a^b f(x)\frac{d[g(x)]}{dx}\,dx = \Big[f(x)g(x)\Big]_a^b - \int_a^b g(x)\frac{df(x)}{dx}\,dx.$

Integrating by use of Formula 7 is called *integration by parts.* Since in Chapter 5 we need to integrate a relatively complex function by parts, we shall illustrate the use of Formula 7 by means of a detailed description of that integration. The first part of the problem is to evaluate the following integral:

$$\int_0^\infty x\lambda e^{-\lambda x}\,dx \qquad \text{for} \qquad \lambda > 0. \tag{A.1}$$

If we let $f(x) = x$ and $dg(x)/dx = \lambda e^{-\lambda x}$, then $df(x)/dx = 1$ and $g(x) = -e^{-\lambda x}$. Using Formula 7,

$$\begin{aligned}\int_0^\infty x\lambda e^{-\lambda x}\,dx &= \Big[(x)(-e^{-\lambda x})\Big]_0^\infty - \int_0^\infty (1)(-e^{-\lambda x})\,dx\\ &= \Big[-xe^{-\lambda x}\Big]_0^\infty - \Big[\frac{1}{\lambda}e^{-\lambda x}\Big]_0^\infty\\ &= (0-0) - \left(0 - \frac{1}{\lambda}\right) = \frac{1}{\lambda}.\end{aligned}$$

The second part of the problem in Chapter 5 is to evaluate

$$\int_0^\infty x^2\lambda e^{-\lambda x}\,dx \qquad \text{for} \quad \lambda > 0.$$

Letting $f(x) = x^2$ and $dg(x)/dx = \lambda e^{-\lambda x}$, then

$$\frac{df(x)}{dx} = 2x \qquad \text{and} \qquad g(x) = -e^{-\lambda x}.$$

Again, applying Formula 7,

$$\int_0^\infty x^2\lambda e^{-\lambda x}\,dx = \Big[-x^2e^{-\lambda x}\Big]_0^\infty - \int_0^\infty -2xe^{-\lambda x}\,dx. \tag{A.2}$$

Note that it will not affect Formula A.2 if we multiply its last term by λ/λ, since λ is a constant. After performing this multiplication and rearranging this last integral, we get:

$$\int_0^\infty x^2\lambda e^{-\lambda x}\,dx = \Big[-x^2e^{-\lambda x}\Big]_0^\infty + \frac{2}{\lambda}\Big[\int_0^\infty x\lambda e^{-\lambda x}\,dx\Big].$$

Now the last term (in brackets) is identical to the integral given in Formula A.1, which we know equals $1/\lambda$. Hence,

$$\begin{aligned}\int_0^\infty x^2\lambda e^{-\lambda x}\,dx &= \Big[-x^2e^{-\lambda x}\Big]_0^\infty + \frac{2}{\lambda}\Big[\frac{1}{\lambda}\Big]\\ &= (0-0) + \frac{2}{\lambda^2} = \frac{2}{\lambda^2}.\end{aligned}$$

Exercises

Integrate the following functions:

A.4. $\int_1^5 4\,dx$ [*Ans.*: 16]

A.5. $\int_0^{10} \frac{x}{2}\,dx$ [*Ans.*: 25]

A.6. $\int_4^{16} \frac{\sqrt{x}}{50}\,dx$ $\left[\textit{Ans.}: \frac{56}{75}\right]$

A.7. $\int_0^{\infty} ke^{-kx}\,dx$ [*Ans.*: 1]

A.8. $\int_2^5 \frac{3}{x}\,dx$ [*Ans.*: $3(\log_e 5 - \log_e 2)$]

A.9. $\int_0^4 (2x^2 + x)\,dx$ [*Ans.*: $50\frac{2}{3}$]

APPENDIX B

TABLES

Table I Binomial Distribution

From Robert Schlaifer, *Analysis of Decisions Under Uncertainty* (New York: McGraw-Hill Book Co., Inc., Preliminary Edition, Volume II, 1967) by specific permission of the President and Fellows of Harvard College, who hold the ocpyright.

The following table gives values of the binomial mass function defined by

$$P(x) = {}_nC_x p^x(1 - p)^{n-x}$$
$$= \frac{n!}{x!(n - x)!} p^x(1 - p)^{n-x}.$$

This is the probability of exactly x successes in n independent Bernoulli trials with probability of success on a single trial equal to p. The values of x at the left of any section are to be used in conjunction with the values of p at the top of that section; the values of x at the right of any section are to be used in conjunction with the values of p at the bottom of that section.

Example: To evaluate $P(x)$ for $n = 5$, $x = 3$, and $p = 0.83$, locate the section of the table for $n = 5$, the column for $p = 0.83$, and the row for $x = 3$, and read

$$P(x) = 0.1652.$$

$n = 1$

x	p	01	02	03	04	05	06	07	08	09	10		
0		9900	9800	9700	9600	9500	9400	9300	9200	9100	9000		1
1		0100	0200	0300	0400	0500	0600	0700	0800	0900	1000		0
		99	98	97	96	95	94	93	92	91	90	p	x

x	p	11	12	13	14	15	16	17	18	19	20		
0		8900	8800	8700	8600	8500	8400	8300	8200	8100	8000		1
1		1100	1200	1300	1400	1500	1600	1700	1800	1900	2000		0
		89	88	87	86	85	84	83	82	81	80	p	x

x	p	21	22	23	24	25	26	27	28	29	30		
0		7900	7800	7700	7600	7500	7400	7300	7200	7100	7000		1
1		2100	2200	2300	2400	2500	2600	2700	2800	2900	3000		0
		79	78	77	76	75	74	73	72	71	70	p	x

x	p	31	32	33	34	35	36	37	38	39	40		
0		6900	6800	6700	6600	6500	6400	6300	6200	6100	6000		1
1		3100	3200	3300	3400	3500	3600	3700	3800	3900	4000		0
		69	68	67	66	65	64	63	62	61	60	p	x

x	p	41	42	43	44	45	46	47	48	49	50		
0		5900	5800	5700	5600	5500	5400	5300	5200	5100	5000		1
1		4100	4200	4300	4400	4500	4600	4700	4800	4900	5000		0
		59	58	57	56	55	54	53	52	51	50	p	x

(Continued)

Table I Binomial Distribution ($n = 2$)

$n = 2$

x	p												
x	p	01	02	03	04	05	06	07	08	09	10		
0		9801	9604	9409	9216	9025	8836	8649	8464	8281	8100		2
1		0198	0392	0582	0768	0950	1128	1302	1472	1638	1800		1
2		0001	0004	0009	0016	0025	0036	0049	0064	0081	0100		0
		99	98	97	96	95	94	93	92	91	90	p	x
x	p	11	12	13	14	15	16	17	18	19	20		
0		7921	7744	7569	7396	7225	7056	6889	6724	6561	6400		2
1		1958	2112	2262	2408	2550	2688	2822	2952	3078	3200		1
2		0121	0144	0169	0196	0225	0256	0289	0324	0361	0400		0
		89	88	87	86	85	84	83	82	81	80	p	x
x	p	21	22	23	24	25	26	27	28	29	30		
0		6241	6084	5929	5776	5625	5476	5329	5184	5041	4900		2
1		3318	3432	3542	3648	3750	3848	3942	4032	4118	4200		1
2		0441	0484	0529	0576	0625	0676	0729	0784	0841	0900		0
		79	78	77	76	75	74	73	72	71	70	p	x
x	p	31	32	33	34	35	36	37	38	39	40		
0		4761	4624	4489	4356	4225	4096	3969	3844	3721	3600		2
1		4278	4352	4422	4488	4550	4608	4662	4712	4758	4800		1
2		0961	1024	1089	1156	1225	1296	1369	1444	1521	1600		0
		69	68	67	66	65	64	63	62	61	60	p	x
x	p	41	42	43	44	45	46	47	48	49	50		
0		3481	3364	3249	3136	3025	2916	2809	2704	2601	2500		2
1		4838	4872	4902	4928	4950	4968	4982	4992	4998	5000		1
2		1681	1764	1849	1936	2025	2116	2209	2304	2401	2500		0
		59	58	57	56	55	54	53	52	51	50	p	x

(*Continued*)

Table I Binomial Distribution ($n = 3$)

$n = 3$

x	p	01	02	03	04	05	06	07	08	09	10		
0		9703	9412	9127	8847	8574	8306	8044	7787	7536	7290		3
1		0294	0576	0847	1106	1354	1590	1816	2031	2236	2430		2
2		0003	0012	0026	0046	0071	0102	0137	0177	0221	0270		1
3		0000	0000	0000	0001	0001	0002	0003	0005	0007	0010		0
		99	98	97	96	95	94	93	92	91	90	p	x

x	p	11	12	13	14	15	16	17	18	19	20		
0		7050	6815	6585	6361	6141	5927	5718	5514	5314	5120		3
1		2614	2788	2952	3106	3251	3387	3513	3631	3740	3840		2
2		0323	0380	0441	0506	0574	0645	0720	0797	0877	0960		1
3		0013	0017	0022	0027	0034	0041	0049	0058	0069	0080		0
		89	88	87	86	85	84	83	82	81	80	p	x

x	p	21	22	23	24	25	26	27	28	29	30		
0		4930	4746	4565	4390	4219	4052	3890	3732	3579	3430		3
1		3932	4014	4091	4159	4219	4271	4316	4355	4386	4410		2
2		1045	1133	1222	1313	1406	1501	1597	1693	1791	1890		1
3		0093	0106	0122	0138	0156	0176	0197	0220	0244	0270		0
		79	78	77	76	75	74	73	72	71	70	p	x

x	p	31	32	33	34	35	36	37	38	39	40		
0		3285	3144	3008	2875	2746	2621	2500	2383	2270	2160		3
1		4428	4439	4444	4443	4436	4424	4406	4382	4354	4320		2
2		1989	2089	2189	2289	2389	2488	2587	2686	2783	2880		1
3		0298	0328	0359	0393	0429	0467	0507	0549	0593	0640		0
		69	68	67	66	65	64	63	62	61	60	p	x

x	p	41	42	43	44	45	46	47	48	49	50		
0		2054	1951	1852	1756	1664	1575	1489	1406	1327	1250		3
1		4282	4239	4191	4140	4084	4024	3961	3894	3823	3750		2
2		2975	3069	3162	3252	3341	3428	3512	3594	3674	3750		1
3		0689	0741	0795	0852	0911	0973	1038	1106	1176	1250		0
		59	58	57	56	55	54	53	52	51	50	p	x

(Continued)

Table I Binomial Distribution ($n = 4$)

$n = 4$

x	p	01	02	03	04	05	06	07	08	09	10		
0		9606	9224	8853	8493	8145	7807	7481	7164	6857	6561		4
1		0388	0753	1095	1416	1715	1993	2252	2492	2713	2916		3
2		0006	0023	0051	0088	0135	0191	0254	0325	0402	0486		2
3		0000	0000	0001	0002	0005	0008	0013	0019	0027	0036		1
4		0000	0000	0000	0000	0000	0000	0000	0000	0001	0001		0
		99	98	97	96	95	94	93	92	91	90	p	x

x	p	11	12	13	14	15	16	17	18	19	20		
0		6274	5997	5729	5470	5220	4979	4746	4521	4305	4096		4
1		3102	3271	3424	3562	3685	3793	3888	3970	4039	4096		3
2		0575	0669	0767	0870	0975	1084	1195	1307	1421	1536		2
3		0047	0061	0076	0094	0115	0138	0163	0191	0222	0256		1
4		0001	0002	0003	0004	0005	0007	0008	0010	0013	0016		0
		89	88	87	86	85	84	83	82	81	80	p	x

x	p	21	22	23	24	25	26	27	28	29	30		
0		3895	3702	3515	3336	3164	2999	2840	2687	2541	2401		4
1		4142	4176	4200	4214	4219	4214	4201	4180	4152	4116		3
2		1651	1767	1882	1996	2109	2221	2331	2439	2544	2646		2
3		0293	0332	0375	0420	0469	0520	0575	0632	0693	0756		1
4		0019	0023	0028	0033	0039	0046	0053	0061	0071	0081		0
		79	78	77	76	75	74	73	72	71	70	p	x

x	p	31	32	33	34	35	36	37	38	39	40		
0		2267	2138	2015	1897	1785	1678	1575	1478	1385	1296		4
1		4074	4025	3970	3910	3845	3775	3701	3623	3541	3456		3
2		2745	2841	2933	3021	3105	3185	3260	3330	3396	3456		2
3		0822	0891	0963	1038	1115	1194	1276	1361	1447	1536		1
4		0092	0105	0119	0134	0150	0168	0187	0209	0231	0256		0
		69	68	67	66	65	64	63	62	61	60	p	x

x	p	41	42	43	44	45	46	47	48	49	50		
0		1212	1132	1056	0983	0915	0850	0789	0731	0677	0625		4
1		3368	3278	3185	3091	2995	2897	2799	2700	2600	2500		3
2		3511	3560	3604	3643	3675	3702	3723	3738	3747	3750		2
3		1627	1719	1813	1908	2005	2102	2201	2300	2400	2500		1
4		0283	0311	0342	0375	0410	0448	0488	0531	0576	0625		0
		59	58	57	56	55	54	53	52	51	50	p	x

(Continued)

Table I Binomial Distribution (n = 5)

$n = 5$

x	p	01	02	03	04	05	06	07	08	09	10		
0		9510	9039	8587	8154	7738	7339	6957	6591	6240	5905		5
1		0480	0922	1328	1699	2036	2342	2618	2866	3086	3280		4
2		0010	0038	0082	0142	0214	0299	0394	0498	0610	0729		3
3		0000	0001	0003	0006	0011	0019	0030	0043	0060	0081		2
4		0000	0000	0000	0000	0000	0001	0001	0002	0003	0004		1
		99	98	97	96	95	94	93	92	91	90	p	x

x	p	11	12	13	14	15	16	17	18	19	20		
0		5584	5277	4984	4704	4437	4182	3939	3707	3487	3277		5
1		3451	3598	3724	3829	3915	3983	4034	4069	4089	4096		4
2		0853	0981	1113	1247	1382	1517	1652	1786	1919	2048		3
3		0105	0134	0166	0203	0244	0289	0338	0392	0450	0512		2
4		0007	0009	0012	0017	0022	0028	0035	0043	0053	0064		1
5		0000	0000	0000	0001	0001	0001	0001	0002	0002	0003		0
		89	88	87	86	85	84	83	82	81	80	p	x

x	p	21	22	23	24	25	26	27	28	29	30		
0		3077	2887	2707	2536	2373	2219	2073	1935	1804	1681		5
1		4090	4072	4043	4003	3955	3898	3834	3762	3685	3601		4
2		2174	2297	2415	2529	2637	2739	2836	2926	3010	3087		3
3		0578	0648	0721	0798	0879	0962	1049	1138	1229	1323		2
4		0077	0091	0108	0126	0146	0169	0194	0221	0251	0283		1
5		0004	0005	0006	0008	0010	0012	0014	0017	0021	0024		0
		79	78	77	76	75	74	73	72	71	70	p	x

x	p	31	32	33	34	35	36	37	38	39	40		
0		1564	1454	1350	1252	1160	1074	0992	0916	0845	0778		5
1		3513	3421	3325	3226	3124	3020	2914	2808	2700	2592		4
2		3157	3220	3275	3323	3364	3397	3423	3441	3452	3456		3
3		1418	1515	1613	1712	1811	1911	2010	2109	2207	2304		2
4		0319	0357	0397	0441	0488	0537	0590	0646	0706	0768		1
5		0029	0034	0039	0045	0053	0060	0069	0079	0090	0102		0
		69	68	67	66	65	64	63	62	61	60	p	x

x	p	41	42	43	44	45	46	47	48	49	50		
0		0715	0656	0602	0551	0503	0459	0418	0380	0345	0313		5
1		2484	2376	2270	2164	2059	1956	1854	1755	1657	1562		4
2		3452	3442	3424	3400	3369	3332	3289	3240	3185	3125		3
3		2399	2492	2583	2671	2757	2838	2916	2990	3060	3125		2
4		0834	0902	0974	1049	1128	1209	1293	1380	1470	1562		1
5		0116	0131	0147	0165	0185	0206	0229	0255	0282	0312		0
		59	58	57	56	55	54	53	52	51	50	p	x

(*Continued*)

Table I Binomial Distribution (n = 6)

$n = 6$

x	p	01	02	03	04	05	06	07	08	09	10		
0		9415	8858	8330	7828	7351	6899	6470	6064	5679	5314		6
1		0571	1085	1546	1957	2321	2642	2922	3164	3370	3543		5
2		0014	0055	0120	0204	0305	0422	0550	0688	0833	0984		4
3		0000	0002	0005	0011	0021	0036	0055	0080	0110	0146		3
4		0000	0000	0000	0000	0001	0002	0003	0005	0008	0012		2
5		0000	0000	0000	0000	0000	0000	0000	0000	0000	0001		1
		99	98	97	96	95	94	93	92	91	90	p	x

x	p	11	12	13	14	15	16	17	18	19	20		
0		4970	4644	4336	4046	3771	3513	3269	3040	2824	2621		6
1		3685	3800	3888	3952	3993	4015	4018	4004	3975	3932		5
2		1139	1295	1452	1608	1762	1912	2057	2197	2331	2458		4
3		0188	0236	0289	0349	0415	0486	0562	0643	0729	0819		3
4		0017	0024	0032	0043	0055	0069	0086	0106	0128	0154		2
5		0001	0001	0002	0003	0004	0005	0007	0009	0012	0015		1
6		0000	0000	0000	0000	0000	0000	0000	0000	0000	0001		0
		89	88	87	86	85	84	83	82	81	80	p	x

x	p	21	22	23	24	25	26	27	28	29	30		
0		2431	2252	2084	1927	1780	1642	1513	1393	1281	1176		6
1		3877	3811	3735	3651	3560	3462	3358	3251	3139	3025		5
2		2577	2687	2789	2882	2966	3041	3105	3160	3206	3241		4
3		0913	1011	1111	1214	1318	1424	1531	1639	1746	1852		3
4		0182	0214	0249	0287	0330	0375	0425	0478	0535	0595		2
5		0019	0024	0030	0036	0044	0053	0063	0074	0087	0102		1
6		0001	0001	0001	0002	0002	0003	0004	0005	0006	0007		0
		79	78	77	76	75	74	73	72	71	70	p	x

x	p	31	32	33	34	35	36	37	38	39	40		
0		1079	0989	0905	0827	0754	0687	0625	0568	0515	0467		6
1		2909	2792	2673	2555	2437	2319	2203	2089	1976	1866		5
2		3267	3284	3292	3290	3280	3261	3235	3201	3159	3110		4
3		1957	2061	2162	2260	2355	2446	2533	2616	2693	2765		3
4		0660	0727	0799	0873	0951	1032	1116	1202	1291	1382		2
5		0119	0137	0157	0180	0205	0232	0262	0295	0330	0369		1
6		0009	0011	0013	0015	0018	0022	0026	0030	0035	0041		0
		69	68	67	66	65	64	63	62	61	60	p	x

x	p	41	42	43	44	45	46	47	48	49	50		
0		0422	0381	0343	0308	0277	0248	0222	0198	0176	0156		6
1		1759	1654	1552	1454	1359	1267	1179	1095	1014	0937		5
2		3055	2994	2928	2856	2780	2699	2615	2527	2436	2344		4
3		2831	2891	2945	2992	3032	3065	3091	3110	3121	3125		3
4		1475	1570	1666	1763	1861	1958	2056	2153	2249	2344		2
5		0410	0455	0503	0554	0609	0667	0729	0795	0864	0937		1
6		0048	0055	0063	0073	0083	0095	0108	0122	0138	0156		0
		59	58	57	56	55	54	53	52	51	50	p	x

(Continued)

Table I Binomial Distribution (n = 7)

$n = 7$

x	p	01	02	03	04	05	06	07	08	09	10		
0		9321	8681	8080	7514	6983	6485	6017	5578	5168	4783		7
1		0659	1240	1749	2192	2573	2897	3170	3396	3578	3720		6
2		0020	0076	0162	0274	0406	0555	0716	0886	1061	1240		5
3		0000	0003	0008	0019	0036	0059	0090	0128	0175	0230		4
4		0000	0000	0000	0001	0002	0004	0007	0011	0017	0026		3
5		0000	0000	0000	0000	0000	0000	0000	0001	0001	0002		2
		99	98	97	96	95	94	93	92	91	90	p	x

x	p	11	12	13	14	15	16	17	18	19	20		
0		4423	4087	3773	3479	3206	2951	2714	2493	2288	2097		7
1		3827	3901	3946	3965	3960	3935	3891	3830	3756	3670		6
2		1419	1596	1769	1936	2097	2248	2391	2523	2643	2753		5
3		0292	0363	0441	0525	0617	0714	0816	0923	1033	1147		4
4		0036	0049	0066	0086	0109	0136	0167	0203	0242	0287		3
5		0003	0004	0006	0008	0012	0016	0021	0027	0034	0043		2
6		0000	0000	0000	0000	0001	0001	0001	0002	0003	0004		1
		89	88	87	86	85	84	83	82	81	80	p	x

x	p	21	22	23	24	25	26	27	28	29	30		
0		1920	1757	1605	1465	1335	1215	1105	1003	0910	0824		7
1		3573	3468	3356	3237	3115	2989	2860	2731	2600	2471		6
2		2850	2935	3007	3067	3115	3150	3174	3186	3186	3177		5
3		1263	1379	1497	1614	1730	1845	1956	2065	2169	2269		4
4		0336	0389	0447	0510	0577	0648	0724	0803	0886	0972		3
5		0054	0066	0080	0097	0115	0137	0161	0187	0217	0250		2
6		0005	0006	0008	0010	0013	0016	0020	0024	0030	0036		1
7		0000	0000	0000	0000	0001	0001	0001	0001	0002	0002		0
		79	78	77	76	75	74	73	72	71	70	p	x

x	p	31	32	33	34	35	36	37	38	39	40		
0		0745	0672	0606	0546	0490	0440	0394	0352	0314	0280		7
1		2342	2215	2090	1967	1848	1732	1619	1511	1407	1306		6
2		3156	3127	3088	3040	2985	2922	2853	2778	2698	2613		5
3		2363	2452	2535	2610	2679	2740	2793	2838	2875	2903		4
4		1062	1154	1248	1345	1442	1541	1640	1739	1838	1935		3
5		0286	0326	0369	0416	0466	0520	0578	0640	0705	0774		2
6		0043	0051	0061	0071	0084	0098	0113	0131	0150	0172		1
7		0003	0003	0004	0005	0006	0008	0009	0011	0014	0016		0
		69	68	67	66	65	64	63	62	61	60	p	x

x	p	41	42	43	44	45	46	47	48	49	50		
0		0249	0221	0195	0173	0152	0134	0117	0103	0090	0078		7
1		1211	1119	1032	0950	0872	0798	0729	0664	0604	0547		6
2		2524	2431	2336	2239	2140	2040	1940	1840	1740	1641		5
3		2923	2934	2937	2932	2918	2897	2867	2830	2786	2734		4
4		2031	2125	2216	2304	2388	2468	2543	2612	2676	2734		3
5		0847	0923	1003	1086	1172	1261	1353	1447	1543	1641		2
6		0196	0223	0252	0284	0320	0358	0400	0445	0494	0547		1
7		0019	0023	0027	0032	0037	0044	0051	0059	0068	0078		0
		59	58	57	56	55	54	53	52	51	50	p	x

(*Continued*)

Table I Binomial Distribution ($n = 8$)

$n = 8$

x	p 01	02	03	04	05	06	07	08	09	10	
0	9227	8508	7837	7214	6634	6096	5596	5132	4703	4305	8
1	0746	1389	1939	2405	2793	3113	3370	3570	3721	3826	7
2	0026	0099	0210	0351	0515	0695	0888	1087	1288	1488	6
3	0001	0004	0013	0029	0054	0089	0134	0189	0255	0331	5
4	0000	0000	0001	0002	0004	0007	0013	0021	0031	0046	4
5	0000	0000	0000	0000	0000	0000	0001	0001	0002	0004	3
	99	98	97	96	95	94	93	92	91	90 p	x

x	p 11	12	13	14	15	16	17	18	19	20	
0	3937	3596	3282	2992	2725	2479	2252	2044	1853	1678	8
1	3892	3923	3923	3897	3847	3777	3691	3590	3477	3355	7
2	1684	1872	2052	2220	2376	2518	2646	2758	2855	2936	6
3	0416	0511	0613	0723	0839	0959	1084	1211	1339	1468	5
4	0064	0087	0115	0147	0185	0228	0277	0332	0393	0459	4
5	0006	0009	0014	0019	0026	0035	0045	0058	0074	0092	3
6	0000	0001	0001	0002	0002	0003	0005	0006	0009	0011	2
7	0000	0000	0000	0000	0000	0000	0000	0000	0001	0001	1
	89	88	87	86	85	84	83	82	81	80 p	x

x	p 21	22	23	24	25	26	27	28	29	30	
0	1517	1370	1236	1113	1001	0899	0806	0722	0646	0576	8
1	3226	3092	2953	2812	2670	2527	2386	2247	2110	1977	7
2	3002	3052	3087	3108	3115	3108	3089	3058	3017	2965	6
3	1596	1722	1844	1963	2076	2184	2285	2379	2464	2541	5
4	0530	0607	0689	0775	0865	0959	1056	1156	1258	1361	4
5	0113	0137	0165	0196	0231	0270	0313	0360	0411	0467	3
6	0015	0019	0025	0031	0038	0047	0058	0070	0084	0100	2
7	0001	0002	0002	0003	0004	0005	0006	0008	0010	0012	1
8	0000	0000	0000	0000	0000	0000	0000	0000	0001	0001	0
	79	78	77	76	75	74	73	72	71	70 p	x

x	p 31	32	33	34	35	36	37	38	39	40	
0	0514	0457	0406	0360	0319	0281	0248	0218	0192	0168	8
1	1847	1721	1600	1484	1373	1267	1166	1071	0981	0896	7
2	2904	2835	2758	2675	2587	2494	2397	2297	2194	2090	6
3	2609	2668	2717	2756	2786	2805	2815	2815	2806	2787	5
4	1465	1569	1673	1775	1875	1973	2067	2157	2242	2322	4
5	0527	0591	0659	0732	0808	0888	0971	1058	1147	1239	3
6	0118	0139	0162	0188	0217	0250	0285	0324	0367	0413	2
7	0015	0019	0023	0028	0033	0040	0048	0057	0067	0079	1
8	0001	0001	0001	0002	0002	0003	0004	0004	0005	0007	0
	69	68	67	66	65	64	63	62	61	60 p	x

x	p 41	42	43	44	45	46	47	48	49	50	
0	0147	0128	0111	0097	0084	0072	0062	0053	0046	0039	8
1	0816	0742	0672	0608	0548	0493	0442	0395	0352	0312	7
2	1985	1880	1776	1672	1569	1469	1371	1275	1183	1094	6
3	2759	2723	2679	2627	2568	2503	2431	2355	2273	2187	5
4	2397	2465	2526	2580	2627	2665	2695	2717	2730	2734	4
5	1332	1428	1525	1622	1719	1816	1912	2006	2098	2187	3
6	0463	0517	0575	0637	0703	0774	0848	0926	1008	1094	2
7	0092	0107	0124	0143	0164	0188	0215	0244	0277	0312	1
8	0008	0010	0012	0014	0017	0020	0024	0028	0033	0039	0
	59	58	57	56	55	54	53	52	51	50 p	x

(*Continued*)

Table I Binomial Distribution (n = 9)

$n = 9$

x	p 01	02	03	04	05	06	07	08	09	10		
0	9135	8337	7602	6925	6302	5730	5204	4722	4279	3874		9
1	0830	1531	2116	2597	2985	3292	3525	3695	3809	3874		8
2	0034	0125	0262	0433	0629	0840	1061	1285	1507	1722		7
3	0001	0006	0019	0042	0077	0125	0186	0261	0348	0446		6
4	0000	0000	0001	0003	0006	0012	0021	0034	0052	0074		5
5	0000	0000	0000	0000	0000	0001	0002	0003	0005	0008		4
6	0000	0000	0000	0000	0000	0000	0000	0000	0000	0001		3
	99	98	97	96	95	94	93	92	91	90	p	x

x	p 11	12	13	14	15	16	17	18	19	20		
0	3504	3165	2855	2573	2316	2082	1869	1676	1501	1342		9
1	3897	3884	3840	3770	3679	3569	3446	3312	2169	3020		8
2	1927	2119	2295	2455	2597	2720	2823	2908	2973	3020		7
3	0556	0674	0800	0933	1069	1209	1349	1489	1627	1762		6
4	0103	0138	0179	0228	0283	0345	0415	0490	0573	0661		5
5	0013	0019	0027	0037	0050	0066	0085	0108	0134	0165		4
6	0001	0002	0003	0004	0006	0008	0012	0016	0021	0028		3
7	0000	0000	0000	0000	0000	0001	0001	0001	0002	0003		2
	89	88	87	86	85	84	83	82	81	80	p	x

x	p 21	22	23	24	25	26	27	28	29	30		
0	1199	1069	0952	0846	0751	0665	0589	0520	0458	0404		9
1	2867	2713	2558	2404	2253	2104	1960	1820	1685	1556		8
2	3049	3061	3056	3037	3003	2957	2899	2831	2754	2668		7
3	1891	2014	2130	2238	2336	2424	2502	2569	2624	2668		6
4	0754	0852	0954	1060	1168	1278	1388	1499	1608	1715		5
5	0200	0240	0285	0335	0389	0449	0513	0583	0657	0735		4
6	0036	0045	0057	0070	0087	0105	0127	0151	0179	0210		3
7	0004	0005	0007	0010	0012	0016	0020	0025	0031	0039		2
8	0000	0000	0001	0001	0001	0001	0002	0002	0003	0004		1
	79	78	77	76	75	74	73	72	71	70	p	x

x	p 31	32	33	34	35	36	37	38	39	40		
0	0355	0311	0272	0238	0207	0180	0156	0135	0117	0101		9
1	1433	1317	1206	1102	1004	0912	0826	0747	0673	0605		8
2	2576	2478	2376	2270	2162	2052	1941	1831	1721	1612		7
3	2701	2721	2731	2729	2716	2693	2660	2618	2567	2508		6
4	1820	1921	2017	2109	2194	2272	2344	2407	2462	2508		5
5	0818	0904	0994	1086	1181	1278	1376	1475	1574	1672		4
6	0245	0284	0326	0373	0424	0479	0539	0603	0671	0743		3
7	0047	0057	0069	0082	0098	0116	0136	0158	0184	0212		2
8	0005	0007	0008	0011	0013	0016	0020	0024	0029	0035		1
9	0000	0000	0000	0001	0001	0001	0001	0002	0002	0003		0
	69	68	67	66	65	64	63	62	61	60	p	x

x	p 41	42	43	44	45	46	47	48	49	50		
0	0087	0074	0064	0054	0046	0039	0033	0028	0023	0020		9
1	0542	0484	0431	0383	0339	0299	0263	0231	0202	0176		8
2	1506	1402	1301	1204	1110	1020	0934	0853	0776	0703		7
3	2442	2369	2291	2207	2119	2027	1933	1837	1739	1641		6
4	2545	2573	2592	2601	2600	2590	2571	2543	2506	2461		5
5	1769	1863	1955	2044	2128	2207	2280	2347	2408	2461		4
6	0819	0900	0983	1070	1160	1253	1348	1445	1542	1641		3
7	0244	0279	0318	0360	0407	0458	0512	0571	0635	0703		2
8	0042	0051	0060	0071	0083	0097	0014	0132	0153	0176		1
9	0003	0004	0005	0006	0008	0009	0011	0014	0016	0020		0
	59	58	57	56	55	54	53	52	51	50	p	x

(*Continued*)

Table I Binomial Distribution ($n = 10$)

$n = 10$

x	p	01	02	03	04	05	06	07	08	09	10		
0		9044	8171	7374	6648	5987	5386	4840	4344	3894	3487		10
1		0914	1667	2281	2770	3151	3438	3643	3777	3851	3874		9
2		0042	0153	0317	0519	0746	0988	1234	1478	1714	1937		8
3		0001	0008	0026	0058	0105	0168	0248	0343	0452	0574		7
4		0000	0000	0001	0004	0010	0019	0033	0052	0078	0112		6
5		0000	0000	0000	0000	0001	0001	0003	0005	0009	0015		5
6		0000	0000	0000	0000	0000	0000	0000	0000	0001	0001		4
		99	98	97	96	95	94	93	92	91	90	p	x

x	p	11	12	13	14	15	16	17	18	19	20		
0		3118	2785	2484	2213	1969	1749	1552	1374	1216	1074		10
1		3854	3798	3712	3603	3474	3331	3178	3017	2852	2684		9
2		2143	2330	2496	2639	2759	2856	2929	2980	3010	3020		8
3		0706	0847	0995	1146	1298	1450	1600	1745	1883	2013		7
4		0153	0202	0260	0326	0401	0483	0573	0670	0773	0881		6
5		0023	0033	0047	0064	0085	0111	0141	0177	0218	0264		5
6		0002	0004	0006	0009	0012	0018	0024	0032	0043	0055		4
7		0000	0000	0000	0001	0001	0002	0003	0004	0006	0008		3
8		0000	0000	0000	0000	0000	0000	0000	0000	0001	0001		2
		89	88	87	86	85	84	83	82	81	80	p	x

x	p	21	22	23	24	25	26	27	28	29	30		
0		0947	0834	0733	0643	0563	0492	0430	0374	0326	0282		10
1		2517	2351	2188	2030	1877	1730	1590	1456	1330	1211		9
2		3011	2984	2942	2885	2816	2735	2646	2548	2444	2335		8
3		2134	2244	2343	2429	2503	2563	2609	2642	2662	2668		7
4		0993	1108	1225	1343	1460	1576	1689	1798	1903	2001		6
5		0317	0375	0439	0509	0584	0664	0750	0839	0933	1029		5
6		0070	0088	0109	0134	0162	0195	0231	0272	0317	0368		4
7		0011	0014	0019	0024	0031	0039	0049	0060	0074	0090		3
8		0001	0002	0002	0003	0004	0005	0007	0009	0011	0014		2
9		0000	0000	0000	0000	0000	0000	0001	0001	0001	0001		1
		79	78	77	76	75	74	73	72	71	70	p	x

x	p	31	32	33	34	35	36	37	38	39	40		
0		0245	0211	0182	0157	0135	0115	0098	0084	0071	0060		10
1		1099	0995	0898	0808	0725	0649	0578	0514	0456	0430		9
2		2222	2107	1990	1873	1757	1642	1529	1419	1312	1209		8
3		2662	2644	2614	2573	2522	2462	2394	2319	2237	2150		7
4		2093	2177	2253	2320	2377	2424	2461	2487	2503	2508		6
5		1128	1229	1332	1434	1536	1636	1734	1829	1920	2007		5
6		0422	0482	0547	0616	0689	0767	0849	0934	1023	1115		4
7		0108	0130	0154	0181	0212	0247	0285	0327	0374	0425		3
8		0018	0023	0028	0035	0043	0052	0063	0075	0090	0106		2
9		0002	0002	0003	0004	0005	0006	0008	0010	0013	0016		1
10		0000	0000	0000	0000	0000	0000	0000	0001	0001	0001		0
		69	68	67	66	65	64	63	62	61	60	p	x

x	p	41	42	43	44	45	46	47	48	49	50		
0		0051	0043	0036	0030	0025	0021	0017	0014	0012	0010		10
1		0355	0312	0273	0238	0207	0180	0155	0133	0114	0098		9
2		1111	1017	0927	0843	0763	0688	0619	0554	0494	0439		8
3		2058	1963	1865	1765	1665	1654	1464	1364	1267	1172		7
4		2503	2488	2462	2427	2384	2331	2271	2204	2130	2051		6
5		2087	2162	2229	2289	2340	2383	2417	2441	2456	2461		5
6		1209	1304	1401	1499	1596	1692	1786	1878	1966	2051		4
7		0480	0540	0604	0673	0746	0824	0905	0991	1080	1172		3
8		0125	0147	0171	0198	0229	0263	0301	0343	0389	0439		2
9		0019	0024	0029	0035	0042	0050	0059	0070	0083	0098		1
10		0001	0002	0002	0003	0003	0004	0005	0006	0008	0010		0
		59	58	57	56	55	54	53	52	51	50	p	x

(Continued)

Table I Binomial Distribution ($n = 20$)

$n = 20$

x	p 01	02	03	04	05	06	07	08	09	10	
0	8179	6676	5438	4420	3585	2901	2342	1887	1516	1216	20
1	1652	2725	3364	3683	3774	3703	3526	3282	3000	2702	19
2	0159	0528	0988	1458	1887	2246	2521	2711	2828	2852	18
3	0010	0065	0183	0364	0596	0860	1139	1414	1672	1901	17
4	0000	0006	0024	0065	0133	0233	0364	0523	0703	0898	16
5	0000	0000	0002	0009	0022	0048	0088	0145	0222	0319	15
6	0000	0000	0000	0001	0003	0008	0017	0032	0055	0089	14
7	0000	0000	0000	0000	0000	0001	0002	0005	0011	0020	13
8	0000	0000	0000	0000	0000	0000	0000	0001	0002	0004	12
9	0000	0000	0000	0000	0000	0000	0000	0000	0000	0001	11
	99	98	97	96	95	94	93	92	91	90 p	x

x	p 11	12	13	14	15	16	17	18	19	20	
0	0972	0776	0617	0490	0388	0306	0241	0189	0148	0115	20
1	2403	2115	1844	1595	1368	1165	0986	0829	0693	0576	19
2	2822	2740	2618	2466	2293	2109	1919	1730	1545	1369	18
3	2093	2242	2347	2409	2428	2410	2358	2278	2175	2054	17
4	1099	1299	1491	1666	1821	1951	2053	2125	2168	2182	16
5	0435	0567	0713	0868	1028	1189	1345	1493	1627	1746	15
6	0134	0193	0266	0353	0454	0566	0689	0819	0954	1091	14
7	0033	0053	0080	0115	0160	0216	0282	0360	0448	0545	13
8	0007	0012	0019	0030	0046	0067	0094	0128	0171	0222	12
9	0001	0002	0004	0007	0011	0017	0026	0038	0053	0074	11
10	0000	0000	0001	0001	0002	0004	0006	0009	0014	0020	10
11	0000	0000	0000	0000	0000	0001	0001	0002	0003	0005	9
12	0000	0000	0000	0000	0000	0000	0000	0000	0001	0001	8
	89	88	87	86	85	84	83	82	81	80 p	x

x	p 21	22	23	24	25	26	27	28	29	30	
0	0090	0069	0054	0041	0032	0024	0016	0014	0011	0008	20
1	0477	0392	0321	0261	0211	0170	0137	0109	0087	0068	19
2	1204	1050	0910	0783	0669	0569	0480	0403	0336	0278	18
3	1920	1777	1631	1484	1339	1199	1065	0940	0823	0716	17
4	2169	2131	2070	1991	1897	1790	1675	1553	1429	1304	16
5	1845	1923	1979	2012	2023	2013	1982	1933	1868	1789	15
6	1226	1356	1478	1589	1686	1768	1833	1879	1907	1916	14
7	0652	0765	0883	1003	1124	1242	1356	1462	1558	1643	13
8	0282	0351	0429	0515	0609	0709	0815	0924	1034	1144	12
9	0100	0132	0171	0217	0271	0332	0402	0479	0563	0654	11
10	0029	0041	0056	0075	0099	0128	0163	0205	0253	0308	10
11	0007	0010	0015	0022	0030	0041	0055	0072	0094	0120	9
12	0001	0002	0003	0005	0008	0011	0015	0021	0029	0039	8
13	0000	0000	0001	0001	0002	0002	0003	0005	0007	0010	7
14	0000	0000	0000	0000	0000	0000	0010	0010	0001	0002	6
	79	78	77	76	75	74	73	72	71	70 p	x

(*Continued*)

Table I Binomial Distribution (n = 20, cont.)

$n = 20$

x \ p	31	32	33	34	35	36	37	38	39	40	
0	0006	0004	0003	0002	0002	0001	0001	0001	0001	0000	20
1	0054	0042	0033	0025	0020	0015	0011	0009	0007	0005	19
2	0229	0188	0153	0124	0100	0080	0064	0050	0040	0031	18
3	0619	0531	0453	0383	0323	0270	0224	0185	0152	0123	17
4	1181	1062	0947	0839	0738	0645	0559	0482	0412	0350	16
5	1698	1599	1493	1384	1272	1161	1051	0945	0843	0746	15
6	1907	1881	1839	1782	1712	1632	1543	1447	1347	1244	14
7	1714	1770	1811	1836	1844	1836	1812	1774	1722	1659	13
8	1251	1354	1450	1537	1614	1678	1730	1767	1790	1797	12
9	0750	0849	0952	1056	1158	1259	1354	1444	1526	1597	11
10	0370	0440	0516	0598	0686	0779	0875	0974	1073	1171	10
11	0151	1188	0231	0280	0336	0398	0467	0542	0624	0710	9
12	0051	0066	0085	0108	0136	0168	0206	0249	0299	0355	8
13	0014	0019	0026	0034	0045	0058	0074	0094	0118	0146	7
14	0003	0005	0006	0009	0012	0016	0022	0029	0038	0049	6
15	0001	0001	0001	0002	0003	0004	0005	0007	0010	0013	5
16	0000	0000	0000	0000	0000	0001	0001	0001	0002	0003	4
	69	68	67	66	65	64	63	62	61	60	p \ x

x \ p	41	42	43	44	45	46	47	48	49	50	
1	0004	0003	0002	0001	0001	0001	0001	0000	0000	0000	19
2	0024	0018	0014	0011	0008	0006	0005	0003	0002	0002	18
3	0100	0080	0064	0051	0040	0031	0024	0019	0014	0011	17
4	0295	0247	0206	0170	0139	0113	0092	0074	0059	0046	16
5	0656	0573	0496	0427	0365	0309	0260	0217	0180	0148	15
6	1140	1037	0936	0839	0746	0658	0577	0501	0432	0370	14
7	1585	1502	1413	1318	1221	1122	1023	0925	0830	0739	13
8	1790	1768	1732	1683	1623	1553	1474	1388	1296	1201	12
9	1658	1707	1742	1763	1771	1763	1742	1708	1661	1602	11
10	1268	1359	1446	1524	1593	1652	1700	1734	1755	1762	10
11	0801	0895	0991	1089	1185	1280	1370	1455	1533	1602	9
12	0417	0486	0561	0642	0727	0818	0911	1007	1105	1201	8
13	0178	0217	0260	0310	0366	0429	0497	0572	0653	0739	7
14	0062	0078	0098	0122	0150	0183	0221	0264	0314	0370	6
15	0017	0023	0030	0038	0049	0062	0078	0098	0121	0148	5
16	0004	0005	0007	0009	0013	0017	0022	0028	0036	0046	4
17	0001	0001	0001	0002	0002	0003	0005	0006	0008	0011	3
18	0000	0000	0000	0000	0000	0000	0001	0001	0001	0002	2
	59	58	57	56	55	54	53	52	51	50	p \ x

(Continued)

Table I Binomial Distribution (n = 50)

n = 50

x p	01	02	03	04	05	06	07	08	09	10	
0	6050	3642	2181	1299	0769	0453	0266	0155	0090	0052	50
1	3056	3716	3372	2706	2025	1447	0999	0672	0443	0286	49
2	0756	1858	2555	2762	2611	2262	1843	1433	1073	0779	48
3	0122	0607	1264	1842	2199	2311	2219	1993	1698	1386	47
4	0015	0145	0459	0902	1360	1733	1963	2037	1973	1809	46
5	0001	0027	0131	0346	0658	1018	1359	1629	1795	1849	45
6	0000	0004	0030	0108	0260	0487	0767	1063	1332	1541	44
7	0000	0001	0006	0028	0086	0195	0363	0581	0828	1076	43
8	0000	0000	0001	0006	0024	0067	0147	0271	0440	0643	42
9	0000	0000	0000	0001	0006	0020	0052	0110	0203	0333	41
10	0000	0000	0000	0000	0001	0005	0016	0039	0082	0152	40
11	0000	0000	0000	0000	0000	0001	0004	0012	0030	0061	39
12	0000	0000	0000	0000	0000	0000	0001	0004	0010	0022	38
13	0000	0000	0000	0000	0000	0000	0000	0001	0003	0007	37
14	0000	0000	0000	0000	0000	0000	0000	0000	0001	0002	36
15	0000	0000	0000	0000	0000	0000	0000	0000	0000	0001	35
	99	98	97	96	95	94	93	92	91	90	p x

x p	11	12	13	14	15	16	17	18	19	20	
0	0029	0017	0009	0005	0003	0002	0001	0000	0000	0000	50
1	0182	0114	0071	0043	0026	0016	0009	0005	0003	0002	49
2	0552	0382	0259	0172	0113	0073	0046	0029	0018	0011	48
3	1091	0833	0619	0449	0319	0222	0151	0102	0067	0044	47
4	1584	1334	1086	0858	0661	0496	0364	0262	0185	0128	46
5	1801	1674	1493	1286	1072	0869	0687	0530	0400	0295	45
6	1670	1712	1674	1570	1419	1242	1055	0872	0703	0554	44
7	1297	1467	1572	1606	1575	1487	1358	1203	1037	0870	43
8	0862	1075	1262	1406	1493	1523	1495	1420	1307	1169	42
9	0497	0684	0880	1068	1230	1353	1429	1454	1431	1364	41
10	0252	0383	0539	0713	0890	1057	1200	1309	1376	1398	40
11	0113	0190	0293	0422	0571	0732	0894	1045	1174	1271	39
12	0045	0084	0142	0223	0328	0453	0595	0745	0895	1033	38
13	0016	0034	0062	0106	0169	0252	0356	0478	0613	0755	37
14	0005	0012	0025	0046	0079	0127	0193	0277	0380	0499	36
15	0002	0004	0009	0018	0033	0058	0095	0146	0214	0299	35
16	0000	0001	0003	0006	0013	0024	0042	0070	0110	0164	34
17	0000	0000	0001	0002	0005	0009	0017	0031	0052	0082	33
18	0000	0000	0000	0001	0001	0003	0007	0012	0022	0037	32
19	0000	0000	0000	0000	0000	0001	0002	0005	0009	0016	31
20	0000	0000	0000	0000	0000	0000	0001	0002	0003	0006	30
21	0000	0000	0000	0000	0000	0000	0000	0000	0001	0002	29
22	0000	0000	0000	0000	0000	0000	0000	0000	0000	0001	28
	89	88	87	86	85	84	83	82	81	80	p x

(Continued)

Table I Binomial Distribution (n = 50, cont.)

	$n = 50$										
x p	21	22	23	24	25	26	27	28	29	30	
1	0001	0001	0000	0000	0000	0000	0000	0000	0000	0000	49
2	0007	0004	0002	0001	0001	0000	0000	0000	0000	0000	48
3	0028	0018	0011	0007	0004	0002	0001	0001	0000	0000	47
4	0088	0059	0039	0025	0016	0010	0006	0004	0002	0001	46
5	0214	0152	0106	0073	0049	0033	0021	0014	0009	0006	45
6	0427	0322	0238	0173	0123	0087	0060	0040	0027	0018	44
7	0713	0571	0447	0344	0259	0191	0139	0099	0069	0048	43
8	1019	0865	0718	0583	0463	0361	0276	0207	0152	0110	42
9	1263	1139	1001	0859	0721	0592	0476	0375	0290	0220	41
10	1377	1317	1226	1113	0985	0852	0721	0598	0485	0386	40
11	1331	1351	1332	1278	1194	1089	0970	0845	0721	0602	39
12	1150	1238	1293	1311	1294	1244	1166	1068	0957	0838	38
13	0894	1021	1129	1210	1261	1277	1261	1215	1142	1050	37
14	0628	0761	0891	1010	1110	1186	1233	1248	1233	1189	36
15	0400	0515	0639	0766	0888	1000	1094	1165	1209	1223	35
16	0233	0318	0417	0529	0648	0769	0885	0991	1080	1147	34
17	0124	0179	0249	0334	0432	0540	0655	0771	0882	0983	33
18	0060	0093	0137	0193	0264	0348	0444	0550	0661	0772	32
19	0027	0044	0069	0103	0148	0206	0277	0360	0454	0558	31
20	0011	0019	0032	0050	0077	0112	0159	0217	0288	0370	30
21	0004	0008	0014	0023	0036	0056	0084	0121	0168	0227	29
22	0001	0003	0005	0009	0016	0026	0041	0062	0090	0128	28
23	0000	0001	0002	0004	0006	0011	0018	0029	0045	0067	27
24	0000	0000	0001	0001	0002	0004	0008	0013	0021	0032	26
25	0000	0000	0000	0000	0001	0002	0003	0005	0009	0014	25
26	0000	0000	0000	0000	0000	0001	0001	0002	0003	0006	24
27	0000	0000	0000	0000	0000	0000	0000	0001	0001	0002	23
28	0000	0000	0000	0000	0000	0000	0000	0000	0000	0001	22
	79	78	77	76	75	74	73	72	71	70	x

(*Continued*)

Table I Binomial Distribution (n = 50, cont.)

$n = 50$

x \ p	31	32	33	34	35	36	37	38	39	40	
4	0001	0000	0000	0000	0000	0000	0000	0000	0000	0000	46
5	0003	0002	0001	0001	0000	0000	0000	0000	0000	0000	45
6	0011	0007	0005	0003	0002	0001	0001	0000	0000	0000	44
7	0032	0022	0014	0009	0006	0004	0002	0001	0001	0000	43
8	0078	0055	0037	0025	0017	0011	0007	0004	0003	0002	42
9	0164	0120	0086	0061	0042	0029	0019	0013	0008	0005	41
10	0301	0231	0174	0128	0093	0066	0046	0032	0022	0014	40
11	0493	0395	0311	0240	0182	0136	0099	0071	0050	0035	39
12	0719	0604	0498	0402	0319	0248	0189	0142	0105	0076	38
13	0944	0831	0717	0606	0502	0408	0325	0255	0195	0147	37
14	1121	1034	0933	0825	0714	0607	0505	0412	0330	0260	36
15	1209	1168	1103	1020	0923	0819	0712	0606	0507	0415	35
16	1188	1202	1189	1149	1088	1008	0914	0813	0709	0606	34
17	1068	1132	1171	1184	1171	1133	1074	0997	0906	0808	33
18	0880	0976	1057	1118	1156	1169	1156	1120	1062	0987	32
19	0666	0774	0877	0970	1048	1107	1144	1156	1144	1109	31
20	0463	0564	0670	0775	0875	0956	1041	1098	1134	1146	30
21	0297	0379	0471	0570	0673	0776	0874	0962	1035	1091	29
22	0176	0235	0306	0387	0478	0575	0676	0777	0873	0959	28
23	0096	0135	0183	0243	0313	0394	0484	0580	0679	0778	27
24	0049	0071	0102	0141	0190	0249	0319	0400	0489	0584	26
25	0023	0035	0052	0075	0106	0146	0195	0255	0325	0405	25
26	0010	0016	0025	0037	0055	0079	0110	0150	0200	0259	24
27	0004	0007	0011	0017	0026	0039	0058	0082	0113	0154	23
28	0001	0003	0004	0007	0012	0018	0028	0041	0060	0084	22
29	0000	0001	0002	0003	0005	0008	0012	0019	0029	0043	21
30	0000	0000	0001	0001	0002	0003	0005	0008	0013	0020	20
31	0000	0000	0000	0000	0001	0001	0002	0003	0005	0009	19
32	0000	0000	0000	0000	0000	0000	0001	0001	0002	0003	18
33	0000	0000	0000	0000	0000	0000	0000	0000	0001	0001	17
	69	68	67	66	65	64	63	62	61	60	p / x

(Continued)

Table I Binomial Distribution (n = 50, cont.)

					$n = 50$						
x p	41	42	43	44	45	46	47	48	49	50	
8	0001	0001	0000	0000	0000	0000	0000	0000	0000	0000	42
9	0083	0002	0001	0001	0000	0000	0000	0000	0000	0000	41
10	0009	0006	0004	0002	0001	0001	0001	0000	0000	0000	40
11	0024	0016	0010	0007	0004	0003	0002	0001	0001	0000	39
12	0054	0037	0026	0017	0011	0007	0005	0003	0002	0001	38
13	0109	0079	0057	0040	0027	0018	0012	0008	0005	0003	37
14	0200	0152	0113	0082	0059	0041	0029	0019	0013	0008	36
15	0334	0264	0204	0155	0116	0085	0061	0043	0030	0020	35
16	0508	0418	0337	0267	0207	0158	0118	0086	0062	0044	34
17	0706	0605	0508	0419	0339	0269	0209	0159	0119	0087	33
18	0899	0803	0703	0604	0508	0420	0340	0270	0210	0160	32
19	1053	0979	0893	0799	0700	0602	0507	0419	0340	0270	31
20	1134	1099	1044	0973	0588	0795	0697	0600	0506	0419	30
21	1126	1137	1126	1092	1030	0967	0884	0791	0695	0598	29
22	1031	1086	1119	1131	1119	1086	1033	0963	0880	0788	28
23	0872	0957	1028	1082	1115	1126	1115	1082	1029	0960	27
24	0682	0780	0872	0956	1026	1079	1112	1124	1112	1080	26
25	0493	0587	0684	0781	0873	0956	1026	1079	1112	1123	25
26	0329	0409	0497	0590	0687	0783	0875	0957	1027	1080	24
27	0203	0263	0333	0412	0500	0593	0690	0786	0877	0960	23
28	0116	0157	0206	0266	0336	0415	0502	0596	0692	0788	22
29	0061	0086	0118	0159	0208	0268	0338	0417	0504	0598	21
30	0030	0044	0062	0087	0119	0160	0210	0270	0339	0419	20
31	0013	0020	0030	0044	0063	0088	0120	0161	0210	0270	19
32	0006	0009	0014	0021	0031	0044	0063	0088	0120	0160	18
33	0002	0003	0006	0009	0014	0021	0031	0044	0063	0087	17
34	0001	0001	0002	0003	0006	0009	0014	0020	0030	0044	16
35	0000	0000	0001	0001	0002	0003	0005	0009	0013	0020	15
36	0000	0000	0000	0000	0001	0001	0002	0003	0005	0006	14
37	0000	0000	0000	0000	0000	0000	0001	0001	0002	0003	13
38	0000	0000	0000	0000	0000	0000	0000	0000	0001	0001	12
	59	58	57	56	55	54	53	52	51	50	p x

(Continued)

Table I Binomial Distribution (n = 100)

$n = 100$

x	p	01	02	03	04	05	06	07	08	09	10		
0		3660	1326	0476	0169	0059	0021	0007	0002	0001	0000		100
1		3697	2707	1471	0703	0312	0131	0053	0021	0008	0003		99
2		1849	2734	2252	1450	0812	0414	0198	0090	0039	0016		98
3		0610	1823	2275	1973	1396	0864	0486	0254	0125	0059		97
4		0149	0902	1706	1994	1781	1338	0888	0536	0301	0159		96
5		0029	0353	1013	1595	1800	1639	1283	0895	0571	0339		95
6		0005	0114	0496	1052	1500	1657	1529	1233	0895	0596		94
7		0001	0031	0206	0589	1060	1420	1545	1440	1188	0889		93
8		0000	0007	0074	0285	0649	1054	1352	1455	1366	1148		92
9		0000	0002	0023	0121	0349	0687	1040	1293	1381	1304		91
10		0000	0000	0007	0046	0167	0399	0712	1024	1243	1319		90
11		0000	0000	0002	0016	0072	0209	0439	0728	1006	1199		89
12		0000	0000	0000	0005	0028	0099	0245	0470	0738	0988		88
13		0000	0000	0000	0001	0010	0043	0125	0276	0494	0743		87
14		0000	0000	0000	0000	0003	0017	0058	0149	0304	0513		86
15		0000	0000	0000	0000	0001	0006	0025	0074	0172	0327		85
16		0000	0000	0000	0000	0000	0002	0010	0034	0090	0193		84
17		0000	0000	0000	0000	0000	0001	0004	0015	0044	0106		83
18		0000	0000	0000	0000	0000	0000	0001	0006	0020	0054		82
19		0000	0000	0000	0000	0000	0000	0000	0002	0009	0026		81
20		0000	0000	0000	0000	0000	0000	0000	0001	0003	0012		80
21		0000	0000	0000	0000	0000	0000	0000	0000	0001	0005		79
22		0000	0000	0000	0000	0000	0000	0000	0000	0000	0002		78
23		0000	0000	0000	0000	0000	0000	0000	0000	0000	0001		77
		99	98	97	96	95	94	93	92	91	90	p	x

(*Continued*)

Table I Binomial Distribution (n = 100, cont.)

	$n = 100$										
x p	11	12	13	14	15	16	17	18	19	20	
1	0001	0000	0000	0000	0000	0000	0000	0000	0000	0000	99
2	0007	0003	0001	0000	0000	0000	0000	0000	0000	0000	98
3	0027	0012	0005	0002	0001	0000	0000	0000	0000	0000	97
4	0080	0038	0018	0008	0003	0001	0001	0000	0000	0000	96
5	0189	0100	0050	0024	0011	0005	0002	0001	0000	0000	95
6	0369	0215	0119	0063	0031	0015	0007	0003	0001	0001	94
7	0613	0394	0238	0137	0075	0039	0020	0009	0004	0002	93
8	0881	0625	0414	0259	0153	0086	0047	0024	0012	0006	92
9	1112	0871	0632	0430	0276	0168	0098	0054	0029	0015	91
10	1251	1080	0860	0637	0444	0292	0182	0108	0062	0034	90
11	1265	1205	1051	0849	0640	0454	0305	0194	0118	0069	89
12	1160	1219	1165	1025	0838	0642	0463	0316	0206	0128	88
13	0970	1125	1179	1130	1001	0827	0642	0470	0327	0216	87
14	0745	0954	1094	1143	1098	0979	0817	0641	0476	0335	86
15	0528	0745	0938	1067	1111	1070	0960	0807	0640	0481	85
16	0347	0540	0744	0922	1041	1082	1044	0941	0798	0638	84
17	1212	0364	0549	0742	0908	1019	1057	1021	0924	0789	83
18	0121	0229	0379	0557	0739	0895	0998	1033	1000	0909	82
19	0064	0135	0244	0391	0563	0736	0882	0979	1012	0981	81
20	0032	0074	0148	0258	0402	0567	0732	0870	0962	0993	80
21	0015	0039	0084	0160	0270	0412	0571	0728	0859	0946	79
22	0007	0019	0045	0094	0171	0282	0420	0574	0724	0849	78
23	0003	0009	0023	0052	0103	0182	0292	0427	0576	0720	77
24	0001	0004	0011	0027	0058	0111	0192	0301	0433	0577	76
25	0000	0002	0005	0013	0031	0064	0119	0201	0309	0439	75
26	0000	0001	0002	0006	0016	0035	0071	0127	0209	0317	74
27	0000	0000	0001	0003	0008	0018	0040	0076	0134	0217	73
28	0000	0000	0000	0001	0004	0009	0021	0044	0082	0141	72
29	0000	0000	0000	0000	0002	0004	0011	0024	0048	0088	71
30	0000	0000	0000	0000	0001	0002	0005	0012	0027	0052	70
31	0000	0000	0000	0000	0000	0001	0002	0006	0014	0029	69
32	0000	0000	0000	0000	0000	0000	0001	0003	0007	0016	68
33	0000	0000	0000	0000	0000	0000	0000	0001	0003	0008	67
34	0000	0000	0000	0000	0000	0000	0000	0001	0002	0004	66
35	0000	0000	0000	0000	0000	0000	0000	0000	0001	0002	65
36	0000	0000	0000	0000	0000	0000	0000	0000	0000	0001	64
	89	88	87	86	85	84	83	82	81	80	p x

(Continued)

Table I Binomial Distribution (n = 100, cont.)

$n = 100$

x	p 21	22	23	24	25	26	27	28	29	30	
7	0001	0000	0000	0000	0000	0000	0000	0000	0000	0000	93
8	0003	0001	0001	0000	0000	0000	0000	0000	0000	0000	92
9	0007	0003	0002	0001	0000	0000	0000	0000	0000	0000	91
10	0018	0009	0004	0002	0001	0000	0000	0000	0000	0000	90
11	0038	0021	0011	0005	0003	0001	0001	0000	0000	0000	89
12	0076	0043	0024	0012	0006	0003	0001	0001	0000	0000	88
13	0136	0082	0048	0027	0014	0007	0004	0002	0001	0000	87
14	0225	0144	0089	0052	0030	0016	0009	0004	0002	0001	86
15	1343	0233	0152	0095	0057	0033	0018	0010	0005	0002	85
16	0484	0350	0241	0159	0100	0061	0035	0020	0011	0006	84
17	0636	0487	0356	0248	0165	0106	0065	0038	0022	0012	83
18	0780	0634	0490	0361	0254	0171	0111	0069	0041	0024	82
19	0895	0772	0631	0492	0365	0259	0177	0115	0072	0044	81
20	0963	0881	0764	0629	0493	0369	0264	0182	0120	0076	80
21	0975	0947	0869	0756	0626	0494	0373	0269	0186	0124	79
22	0931	0959	0932	0858	0749	0623	0495	0376	0273	0190	78
23	0839	0917	0944	0919	0847	0743	0621	0495	0378	0277	77
24	0716	0830	0905	0931	0906	0837	0736	0618	0496	0380	76
25	0578	0712	0822	0893	0918	0894	0828	0731	0615	0496	75
26	0444	0579	0708	0814	0883	0906	0883	0819	0725	0613	74
27	0323	0448	0580	0704	0806	0873	0896	0873	0812	0720	73
28	0224	0329	0451	0580	0701	0799	0864	0886	0864	0804	72
29	0148	0231	0335	0455	0580	0697	0793	0855	0876	0856	71
30	0093	0154	0237	0340	0458	0580	0694	0787	0847	0868	70
31	0056	0098	0160	0242	0344	0460	0580	0691	0781	0840	69
32	0032	0060	0103	0165	0248	0349	0462	0579	0688	0776	68
33	0018	0035	0063	0107	0170	0252	0352	0464	0579	0685	67
34	0009	0019	0037	0067	0112	0175	0257	0356	0466	0579	66
35	0005	0010	0021	0040	0070	0116	0179	0261	0359	0468	65
36	0002	0005	0011	0023	0042	0073	0120	0183	0265	0362	64
37	0001	0003	0006	0012	0024	0045	0077	0123	0187	0268	63
38	0000	0001	0003	0006	0013	0026	0047	0079	0127	0191	62
39	0000	0001	0001	0003	0007	0015	0028	0049	0082	0130	61
40	0000	0000	0001	0002	0004	0008	0016	0029	0051	0085	60
41	0000	0000	0000	0001	0002	0004	0008	0017	0031	0053	59
42	0000	0000	0000	0000	0001	0002	0004	0009	0018	0032	58
43	0000	0000	0000	0000	0000	0001	0002	0005	0010	0019	57
44	0000	0000	0000	0000	0000	0000	0001	0002	0005	0010	56
45	0000	0000	0000	0000	0000	0000	0000	0001	0003	0005	55
46	0000	0000	0000	0000	0000	0000	0000	0001	0001	0003	54
47	0000	0000	0000	0000	0000	0000	0000	0000	0001	0001	53
48	0000	0000	0000	0000	0000	0000	0000	0000	0000	0001	52
	79	78	77	76	75	74	73	72	71	70 p	x

(*Continued*)

Table I Binomial Distribution (n = 100, cont.)

x	p = 31	32	33	34	35	36	37	38	39	40	
15	0001	0001	0000	0000	0000	0000	0000	0000	0000	0000	85
16	0003	0001	0001	0000	0000	0000	0000	0000	0000	0000	84
17	0006	0003	0002	0001	0000	0000	0000	0000	0000	0000	83
18	0013	0007	0004	0002	0001	0000	0000	0000	0000	0000	82
19	0025	0014	0008	0004	0002	0001	0000	0000	0000	0000	81
20	0046	0027	0015	0008	0004	0002	0001	0001	0000	0000	80
21	0079	0049	0029	0016	0009	0005	0002	0001	0001	0000	79
22	0127	0082	0051	0030	0017	0010	0005	0003	0001	0001	78
23	0194	0131	0085	0053	0032	0018	0010	0006	0003	0001	77
24	0280	0198	0134	0088	0055	0033	0019	0011	0006	0003	76
25	0382	0283	0201	0137	0090	0057	0035	0020	0012	0006	75
26	0496	0384	0286	0204	0140	0092	0059	0036	0021	0012	74
27	0610	0495	0386	0288	0207	0143	0095	0060	0037	0022	73
28	0715	0608	0495	0387	0290	0209	0145	0097	0062	0038	72
29	0797	0710	0605	0495	0388	0292	0211	0147	0098	0063	71
30	0848	0791	0706	0603	0494	0389	0294	0213	0149	0100	70
31	0860	0840	0785	0702	0601	0494	0389	0295	0215	0151	69
32	0833	0853	0833	0779	0698	0599	0493	0390	0296	0217	68
33	0771	0827	0846	0827	0774	0694	0597	0493	0390	0297	67
34	0683	0767	0821	0840	0821	0769	0691	0595	0492	0391	66
35	0578	0680	0763	0816	0834	0816	0765	0688	0593	0491	65
36	0469	0578	0678	0759	0811	0829	0811	0761	0685	0591	64
37	0365	0471	0578	0676	0755	0806	0824	0807	0757	0682	63
38	0272	0367	0472	0577	0674	0752	0802	0820	0803	0754	62
39	0194	0275	0369	0473	0577	0672	0749	0799	0816	0799	61
40	0133	0197	0277	0372	0474	0577	0671	0746	0795	0812	60
41	0087	0136	0200	0280	0373	0475	0577	0670	0744	0792	59
42	0055	0090	0138	0203	0282	0375	0476	0576	0668	0742	58
43	0033	0057	0092	0141	0205	0285	0377	0477	0576	0667	57
44	0019	0035	0059	0094	0143	0207	0287	0378	0477	0576	56
45	0011	0020	0036	0060	0096	0045	0210	0289	0380	0478	55
46	0006	0011	0021	0037	0062	0098	0147	0211	0290	0381	54
47	0003	0006	0012	0022	0038	0063	0099	0149	0213	0292	53
48	0001	0003	0007	0012	0023	0039	0064	0101	0151	0215	52
49	0001	0002	0003	0007	0013	0023	0040	0066	0102	0152	51
50	0000	0001	0002	0004	0007	0013	0024	0041	0067	0103	50
51	0000	0000	0001	0002	0004	0007	0014	0025	0042	0068	49
52	0000	0000	0000	0001	0002	0004	0008	0014	0025	0042	48
53	0000	0000	0000	0000	0001	0002	0004	0008	0015	0026	47
54	0000	0000	0000	0000	0000	0001	0002	0004	0008	0015	46
55	0000	0000	0000	0000	0000	0000	0001	0002	0004	0008	45
56	0000	0000	0000	0000	0000	0000	0000	0001	0002	0004	44
57	0000	0000	0000	0000	0000	0000	0000	0001	0001	0002	43
58	0000	0000	0000	0000	0000	0000	0000	0000	0001	0001	42
59	0000	0000	0000	0000	0000	0000	0000	0000	0000	0001	41
	69	68	67	66	65	64	63	62	61	60 p	x

n = 100

(Continued)

Table I Binomial Distribution (n = 100, cont.)

	$n = 100$										
x p	41	42	43	44	45	46	47	48	49	50	
23	0001	0000	0000	0000	0000	0000	0000	0000	0000	0000	77
24	0002	0001	0000	0000	0000	0000	0000	0000	0000	0000	76
25	0003	0002	0001	0000	0000	0000	0000	0000	0000	0000	75
26	0007	0003	0002	0001	0000	0000	0000	0000	0000	0000	74
27	0013	0007	0004	0002	0001	0000	0000	0000	0000	0000	73
28	0023	0013	0007	0004	0002	0001	0000	0000	0000	0000	72
29	0039	0024	0014	0008	0004	0002	0001	0000	0000	0000	71
30	0065	0040	0024	0014	0008	0004	0002	0001	0001	0000	70
31	0102	0066	0041	0025	0014	0008	0004	0002	0001	0001	69
32	0152	0103	0067	0042	0025	0015	0008	0004	0002	0001	68
33	0218	0154	0104	0068	0043	0026	0015	0008	0004	0002	67
34	0298	0219	0155	0105	0069	0043	0026	0015	0009	0005	66
35	0391	0299	0220	0156	0106	0069	0044	0026	0015	0009	65
36	0491	0391	0300	0221	0157	0107	0070	0044	0027	0016	64
37	0590	0490	0391	0300	0222	0157	0107	0070	0044	0027	63
38	0680	0588	0489	0391	0301	0222	0158	0108	0071	0045	62
39	0751	0677	0587	0489	0391	0301	0223	0158	0108	0071	61
40	0796	0748	0675	0586	0488	0391	0301	0223	0159	0108	60
41	0809	0793	0745	0673	0584	0487	0391	0301	0223	0159	59
42	0790	0806	0790	0743	0672	0583	0487	0390	0301	0223	58
43	0740	0787	0804	0788	0741	0670	0582	0486	0390	0301	57
44	0666	0739	0785	0802	0786	0739	0669	0581	0485	0390	56
45	0576	0666	0737	0784	0800	0784	0738	0668	0580	0485	55
46	0479	0576	0665	0736	0782	0798	0783	0737	0667	0580	54
47	0382	0480	0576	0065	0736	0781	0797	0781	0736	0666	53
	59	58	57	56	55	54	53	52	51	50	p x

(*Continued*)

Table I Binomial Distribution (n = 100, cont.)

x	p 41	42	43	44	45	46	47	48	49	50	
					n = 100						
48	0293	0383	0480	0577	0665	0735	0781	0797	0781	0735	52
49	0216	0295	0384	0481	0577	0664	0735	0780	0796	0780	51
50	0153	0218	0296	0385	0482	0577	0665	0735	0780	0796	50
51	0104	0155	0219	0297	0386	0482	0578	0665	0735	0780	49
52	0068	0105	0156	0220	0298	0387	0483	0578	0665	0735	48
53	0043	0069	0106	0156	0221	0299	0388	0483	0579	0666	47
54	0026	0044	0070	0107	0157	0221	0299	0388	0484	0580	46
55	0015	0026	0044	0070	0108	0158	0222	0300	0389	0485	45
56	0008	0015	0027	0044	0071	0108	0158	0222	0300	0390	44
57	0005	0009	0016	0027	0045	0071	0108	0158	0223	0301	43
58	0002	0005	0009	0016	0027	0045	0071	0108	0159	0223	42
59	0001	0002	0005	0009	0016	0027	0045	0071	0109	0159	41
60	0001	0001	0002	0005	0009	0016	0027	0045	0071	0108	40
61	0000	0001	0001	0002	0005	0009	0016	0027	0045	0071	39
62	0000	0000	0001	0001	0002	0005	0009	0016	0027	0045	38
63	0000	0000	0000	0001	0001	0002	0005	0009	0016	0027	37
64	0000	0000	0000	0000	0001	0001	0002	0005	0009	0016	36
65	0000	0000	0000	0000	0000	0001	0001	0002	0005	0009	35
66	0000	0000	0000	0000	0000	0000	0001	0001	0002	0005	34
67	0000	0000	0000	0000	0000	0000	0000	0001	0001	0002	33
68	0000	0000	0000	0000	0000	0000	0000	0000	0001	0001	32
69	0000	0000	0000	0000	0000	0000	0000	0000	0000	0001	31
	59	58	57	56	55	54	53	52	51	50 p	x

Table II Poisson Distribution ($\lambda = 0.1$ to $\lambda = 20$)

The following table gives the probability of exactly x successes, for various values of λ, as defined by the Poisson mass function.

$$P(x) = \frac{e^{-\lambda}\lambda^x}{x!}$$

Examples: If $\lambda = 1.5$, then $P(2) = 0.2510$, $P(3) = 0.1255$.

Poisson Probabilities

($\lambda = 0.1$ to $\lambda = 2.0$)

x	λ = 0.1	0.2	0.3	0.4	0.5	0.6	0.7	0.8	0.9	1.0
0	.9048	.8187	.7408	.6703	.6065	.5488	.4966	.4493	.4066	.3679
1	.0905	.1637	.2222	.2681	.3033	.3293	.3476	.3595	.3659	.3679
2	.0045	.0164	.0333	.0536	.0758	.0988	.1217	.1438	.1647	.1839
3	.0002	.0011	.0033	.0072	.0126	.0198	.0284	.0383	.0494	.0613
4	.0000	.0001	.0002	.0007	.0016	.0030	.0050	.0077	.0111	.0153
5	.0000	.0000	.0000	.0001	.0002	.0004	.0007	.0012	.0020	.0031
6	.0000	.0000	.0000	.0000	.0000	.0000	.0001	.0002	.0003	.0005
7	.0000	.0000	.0000	.0000	.0000	.0000	.0000	.0000	.0000	.0001

x	λ = 1.1	1.2	1.3	1.4	1.5	1.6	1.7	1.8	1.9	2.0
0	.3329	.3012	.2725	.2466	.2231	.2019	.1827	.1653	.1496	.1353
1	.3662	.3614	.3543	.3452	.3347	.3230	.3106	.2975	.2842	.2707
2	.2014	.2169	.2303	.2417	.2510	.2584	.2640	.2678	.2700	.2707
3	.0738	.0867	.0998	.1128	.1255	.1378	.1496	.1607	.1710	.1804
4	.0203	.0260	.0324	.0395	.0471	.0551	.0636	.0723	.0812	.0902
5	.0045	.0062	.0084	.0111	.0141	.0176	.0216	.0260	.0309	.0361
6	.0008	.0012	.0018	.0026	.0035	.0047	.0061	0078	.0098	.0120
7	.0001	.0002	.0003	.0005	.0008	.0011	.0015	.0020	.0027	.0034
8	.0000	.0000	.0001	.0001	.0001	.0002	.0003	.0005	.0006	.0009
9	.0000	.0000	.0000	.0000	.0000	.0000	.0001	.0001	.0001	.0002

(*Continued*)

Table II Poisson Distribution (λ = 2.1 to λ = 4.0)

x	2.1	2.2	2.3	2.4	2.5	2.6	2.7	2.8	2.9	3.0
0	.1225	.1108	.1003	.0907	.0821	.0743	.0672	.0608	.0550	.0498
1	.2572	.2438	.2306	.2177	.2052	.1931	.1815	.1703	.1596	.1494
2	.2700	.2681	.2652	.2613	.2565	.2510	.2450	.2384	.2314	.2240
3	.1890	.1966	.2033	.2090	.2138	.2176	.2205	.2225	.2237	.2240
4	.0992	.1082	.1169	.1254	.1336	.1414	.1488	.1557	.1622	.1680
5	.0417	.0476	.0538	.0602	.0668	.0735	.0804	.0872	.0940	.1008
6	.0146	.0174	.0206	.0241	.0278	.0319	.0362	.0407	.0455	.0504
7	.0044	.0055	.0068	.0083	.0099	.0118	.0139	.0163	.0188	.0216
8	.0011	.0015	.0019	.0025	.0031	.0038	.0047	.0057	.0068	.0081
9	.0003	.0004	.0005	.0007	.0009	.0011	.0014	.0018	.0022	.0027
10	.0001	.0001	.0001	.0002	.0002	.0003	.0004	.0005	.0006	.0008
11	.0000	.0000	.0000	.0000	.0000	.0001	.0001	.0001	.0002	.0002
12	.0000	.0000	.0000	.0000	.0000	.0000	.0000	.0000	.0000	.0001

x	3.1	3.2	3.3	3.4	3.5	3.6	3.7	3.8	3.9	4.0
0	.0450	.0408	.0369	.0334	.0302	.0273	.0247	.0224	.0202	.0183
1	.1397	.1304	.1217	.1135	.1057	.0984	.0915	.0850	.0789	.0733
2	.2165	.2087	.2008	.1929	.1850	.1771	.1692	.1615	.1539	.1465
3	.2237	.2226	.2209	.2186	.2158	.2125	.2087	.2046	.2001	.1954
4	.1734	.1781	.1823	.1858	.1888	.1912	.1931	.1944	.1951	.1954
5	.1075	.1140	.1203	.1264	.1322	.1377	.1429	.1477	.1522	.1563
6	.0555	.0608	.0662	.0716	.0771	.0826	.0881	.0936	.0989	.1042
7	.0246	.0278	.0312	.0348	.0385	.0425	.0466	.0508	.0551	.0595
8	.0095	.0111	.0129	.0148	.0169	.0191	.0215	.0241	.0269	.0298
9	.0033	.0040	.0047	.0056	.0066	.0076	.0089	.0102	.0116	.0132
10	.0010	.0013	.0016	.0019	.0023	.0028	.0033	.0039	.0045	.0053
11	.0003	.0004	.0005	.0006	.0007	.0009	.0011	.0013	.0016	.0019
12	.0001	.0001	.0001	.0002	.0002	.0003	.0003	.0004	.0005	.0006
13	.0000	.0000	.0000	.0000	.0001	.0001	.0001	.0001	.0002	.0002
14	.0000	.0000	.0000	.0000	.0000	.0000	.0000	.0000	.0000	.0001

(Continued)

Table II Poisson Distribution (λ = 4.1 to λ = 6.0)

x	λ 4.1	4.2	4.3	4.4	4.5	4.6	4.7	4.8	4.9	5.0
0	.0166	.0150	.0136	.0123	.0111	.0101	.0091	.0082	.0074	.0067
1	.0679	.0630	.0583	.0540	.0500	.0462	.0427	.0395	.0365	.0337
2	.1393	.1323	.1254	.1188	.1125	.1063	.1005	.0948	.0894	.0842
3	.1904	.1852	.1798	.1743	.1687	.1631	.1574	.1517	.1460	.1404
4	.1951	.1944	.1933	.1917	.1898	.1875	.1849	.1820	.1789	.1755
5	.1600	.1633	.1662	.1687	.1708	.1725	.1738	.1747	.1753	.1755
6	.1093	.1143	.1191	.1237	.1281	.1323	.1362	.1398	.1432	.1462
7	.0640	.0686	.0732	.0778	.0824	.0869	.0914	.0959	.1002	.1044
8	.0328	.0360	.0393	.0428	.0463	.0500	.0537	.0575	.0614	.0653
9	.0150	.0168	.0188	.0209	.0232	.0255	.0280	.0307	.0334	.0363
10	.0061	.0071	.0081	.0092	.0104	.0118	.0132	.0147	.0164	.0181
11	.0023	.0027	.0032	.0037	.0043	.0049	.0056	.0064	.0073	.0082
12	.0008	.0009	.0011	.0014	.0016	.0019	.0022	.0026	.0030	.0034
13	.0002	.0003	.0004	.0005	.0006	.0007	.0008	.0009	.0011	.0013
14	.0001	.0001	.0001	.0001	.0002	.0002	.0003	.0003	.0004	.0005
15	.0000	.0000	.0000	.0000	.0001	.0001	.0001	.0001	.0001	.0002

x	λ 5.1	5.2	5.3	5.4	5.5	5.6	5.7	5.8	5.9	6.0
0	.0061	.0055	.0050	.0045	.0041	.0037	.0033	.0030	.0027	.0025
1	.0311	.0287	.0265	.0244	.0225	.0207	.0191	.0176	.0162	.0149
2	.0793	.0746	.0701	.0659	.0618	.0580	.0544	.0509	.0477	.0446
3	.1348	.1293	.1239	.1185	.1133	.1082	.1033	.0985	.0938	.0892
4	.1719	.1681	.1641	.1600	.1558	.1515	.1472	.1428	.1383	.1339
5	.1753	.1748	.1740	.1728	.1714	.1697	.1678	.1620	.1632	.1606
6	.1490	.1515	.1537	.1555	.1571	.1584	.1594	.1656	.1605	.1606
7	.1086	.1125	.1163	.1200	.1234	.1267	.1298	.1301	.1353	.1377
8	.0692	.0731	.0771	.0810	.0849	.0887	.0925	.0926	.0998	.1033
9	.0392	.0423	0.454	.0486	.0519	.0552	.0586	.0662	.0654	.0688
10	.0200	.0220	.0241	.0262	.0285	.0309	.0334	.0359	.0386	.0413
11	.0093	.0104	.0116	.0129	.0143	.0157	.0173	.0190	.0207	.0225
12	.0039	.0045	.0051	.0058	.0065	.0073	.0082	.0092	.0102	.0113
13	.0015	.0018	.0021	.0024	.0028	.0032	.0036	.0041	.0046	.0052
14	.0006	.0007	.0008	.0009	.0011	.0013	.0015	.0017	.0019	.0022
15	.0002	.0002	.0003	.0003	.0004	.0005	.0006	.0007	.0008	.0009
16	.0001	.0001	.0001	.0001	.0001	.0002	.0002	.0002	.0003	.0003
17	.0000	.0000	.0000	.0000	.0000	.0001	.0001	.0001	.0001	.0001

(*Continued*)

Table II Poisson Distribution (λ = 6.1 to λ = 8.0)

	λ									
x	6.1	6.2	6.3	6.4	6.5	6.6	6.7	6.8	6.9	7.0
0	.0022	.0020	.0018	.0017	.0015	.0014	.0012	.0011	.0010	.0009
1	.0137	.0126	.0116	.0106	.0098	.0090	.0082	.0076	.0070	.0064
2	.0417	.0390	.0364	.0340	.0318	.0296	.0276	.0258	.0240	.0223
3	.0848	.0806	.0765	.0726	.0688	.0652	.0617	.0584	.0552	.0521
4	.1294	.1249	.1205	.1162	.1118	.1076	.1034	.0992	.0952	.0912
5	.1579	.1549	.1519	.1487	.1454	.1420	.1385	.1349	.1314	.1277
6	.1605	.1601	.1595	.1586	.1575	.1562	.1546	.1529	.1511	.1490
7	.1399	.1418	.1435	.1450	.1462	.1472	.1480	.1486	.1489	.1490
8	.1066	.1099	.1130	.1160	.1188	.1215	.1240	.1263	.1284	.1304
9	.0723	.0757	.0791	.0825	.0858	.0891	.0923	.0954	.0985	.1014
10	.0441	.0469	.0498	.0528	.0558	.0588	.0618	.0649	.0679	.0710
11	.0245	.0265	.0285	.0307	.0330	.0353	.0377	.0401	.0426	.0452
12	.0124	.0137	.0150	.0164	.0179	.0194	.0210	.0227	.0245	.0264
13	.0058	0.065	.0073	.0081	.0089	.0098	.0108	.0119	.0130	.0142
14	.0025	.0029	.0033	.0037	.0041	.0046	.0052	.0058	.0064	.0071
15	.0010	.0012	.0014	.0016	.0018	.0020	.0023	.0026	.0029	.0033
16	.0004	.0005	.0005	.0006	.0007	.0008	.0010	.0011	.0013	.0014
17	.0001	.0002	.0002	.0002	.0003	.0003	.0004	.0004	.0005	.0006
18	.0000	.0001	.0001	.0001	.0001	.0001	.0001	.0002	.0002	.0002
19	.0000	.0000	.0000	.0000	.0000	.0000	.0000	.0001	.0001	.0001

	λ									
x	7.1	7.2	7.3	7.4	7.5	7.6	7.7	7.8	7.9	8.0
0	.0008	.0007	.0007	.0006	.0006	.0005	.0005	.0004	.0004	.0003
1	.0059	.0054	.0049	.0045	.0041	.0038	.0035	.0032	.0029	.0027
2	.0208	.0194	.0180	.0167	.0156	.0145	.0134	.0125	.0116	.0107
3	.0492	.0464	.0438	.0413	.0389	.0366	.0345	.0324	.0305	.0286
4	.0874	.0836	.0799	.0764	.0729	.0696	.0663	.0632.	0602	.0573
5	.1241	.1204	.1167	.1130	.1094	.1057	.1021	.0986	.0951	.0916
6	.1468	.1445	.1420	.1394	.1367	.1339	.1311	.1282	.1252	.1221
7	.1489	.1486	.1481	.1474	.1465	.1454	.1442	.1428	.1413	.1396
8	.1321	.1337	.1351	.1363	.1373	.1382	.1388	.1392	.1395	.1396
9	.1042	.1070	.1096	.1121	.1144	.1167	.1187	.1207	.1224	.1241
10	.0740	.0770	.0800	.0829	.0858	.0887	.0914	.0941	.0967	.0993
11	.0478	.0504	.0531	.0558	.0585	.0613	.0640	.0667	.0695	.0722
12	.0283	.0303	.0323	.0344	.0366	.0388	.0411	.0434	.0457	.0481
13	.0154	.0168	.0181	.0196	.0211	.0227	.0243	.0260	.0278	.0296
14	.0078	.0086	.0095	.0104	.0113	.0123	.0134	.0145	.0157	.0169
15	.0037	.0041	.0046	.0051	.0057	.0062	.0069	.0075	.0083	.0090
16	.0016	.0019	.0021	.0024	.0026	.0030	.0033	.0037	.0041	.0045
17	.0007	.0008	.0009	.0010	.0012	.0013	.0015	.0017	.1119	.0021
18	.0003	.0003	.0004	.0004	.0005	.0006	.0006	.0007	.0008	.0009
19	.0001	.0001	.0001	.0002	.0002	.0002	.0003	.0003	.0003	.0004
20	.0000	.0000	.0001	.0001	.0001	.0001	.0001	.0001	.0001	.0002
21	.0000	.0000	.0000	.0000	.0000	.0000	.0000	.0000	.0001	.0001

(Continued)

Table II Poisson Distribution (λ = 8.1 to λ = 10)

	λ									
x	8.1	8.2	8.3	8.4	8.5	8.6	8.7	8.8	8.9	9.0
0	.0003	.0003	.0002	.0002	.0002	.0002	.0002	.0002	.0001	.0001
1	.0025	.0023	.0021	.0019	.0017	.0016	.0014	.0013	.0012	.0011
2	.0100	.0092	.0086	.0079	.0074	.0068	.0063	.0058	.0054	.0050
3	.0269	.0252	.0237	.0222	.0208	.0195	.0183	.0171	.0160	.0150
4	.0544	.0517	.0491	.0466	.0443	.0420	.0398	.0377	.0357	.0337
5	.0882	.0849	.0816	.0784	.0752	.0722	.0692	.0663	.0635	.0607
6	.1191	.1160	.1128	.1097	.1066	.1034	.1003	.0972	.0941	.0911
7	.1378	.1358	.1338	.1317	.1294	.1271	.1247	.1222	.1197	.1171
8	.1395	.1392	.1388	.1382	.1375	.1366	.1356	.1344	.1332	.1318
9	.1256	.1269	.1280	.1290	.1299	.1306	.1311	.1315	.1317	.1318
10	.1017	.1040	.1063	.1084	.1104	.1123	.1140	.1157	.1172	.1186
11	.0749	.0776	.0802	.0828	.0853	.0878	.0902	.0925	.0948	.0970
12	.0505	.0530	.0555	.0579	.0604	.0629	.0654	.0679	.0703	.0728
13	.0315	.0334	.0354	.0374	.0395	.0416	.0438	.0459	.0481	.0504
14	.0182	.0196	.0210	.0225	.0240	.0256	.0272	.0289	.0306	.0324
15	.0098	.0107	.0116	.0126	.0136	.0147	.0158	.0169	.0182	.0194
16	.0050	.0055	.0060	.0066	.0072	.0079	.0086	.0093	.0101	.0109
17	.0024	.0026	.0029	.0033	.0036	.0040	.0044	.0048	.0053	.0058
18	.0011	.0012	.0014	.0015	.0017	.0019	.0021	.0024	.0026	.0029
19	.0005	.0005	.0006	.0007	.0008	.0009	.0010	.0011	.0012	.0014
20	.0002	.0002	.0002	.0003	.0003	.0004	.0004	.0005	.0005	.0006
21	.0001	.0001	.0001	.0001	.0001	.0002	.0002	.0002	.0002	.0003
22	.0000	.0000	.0000	.0000	.0001	.0001	.0001	.0001	.0001	.0001

	λ									
x	9.1	9.2	9.3	9.4	9.5	9.6	9.7	9.8	9.9	10
0	.0001	.0001	.0001	.0001	.0001	.0001	.0001	.0001	.0001	.0000
1	.0010	.0009	.0009	.0008	.0007	.0007	.0006	.0005	.0005	.0005
2	.0046	.0043	.0040	.0037	.0034	.0031	.0029	.0027	.0025	.0023
3	.0140	.0131	.0123	.0115	.0107	.0100	.0093	.0087	.0081	.0076
4	.0319	.0302	.0285	.0269	.0254	.0240	.0226	.0213	.0201	.0189
5	.0581	.0555	.0530	.0506	.0483	.0460	.0439	.0418	.0398	.0378
6	.0881	.0851	.0822	.0793	.0764	.0736	.0709	.0682	.0656	.0631
7	.1145	.1118	.1091	.1064	.1037	.1010	.0982	.0955	.0928	.0901
8	.1302	.1286	.1269	.1251	.1232	.1212	.1191	.1170	.1148	.1126
9	.1317	.1315	.1311	.1306	.1300	.1293	.1284	.1274	.1263	.1251
10	.1198	.1210	.1219	.1228	.1235	.1241	.1245	.1249	.1250	.1251
11	.0991	.1012	.1031	.1049	.1067	.1083	.1098	.1112	.1125	.1137
12	.0752	.0776	.0799	.0822	.0844	.0866	.0888	.0908	.0928	.0948
13	.0526	.0549	.0572	.0594	.0617	.0640	.0662	.0685	.0707	.0729
14	.0342	.0361	.0380	.0399	.0419	.0439	.0459	.0479	.0500	.0521
15	.0208	.0221	.0235	.0250	.0265	.0281	.0297	.0313	.0330	.0347
16	.0118	.0127	.0137	.0147	.0157	.0168	.0180	.0192	.0204	.0217
17	.0063	.0069	.0075	.0081	.0088	.0095	.0103	.0111	.0119	.0128
18	.0032	.0035	.0039	.0042	.0046	.0051	.0055	.0060	.0065	.0071
19	.0015	.0017	.0019	.0021	.0023	.0026	.0028	.0031	.0034	.0037
20	.0007	.0008	.0009	.0010	.0011	.0012	.0014	.0015	.0017	.0019
21	.0003	.0003	.0004	.0004	.0005	.0006	.0006	.0007	.0008	.0009
22	.0001	.0001	.0002	.0002	.0002	.0002	.0003	.0003	.0004	.0004
23	.0000	.0001	.0001	.0001	.0001	.0001	.0001	.0001	.0002	.0002
24	.0000	.0000	.0000	.0000	.0000	.0000	.0000	.0001	.0001	.0001

(Continued)

Table II Poisson Distribution (λ = 11 to λ = 20)

	λ									
x	11	12	13	14	15	16	17	18	19	20
0	.0000	.0000	.0000	.0000	.0000	.0000	.0000	.0000	.0000	.0000
1	.0002	.0001	.0000	.0000	.0000	.0000	.0000	.0000	.0000	.0000
2	.0010	.0004	.0002	.0001	.0000	.0000	.0000	.0000	.0000	.0000
3	.0037	.0018	.0008	.0004	.0002	.0001	.0000	.0000	.0000	.0000
4	.0102	.0053	.0027	.0013	.0006	.0003	.0001	.0001	.0000	.0000
5	.0224	.0127	.0070	.0037	.0019	.0010	.0005	.0002	.0001	.0001
6	.0411	.0255	.0152	.0087	.0048	.0026	.0014	.0007	.0004	.0002
7	.0646	.0437	.0281	.0174	.0104	.0060	.0034	.0018	.0010	.0005
8	.0888	.0655	.0457	.0304	.0194	.0120	.0072	.0042	.0024	.0013
9	.1085	.0874	.0661	.0473	.0324	.0213	.0135	.0083	.0050	.0029
10	.1194	.1048	.0859	.0063	.0486	.0341	.0230	.0150	.0095	.0058
11	.1194	.1144	.1015	.0844	.0663	.0496	.0355	.0245	.0164	.0106
12	.1094	.1144	.1099	.0984	.0829	.0661	.0504	.0368	.0259	.0176
13	.0926	.1056	.1099	.1060	.0956	.0814	.0658	.0509	.0378	.0271
14	.0728	.0905	.1021	.1060	.1024	.0930	.0800	.0655	.0514	.0387
15	.0534	.0724	.0885	.0989	.1024	.0992	.0906	.0786	.0650	.0516
16	.0367	.0543	.0719	.0866	.0960	.0992	.0963	.0884	.0772	.0646
17	.0237	.0383	.0550	.0713	.0847	.0934	.0963	.0936	.0863	.0760
18	.0145	.0256	.0397	.0554	.0706	.0830	.0909	.0936	.0911	.0844
19	.0084	.0161	.0272	.0409	.0557	.0699	.0814	.0887	.0911	.0888
20	.0046	.0097	.0177	.0286	.0418	.0559	.0692	.0798	.0866	.0888
21	.0024	.0055	.0109	.0191	.0299	.0426	.0560	.0684	.0783	.0846
22	.0012	.0030	.0065	.0121	.0204	.0310	.0433	.0560	.0676	.0769
23	.0006	.0016	.0037	.0074	.0133	.0216	.0320	.0438	.0559	.0669
24	.0003	.0008	.0020	.0043	.0083	.0144	.0226	.0328	.0442	.0557
25	.0001	.0004	.0010	.0024	.0050	.0092	.0154	.0237	.0336	.0446
26	.0000	.0002	.0005	.0013	.0029	.0057	.0101	.0164	.0246	.0343
27	.0000	.0001	.0002	.0007	.0016	.0034	.0063	.0109	.0173	.0254
28	.0000	.0000	.0001	.0003	.0009	.0019	.0038	.0070	.0117	.0181
29	.0000	.0000	.0001	.0002	.0004	.0011	.0023	.0044	.0077	.0125
30	.0000	.0000	.0000	.0001	.0002	.0006	.0013	.0026	.0049	.0083
31	.0000	.0000	.0000	.0000	.0001	.0003	.0007	.0015	.0030	.0054
32	.0000	.0000	.0000	.0000	.0001	.0001	.0004	.0009	.0018	.0034
33	.0000	.0000	.0000	.0000	.0000	.0001	.0002	.0005	.0010	.0020
34	.0000	.0000	.0000	.0000	.0000	.0000	.0001	.0002	.0006	.0012
35	.0000	.0000	.0000	.0000	.0000	.0000	.0000	.0001	.0003	.0007
36	.0000	.0000	.0000	.0000	.0000	.0000	.0000	.0001	.0002	.0004
37	.0000	.0000	.0000	.0000	.0000	.0000	.0000	.0000	.0001	.0002
38	.0000	.0000	.0000	.0000	.0000	.0000	.0000	.0000	.0000	.0001
39	.0000	.0000	.0000	.0000	.0000	.0000	.0000	.0000	.0000	.0001

Table III Cumulative Normal Distribution

$$F(z) = \int_{-\infty}^{z} \frac{1}{\sqrt{2\pi}} e^{-z^2/2}\, dz \quad \textit{Example:}$$

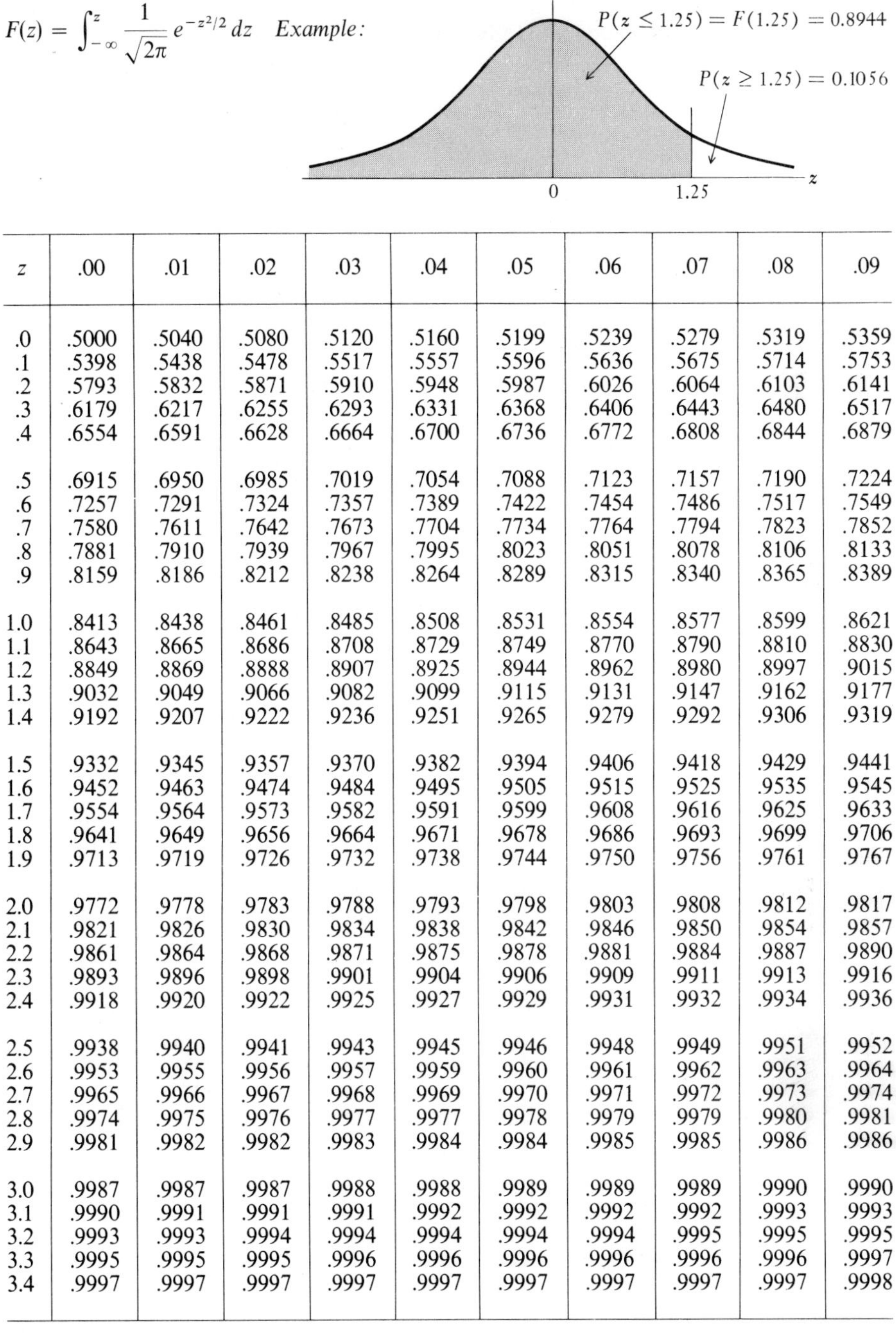

z	.00	.01	.02	.03	.04	.05	.06	.07	.08	.09
.0	.5000	.5040	.5080	.5120	.5160	.5199	.5239	.5279	.5319	.5359
.1	.5398	.5438	.5478	.5517	.5557	.5596	.5636	.5675	.5714	.5753
.2	.5793	.5832	.5871	.5910	.5948	.5987	.6026	.6064	.6103	.6141
.3	.6179	.6217	.6255	.6293	.6331	.6368	.6406	.6443	.6480	.6517
.4	.6554	.6591	.6628	.6664	.6700	.6736	.6772	.6808	.6844	.6879
.5	.6915	.6950	.6985	.7019	.7054	.7088	.7123	.7157	.7190	.7224
.6	.7257	.7291	.7324	.7357	.7389	.7422	.7454	.7486	.7517	.7549
.7	.7580	.7611	.7642	.7673	.7704	.7734	.7764	.7794	.7823	.7852
.8	.7881	.7910	.7939	.7967	.7995	.8023	.8051	.8078	.8106	.8133
.9	.8159	.8186	.8212	.8238	.8264	.8289	.8315	.8340	.8365	.8389
1.0	.8413	.8438	.8461	.8485	.8508	.8531	.8554	.8577	.8599	.8621
1.1	.8643	.8665	.8686	.8708	.8729	.8749	.8770	.8790	.8810	.8830
1.2	.8849	.8869	.8888	.8907	.8925	.8944	.8962	.8980	.8997	.9015
1.3	.9032	.9049	.9066	.9082	.9099	.9115	.9131	.9147	.9162	.9177
1.4	.9192	.9207	.9222	.9236	.9251	.9265	.9279	.9292	.9306	.9319
1.5	.9332	.9345	.9357	.9370	.9382	.9394	.9406	.9418	.9429	.9441
1.6	.9452	.9463	.9474	.9484	.9495	.9505	.9515	.9525	.9535	.9545
1.7	.9554	.9564	.9573	.9582	.9591	.9599	.9608	.9616	.9625	.9633
1.8	.9641	.9649	.9656	.9664	.9671	.9678	.9686	.9693	.9699	.9706
1.9	.9713	.9719	.9726	.9732	.9738	.9744	.9750	.9756	.9761	.9767
2.0	.9772	.9778	.9783	.9788	.9793	.9798	.9803	.9808	.9812	.9817
2.1	.9821	.9826	.9830	.9834	.9838	.9842	.9846	.9850	.9854	.9857
2.2	.9861	.9864	.9868	.9871	.9875	.9878	.9881	.9884	.9887	.9890
2.3	.9893	.9896	.9898	.9901	.9904	.9906	.9909	.9911	.9913	.9916
2.4	.9918	.9920	.9922	.9925	.9927	.9929	.9931	.9932	.9934	.9936
2.5	.9938	.9940	.9941	.9943	.9945	.9946	.9948	.9949	.9951	.9952
2.6	.9953	.9955	.9956	.9957	.9959	.9960	.9961	.9962	.9963	.9964
2.7	.9965	.9966	.9967	.9968	.9969	.9970	.9971	.9972	.9973	.9974
2.8	.9974	.9975	.9976	.9977	.9977	.9978	.9979	.9979	.9980	.9981
2.9	.9981	.9982	.9982	.9983	.9984	.9984	.9985	.9985	.9986	.9986
3.0	.9987	.9987	.9987	.9988	.9988	.9989	.9989	.9989	.9990	.9990
3.1	.9990	.9991	.9991	.9991	.9992	.9992	.9992	.9992	.9993	.9993
3.2	.9993	.9993	.9994	.9994	.9994	.9994	.9994	.9995	.9995	.9995
3.3	.9995	.9995	.9995	.9996	.9996	.9996	.9996	.9996	.9996	.9997
3.4	.9997	.9997	.9997	.9997	.9997	.9997	.9997	.9997	.9997	.9998

Table IV Exponential Distribution

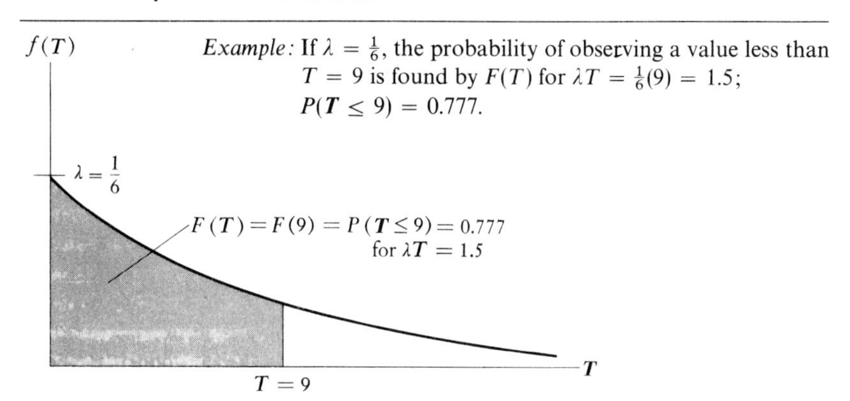

λT	$F(T)$	λT	$F(T)$	λT	$F(T)$	λT	$F(T)$
0.0	0.000	2.5	0.918	5.0	0.9933	7.5	0.99945
0.1	0.095	2.6	0.926	5.1	0.9939	7.6	0.99950
0.2	0.181	2.7	0.933	5.2	0.9945	7.7	0.99955
0.3	0.259	2.8	0.939	5.3	0.9950	7.8	0.99959
0.4	0.330	2.9	0.945	5.4	0.9955	7.9	0.99963
0.5	0.393	3.0	0.950	5.5	0.9959	8.0	0.99966
0.6	0.451	3.1	0.955	5.6	0.9963	8.1	0.99970
0.7	0.503	3.2	0.959	5.7	0.9967	8.2	0.99972
0.8	0.551	3.3	0.963	5.8	0.9970	8.3	0.99975
0.9	0.593	3.4	0.967	5.9	0.9973	8.4	0.99978
1.0	0.632	3.5	0.970	6.0	0.9975	8.5	0.99980
1.1	0.667	3.6	0.973	6.1	0.9978	8.6	0.99982
1.2	0.699	3.7	0.975	6.2	0.9980	8.7	0.99983
1.3	0.727	3.8	0.978	6.3	0.9982	8.8	0.99985
1.4	0.753	3.9	0.980	6.4	0.9983	8.9	0.99986
1.5	0.777	4.0	0.982	6.5	0.9985	9.0	0.99989
1.6	0.798	4.1	0.983	6.6	0.9986	9.1	0.99989
1.7	0.817	4.2	0.985	6.7	0.9988	9.2	0.99990
1.8	0.835	4.3	0.986	6.8	0.9989	9.3	0.99991
1.9	0.850	4.4	0.988	6.9	0.9990	9.4	0.99992
2.0	0.865	4.5	0.989	7.0	0.9991	9.5	0.99992
2.1	0.878	4.6	0.990	7.1	0.9992	9.6	0.99993
2.2	0.889	4.7	0.991	7.2	0.9993	9.7	0.99994
2.3	0.900	4.8	0.992	7.3	0.9993	9.8	0.99994
2.4	0.909	4.9	0.993	7.4	0.9993	9.9	0.99995

Table V Critical Values of the χ^2 Distribution

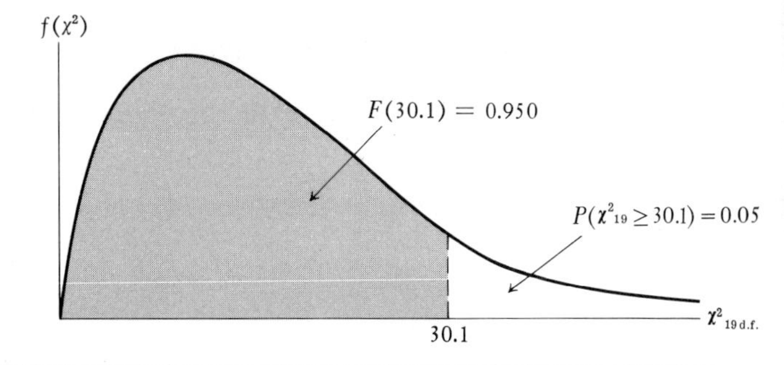

$$F(\chi^2) = \int_0^{\chi^2} \frac{\chi^{(v-2)/2} e^{-\chi/2}\, d\chi}{2^{v/2}[(v - 2/2)]}$$

Example: $P(\chi^2_{19} \leq 30.1)$ for d.f. = 19
$P(\chi^2 \leq 30.1) = F(30.1) = 0.950$

ν \ F	.005	.010	.025	.050	.100	.900	.950	.975	.990	.995
1	$.0^4393$	$.0^3157$	$.0^3982$	$.0^2393$	.0158	2.71	3.84	5.02	6.63	7.88
2	.0100	.0201	.0506	.103	.211	4.61	5.99	7.38	9.21	10.6
3	.0717	.115	.216	.352	.584	6.25	7.81	9.35	11.3	12.8
4	.207	.297	.484	.711	1.06	7.78	9.49	11.1	13.3	14.9
5	.412	.554	.831	1.15	1.61	9.24	11.1	12.8	15.1	16.7
6	.676	.872	1.24	1.64	2.20	10.6	12.6	14.4	16.8	18.5
7	.989	1.24	1.69	2.17	2.83	12.0	14.1	16.0	18.5	20.3
8	1.34	1.65	2.18	2.73	3.49	13.4	15.5	17.5	20.1	22.0
9	1.73	2.09	2.70	3.33	4.17	14.7	16.9	19.0	21.7	23.6
10	2.16	2.56	3.25	3.94	4.87	16.0	18.3	20.5	23.2	25.2
11	2.60	3.05	3.82	4.57	5.58	17.3	19.7	21.9	24.7	26.8
12	3.07	3.57	4.40	5.23	6.30	18.5	21.0	23.3	26.2	28.3
13	3.57	4.11	5.01	5.89	7.04	19.8	22.4	24.7	27.7	29.8
14	4.07	4.66	5.63	6.57	7.79	21.1	23.7	26.1	29.1	31.3
15	4.60	5.23	6.26	7.26	8.55	22.3	25.0	27.5	30.6	32.8
16	5.14	5.81	6.91	7.96	9.31	23.5	26.3	28.8	32.0	34.3
17	5.70	6.41	7.56	8.67	10.1	24.8	27.6	30.2	33.4	35.7
18	6.26	7.01	8.23	9.39	10.9	26.0	28.9	31.5	34.8	37.2
19	6.84	7.63	8.91	10.1	11.7	27.2	30.1	32.9	36.2	38.6
20	7.43	8.26	9.59	10.9	12.4	28.4	31.4	34.2	37.6	40.0
21	8.03	8.90	10.3	11.6	13.2	29.6	32.7	35.5	38.9	41.4
22	8.64	9.54	11.0	12.3	14.0	30.8	33.9	36.8	40.3	42.8
23	9.26	10.2	11.7	13.1	14.8	32.0	35.2	38.1	41.6	44.2
24	9.89	10.9	12.4	13.8	15.7	33.2	36.4	39.4	43.0	45.6
25	10.5	11.5	13.1	14.6	16.5	34.4	37.7	40.6	44.3	46.9
26	11.2	12.2	13.8	15.4	17.3	35.6	38.9	41.9	45.6	48.3
27	11.8	12.9	14.6	16.2	18.1	36.7	40.1	43.2	47.0	49.6
28	12.5	13.6	15.3	16.9	18.9	37.9	41.3	44.5	48.3	51.0
29	13.1	14.3	16.0	17.7	19.8	39.1	42.6	45.7	49.6	52.3
30	13.8	15.0	16.8	18.5	20.6	40.3	43.8	47.0	50.9	53.7
z_α	−2.58	−2.33	−1.96	−1.64	−1.28	+1.28	+1.64	+1.96	+2.33	+2.58

NOTE: For $v > 30$ (i.e., for more than 30 degrees of freedom) take

$$\chi^2 = v\left[1 - \frac{2}{9v} + z_\alpha\sqrt{\frac{2}{9v}}\right]^2 \quad \text{or} \quad \chi^2 = \tfrac{1}{2}[z_\alpha + \sqrt{(2v-1)}]^2$$

according to the degree of accuracy required. z_α is the standarized normal deviate corresponding to the α level of significance, and is shown in the bottom line of the table.

This table is abridged from "Tables of percentage points of the incomplete beta function and of the chi-square distribution," *Biometrika*, Vol. 32 (1941). Reprinted with permission of its author, Catherine M. Thompson, and the editor of *Biometrika*.

Table VI Random Digits

From RAND Corporation, *A Million Random Digits*. By permission.

07018	31172	12572	23968	55216	85366	56223	09300	94564	18172
52444	65625	97918	46794	62370	59344	20149	17596	51669	47429
72161	57299	87521	44351	99981	55008	93371	60620	66662	27036
17918	75071	91057	46829	47992	26797	64423	42379	91676	75127
13623	76165	43195	50205	75736	77473	07268	31330	07337	55901
27426	97534	89707	97453	90836	78967	00704	85734	21776	85764
96039	21338	88169	69530	53300	29895	71507	28517	77761	17244
68282	98888	25545	69406	29470	46476	54562	79373	72993	98998
54262	21477	33097	48125	92982	98382	11265	25366	06636	25349
66290	27544	72780	91384	47296	54892	59168	83951	91075	04724
53348	39044	04072	62210	01209	43999	54952	68699	31912	09317
34482	42758	40128	48436	30254	50029	19016	56837	05206	33851
99268	98715	07545	27317	52459	75366	43688	27460	65145	65429
95342	97178	10401	31615	95784	77026	33087	65961	10056	72834
38556	60373	77935	64608	28949	94764	45312	71171	15400	72182
39159	04795	51163	84475	60722	35268	05044	56420	39214	89822
41786	18169	96649	92406	42773	23672	37333	85734	99886	81200
95627	30768	30607	89023	60730	31519	53462	90489	81693	17849
98738	15548	42263	79489	85118	97073	01574	57310	59375	54417
75214	61575	27805	21930	94726	39454	19616	72239	93791	22610
73904	89123	19271	15792	72675	62175	48746	56084	54029	22296
33329	08896	94662	05781	59187	53284	28024	45421	37956	14252
66364	94799	62211	37539	80172	43269	91133	05562	82385	91760
68349	16984	86532	96186	53893	48268	82821	19526	63257	14288
19193	99621	66899	12351	72438	99839	24228	32079	53517	18558
09237	23489	19172	80439	76263	98918	59330	20121	89779	58862
11007	77008	27646	82072	28048	41589	70883	72035	81800	50296
60622	25875	26446	25738	32962	24266	26814	01194	48587	93319
79973	26895	65304	34978	43053	28951	22676	05303	39725	60054
71080	74487	83196	61939	05045	20405	69321	80823	20905	68727
09923	36773	21247	54735	68996	16937	18134	51873	10973	77090
63094	85087	94186	67793	18178	82224	17069	87880	54945	73489
19806	76028	54285	90845	35464	68076	15868	70063	26794	81386
17295	78454	21700	12301	88832	96796	59341	16136	01803	17537
59338	61051	97260	89829	69121	86547	62195	72492	33536	60137

Table VII Critical Values of the t-Distribution

$$F(t) = \int_{-\infty}^{t} \frac{\left(\frac{\nu - 1}{2}\right)!}{\left(\frac{\nu - 2}{2}\right)! \sqrt{\pi n} \left(1 + \frac{t^2}{\nu}\right)^{(\nu+1)/2}} \, dt$$

$f(t)$

$F(t) = P(t_{19} \leq 2.093) = 0.975$

0.025

0 2.093 $t_{19 \text{ d.f.}}$

Example: $n = 20$, $\nu = 19$

ν \ F	.75	.90	.95	.975	.99	.995	.9995
1	1.000	3.078	6.314	12.706	31.821	63.657	636.619
2	.816	1.886	2.920	4.303	6.965	9.925	31.598
3	.765	1.638	2.353	3.182	4.541	5.841	12.941
4	.741	1.533	2.132	2.776	3.747	4.604	8.610
5	.727	1.476	2.015	2.571	3.365	4.032	6.859
6	.718	1.440	1.943	2.447	3.143	3.707	5.959
7	.711	1.415	1.895	2.365	2.998	3.499	5.405
8	.706	1.397	1.860	2.306	2.896	3.355	5.041
9	.703	1.383	1.833	2.262	2.821	3.250	4.781
10	.700	1.372	1.812	2.228	2.764	3.169	4.587
11	.697	1.363	1.796	2.201	2.718	3.106	4.437
12	.695	1.356	1.782	2.179	2.681	3.055	4.318
13	.694	1.350	1.771	2.160	2.650	3.012	4.221
14	.692	1.345	1.761	2.145	2.624	2.977	4.140
15	.691	1.341	1.753	2.131	2.602	2.947	4.073
16	.690	1.337	1.746	2.120	2.583	2.921	4.015
17	.689	1.333	1.740	2.110	2.567	2.898	3.965
18	.688	1.330	1.734	2.101	2.552	2.878	3.922
19	.688	1.328	1.729	2.093	2.539	2.861	3.883
20	.687	1.325	1.725	2.086	2.528	2.845	3.850
21	.686	1.323	1.721	2.080	2.518	2.831	3.819
22	.686	1.321	1.717	2.074	2.508	2.819	3.792
23	.685	1.319	1.714	2.069	2.500	2.807	3.767
24	.685	1.318	1.711	2.064	2.492	2.797	3.745
25	.684	1.316	1.708	2.060	2.485	2.787	3.725
26	.684	1.315	1.706	2.056	2.479	2.779	3.707
27	.684	1.314	1.703	2.052	2.473	2.771	3.690
28	.683	1.313	1.701	2.048	2.467	2.763	3.674
29	.683	1.311	1.699	2.045	2.462	2.756	3.659
30	.683	1.310	1.697	2.042	2.457	2.750	3.646
40	.681	1.303	1.684	2.021	2.423	2.704	3.551
60	.679	1.296	1.671	2.000	2.390	2.660	3.460
120	.677	1.289	1.658	1.980	2.358	2.617	3.373
∞	.674	1.282	1.645	1.960	2.326	2.576	3.291

This table is abridged from Table III of Fisher & Yates: *Statistical Tables for Biological, Agricultural and Medical Research*, published by Longman Group Ltd. London (previously published by Oliver & Boyd Ltd. Edinburgh) and by permission of the authors and publishers.

Table VIII(a) Critical Values of the F-Distribution ($\alpha = 0.05$)

Tables VIII(a), (b), and (c) from M. Merrington and C. M. Thompson, "Tables of percentage points of the inverted beta (F) distribution." *Biometrika*, Vol. 33 (1943) by permission of the *Biometrika* Trustees.

The following table gives the critical values of the $\boldsymbol{F}$-distribution for $\alpha = 0.05$. This probability represents the area exceeding the value of $F_{0.05,\, \nu_1,\, \nu_2}$, as shown by the shaded area in the figure below.

Examples: If $\nu_1 = 15$ (d.f. for the numerator), and $\nu_2 = 20$, then the critical value cutting off 0.05 is 2.20.

$$P(F \geq 2.20) = 0.05,$$

$$P(F \leq 2.20) = 0.95.$$

Values of $F_{0.05,\, \nu_1,\, \nu_2}$

ν_1 = Degrees of freedom for numerator

ν_2 = Degrees of freedom for denominator

ν_2 \ ν_1	1	2	3	4	5	6	7	8	9	10	12	15	20	24	30	40	60	120	∞
1	161	200	216	225	230	234	237	239	241	242	244	246	248	249	250	251	252	253	254
2	18.5	19.0	19.2	19.2	19.3	19.3	19.4	19.4	19.4	19.4	19.4	19.4	19.4	19.5	19.5	19.5	19.5	19.5	19.5
3	10.1	9.55	9.28	9.12	9.01	8.94	8.89	8.85	8.81	8.79	8.74	8.70	8.66	8.64	8.62	8.59	8.57	8.55	8.53
4	7.71	6.94	6.59	6.39	6.26	6.16	6.09	6.04	6.00	5.96	5.91	5.86	5.80	5.77	5.75	5.72	5.69	5.66	5.63
5	6.61	5.79	5.41	5.19	5.05	4.95	4.88	4.82	4.77	4.74	4.68	4.62	4.56	4.53	4.50	4.46	4.43	4.40	4.37
6	5.99	5.14	4.76	4.53	4.39	4.28	4.21	4.15	4.10	4.06	4.00	3.94	3.87	3.84	3.81	3.77	3.74	3.70	3.67
7	5.59	4.74	4.35	4.12	3.97	3.87	3.79	3.73	3.68	3.64	3.57	3.51	3.44	3.41	3.38	3.34	3.30	3.27	3.23
8	5.32	4.46	4.07	3.84	3.69	3.58	3.50	3.44	3.39	3.35	3.28	3.22	3.15	3.12	3.08	3.04	3.01	2.97	2.93
9	5.12	4.26	3.86	3.63	3.48	3.37	3.29	3.23	3.18	3.14	3.07	3.01	2.94	2.90	2.86	2.83	2.79	2.75	2.71
10	4.96	4.10	3.71	3.48	3.33	3.22	3.14	3.07	3.02	2.98	2.91	2.85	2.77	2.74	2.70	2.66	2.62	2.58	2.54
11	4.84	3.98	3.59	3.36	3.20	3.09	3.01	2.95	2.90	2.85	2.79	2.72	2.65	2.61	2.57	2.53	2.49	2.45	2.40
12	4.75	3.89	3.49	3.26	3.11	3.00	2.91	2.85	2.80	2.75	2.69	2.62	2.54	2.51	2.47	2.43	2.38	2.34	2.30
13	4.67	3.81	3.41	3.18	3.03	2.92	2.83	2.77	2.71	2.67	2.60	2.53	2.46	2.42	2.38	2.34	2.30	2.25	2.21
14	4.60	3.74	3.34	3.11	2.96	2.85	2.76	2.70	2.65	2.60	2.53	2.46	2.39	2.35	2.31	2.27	2.22	2.18	2.13
15	4.54	3.68	3.29	3.06	2.90	2.79	2.71	2.64	2.59	2.54	2.48	2.40	2.33	2.29	2.25	2.20	2.16	2.11	2.07
16	4.49	3.63	3.24	3.01	2.85	2.74	2.66	2.59	2.54	2.49	2.42	2.35	2.28	2.24	2.19	2.15	2.11	2.06	2.01
17	4.45	3.59	3.20	2.96	2.81	2.70	2.61	2.55	2.49	2.45	2.38	2.31	2.23	2.19	2.15	2.10	2.06	2.01	1.96
18	4.41	3.55	3.16	2.93	2.77	2.66	2.58	2.51	2.46	2.41	2.34	2.27	2.19	2.15	2.11	2.06	2.02	1.97	1.92
19	4.38	3.52	3.13	2.90	2.74	2.63	2.54	2.48	2.42	2.38	2.31	2.23	2.16	2.11	2.07	2.03	1.98	1.93	1.88
20	4.35	3.49	3.10	2.87	2.71	2.60	2.51	2.45	2.39	2.35	2.28	2.20	2.12	2.08	2.04	1.99	1.95	1.90	1.84
21	4.32	3.47	3.07	2.84	2.68	2.57	2.49	2.42	2.37	2.32	2.25	2.18	2.10	2.05	2.01	1.96	1.92	1.87	1.81
22	4.30	3.44	3.05	2.82	2.66	2.55	2.46	2.40	2.34	2.30	2.23	2.15	2.07	2.03	1.98	1.94	1.89	1.84	1.78
23	4.28	3.42	3.03	2.80	2.64	2.53	2.44	2.37	2.32	2.27	2.20	2.13	2.05	2.01	1.96	1.91	1.86	1.81	1.76
24	4.26	3.40	3.01	2.78	2.62	2.51	2.42	2.36	2.30	2.25	2.18	2.11	2.03	1.98	1.94	1.89	1.84	1.79	1.73
25	4.24	3.39	2.99	2.76	2.60	2.49	2.40	2.34	2.28	2.24	2.16	2.09	2.01	1.96	1.92	1.87	1.82	1.77	1.71
30	4.17	3.32	2.92	2.69	2.53	2.42	2.33	2.27	2.21	2.16	2.09	2.01	1.93	1.89	1.84	1.79	1.74	1.68	1.62
40	4.08	3.23	2.84	2.61	2.45	2.34	2.25	2.18	2.12	2.08	2.00	1.92	1.84	1.79	1.74	1.69	1.64	1.58	1.51
60	4.00	3.15	2.76	2.53	2.37	2.25	2.17	2.10	2.04	1.99	1.92	1.84	1.75	1.70	1.65	1.59	1.53	1.47	1.39
120	3.92	3.07	2.68	2.45	2.29	2.18	2.09	2.02	1.96	1.91	1.83	1.75	1.66	1.61	1.55	1.50	1.43	1.35	1.25
∞	3.84	3.00	2.60	2.37	2.21	2.10	2.01	1.94	1.88	1.83	1.75	1.67	1.57	1.52	1.46	1.39	1.32	1.22	1.00

Table VIII(b) Critical Values of the F-Distribution ($\alpha = 0.025$)

The following table gives the critical values of the F-distribution for $\alpha = 0.025$. This probability represents the area exceeding the value of $F_{0.025, \nu_1, \nu_2}$, as shown by the shaded area in the figure below.

Examples: If $\nu_1 = 15$ (representing the greater mean square), and $\nu_2 = 20$, then the critical value for $\alpha = 0.025$ is 2.570.

$$P(F \geq 2.570) = 0.025,$$

$$P(F \leq 2.570) = 0.975.$$

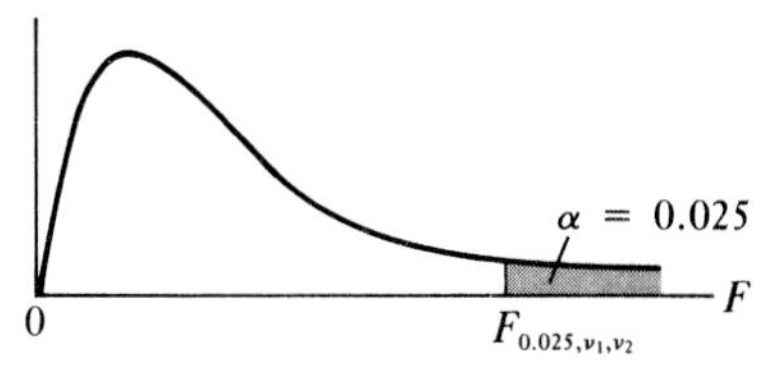

VALUES OF $F_{0.025, \nu_1, \nu_2}$

ν_1 = Degrees of freedom for numerator

ν_2 = Degrees of freedom for denominator

ν_2	1	2	3	4	5	6	7	8	9	10	12	15	20	24	30	40	60	120	∞
1	647.8	799.5	864.2	899.6	921.8	937.1	948.2	956.7	963.3	968.6	976.7	984.9	993.1	997.2	1001	1006	1010	1014	1018
2	38.51	39.00	39.17	39.25	39.30	39.33	39.36	39.37	39.39	39.40	39.41	39.43	39.45	39.46	39.46	39.47	39.48	39.49	39.50
3	17.44	16.04	15.44	15.10	14.88	14.73	14.62	14.54	14.47	14.42	14.34	14.25	14.17	14.12	14.08	14.04	13.99	13.95	13.90
4	12.22	10.65	9.98	9.60	9.36	9.20	9.07	8.98	8.90	8.84	8.75	8.66	8.56	8.51	8.46	8.41	8.36	8.31	8.26
5	10.01	8.43	7.76	7.39	7.15	6.98	6.85	6.76	6.68	6.62	6.52	6.43	6.33	6.28	6.23	6.18	6.12	6.07	6.02
6	8.81	7.26	6.60	6.23	5.99	5.82	5.70	5.60	5.52	5.46	5.37	5.27	5.17	5.12	5.07	5.01	4.96	4.90	4.85
7	8.07	6.54	5.89	5.52	5.29	5.21	4.99	4.90	4.82	4.76	4.67	4.57	4.47	4.42	4.36	4.31	4.25	4.20	4.14
8	7.57	6.06	5.42	5.05	4.82	4.65	4.53	4.43	4.36	4.30	4.20	4.10	4.00	3.95	3.89	3.84	3.78	3.73	3.67
9	7.21	5.71	5.08	4.72	4.48	4.32	4.20	4.10	4.03	3.96	3.87	3.77	3.67	3.61	3.56	3.51	3.45	3.39	3.33
10	6.94	5.46	4.83	4.47	4.24	4.07	3.95	3.85	3.78	3.72	3.62	3.52	3.42	3.37	3.31	3.26	3.20	3.14	3.08
11	6.72	5.26	4.63	4.28	4.04	3.88	3.76	3.66	3.59	3.53	3.43	3.33	3.23	3.17	3.12	3.06	3.00	2.94	2.88
12	6.55	5.10	4.47	4.12	3.89	3.73	3.61	3.51	3.44	3.37	3.28	3.18	3.07	3.02	2.96	2.91	2.85	2.79	2.72
13	6.41	4.97	4.35	4.00	3.77	3.60	3.48	3.39	3.31	3.25	3.15	3.05	2.95	2.89	2.84	2.78	2.72	2.66	2.60
14	6.30	4.86	4.24	3.89	3.66	3.50	3.38	3.29	3.21	3.15	3.05	2.95	2.84	2.79	2.73	2.67	2.61	2.55	2.49
15	6.20	4.77	4.15	3.80	3.58	3.41	3.29	3.20	3.12	3.06	2.96	2.86	2.76	2.70	2.64	2.59	2.52	2.46	2.40
16	6.12	4.69	4.08	3.73	3.50	3.34	3.22	3.12	3.05	2.99	2.89	2.79	2.68	2.63	2.57	2.51	2.45	2.38	2.32
17	6.04	4.62	4.01	3.66	3.44	3.28	3.16	3.06	2.98	2.92	2.82	2.72	2.62	2.56	2.50	2.44	2.38	2.32	2.25
18	5.98	4.56	3.95	3.61	3.38	3.22	3.10	3.01	2.93	2.87	2.77	2.67	2.56	2.50	2.44	2.38	2.32	2.26	2.19
19	5.92	4.51	3.90	3.56	3.33	3.17	3.05	2.96	2.88	2.82	2.72	2.62	2.51	2.45	2.39	2.33	2.27	2.20	2.13
20	5.87	4.46	3.86	3.51	3.29	3.13	3.01	2.91	2.84	2.77	2.68	2.57	2.46	2.41	2.35	2.29	2.22	2.16	2.09
21	5.83	4.42	3.82	3.48	3.25	3.09	2.97	2.87	2.80	2.73	2.64	2.53	2.42	2.37	2.31	2.25	2.18	2.11	2.04
22	5.79	4.38	3.78	3.44	3.22	3.05	2.93	2.84	2.76	2.70	2.60	2.50	2.39	2.33	2.27	2.21	2.14	2.08	2.00
23	5.75	4.35	3.75	3.41	3.18	3.02	2.90	2.81	2.73	2.67	2.57	2.47	2.36	2.30	2.24	2.18	2.11	2.04	1.97
24	5.72	4.32	3.72	3.38	3.15	2.99	2.87	2.78	2.70	2.54	2.54	2.44	2.33	2.27	2.21	2.15	2.08	2.01	1.94
25	5.69	4.29	3.69	3.35	3.13	2.97	2.85	2.75	2.68	2.61	2.51	2.41	2.30	2.24	2.18	2.12	2.05	1.98	1.91
26	5.66	4.27	3.67	3.33	3.10	2.94	2.82	2.73	2.65	2.59	2.49	2.39	2.28	2.22	2.16	2.09	2.03	1.95	1.88
27	5.63	4.24	3.65	3.31	3.08	2.92	2.80	2.71	2.63	2.57	2.47	2.36	2.25	2.19	2.13	2.07	2.00	1.93	1.85
28	5.61	4.22	3.63	3.29	3.06	2.90	2.78	2.69	2.61	2.55	2.45	2.34	2.23	2.17	2.11	2.05	1.98	1.91	1.83
29	5.59	4.20	3.61	3.27	3.04	2.88	2.76	2.67	2.59	2.53	2.43	2.32	2.21	2.15	2.09	2.03	1.96	1.89	1.81
30	5.57	4.18	3.59	3.25	3.03	2.87	2.75	2.65	2.57	2.51	2.41	2.31	2.20	2.14	2.07	2.01	1.94	1.87	1.79
40	5.42	4.05	3.46	3.13	2.90	2.74	2.62	2.53	2.45	2.39	2.29	2.18	2.07	2.01	1.94	1.88	1.80	1.72	1.64
60	5.29	3.93	3.34	3.01	2.79	2.63	2.51	2.41	2.33	2.27	2.17	2.06	1.94	1.88	1.82	1.74	1.67	1.58	1.48
120	5.15	3.80	3.23	2.89	2.67	2.52	2.39	2.30	2.22	2.16	2.05	1.94	1.82	1.76	1.69	1.61	1.53	1.43	1.31
∞	5.02	3.69	3.12	2.79	2.57	2.41	2.29	2.19	2.11	2.05	1.94	1.83	1.71	1.64	1.57	1.48	1.39	1.27	1.00

Table VIII(c) Critical Values of the F-Distribution ($\alpha = 0.01$)

The following table gives the critical values of the $\boldsymbol{F}$-distribution for $\alpha = 0.01$. This probability represents the area exceeding the value of $F_{0.01, v_1, v_2}$, as shown by the shaded area in the figure below.

Examples: If $v_1 = 15$ (representing the greater mean square), and $v_2 = 20$, then the critical value for $\alpha = 0.01$ is 3.09.

$$P(F \geq 3.09) = 0.01,$$
$$P(F \leq 3.09) = 0.99.$$

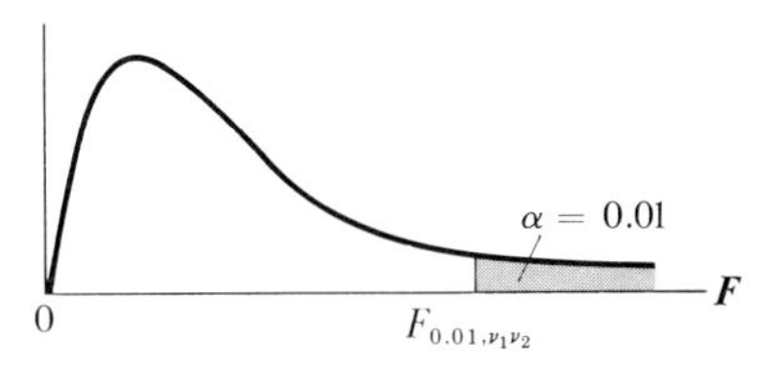

VALUES OF $F_{0.01, v_1, v_2}$

v_1 = Degrees of freedom for numerator

v_2 = Degrees of freedom for denominator

v_2	1	2	3	4	5	6	7	8	9	10	12	15	20	24	30	40	60	120	∞
1	4,052	5,000	5,403	5,625	5,764	5,859	5,928	5,982	6,023	6,056	6,106	6,157	6,209	6,235	6,261	6,287	6,313	6,339	6,366
2	98.5	99.0	99.2	99.2	99.3	99.3	99.4	99.4	99.4	99.4	99.4	99.4	99.4	99.5	99.5	99.5	99.5	99.5	99.5
3	34.1	30.8	29.5	28.7	28.2	27.9	27.7	27.5	27.3	27.2	27.1	26.9	26.7	26.6	26.5	26.4	26.3	26.2	26.1
4	21.2	18.0	16.7	16.0	15.5	15.2	15.0	14.8	14.7	14.5	14.4	14.2	14.0	13.9	13.8	13.7	13.7	13.6	13.5
5	16.3	13.3	12.1	11.4	11.0	10.7	10.5	10.3	10.2	10.1	9.89	9.72	9.55	9.47	9.38	9.29	9.20	9.11	9.02
6	13.7	10.9	9.78	9.15	8.75	8.47	8.26	8.10	7.98	7.87	7.72	7.56	7.40	7.31	7.23	7.14	7.06	6.97	6.88
7	12.2	9.55	8.45	7.85	7.46	7.19	6.99	6.84	6.72	6.62	6.47	6.31	6.16	6.07	5.99	5.91	5.82	5.74	5.65
8	11.3	8.65	7.59	7.01	6.63	6.37	6.18	6.03	5.91	5.81	5.67	5.52	5.36	5.28	5.20	5.12	5.03	4.95	4.86
9	10.6	8.02	6.99	6.42	6.06	5.80	5.61	5.47	5.35	5.26	5.11	4.96	4.81	4.73	4.65	4.57	4.48	4.40	4.31
10	10.0	7.56	6.55	5.99	5.64	5.39	5.20	5.06	4.94	4.85	4.71	4.56	4.41	4.33	4.25	4.17	4.08	4.00	3.91
11	9.65	7.21	6.22	5.67	5.32	5.07	4.89	4.74	4.63	4.54	4.40	4.25	4.10	4.02	3.94	3.86	3.78	3.69	3.60
12	9.33	6.93	5.95	5.41	5.06	4.82	4.64	4.50	4.39	4.30	4.16	4.01	3.86	3.78	3.70	3.62	3.54	3.45	3.36
13	9.07	6.70	5.74	5.21	4.86	4.62	4.44	4.30	4.19	4.10	3.96	3.82	3.66	3.59	3.51	3.43	3.34	3.25	3.17
14	8.86	6.51	5.56	5.04	4.70	4.46	4.28	4.14	4.03	3.94	3.80	3.66	3.51	3.43	3.35	3.27	3.18	3.09	3.00
15	8.68	6.36	5.42	4.89	4.56	4.32	4.14	4.00	3.89	3.80	3.67	3.52	3.37	3.29	3.21	3.13	3.05	2.96	2.87
16	8.53	6.23	5.29	4.77	4.44	4.20	4.03	3.89	3.78	3.69	3.55	3.41	3.26	3.18	3.10	3.02	2.93	2.84	2.75
17	8.40	6.11	5.19	4.67	4.34	4.10	3.93	3.79	3.68	3.59	3.46	3.31	3.16	3.08	3.00	2.92	2.83	2.75	2.65
18	8.29	6.01	5.09	4.58	4.25	4.01	3.84	3.71	3.60	3.51	3.37	3.23	3.08	3.00	2.92	2.84	2.75	2.66	2.57
19	8.19	5.93	5.01	4.50	4.17	3.94	3.77	3.63	3.52	3.43	3.30	3.15	3.00	2.92	2.84	2.76	2.67	2.58	2.49
20	8.10	5.85	4.94	4.43	4.10	3.87	3.70	3.56	3.46	3.37	3.23	3.09	2.94	2.86	2.78	2.69	2.61	2.52	2.42
21	8.02	5.78	4.87	3.37	4.04	3.81	3.64	3.51	3.40	3.31	3.17	3.03	2.88	2.80	2.72	2.64	2.55	2.46	2.36
22	7.95	5.72	4.82	4.31	3.99	3.76	3.59	3.45	3.35	3.26	3.12	2.98	2.83	2.75	2.67	2.58	2.50	2.40	2.31
23	7.88	5.66	4.76	4.26	3.94	3.71	3.54	3.41	3.30	3.21	3.07	2.93	2.78	2.70	2.62	2.54	2.45	2.35	2.26
24	7.82	5.61	4.72	4.22	3.90	3.67	3.50	3.36	3.26	3.17	3.03	2.89	2.74	2.66	2.58	2.49	2.40	2.31	2.21
25	7.77	5.57	4.68	4.18	3.86	3.63	3.46	3.32	3.22	3.13	2.99	2.85	2.70	2.62	2.53	2.45	2.36	2.27	2.17
30	7.56	5.39	4.51	4.02	3.70	3.47	3.30	3.17	3.07	2.98	2.84	2.70	2.55	2.47	2.39	2.30	2.21	2.11	2.01
40	7.31	5.18	4.31	3.83	3.51	3.29	3.12	2.99	2.89	2.80	2.66	2.52	2.37	2.29	2.20	2.11	2.02	1.92	1.80
60	7.08	4.98	4.13	3.65	3.34	3.12	2.95	2.82	2.72	2.63	2.50	2.35	2.20	2.12	2.03	1.94	1.84	1.73	1.60
120	6.85	4.79	3.95	3.48	3.17	2.96	2.79	2.66	2.56	2.47	2.34	2.19	2.03	1.95	1.86	1.76	1.66	1.53	1.38
∞	6.63	4.61	3.78	3.32	3.02	2.80	2.64	2.51	2.41	2.32	2.18	2.04	1.88	1.79	1.70	1.59	1.47	1.32	1.00

Table IX Values of

$$z = \frac{1}{2}\ln\frac{1+r}{1-r}$$

(For negative values of r put a minus sign in front of the tabled numbers.)

r	0.00	0.01	0.02	0.03	0.04	0.05	0.06	0.07	0.08	0.09
0.0	0.0000	.01000	.02000	.03001	.04002	.05004	.06007	.07012	.08017	.09024
0.1	.10034	.11045	.12058	.13074	.14093	.15114	.16139	.17167	.18198	.19234
0.2	.20273	.21317	.22366	.23419	.24477	.25541	.26611	.27686	.28768	.29857
0.3	.30952	.32055	.33165	.34283	.35409	.36544	.37689	.38842	.40006	.41180
0.4	.42365	.43561	.44769	.45990	.47223	.48470	.49731	.51007	.52298	.53606
0.5	.54931	.56273	.57634	.59014	.60415	.61838	.63283	.64752	.66246	.67767
0.6	.69315	.70892	.72500	.74142	.75817	.77530	.79281	.81074	.82911	.84795
0.7	.86730	.88718	.90764	.92873	.95048	.97295	.99621	1.02033	1.04537	1.07143
0.8	1.09861	1.12703	1.15682	1.18813	1.22117	1.25615	1.29334	1.33308	1.37577	1.42192
0.9	1.47222	1.52752	1.58902	1.65839	1.73805	1.83178	1.94591	1.09229	2.29756	2.64665

Table X The Durbin–Watson *d*-Statistic

Significance points of d_L and d_U: 5%

n	$m = 1$		$m = 2$		$m = 3$		$m = 4$		$m = 5$	
	d_L	d_U	d_L	d_U	d_L	d_U	d_L	d_U	d_L	d_U
15	1.08	1.36	0.95	1.54	0.82	1.75	0.69	1.97	0.56	2.21
16	1.10	1.37	0.98	1.54	0.86	1.73	0.74	1.93	0.62	2.15
17	1.13	1.38	1.02	1.54	0.90	1.71	0.78	1.90	0.67	2.10
18	1.16	1.39	1.05	1.53	0.93	1.69	0.82	1.87	0.71	2.06
19	1.18	1.40	1.08	1.53	0.97	1.68	0.86	1.85	0.75	2.02
20	1.20	1.41	1.10	1.54	1.00	1.68	0.90	1.83	0.79	1.99
21	1.22	1.42	1.13	1.54	1.03	1.67	0.93	1.81	0.83	1.96
22	1.24	1.43	1.15	1.54	1.05	1.66	0.96	1.80	0.86	1.94
23	1.26	1.44	1.17	1.54	1.08	1.66	0.99	1.79	0.90	1.92
24	1.27	1.45	1.19	1.55	1.10	1.66	1.01	1.78	0.93	1.90
25	1.29	1.45	1.21	1.55	1.12	1.66	1.04	1.77	0.95	1.89
26	1.30	1.46	1.22	1.55	1.14	1.65	1.06	1.76	0.98	1.88
27	1.32	1.47	1.24	1.56	1.16	1.65	1.08	1.76	1.01	1.86
28	1.33	1.48	1.26	1.56	1.18	1.65	1.10	1.75	1.03	1.85
29	1.34	1.48	1.27	1.56	1.20	1.65	1.12	1.74	1.05	1.84
30	1.35	1.49	1.28	1.57	1.21	1.65	1.14	1.74	1.07	1.83
31	1.36	1.50	1.30	1.57	1.23	1.65	1.16	1.74	1.09	1.83
32	1.37	1.50	1.31	1.57	1.24	1.65	1.18	1.73	1.11	1.82
33	1.38	1.51	1.32	1.58	1.26	1.65	1.19	1.73	1.13	1.81
34	1.39	1.51	1.33	1.58	1.27	1.65	1.21	1.73	1.15	1.81
35	1.40	1.52	1.34	1.58	1.28	1.65	1.22	1.73	1.16	1.80
36	1.41	1.52	1.35	1.59	1.29	1.65	1.24	1.73	1.18	1.80
37	1.42	1.53	1.36	1.59	1.31	1.66	1.25	1.72	1.19	1.80
38	1.43	1.54	1.37	1.59	1.32	1.66	1.26	1.72	1.21	1.79
39	1.43	1.54	1.38	1.60	1.33	1.66	1.27	1.72	1.22	1.79
40	1.44	1.54	1.39	1.60	1.34	1.66	1.29	1.72	1.23	1.79
45	1.48	1.57	1.43	1.62	1.38	1.67	1.34	1.72	1.29	1.78
50	1.50	1.59	1.46	1.63	1.42	1.67	1.38	1.72	1.34	1.77
55	1.53	1.60	1.49	1.64	1.45	1.68	1.41	1.72	1.38	1.77
60	1.55	1.62	1.51	1.65	1.48	1.69	1.44	1.73	1.41	1.77
65	1.57	1.63	1.54	1.66	1.50	1.70	1.47	1.73	1.44	1.77
70	1.58	1.64	1.55	1.67	1.52	1.70	1.49	1.74	1.46	1.77
75	1.60	1.65	1.57	1.68	1.54	1.71	1.51	1.74	1.49	1.77
80	1.61	1.66	1.59	1.69	1.56	1.72	1.53	1.74	1.51	1.77
85	1.62	1.67	1.60	1.70	1.57	1.72	1.55	1.75	1.52	1.77
90	1.63	1.68	1.61	1.70	1.59	1.73	1.57	1.75	1.54	1.78
95	1.64	1.69	1.62	1.71	1.60	1.73	1.58	1.75	1.56	1.78
100	1.65	1.69	1.63	1.72	1.61	1.74	1.59	1.76	1.57	1.78

Table X (continued)

Significance points of d_L and d_U: 1%

n	$m = 1$		$m = 2$		$m = 3$		$m = 4$		$m = 5$	
	d_L	d_U	d_L	d_U	d_L	d_U	d_L	d_U	d_L	d_U
15	0.81	1.07	0.70	1.25	0.59	1.46	0.49	1.70	0.39	1.96
16	0.84	1.09	0.74	1.25	0.63	1.44	0.53	1.66	0.44	1.90
17	0.87	1.10	0.77	1.25	0.67	1.43	0.57	1.63	0.48	1.85
18	0.90	1.12	0.80	1.26	0.71	1.42	0.61	1.60	0.52	1.80
19	0.93	1.13	0.83	1.26	0.74	1.41	0.65	1.58	0.56	1.77
20	0.95	1.15	0.86	1.27	0.77	1.41	0.68	1.57	0.60	1.74
21	0.97	1.16	0.89	1.27	0.80	1.41	0.72	1.55	0.63	1.71
22	1.00	1.17	0.91	1.28	0.83	1.40	0.75	1.54	0.66	1.69
23	1.02	1.19	0.94	1.29	0.86	1.40	0.77	1.53	0.70	1.67
24	1.04	1.20	0.96	1.30	0.88	1.41	0.80	1.53	0.72	1.66
25	1.05	1.21	0.98	1.30	0.90	1.41	0.83	1.52	0.75	1.65
26	1.07	1.22	1.00	1.31	0.93	1.41	0.85	1.52	0.78	1.64
27	1.09	1.23	1.02	1.32	0.95	1.41	0.88	1.51	0.81	1.63
28	1.10	1.24	1.04	1.32	0.97	1.41	0.90	1.51	0.83	1.62
29	1.12	1.25	1.05	1.33	0.99	1.42	0.92	1.51	0.85	1.61
30	1.13	1.26	1.07	1.34	1.01	1.42	0.94	1.51	0.88	1.61
31	1.15	1.27	1.08	1.34	1.02	1.42	0.96	1.51	0.90	1.60
32	1.16	1.28	1.10	1.35	1.04	1.43	0.98	1.51	0.92	1.60
33	1.17	1.29	1.11	1.36	1.05	1.43	1.00	1.51	0.94	1.59
34	1.18	1.30	1.13	1.36	1.07	1.43	1.01	1.51	0.95	1.59
35	1.19	1.31	1.14	1.37	1.08	1.44	1.03	1.51	0.97	1.59
36	1.21	1.32	1.15	1.38	1.10	1.44	1.04	1.51	0.99	1.59
37	1.22	1.32	1.16	1.38	1.11	1.45	1.06	1.51	1.00	1.59
38	1.23	1.33	1.18	1.39	1.12	1.45	1.07	1.52	1.02	1.58
39	1.24	1.34	1.19	1.39	1.14	1.45	1.09	1.52	1.03	1.58
40	1.25	1.34	1.20	1.40	1.15	1.46	1.10	1.52	1.05	1.58
45	1.29	1.38	1.24	1.42	1.20	1.48	1.16	1.53	1.11	1.58
50	1.32	1.40	1.28	1.45	1.24	1.49	1.20	1.54	1.16	1.59
55	1.36	1.43	1.32	1.47	1.28	1.51	1.25	1.55	1.21	1.59
60	1.38	1.45	1.35	1.48	1.32	1.52	1.28	1.56	1.25	1.60
65	1.41	1.47	1.38	1.50	1.35	1.53	1.31	1.57	1.28	1.61
70	1.43	1.49	1.40	1.52	1.37	1.55	1.34	1.58	1.31	1.61
75	1.45	1.50	1.42	1.53	1.39	1.56	1.37	1.59	1.34	1.62
80	1.47	1.52	1.44	1.54	1.42	1.57	1.39	1.60	1.36	1.62
85	1.48	1.53	1.46	1.55	1.43	1.58	1.41	1.60	1.39	1.63
90	1.50	1.54	1.47	1.56	1.45	1.59	1.43	1.61	1.41	1.64
95	1.51	1.55	1.49	1.57	1.47	1.60	1.45	1.62	1.42	1.64
100	1.52	1.56	1.50	1.58	1.48	1.60	1.46	1.63	1.44	1.65

SOURCE: Reproduced by permission of the editor and authors, from J. Durbin and G. S. Watson, "Testing for serial correlation in least squares regression, (II)," *Biometrika*, **38**, 1951, pp. 159–178.

Table XI Critical Values of r in the Runs Test

Given in the body of Table XI are various critical values of r for various values of n_1 and n_2. For the Wald–Wolfowitz two-sample runs test, any value of r which is equal to or smaller than that shown in Table XI is significant at the 0.05 level.

n_1 \ n_2	2	3	4	5	6	7	8	9	10	11	12	13	14	15	16	17	18	19	20
2											2	2	2	2	2	2	2	2	2
3					2	2	2	2	2	2	2	2	2	3	3	3	3	3	3
4				2	2	2	3	3	3	3	3	3	3	3	4	4	4	4	4
5			2	2	3	3	3	3	3	4	4	4	4	4	4	4	5	5	5
6		2	2	3	3	3	3	4	4	4	4	5	5	5	5	5	5	6	6
7		2	2	3	3	3	4	4	5	5	5	5	5	6	6	6	6	6	6
8		2	3	3	3	4	4	5	5	5	6	6	6	6	6	7	7	7	7
9		2	3	3	4	4	5	5	5	6	6	6	7	7	7	7	8	8	8
10		2	3	3	4	5	5	5	6	6	7	7	7	7	8	8	8	8	9
11		2	3	4	4	5	5	6	6	7	7	7	8	8	8	9	9	9	9
12	2	2	3	4	4	5	6	6	7	7	7	8	8	8	9	9	9	10	10
13	2	2	3	4	5	5	6	6	7	7	8	8	9	9	9	10	10	10	10
14	2	2	3	4	5	5	6	7	7	8	8	9	9	9	10	10	10	11	11
15	2	3	3	4	5	6	6	7	7	8	8	9	9	10	10	11	11	11	12
16	2	3	4	4	5	6	6	7	8	8	9	9	10	10	11	11	11	12	12
17	2	3	4	4	5	6	7	7	8	9	9	10	10	11	11	11	12	12	13
18	2	3	4	5	5	6	7	8	8	9	9	10	10	11	11	12	12	13	13
19	2	3	4	5	6	6	7	8	8	9	10	10	11	11	12	12	13	13	13
20	2	3	4	5	6	6	7	8	9	9	10	10	11	12	12	13	13	13	14

Adapted from Swed, Frieda S., and Eisenhart, C. 1943. Tables for testing randomness of grouping in a sequence of alternatives. *Ann. Math. Statist.*, **14**, 83–86, with the kind permission of the authors and publisher.

Table XII Critical Values of T for the Wilcoxon Matched-Pairs Signed-Ranks Test

n	Level of significance for one-tailed test		
	.025	.01	.005
	Level of significance for two-tailed test		
	.05	.02	.01
6	0	—	—
7	2	0	—
8	4	2	0
9	6	3	2
10	8	5	3
11	11	7	5
12	14	10	7
13	17	13	10
14	21	16	13
15	25	20	16
16	30	24	20
17	35	28	23
18	40	33	28
19	46	38	32
20	52	43	38
21	59	49	43
22	66	56	49
23	73	62	55
24	81	69	61
25	89	77	68

Adapted from Table I of Wilcoxon, F. 1949. *Some rapid approximate statistical procedures.* New York: American Cyanamid Company, p. 13, with the kind permission of the author and publisher.

Table XIII Critical Values of *D* in the Kolmogorov–Smirnov One-Sample Test

Sample size (n)	Level of significance for D = maximum $\mid F_i - S_i \mid$				
	.20	.15	.10	.05	.01
1	.900	.925	.950	.975	.995
2	.684	.726	.776	.842	.929
3	.565	.597	.642	.708	.828
4	.494	.525	.564	.624	.733
5	.446	.474	.510	.565	.669
6	.410	.436	.470	.521	.618
7	.381	.405	.438	.486	.577
8	.358	.381	.411	.457	.543
9	.339	.360	.388	.432	.514
10	.322	.342	.368	.410	.490
11	.307	.326	.352	.391	.468
12	.295	.313	.338	.375	.450
13	.284	.302	.325	.361	.433
14	.274	.292	.314	.349	.418
15	.266	.283	.304	.338	.404
16	.258	.274	.295	.328	.392
17	.250	.266	.286	.318	.381
18	.244	.259	.278	.309	.371
19	.237	.252	.272	.301	.363
20	.231	.246	.264	.294	.356
25	.21	.22	.24	.27	.32
30	.19	.20	.22	.24	.29
35	.18	.19	.21	.23	.27
Over 35	$\frac{1.07}{\sqrt{n}}$	$\frac{1.14}{\sqrt{n}}$	$\frac{1.22}{\sqrt{n}}$	$\frac{1.36}{\sqrt{n}}$	$\frac{1.63}{\sqrt{n}}$

Adapted from Massey, F. J., Jr. 1951. The Kolmogorov–Smirnov test for goodness of fit. *J. Amer. Statist. Ass.*, **46**, 70, with the kind permission of the author and publisher.

ANSWERS TO SELECTED PROBLEMS

Chapter 1

1.3 30%

1.5 d) approximately 28.5 e) midpoint 25 f) $\mu = 30.88$

1.7 a) 125; index increased b) 125; index decreased

1.9 a) $\mu = 21.2$ b) mode = 20.43; median = 20.43

1.13 $\mu = 26.4$

1.15 $\sigma^2 = 212.68$

1.17 a) $\sigma^2 = 0.86$ c) 80% and 94%

1.19 a) $\mu = 10.25$ b) $\sigma^2 = 28.69$

1.21 a) guess $\mu = 480, \sigma = 100$ b) 68%, 16%

1.23 a) $\mu = 6.97; \sigma^2 = 35.904$

1.25 $\mu = 81.91; \sigma^2 = 137.085$

1.27 a) $q_1 = 1892, q_2 = 1971, q_3 = 2041$ b) 1999 c) 2117

1.31 a) $\mu = 15{,}856.3$; median $= 15{,}900$; $\sigma^2 = 992{,}500$; interquartile range $= 15{,}025$
b) mean < median slightly, therefore slightly skewed to the left.
c) $x \geq 15{,}000$ is $1 - 0.375 = 0.625$; $x < 16{,}001$ is 0.5625 d) $\mu = 15{,}781$

1.33 a) $\mu = \$28{,}986.50$; $\sigma^2 = 4{,}818.14$; median $= \$27{,}885$

1.41 median $= \$19{,}875$

1.43 b) mode $= 10$, median $= 35.6$, mean $= 47.6$ c) $\mu \pm 1\sigma = 73\%$; $\mu \pm 2\sigma = 93\%$
d) 63.02 customers

1.47 d) median

1.49 $\mu = 1.88, \sigma^2 = 0.92$

1.51 a) second has slightly higher relative variability
b) variability among men slightly higher
c) #1 most risky, #3 least risky; #1 has highest average return

Chapter 2

2.3 a) discrete and finite b) discrete and infinite c) discrete and finite
d) continuous and infinite e) continuous and infinite

2.5 number of winners $= 1501$; $P(\text{win}) = 0.03002$

2.7 $P(\text{face card}) = 12/52$

2.9 a) objective (if the number of participants is indicated)
b) objective c) subjective d) subjective e) subjective

2.11 a)

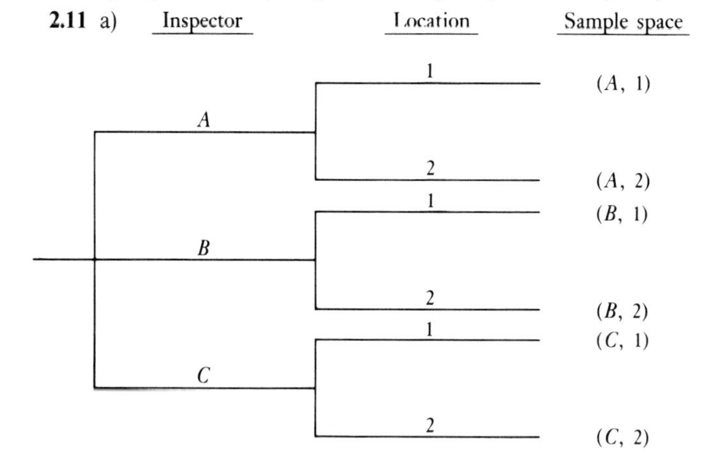

$P(\text{each sample point}) = 1/6$

b) $P(A, 1) = 1/6$

2.13 a) 10,000 students b) $P(\text{woman}) = 0.45$ c) $P(\text{senior}) = 0.15$
d) $P(\text{sophomore} \cap \text{male}) = 0.12$; $P(\text{sophomore} \mid \text{male}) = 0.22$

2.15 a) $P(\text{cash adv.}) = 0.2073$; $P(\overline{\text{cash adv.}}) = 0.7927$ b) 0.07 c) 0.11
d) 1707/10000 e) 8305/10000

2.17 $P(\text{male} \cap >21) = 0.122$; $P(\text{male} \cup \text{female}) = 1.0$

2.19 a) $P(3 \text{ def.}) = 0.000125$

2.21 a) No; $P(>1 \text{ wrong}) = 0.984$ b) $P(3 \text{ incorrect answers}) = 0.422$

2.25 a) $P(\text{profit}) = 3/8$ b) $P(\text{fav.}|\text{profit}) = 4/5$

2.27 0.625

2.29 $P(\text{male}|\text{senior}) = 0.483$

2.31 a) $P(\text{cash adv.}) = 0.2073$ b) $P(\text{balance due} < 100|\text{cash adv.}) = 0.1105$

2.33 a) $P(\text{none}|A_3) = 1/3$ b) not independent

2.35 $P(R|H) = 1/3$; $P(U|H) = 2/3$

2.37 10

2.39 a) 35 b) 840 c) 4/7

2.41 8 possibilities

2.43 *ABC, ACB, BCA, BAC, CAB, CBA*

2.45 a) 3,628,800 b) 120 c) 720

2.47 On the average it would take 1.15 million years to find the correct code. This should be sufficient!

2.49 $P(\text{at least } 3) = 0.643$

2.51 Multinomial distribution: $\text{prob} = \dfrac{7!}{2!4!1!}\left(\dfrac{1}{2}\right)^2\left(\dfrac{1}{4}\right)^4\left(\dfrac{1}{52}\right)^1$

2.53 a) number of sample points = 3,628,800; $P(\text{one outcome}) = 1/3{,}628{,}800$
b) 0.325

2.55 a) $\frac{1}{4}$ b) $\frac{1}{2}$ c) $\frac{1}{3}$

2.57 0.507

2.61 $P(\text{at least } 1) = \frac{3}{5}$

2.63 0.5177 vs. 0.4914

Chapter 3

3.3 b) $E[\mathbf{x}] = 2.4$; $V[\mathbf{x}] = 1.44$
c) $\sigma = 1.2$ so $\mu \pm \sigma$ includes 60% of the values.

3.5 b) $E[\mathbf{x}] = 1.6$; c) $V[\mathbf{x}] = 17.84$

3.7 a) The unskilled workers are 9/14 of the total number of workers.
b) $E[\mathbf{x}] = 2.571$; $V[\mathbf{x}] = 0.39$

3.9 $E[\mathbf{x}] = 2.57$; $(E[\mathbf{x}])^2 = 6.61 \neq E[\mathbf{x}^2] = 7.0$

3.11 b) $P(\mathbf{x} \leq \frac{1}{2}) = \frac{1}{4}$; $F(1) = P(\mathbf{x} \leq 1) = 1$

3.13 a) $F(0) = 0$; $F(\frac{1}{2}) = \frac{1}{4}$; $F(0.707) = \frac{1}{2}$; $F(1) = 1$; $F(3.7) = 1$
b) $F(x) = x^2$ for $0 \leq x \leq 1$; $F(x) = 0$ for $x \leq 0$; $F(x) = 1.0$ for $x \geq 1.0$

3.15 b) $P(\mathbf{d} > 45) = 0.75$; $F(50) = 0.444$ c) $E[\mathbf{d}] = 50$

3.17 $F(x) = \dfrac{4}{3}x - \dfrac{x^3}{3}$; $F\left(\dfrac{1}{3}\right) = \dfrac{35}{81}$

3.19 b) $F(x) = x^3$; $F(\frac{1}{4}) = 0.0156$

3.21 a) $E[x] = 20{,}000$; $E[y] = 20{,}000$ c) $\sigma_x^2 = 2 \times 10^8$; $\sigma_y^2 = 0.667 \times 10^8$

3.23 a) $E[y] = 32$; $V[y] = 36$

b) 1. $E[x \cdot y] = E[x]E[y] = 50$; $E[x + 2y] = E[x] + E[2y] = 5 + 20 = 25$
$E[13 - 2x] = E[13] - E[2x] = 13 - 10 = 3$

2. $V[x - y] = V[x] + V[y] = 9 + 25 = 34$;
$V[x + 2y] = V[x] + 4V[y] = 9 + 100 = 109$
$V[13 - 2x] = V[13] + 4V[x] = 0 + 36 = 36$

3. $C[x, y] = 0$ since x and y are independent.

3.25 a) $P_x(2) = 0.40$; $P_y(2) = 0.40$; $P_{x|y}(3|2) = \frac{1}{4}$

b) Dependent, since $P_x(3) = 0.40 \neq P(x = 3|y = 2)$

c) $E[x] = 2.1$; $E[y] = 2.2$ d) $E[x \cdot y] = 4.8$; $E[x + y] = 4.3$

e) $C[x, y] = 0.18$ f) $E[5x - 3y] = 3.9$ g) $V[5x - 3y] = 14.39$

3.29 a) The table below shows the cumulative joint mass function.

		x: 1	5	10
	1	1/20	3/20	6/20
y	2	5/20	7/20	13/20
	3	7/20	13/20	1.0

b) $P_x(1) = 7/20$; $P_x(5) = 6/20$; $P_x(10) = 7/20$; $P_y(1) = 6/20$; $P_y(2) = 7/20$;
$P_y(3) = 7/20$ c) $E[x] = 107/20$; $E[y] = 41/20$ d) $P(x = 10|y = 1) = \frac{1}{2}$

e) Not independent, since $P(x = 10) = 7/20 \neq P(x = 10|y = 1)$.

3.31 a) $F(x_1, x_2) = \frac{1}{4}(x_1^2x_2 + 3x_1x_2^2)$; $F(\frac{1}{2}, \frac{1}{2}) = \frac{1}{8}$

c) $P(0 \leq x_1 \leq \frac{1}{2}, \frac{1}{2} \leq x_2 \leq 1) = \frac{5}{16}$ d) $f(x_1) = \frac{1}{2}x_1 + \frac{3}{4}$
$f(x_2) = \frac{1}{4} + \frac{3}{2}x_2$

e) $F(x_2|x_1) = \dfrac{\frac{1}{2}x_1 + \frac{3}{2}x_2}{\frac{1}{2}x_1 + \frac{3}{4}}$ f) $P(0 \leq x_2 \leq \frac{3}{4}|x_1 = 0) = 9/16$

3.33 b) $E[x] = 2.0$; $V[x] = 1.0$ c) $E[y] = 0$; $V[y] = 4.0$

d) $E[y] = 0$; $V[y] = 4$

3.35 b) $P(y = 0|x = 3) = P(y = 1|x = 3) = P(y = 2|x = 3) = 0$
$P(y = 3|x = 3) = P(y = 4|x = 3) = P(y = 5|x = 3) = P(y = 6|x = 3) = \frac{1}{4}$

c) $\sum yP(y) = 4.75$

3.39 To A: = \$2.50; to B: = \$14.17; to C: = \$6.87; to D: = \$1.46

3.45 a) $E[x] = 6.4$; $E[y] = 6.0$; $V[x] = 2.24$; $V[y] = 3.2$

b) $E[0.5x + 0.5y] = 6.2$; $V[0.5x + 0.5y] = 0.16$

Chapter 4

4.1 $P(x \leq 4) = 0.9590$

4.3 $P(x = 2) = 0.1536$

4.5 b) $P(x = 7) = 0.0079$; $P(x \geq 7) = 0.0086$ c) $\mu = 3.2$, $\sigma^2 = 1.92$

4.7 a) $E[x] = 5$; $V[x] = 4.75$ b) $P(x \leq 1|n = 100,\ p = 0.05) = 0.0371$ from Table I.

4.9 a) Average = 47.5 days; $V[x] = 2.375$. Using $\mu \pm 4\sigma$, 41.34 to 53.66

4.11 Average = 25; \$1250. Using $\mu \pm 4\sigma$, 42.32 cars, \$2116 maximum

4.13 b) $P(x \leq 3 | n = 10, p = 0.60) = 0.0548$.

4.15 $P\left(\frac{x}{n} \leq 0.12 | n = 100, p = 0.20\right) = P(x \leq 12 | n = 100, p = 0.20) = 0.0255$

4.17 a) $P(3w, 3m) = 0.4762$ b) $P(4w, 2m) + P(5w, 1m) = 0.2619$

4.21 $\frac{{}_4C_3 \cdot {}_{16}C_1}{{}_{20}C_4} = 0.0132$

4.23 $n = 4$; probability of being a winner is $p = 4/20 = 0.20$; $P(x = 3) = 0.0256$ (from Table I, $n = 4$, $p = 0.20$). (*Note:* Approximation is not very good to the answer to Exercise 4.21 ($p = 0.0132$) since $n = 4$ is 20% of $N = 20$.)

4.25 a) $P(7D, 1R) = \binom{55}{7}\binom{45}{1}\Big/\binom{100}{8}$ b) $P(x = 1 | n = 8, p = 0.45) = 0.0548$

c) $P(x \geq 5 | n = 8, p = 0.45) = 0.2603$

4.27 0.4438

4.29 a) $P(x = 10) = 0.1251$ b) $P(x > 12) = 0.208$ c) $P(x < 6) = 0.0671$

4.31 $P(x \leq 1) = P(0) + P(1) = 0.1353 + 0.2707 = 0.4060$

4.35 $P(x = 5) = 0.0005$ (binomial); $P(x = 5) = 0.0012$ (Poisson), $\lambda = 0.80$; $V[\text{binomial}] = 0.736$; $V[\text{Poisson}] = 0.80$

4.37 a) $\mu \pm 1\sigma = 0.9197$; $\mu \pm 2\sigma = 0.9810$ from Table II b) $\lambda = 4$; $\mu \pm \sigma = 0.7978$; $\mu \pm 2\sigma = 0.9787$; $\lambda = 16$; $\mu \pm \sigma = 0.7411$; $\mu \pm 2\sigma = 0.9677$

4.39 a) $P(x \geq 3) = 0.0258$ b) Three or more defectives will occur so seldom (2.58%) when $p = 0.10$ that if 3 defectives are observed, action should be taken.

4.45 a) 29% b) 0.525 c) $P(1 < x < 4) = 0.2560$

4.47 d) Poisson

4.49 a) 0.3668 b) 12.5 c) 18.75

Chapter 5

5.5 a) $P(z \leq 1.645) = 0.95$; $1.645 = (a - 42)/2 \rightarrow a = 45.29$
b) $P(z \geq -1.645) = 0.95$; $-1.645 = (b - 42)/2 \rightarrow b = 38.71$
c) $P(38.08 \leq x \leq 45.92) = 0.95$

5.7 $P(x \leq a) = 0.99$; $P(z \leq 2.33) = 0.99$; $(a - 10)/2 = 2.33$; $a = 14.66$ days

5.9 a) $P(x \geq 16) = 0.1587$; $P(x \geq 24) = 0.0013$; $P(8 \leq x \leq 16) = 0.6826$
b) $a = 4$; $b = 20$

5.11 a) $21,564; $20,350
b) $P(15{,}080 \leq x \leq 22{,}920) = 0.9500$

5.13 a) $P(x \geq 8{,}000) = 0.0918$ b) Break-even attendance is 4737.

5.15 The GMAT score is better because the z-value is larger (1.25 vs. 1.20)

5.17 $P(z \geq 3.89) \cong 0$

5.19 $P(z \leq -2.13) = 0.0166$

5.21 b) $\mu = 1/\lambda = 1/3$; $\sigma^2 = 1/\lambda^2 = 1/9$ c) 86.5% and 95.0%

5.23 a) $\lambda = 2$ b) $\sigma^2 = 1/4$ c) $P(0 \leq T \leq 1) = 0.865$ d) $P(T \geq 2.0) = 0.018$

5.25 a) $P(x = 10) = 0.1251$ b) $P(T \geq 1/10) = 0.368$ c) $P(3 \leq T \leq 6) = 0.239$

5.27 b) $P(0 \le T \le 20) = 0.330$; $P(20 \le T \le 60) = 0.369$; $P(T > 60) = 0.301$
c) $\mu = (1/\lambda) = 50$; $\sigma^2 = (1/\lambda)^2 = 2500$; $\sigma = 50$; median $= 35$

5.29 $P(2.18 \le \chi_8^2 \le 17.5) = 0.95$

5.35 c) $0.05 > P(\chi^2 \ge 16) > 0.025$

Chapter 6

6.9 $\bar{x} = 42.17$; $s = \sqrt{10.97} = 3.31$

6.11 a) $\bar{x} = 55.8$; $s^2 = 107.51$
b) median $= 55$; mode $= 55.70$; range $= |70 - 40| = 30$
c) $\bar{x} \pm s = 55.8 \pm 10.4 = 45.4$ to 66.2; 60% of data; $\bar{x} \pm 2s = 55.8 \pm 2(10.4) = 35.0$ to 76.6; 100% of data

6.13 $\bar{x} = (1/n)\sum xf = (1/100)(80) = 0.80$. $s^2 = 0.8687$; $s = 0.932$

6.15 a) $\bar{x} = (1/n)\sum xf = 222.75$

6.17 a) $\bar{x} = 105$; b) $s^2 = [1/(n - 1)]\sum(x - \bar{x})^2 = (1/3)(20) = 6.667$; $s = 2.582$
c) more suspicious since $\bar{x} = 105$ is large relative to $\mu = 96$
d) range $= |108 - 102| = 6$ e) range considers only highest and lowest values

6.21 a) attribute sampling

6.25 a) $P(\bar{\mathbf{x}} \ge 5) = P(\mathbf{t} \ge 2.44)$; $0.01 > P(\mathbf{t} \ge 2.44) > 0.005$; $P(\mathbf{x} \le 5) = P(\mathbf{t} \le -2.93)$, which is less than 0.005

6.27 a) $P(\mathbf{x} \ge 35) = 0.0475$; $P(25 \le \mathbf{x} \le 35) = 0.9050$
b) $P(\bar{\mathbf{x}} > 31) = 0.0228$; $P(\bar{\mathbf{x}} < 30.5) = 0.8413$; $P(29 \le \bar{\mathbf{x}} \le 31) = 0.9544$

6.29 b) $\mu = (1/10)[0 + 1 + 2 + \cdots + 9] = (1/10)(45) = 4.5$

6.31 a) $\sigma_{\bar{x}} = (0.10/8)\sqrt{\dfrac{400 - 64}{399}} = 0.0115$

$$P(\bar{\mathbf{x}} \le 19.95) = P\left(\mathbf{z} \le \frac{19.95 - 20}{0.0115}\right) = P(\mathbf{z} \le -4.35).$$ This probability is $\cong 0$.

6.33 Distribution will be normal with a mean of $\mu = 5$ and $\sigma^2 = \sigma^2/n = (1/25)/100 = 1/2500$.

6.35 b) $\sigma_{\bar{x}} = \sqrt{[2500/25][75/99]} = 8.706$

6.37 a) $0.05 < P(\bar{\mathbf{x}} \le 96 | \mu = 110) < 0.10$; $0.05 < P(\bar{\mathbf{x}} \ge 96 | \mu = 80) < 0.10$
b) $P(\bar{\mathbf{x}} \le 96 | \mu = 110) = 0.0228$; $P(\bar{\mathbf{x}} \le 96 | \mu = 100) = 0.2119$

6.39 a) $\bar{x} = 122.75$; $s^2 = 1415.34$
b) $P(\mathbf{t} \ge 0.73)$ is between 0.25 and 0.10 (check $\nu = 60$, $\nu = 120$).
c) No; the sample result is not very unexpected if $\mu_x = 120$.

6.41 a) With $s = 6$, $n = 9$, d.f. $= 8$; $0.025 < P(\bar{\mathbf{x}} \ge 54) = P(\mathbf{t} \ge 2) < 0.05$
$0.005 < P(\bar{\mathbf{x}} \le 44) = P(\mathbf{t} \le -3) < 0.01$
$P(45 \le \bar{\mathbf{x}} \le 55) = P(-2.5 \le \mathbf{t} \le 2.5) \cong 1.0$
b) All t-values double; first two probabilities $\cong 0$; third probability $\cong 1.0$.

6.43 $\bar{x} = 282.50$; $s = 10.61$; $t_c = -17.5/(10.61/\sqrt{2}) = -2.33$; $P(\mathbf{t} \le -2.33) > 0.10$ for 1 d.f.

6.45 $P(\chi^2 \ge 5.99) = 0.05$; $P\left(\dfrac{s^2}{\sigma^2} \ge 3.0\right) = 0.05$

6.47 $P(s^2 \leq 36) = P(\chi^2 \leq 5.76) > 0.10$

6.49 a) $E[\chi^2] = v = n - 1 = 24$; $V[\chi^2] = 2v = 48$
b) $P(\chi^2 \geq 43.0) = 0.01$; $P(\chi^2 \geq 33.2) = 0.10$; $P(\chi^2 \leq 9.89) = 0.005$

6.51 $F_c = 100/250 = 0.400$; $P(F \leq 0.400) \cong 0.05$

6.53 $F_c = 0.533$; $0.05 > P(F \leq 0.533) > 0.025$

6.55 $F_c = 0.694$; $P(F \leq 0.694) > 0.05$

6.63 a) Providing no known method for making statistical inferences about the population solely from the results of the sample.

6.65 b) Greater than $250

Chapter 7

7.3 Each of the estimators is unbiased.

7.5 a) $\mu = 10$; $\sigma^2 = 25$; $\sigma = 5$ c) $E[\bar{x}] = 10$ d) $E[s^2] = 25$
e) $E[s] = 4.329$. (Note: $E[s] = 4.329 \neq \sigma = 5$.) f) $E[md] = 10$; median is unbiased.

7.7 mean of $\bar{x}$'s is 10.0; variance of means = 12.5; variance of means in Problem 7.6 was 8.34; variance of $\bar{x}$'s is decreasing, as property of consistency says it should.

7.9 a) x/n is unbiased; b) $V[x/n] = 0.06$ and 0.048. The property of consistency is supported.

7.11 $E[s^2] = (1/16)(52) = 6.5$, which equals variance of population;
$E[s] = (1/16)(31.11) = 1.94$, which does not equal $\sigma = \sqrt{6.5} = 2.55$.

7.13 $E[x/n] = (1/n)E[x] = (1/n)(np) = p$

7.15 a) $E[\bar{x}] = \lambda$ b) relative efficiency $= \dfrac{1/2\lambda}{5/9\lambda} = 9/10$

7.17 $E[s^2] = 120/12 = 10.0$; $\sigma^2 = 7.5$, $E[s^2] \neq \sigma^2$

7.21 a) $\bar{x} + t_{0.025,\,24}(s/\sqrt{n}) = 900 \pm 2.064(150/5) = 838.08 \leq \mu \leq 961.92$
b) $\bar{x} \pm z_{0.025}(s/\sqrt{n}) = 900 + 1.96(150/5) = 841.20 \leq \mu \leq 958.80$
Second interval is a bit shorter, although not by much.

7.23 90% C.I. for μ, $\bar{x} - 1.645(\sigma/\sqrt{n}) \leq \mu \leq \bar{x} + 1.645(\sigma/\sqrt{n})$
$24.3 - 1.645(6/\sqrt{16}) \leq \mu \leq 24.3 + 1.645(6/\sqrt{16})$ or $21.833 \leq \mu \leq 26.768$

7.25 $0.645 \leq \mu \leq 7.335$

7.27 $n = 385$; cost $= 385(0.35) = \$134.75$

7.29 a) [52,262.5, 69,737.5] b) $n =$ 3,394 days

7.31 $n = 663$, and $20n = N = 13{,}260$

7.33 $50.94 \leq \sigma^2 \leq 142.5$

7.35 a) $\$24{,}505 \leq \mu \leq \$31{,}495$ b) $n = 543$

7.37 $n = 332$; $N = 2n = 2(332) = 664$

7.41 The distribution of $\bar{x}$ would approximate a normal, but not closely due to the skewness. $P(\bar{x} \geq 2.75) = 0.0228$; 99% confidence interval: $2.29 \leq \mu \leq 2.81$.

7.43 The amount of bias is $(1/n)\sigma^2$.

7.47 Correct answer is (d), specified confidence interval.

7.49 Correct answer is (a), 72. Note, in Formula (7.7), that n is inversely related to the *square of D*. Thus, if D is doubled, then n will decrease by a factor of 4.

7.51 Correct answer is (d); between 240 and 560 are delinquent.

7.53 $\lambda = \sum x_i/n$

Chapter 8

8.3 a) $H_0:\mu = \$5000$ vs. $H_a:\mu < \$5000$
Type I: reject $\mu = \$5000$ when μ is, in fact, equal to \$5000
Type II: accept $\mu = \$5000$ when μ is, in fact, less than \$5000
b) Use $z = (\bar{x} - \mu_0)/(\sigma/\sqrt{n})$; c) $\bar{x} \geq \$4588$ is acceptance region.
d) $\bar{x} = 4650$; since \$4650 does not lie in the critical region, we must accept H_0. Report $P(z \leq -1.40) = 0.0808$.
e) $2P(z \leq -1.40) = 2(0.0808) = 0.1616$
f) It is not very random to select four consecutive Saturdays.

8.5 $H_0:\mu = \$12{,}000$ vs. $H_a:\mu < \$12{,}000$; $z_c = -2.50$, reject H_0.

8.7 a) Use $H_0:\mu \leq 70{,}000$, with the manufacturer's claim represented by H_a. Manufacturer would prefer $H_0:\mu \geq 70{,}001$. The alternative to $H_0:\mu \leq 70{,}000$ should be $H_a:\mu > 70{,}000$.
Type I Error: conclude the tire lasts $> 70{,}000$ when it lasts $\leq 70{,}000$.
Type II Error: concluding the tire lasts 70,000 or less when it lasts $> 70{,}000$.
b) C.R. = 71,645; z critical region is 1.645.
c) Reject H_a and accept H_0; $P(\bar{x} \geq 71{,}250) = P(z \geq 1.250) = 0.1050$.

8.9 a) $H_0:\mu = 14{,}500$ vs. $H_a:\mu < 14{,}500$
b) $z_{0.05} = -1.645$, which means the critical region is $z \leq -1.645$ or $\bar{x} \leq 14{,}467.10$
c) Reject H_0 if $\bar{x} = \$14{,}300$, since this value falls in the critical region.
d) $P(\bar{x} \leq 14{,}300) \cong 0$

8.11 a) $H_0:\mu \leq 50$ vs $H_a:\mu > 50$
b) $z = [(\bar{x} - \mu_0)/(\sigma/\sqrt{n})]$, where $\mu_0 = 50$ and $\sigma = 3$, $z_c = 1.333$; accept H_0, reject shipment
c) $1.645[(\bar{x} - \mu_0)/(\sigma/\sqrt{n})] = (\bar{x} - 50)/(3/\sqrt{64})$ (solve for $\bar{x}$); therefore, $\bar{x} = 50.617$ hours.

8.13 $H_0:p = 1/2$ vs. $H_a:p \neq 1/2$, where p is the proportion who prefer the Chrysler car. Could use the alternative $H_a:p > 1/2$.

8.15 To calculate β we first need to determine the critical region (from Problem 8.2) in terms of $\bar{x}$. In terms of z, the critical region was $z \leq -2.33$. Solve for $\bar{x}$, assuming μ_0 is true; thus the acceptance region is $\bar{x} \geq 23.94$. $\beta = P(\text{rejecting } H_a | H_a \text{ is true}) = P(\bar{x} \geq 23.94 | \mu_a = 23.90) = 0.0418$;
Power $= 1 - \beta = 0.9582$.

8.17 First, we need to find the acceptance region for Problem 8.8 in terms of $\bar{x}$. In terms of z, the acceptance region was $-2.575 \leq z \leq 2.575$; $0.997 \leq \bar{x} \leq 1.203$. $\beta(0.90) = 0.0075$; $\beta(0.95) = 0.1200$; $\beta(1.00) = 0.5319$; $\beta(1.05) = 0.9082$; $\beta(1.10) = 0.9899$

8.19 a) $\alpha = 0.50$; $\beta = 0.75$ b) $\alpha = 0.75$; $\beta = 0.625$
c) Second more appropriate, since β is lower.

8.21 a) reject H_0 b) $\alpha = P(\bar{x} \leq 14) = P(z \leq -2.40) = 0.0082$
c) $P(z \leq -2.50) = 0.0062$
d) $z = -2.575$; $\mu_0 - 2.575(\sigma/\sqrt{n}) = 20 - 2.575(2.5)$; critical region: $\bar{x} \leq 13.56$

8.23 $H_0: \mu \geq 8.0$ vs. $H_a: \mu < 8.0$; $\bar{x} = (1/n)\sum x_i = 6$; $s = \sqrt{2} = 1.414$. Critical region is $t \leq 3.365$; accept H_a

8.25 a) $H_0: \mu_1 \geq \mu_2$ vs. $H_a: \mu_1 < \mu_2$. We must adjust H_0 to $H_0: \mu_1 - \mu_2 = 0$
d) $z_c = -3$; reject H_0

8.27 $H_0: \mu_1 - \mu_2 = 0$ vs. $H_a: \mu_1 - \mu_2 \geq 0$
If σ_1^2 and σ_2^2 can be assumed to be equal, then t test is appropriate. $t_c = 2.677$. Since 2.677 exceeds critical value of 2.583, we reject H_0.

8.29 $\bar{x}_1 = 31.025$; $s_1^2 = 0.0758$; $\bar{x}_2 = 31.325$; $s_2^2 = 0.0492$; $t_c = -1.695$
Since $t_c = -1.695$ is not less than the critical value of $-t_{0.005,\,6} = -3.707$, accept H_0.

8.31 $H_0: \mu_B - \mu_A = 1$ vs. $H_a: \mu_B - \mu_A > 1$; $\bar{D} = 2.5$, $s_D = 2.0$; $t_c = (2.5 - 1)/(2/\sqrt{4}) = 1.5$; do not reject H_0.

8.33 $H_0: \mu_1 - \mu_2 = 0$ vs. $H_a: \mu_1 - \mu_2 \neq 0$; $\bar{D} = 10$, $s_D = 4$, $\alpha = 0.01$, $n = 4$, $t_c = (10 - 0)/(4/\sqrt{4}) = 5.0$; accept H_0.

8.35 a) Since σ_1^2 and σ_2^2 are unknown, we use the t-distribution, although a normal approximation could be used because n_1 and n_2 are large. $t_c = -1.85$; $0.10 > 2P(\mathbf{t} \leq -1.85) > 0.05$. For the one-sided test, reject H_0 if α is 0.05 or larger.

8.37 We now need to use the finite population correction factor in the denominator of the formula for z_c; $z_c = -2.71$. $P(\mathbf{z} \leq -2.71) = 0.0068$

8.39 a) $H_0: \mu = 5000$ vs. $H_a: \mu < 5000$ b) Use $\mathbf{t} = (\bar{\mathbf{x}} - \mu_0)/(\mathbf{s}/\sqrt{n})$
c) The critical region is $t \leq -2.353$; $\bar{x} = \$4650$; $s = 443.47$; $t_c = -1.578$;
d) reject H_a. e) $0.25 > P(\mathbf{t} < -1.578) > 0.10$

8.41 $H_0: \mu = 31.5$ vs. $H_a: \mu < 31.5$, using $\alpha = 0.01$; $t_c = -3.667$; reject H_0.

8.43 Optimal C.R. is $x \geq 19$ (try 18 and 20); $\alpha = 0.0046$; $\beta = 0.0630$
$E.C.(0.10) = \$2.30$; $E.C.(0.25) = 12.60$; $T.E.C.(x \geq 19) = \$30.39$.
$T.E.C.(x \geq 18) = \$30.76$; $T.E.C.(x \geq 20) = \$31.67$; total cost is more for $n = 100$ (\$30.39) than for $n = 50$(\$28.39).

8.45 a)

x	$p = 0.10$	x	$p = 0.25$
6	0.0089	0	0.0032
7	0.0020	1	0.0211
8	0.0004	2	0.1669
9	0.0001	3	0.1339
	α=0.0114	4	0.1897
		5	0.2023
			β=0.7171

b) $E[\text{cost}]$ given $p = 0.10$ is $0.0114(500) = \$5.70$
$E[\text{cost}]$ given $p = 0.25$ is $0.7171(200) = \$143.42$
c) Total expected cost = \$43.03
d) $x \geq 5$ is the best critical region

8.47 a) $\alpha = P(\mathbf{x} \geq 31 \mid n = 50, p = 0.50) = 0.0673$;
$\beta = P(\mathbf{x} < 31 \mid n = 50, p = 0.75) = 0.0138$
b) Type I error: Conclude that $p = 0.75$ when $p = 0.50$;
Chrysler might falsely advertise the popularity of its car.
Type II error: Conclude that $p = 0.50$ when $p = 0.75$;
Chrysler might underestimate the popularity of its car.

c) No; the more serious error should have the lower probability. As it is now, $\alpha > \beta$.
d) Use $x \geq 33$; $\alpha = 0.0815$ and $\beta = 0.0834$

8.49 $\alpha = P(x \geq 30.5) = 0.606$; $\beta = P(x \leq 30.5) = 0.0110$

8.51 $H_0{:}p = 0.20$ vs. $H_a{:}p > 0.20$, $x/n = 16/64$, $\alpha = 0.05$; $z_{0.05} = 1.645$; $z_c = 1.0$
Since z_c is less than $z_{0.05} = 1.645$, $H_0{:}p = 0.20$ cannot be rejected.

8.53 $H_0{:}p_1 - p_2 = 0$ vs. $H_a{:}p_1 - p_2 \neq 0$; $z_c = 1.715$; do not reject H_0 since 1.715 is not in the critical region.

8.55 $H_0{:}p_2 - p_1 = 0.10$ vs. $H_a{:}p_2 - p_1 > 0.10$; $z_c = 0.384$. Do not reject H_0.

8.61 a) $H_0{:}\sigma_1^2 = \sigma_2^2$ vs. $H_a{:}\sigma_1^2 \neq \sigma_2^2$. Use the F-test statistic. b) $F_c = 0.5625$.
$F_{(\text{lower critical value, } 60, 30)} = 1/1.65 = 0.6061$. Since $F_c = 0.5625$ is less than the critical value of 0.6061, we accept $H_a{:}\sigma_1^2 \neq \sigma_2^2$.

8.63 Change hypotheses to: $H_0{:}\sigma^2 = 25$ vs. $H_a{:}\sigma^2 > 25$; $\bar{x} = 17$; $s^2 = 43.5$
$\chi_c^2 = 6.96$. Do not reject H_0; report $p > 0.10$.

8.65 Given $\mu = 17{,}500$, $\alpha = 0.05$, $\sigma = \$1000$, a test is to be constructed so that the distance from 17,500 to 17,600 is 1.645 standard deviations. Thus: $100 = 1.645\sigma/\sqrt{n} = 1.645(1000)/\sqrt{n}$; $\sqrt{n} = 1.645(1000)/100 = 16.45$; $n = 270.6$. The firm must survey 271 persons.

8.67 a) $H_0{:}p = 0.50$, $H_a{:}p > 0.50$. Given $n = 100$ and $x = 60$. Using Table I, $P(x \geq 60 \mid n = 100, p = 0.50) = 0.0285$. Since this p-value is quite small, reject H_0. Conclude that more than 50% of high-income families save more than 10%.
b) Can't say without comparison to sample of average-income families, using the two-sample methods in this chapter.

8.69 $t_c = -4.08$; reject H_0 and conclude new technique better.

8.73 a) $\bar{x} = 81.0$; $s^2 = 25.56$ b) $[72.821, 89.179]$ c) $[0, 0.5257]$
d) $t_c = 2.502$; reject H_0; $0.025 > P(t \geq 2.502) > 0.01$ e) $\beta = 0.2776$
f) $\chi^2 = 6.434$ g) $n = 192.08$ or 193 observations
h) $z_c = -1.897$; $P(z \leq -1.90) = 0.0287$; reject H_0

Chapter 9

9.3 No. By the "luck of the draw," you lost \$1. This is a bad *outcome*, but that doesn't mean that it was necessarily a bad *decision* to risk this possible loss of \$1 for a 50–50 chance at winning \$10.

9.7

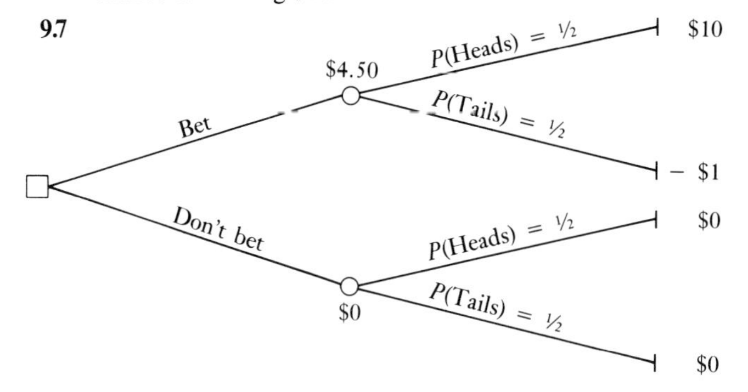

Maximin: Don't bet. Maximax: Bet. EMV: Bet.

9.9 a)

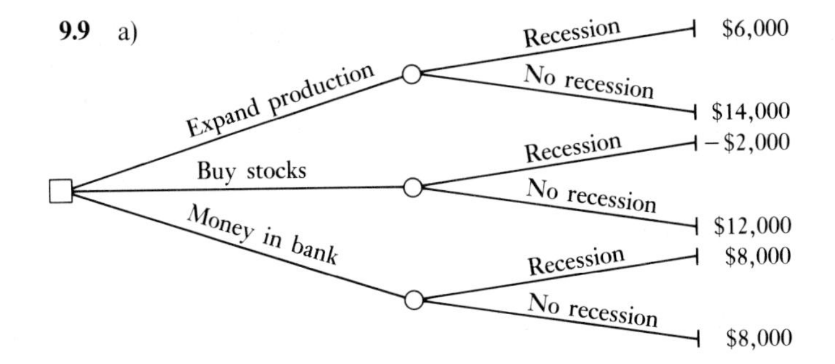

b) Put the money in the bank.

9.13 a) 0.05 b) 0.0087 c) 0.1059

9.15 a) 0.0667 b) 0.7846 c) 0.0219

9.17 EMV(Stock 100) = 7.2; EMV(Stock 50) = 7.8; EMV(Don't stock) = 0; stock 50. EMV(Perfect info.) = 11.6; EVPI = 3.8.

9.19 EMV(Expand) = \$6000; EMV(Buy stocks) = \$6400; EMV(Bank) = \$8000; put money in bank.
EMV(Perfect info.) = \$11,600; EVPI = \$3600.

9.21 a)

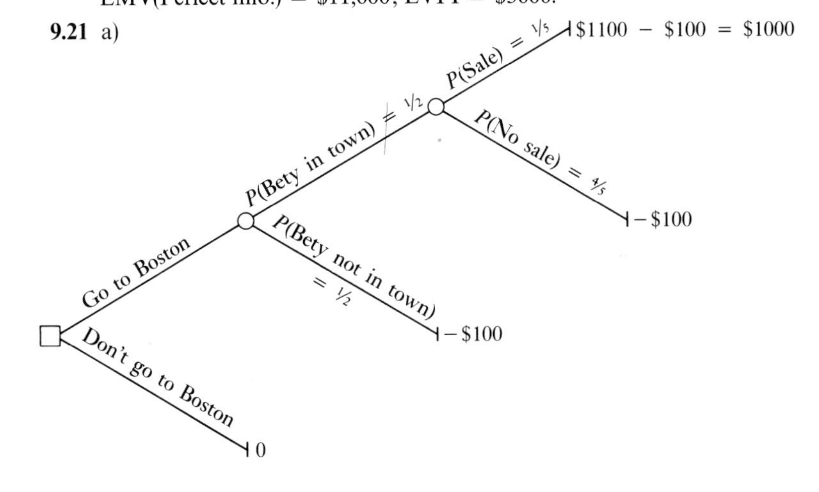

EMV(Go) = \$10; EMV(Don't go) = 0; to maximize, fly.

b) Joe values right arm at \$50 c) EVSI = \$26; ENGS = \$16.

9.23 a) EMV(Ship) = 17; EMV(Repair) = 10
b) Repair coat, EMV = 10 c) ship coat, EMV = 19.28
d) EMV(Sample info.) = 17.70 e) EVSI = 0.70; ENGS = −0.30.

9.25 P(new product | no new plant) = 0.30. If plant is not being built, choose "no advertising." P(new plant) = 0.60.
EMV(Sample info.) = 460,000; EVSI = 20,000.

9.27 EVSI = 0.74 when $n = 1$, 0.84 when $n = 2$, and 0.99 when $n = 3$.
ENGS = 0.59 when $n = 1$, 0.54 when $n = 2$, and 0.54 when $n = 3$.

9.29 a)

Sample Outcome	p	Prior Prob.	Likelihood	Prior Prob. × Likelihood	Post. Prob.
No purchases	0.1	0.2	0.81	0.162	162/562
	0.2	0.3	0.64	0.192	192/562
	0.3	0.3	0.49	0.147	147/562
	0.4	0.1	0.36	0.036	36/562
	0.5	0.1	0.25	0.025	25/562
				0.562	
One purchase	0.1	0.2	0.18	0.036	36/356
	0.2	0.3	0.32	0.096	96/356
	0.3	0.3	0.42	0.126	126/356
	0.4	0.1	0.48	0.048	48/356
	0.5	0.1	0.50	0.050	50/356
				0.356	
Two purchases	0.1	0.2	0.01	0.002	2/82
	0.2	0.3	0.04	0.012	12/82
	0.3	0.3	0.09	0.027	27/82
	0.4	0.1	0.16	0.016	16/82
	0.5	0.1	0.25	0.025	25/82
				0.082	

b)

	No Purchases	One Purchase	Two Purchases
EMV(Stock 100)	2.76	11.28*	19.90*
EMV(Stock 50)	5.77*	9.87	12.73
EMV(Don't stock)	0	0	0

(Optimal actions are starred.)

c) EMV(Sample information) = 8.89; EVSI = 1.09.

9.31 a)

Sample	θ	Prior	Likelihood	Prior x Lik.	Posterior
Favorable	θ_1	0.30	0.90	0.27	0.5625
	θ_2	0.70	0.30	0.21	0.4375
				0.48	
Unfavorable	θ_1	0.30	0.10	0.03	0.0577
	θ_2	0.70	0.70	0.49	0.9423
				0.52	

Sample	EMV(Mkt. | Sample)	EMV(Don't mkt. | Sample)
Favorable	150,000	0
Unfavorable	−253,840	0

EMV(Sample information) = 72,000; EVSI = 72,000.

b)

Sample	$P(\theta_1 \mid \text{Sample})$	$P(\theta_2 \mid \text{Sample})$	EMV(Mkt. | Sample)	$P(\text{Sample})$
F_A, F_B	0.9536	0.0464	462,880	0.151
F_A, N_B	0.3913	0.6087	13,040	0.046
F_A, U_B	0.3396	0.6604	−28,320	0.053
N_A, F_B	0.8372	0.1628	369,760	0.086
N_A, N_B	0.1385	0.8615	−189,200	0.065
N_A, U_B	0.1139	0.8861	−208,880	0.079
U_A, F_B	0.3288	0.6712	−36,960	0.073
U_A, N_B	0.0151	0.9849	−287,920	0.199
U_A, U_B	0.0121	0.9879	−290,320	0.248

$$\text{EMV(Sample information)} = 462{,}880(0.151) + 13{,}040(0.046) + 369{,}760(0.086) = 102{,}300.$$

9.33 a) EVSI = 1717, ENGS = 0.71 b) EVSI = 2.51, ENGS = 0.51; Sample of $n = 1$ is better than $n = 2$.

9.35 $$\left.\begin{array}{l} m - 0.67\sqrt{v} = 160 \\ m + 0.25\sqrt{v} = 180 \end{array}\right\} \Rightarrow \begin{array}{l} m = 174.6 \\ v = 471 \end{array}$$

9.37 We have $3.5 = m - 1.5\sqrt{v}$ and $6 = m + \sqrt{v}$. Solving yields $m = 5$ and $v = 1$. The posterior distribution is normal with

$$M = \frac{(5/1) + (1.2/4)}{(1/1) + (1/1.2)} = 2.89 \text{ and variance } V = \frac{1}{(1/1) + (1/1.2)} = 0.55.$$

9.39 Here $k_u = 20{,}000$ and $k_o = 8000$. Thus, we want the

$$\frac{100k_u}{k_u + k_o} = \frac{100(20{,}000)}{28{,}000} = 71.43 \text{ percentile, which is 8 houses.}$$

9.41 Here we want the 16.67 percentile, which is 762 gallons.

9.47 Utility table:

	Recession	**No recession**	**Expected utility**
Expand production	−12	19	9.70
Buy stocks	−4	17	10.70
Money in bank	13	13	13.00

The company should put the money in the bank.

9.49 In Problem 9.10 it takes a probability of recession of 0.30 to make the bank the best choice. With the risk-averse utility function in Problem 9.48, this cutoff point is reduced from 0.30 to 0.1935. The bank is the least risky choice, and with the risk-averse utility function, it takes a smaller probability of a recession to entice the company to take a riskier action.

9.51 EMV(Std. bid—Airport) = 0.40(2,000,000) + 0.60(−500,000) = 500,000
EMV(Lower bid—Airport) = 0.70(1,000,000) + 0.30(−500,000) = 550,000
EMV(Std. bid—Dam) = 0.30(4,000,000) + 0.70(−1,000,000) = 500,000
EMV(Lower bid—Dam) = 0.50(2,000,000) + 0.50(−1,000,000) = 500,000
Use the lower bid for the airport.

9.53

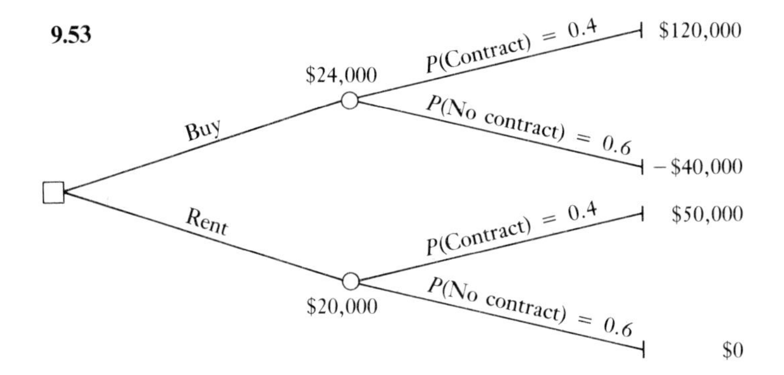

The contractor should buy the equipment.

9.55 The likelihoods are binomial probabilities, $P(R = 9 \mid n = 15, p)$.

p	Prior Prob.	Likelihood	Prior Prob. × Likelihood	Posterior Prob.
0.4	0.2	0.0612	0.01224	0.078
0.5	0.4	0.1527	0.06108	0.392
0.6	0.4	0.2066	0.08264	0.530
			0.15596	

9.57 EMV(Perfect information) = 48,000; EVPI = 24,000.

9.59 Two flips:

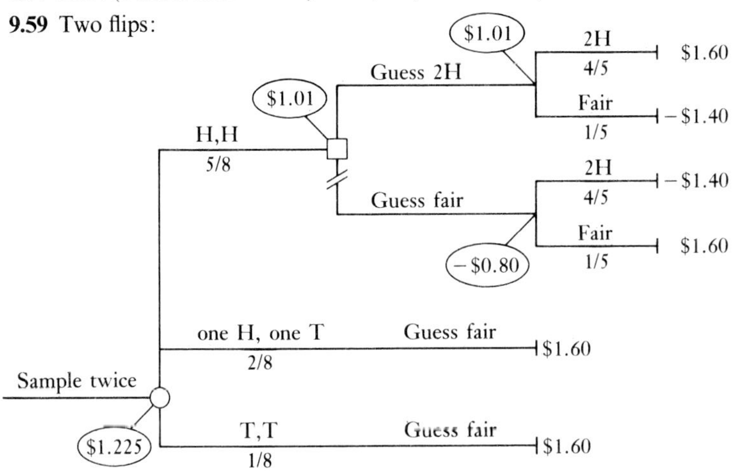

ENGS = 1.225 – 0.50 = $0.7125; EVSI = $1.125.

9.61 ENGS = $1.25 – 0.50 = $0.775; EVSI = 0.775 + 0.35 = $0.1125. (*Note:* Average sample cost = $0.35, since $\frac{1}{4}$ of time you will not take second sample.)

9.63 a) EMV(Guess 70*R* – 30*B*) = 1.20; EMV(Guess 70*B* – 30*R*) = 3.60.
b) EMV(Perfect information) = 8.40; EVSI = 8.40 – 3.60 = 4.80.

c)

	Red	Blue	
EMV(Guess 70R − 30B)	2.87	−0.22	EMV(Sample info.) = 4.80
EMV(Guess 70B − 30R)	0.26	6.44	EVSI = 1.20

d)

	3 Red	2 Red, 1 Blue	1 Red, 2 Blue	3 Blue
EMV(Guess 70R − 30B)	5.15	2.87	−0.22	−1.60
EMV(Guess 70B − 30R)	−4.30	0.26	6.44	9.20

EMV(Sample info.) = 5.81; EVSI = 2.21.

9.65 The posterior distribution is normal with $M = 5.87$ and $V = 0.1829$.

9.67 Here $k_u = 10$ and $k_o = 15$, so we want the 40th percentile, which is 11,250 processors.

9.69 b) EU(Investment) = 0(0.2) + 0.45(0.1) + 66(0.1) + 0.82(0.4) + 1.00(0.2) = 0.639. EU(No investment) = $U(0)$ = 0.60. He should make the investment.

9.71 ENGS = 60,199; EVSI = 65,199. (Your answer may differ because of rounding.)

Chapter 10

10.5 $\hat{y} = 90 + x$; for $x = 71$, $\hat{y} = 161$ pounds   b) $\hat{y} = 68.35$

10.7 b) $\hat{y} = 1.3 + 0.90x$; substituting $(\bar{x}, \bar{y}) = (3, 4)$ into the regression line, $4 = 1.3 + 0.9(3) = 1.3 + 2.7 = 4$

10.9 $\hat{y} = -920 + 0.03x$; $\hat{y} = -920 + 0.03(300{,}000) = 8{,}080$ for $x = 300{,}000$

10.11 $\hat{y} = 11 + 0.01x$

10.13 $\hat{y} = 95.84 - 4.8x$

10.15 a) $\hat{y} = 68.42 - 2.69x$
c) $s_{xy} = (1/(n-1))\sum(x - \bar{x})(y - \bar{y}) = (1/4)(-350) = -87.5$

d)

x	$\hat{y} = 68.42 - 2.69x$	y	$(y - \hat{y})$	
25	1.17	3	1.83	
20	14.62	9	−5.62	
10	41.52	40	−1.52	$\sum(y - \hat{y}) = 0$
20	14.62	15	0.38	
15	28.07	33	4.93	

10.17 $\hat{y} = 421.79 + 665.29x$

10.21 a) $\hat{y} = 75 - 5x$ b) $\hat{y} = 75 - 5(6) = 45$

10.25 a) $s_b = 0.50$; $t = 2.450$
$0.02 \geq 2P(t \geq 2.450) \geq 0.01$ for 60 d.f.; accept $H_a: \beta \neq 0$ at $\alpha = 0.02$
b) $0.225 \leq \beta \leq 2.225$
c) SSR = 54; MSR = 54; MSE = 9; $F = 6.0$; $F = (2.45)^2 = 6.0$
$0.025 > P(F \geq 6.00) > 0.01$; $0.02 > 2P(t \geq 2.45) > 0.01$ for 60 d.f. and a two-sided test. Thus, both tests reject H_0 for $\alpha = 0.025$ and not for $\alpha = 0.01$.

10.27 a) $\hat{y} = 68.114 + 1.486x$; for $x = 7$, $\hat{y} = 78.516$

10.29 a) $t = 4$, $t^2 = 4^2 = 16$, so consistent   b) $0.475 \leq \beta \leq 1.445$
c) [$29,153, 30,447]   d) [29,403, 30,197]

10.31 a) $\hat{y} = 26 + (3/4)(60) = 71$, $71/9 = 7.89$ maids
Thus, employ 8 maids.
b) $\hat{y} = 53$; total deviation: $y - \bar{y} = 55 - 60 = -5$;
Explained deviation: $\hat{y} - \bar{y} = 53 - 60 = -7$; Unexplained deviation:
$y - \hat{y} = 55 - 53 = 2$

10.33 $t = 2.064$; do not reject $H_0\colon \beta = 0$

10.35 $s_f = 1.032$; *For 1981*: [18.66, 23.42]; *for 1985*: [20.87, 26.09]

10.37 a) $\hat{y} = 1{,}386{,}969.3 + 0.533\,pop.$ b) $t = 10.5$; reject H_0
c) $F = 111.470$ (agrees with t-value)
d) $\hat{y} = 1{,}386{,}969.3 + 0.533(19{,}660{,}000) = 11{,}865{,}749$

10.41 a) Divide each x-value by 100 to make calculations easier.
$\hat{y} = 25.158 - 0.133x$; for $x = 43$, $\hat{y} = 19.439$
b) $t = 2.608$; do not reject $H_0\colon \beta = 0$

10.45 One should be *very careful* in making "cause and effect" inferences, but they need not be avoided.

10.47 a) $\hat{y} = (0.517)(1.491)^x$
b) Errors: $y - \hat{y}$: 0.03, −0.05, −0.01, 0.05, −0.01, 0.02, 0.03

Chapter 11

11.1 a) SSE = 7000 b) $r^2 = 0.64$; therefore 36% is unexplained
c) SST = 19,444

11.3 $\hat{y} = 20 + 2x$; SSR/SST = 200/242 = 0.826

11.5 a) $r = 0.866$
b) $s_{xy} = 1.50$; $(0.866)^2 = 0.75$. Thus, 75% is explained.

11.7 a) $r = 0.9736$ b) $t = 7.386$. Reject $H_0\colon \rho = 0$

11.11 a) $r = 0.301$ b) $t = 2.450$ c) 9.09% is explained

11.13 a) $r^2 = \text{SSR/SST} = 0.11 \Rightarrow 11\%$ explained
b) $t = \dfrac{r\sqrt{n-2}}{\sqrt{1-r^2}} = \dfrac{0.332\sqrt{24}}{\sqrt{1-0.11}} = 1.72$; do not reject H_0 at $\alpha = 0.05$.
c) $0.10 \geq 2P(t \geq 1.72) > 0.05$
d) $F = \text{MSR/MSE} = \dfrac{\text{SSR}/1}{\text{SSE}/24} = \dfrac{13.2/1}{106.8/24} = 2.966$
Since $1.72^2 = t^2 = 2.966 = F$, answers are consistent.

11.15 a) $s_e = \sqrt{\text{SSE}/(n-2)} = \sqrt{21312/15} = \sqrt{1420.8} = 37.7$
$r = \sqrt{\text{SSR/SST}} = \sqrt{7104/28{,}416} = -0.50$ (minus because covariance is negative)
b) $t_c = (r\sqrt{n-2})/(\sqrt{1-r^2}) = (-0.5\sqrt{15})/(\sqrt{1-0.25}) = -2.24$; since $-t_{0.05,\,15} = -1.753$, reject H_0.

11.17 The computations are easier if we divide each x- and y-value by 1000.
a) $r = \dfrac{s_{xy}}{s_x s_y} = \dfrac{18907}{(795)(26)} = 0.914$

b) $t = \dfrac{r\sqrt{n-2}}{\sqrt{1-r^2}} = \dfrac{0.914\sqrt{8}}{\sqrt{1-(0.914)^2}} = 6.37$; reject H_0. The transformations do not affect r.

11.19 a) $z = \dfrac{z_r - \mu_r}{\sigma_r}$, where

$$z_r = \frac{1}{2}\ln\left(\frac{1+r}{1-r}\right) = \frac{1}{2}\ln\left(\frac{1+0.33}{1-0.33}\right) = 0.3428$$

$$\mu_r = \frac{1}{2}\ln\left(\frac{1+\rho}{1-\rho}\right) = \frac{1}{2}\ln\left(\frac{1+0.05}{1-0.05}\right) = 0.0500 \text{ (from Table IX)}$$

and

$$\sigma_r = \frac{1}{\sqrt{n-3}} = \frac{1}{\sqrt{23}} = 0.2085;\ z = \frac{0.3428 - 0.0500}{0.02085} = 1.40.$$

$P(z \geq 1.40) = 0.0808 \rightarrow$ do not reject H_0 unless $\alpha \geq 0.0808$

b) $P(z \geq 1.13) = 0.1298$

11.21 a) $\hat{y} = 20.0 + 2.0$ ADEXP; $r = 0.909$; $r^2 = 0.826$; $t_c = 4.364$; reject H_0

b) $\hat{y} = 20.0 + 2(8) = 36.0$ or \$36,000

11.23 a) $r = 0.92$

b) $t_c = 6.835$. Since $t_{0.05,\,8} = 1.860$, we reject H_0 and accept $H_a\!: p > 0$.

11.25 $H_0\!: \beta = 0$ vs. $H_a\!: \beta > 0$; $t_c = (b - \beta_0)/s_b = (6 - 0)/1.5 = 4.0$
Since $t_{0.05,\,10} = 1.812$, we reject H_0 and accept H_a.

11.27 a) $r^2 = \text{SSR/SST} = 13.2/120 = 0.11 \Rightarrow 11\%$ explained; $r = \sqrt{0.11} = 0.3316$

b) $t_c = (r\sqrt{n-2})/(\sqrt{1-r^2}) = (0.3316\sqrt{24})/\sqrt{(1-0.11)} = 1.72$
(Note that sign of r is $+$, because problem states that slope is positive.)

c) $F = (\text{SSR}/1)/(\text{SSE}/24) = (13.2/1)/(106.8/24) = 2.96$. Since $2.96 = (1.72)^2$, the values are consistent.

11.29 $s_x = 223.60$; $s_y = 1581.14$; $s_{xy} = 325{,}000$; $r = 0.919$

11.31 $r^2 = (0.917)^2$ for Problem 10.40; $r^2 = (0.964)^2$ using $\sqrt{x}$; $\sqrt{x}$ is better.

Chapter 12

12.5 a) $\hat{y} = 1.03145 + 2.1473x - 0.3717z$ b) $\hat{y} = 3.32315$ c) $s_e = 0.495$

12.9 a) For each one-unit change in x_2, $\hat{y}$ will change by 6 units, assuming that x_1 and x_3 are held constant.

b) $t_1 = 3.33$; $t_2 = 1.2$; $t_3 = -5$; x_3 is most significant.

c) If x_2 is eliminated, SSE will increase, R^2 will decrease, and a, b_1 and b_3 will change. We can't tell what will happen to s_e, since the d.f. change.

d) $\hat{y} = 28$

12.13 a) $\hat{y} = 5.39 + 0.7905x_1 + 0.5323x_2$

b) b_1: for a given Test II score, if a Test I score increases by 1 point, the final rating will increase by 0.79 points. If x_1 had been the only independent variable, then the value of b_1 would not have been 0.79, since simple linear regression would ignore the effect of Test II on the final rating.

c) SSR $= 700.37$; SST $= 770.4$; SSE $= 70.03$

d) $F_c = \dfrac{\text{SSR}/m}{\text{SSE}/(n - m - 1)} = \dfrac{700.37/2}{70.4/7} = 35.00.$

Since this calculated value exceeds the critical value for $\alpha = 0.05$, we reject $H_0: \rho = 0$.

12.15 a) $\hat{y} = 35.769 + 4.767x$; $R^2 = 0.389$

b) $\hat{y} = -152.16 + 60.738x - 3.968x^2$; $R^2 = 0.840$

c) Model 2 provides a better fit, but must be interpreted with caution because x and x^2 are so highly correlated (0.9958). Model 2 could be worse if it makes no intuitive sense.

12.17 a) Seems to be a parabolic relationship. b) $\hat{y} = 154.9 - 36.6x + 3.036x^2$

c)

y	$\hat{y}$	$y - \hat{y}$	y	$\hat{y}$	$y - \hat{y}$
100	121.3	−21.3	35	44.6	−9.6
120	93.8	26.2	60	47.5	12.5
75	72.4	2.6	75	56.4	18.6
50	57.1	−7.1	80	71.4	8.6
40	47.8	−7.8	70	92.5	−22.5

12.21 b) $t = -4.7$, so conclude that formal character of constitution *does* affect the proportion of revenues the central government obtains.

c) $F = 23.21$; reject $H_0: \rho = 0$, and conclude model is significant.

d) The value of d has no meaningful interpretation, since the sample is a cross section. The order of the observations, and thus the value of d, is arbitrary.

12.23 a) $r^2_{yx_4 \cdot x_1x_2x_3} = \dfrac{\text{Extra explained by } x_4}{\text{SSE}(x_1x_2x_3)} = 0.15$ b) $F_c = 2.368$; accept H_0

12.25 a) $R^2 = \dfrac{\text{SSR}}{\text{SST}} = 1 - \dfrac{150}{600} = 0.75$ b) $r^2_{yx_3 \cdot x_1x_2} = \dfrac{450 - 350}{600 - 350} = \dfrac{100}{250} = 0.4$

12.27 $R^2_{y \cdot x_1x_2x_3} = 0.80$; $r^2_{yx_1 \cdot x_2x_3} = \dfrac{400}{600} = 0.67$

12.29 a) $R^2_{y \cdot xzw} = 0.90$ b) $r^2_{y,x \cdot zw} = 0.75$

12.31 a) $\hat{y} = -4.50 + 2x_1 + 2.833x_2$ b) SST = 166; SSR = 156.167; SSE = 9.833; $s_e = 2.217$; $R^2 = 0.9407$

c) $r^2_{yx_1 \cdot x_2} = 0.6197$; $r^2_{yx_2 \cdot x_1} = 0.662$

d) $F_c = 15.88$; since F_c does not exceed $F_{(2,2)}$, we reject H_a, accept the null hypothesis that $\rho = 0$, and conclude that x_1 and x_2 do not affect y.

e) ANOVA table

	SS	d.f.	MS
Reg.	156.167	2	78.083
Error	9.833	2	4.916
Total	166	4	

$F = \dfrac{78.083}{4.916} = 15.88$

12.33 a) $t_1 = 28.27$; $t_2 = 10.36$; x_1 is more important

b) $y_{12} = 211.35$; $e_{12} = 8.65$

c) 2 d.f. in the numerator; 9 d.f. in the denominator

d) $R^2 = (0.955)^2 = 0.99$

g) No problem of multicollinearity since $r_{x_1x_2}$ is very close to zero.
h) The prediction using $x_1 = 10$ and $x_2 = 12$ would be expected to be more accurate since these two values are closer to $\bar{x}_1 = 12.58$ and $\bar{x}_2 = 11.17$ than $x_1 = 20$, $x_2 = 30$.

12.35 a) $\hat{y} = 17.321 + 4.539x_2 - 2.317x_4$ b) $s_e = 2.24$ c) $R = 0.485$
d) $\beta_4{:}t = 14.758$; reject H_0. $\beta_2{:}t = 4.024$; reject H_0.
e) $F = \text{MSR}/\text{MSE} = 50/5 = 10$ with 2.65 d.f.; reject H_0.
f) x_1 will enter next because its partial correlation is higher.

12.37 a) $\hat{y} = 0.3646$ POP + 540,354.8 URATE + 612,216.3 COMP + 21.989 ADEXP − 1,790,782.5 [Note: each POP value divided by 1,000]
$= 0.3646(19660) + 540{,}354.8(6.4) + 612{,}216.3(2) + 21.989(35{,}000) - 1{,}790{,}782.5$
$= 10{,}829{,}334$.

12.39 a) No; for each additional variable, R will either stay the same, or increase. [The adjusted R value may decrease, however.]
b) The value of F may increase or decrease.
c) The value of $\sum(y - \hat{y})^2$ may increase or stay the same because additional variables can never make $\hat{y}$ a worse predictor of y. The value of s_e could increase or decrease because any change in SSE will tend to decrease s_e while any change in $(n - m - 1)$ will tend to increase s_e.

12.45 The max EMV occurs at a bid of 99.10 if bid prices are in intervals of 0.05.

Chapter 13

13.3 a) \$300/year $(=b)$
b) $x = 11$; $\hat{y}_{1981} = 19{,}800$; income is \$800 below trend
d) \$476.25 and \$469.23 e) $\hat{y} = 82.89 + 0.98x$ f) $\hat{y} = 331.568 + 3.936x$

13.5 b) $\hat{y} = 330 + 16.25x$ c) $\hat{y} = 330 + 15.47x$

13.7 a) $\hat{y} = (5.715)(0.682)^{0.298^{x'}}$ b) $\hat{y} = 0.1842 + (0.0722)(0.1038)^{x'}$
c) From the values of $\hat{y}$(Gompertz) and $\hat{y}$(logistic) we see that the two fits are about the same.

13.9

Year	1974	1975	1976	1977	1978	1979	1980	1981	1982
Sales	18	20	22	19	21	24	21	23	27
M.A.		20	20.33	20.67	21.33	22	22.67	23.67	

13.11 $\hat{y} = 50 + 5.5x$; $\hat{y}_{1980} = 50 + 5.5(2) = 61.0$

$$\text{M.A.}_{1980} = \frac{51 + 68 + 65}{3} = \frac{184}{3} = 61.33$$

13.13 Since $\hat{y} = 3600 + 480x$ is for annual trend, to get an equation for *months*, divide b by $12 \rightarrow \hat{y} = 3600 + 40x$. Since $x = 0$ for October 1980, $x = 42$ for April 1984, and $\hat{y} = 3600 + 40(42) = 5280$ is the trend value for April 1984. Seasonally adjusted figure is $5280 \times 0.80 = 4224$.

13.19 Assume prediction for 1979 was the observed sales in 1978
$\hat{\hat{S}}_{79} = 19{,}000$. Then
$\hat{\hat{S}}_{83} = aS_{82} + a(1 - a)S_{81} + a(1 - a)^2S_{80} + (1 - a)^3[aS_{79} + (1 - a)\hat{\hat{S}}_{79}]$
$= \frac{2}{3}(28{,}000) + \frac{2}{3}(1/3)(26{,}000) + \frac{2}{3}(1/3)^2(24{,}000)$
$+ (\frac{1}{2})^3[\frac{2}{3}(23{,}000) + \frac{1}{3}(19{,}000)] = 27{,}025$

13.21 c) Seasonally adjusted for January = 10,100,000
Seasonally adjusted for August = 9,500,000
Percentage change = −5.94%

13.23 b) $\hat{y} = 24.52 - 0.085x$

13.25 $\dfrac{\$3.5 \text{ million}}{0.95} = \$3.68$ million in 1977 dollars

$\dfrac{\$4.5 \text{ million}}{1.40} = \$3.21$ million in 1977 dollars

13.27 $\dfrac{212/130}{200/125} = 1.019 \rightarrow 1.9\%$ growth

13.29 $\dfrac{20{,}000}{120}(100) = \$16{,}667$ real income; change in real income is $16,667 − 17,000 = −$333

13.35 a) $I_n = \dfrac{\sum P_n Q_0}{\sum P_0 Q_0} \times 100 = \dfrac{\sum P_{60} Q_{50}}{\sum P_{50} Q_{50}} \times 100 = \dfrac{15}{10} \times 100 = 150.0$

b) $1750 in 1960 is worth the equivalent of (1750/150) × 100 = $1166.67. Thus, $1200 in 1950 bought more to eat.

13.39 $y = T \times S \times C \times I$ (when S and C are included, include them as index/100). $y = 55{,}000$, $T = 44{,}000$, $S = 95$, $C = 119$; so I index = 110.5.

13.41 $\hat{y} = 37.50 + 6x$, where x is in years. To represent trend equation for months, divide b by 12: $\hat{y} = 37.50 + 0.5x$, where $x = 0$ now represents July 1, 1977. Then March 1, 1979 $\rightarrow x = 20$. So to move the origin to March 1, 1979: $\hat{y} = 37.5 + \frac{1}{2}(x + 20) = 37.50 + 10 + \frac{1}{2}x = 47.50 + \frac{1}{2}x$.

13.43 a) $\hat{y} = 4 + 1.9x$; x units = 1 year; $x = 0$ at mid-1980

b) $s_e = \sqrt{\dfrac{\sum e^2}{n-2}} = \sqrt{\dfrac{3.9}{3}} = \sqrt{1.3} = 1.14$ c) $\hat{y}_{1984} = 4 + 1.9(4) = 11.7$

d) total deviation $= y - \bar{y} = -2$; explained $= -1.9$; unexplained $= -0.1$

13.45 a) $\log \hat{y} = 2.8485 + 0.3784x$ or $\hat{y} = (706)(2.39)^x$

b) for 1984 ($x = 4$): $\log \hat{y} = 4.3621$; $\hat{y} = 23{,}016$
y: 100, 400, 700, 1600, 3900
$\hat{y}$: 123.1, 295, 706, 1686, 4030

c) $\hat{y} = -38.46 + (738.46)(5\frac{1}{3})^x$
y: 100, 400, 700, 1600, 3900
$\hat{y}$: 100, 281, 700, 1666, 3900

13.47 a) $\hat{y} = 90{,}000 + 5000x$, x units = 1 quarter
$x = 0$ at mid-first quarter, 1981, so that
$x = 1$ represents 2nd quarter, $x = 2$ represents 3rd quarter, and
$x = 3$ represents 4th quarter

b) 1st, 100,000/0.8 = 125,000; 2nd, 150,000/1.3 = 115,385;
3rd, 120,000/1.1 = 109,091; 4th, 110,000/0.8 = 137,500.

c) $\hat{y} = 90{,}000 + 5000(9) = 135{,}000$; $135{,}000(1.3) = 175{,}500$

13.49 a) average monthly ratio is 1143/12 = 95.25
$S_{\text{Mar}} = 100/95.25 = 105$, $S_{\text{Apr}} = 110/95.25 = 115.5$, $S_{\text{Aug}} = 72/95.25 = 75.6$

b) \$10 million average/month: \$10 million × 1.155 = \$11.55 million for Apr., \$10 million × 0.775 = \$7.56 million for Aug.

13.51 a) Using least-squares method, $\hat{y} = 4.028 + 0.139x$
b) $r = 0.14$; c) $r^2 = 0.019$ or 1.9% of variation explained

d)

Year	y	Trend Value	Year	y	Trend Value
1971	2	2.8	1976	3	5.3
1972	3	3.3	1977	5	5.8
1973	6	3.8	1978	7	6.3
1974	5	4.3	1979	8	6.8
1975	4	4.8	1980	7	7.3

e) No. This is annual data.
f) Moving average 1973 = 4.67; moving average 1976 = 8.67

Chapter 14

14.3 The sample means are $\bar{y}_1 = 59/5 = 11.8$, $\bar{y}_2 = 44/5 = 8.8$, $\bar{y}_3 = 61/5 = 12.2$, and $\bar{y}_4 = 43/5 = 8.6$. The grand sample mean is $\bar{y} = 207/20 = 10.35$. The estimated treatment effects are 1.45, −1.55, 1.85, −1.75.

14.5 For the Southeast, $\bar{y}_1 = 893/9 = 99.22$; for the Southwest, $\bar{y}_2 = 1065/10 = 106.50$; for the Northeast, $\bar{y}_3 = 1038/10 = 103.80$; for the Northwest, $\bar{y}_4 = 512/5 = 102.40$; for the Midwest, $\bar{y}_5 = 785/7 = 112.14$; and for the Far West, $\bar{y}_6 = 633/6 = 105.50$. The estimate for the entire U.S. is the grand sample mean, $\bar{y} = 4926/47 = 104.81$. The estimated treatment effects are −5.59, 1.69, −1.01, −2.41, 7.33, 0.69.

14.7

Source of Variation	SS	d.f.	MS
Between brands	224.8	2	112.40
Within brands	118.8	15	7.92
Total	343.6	17	

$F = \dfrac{112.40}{7.92} = 14.19$. This is significant at the $\alpha = 0.01$ level.

14.9

	Case 1	Case 2
SS total	112	112
SS between	0	112
SS within	112	0

These are extreme cases in the sense that SS between is zero in Case 1 and SS within is zero in Case 2.

14.13

Source of Variation	SS	d.f.	MS
Between machines	32,396	3	10,799
Within machines	12,178	32	381
Total	44,574	35	

$F = 28.34$

This is much larger than the critical value for F when $\alpha = 0.01$, so we reject the hypothesis of equality among the four population means.

14.17 *Programs*

1–2: 1.45 + 1.55 ± 3.18 = 3.00 ± 3.18 = (−0.18, 6.18).

1–3: 1.45 − 1.85 ± 3.18 = −0.40 ± 3.18 = (−3.58, 2.78).

1–4: 1.45 + 1.75 ± 3.18 = 3.20 ± 3.18 = (0.02, 6.38).

2–3: −1.55 − 1.85 ± 3.18 = −3.40 ± 3.18 = (−6.58, −0.22).

2–4: −1.55 + 1.75 ± 3.18 = 0.20 ± 3.18 = (−2.98, 3.38).

3–4: 1.85 + 1.75 ± 3.18 = 0.10 ± 3.18 = (−3.08, 3.28).

14.19 The estimated treatment effects are: 37.25, 21.28, −26.64, −32.20. The simultaneous confidence intervals are:

A–B: 37.25 − 21.58 ± 27.14 = (−11.47, 42.81).

A–C: 37.25 + 26.64 ± 27.14 = (36.75, 91.03).

A–D: 37.25 + 32.20 ± 27.14 = (42.31, 96.59).

B–C: 21.58 + 26.64 ± 27.14 = (21.08, 75.36).

B–D: 21.58 + 32.20 ± 27.14 = (26.64, 80.92).

C–D: −26.64 + 32.20 ± 27.14 = (−21.58, 32.70).

14.23 The variation within each group (that is, each combination of treatments) reflects chance fluctuations.

14.27 The totals (not means) are:

	A	*B*	*C*	
Asphalt	152	163	137	452
Concrete	159	182	137	478
Gravel	135	129	139	403
	446	474	413	1333 = grand sum

C.F. = $(1333)^2/36$ = 49,358; SS Total = 673; SS Columns = 155; SS Rows = 242; SS Error = 80; SS Interaction = 196

Source of Variation	SS	d.f.	MS	*F*
Columns	155	2	77.5	26.18
Rows	242	2	121.0	40.88
Interaction	196	4	49.0	16.55
Error	80	27	2.96	
Total	673	35		

The column effects, row effects, and interaction effects are all significant at the $\alpha = 0.05$ level.

14.29 Brand *A* = 0.14; Brand *B* = 2.47; Brand *C* = −2.61
Asphalt = 0.64; Concrete = 2.81; Gravel = −3.44
Brand *A*, Asphalt = 0.19; Brand *A*, Concrete = −0.22; Brand *A*, Gravel = 0.03
Brand *B*, Asphalt = 0.61; Brand *B*, Concrete = 3.19; Brand *B*, Gravel = −3.81
Brand *C*, Asphalt = −0.81; Brand *C*, Concrete = −2.97; Brand *C*, Gravel = 3.78

14.31 Sums:

	Center	Lake	West	North	
Drug	10.03	10.87	9.58	10.49	40.97
Discount	9.20	9.35	9.19	9.36	37.10
Grocery	10.84	10.75	10.68	11.31	43.58
	30.07	30.97	29.45	31.16	121.65

C.F. = $(121.65)^2 48$ = 308.3067; SS Total = 1.8684; SS Columns = 0.1595
SS Rows = 1.3288; SS Error = 0.2378; SS Interaction = 0.1423

Source of Variation	SS	d.f.	MS	F
Columns	0.1595	3	0.0532	8.06
Rows	1.3288	2	0.6644	100.67
Interaction	0.1423	6	0.0237	3.59
Error	0.2378	36	0.0066	
Total	1.8684	47		

The 3 F-tests all yield significant results, and the estimated effects are as follows. Center City = −0.02; Lakefront = 0.05; West Side = −0.08; North suburbs = 0.07; Drug Store = 0.03; Discount Store = −0.21; Grocery Store = 0.19; Center, Drug = −0.03; Center, Discount = 0.00; Center, Grocery = 0.01; Lake, Drug = 0.11; Lake, Discount = −0.03; Lake, Grocery = −0.08; West, Drug = −0.08; West, Discount = 0.06; West, Grocery = 0.03; North, Drug = −0.01; North, Discount = −0.05; North, Grocery = 0.04

14.39

Source of Variation	SS	d.f.	MS	F
Linear regression	54	1	54	6
Error	540	60	9	
Total	594	61		

The linear regression effect is significant at the 0.05 level but not at the 0.01 level.

14.43 a)

Source of Variation	SS	d.f.	MS	F
Between sections	1409	2	704.5	12.74
Within sections	1825	33	55.3	
Total	3236	35		

The value of F is much larger than the critical value for $\alpha = 0.01$. There are significant differences between the sections.

b) $\hat{\tau}_1 = 67.25 - 69.81 = -2.56$, $\hat{\tau}_2 = 63.75 - 69.81 = -6.06$, and $\hat{\tau}_3 = 78.42 - 69.81 = 8.61$.

$$\sqrt{(J-1)F(\text{MSWithin})\left(\frac{1}{n_k} + \frac{1}{n_l}\right)} = \sqrt{2(3.29)(55.3)\left(\frac{1}{12} + \frac{1}{12}\right)} = 7.79.$$

Intervals: 1–2: $-2.56 + 6.06 \pm 7.79 = (-4.29, 11.29)$.
1–3: $-2.56 - 8.61 \pm 7.79 = (-18.96, -3.38)$.
2–3: $-6.06 - 8.61 \pm 7.79 = (-22.46, -6.88)$.

The differences between Sections 1 and 3 and between Sections 2 and 3 are significant.

14.45

Source of Variation	SS	d.f.	MS	F
Between	69	2	34.5	6.05
Within	119	21	5.7	
Total	188	23		

The critical values of F for $\alpha = 0.05$ and $\alpha = 0.01$ are 3.47 and 5.78. The means differ significantly. $\hat{\tau}_1 = 2.38$, $\hat{\tau}_2 = -1.50$, $\hat{\tau}_3 = -0.87$, and $\sqrt{2(3.47)(5.7)(\frac{1}{8} + \frac{1}{8})} = 3.14$.

Intervals: Union A–Union B: $2.38 + 1.50 \pm 3.14 = (0.74, 7.02)$.
Union A–Nonunion: $2.38 + 0.87 \pm 3.14 = (0.11, 6.39)$.
Union B–Nonunion: $-1.50 + 0.87 \pm 3.14 = (-3.77, 2.51)$.

Union A and Union B are significantly different, as are Union A and Nonunion. However, Union B and Nonunion are not significantly different.

14.47 a) $\bar{y}_1 = 51.3/8 = 6.41$, $\bar{y}_2 = 48.8/6 = 8.13$, $\bar{y}_3 = 28.6/5 = 5.72$, and $\bar{y} = 128.7/19 = 6.77$. The estimated treatment effects are $\hat{\tau}_1 = 6.41 - 6.77 = -0.36$, $\hat{\tau}_2 = 8.13 - 6.77 = 1.36$, and $\hat{\tau}_3 = 5.72 - 6.77 = -1.05$.

b)

Source of Variation	SS	d.f.	MS	F
Between	17.7	2	8.85	22.12
Within	6.4	16	0.40	
Total	24.1	18		

c) Yes, the costs are different at the 0.05 level of significance.

14.49

Source of Variation	SS	d.f.	MS	F
Programs	60	3	20.00	7.49
Age	10	1	10.00	3.75
Interaction	2	3	0.67	0.25
Error	64	24	2.67	
Total	136	31		

At the $\alpha = 0.01$ level, only the program effects are significant.

14.51 Yes. The F-value is much higher than the critical value for $\alpha = 0.05$ with 1 and 28 degrees of freedom.

14.53 Yes, $F = 51.05$ is much larger than the critical value, 2.76.

14.55 The effect of the level of extroversion is significant, but education effects and interaction effects are not significant. The extroverts are willing to pay up to \$15.89 on average, compared with \$12.67 for the ambiverts and \$7.11 for the introverts.

Chapter 15

15.3 a) interval b) nominal c) ratio d) nominal

15.5 $E[U] = 800$; $\sigma_U^2 = 10{,}800$; $z = -4.80$; $P(z \leq -4.80) \cong 0$

15.7 7 −'s and 3 +'s, $n = 10$, $x = 7$, $p = 0.50$; $2P(\mathbf{x} \geq 7) = 0.3438$ (from Table I) for a two-tailed test. Do not reject H_0.

15.9 1st–2nd; 4 plus signs, 1 minus, $2P(\mathbf{x} \geq 4) = 0.3748$ (from Table I). Do not reject H_0.

15.11 Probability of 1 or less accidents when $p = 0.05$ and $n = 100$ is 0.0371, from Table I. Hence, accept $H_a: p = 0.05$ at $\alpha = 0.0371$ or higher.

15.13 $H_0: p = 0.50$ vs. $H_a: p \neq 0.50$, where $p =$ proportion of population who think the president is less effective in the second evaluation. 60% of 400 is 240 people.

$$z = \frac{x - np_0}{\sqrt{np_0 q_0}} = \frac{240 - 200}{\sqrt{400(0.50)(0.50)}} = 4.00 > 1.96; \text{ reject } H_0$$

15.15 a) nominal b) nominal c) nominal d) ordinal e) ordinal f) ordinal g) nominal h) ordinal

15.17 a) $H_0: E_{\text{purple}} = 696\frac{3}{4}$, $E_{\text{white}} = 232\frac{1}{4}$
H_a: The population frequencies are different from these.
$\chi_c^2 = 0.3907$; do not reject H_0.
b) normal approximation; $z_c = 0.633$; accept H_0: the proportion of purple flowers is 3/4.

15.19 H_0: the manager's proportions correct; H_a: manager's proportions incorrect. $\chi_c^2 = 11$; accept H_0

15.21

x	Cum. Observed	Cum. Theoretical ($\lambda = 7$)	$\lvert F - S \rvert$
0	144/280 = 0.5143	0.4966	0.0177
1	235/280 = 0.8393	0.8442	0.0049
2	267/280 = 0.9536	0.9659	0.0123
3	278/280 = 0.9929	0.9943	0.0014
≥ 4	280/280 = 1.0000	1.0000	0

From Table XIII, do not reject H_0.

15.23 a) H_0: There is no difference in the time of birth between males and females.

b) $$\chi^2 = \frac{(10 - 5)^2}{10} + \frac{(7.5 - 5)^2}{7.5} + \cdots + \frac{(12.5 - 15)^2}{12.5} = 7.66$$

$0.025 \geq P(\chi^2 \geq 7.66) \geq 0.01$; reject H_0 at $\alpha = 0.025$, d.f. $= 2$.

15.25 a) H_0: There is no difference in deaths among the experimental and control groups.
b) $\chi^2 = 8.0$; $P(\chi^2 \geq 8) \leq 0.005$; reject H_0 at $\alpha = 0.005$

15.27 $C = \sqrt{0.39071(929.3907)} = 0.0205$

15.29 $r_s = 0.60$; $\tau = 0.40$

INDEX

(Continued from front endpapers)

Chapter 7

α	probability of making an error (alpha)
C.I.	confidence interval
D	maximum difference between point estimate and the population parameter

Chapter 8

H_0 and H_a	null and alternative hypotheses
α and β	probability of Type I (alpha) and Type II (beta) errors
μ_0	value of μ specified by H_0
z_c or t_c	calculated value of z or t
z_α	value of z-distribution having α of the area beyond it
d_i	difference score between ith pair of the matched pairs
Δ	average difference between matched pairs in population
s_D	standard deviation of difference scores

Chapter 9

EMV	expected monetary value
EVPI	expected value of perfect information
ENGS	expected net gain from sampling
EVSI	expected value of sample information
$f(\theta)$	prior density function of θ
m, v	mean and variance of prior distribution
M, V	mean and variance of posterior distribution
EU	expected utility
$U(M)$	utility for money

Chapter 10

$\mu_{y \cdot x}$	mean of y values given an x value
α, β	intercept and slope of a population regression line
ϵ_i	error or residual for the ith population value
y_i	ith value in the population or sample
$\hat{y}$	least squares estimate of $\mu_{y \cdot x}$
a, b	least squares estimate of α and β
e_i	sample estimate of ϵ_i
s_e	standard error of the estimate
s_b	standard error of the coefficient b
BLUE	best linear unbiased estimate
$s_f, s_{\bar{y}}$	standard error of the forecast, and of $\mu_{y \cdot x}$
x_g	a given value of the variable x
MSR, MSE	mean square regression, mean square error